Dynamic Stereochemistry of Chiral Compounds
Principles and Applications

Dynamic Stereochemistry of Chiral Compounds

Principles and Applications

Christian Wolf
Department of Chemistry, Georgetown University, Washington, DC, USA

RSCPublishing

ISBN: 978-0-85404-246-3

A catalogue record for this book is available from the British Library

Published by The Royal Society of Chemistry,
Thomas Graham House, Science Park, Milton Road,
Cambridge CB4 0WF, UK

Registered Charity Number 207890

For further information see our web site at www.rsc.org

Preface

Since the pioneering work of Pasteur, Le Bel and van't Hoff stereochemistry has evolved to a multi-faceted and interdisciplinary field that continues to grow at an exponential rate. Today, dynamic stereochemistry plays a fundamental role across the chemical sciences, ranging from asymmetric synthesis to drug discovery and nanomaterials. The immense interest and activity in these areas have led to the development of new methodologies and research directions in recent years. Chirality plays a pivotal role in efforts to control molecular motion and has paved the way for microscopic propellers, gears, switches, motors and other technomimetic devices. The concepts of Euclidian chirality have been extended to topological chirality of fascinating mechanically interlocked assemblies including rotaxanes, catenanes, and even molecular knots or pretzelanes. The impressive advance of asymmetric synthesis has been accompanied by significant progress in stereochemical analysis. Mechanistic insights into isomerization reactions and information about the conformational and configurational stability of chiral compounds are indispensable for today's chemist, and new techniques such as dynamic chromatography and stopped-flow procedures that complement chiroptical and NMR spectroscopic methods have been established. A profound understanding of the stability of chiral target compounds, intermediates and starting materials to racemization and diastereomerization is indispensable for planning an efficient synthetic route. In many cases, interconversion of stereoisomers compromises the efficacy of asymmetric synthesis and ultimately results in the loss of stereochemical purity, but it can also be advantageous. While the usefulness and scope of asymmetric transformations of the first and second kind have been known for a long time, dynamic kinetic resolution, dynamic kinetic asymmetric transformation and dynamic thermodynamic resolution have become powerful synthetic alternatives. Many strategies that afford excellent control of stereolabile substrates and reaction intermediates have been developed and are nowadays routinely employed in the synthesis of a wide range of chiral compounds including natural products. Asymmetric synthesis of complex target compounds generally entails incorporation of several chiral elements in addition to strategic carbon-carbon bond formation. Once molecular chirality has been established it is often necessary to further manipulate it. Numerous methods for selective translocation of a chiral element along an existing carbon framework or interconversion of elements of chirality without loss of stereochemical purity have been introduced and provide invaluable synthetic prospects. Both, the progress and the diversity of stereodynamic chemistry, in particular asymmetric synthesis, are unrivaled and constantly fueled by an enormous amount of new scientific contributions.

Over the last twenty years, several excellent books about asymmetric synthesis have appeared. Traditionally, asymmetric synthesis is discussed based on (1) reaction types, for example aldol reactions, hydroboration, epoxidation, dihydroxylation, hydrogenation, and so on, or (2) emerging concepts such as organocatalysis, biomimetic methods and phase-transfer catalysis. In addition, many reviews on broadly applicable asymmetric catalysts and methods can be found in the literature. While there is no need to duplicate these books and articles, I have felt that a conceptually different textbook that embraces asymmetric synthesis, interconversion of chiral compounds, analytical methods suitable for the study of racemization and diastereomerization reactions, as well as a discussion of topologically chiral assemblies and molecular propellers, switches and motors in the context of dynamic stereochemistry would be very helpful for both teaching and research. Stereoselective synthesis, analysis and stereodynamic properties and applications of chiral compounds are now combined in one text. The book is aimed at graduate students and is intended to serve as a guide for researchers with an interest in synthetic, analytical and mechanistic aspects of dynamic stereochemistry of chiral compounds.

This book is organized into three parts that contain nine chapters and a glossary of important stereochemical terms and definitions. The first chapter provides an introduction to the significance and interdisciplinary character of dynamic stereochemistry. Chapter 2 covers principles, terminology and nomenclature of stereochemistry with an emphasis on Euclidian and topological chirality. The reader is familiarized with stereodynamic properties of chiral compounds and the relative contributions of interconverting configurational and conformational isomers to selectivity, reactivity and chiral recognition. Racemization, enantiomerization and diastereomerization including mutarotation and epimerization are discussed in Chapter 3. Mathematical treatments of reversible and irreversible isomerization kinetics are presented and the wealth of racemization and diastereomerization mechanisms is reviewed separately for each class of compounds to provide a systematic overview. Numerous examples of compounds with individual energy barriers to enantioconversion are given to highlight steric and electronic effects. The principles and scope of analytical techniques that are commonly used to determine the conformational and configurational stability of chiral compounds are discussed in Chapter 4. Special emphasis is given to chiroptical methods, variable-temperature NMR spectroscopy, dynamic chromatography, and stopped-flow analysis.

The second part of this book focuses on asymmetric synthesis. Chapter 5 introduces the reader to the principles of asymmetric synthesis and outlines basic concepts of stereodifferentiation and reaction control. Chapter 6 covers asymmetric synthesis with chiral organolithium compounds and atroposelective synthesis of biaryls and nonbiaryls. This discussion leads to other strategies that are aimed at manipulation of chirality without concomitant racemization or diastereomerization, for example chirality transfer and interconversion of chiral elements during pericyclic rearrangements, S_N2' and S_E2' displacements, and self-regeneration of stereogenicity with temporary or transient chiral intermediates. Synthetic methods that utilize relays and stereodynamic catalysts for amplification of chirality and asymmetric induction link the above topics to the following chapter. Strategies that incorporate stereolabile chiral compounds and intermediates into asymmetric synthesis under thermodynamic (asymmetric transformations of the first and second kind, dynamic thermodynamic resolution) or kinetic reaction control (dynamic kinetic resolution and dynamic kinetic asymmetric transformation) are presented in Chapter 7. Because of the considerable overlap and relevance to some of the above methodologies, kinetic resolutions are also covered in detail. The scope and application spectrum of asymmetric reactions and concepts presented in Chapters 5 to 7 are highlighted with many examples and the stereochemical outcome is explained with a mechanistic rationale including transition state structures whenever possible.

The third part of this book comprises stereodynamic devices, manipulation of molecular motion, and the chemistry of topologically chiral assemblies. Chapter 8 examines the central role that chirality plays in the design of molecular propellers, bevel gears, brakes, switches, sensors, and motors. The synthesis, chirality and stereodynamics of catenanes, rotaxanes and other mechanically interlocked compounds are presented in Chapter 9.

A single monograph can not comprehensively cover the enormous scope and the many facets of dynamic stereochemistry. Topics such as switching and amplification of chirality in polymers, gels and liquid crystals had to be excluded due to limitations of space and time. In writing this book during the last two and a half years, I have tried to adhere to well-defined and established stereochemical terminology and emphasized important limitations and conflicting definitions in the text. To assist the reader, a detailed glossary of stereochemical definitions and terms is included at the end of the book, and a list of abbreviations and acronyms is provided at the beginning. All topics are extensively referenced and the principal researcher is frequently named in the text to encourage further reading and to facilitate additional literature search.

I would like to thank my colleagues, in particular Professors William H. Pirkle and the late Wilfried A. König, and my students for continuing inspiration and helpful discussions. I wish to thank Thomas J. Nguyen for the technical drawings illustrating the conceptual linkage between macroscopic mechanical devices and their molecular analogs. And I am particularly grateful to my wife Julia for her patience, encouragement and understanding during the endless hours involved in writing this book.

Christian Wolf
Washington, DC

This book is dedicated to
my wife Julia

Contents

Abbreviations and Acronyms

a	axial
a′	pseudoaxial
AAA	asymmetric allylic alkylation
Ab	antibody
Ac	acetyl
ac	anticlinal
acac	acetylacetonate
ACN	acetonitrile
AD	asymmetric dihydroxylation
Ad	adamantyl
AE	asymmetric epoxidation
AIBN	azobisisobutyronitrile
Ala	alanine
ap	antiperiplanar
aq	aqueous
Ar	aryl
Asp	aspartic acid
atm	atmospheric pressure
ATP	adenosine triphosphate
$b_{0.5}$	peak width at half height
Bc	butyryl
BHT	2,6-di-*tert*-butyl-4-methylphenol
Bn	benzyl
Boc	*t*-butoxycarbonyl, *t*-BuOC(O)-
BPDM	benzphetamin *N*-demethylase
Bu	*n*-butyl
Bz	benzoyl
°C	degree Celsius
c	concentration; conversion
CAL	*Candida antarctica* lipase
CAN	cerium ammonium nitrate
Cat	catalyst
Cb	*N,N*-diisopropylcarbamoyl, $(i\text{-Pr})_2\text{NC(O)}$-
Cby	2,2,4,4-tetramethyloxazolidine-3-carbonyl

Cbz	carbobenzyloxy, BnOC(O)-
CD	circular dichroism
CDA	chiral derivatizing agent
CEC	capillary electrochromatography
CIDR	crystallization-induced dynamic resolution
CIP	Cahn–Ingold–Prelog
CLA	chiral Lewis acid; complete line shape analysis
cod	1,5-cyclooctadiene
Cp	cyclopentadienyl
CPE	circular polarization of emission
CPL	circularly polarized light
CP-MAS	cross polarization magic angle spinning
18-crown-6	1,4,7,10,13,16-hexaoxacyclooctadecane
CSA	chiral solvating agent
CSP	chiral stationary phase
CSR	chiral shift reagent
Cy, c-C_6H_{11}	cyclohexyl
CZE	capillary zone electrophoresis
D	sodium D-line (589 nm); absolute configuration (Fischer–Rosanoff convention)
d	day(s)
DABCO	1,4-diazabicyclo[2.2.2]octane
dba	dibenzylidene acetone
DBB	4,4′-di-*tert*-butylbiphenyl
DBN	1,5-diazabicyclo[4.3.0]non-5-ene
DBU	1,8-diazabicyclo[5.4.0]undec-7-ene
DCC	dicyclohexyl carbodiimide
DDQ	2,3-dichloro-5,6-dicyano-1,4-benzoquinone
de	diastereomeric excess
DEAD	diethyl azodicarboxylate
DGC	dynamic chromatography
DHPLC	dynamic high performance liquid chromatography
DIAD	diisopropyl azodicarboxylate
DIEA	*N,N*-diisopropylethylamine
DKR	dynamic kinetic resolution
dm	decimeter
DMAD	dimethyl azodicarboxylate
DMAP	4-dimethylaminopyridine
DME	1,2-dimethoxyethane
DMEKC	dynamic micellar electrokinetic chromatography
DMF	*N,N*-dimethyl formamide
DMSO	dimethyl sulfoxide
DNA	deoxyribonucleic acid
DNB	3,5-dinitrobenzoyl
DNMR	dynamic nuclear magnetic resonance
DOPA	3,4-dihydroxyphenylalanine
dppf	1,1′-bis(diphenylphosphino)ferrocene
dr	diastereomeric ratio
DSFC	dynamic supercritical fluid chromatography
DSubFC	dynamic subcritical fluid chromatography
DTR	dynamic thermodynamic resolution

DYKAT	dynamic kinetic asymmetric transformation
E	electrophile
E	enzymatic enantioselectivity factor
(*E*)	entgegen, relative configuration
e	equatorial
e′	pseudoequatorial
ECCD	exciton-coupled circular dichroism
ee	enantiomeric excess
ee$_{pss}$	enantiomeric excess at photostationary state
e.g.	*exempli gratia*, for example
en	ethylenediamine
enant	enantiomerization
epi	epimer; epimerization
equiv	equivalent
er	enantiomeric ratio
EROD	ethoxyresorufin-*O*-deethylase
ESR	electron spin resonance
Et	ethyl
et al.	*et alii*, and others
EXSY	NMR exchange spectroscopy
FID	flame ionization detection
Fmoc	fluorenyl-9-methoxycarbonyl
G	Gibbs free energy
$G^{\neq}$	Gibbs activation energy
g	gram
g	anisotropy factor
GC	gas chromatography
Gly	glycine
H	enthalpy
$H^{\neq}$	activation enthalpy
h	hour(s); Planck's constant
HCH	hexachlorocyclohexane
hfc	heptafluorobutyrylcamphorato
HKR	hydrolytic kinetic resolution
hν	irradiation of light
HOMO	highest occupied molecular orbital
HPLC	high performance liquid chromatography
HSA	human serum albumin
i-Bu	isobutyl
i.e.	*id est*, that is
i-Pr	isopropyl
IR	infrared
J	joule
J	coupling constant
K	Kelvin
K	equilibration constant
k	kilo
k	rate constant
k$_B$	Boltzmann's constant
KHMDS	potassium hexamethyldisilazide

KR	kinetic resolution
L	liter; ligand; absolute configuration (Fischer-Rosanoff convention)
l	length
LASER	light amplification by stimulated emission of radiation
LDA	lithium diisopropylamide
Leu	leucine
LHMDS	lithium hexamethyldisilazide
ln	natural logarithm
LTMP	lithium tetramethylpiperidide
LUMO	lowest unoccupied molecular orbital
Lys	lysine
M	molar
(*M*)	denotes left-handed helicity (CIP convention)
m	milli
m	*meta*
MC	3-methylcholanthrene
m-CPBA	3-chloroperbenzoic acid
Me	methyl
MEKC	micellar electrokinetic chromatography
Mes	mesityl
Met	methionine
min	minute(s)
MOM	methoxymethyl
MP	mobile phase
MS	mass spectrometry
Ms	methylsulfonyl, CH_3SO_2-
MTBE	methyl *tert*-butyl ether
N	0,1,2,3 . . .
n	nano
n	1,2,3 . . .
n_A	population of state *A*
NADH	nicotinamide adenine dinucleotide
NADPH	nicotinamide adenine dinucleotide phosphate
NBS	*N*-bromosuccinimide
Nf	nonaflate, $C_4F_9SO_2-$
NLE	nonlinear effect
nm	nanometer
NMO	*N*-methylmorpholine *N*-oxide
NMR	nuclear magnetic resonance
NOESY	nuclear Overhauser effect spectroscopy
NSAID	nonsteroidal anti-inflammatory drug
Nu	nucleophile
o	*ortho*
obs	observed
ORD	optical rotary dispersion
Ox	oxidation
(*P*)	denotes right-handed helicity (CIP convention)
p	*para*
PB	phenobarbital
PCB	polychlorinated biphenyl

PCL	*Pseudomonas cepacia* lipase
Ph	phenyl
pH	$-\log_{10}[\mathrm{H_3O^+}]$
Phe	phenylalanine
Phg	phenylglycine
PhMe	toluene
Piv	trimethylacetyl
PKR	parallel kinetic resolution
PLP	pyridoxal-5'-phosphate
PM3	parametric method 3 (semi-empirical molecular modeling software)
PMB	*para*-methoxybenzyl
PPAR	peroxisome proliferator-activated receptor
ppm	parts per million
Pr	*n*-propyl
PS	lipase from *Pseudomonas stutzeri*
py	pyridine
R	alkyl; universal gas constant
(*R*)	*rectus*, denotes absolute configuration (CIP convention)
rac	racemic; racemization
Red	reduction
RNA	ribonucleic acid
S	entropy
(*S*)	*sinister*, denotes absolute configuration (CIP convention)
$S^{\neq}$	activation entropy
s	second(s)
s	enantioselectivity factor
s-Bu	*sec*-butyl
sc	synclinal
S_E	electrophilic substitution
SFC	supercritical fluid chromatography
sia	3-methyl-2-butyl
SMB	simulated moving bed chromatography
S_N	nucleophilic substitution
S_Ni	intramolecular nucleophilic substitution
SP	stationary phase
sp	synperiplanar
SRS	self-regeneration of stereocenters
SubFC	subcritical fluid chromatography
T	temperature
T_c	coalescence temperature
t	time
TBAF	tetrabutylammonium fluoride
TBDMS	*tert*-butyldimethylsilyl
TBDPS	*tert*-butyldiphenylsilyl
TBHP	*tert*-butyl hydroperoxide
t-Boc	*t*-butoxycarbonyl, *t*-BuOC(O)-
TBS	tributylsilyl
t-Bu	*tert*-butyl
TCDD	2,3,7,8-tetrachlorodibenzo-*p*-dioxin
Tf	triflate, $\mathrm{CF_3SO_2}$-

TFA	trifluoroacetic acid; trifluoroacetyl
TFAA	trifluoroacetic anhydride
tfc	trifluoroacetylcamphorato
THF	tetrahydrofuran
THP	tetrahydropyran
TIPS	triisopropylsilyl
tipyl	2,4,6-triisopropylphenyl
TLC	thin layer chromatography
TMEDA	N,N,N',N'-tetramethylethylenediamine
TMS	trimethylsilyl
Tol	tolyl
Tr	trityl, Ph_3C-
TR-CPL	time-resolved circularly polarized luminescence
TS	transition state
Ts	tosyl, $4\text{-}CH_3C_6H_4SO_2$-
TTF	tetrathiafulvalene
Tyr	tyrosine
URO	uroporphyrinogen
UV	ultraviolet
UV-vis	ultraviolet-visible
V	volt
VCD	vibrational circular dichroism
VE	valence electron
(Z)	*zusammen*, relative configuration
α	denotes anomer; *C*-2
α	rotation angle; chromatographic enantioselectivity factor
α,ω	denotes chain termini
β	denotes anomer; *C*-3
γ	*C*-4
Δ	heat; difference; right-handed complex
δ	*C*-5
ε	ellipticity
θ	molar ellipticity
κ	transmission coefficient
Λ	left-handed complex
λ	wavelength
μ	micro
ν	frequency
$\tau_{1/2}$	half-life time
Φ	quantum yield for photoracemization
χ	mole fraction
(+)	dextrorotatory
(−)	levorotatory

Structures and Acronyms of Chiral Ligands

BINAP   BINOL   BIPHEB   BIPHOS   BOX   CBS oxazaborolidine

DABN   DAIB   DET   DIPT   DIOP

DPEN   DUPHOS   (DHQD)$_2$-PHAL   (DHQD)$_2$-AQN

hfc   MTPA   NOBIN   PPFA   salen

SEGPHOS   sparteine   SYNPHOS   TADDOL   tfc   TSDPEN

CHAPTER 1

Introduction

In 1848, Louis Pasteur, one of the pioneering stereochemists, recognized the omnipresence and significance of chirality, which prompted his famous statement that the universe is chiral (*l'univers est dissymétrique*).[1] Today, we know that chirality can indeed be encountered at all levels in nature – in the form of the elementary particle known as the helical neutrino, inherently chiral proteins, carbohydrates and DNA, or helical bacteria, plants and sea shells. Pasteur realized that chiral objects exist as a pair of enantiomorphous mirror images that are nonsuperimposable and related to each other like a right-handed and left-handed glove. At the molecular level, chirality gives rise to enantiomers that can exhibit strikingly different chemical and physical properties in a chiral environment. Many biologically active compounds, for example pharmaceuticals, agrochemicals, flavors, fragrances, and nutrients, are chiral, and more than 50% of today's top-selling drugs including Lipitor (cholesterol reducer, global sales in 2004: $12.0 billion), Zocor (cholesterol reducer, $5.9 billion), Plavix (antithrombic, $5.0 billion) and Nexium (antiulcerant, $4.8 billion) are sold as single enantiomers, Figure 1.1.

The increasing demand for enantiopure chemicals has been accompanied by significant progress in asymmetric synthesis[2–8] and catalysis,[9–14] and by the development of analytical techniques for the determination of the stereochemical purity of chiral compounds. Stereoselective analysis usually entails chiroptical measurements,[15] NMR and mass spectroscopic methods,[16–18] electrophoresis,[19] chiral chromatography[20] or UV and fluorescence sensing assays,[21–27] and it can provide invaluable information about the stability of chiral compounds to racemization and diastereomerization. A renowned example of a chiral drug that undergoes fast enantioconversion under physiological conditions is thalidomide (Thalidomid, Contergan) which was prescribed to pregnant women in the 1960s to alleviate morning sickness. One of the enantiomeric forms of thalidomide does indeed have sedative and antinausea effects, but the other enantiomer is a potent teratogen.[i] The racemic drug was approved in Europe for the treatment of pregnant women suffering from nausea and its use caused severe birth defects. Even formulation of the pure nontoxic (*R*)-enantiomer of thalidomide would have been unsafe because racemization takes place *in vivo* and the teratogenic (*S*)-enantiomer is rapidly generated in the human body, Scheme 1.1.[ii]

Since the thalidomide tragedy, the significance of the stereochemical integrity of biologically active compounds has received increasing attention and the investigation of the stereodynamic

---

[i] Teratogenic agents interfere with embryonic development and cause congenital malformations (birth defects) in babies.

[ii] Despite its inherent toxicity and stereochemical instability, thalidomide has recently been approved by the US Food and Drug Administration under a specially restricted distribution program for cancer therapy and for the treatment of the painful disfiguring skin sores associated with leprosy.

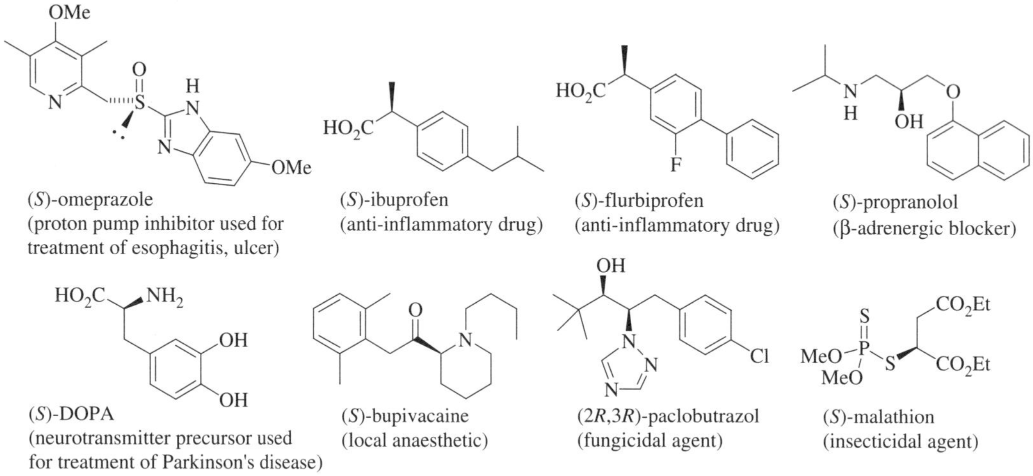

Figure 1.1 Structures of chiral pharmaceuticals and agrochemicals.

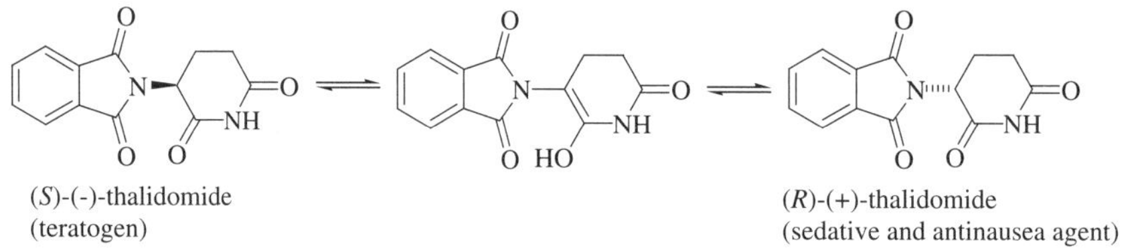

Scheme 1.1 Interconversion of the enantiomers of thalidomide.

properties of chiral molecules has become an integral part of modern drug development. For example, a variety of 2-arylpropionic acids including ibuprofen (Advil), naproxen (Aleve), keto-profen (Oruvail), and flurbiprofen (Ansaid) has found widespread use as pain relievers and nonsteroidal anti-inflammatory drugs (NSAIDs). The anti-inflammatory activity of these profens resides primarily with the (*S*)-enantiomer.[iii] The enantiomers of flurbiprofen possess different pharmacokinetic properties and show substantial racemization under physiological conditions. Although (*S*)-naproxen is the only profen that was originally marketed in enantiopure form, *in vivo* interconversion of the enantiomers of NSAIDs is an important issue in preclinical pharmacological and toxicological studies.[28] Another noteworthy example that underscores the significance of racemization of chiral drugs is the potent gastric acid secretion inhibitor omeprazole (Prilosec) which has been used in its racemic form for the treatment of esophagitis and ulcer.[29] The active form of this prodrug is an achiral sulfenamide derivative, and one could conclude that the presence of a chiral center in omeprazole does not affect its pharmacological activity. Nevertheless, the enantiomers of omeprazole have strikingly different pharmacokinetic profiles. The (*S*)-enantiomer of omeprazole affords higher bioavailability in humans because the (*R*)-form is more readily metabolized in the liver. To enhance the potency of this blockbuster drug, the pharmaceutical industry has launched esomeprazole (Nexium), the pure (*S*)-enantiomer of omeprazole.

The unique structure and mode of action of the antibiotic glycopeptide vancomycin (Vancocin) has fascinated synthetic and medicinal chemists alike. The antibacterial activity of vancomycin, *i.e.*,

---

[iii] Some (*S*)-profens have distinct ulcerogenic properties that are increased in the presence of the (*R*)-enantiomer.

its ability to inhibit cell wall biosynthesis, stems from its rigid macrocyclic structure which is crucial for effective recognition and binding to the terminal D-alanine-D-alanine sequence of bacterial peptides. The cup-shaped structure of vancomycin is a consequence of the chirality of the amino acids in conjunction with the chiral axis of the actinoidinic acid moiety consisting of aryl rings *A* and *B* and the planar chirality of aryl ethers *C* and *E*, Scheme 1.2. The stereodynamics of these chiral elements have played a pivotal role in the total synthesis of the vancomycin aglycon (the part of the glycopeptide lacking carbohydrate units).[30–33] An intriguing example is Evans' synthesis of vancomycin from a bicyclic precursor bearing an unnatural (*M*)-actinoidinic acid moiety. Evans and coworkers recognized that the axial chirality of the atropisomeric actinoidinic acid unit is controlled by the global aglycon architecture, rather than by individual proximate stereogenic elements. They also realized that rotation about the central *C–C* bond between aryl rings *A* and *B* proceeds at ambient temperature and favors formation of the (*P*)-conformer with high selectivity. In fact, thermal equilibration at 55 °C results in atropisomerselective interconversion of the two diastereomeric tetrapeptides and establishes the desired axial chirality with 90% diastereomeric excess. This thermodynamically controlled transformation significantly enhances the overall efficiency of the total synthesis of vancomycin, Scheme 1.2. Similarly, Boger and others exploited thermally controlled atropisomerization reactions which have been strategically incorporated into several synthetic routes towards vancomycin and other complex antibiotics including teicoplanin.

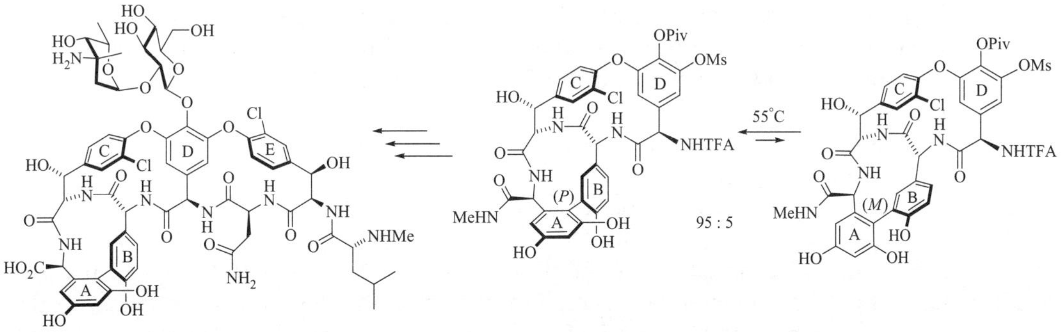

Scheme 1.2   Structure of vancomycin (left), and stereoisomer interconversion favoring the formation of the natural (*P*)-diastereoisomer of a tetrapeptide precursor (right).

The unique stereodynamics of chiral compounds have paved the way to artificial machines and other molecular devices that lie at the interface of chemistry, engineering, physics, and molecular biology. The design of gears, rotors, switches, scissors, brakes, shuttles, turnstiles, and even motors showing unidirectional motion has certainly been inspired by the coordinated movement in biological systems such as muscle fibers, flagella and cilia.[34–36] Feringa and coworkers developed light-driven molecular motors derived from sterically overcrowded chiral alkenes exhibiting thermal bistability and nondestructive read-out.[37,38] The subtle interplay between the chiral center and the inherently helical conformation of this type of molecular motor provides control of the rotation about the carbon–carbon double bond in a series of thermally and photochemically initiated isomerization steps. In other words, both chirality and conformational flexibility are essential for unidirectional motion of the rotor around the stator, Scheme 1.3. The first isomerization step requires irradiation of UV light to a solution of the stable (3′*S*)-(*P*)-*trans*-form of the chiral alkene. This generates a photostationary state favoring 69.4% of the unstable (3′*S*)-(*M*)-*cis*-form that thermally relaxes to 94% of the stable (3′*S*)-(*P*)-*cis*-isomer in the second step. The same concept is exploited for photoisomerization of the (3′*S*)-(*P*)-*cis*-alkene to 48.9% of the (3′*S*)-(*M*)-*trans*-isomer

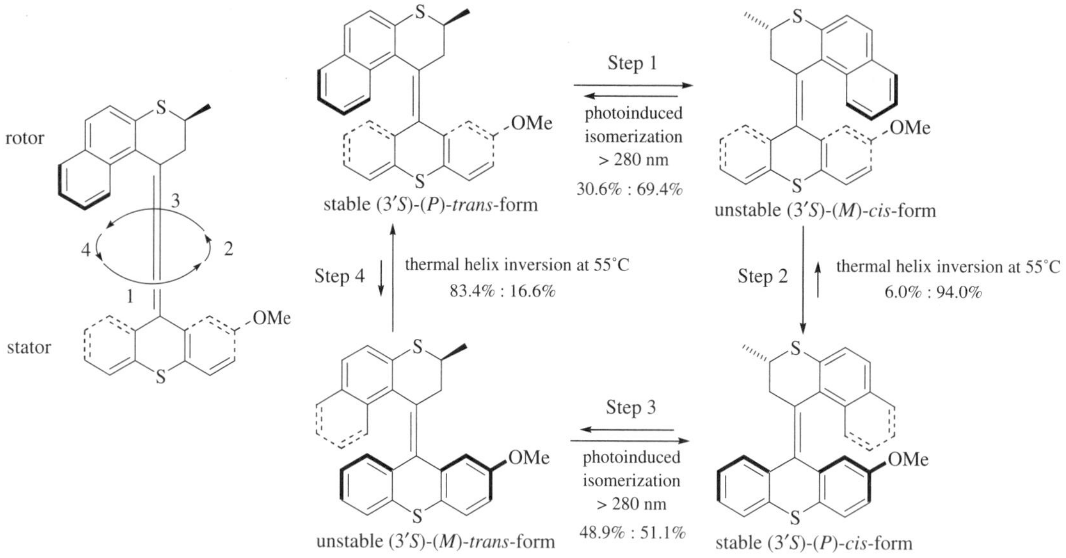

Scheme 1.3   Four isomerization steps of a unidirectional motor.

in the third step which is followed by thermal helix inversion to regenerate 83.4% of the (3′S)-(P)-*trans*-isomer in the final step. Since the interconverting isomers exist as mixtures obeying Boltzmann distributions and reversible reaction kinetics, this molecular motor is not rotating in an exclusively monodirectional sense. However, the discrete and synergetic photochemical and thermal isomerization reactions of the diastereomeric mixtures result in an overall clockwise motion observed from the stator.

Since the discovery of the ubiquity of chirality by Pasteur, stereochemistry has undoubtedly emerged as one of the most important and fascinating areas within the chemical sciences. Stereochemistry embraces a broad variety of closely intertwined static and dynamic aspects that are all related to the three-dimensional structure of molecules. While static stereochemistry deals with the spatial arrangement of atoms in molecules and the corresponding chemical and physical properties, dynamic stereochemistry emphasizes structural change and comprises asymmetric reactions as well as interconversion of configurational and conformational isomers. The few examples outlined above highlight both the significance of chirality and the fundamental role that dynamic stereochemistry plays in modern chemistry, spanning multiple disciplines from asymmetric synthesis and drug discovery to material sciences. The principles and applications of stereodynamic chemistry of chiral compounds are discussed in the chapters following.

REFERENCES

1. Pasteur, L. *Ann. Chim. Physique* **1848**, *24*, 442-459.
2. Helmchen, G.; Hoffmann, R. W.; Mulzer, J.; Schaumann, E. *Stereoselective Synthesis* in *Methods of Organic Chemistry*, Houben-Weyl, Vol. 21a-21f, 4th ed., Thieme, Stuttgart, 1995.
3. Gawley, R. E.; Aubé, J. *Principles of Asymmetric Synthesis*, Tetrahedron Organic Chemistry Series, Elsevier, New York, 1996.
4. Ho, T.-L. *Stereoselectivity in Synthesis*, Wiley-VCH, New York, 1999.
5. Lin, G.-Q.; Li, Y.-M.; Chan, A. S. C. *Principles and Applications of Asymmetric Synthesis*, Wiley-VCH, New York, 2001.
6. Sharpless, K. B. *Angew. Chem., Int. Ed.* **2002**, *41*, 2024-2032.

7. Song, C. E.; Lee, S.-G. *Chem. Rev.* **2002**, *102*, 3495-3524.
8. Liu, M.; Sibi, M. P. *Tetrahedron* **2002**, *58*, 7991-8035.
9. Jonathan, M. J. W. *Catalysis in Asymmetric Synthesis*, Sheffield Academic Press, Sheffield, 1999.
10. Ojima, I. *Catalytic Asymmetric Synthesis*, 2nd ed., Wiley-VCH, New York, 2000.
11. Noyori, R. *Angew. Chem., Int. Ed.* **2002**, *41*, 2008-2022.
12. Shibasaki, M.; Yoshikawa, N. *Chem. Rev.* **2002**, *102*, 2187-2209.
13. Inanaga, J.; Furuno, H.; Hayano, T. *Chem. Rev.* **2002**, *102*, 2211-2225.
14. Mikami, K.; Terada, M.; Matsuzawa, H. *Angew. Chem., Int. Ed.* **2002**, *41*, 3554-3571.
15. Ding, K.; Shii, A.; Mikami, K. *Angew. Chem., Int. Ed.* **1999**, *38*, 497-501.
16. Guo, J.; Wu, J.; Siuzdak, G.; Finn, M. G. *Angew. Chem., Int. Ed.* **1999**, *38*, 1755-1758.
17. Evans, M. A.; Morken, J. P. *J. Am. Chem. Soc.* **2002**, *124*, 9020-9021.
18. Markert, C.; Pfaltz, A. *Angew. Chem., Int. Ed.* **2004**, *43*, 2498-2500.
19. Reetz, M. T.; Kuhling, K. M.; Deege, A.; Hinrichs, H.; Belder, D. *Angew. Chem., Int. Ed.* **2000**, *39*, 3891-3893.
20. Eliel, E. L.; Wilen, S. H. *Stereochemistry of Organic Compounds*, John Wiley & Sons, New York, 1994, pp. 214-274.
21. Pu, L. *Chem. Rev.* **2004**, *104*, 1687-1716.
22. Mei, X.; Wolf, C. *Chem. Commun.* **2004**, 2078-2079.
23. Zhao, J.; Fyles, T. M.; James, T. D. *Angew. Chem., Int. Ed.* **2004**, *43*, 3461-3464.
24. Eelkema, R.; van Delden, R. A.; Feringa, B. L. *Angew. Chem., Int. Ed.* **2004**, *43*, 5013-5016.
25. Mei, X.; Wolf, C. *J. Am. Chem. Soc.* **2004**, *126*, 14736-14737.
26. Tumambac, G. E.; Wolf, C. *Org. Lett.* **2005**, *7*, 4045-4048.
27. Folmer-Andersen, J. F.; Lynch, V. M.; Anslyn, E. V. *J. Am. Chem. Soc.* **2005**, *127*, 7986-7987.
28. Leipold, D. D.; Kantoci, D.; Murray Jr., E. D.; Quiggle, D. D.; Wechter, W. J. *Chirality* **2004**, *16*, 379-387.
29. Lindberg, P.; Braendstroem, A.; Wallmark, B.; Mattsson, H.; Rikner, L.; Hoffman, K.-J. *Med. Res. Rev.* **1990**, *10*, 1-54.
30. Evans, D. A.; Wood, M. R.; Trotter, B. W.; Richardson, T. I.; Barrow, J. C.; Katz, J. L. *Angew. Chem., Int. Ed.* **1998**, *37*, 2700-2704.
31. Evans, D. A.; Dinsmore, C. J.; Watson, P. S.; Wood, M. R.; Richardson, T. I.; Trotter, B. W.; Katz, J. L. *Angew. Chem., Int. Ed.* **1998**, *37*, 2704-3708.
32. Boger, D. L.; Miyazaki, S.; Kim, S. H.; Wu, J. H.; Loiseleur, O.; Castle, S. L. *J. Am. Chem. Soc.* **1999**, *121*, 3226-3227.
33. Boger, D. L.; Miyazaki, S.; Kim, S. H.; Wu, J. H.; Castle, S. L.; Loiseleur, O.; Jin, Q. *J. Am. Chem. Soc.* **1999**, *121*, 10004-10011.
34. Kelly, T. R.; DeSilva, H.; Silva, R. A. *Nature* **1999**, *401*, 150-152.
35. Koumura, N.; Zijlstra, R. W. J.; van Delden, R. A.; Harada, N.; Feringa, B. L. *Nature* **1999**, *401*, 152-155.
36. Leigh, D. A.; Wong, J. K. Y.; Dehez, F.; Zerbetto, F. *Nature* **2003**, *424*, 174-179.
37. Koumura, N.; Geertsema, E. M.; Meetsma, A.; Feringa, B. L. *J. Am. Chem. Soc.* **2000**, *122*, 12005-12006.
38. Van Delden, R. A.; ter Wiel, M. K. J.; de Jong, H.; Meetsma, A.; Feringa, B. L. *Org. Biomol. Chem.* **2004**, *2*, 1531-1541.

Principles of Chirality and Dynamic Stereochemistry

Stereochemistry has evolved into a multi-faceted discipline, and numerous concepts and descriptors have been introduced since the discovery and analysis of chiral organic compounds by van't Hoff, Le Bel and Pasteur in the nineteenth century. Unfortunately, the stereochemical terminology used in the literature is not always well defined, and in some cases it can be ambiguous. The principles of stereoisomerism and chirality and the relationship between reactivity and stereodynamics of chiral compounds are discussed in this chapter and complemented with brief definitions in the glossary.

2.1 STEREOCHEMISTRY OF CHIRAL COMPOUNDS

Stereoisomeric compounds exhibit the same constitution, *i.e.*, molecular formula and connectivity of atoms, but have a different spatial arrangement. Stereoisomers are classified into enantiomers and diastereoisomers. Chiral molecules that have the relationship of mirror images are called enantiomers. All other stereoisomers are diastereoisomers, Figure 2.1. Enantiomers have the same chemical and physical properties when present in an achiral environment. But they show different chiroptical behavior and cause a clockwise (dextrorotatory (+)-form) or counterclockwise (levorotatory (−)-form) rotation of the plane of linearly polarized light by equal amounts.[i] Enantiomers may have distinguishable properties in a chiral environment if they experience diastereomeric interactions. For example, the enantiomers of chiral drugs such as omeprazole, ibuprofen and DOPA exhibit different pharmacological and pharmacokinetic activities because they interact with enzymes and receptors consisting of amino acids and other chiral biomolecules. By contrast, diastereoisomers usually have different chemical and physical properties in both achiral and chiral environments. Many chiral compounds possess a tetrahedral carbon atom bearing four different substituents, Figure 2.2. The presence of one chiral carbon center in a molecule gives rise to a pair of two enantiomers. Other important stereogenic centers consist of a nitrogen, phosphorus, boron, silicon or sulfur atom carrying nonidentical substituents, Figure 2.3.

Enantiomers and diastereoisomers can be chiral due to the presence of an asymmetric center, but this is not always the case. A compound is chiral if it is not superimposable on its mirror image. This unequivocal definition provides a necessary and sufficient condition for chirality. Chiral

---

[i] The absolute configuration of a chiral compound, and the assignment of its three-dimensional structure based on CIP rules (descriptors *R,S*) must not be confused with its ability to rotate the plane of monochromatic linearly polarized light in a clockwise or counterclockwise direction (descriptors +,−).

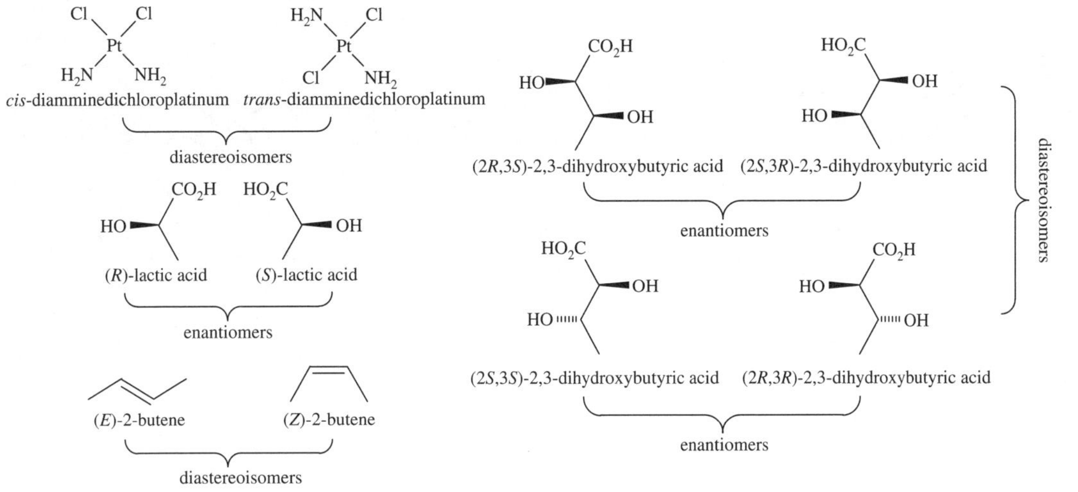

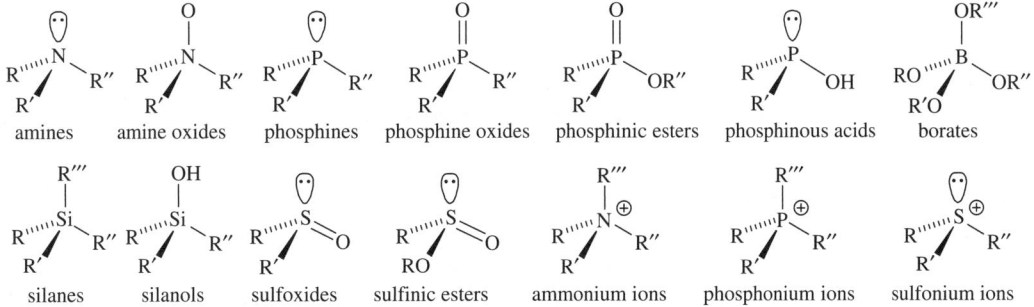

Figure 2.1 Examples of enantiomers and diastereoisomers.

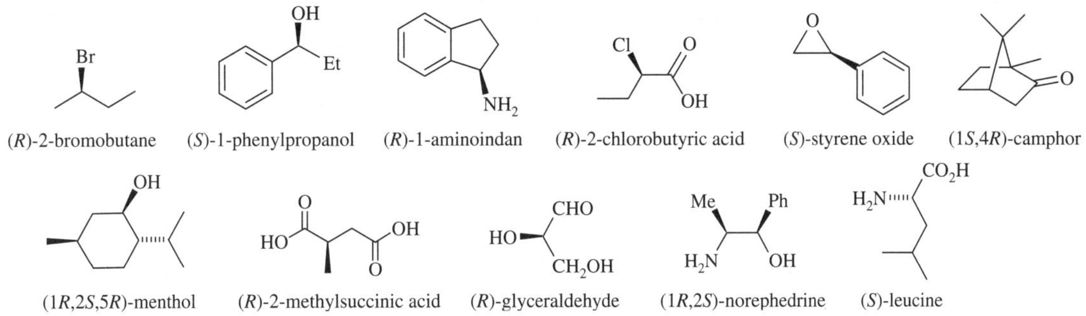

(Figure 2.1 content)

cis-diamminedichloroplatinum   *trans*-diamminedichloroplatinum

diastereoisomers

(R)-lactic acid   (S)-lactic acid

enantiomers

(E)-2-butene   (Z)-2-butene

diastereoisomers

(2R,3S)-2,3-dihydroxybutyric acid   (2S,3R)-2,3-dihydroxybutyric acid

enantiomers

(2S,3S)-2,3-dihydroxybutyric acid   (2R,3R)-2,3-dihydroxybutyric acid

enantiomers

diastereoisomers

Figure 2.2 Chiral compounds with asymmetric carbon atoms.

(R)-2-bromobutane   (S)-1-phenylpropanol   (R)-1-aminoindan   (R)-2-chlorobutyric acid   (S)-styrene oxide   (1S,4R)-camphor

(1R,2S,5R)-menthol   (R)-2-methylsuccinic acid   (R)-glyceraldehyde   (1R,2S)-norephedrine   (S)-leucine

Figure 2.3 Chiral compounds with stereogenic centers others than carbon atoms.

amines   amine oxides   phosphines   phosphine oxides   phosphinic esters   phosphinous acids   borates

silanes   silanols   sulfoxides   sulfinic esters   ammonium ions   phosphonium ions   sulfonium ions

molecules may possess a rotation axis, but no symmetry plane, rotation-reflection axis and inversion center. This criterion may be fulfilled by the presence of an asymmetric center or other elements of chirality. Many organic compounds lack asymmetric atoms but possess axial, planar or helical chirality. For example, substituted allenes, biphenyls, spiranes, and alkylidenecycloalkanes have a chiral axis and exist in the form of two enantiomers. Certain ansa compounds,

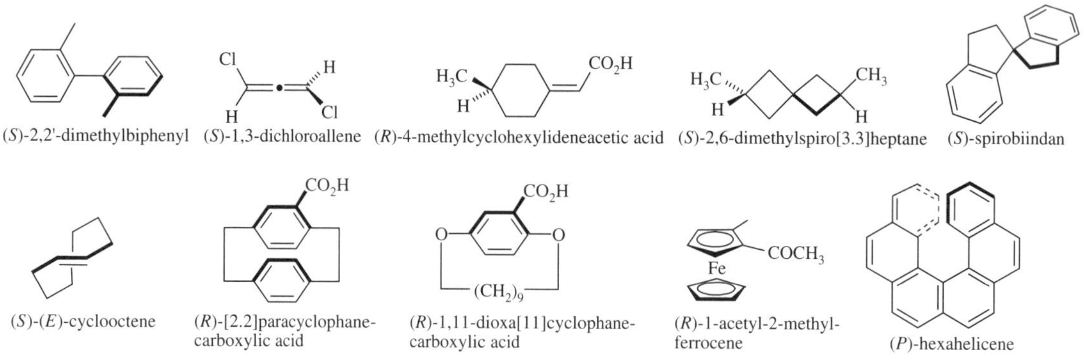

Figure 2.4 Molecules with a chiral axis, plane or helix.

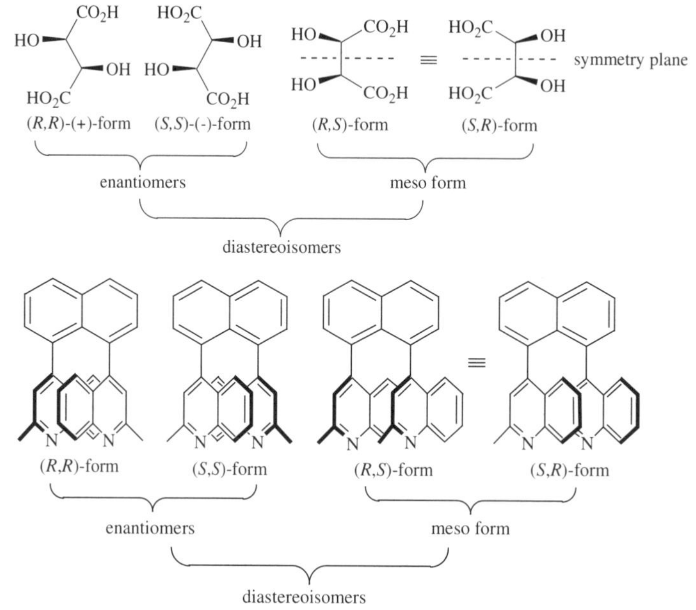

Figure 2.5 The stereoisomers of tartaric acid (top) and 1,8-bis(2'-methyl-4'-quinolyl)naphthalene (bottom).

paracyclophanes, *trans*-cycloalkenes with a twisted double bond, and metallocene complexes display a chiral plane, and helicenes afford a chiral helical structure, Figure 2.4.

Chiral compounds may have one or more than one stereogenic center, axis, plane or helix. In most cases, each element of chirality gives rise to two distinct stereoisomers, and one can expect 2^n stereoisomers for n chiral elements. However, the existence of a symmetry plane, rotation-reflection axis or inversion center results in degeneracies. A compound that possesses at least one of these symmetry elements is achiral because it is superimposable on its mirror image. Tartaric acid has two chiral centers and should exhibit two diastereoisomeric pairs of enantiomers. In fact, it exists only in the form of two enantiomeric (*R,R*)- and (*S,S*)-isomers and a third diastereoisomeric form that has a symmetry plane and is achiral because its mirror images are superimposable and therefore identical, Figure 2.5. Similarly, 1,8-bis(2'-methyl-4'-quinolyl)naphthalene has two chiral axes but affords three stereoisomers: two enantiomers and one meso form. These examples clearly demonstrate that the presence of a stereogenic element in a molecule such as a tetrahedral carbon atom bearing four different substituents or a chiral axis does not necessarily render the molecule chiral.

A discussion of the stereochemistry of chiral compounds requires a clear understanding of the relevant nomenclature. The Cahn–Ingold–Prelog (CIP) rules are used to give an unambiguous description of the absolute configuration of enantiomers and diastereoisomers.[1,2] Because of its importance, the CIP convention and the descriptors *R*, *S*, *M*, and *P* are explained below in some detail, whereas definitions of other stereochemical terms can be found in the glossary. The descriptors *R*, *S*, *M*, and *P* are applicable to all kinds of chiral elements. First, the priority of the atoms attached to the chiral element is assigned based on the following sequence rules:

1. Atoms with a higher atomic number have a higher rank than atoms with a lower atomic number. Note that the substituent with the lowest priority at a tetrahedral carbon atom may be a hydrogen atom or a lone electron pair at a three-coordinate chiral heteroatom such as nitrogen, phosphorus or sulfur. Isotopes with higher atomic mass number precede isotopes with lower mass number, *e.g.*, $D > H$ and $^{13}C > {}^{12}C$.

2. When two or more substituents are the same element, one has to compare atoms in the next sphere until the priority of all groups has been established. It is important that all substituents in a given sphere must be evaluated before one proceeds further away from the core of the chiral element to the next array of atoms. Comparison of a dimethyl acetal moiety with an ethoxy group shows that the former has a higher priority because it carries two oxygens, in contrast to the ether moiety which bears only one oxygen atom. In other words, all atoms in one sphere must be fully analyzed before one continues to the next sphere. In the case of equal priority of a given sphere, one always proceeds by following the branch of the highest priority until a rank can finally be assigned, Figure 2.6.

3. The rank of substituents carrying multiple bonds is established based on the principle of ligand complementation. Multiple bonds are complemented by phantom atoms through multiple representation of atoms that are doubly or triply bonded. An organometallic π-complex is treated the same way. A metal carrying an alkene or alkine ligand is hypothetically replaced by one (two) metal(s) attached to each end of the ligand, *e.g.*, a palladium π-complex bearing an olefinic group –CH=CHR is represented by –C(Pd)H–C(Pd)HR. The rule of ligand complementation is also applicable to polydentate and cyclic ligands. A cyclic structure is disconnected and complemented with a phantom atom at the bonds where it reduplicates itself. For example, the priority of the substituted pyrrolidine shown in Figure 2.7 is assigned by cutting the ring into two hypothetical branches with peripheral phantom atoms in addition to the real branches. Disconnection at *A* gives –CH₂NHCH₂CH₂(C) having the highest priority because –CH₂NHCH₂CH₃, the group that is assigned second priority, carries a peripheral hydrogen which has a lower priority than a phantom carbon. Disconnection at *B* results in the formation of –CH₂CH₂NHCH₂(C), which has the lowest rank. This case thus also illustrates that, for a given element, a real atom has precedence over a phantom atom: –CH₂CH₂NHCH₂CH₃ > –CH₂CH₂NHCH₂(C).

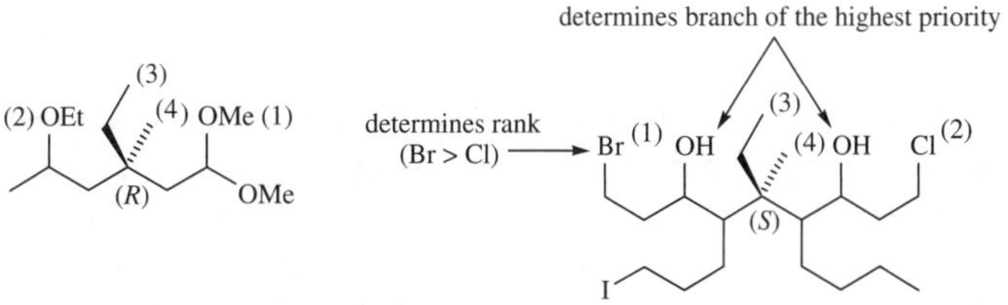

Figure 2.6 Comparison of CIP spheres and determination of the branch of the highest priority.

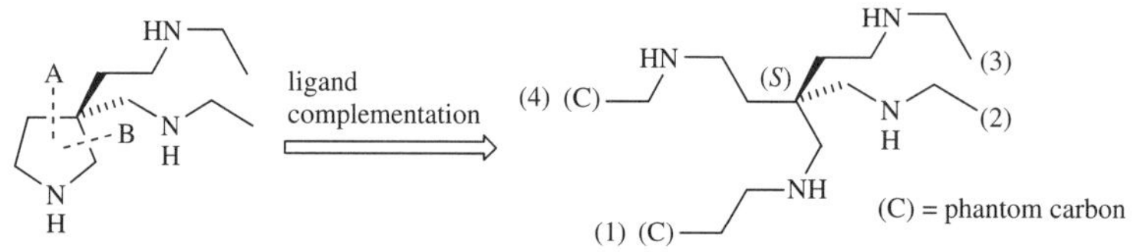

Figure 2.7 Ligand complementation in cyclic compounds.

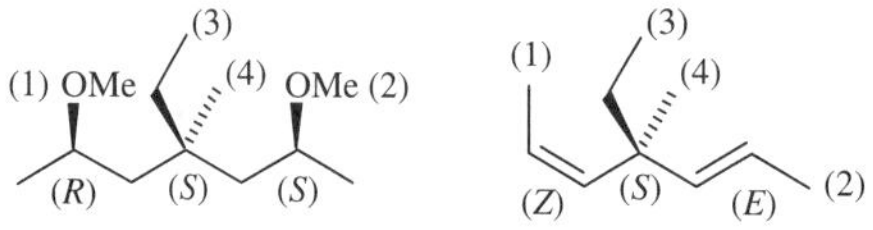

Figure 2.8 CIP rules for stereoisomeric substituents.

4. If ligands possess a different stereochemical environment and can not be distinguished based on the above, the following rule applies: *cis*-configuration precedes *trans*-configuration, $Z > E$, $R > S$, $M > P$, Figure 2.8.

The CIP rules described above give the following priority of common functional groups:

$-I$ > $-Br$ > $-Cl$ > $-SO_3H$ > $-SO_2R$ > $-SR$ > $-SH$ > $-P(O)R_2$ > $-PR_2$ > $-F$ > $-OR$ > $-NO_2$ > $-NR_2$ > $-NH_2$ > $-CO_2R$ > $-CO_2H$ > $-C(O)R$ > $-CHO$ > $-CH(OR)_2$ > $-CH(OH)_2$ > $-CH_2OR$ > $-CH_2OH$ > $-CN$ > $-CH_2NH_2$ > *o*-tolyl > *m*-tolyl > *p*-tolyl > $-C_6H_5$ > $-C\equiv CH$ > *t*-Bu > $-CH(C_6H_{13})_2$ > cyclohexyl > vinyl > *i*-Pr > benzyl > allyl > *n*-alkyl > $-H$.

Once the priority of the substituents or groups around the chiral center is established, the molecule is viewed in such a way that the moiety with the lowest priority is placed behind the central atom. The three remaining substituents or groups are thus oriented towards the viewer and afford a tripodal arrangement. If the sense of direction following the established priority of these three substituents describes a clockwise rotation, (*R*)-configuration (for Latin *rectus*, right) is assigned. If the rotation is counterclockwise, the chiral center has (*S*)-configuration (for Latin *sinister*, left), Figure 2.9.

This nomenclature can also be applied to compounds possessing a chiral axis or plane. In both cases the priority of substituents must be assigned with the help of an additional CIP rule, which specifies that the groups on an arbitrarily chosen side of a chiral axis or plane always precede substituents on the other side of the stereogenic element. It is helpful to view the chiral axis of biphenyls, allenes and spiro compounds as a stretched tetrahedron. Because of the inherently lower symmetry of a stretched tetrahedron compared to a regular tetrahedron, it does not require the presence of four different substituents to constitute a chiral axis. For example, 2,2′-dimethyl-biphenyl is chiral. The symmetric substitution pattern gives rise to a C_2-symmetric structure, but incorporation of the same substituents to a tetrahedral carbon generates an achiral compound with a symmetry plane. To assign the CIP descriptor to an axially chiral compound, one can choose either side of the chiral axis to determine the first two priorities. This side always has precedence over the opposite side. Then the remaining ranks are assigned on the side of lower priority, and the clockwise (R_a) or counterclockwise (S_a) sense of rotation is established. Some classes of compounds, including (*E*)-cycloalkenes, cyclophanes and metallocenes, display a planar aryl or alkene group that may have one or several substituents residing outside that plane. A substitution pattern that destroys a perpendicular symmetry plane or an inversion center establishes planar chirality.

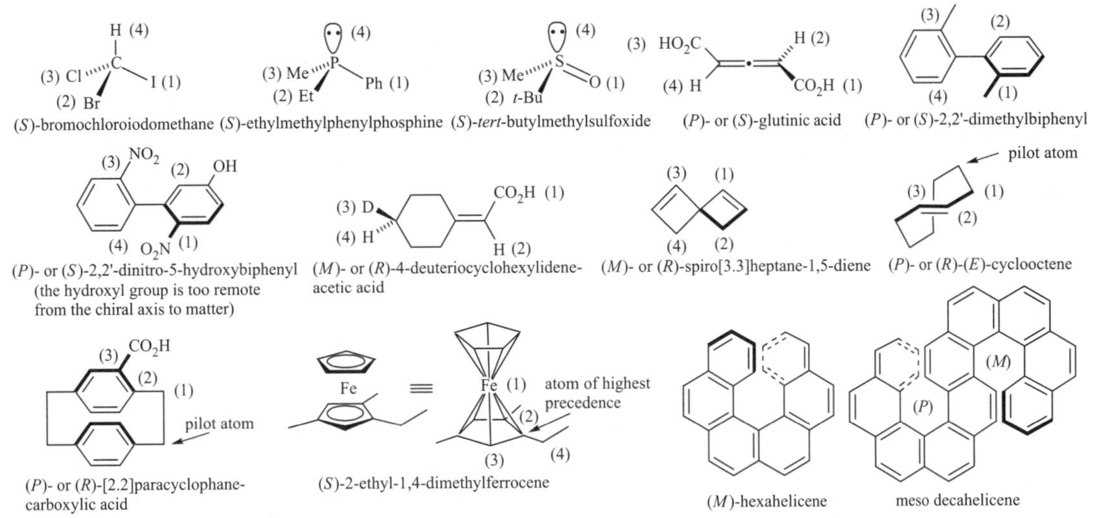

Figure 2.9 Description of axial, planar and helical chirality based on CIP rules. Chiral axes of symmetrically substituted biaryls, allenes and spiro compounds are arbitrarily viewed from the right-hand side to assign CIP priorities.

To assign CIP descriptors for molecules with a chiral plane, one views the plane from the so-called pilot atom which is the out-of-plane atom closest to the stereogenic element. The arrangement of the next three subsequent atoms located in the chiral plane determined according to the CIP rules will either describe a clockwise (R_p) or a counterclockwise (S_p) rotation.[ii] Although metallocenes exhibit planar chirality, they are conventionally treated as molecules with a chiral center. First, bonds between the π-donor and the metal atom are complemented as described above and the atom of the highest priority is determined. The asymmetric environment of this tetrahedral atom is then analyzed using the CIP rules to assign the absolute configuration, Figure 2.9.

As mentioned above, molecules with a helical shape may also be chiral. The sense of helicity is described as right-handed (P) if it resembles a screw that rotates clockwise away from the viewer, and left-handed (M) if the rotation is counterclockwise. While hexahelicene exists in the form of two enantiomers, decahelicene has two stereogenic elements, *i.e.*, two helical turns, which gives rise to a pair of (P,P)- and (M,M)-enantiomers and a meso (M,P)-isomer having an inversion center, Figure 2.9. The sense of chirality in molecules with a stereogenic axis or plane may also be characterized using descriptors M and P. For axially chiral compounds, only the substituents of the highest CIP rank on each side of the stereogenic axis (denoted 1 and 3 in Figure 2.9) are considered. After determination of these two substituents, one views the molecule along the chiral axis from an arbitrarily chosen side. The helical descriptors are then assigned through rotation of the front substituent in direction to the substituent located at the rear of the axis. The descriptor P denotes a clockwise rotation, whereas M is assigned in the case of a counterclockwise rotation. In order to assign helical descriptors to a chiral plane, one looks from the pilot atom at the subsequent atoms using the CIP convention. Again, the descriptor P refers to a clockwise rotation and M to a counterclockwise movement. Application of the two conventions to axially chiral compounds reveals that descriptors R_a and M, as well as S_a and P, correspond to each other. In the case of planar chirality, the descriptors R_p and P, and S_p and M, respectively, are synonymous.

[ii] The use of suffixes *a* and *p* to indicate axial and planar chirality is optional.

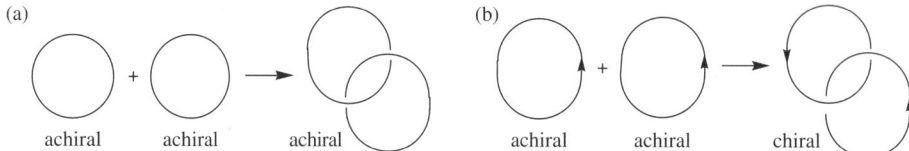

Figure 2.10 Achiral (a) and chiral (b) catenanes devoid of classical elements of chirality.

In addition to classical or Euclidian elements of chirality, introduction of directionality to mechanically interlocked molecules can generate so-called topological chirality. Examples of nonclassical chirality can be encountered in catenanes and rotaxanes. Catenanes contain two or more interlocked rings, and rotaxanes consist of a molecule that is threaded through at least one cyclic molecule which is trapped on the rod by bulky terminal groups. Intertwining of two achiral rings often affords an achiral catenane, but a chiral assembly existing in the form of two mirror images is obtained if the two achiral rings possess directionality, Figure 2.10.[3,4] Similarly, incorporation of directionality into both axle and wheel of a rotaxane generates topological chirality. In this case, the wheel attains either a clockwise or a counterclockwise orientation with respect to the axle, resulting in cycloenantiomerism, *vide infra*.[5] Topologically chiral compounds may lack classical elements of chirality but still exist as a pair of enantiomers.

The assignment of the absolute configuration of topologically chiral molecules has been under debate for some time, and different procedures utilizing CIP rules have been proposed.[6,7] The chirality of catenanes and rotaxanes composed of oriented ring and thread components can be described through a combination of axial and polar vectors generating a right-handed or a left-handed screw. The directionality of the interlocked units is assigned based on CIP priority rules. In both ring and thread of a rotaxane, the direction from the atom of the highest priority to the atom of the second highest priority following the shortest distance is determined to define (a) an axial vector depicting the orientation of the thread and (b) a polar vector describing the orientation of the ring. If the vector combination represents the motion of a right-handed screw, (*R*)-configuration is assigned, whereas the (*S*)-enantiomer resembles a left-handed screw, Figure 2.11. [2]Rotaxanes consisting of only one oriented component are of course achiral. However, the presence of at least one directional ring on a [3]rotaxane exhibiting a nondirectional thread gives rise to topological chirality, Figure 2.12. The concept used for the assignment of the absolute configuration outlined for chiral [2]rotaxanes can also be applied to cyclostereoisomeric [3]rotaxanes that are chiral only by virtue of directionality of at least one ring component. In this case, one views the oriented wheel which absolute configuration is under consideration from the direction of the other wheel. The part of the thread bearing the second wheel has a higher priority than the part carrying the wheel under consideration. For the determination of the configuration at ring *B* of the [3]rotaxane shown in Figure 2.12, the side of the nondirectional axle bearing ring *A* receives a higher priority to artificially assign an axial vector. Combination of this axial vector and the polar vector of ring *B* then establishes (*R*)-configuration. Applying the same method to determine the chirality at ring *A* confirms that the [3]rotaxane possesses (*R,R*)-configuration.

To apply the same concept to catenanes, one describes the orientation of one ring with a polar vector while the other is viewed as a local thread and assigned an axial vector. The ring used to determine the direction of the axial and polar vector, respectively, can be chosen arbitrarily. For example, assignment of the axial vector based on the local directionality of ring *A* and the polar vector based on the orientation of ring *B* affords (*R*)-configuration for the [2]catenane shown in Figure 2.13.

The concept of cyclostereoisomerism resulting from ring directionality and the distribution pattern of chiral but otherwise identical building blocks in a cyclic molecule, for example (*R*)- and (*S*)-enantiomers of alanine in cyclohexaalanine, was first introduced by Prelog and Gerlach.[8,9] The

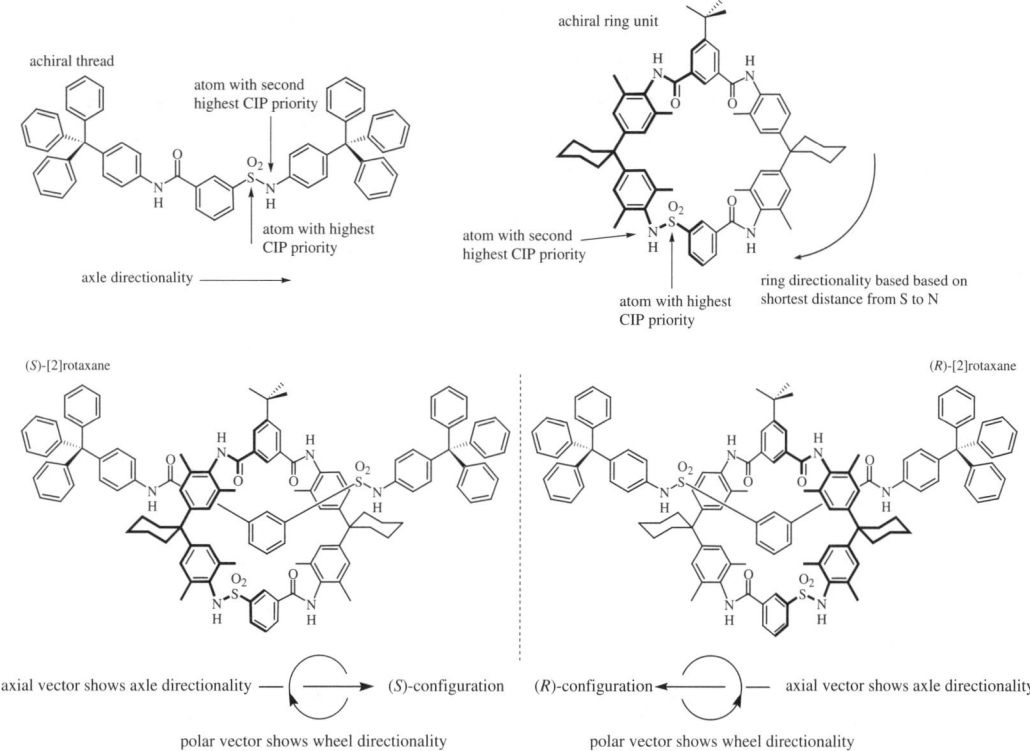

Figure 2.11 Assignment of the absolute configuration of the enantiomers of a [2]rotaxane.

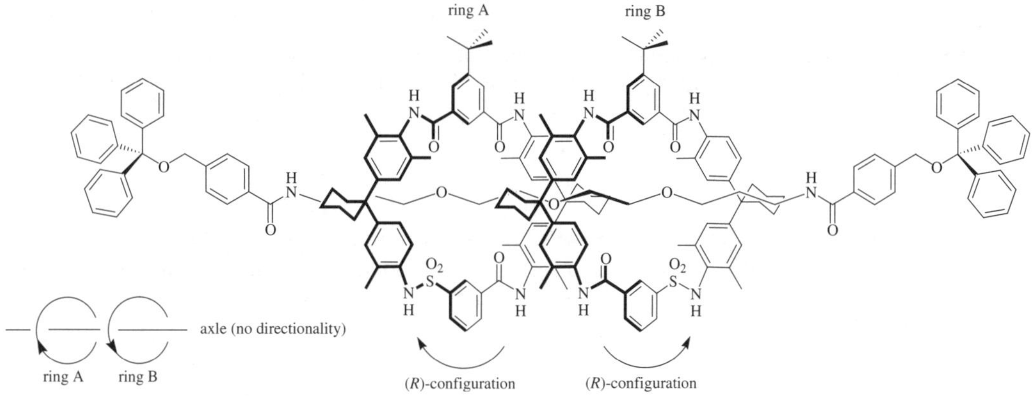

Figure 2.12 Structure of an (*R*,*R*)-[3]rotaxane.

amino acid sequence and connectivity of stereogenic centers exhibiting both (*R*)- and (*S*)-configuration in cyclopeptides give rise to cycloenantiomers and cyclodiastereoisomers, Figure 2.14. The presence of ring directionality in cyclopeptides consisting of both enantiomers of one amino acid does not change the overall number of possible stereoisomers, *i.e.*, a cyclic peptide exists in as many stereoisomers as the open chain form. Mislow therefore suggested that the term cyclostereoisomerism should be restricted to cases in which ring directionality generates additional stereoisomers, and demonstrated that this can be achieved with hexaisopropylbenzene derivatives. For instance,

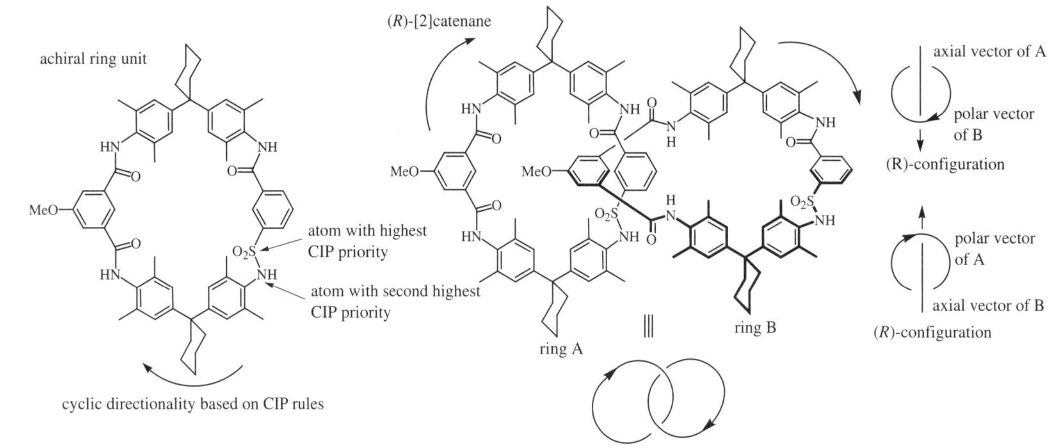

Figure 2.13   Assignment of the absolute configuration of a [2]catenane based on CIP rules.

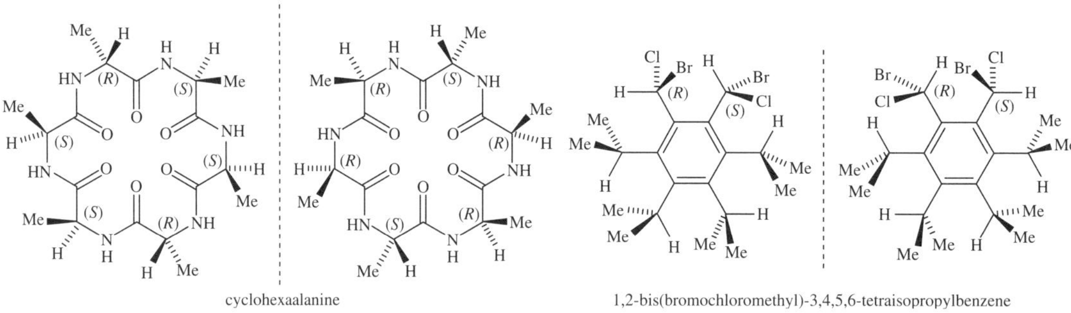

Figure 2.14   Structures of cycloenantiomeric cyclohexaalanine (left) and 1,2-bis(bromochloromethyl)-3,4,5,6-tetraisopropylbenzene (right). Note that the two molecules of cyclohexaalanine differ on account of the directional sense of the peptide bonds, *i.e.*, $R \rightarrow S \rightarrow S \rightarrow R \rightarrow S \rightarrow S$ versus $S \leftarrow R \leftarrow R \leftarrow S \leftarrow R \leftarrow R$.

the directionality of 1,2-bis(bromochloromethyl)-3,4,5,6-tetraisopropylbenzene is due to static gearing between the aryl substituents, which effectively increases the number of possible stereo-isomers, see Chapter 8.7.[10–12] The same situation arises with cyclostereoisomeric rotaxanes. Removal of directionality in either the axle or the wheel in the chiral rotaxane shown in Figure 2.11 would eliminate topological chirality and thus reduce the number of stereoisomers; as a result, one would observe only a single achiral [2]rotaxane instead of two enantiomers.

By analogy with topologically chiral rotaxanes and catenanes consisting of interlocked achiral components, metal complexes containing two or more chelating ligands can display helical chirality, even if all the individual components are achiral. For example, tris(ethylenediamine)cobalt(III), $[Co(en)_3]^{3+}$, is a chiral octahedral complex that exists in the form of two enantiomers. The assignment of the absolute configuration of octahedral tris-chelate complexes is determined by the sense of the helical coordination sphere generated by the ligands. The view along one of the three-fold rotation axes of the octahedral complex reveals a sense of clockwise rotation similar to a right-handed screw in a Δ-complex, whereas the enantiomeric Λ-complex resembles a left-handed screw with a counterclockwise helical sense, Figure 2.15.

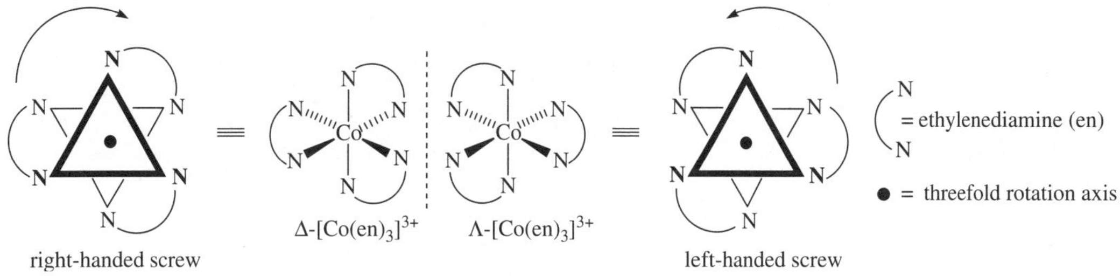

Figure 2.15 Assignment of the absolute configuration of chiral metal complexes.

2.2 DYNAMIC STEREOCHEMISTRY OF CYCLIC AND ACYCLIC CHIRAL COMPOUNDS

Dynamic stereochemistry deals with reversible and irreversible changes of the three-dimensional structure of molecules, for example during thermal interconversion of conformational isomers, photochemical isomerization, and acid- or base-catalyzed racemization. Stereodynamic properties of chiral compounds are critical to the success of asymmetric synthesis including dynamic kinetic resolution and chirality transfer processes, and they have been exploited in a wide range of applications involving chiral amplification with relays and switches or correlated motion in propellers, gears and motors. The diversity and complexity of dynamic stereochemistry becomes evident through a discussion of conformational and configurational isomerism.

Stereoisomers can be classified into configurational and conformational isomers, which are sometimes distinguished based on their stereochemical stability. Interconversion of configurational isomers often requires bond breaking, whereas conformational isomers usually (but not exclusively) isomerize without bond breaking. Accordingly, a relatively high energy barrier to racemization or diastereomerization is commonly associated with configurational isomerism. The thermodynamic stability and bond breaking criteria stress the physical organic properties of stereoisomers: the isolable enantiomers of bromochlorofluoroiodomethane, Tröger's base, and butylmethyl-phenylphosphine are considered to be configurational isomers because they are stable to racemization at room temperature. By contrast, the enantiomers of methylethylpropylamine are often viewed as conformational isomers because they rapidly interconvert through nitrogen inversion, Figure 2.16. Similarly, the enantiomers of axially chiral biphenyls can either be classified as configurational or conformational isomers. For instance, 2,2'-di-*tert*-butylbiphenyl has a high energy barrier to racemization which is observed only at temperatures above 150 °C but the enantiomers of 3,3'-di-*tert*-butylbiphenyl can not be isolated at room temperature due to rapid rotation about the chiral axis. Although the isolation and stability criteria are practical, these examples demonstrate that the categorization of structurally related molecules belonging to the same class of compounds into conformational and configurational isomers is somewhat arbitrary. The ambivalence of this definition is even more evident in the case of 2,2'-diiodobiphenyl, which has a rotational energy barrier of 97 kJ/mol.[13] Chromatographic isolation of the enantiomers, which slowly racemize at room temperature, is possible if the process is not too time-consuming.[iii] However, a clear definition of the isolation conditions, such as temperature and time scale, that should be applied to distinguish configurational from conformational isomers does not exist. Biaryls may or may not be stable to racemization at room temperature, and enantioconversion can proceed via rotation about the chiral axis or bond breaking. For example, the enantiomers of

---

[iii] The enantiomers of 2,2'-diiodobiphenyl show on-column racemization during HPLC separation on triacetyl cellulose at ambient temperatures. Pure enantiomers of 2,2'-diiodobiphenyl can be isolated if the chromatographic process is rapid, but racemization occurs upon standing or during HPLC if the separation process is slow.

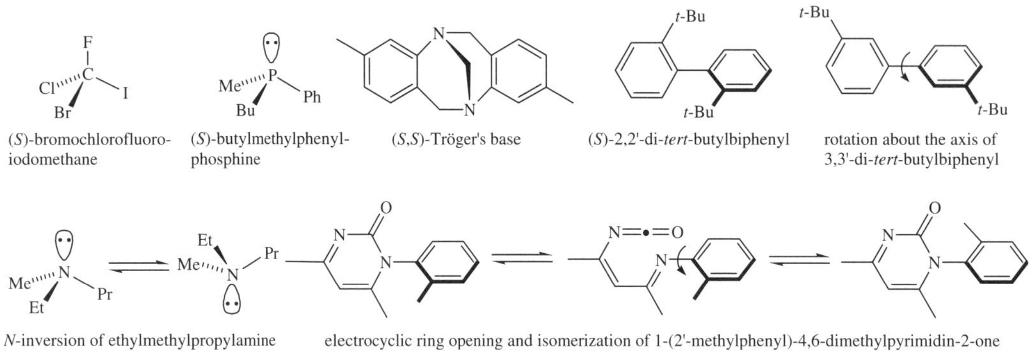

Figure 2.16 Chiral compounds exhibiting conformational or configurational isomerism and racemization mechanisms.

1-arylpyrimidine-2-thiones and their oxygen analogs can be conveniently isolated by chiral chromatography but undergo racemization during [3,3]-electrocyclic rearrangement involving ring opening and bond breaking at elevated temperature.[14,15] Clearly, the stability and bond-breaking criteria are vague and sometimes incompatible.

Alternatively, configurational isomers have been defined as stereoisomers possessing different bond angles while conformers differ in torsion angles. It should be noted that bond and torsion angles can vary in amplitude and sign. For example, the enantiomers of butyl-methylphenylphosphine display bond angles of the same value, albeit with opposite sign. As the relative stability becomes irrelevant, the enantiomers of compounds with a chiral center, including Tröger's base and ethylmethylpropylamine, are considered configurational isomers. All axially chiral biaryls mentioned above afford conformational enantiomers, irrespective of their rotational energy barrier or isomerization mechanism. One shortcoming of this classification is that (*E*)- and (*Z*)-alkenes, or the enantiomers of allenes and (*E*)-cycloalkenes, having a chiral axis or plane may thus be regarded as conformational isomers. This seems somewhat counterintuitive to many chemists. Both definitions are commonly used in the literature to classify stereoisomers. Since the stability criterion and the corresponding isolation conditions are not clearly defined, and are problematic in some cases, the distinction between configurational and conformational isomerism in the present book is based on differences in bond and torsion angles, respectively.

Organic compounds often exist as a mixture of constantly interconverting conformational isomers. In many cases, achiral compounds populate both achiral and racemic chiral conformations at equilibrium. As has been discussed above for tartaric acid, 2,3-diaminobutane has two chiral centers and forms three configurational isomers, *i.e.*, a meso (2*R*,3*S*)-isomer and a pair of enantiomers with (2*R*,3*R*)- and (2*S*,3*S*)-configuration. A comparison of the chemical and physical properties of acyclic configurational stereoisomers requires analysis of the corresponding mixtures of fluxional conformers. A closer look at the three-dimensional structure of meso 2,3-diamino-butane reveals that it adopts three conformational isomers, with torsion angles varying in either magnitude or sign. The meso isomer can populate an achiral antiperiplanar and two enantiomeric synclinal conformers that rapidly rotate about the central carbon–carbon bond at room temperature, while (2*R*,3*R*)-2,3-diaminobutane and its enantiomer are inherently chiral and exist as a mixture of three interconverting diastereomeric rotamers, Figure 2.17. The ratio of the achiral to the two equienergetic chiral conformations of meso 2,3-diaminobutane can be expected to change with temperature and to vary in the gas, liquid and solid state, but the mixture is always racemic in the absence of a chiral bias. It is important to realize that facile rotation about the carbon–carbon bond in compounds such as meso 2,3-diaminobutane generates enantiomeric conformational isomers that become diastereomeric in a chiral environment; in other words, the presence of a

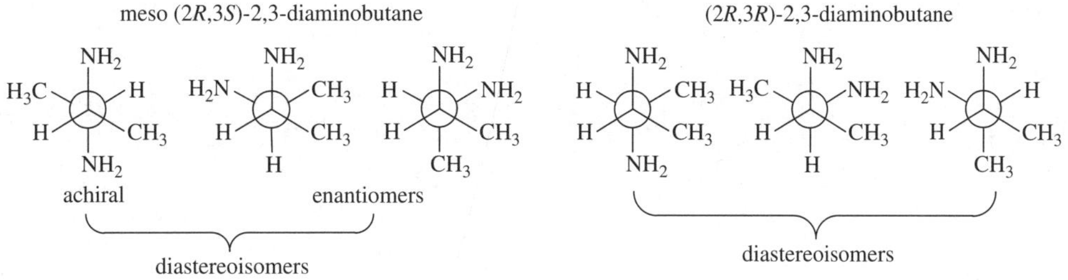

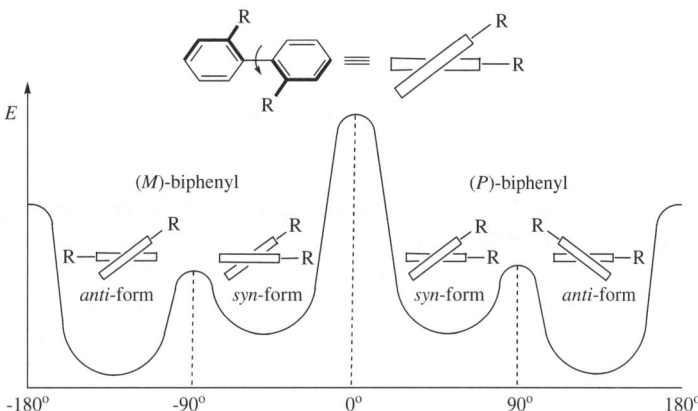

meso (2R,3S)-2,3-diaminobutane

(2R,3R)-2,3-diaminobutane

achiral   enantiomers

diastereoisomers   diastereoisomers

Figure 2.17 Newman projection of the conformational isomers of (2R,3S)- and (2R,3R)-2,3-diaminobutane.

Figure 2.18 Energy profile of 2,2′-disubstituted biphenyls.

nonracemic chiral additive can disturb the equilibrium between the enantiomers. Chelating interactions of meso (2R,3S)-2,3-diaminobutane with an achiral compound, for example maleic acid, selectively stabilizes and thus favors synclinal conformations over the antiperiplanar rotamer, but it does not change the racemic equilibrium. By contrast, hydrogen bond interactions or salt formation with a chiral acid such as (2R,3R)-tartaric acid render the enantiomeric conformers of meso 2,3-diaminobutane diastereomeric and selectively favor one over the other.

Axially chiral biphenyls undergo racemization via rotation about the pivotal aryl–aryl bond, see Chapter 3.3. In the ground state conformation, the aryl planes are neither coplanar nor orthogonal due to a compromise between resonance stabilization and steric interactions. While coplanarity of the aromatic rings would maximize resonance stabilization, it would also cause severe repulsion between *ortho*-substituents. The steric interactions between *ortho*-substituents are minimized in an orthogonal geometry, but at the expense of π-electron overlap which decreases with the cosine of the torsion angle. As a result, the enantiomers of 2,2′-disubstituted biaryls populate two diastereomeric *syn*- and *anti*-conformations, with torsion angles in solution usually ranging from 30 to 80°, and 100 to 150°, respectively. In most cases, the *anti*-form is thermodynamically more stable but 2,2′-dihalobiphenyls have been reported to prefer the *syn*-conformation.[16,17] A typical energy profile of 2,2′-disubstituted biphenyls is shown in Figure 2.18.[iv]

It is therefore not surprising that conformational equilibria of acyclic and cyclic compounds affect the stereochemical outcome of asymmetric transformations, see Chapters 6.5 and 6.6. A well

---

[iv] Incorporation of sterically demanding substituents into the *ortho*-positions of biphenyl significantly increases the energy barrier to rotation and favors a near-orthogonal ground state geometry.

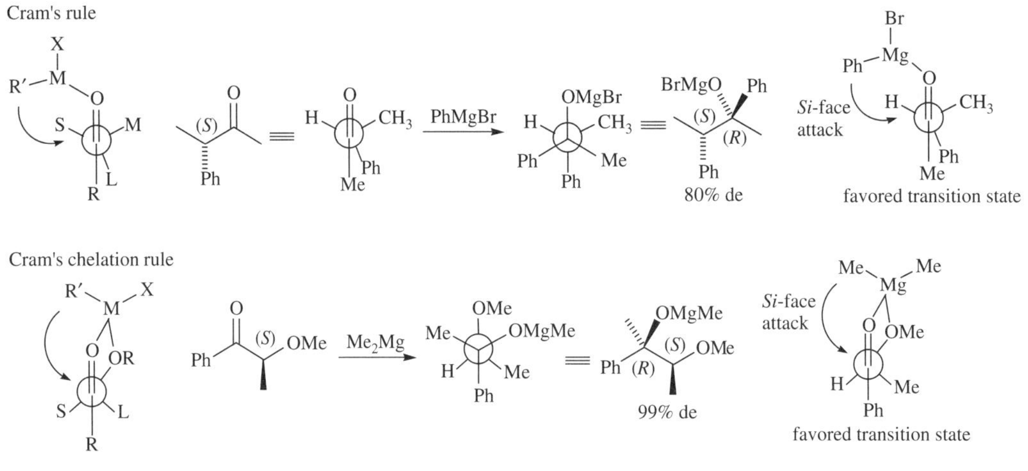

Scheme 2.1   Rationalization of the stereoselectivity of 1,2-nucleophilic additions to chiral ketones, based on Cram's rule.

known example is the asymmetric 1,2-addition of organometallic reagents and hydrides to carbonyl compounds possessing an adjacent chiral center. The diastereofacial selectivity of a nucleophilic addition to a chiral ketone depends on the conformational dynamics of the substrate, and a variety of models that attempt to predict the selectivity of such 1,2-asymmetric inductions has been developed. A powerful rationale was introduced by Cram.[18] Assuming a noncatalytic, kinetically controlled reaction, and disregarding electronic effects and the Curtin–Hammett principle, Cram reasoned that the stereoselectivity of nucleophilic attacks on acyclic ketones and aldehydes could be rationalized by considering a single substrate conformation. He postulated that the substrate resides throughout the reaction in a preferred conformation having the largest substituent L attached to the chiral center antiperiplanar to the adjacent carbonyl function. The organometallic reagent then approaches the carbonyl group from the less sterically hindered side with the smallest substituent S because the medium-sized substituent M affords greater steric hindrance, Scheme 2.1. The diastereoselectivity of the Grignard reaction between phenylmagnesium bromide and (S)-3-phenylbutan-2-one can be rationalized by looking at the Newman projection of the favored transition state. Since the proton provides less steric hindrance than the methyl substituent, phenylation occurs preferentially from the *Si*-face, producing (2R,3S)-2,3-diphenylbutan-2-ol in 80% diastereomeric excess. This simple but powerful rationale was later extended to the so-called Cram chelation model to accurately predict the selectivity of a nucleophilic attack on chiral substrates that possess a stereocenter bearing a heteroatom in α- or β-position to the carbonyl function.[19] Cram and others showed that α-alkoxy ketones and Grignard reagents form a cyclic structure having the alkoxy group synperiplanar to the carbonyl function, Scheme 2.1. The restricted conformational freedom leads to a highly diastereoselective transition state in which the nucleophile is delivered from the less sterically hindered face exhibiting the smaller ligand S. In contrast, attack from the opposite side is impeded by the presence of the larger ligand L. For example, formation of an intermediate dimethylmagnesium complex of (S)-2-methoxy-1-phenyl-propanone effectively locks the substrate into a single conformer and thus restricts the number of possible transition states, which increases stereoselectivity. Comparison of the diastereotopic faces of the five-membered chelate explains why the nucleophilic attack on the carbonyl moiety preferentially occurs from the *Si*-face, yielding (2R,3S)-3-methoxy-2-phenylbutan-2-ol in excellent diastereomeric excess.[20]

The preceding discussion, and additional examples given in Chapters 5, 6 and 7, emphasize the importance of careful stereodynamic analysis of chiral substrates, reagents and catalysts for prediction or interpretation of asymmetric induction. The mechanism of Grignard reactions of α- and β-alkoxy carbonyls is further complicated by Schlenk equilibria, competition between acyclic and cyclic reaction pathways and solvation effects. Cram's model correctly predicts the stereochemical outcome of many diastereoface-differentiating additions, except for reactions involving α-halogenated ketones and α-substituted cyclohexanones. To overcome these limitations, the simple rationale introduced by Cram has been further refined by Karabatsos,[21] Felkin,[22,23] Anh,[24] and Heathcock,[25,26] accounting for the Bürgi–Dunitz trajectory,[27] as well as for complicated conformational equilibria, Hammond's postulate,[28] Curtin–Hammett kinetics,[29] and competing steric and electronic effects. Nevertheless, Cram's model is still commonly used to predict 1,2-asymmetric induction in nucleophilic additions to aldehydes, ketones, imines, and their derivatives.

The stereodynamics of chiral compounds can afford sophisticated relays and molecular devices, including propellers and motors mimicking nature, see Chapters 6.5.4, 8.1 and 8.11. Allosteric regulation and homotrophic or heterotrophic cooperativity of enzymes is controlled by reversible binding of molecular activators and inhibitors. The binding of a so-called effector induces a change in the three-dimensional enzyme structure, which ultimately affects a distant active site, *i.e.*, conformational communication in proteins is the basis for remote reaction control. The idea of stereoselective communication and remote chiral induction is fascinating to synthetic chemists and stereochemists alike. An impressive example of stereodynamic control resulting in highly selective chiral induction transmitted through 22 bonds of a conformationally responsive relay has been accomplished by Clayden and coworkers.[30] Conformational analysis of an array of three rigid amide-substituted xanthenes confirmed that the tertiary amide groups prefer *anti*-conformation to minimize the overall dipole moment and steric repulsion, Scheme 2.2. Introduction of an ephedrine-derived (*S*)-oxazolidine residue at one terminus of a tris(arenedicarboxamide) framework consisting of three xanthene rings induces a right-handed twist (*P*-helicity) on the first two amide groups attached to the neighboring arenedicarboxamide. This local conformational induction then propagates by a domino effect through the contiguous xanthenedicarboxamide moieties showing either (*M*)- or (*P*)-helicity. As each of the six amide groups adopts an *anti*-conformational orientation relative to the previous one, the central chirality of the oxazolidine ring triggers formation of a

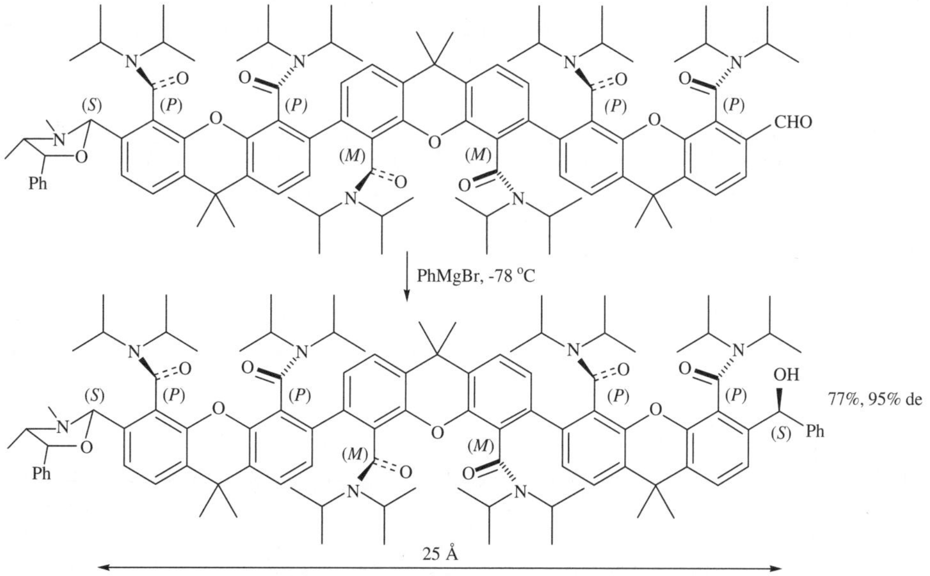

Scheme 2.2 Remote stereocontrol and asymmetric induction via conformational communication along a tris(xanthenedicarboxamide) relay.

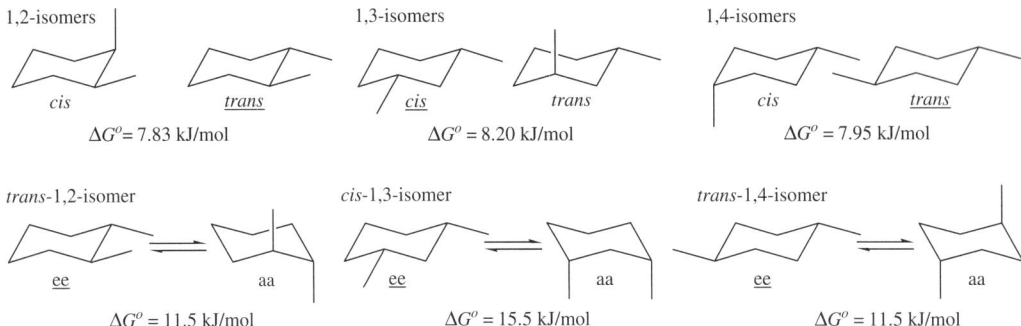

Figure 2.19 Relative stability of configurational (top) and conformational isomers (bottom) of dimethyl-cyclohexanes. The more stable isomer is underlined.

robust atropisomeric array and controls the outcome of diastereoselective reactions at the remote aldehyde group located at the other terminus. Grignard reaction with phenylmagnesium bromide at –78 °C gives the corresponding secondary (*S*)-alcohol in 77% yield and 95% de as a consequence of [1,23]-asymmetric induction over a distance of 2.5 nm.

The high degree of rotational freedom inherent to acyclic molecules allows effective minimization of steric and torsional strain. Cyclic molecules experience less conformational flexibility to balance both factors. The stereodynamic behavior of cyclic molecules is more restricted than that of acyclic compounds, and is controlled by torsion (Pitzer) and angle (Baeyer) strain, and other stabilizing or destabilizing forces such as hydrogen bonding, dipole–dipole interactions and 1,3-syndiaxial or 1,4-transannular repulsion. Three-membered and a few four-membered rings are flat while the ring geometry of all other cycloalkanes is nonplanar and usually quite fluxional. Puckering of four-membered rings is common and reduces torsion strain, albeit at the expense of some angle strain. Five- and six-membered rings form envelope and chair conformations to balance torsion and angle strain. As a consequence, substituents can occupy equatorial or axial positions, and this generates a variety of stereoisomers.

Monosubstitued cyclohexanes are achiral but exist as a mixture of two diastereomeric chair conformations that rapidly interconvert at room temperature via ring inversion. Isomerization of one chair to the other proceeds via transient boat and twist conformations, converting axial into equatorial positions and *vice versa.*[31] In general, the chair bearing the substituent in equatorial position is thermodynamically favored. Disubstituted cyclohexanes can have *cis*- or *trans*-configuration. In the case of *cis*-1,2-, *trans*-1,3-, and *cis*-1,4-cyclohexanes, one substituent occupies an axial and the other an equatorial position. By contrast, the corresponding *trans*-1,2-, *cis*-1,3-, and *trans*-1,4-disubstitued isomers interconvert between diaxial and diequatorial chair conformations, the latter being thermodynamically favored, with the exception of some diaxial *cis*-1,3-disubstituted cyclohexanes that are stabilized by intramolecular interactions such as hydrogen bonding and *trans*-1,2-dihalocyclohexanes which preferentially populate the diaxial conformation, Figure 2.19. In particular, intramolecular hydrogen bonding can have a dramatic effect on the relative stability of conformational isomers. Spectroscopic analysis of the stereodynamics of methyl-β-D-2,4-di-*O*-pyrenecarbonylxylopyranoside by Yuasa and coworkers revealed that the $^4C_1$-conformer of this carbohydrate exhibiting all four substituents in equatorial positions is favored over the $^1C_4$-conformation in DMSO and methanol, Scheme 2.3.[32] However, intramolecular hydrogen bonding between the free hydroxyl and the methoxy group stabilizes the $^1C_4$-conformation and excimer formation in chloroform and other aprotic solvents. The ring flip is therefore controlled by intramolecular hydrogen bonding and the presence or absence of solvents that interfere with this interaction.

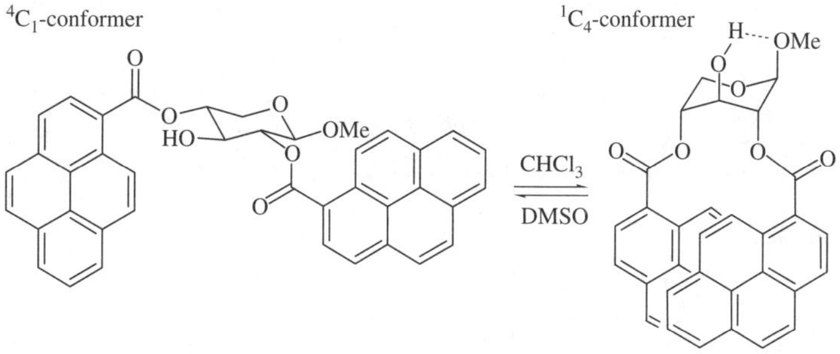

Scheme 2.3   Ring inversion of methyl-β-D-2,4-di-*O*-pyrenecarbonylxylopyranoside.

All *trans*-1,2- and *trans*-1,3-disubstituted cyclohexanes are chiral, even when the two substituents are identical. In these cases, ring flipping does not result in racemization and the enantiomers are isolable at room temperature. For example, ring inversion of the diequatorial isomer of *trans*-1,2-diaminocyclohexane generates the corresponding diaxial conformer but it does not produce a mirror image, Figure 2.20. The situation is different in *cis*-1,2- and *cis*-1,3-disubstituted cyclohexanes. Although *cis*-1,2-diaminocyclohexane is inherently chiral, the enantiomeric conformers can not be isolated at room temperature due to rapid interconversion via chair inversion. The diaxial and diequatorial isomers of *cis*-1,3-diaminocyclohexane possess a symmetry plane and are achiral.[v] Introduction of different substituents into 1,2- and 1,3-disubstituted cyclohexanes provides a mixture of separable stereoisomers, *e.g.*, *cis*- and *trans*-1-amino-2-methylcyclohexane exist as two pairs of noninterconverting enantiomers. The conformations of *cis*- and *trans*-1,4-disubstituted cyclohexanes have a plane of symmetry and are generally achiral.

Nucleophilic additions to 4-substituted cyclohexanones often proceed with low selectivity because of the conformational flexibility of the six-membered ring and weak stereoselective induction from the remote substituent. However, the sterically demanding *tert*-butyl group prefers to reside in the equatorial position and effectively locks 4-*tert*-butylcyclohexanone into one conformation, which in most cases improves diastereoselectivity. Since the contribution from the less stable cyclohexanone conformer having the *tert*-butyl group in the axial position is negligible, the prediction of diastereoselective reactions of 4-*tert*-butylcyclohexanone is simplified and requires stereochemical analysis of only a single conformer. Most nucleophiles including Grignard and organolithium reagents approach the carbonyl group at an angle close to 109° (Bürgi–Dunitz trajectory) from the site opposite to the bulky *tert*-butyl group. Reetz and Stanchev obtained the axial carbinol via organiron(ii)-mediated diastereofacial addition of butylmagnesium chloride in 76% de. This is generally attributed to a less shielded equatorial attack on the carbonyl of the predominant conformation of 4-*tert*-butylcyclohexanone, Scheme 2.4.[33] By contrast, very small nucleophiles such as hydrides or acetylides experience less steric interaction with the remote 4-*tert*-butyl group and favor the axial approach to avoid local interactions with the axial hydrogens at *C*-2 and *C*-6. Reduction of 4-*tert*-butylcyclohexanone with lithium aluminum hydride proceeds preferentially via axial attack and produces the equatorial alcohol as the major product.

Incorporation of small groups into the α-position of cyclohexanone does not change the preference of lithium aluminum hydride for an axial attack. Computation of the two conformations of (*R*)-2-methylcyclohexanone shows that the methyl group does not strongly interfere with an axial approach. The predominant formation of (1*R*,2*R*)-*trans*-2-methylcyclohexan-1-ol can be attributed to a relatively rapid hydride addition to the major conformer of (*R*)-2-methylcyclohexanone having

---

[v]The same considerations apply to cyclopentanes, *i.e.*, *trans*-1,2- and *trans*-1,3-disubstituted cyclopentanes are chiral, irrespective of the structure of the substituents. Similarly, *trans*-1,2-cyclobutanes and *trans*-1,2-cyclopropanes are inherently chiral, whereas both *cis*- and *trans*-1,3-cyclobutanes exhibiting identical substituents are achiral.

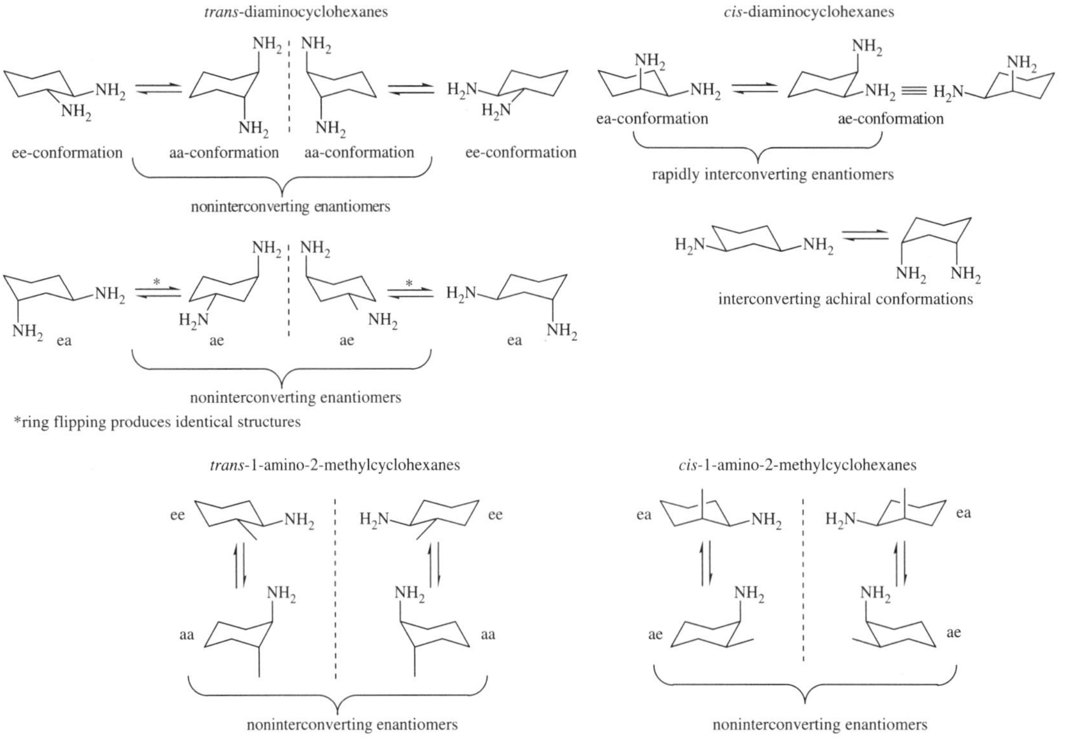

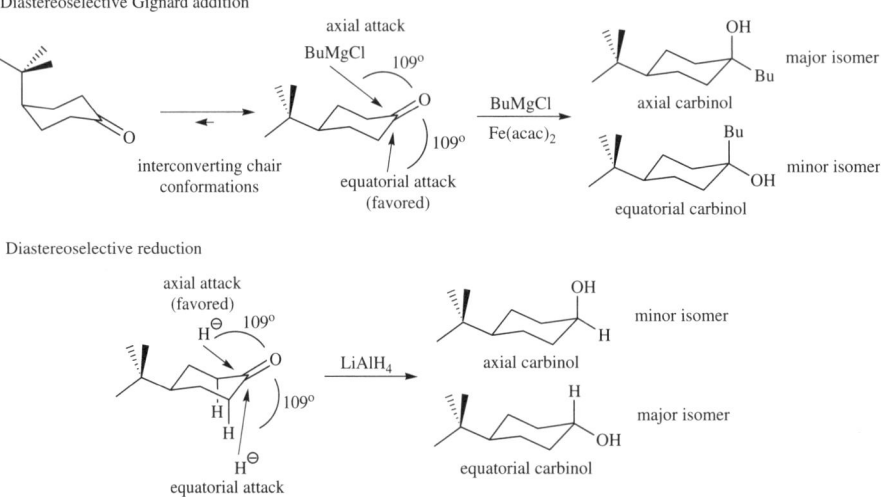

Figure 2.20   Conformations and chirality of disubstituted cyclohexanes.

Scheme 2.4   Stereoselective Grignard reaction and reduction of 4-*tert*-butylcyclohexanone with LiAlH$_4$.

the methyl group in the equatorial position, Scheme 2.5. An increase in the steric bulk of the α-substituent further locks the ketone into the equatorial conformation and results in enhanced steric hindrance to an axial attack. In the case of 2-*tert*-butylcyclohexanone, these interactions exceed the repulsion between the approaching hydride and the axial hydrogens at *C*-2 and *C*-6, and

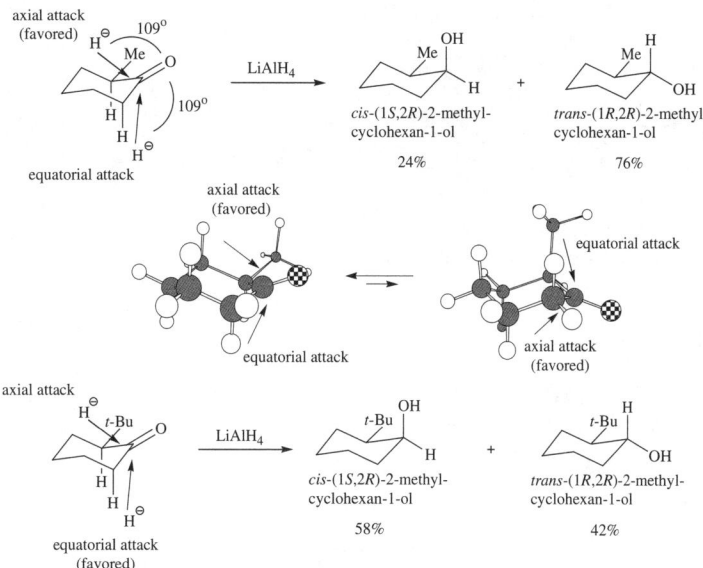

Scheme 2.5 Diastereoselective reduction of (*R*)-2-alkylcyclohexanones.

reduction with LiAlH$_4$ affords 2-*tert*-butylcyclohexanol as a 58:42 diastereomeric mixture in favor of the *cis*-isomer.[34]

An interesting example that underscores the significance of the relative population and inter-conversion of configurational isomers is the diastereoselective *N*-oxidation of *N*-methyl-4-*tert*-butylpiperidine, Scheme 2.6.[35,36] As a result of facile *N*-inversion, this compound exists as a mixture of two rapidly interconverting isomers having either an equatorial or an axial methyl group. The piperidine conformer that bears the methyl group in the equatorial position is approximately 60 times more thermodynamically stable than the axial conformer due to 1,3-syndiaxial repulsion between the axial methyl group and two axial hydrogens. Reaction with hydrogen peroxide proceeds more quickly with the minor diastereoisomer because it provides a more accessible equatorial position for oxidation. Although *N*-inversion is a rapid process, the relative population of the conformational piperidine isomers has a major effect on the stereochemical outcome of the oxidation. The reported oxidation product ratio of 19:1 in favor of the axial *N*-oxide can be attributed to competing nitrogen inversion and diffusion-controlled diastereoselective oxidation.[vi]

Cyclohexenes exist in two half-chair conformations that rapidly isomerize at room temperature. Because the double bond and the two adjacent saturated carbon atoms *C*-3 and *C*-6 are in one plane, allylic substituents reside in pseudoequatorial and pseudoaxial positions while carbons *C*-4 and *C*-5 generate truly axial and equatorial positions. Similar to cyclohexanes, substituents preferentially occupy equatorial positions, although the energetic difference between equatorial and axial conformers is less pronounced in cyclohexenes as a consequence of the reduced steric repulsion experienced by axial substituents. In cyclohexene rings, axial groups participate in only one 1,3-synpseudoaxial interaction, whereas two 1,3-syndiaxial interactions with truly axial hydrogens destabilize the axial conformation in cyclohexanes, Figure 2.21. For the same reason, the general preference of allylic substituents for a pseudoequatorial orientation is less pronounced than the difference in the relative stability of axial and equatorial positions in saturated rings. Electronegative substituents such as halides or amino, amido and alkoxy groups favor the pseudoaxial position.[37] In accordance to the anomeric effect observed with carbohydrates, the

---

[vi] Because both *N*-inversion and oxidation are rapid processes with similar reaction rates, the Curtin–Hammett principle does not apply.

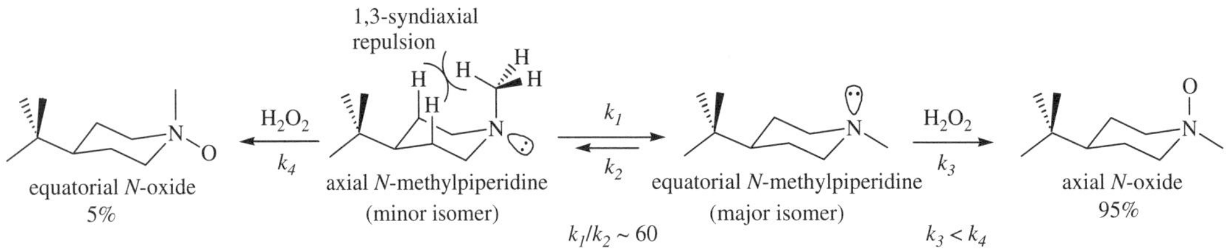

Scheme 2.6   Oxidation of *N*-methyl-4-*tert*-butylpiperidine.

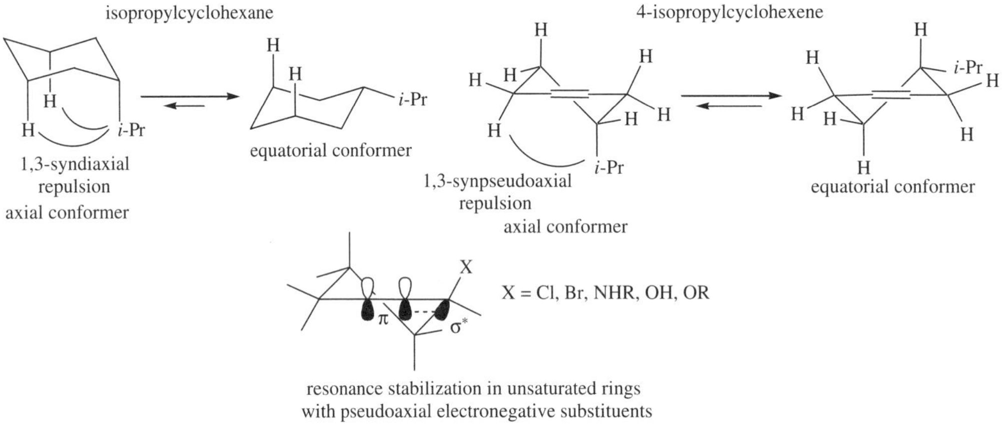

Figure 2.21   Substituent effects on the relative stability of cyclohexane and cyclohexene conformations.

preference for the pseudoaxial position has been explained with resonance stabilization due to overlap of the π-system of the double bond and an antibonding orbital of the allylic σ-bond.

The relative stability of the half-chair conformation of disubstituted cyclohexenes or tetralin (1,2,3,4-tetrahydronaphthalene) derivatives depends on a variety of intramolecular interactions in addition to torsion and angle strain. Most important are resonance between the double bond and the pseudoaxial allylic σ-bond, synpseudoaxial repulsion, dipole–dipole interactions, and hydrogen bonding. For example, incorporation of a methyl group into the saturated moiety of 4-(3′,5′-dinitrobenzamido)-1,2,3,4-tetrahydrophenanthrene at carbons *C*-1, *C*-2 and *C*-3 alters the relative stability of the conformer having the benzylic amido group in pseudoaxial position. Wolf and Pirkle studied the conformational preference and rigidity of disubstituted 4-(3′,5′-dinitrobenz-amido)-1,2,3,4-tetrahydrophenanthrene derivatives which are structural analogs of the so-called Whelk-O selector.[38] Initially developed to separate the enantiomers of naproxen and other profens, this selector forms a chiral cleft which appears to be crucial for enantioselective recognition.[vii] It is assumed that only one enantiomer of a given racemate can diffuse into the cleft to undergo *simultaneous* hydrogen bonding to the amide function and face-to-face and face-to-edge inter-actions with the aromatic moieties while maintaining a heavily populated low energy conformation, Figure 2.22.[39–44]

The propensity of the DNB group to occupy the pseudoaxial position in 4-(3′,5′-dinitrobenz-amido)-1,2,3,4-tetrahydrophenanthrenes was evaluated based on chromatographic structure-activity relationships. Chiral HPLC separation of the racemic Whelk-O analogs shown in Table 2.1 on an

---

[vii] The Whelk-O selector has been used extensively for chromatographic enantioseparation of compounds exhibiting a stereogenic center, axis or plane in close proximity to a hydrogen bond acceptor (usually a carbonyl group).

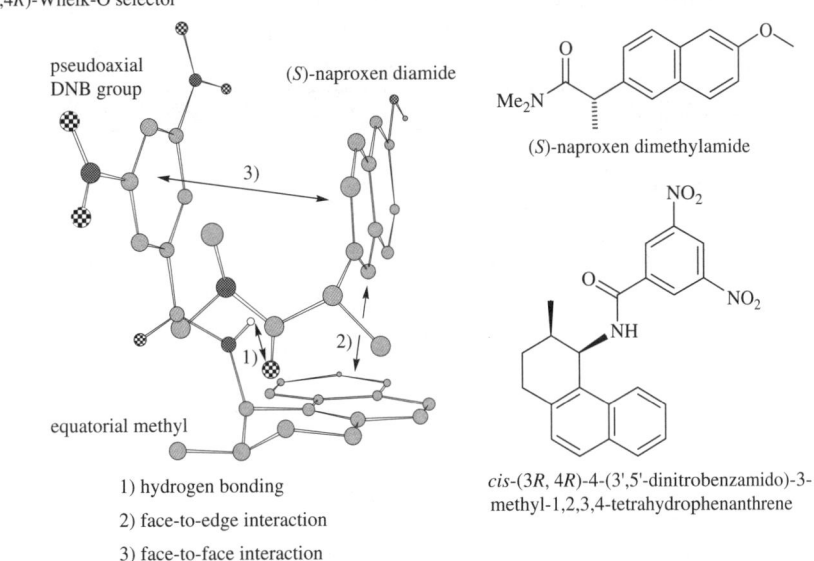

(S)-naproxen dimethylamide

cis-(3R, 4R)-4-(3',5'-dinitrobenzamido)-3-methyl-1,2,3,4-tetrahydrophenanthrene

1) hydrogen bonding
2) face-to-edge interaction
3) face-to-face interaction

Figure 2.22   Chiral cleft and favored conformation of the Whelk-O selector participating in simultaneous hydrogen bonding, face-to-face and face-to-edge interactions with (S)-naproxen dimethylamide. Hydrogens bonded to carbon atoms are not shown. [Reproduced with permission from *Tetrahedron* **2002**, *58*, 3597–3603.]

(S)-naproxen amide-derived chiral stationary phase allowed determination of the difference in the Gibbs free energy, $\Delta\Delta G°$, of transient diastereomeric complexes according to Equation 2.1.

$$\Delta\Delta G° = -RT \ln \alpha \qquad (2.1)$$

where α is the chromatographic separation factor, and $\Delta G°$ is the Gibbs free energy of the formation of transient diastereomeric complexes.

   These energetic differences can be correlated to the influence of the methyl group on the relative stability of the Whelk-O conformers. Stabilization of a cleft-like structure bearing a pseudoaxial DNB group results in high enantioselectivity because (S)-naproxen fits perfectly into the chiral pocket of one of the chromatographed Whelk-O enantiomers, as depicted in Figure 2.22. In contrast, enantioselective recognition is diminished when the Whelk-O enantiomers are likely to adopt a structure with a pseudoequatorial DNB group. In this case, the difference in the Gibbs free energy, $\Delta\Delta G°$, of the transient diastereomeric complexes formed between the CSP and the enantiomeric solutes is reduced as none of them can accommodate simultaneous hydrogen bonding, face-to-face and face-to-edge interactions. Introduction of a methyl group into positions *C*-1, *C*-2 and *C*-3 of the 4-amidotetrahydrophenanthrene ring yields three pairs of diastereoisomers having either *cis*- or *trans*-configuration, Table 2.1. In order to understand the chiral recognition mechanism, one has to remember that the tetrahydrophenanthrene ring of these Whelk-O analogs resembles a cyclohexene or tetralin structure and that the benzylic electron-withdrawing benzamido group generally prefers the pseudoaxial position.

   The difference in the Gibbs free energy between the transient diastereomeric complexes formed by the enantiomers of 4-(3',5'-dinitrobenzamido)-1,2,3,4-tetrahydrophenanthrene and the (S)-naproxen-derived chiral stationary phase was determined as 3.87 kJ/mol. The high selectivity has been attributed to extensive population of the conformer that has the benzamido substituent in

Table 2.1 Energetic differences of transient diastereomeric complexes of Whelk-O analogs and immobilized (*S*)-naproxen diamide.

Whelk-O derivative	Relative configuration	$\Delta\Delta G°$ [kJ/mol]
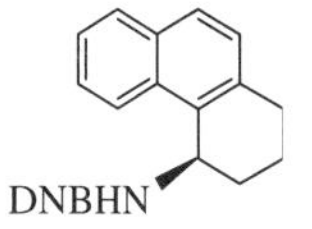 DNBHN	/	3.87
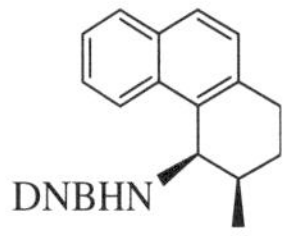 DNBHN	*cis*-3,4-	4.05
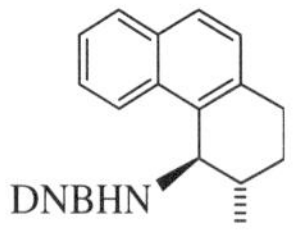 DNBHN	*trans*-3,4-	3.77
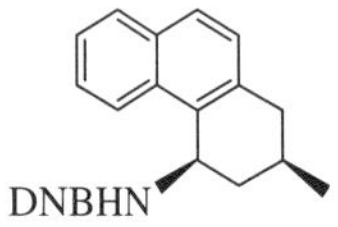 DNBHN	*cis*-2,4-	1.52
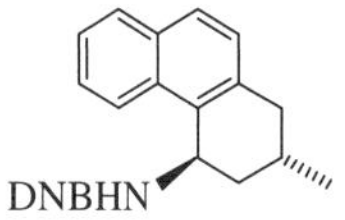 DNBHN	*trans*-2,4-	3.02
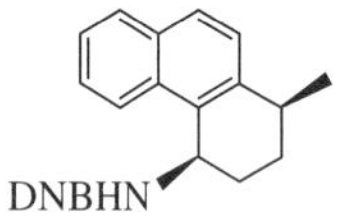 DNBHN	*cis*-1,4-	3.56
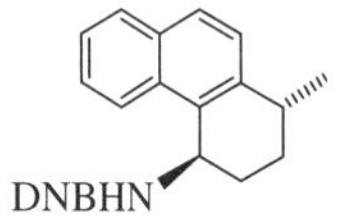 DNBHN	*trans*-1,4-	1.92

Only one enantiomer of the racemic Whelk-O analogs is shown.
DNB = 3,5-dinitrobenzoyl group.

pseudoaxial position, thus forming a chiral cleft exhibiting the 3,5-DNB group perpendicular to the tetrahydrophenanthrene plane, which is a prerequisite for effective chiral recognition. Incorporation of a methyl group into the adjacent position provides *cis*- and *trans*-4-(3′,5′-dinitrobenzamido)-3-methyl-1,2,3,4-tetrahydrophenanthrene. The *cis*-configuration preferentially populates the conformation with a pseudoaxial DNB group and an equatorial methyl group over the conformer having the DNB group in pseudoequatorial and the methyl substituent in axial position. The *cis*-3,4-disubstituted Whelk-O analog therefore strongly favors the cleft-like conformation and undergoes highly enantioselective interactions with (*S*)-naproxen exceeding 4.0 kJ/mol. By contrast, the cleft-like conformation is less heavily populated in the *trans*-3,4-disubstituted Whelk-O derivative because it bears an axial methyl group. The stereoselectivity observed with the *trans*-isomer is therefore reduced to 3.77 kJ/mol. The enantiomers of *trans*-4-(3′,5′-dinitrobenzamido)-2-methyl-1,2,3,4-tetrahydrophenanthrene can afford a conformation having the two substituents in the preferred pseudoaxial and equatorial orientations, whereas the *cis*-2,4-isomer must place the methyl group in axial position in order to form the cleft with a pseudoaxial DNB group. The cleft-like conformer is therefore more heavily populated by the *trans*-2,4-isomer than by the *cis*-derivative. Accordingly, (*S*)-naproxen effectively differentiates between the enantiomers of the *trans*-isomer ($\Delta\Delta G° = 3.02$ kJ/mol) but enantioselective separation and recognition of the *cis*-diastereomer is diminished (1.52 kJ/mol). Finally, *cis*-4-(3′,5′-dinitrobenzamido)-1-methyl-1,2,3,4-tetrahydrophenanthrene shows superior enantioselectivity ($\Delta\Delta G° = 3.56$ kJ/mol) over the *trans*-disubstituted analog (1.92 kJ/mol) because the cleft-like conformer is destabilized in the latter due to synpseudoaxial interactions experienced by the methyl group. Clearly, the relative stability and stereoselectivity of the cleft-like Whelk-O conformers originate from an energetic compromise between the tendency of the DNB group to occupy a pseudoaxial position and the preference of the methyl group for an equatorial or pseudoequatorial orientation.

REFERENCES

1. Cahn, R. S.; Ingold, C. K.; Prelog, V. *Angew. Chem., Int. Ed. Engl.* **1966**, *5*, 385-415.
2. Prelog, V.; Helmchen, G. *Angew. Chem., Int. Ed. Engl.* **1982**, *21*, 567-583.
3. Nierengarten, J.-F.; Dietrich-Buchecker, C. O.; Sauvage, J.-P. *J. Am. Chem. Soc.* **1994**, *116*, 375-376..
4. Chambron, J.-C.; Sauvage, J.-P. *J. Am. Chem. Soc.* **1997**, *119*, 9558-9559.
5. Jäger, R.; Händel, M.; Harren, J.; Rissanen, K.; Vögtle, F. *Liebigs Ann.* **1996**, 1201-1207.
6. Tauber J. J. *Res. Nat. Bur. Stand. Sect. A.* **1963**, *67*, 591-599.
7. Reuter, C.; Mohry, A.; Sobanski, A.; Vögtle, F. *Chem. Eur. J.* **2000**, *6*, 1674-1682.
8. Prelog, V.; Gerlach, H. *Helv. Chim. Acta* **1964**, *47*, 2288-2294.
9. Gerlach, H.; Owtschinnikow, J. A.; Prelog, V. *Helv. Chim. Acta* **1964**, *47*, 2294-2302.
10. Mislow, K. *Chimia* **1986**, *40*, 395-402.
11. Chorev, M.; Goodman, M. *Acc. Chem. Res.* **1992**, *25*, 266-272.
12. Eliel, E. L; Wilen, S. H. *Stereochemistry of Organic Compounds*, Wiley, New York, 1994, pp. 1176-1181.
13. Wolf, C.; König, W. A.; Roussel, C. *Chirality* **1995**, *7*, 610-611.
14. Roussel, C.; Adjimi, M.; Chemlal, A.; Djafri, A. *J. Org. Chem.* **1988**, *53*, 5076-5080.
15. Newell, L. M.; Sekhar, V. C.; DeVries, K. M.; Staigers, T. L.; Finneman. J. I. *J. Chem. Soc., Perkin Trans. 2* **2001**, 961-963.
16. Roberts, R. M. G. *Magn. Reson. Chem.* **1985**, *23*, 52-54.
17. Dynes, J. J.; Baudais, F. L.; Boyd, R. K. *Can. J. Chem.* **1985**, *63*, 1292-1299.
18. Cram, D. J.; Elhafez, F. A. A. *J. Am. Chem. Soc.* **1952**, *74*, 5828-5835.
19. Cram, D. J.; Kopecky, K. R. *J. Am. Chem. Soc.* **1959**, *81*, 2748-2755.
20. Chen, X.; Hortelano, E. R.; Eliel, E. L.; Frye, S. V. *J. Am. Chem. Soc.* **1992**, *114*, 1778-1784.
21. Karabatsos, G. J. *J. Am. Chem. Soc.* **1967**, *89*, 1367-1371.
22. Cherest, M.; Felkin, H.; Prudent, N. *Tetrahedron Lett.* **1968**, *9*, 2199-2204.
23. Cherest, M.; Felkin, H. *Tetrahedron Lett.* **1968**, *9*, 2205-2208.
24. Anh, N. T.; Eisenstein, O. *Nouv. J. Chimie* **1977**, *1*, 61-70.
25. Heathcock, C. H.; Flippin, L. A. *J. Am. Chem. Soc.* **1983**, *105*, 1667-1668.
26. Lodge, E. P.; Heathcock, C. H. *J. Am. Chem. Soc.* **1987**, *109*, 3353-3361.
27. Bürgi, H. B.; Dunitz, J. D.; Schefter, E. *J. Am. Chem. Soc.* **1973**, *95*, 5065-5067.
28. Hammond, G. S. *J. Am. Chem. Soc.* **1955**, *77*, 334-338.
29. Curtin, D. Y. *Rec. Chem. Progr.* **1954**, *15*, 111-128.
30. Clayden, J.; Lund, A.; Vallverdu, L.; Helliwell, M. *Nature* **2004**, *431*, 966-971.

31. Eliel, E. L; Wilen, S. H. *Stereochemistry of Organic Compounds*, Wiley, New York, 1994, pp. 686-725.

32. Yuasa, H.; Miyagawa, N.; Izumi, T.; Nakatani, M.; Izumi, M.; Hashimoto, H. *Org. Lett.* **2004**, *6*, 1489-1492.

33. Reetz, M.; Stanchev, S. *J. Chem. Soc., Chem. Commun.* **1993**, 328-330.

34. Ashby, E. C.; Sevenair, J. P.; Dobbs, F. R. *J. Org. Chem.* **1971**, *36*, 197-199.

35. Shvo, Y.; Kaufman, E. D. *Tetrahedron*, **1972**, *28*, 573-580.

36. Crowley, P. J.; Robinson, M. J. T.; Ward, M. G. *Tetrahedron* **1977**, *33*, 915-925.

37. Lessard, J.; Tan, P. V. M.; Martino, R.; Saunders, J. K. *Can. J. Chem.* **1997**, *55*, 1015-1023.

38. Wolf, C.; Pirkle, W. H. *Tetrahedron* **2002**, *58*, 3597-3603.

39. Pirkle, W. H.; Welch, C. J.; Lamm, B. *J. Org. Chem.* **1992**, *57*, 3854-3860.

40. Wolf, C.; Spence, P. L.; Pirkle, W. H.; Derrico, E. M.; Cavender, D. M.; Rozing, G. P. *J. Chromatogr. A* **1997**, *782*, 175-179.

41. Wolf, C.; Pirkle, W. H. *J. Chromatogr. A* **1997**, *785*, 173-178.

42. Wolf, C.; Pirkle, W. H. *J. Chromatogr. A* **1998**, *799*, 177-184.

43. Wolf, C.; Spence, P. L.; Pirkle, W. H.; Cavender, D. M.; Derrico, E. M. *Electrophoresis* **2000**, *21*, 917-924.

44. Wolf, C.; Pranatharthiharan, L.; Volpe, E. C. *J. Org. Chem.* **2003**, *68*, 3287-3290.

CHAPTER 3

Racemization, Enantiomerization and Diastereomerization

The development of mechanistic insights into isomerization reactions, and information about the conformational and configurational stability of chiral compounds under various conditions, are indispensable for today's chemist. Racemization, enantiomerization and diastereomerization of chiral compounds play a key role across the chemical sciences and provide multiple opportunities for asymmetric synthesis, the design of microscopic motors and machines, drug discovery and medical diagnosis. A basic understanding of the properties and reactions of chiral compounds requires in-depth analysis of the stereodynamics of coexisting isomers. Compounds that exist as a mixture of rapidly interconverting stereoisomers are often conveniently, albeit inaccurately, viewed as one averaged structure. A closer look reveals that organic compounds are generally fluxional and adopt more than one conformation or configuration. For example, amines that have three different substituents at the stereogenic nitrogen atom exist as a pair of configurationally unstable enantiomers, and monosubstituted cyclohexanes populate two diastereomeric chair conformations (and negligible twist conformations) having the substituent in either equatorial or axial position. Because of the inherently low energy barrier to interconversion, it is often difficult to distinguish between the individual isomers of stereolabile compounds. The energy barrier to pyramidal inversion of ethylmethylisopropylamine is 31.4 kJ/mol, and the activation energy required for ring flipping of the isomers of chlorocyclohexane is about 44 kJ/mol, Scheme 3.1. Both processes, racemization of ethylmethylisopropylamine and diastereomerization of chlorocyclohexane, are fast and only one averaged species is observed by NMR spectroscopy at room temperature. One can distinguish between diastereotopic proton signals of the individual isomers of monosubstituted cyclohexanes or chiral amines in the presence of a chiral shift reagent under cryogenic conditions when the rates of interconversion are relatively slow with respect to the NMR time scale. Alternatively, faster methods such as time-resolved microwave and IR spectroscopy generate an instantaneous rather than a time-averaged spectrum and can be used to study the structures and properties of stereolabile isomers even at room temperature.

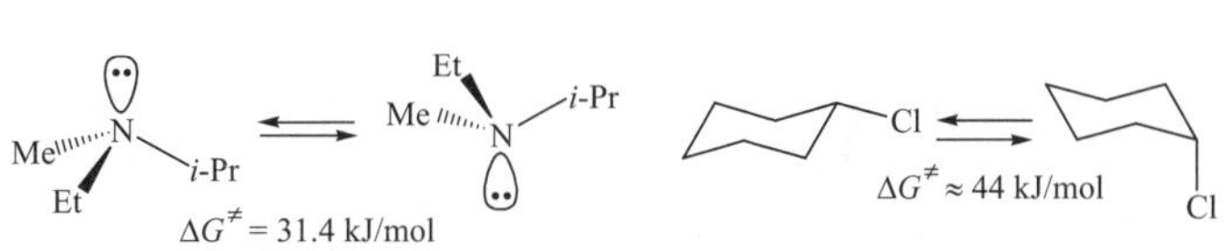

Scheme 3.1 Interconversion of the stereoisomers of ethylmethylisopropylamine and chlorocyclohexane.

3.1 CLASSIFICATION OF ISOMERIZATION REACTIONS OF CHIRAL COMPOUNDS

Stereomutations of conformationally and configurationally unstable chiral molecules can be treated as a macroscopic process, in which an enantiopure or diastereomerically pure compound is transformed by thermal equilibration either to a racemate or to a diastereoisomeric mixture. Alternatively, isomerization reactions can be viewed at the molecular level. The macroscopic change is by definition irreversible, whereas the microscopic view takes into account the individual rate constants of reversible reactions, Figure 3.1.[1] Racemization is an irreversible process that is completed when 50% of an enantiopure compound has been converted to the other enantiomer. The corresponding half-life time is the time during which the enantiomeric excess of a chiral mixture has been reduced to half its initial value, for example from 100% ee to 50% ee. At the microscopic level, the same process is perceived as a reversible enantiomerization reaction with a different rate constant $k_{enant} = 0.5\ k_{rac}$. This process is completed when all molecules have been converted to the other enantiomer. The half-life time of enantiomerization is the time required for 50% interconversion, *i.e.*, a change from 100% ee to 0% ee. Enantiomerization is usually reversible and leads to the formation of a racemate but many unidirectional processes resulting in complete enantioconversion are known. For example, (*R*)-profens are enantioselectively and irreversibly transformed to the (*S*)-enantiomer by coenzyme A conjugate.[2,3] Such unidirectionality can also be achieved if enantioconversion is coupled with stereoselective removal of one enantiomer from the equilibrium, for instance by dynamic kinetic resolution or asymmetric transformation of the second kind. Monitoring of the macroscopic change of a nonracemic mixture of stereolabile enantiomers using polarimetry or circular dichroism spectroscopy provides the rate of racemization, while other methods such as dynamic chromatography or proton–deuterium substitution analysis by ^{1}H NMR spectroscopy afford enantiomerization rates. Since enantiomerization and racemization kinetics are frequently employed to describe the same process, it is important to distinguish between the different mathematical treatment and the corresponding rate constants, k_{enant} and k_{rac}, respectively.

Kinetic analysis of diastereomerization is more complex than the study of racemization or enantiomerization because the rate constant for the transformation of one diastereomer to another is different from the rate constant of the reverse reaction. Diastereomers possess different thermodynamic stabilities and an equimolar ratio is not usually obtained at equilibrium. Epimerization is a special case of diastereomerization, as it refers solely to the interconversion of epimers.[i] The macroscopic analogy to epimerization is mutarotation which describes the irreversible change in the optical rotation of an epimeric mixture until equilibrium is reached.

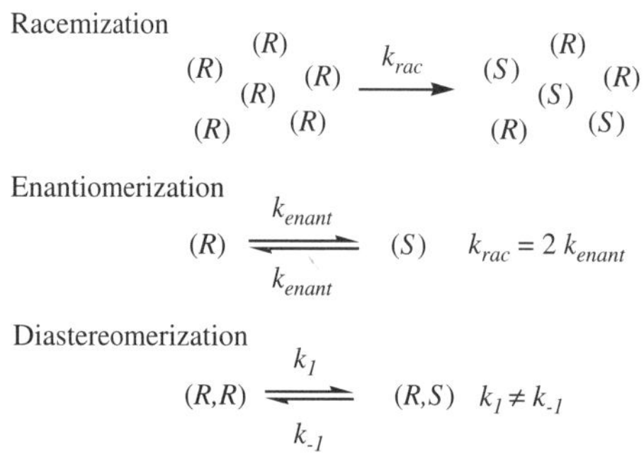

Figure 3.1 Classification of isomerization reactions of chiral compounds.

---

[i]Epimers are defined as diastereoisomers that differ in only one configuration of two or more elements of chirality.

3.1.1 Racemization

Racemization is irreversible, and is completed when the thermodynamic equilibrium (ee = 0%) is reached. The driving force for racemization is the increase in entropy, provided one can disregard enthalpic contributions from heterochiral interactions that can occur in racemic mixtures in addition to strictly homochiral interactions in enantiopure solutions. The entropic driving force for racemization at 25 °C can be estimated as $\Delta G° \approx -RT \ln 2$ which corresponds to $-1.7\,kJ/mol$. Racemization is treated according to irreversible first-order kinetics:

$$\frac{dx}{dt} = k_{rac}\,(R_0 - x) \tag{3.1}$$

where R_0 is the initial concentration of the (R)-enantiomer; $x = R_0 - R,S$ (concentration of the racemate at time t); and k_{rac} is the rate constant of racemization.

$$\int_0^x \frac{dx}{R_0 - x} = -k_{rac} \int_0^t dt \tag{3.2}$$

Integration of both sides of the equation gives:

$$\ln\left(\frac{R_0}{R_0 - x}\right) = k_{rac}\,t \tag{3.3}$$

The half-life of racemization, $\tau_{1/2}$, can be calculated using the rate constant of enantiomerization k_{enant} (assuming $S_0 = 0$ at $t = 0$):[ii]

$$\ln\left(\frac{x_{eq}}{x_{eq} - x}\right) = \ln\left(\frac{R_0}{2R - R_0}\right) = \ln\left(\frac{R + S}{R - S}\right) = 2k_{enant}\,t \tag{3.4}$$

where $R_0 = R + S$.

At 50% ee, the equation becomes:

$$\tau_{1/2} = \frac{\ln 2}{2k_{enant}} \quad \text{or} \quad \tau_{1/2} = \frac{\ln 2}{k_{rac}} \tag{3.5}$$

Both detection and control of racemization are important tasks in asymmetric synthesis because it can affect the stereochemical outcome of a reaction and compromise the stereoisomeric purity of the product. Similarly, a basic understanding of the stereochemical integrity of a chiral drug is crucial for ruling out undesirable racemization which may potentially occur under physiological conditions or during long time storage. Nevertheless, racemization can be utilized as a powerful tool in obtaining enantiopure products from racemic starting materials via dynamic kinetic resolution (DKR). In other words, it is sometimes important to prevent enantioconversion of a chiral compound while in other cases it becomes advantageous if one can identify conditions and catalysts that facilitate racemization. A close look at the solvent-dependent enolization of 2-benzoyl cyclohexanone and the palladium-catalyzed interconversion of axially chiral allenes reflects these opposite perspectives on racemization. Chiral drugs that possess an asymmetric center carrying a proton next to a carbonyl moiety can racemize via base- or acid-catalyzed keto/enol-tautomerization. This is important because stereoisomers are likely to have different pharmacological and pharmacokinetic properties. In particular, β-diketones and β-keto esters such as 2-benzoyl

---

[ii] The half-life of racemization is the time it takes to reduce 100% ee to 50% ee.

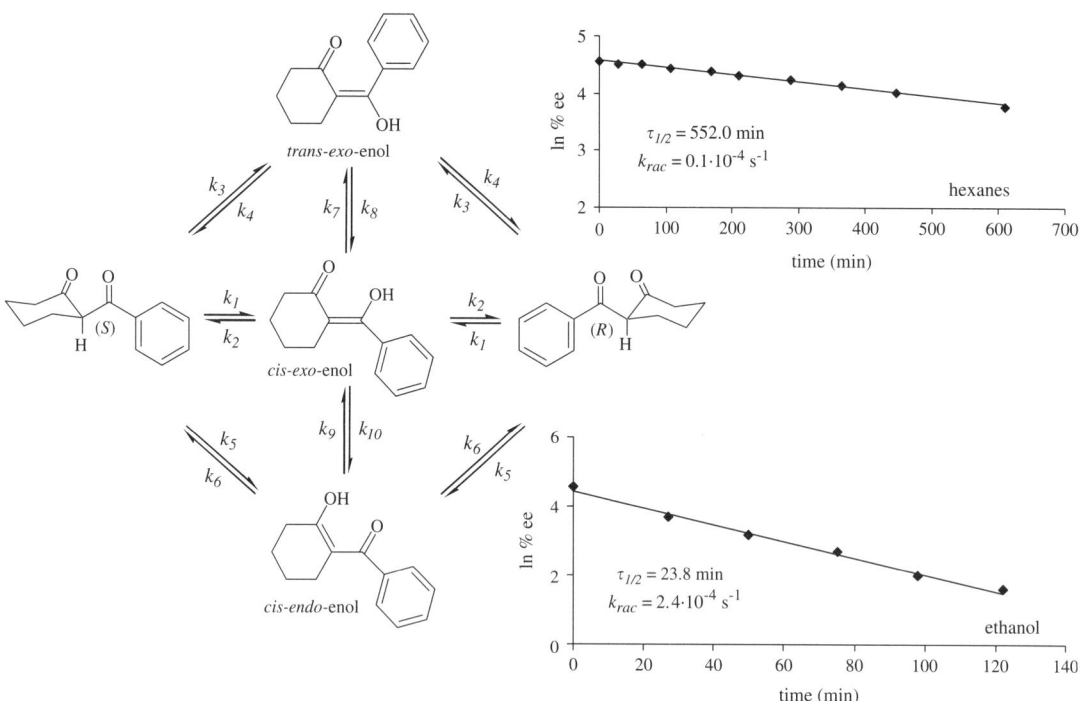

Scheme 3.2   Enantioconversion pathways of 2-benzoyl cyclohexanone and analysis of the racemization kinetics in hexanes and ethanol at 66.0 °C. [Modified with permission from *Chirality* **2005**, *17*, 171-176.]

cyclohexanone are prone to enolization, even in the absence of base or acid. Campbell and Gilow showed that 2-benzoyl cyclanones mainly exist in the keto form, but tautomerize to some extent in protic solvents. They determined the keto/enol ratio of 2-benzoyl cyclohexanone in methanol as 97:3.[4,5] Protic solvents such as ethanol stabilize the three possible intermediate enol forms through hydrogen bonding and thus affect the thermodynamics and kinetics of the tautomerization. Tumambac and Wolf isolated enantiopure 2-benzoyl cyclohexanone by chiral HPLC and studied the configurational stability in hexanes and ethanol at 66.0 °C.[6] Enolization of β-keto esters is enthalpically favored in protic solvents, although this effect is somewhat reduced because the disruption of the intramolecular enol hydrogen bond and subsequent formation of intermolecular hydrogen bonds with the solvent is associated with negative entropy contributions. The inter-conversion of the enantiomers of 2-benzoyl cyclohexanone is further complicated by the fact that the intermediate *cis-exo*-enol form undergoes *exo/endo*-isomerization to the *trans-exo*-enol isomer, Scheme 3.2.[iii] The racemization can therefore proceed via parallel reaction pathways that are linked through complex temperature- and solvent-dependent equilibria. Comparison of the racemization rate constants, and the corresponding half-life times in ethanol and hexanes, reveals that the configurational stability of 2-benzoyl cyclohexanone is highly solvent-dependent. Protic solvents favor formation of the intermediate enol tautomers and subsequent racemization which has been confirmed by NMR spectroscopy. The results demonstrate that enantiopure β-diketones are potentially useful building blocks for asymmetric synthesis, as long as the choice of solvent is taken into consideration. While the stereochemical stability in alkanes and other aprotic solvents

---

[iii] The *cis-exo*-enol form has an extended conjugated system with an exocyclic double bond, and is expected to be more stable than the *trans*-isomer and the enol form having an endocyclic double bond. Only the *cis*-enols undergo intramolecular hydrogen bonding which is probably disrupted in protic solvents. The relative stability of the enol intermediates therefore depends on the nature of the solvent.

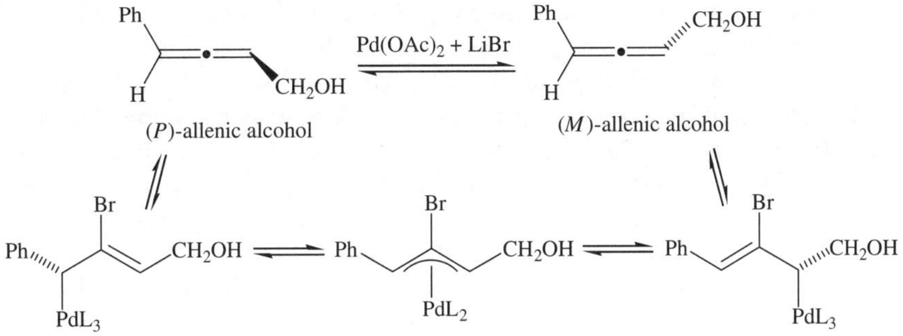

Scheme 3.3 Palladium-catalyzed allene racemization.

Table 3.1 Racemization half-lives.

Allene	$\tau_{1/2}$
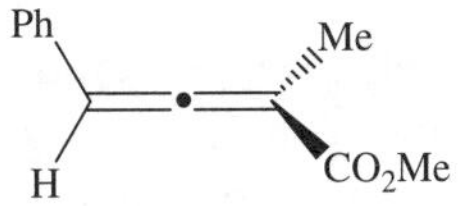	< 1 min
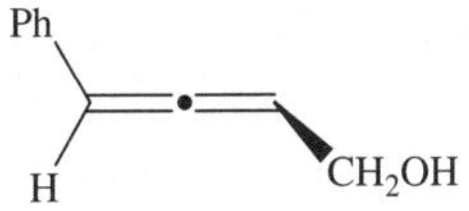	15 min
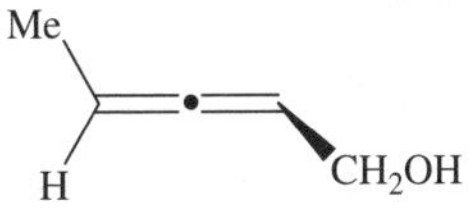	3.3 h
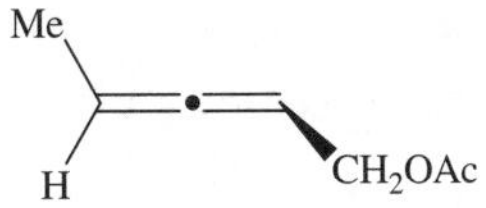	21 min
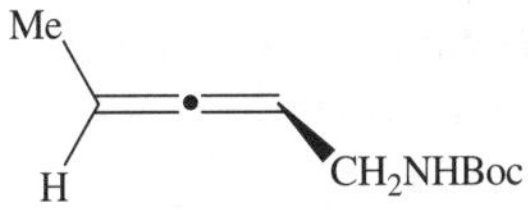	30 min
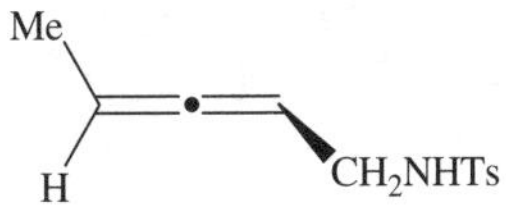	2.25 h

might be a fundamental prerequisite to many asymmetric reactions, rapid racemization of β-diketones and β-keto esters in alcohols can be exploited by dynamic kinetic resolution and other stereodynamic synthetic methods, see Chapter 3.1.2.

From a synthetic standpoint, fast enantioconversion is attractive if it can be coupled with highly stereoselective transformation of one of the stereolabile enantiomers. In this case, the racemic starting material undergoes deracemization and is subsequently converted to an enantiopure product. An attractive feature of such a dynamic kinetic resolution is that it allows quantitative conversion of a racemate to a single enantiomer. The discovery of facile racemization pathways for synthetically versatile building blocks such as allenes is therefore quite important.

Axially chiral allenes are often produced as racemic mixtures. The separation of allenic enantiomers can be accomplished by preferential crystallization of diastereomeric salts or by chiral chromatography, but this entails additional purification steps. A further significant disadvantage of traditional enantioseparation is the inherently low atom economy. The maximum yield of the desired enantiomer is only 50%, and in most cases the other 50% has to be discarded or in some way recycled. Horváth and Bäckvall studied the Pd(II)-catalyzed racemization of axially chiral allenes as a potential entry to dynamic kinetic resolution.[7] They found that catalytic amounts of Pd(OAc)$_2$ facilitate racemization of allenes exhibiting various functional groups through a bromopalladation/debromopalladation sequence in the presence of LiBr, Scheme 3.3. The racemization proceeds via reversible formation of an achiral π-allyl palladium complex. Allenic esters, amines and alcohols racemize with half-life times, $\tau_{1/2}$, ranging from less than one minute to a few hours, Table 3.1. The combination of Pd(II)-controlled racemization with stereoselective transformation of one of the interconverting allenic enantiomers may ultimately lead to an enantioconvergent process with high atom economy, utilizing 100% of the starting materials, see Chapter 7.4.2 for examples.

3.1.2   Enantiomerization

Enantiomerization and racemization are closely related and correspond to microscopically reversible and macroscopically irreversible processes, respectively. The mathematical treatment of enantiomerization and racemization is somewhat arbitrary and depends on the choice of method. A polarimetric study provides the rate of racemization, while dynamic chromatography can be used to monitor directly the reversible enantiomerization processes of the same reaction. The principles and value of enantiomerization and deracemization become evident in a discussion of dynamic kinetic resolution (DKR) and crystallization-induced dynamic resolution (CIDR), *vide infra*. The kinetics for the reversible first-order interconversion of enantiomers can be simplified because the rate constants for the forward and backward reaction are identical:

$$\frac{dx}{dt} = -k_2 \left(S_0 + x \right) + k_1 \left(R_0 - x \right) \tag{3.6}$$

where R_0 is the initial concentration of the (R)-enantiomer; S_0 is the initial concentration of the (S)-enantiomer; $x = R_0 - R$, where R is the enantiomer concentration at time t; and $k_2 = k_1 = k_{enant}$ (rate constant of enantiomerization).

$$\frac{dx}{dt} = -k_{enant} \left\{ \left(S_0 + x \right) - \left(R_0 - x \right) \right\} \tag{3.7}$$

$$= k_{enant} \left(R_0 - S_0 - 2x \right) \tag{3.8}$$

$$= 2 k_{enant} \left(\frac{R_0 - S_0}{2} - x \right) \tag{3.9}$$

At equilibrium, $\frac{dx}{dt} = 0$; $x_{eq} = \frac{R_0 - S_0}{2}$
   It follows that:

$$\frac{dx}{dt} = 2 k_{enant} \left(x_{eq} - x \right) \tag{3.10}$$

and

$$\int_0^x \frac{dx}{x_{eq} - x} = 2\,k_{enant} \int_0^t dt \qquad (3.11)$$

Integration gives:

$$\ln\left(\frac{x_{eq}}{x_{eq} - x}\right) = \ln\left(\frac{R_0 - R_{eq}}{R - R_{eq}}\right) = 2\,k_{enant}\,t \qquad (3.12)$$

Dynamic kinetic resolution of rapidly interconverting enantiomers can be achieved if one of the enantiomers undergoes a selective reaction in the presence of a chiral catalyst or reagent. In this case, unidirectional conversion of the racemate to a single enantiomer is driven by concomitant irreversible enantioselective consumption. Noyori and coworkers observed that chiral β-keto esters are stereolabile under conditions that allow (BINAP)RuBr$_2$-catalyzed asymmetric hydrogenation to β-hydroxy esters.[8] The combination of rapid enantioconversion and stereoselective hydrogenation results in complete deracemization of β-keto carboxylic esters because one enantiomer is selectively recognized and significantly more rapidly reduced by the chiral ruthenium complex. For example, dynamic kinetic resolution of racemic methyl 2-oxo-cyclopentanecarboxylate, involving hydrogenation in the presence of [(*M*)-BINAP]RuBr$_2$, produces the corresponding (1*R*,2*R*)-*trans*-β-hydroxymethylester with high *anti*-diastereoselectivity in 99% yield and 92% ee. The racemic β-keto ester is thus converted to a single stereosiomer by a convenient one-pot procedure, Scheme 3.4.

Preparative separation of enantiomers is usually achieved by chiral chromatography or selective crystallization of diastereomeric salts. Chiral carboxylic acids and amino acids are often separated into enantiomers by formation of diastereoisomers with enantiopure resolving agents such as alkaloids. A major drawback to this approach is that the maximum yield is limited to only 50%. Coupling of such a resolution process with enantiomerization can result in complete transformation of a racemic mixture to one enantiomer. The thermodynamically controlled conversion of a mixture of stereoisomers to an enriched or pure single stereoisomer followed by physical separation based on crystallization or extraction is called asymmetric transformation of the second kind or crystallization-induced dynamic resolution. Through careful optimization of temperature, solvent

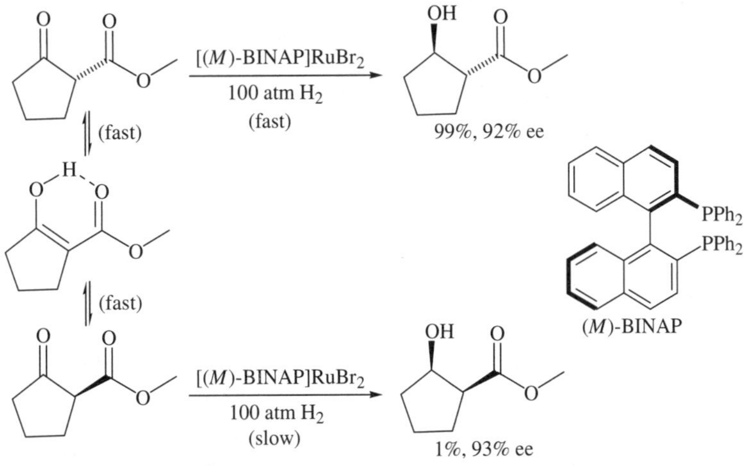

Scheme 3.4 Stereoselective hydrogenation via dynamic kinetic resolution.

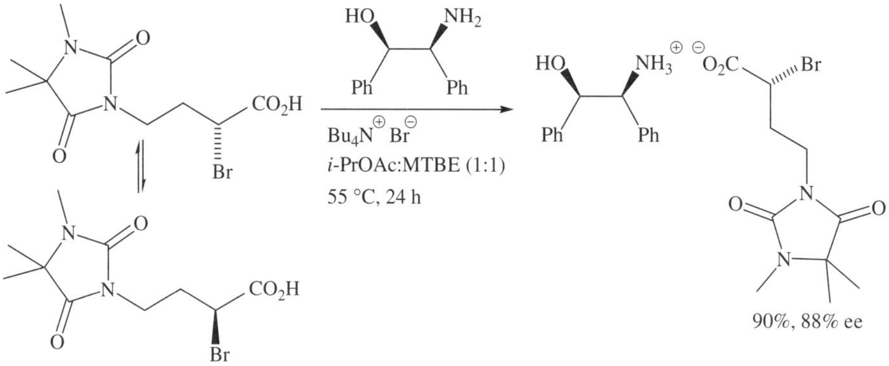

Scheme 3.5 Crystallization-induced dynamic resolution of α-substituted carboxylic acids.

composition, concentration of the substrate and additives, and choice of a chiral resolving agent, Kiau and coworkers developed a process that allows deracemization of stereolabile α-substituted carboxylic acids due to selective precipitation of diastereomeric ammonium salts.[9] For instance, crystallization-induced dynamic resolution of racemic 2-bromo-3,4,4-trimethyl-2,5-dioxo-1-imida-zolidinebutanoic acid in the presence of (1*R*,2*S*)-2-amino-1,2-diphenylethanol and catalytic amounts of tetrabutylammonium bromide furnishes the (*R*)- carboxylic acid in 88% ee, Scheme 3.5.

3.1.3 Diastereomerization

The kinetics of enantiomerization and racemization reactions that proceed in a single step, *i.e.*, through one transition state and without formation of intermediates, can be fully described with a single rate constant. The mathematical treatment of diastereomerization is inherently more complex because the rate constants for the forward and reverse reaction are not identical. Since the interconverting stereoisomers have different thermodynamic stabilities, the study of diastereomerization reactions requires determination of at least two rate constants whose ratio is the equilibrium constant, K_{eq}.

An interesting case arises from the reversible interconversion of enantiomeric 1,8-diarylnaph-thalenes, which requires two consecutive diastereomerization steps involving an achiral interme-diate. Crystallographic analysis and computational studies have shown that this class of compounds affords highly congested structures with cofacial aryl groups residing almost perfectly perpendicular to the naphthalene ring. Substitution of the *peri*-rings gives rise to three stereoiso-mers: two C_2-symmetric and therefore chiral *anti*-isomers, and one achiral *syn*-isomer. Rotation of either ring about the aryl–naphthalene bond results in diastereomerization. Tumambac and Wolf were able to isolate enantiopure *anti*-isomers of a 1,8-diquinolylnaphthalene that is stable to isomerization at 25 °C. Enantioconversion occurs at higher temperatures via two subsequent diastereomerization steps, Scheme 3.6.[10] At equilibrium, the ratio of the two enantiomeric *anti*-conformers to the diastereomeric *syn*-conformer of 1,8-bis(2'-phenyl-4'-quinolyl)naphthalene is 1.2:1, which corresponds to a difference in the Gibbs free energy of the *anti*- and *syn*-isomers, ΔG, of 1.6 kJ/mol according to the Boltzmann equation:[iv]

$$\frac{n_A}{2\,n_B} = \mathrm{e}^{-\Delta G/RT} \tag{3.13}$$

where n_A and n_B are the population of states A and B; and $\Delta G = G_A - G_B$.

[iv] The factor of 2 accounts for the coexistence of two enantiomeric *anti*-isomers. Equilibration of the diquinolylnaphthalene was performed at 97.8 °C.

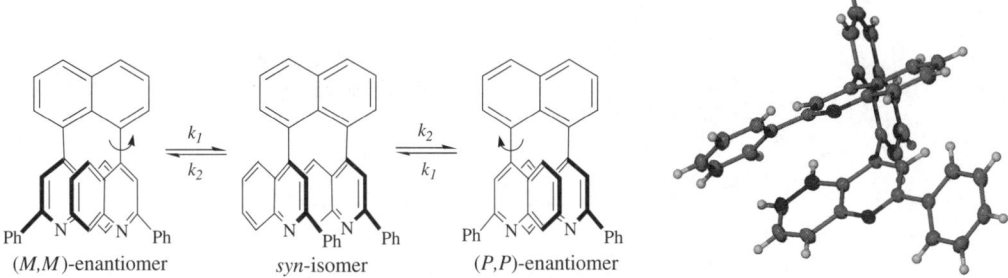

Scheme 3.6   Isomerization of the stereoisomers of 1,8-bis(2′-phenyl-4′-quinolyl)naphthalene and the crystal structure of the *anti*-isomer. [Reproduced with permission from *J. Org. Chem.* **2005**, *70*, 2930-2938.]

The three stereoisomers were separated by preparative chiral HPLC. The conversion of one highly enantioenriched *anti*-isomer to the diastereomeric *syn*-isomer, and subsequently to the opposite enantiomer, occured at 97.8 °C, and was monitored by chiral HPLC at room temperature. The mathematical treatment of the reversible first-order interconversion kinetics can be fully described by the two rate constants k_1 and k_2, since the equienergetic *anti*-isomers form a single meso intermediate and experience the same rotational energy barrier. For the same reason, isomerization of the meso isomer to either enantiomer proceeds at identical rates via equienergetic transition states:

$$\frac{dx}{dt} = -2\,k_2\,(S_0 + x) + 2\,k_1\,(A_0 - x) \qquad (3.14)$$

where A_0 is the initial concentration of the *anti*-isomer and S_0 is the initial concentration of the *syn*-isomer; $x = A_0 - A$ (concentration of the *anti*-isomer at time t); k_1 is the rate constant for the *anti*-to-*syn*-interconversion; and k_2 is the rate constant for the *syn*-to-*anti*-interconversion.

$$\frac{dx}{dt} = -2\,k_2\,S_0 - 2\,k_2\,x + 2\,k_1\,A_0 - 2\,k_1\,x \qquad (3.15)$$

$$= 2\,\{(k_1\,A_0 - k_2\,S_0) - (k_1 + k_2)\}x \qquad (3.16)$$

$$= 2(k_1 + k_2)\left(\frac{k_1\,A_0 - k_2\,S_0}{k_1 + k_2} - x\right) \qquad (3.17)$$

At equilibrium $\frac{dx}{dt} = 0$ and $x_{eq} = \frac{k_1\,A_0 - k_2\,S_0}{k_1 + k_2} = A_0 - A_{eq}$

$$\text{Hence: } \frac{dx}{dt} = 2\,(k_1 + k_2)\,(x_{eq} - x) \qquad (3.18)$$

$$\int_0^x \frac{dx}{(x_{eq} - x)} = \int_0^t 2\,(k_1 + k_2)\,dt \qquad (3.19)$$

Integration gives:

$$\ln\left(\frac{x_{eq}}{x_{eq} - x}\right) = 2(k_1 + k_2)\,t \qquad (3.20)$$

The rate constants for *anti*-to-*syn*-isomerization, k_1, and for *syn*-to-*anti*-interconversion, k_2, were determined as $9.10 \cdot 10^{-5}\,\text{s}^{-1}$ and $1.09 \cdot 10^{-4}\,\text{s}^{-1}$, respectively, Figure 3.2. The rotational energy barrier of 1,8-bis(2′-phenyl-4′-quinolyl)naphthalene was calculated as 122.4 ± 0.2 (121.8) kJ/mol for the conversion of the *anti(syn)*- to the *syn(anti)*-isomer, using the equilibrium constant $K = k_1/k_2$ and the Eyring equation.

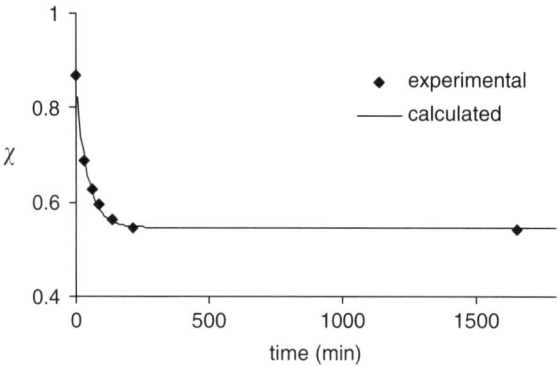

Figure 3.2 Plot of the mole fraction, χ, of *anti*-1,8-bis(2′-phenyl-4′-quinolyl)naphthalene *versus* time monitored at 97.8 °C, and curve fit based on reversible first-order isomerization kinetics. [Reproduced with permission from *J. Org. Chem.* **2005**, *70*, 2930-2938.]

The enol formation of 2-benzoyl cyclohexanone, discussed in Chapter 3.1.1, generates achiral intermediates and results in irreversible racemization. Although enolization of carbonyl compounds bearing an acidic proton at an asymmetric carbon in α-position is often undesirable, it can be useful synthetically if it allows selective isomerization towards an otherwise elusive stereoisomer. Stereoselective aldol reactions have been intensively studied and exploited for asymmetric synthesis of a number of compounds. However, the preparation of some aldol stereoisomers still remains a major synthetic challenge, and can only be accomplished through thermodynamically controlled diastereomerization of β-hydroxy aldehydes or ketones. This is possible because aldol products are prone to retro aldol reaction and isomerization via enolization in the presence of base. Ward and coworkers have shown that certain aldol adducts undergo effective *syn/anti*-isomerization in the presence of imidazole or 4-dimethylaminopyridine, and this constitutes a useful entry to challenging aldol stereoisomers.[11] Enolization of β-hydroxy ketones derived from tetrahydro-4*H*-thiopyran-4-one proceeds under mild conditions, excluding by-products from competing elimination and retro aldol reactions. For example, the MgBr$_2$-promoted aldol reaction of tetrahydrothiopyran-4-one-derived trimethylsilyl enol ethers and structurally similar carboxaldehydes generally favors formation of the *syn*-isomer. Imidazole-catalyzed tautomerization of the *syn*-isomer generates a chiral enol intermediate that forms an equilibrium with both diastereomeric aldol adducts, Scheme 3.7. The diastereomerization yields a *syn/anti* ratio of 1.6:1 at room temperature. Repetitive separation of the diastereomers and isomerization of the recycled *syn*-isomer thus provides access to the *anti*-diastereoisomer which can not be prepared otherwise. The equilibration process is highly solvent-dependent and half-life times range from 12 to 160 hours, Table 3.2. The reversible isomerization of the *syn*-isomer involves a slow enolization step followed by rapid forward and reverse ketonization that either forms the *anti*-isomer or regenerates the *syn*-adduct. The observable isomerization rate constant, k_{obs}, and the corresponding half-life time, $\tau_{1/2}$, are governed by four different rate constants, and describe the macroscopically irreversible transformation of the pure or diastereomerically enriched *syn*-isomer to the final *syn/anti*-mixture at equilibrium:

$$k_{obs} = \frac{\ln 2}{\tau_{1/2}} \tag{3.21}$$

where k_{obs} is the first-order rate constant of diastereomerization.[v]

---

[v] Determination of the overall diastereomerization rate and the equilibrium constant does not provide the individual rate constants for the enolization and ketonization processes.

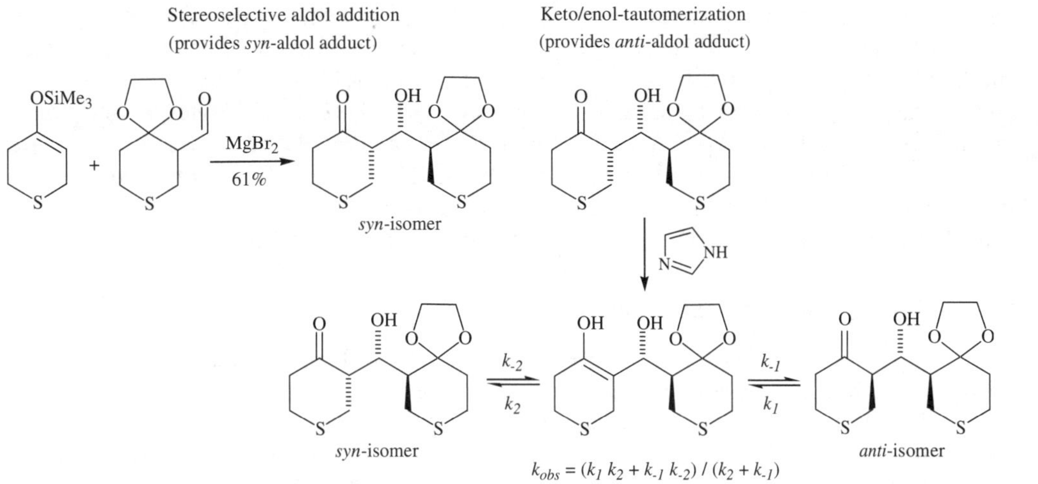

Scheme 3.7 Stereoselective aldol reaction and enolization of β-hydroxy ketone-derived tetrahydro-4*H*-thiopyran-4-ones.

Table 3.2 Solvent effects on the *syn/anti* ratio and diastereomization of β-hydroxy ketones.

Solvent	$K_{eq}(syn/anti)$	$k_{obs}\ [10^{-2}\ h^{-1}]$	$\tau_{1/2}\ [h]$
CDCl$_3$	1.8:1	5.9	12
CD$_2$Cl$_2$	1.9:1	2.9	24
acetone-d$_6$	1.6:1	0.55	130
DMF-d$_7$	2.1:1	0.43	160
C$_6$D$_6$	1.5:1	5.3	13

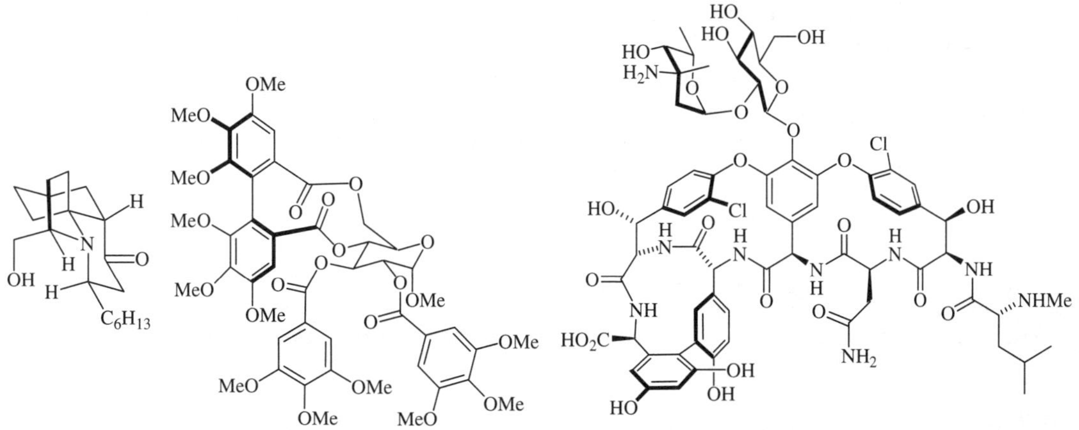

Figure 3.3 Structures of (+)-cylindricene C, permethyl tellimagrandin I and vancomycin (from left to right).

Diastereomerization under thermodynamic control, *i.e.*, asymmetric transformation of the first kind, has been used as a powerful strategy for asymmetric synthesis of complex chiral molecules. For example, Hsung *et al.* reported the total synthesis of the alkaloid (+)-cylindricine C in eight steps including diastereomerization of an aza-tricyclic precursor, Figure 3.3.[12] Meyers and coworkers combined diastereoselective Ullmann coupling and concomitant diastereomerization towards an atropisomeric (S)-hexamethoxybiphenyl derivative which was successfully converted to permethyl tellimagrandin I,[13,14] and several other groups have incorporated asymmetric transformations of intermediate stereolabile diastereomers into ingenious multi-step syntheses of vancomycin.[15–18]

3.1.4 Epimerization and Mutarotation

Epimerization and mutarotation are generally associated with interconversion of carbohydrate diastereomers. Many saccharides, such as pentoses and hexoses, exist as a mixture of an open chain form and two diastereomeric cyclic structures exhibiting an additional chiral carbon center. The diastereomers that are formed through intramolecular hemiacetal or hemiketal formation are referred to as anomers, and are more stable than the open chain form which is often present in trace amounts. Formation of hemiacetals and hemiketals is generally reversible and the isomers interconvert rapidly in aqueous solution. If the pure anomers have a distinct specific optical rotation, the interconversion process can be monitored by polarimetry. The specific rotation of pure α-D-glucopyranose in water is +112° but the value decreases over time to +52.7°. The anomeric β-D-glucopyranose has a specific rotation of +18.7° which increases to the same value in water. In both cases, the change in the optical rotation of the aqueous mixture continues until equilibrium is reached. This macroscopic observation is called mutarotation. Since anomers are diastereomeric, the equilibrium mixture does not usually contain equal amounts of the two isomers. For glucose, the equilibrium mixture consists of 64% of the β- and 36% of the α-anomer because the former displays all substituents in the equatorial position and is therefore thermodynamically more stable. The mutarotation of glucose is a consequence of the interconversion of two pyranose forms via an open chain intermediate but other carbohydrates undergo more complex isomerizations. For instance, fructose equilibrates between its open chain form and the anomers of fructopyranose and fructo-furanose. Mutarotation often complicates the chromatographic purification of carbohydrates as it competes with the separation process. If mutarotation proceeds within the chromatographic time scale, it is likely to cause band broadening or peak splitting. To achieve resolution of anomeric carbohydrates, the separation process must be much faster than the interconversion which can occur within a few minutes in an aqueous mobile phase. Although mutarotation can impede chromato-graphic separation, it can be exploited for analytical purposes. For example, computer simulation of elution profiles that originate from simultaneous separation and interconversion of anomers during liquid chromatography can be used to study the mutarotation kinetics, see Chapter 4.3.[19]

The mutarotation of (+)-α-D-glucose in water can easily be monitored by polarimetry at room temperature and is treated similarly to enantiomerization and racemization reactions. Muta-rotation is a special case of diastereomerization and two different rate constants, k_1 and k_2, have to be considered (Scheme 3.8). Application of Equation 3.12 to a reversible first-order reaction, with $k_1 \neq k_2$, gives:

$$\ln\left(\frac{x_\infty}{x_\infty - x}\right) = \ln\left(\frac{C_{\alpha_0} - C_{\alpha_\infty}}{C_{\alpha_t} - C_{\alpha_\infty}}\right) = (k_1 + k_2)\, t \qquad (3.22)$$

Since the term $\ln\left(C_{\alpha_0} - C_{\alpha_\infty}\right)$ is a constant, the equation can be simplified to:

$$\ln\left(C_{\alpha_t} - C_{\alpha_\infty}\right) = -(k_1 + k_2)\, t + C \qquad (3.23)$$

where C_{α_0} is the concentration of the α-anomer at $t = 0$; C_{α_t} is the concentration of the α-anomer at t; C_{α_∞} is the concentration of the α-anomer at $t = \infty$ (at equilibrium); k_1 is the rate constant for the

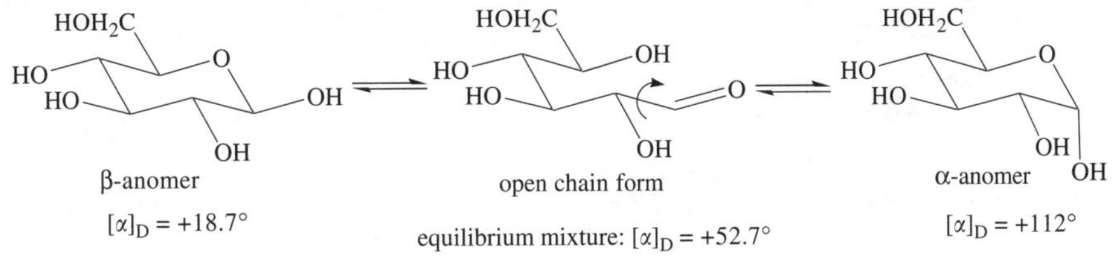

Scheme 3.8 Mutarotation of D-glucose.

α-to-β isomerization; k_2 is the rate constant for the β-to-α isomerization; and $C = \ln\left(C_{\alpha_0} - C_{\alpha_\infty}\right)$.

$$[\alpha]_D^T = \frac{\alpha}{c\,l} \tag{3.24}$$

The specific rotation is defined as :

where $[\alpha]$ is the specific rotation measured with monochromatic light of wavelength λ (usually the sodium D line at 589 nm) at temperature T; α is the observed rotation; l is the path length in dm; and c is the concentration in g/mL.

In general, the rotation of an anomeric mixture at t, α_t, is given by:

$$\alpha_t = [\alpha]_\alpha \, l \, C_{\alpha_t} + [\alpha]_\beta \, l \, C_{\beta_t} = [\alpha]_\alpha \, l \, C_{\alpha_t} + [\alpha]_\beta \, l \left(C_{\alpha_0} - C_{\alpha_t}\right) \tag{3.25}$$

Rearrangement of Equation 3.25 provides C_{α_t}:

$$C_{\alpha_t} = \frac{\alpha_t - [\alpha]_\beta \, l \, C_{\alpha_0}}{\left([\alpha]_\alpha - [\alpha]_\beta\right) l} \tag{3.26}$$

and at $t = \infty$:

$$C_{\alpha\infty} = \frac{\alpha_\infty - [\alpha]_\beta \, l \, C_{\alpha_0}}{\left([\alpha]_\alpha - [\alpha]_\beta\right) l} \tag{3.27}$$

where α_∞ is the observed optical rotation at $t = \infty$ (at equilibrium); $[\alpha]_\alpha$ is the specific rotation of the α-anomer; and $[\alpha]_\beta$ is the specific rotation of the β-anomer.

Combination of these equations gives:

$$C_{\alpha_t} - C_{\alpha\infty} = \frac{\alpha_t - [\alpha]_\beta \, l \, C_{\alpha_0}}{\left([\alpha]_\alpha - [\alpha]_\beta\right) l} - \frac{\alpha_\infty - [\alpha]_\beta \, l \, C_{\alpha_0}}{\left([\alpha]_\alpha - [\alpha]_\beta\right) l} \tag{3.28}$$

$$C_{\alpha_t} - C_{\alpha\infty} = \frac{\alpha_t - \alpha_\infty}{\left([\alpha]_\alpha - [\alpha]_\beta\right) l} \tag{3.29}$$

$$\ln\left(C_{\alpha_t} - C_{\alpha\infty}\right) = \ln\frac{\alpha_t - \alpha_\infty}{\left([\alpha]_\alpha - [\alpha]_\beta\right) l} = \ln\left(\alpha_t - \alpha_\infty\right) - \ln\left([\alpha]_\alpha - [\alpha]_\beta\right) l \tag{3.30}$$

$$= \ln\left(\alpha_t - \alpha_\infty\right) + C \tag{3.31}$$

The equation finally becomes:

$$\ln\left(\alpha_t - \alpha_\infty\right) = -\left(k_1 + k_2\right) t + C \tag{3.32}$$

where C is a constant.

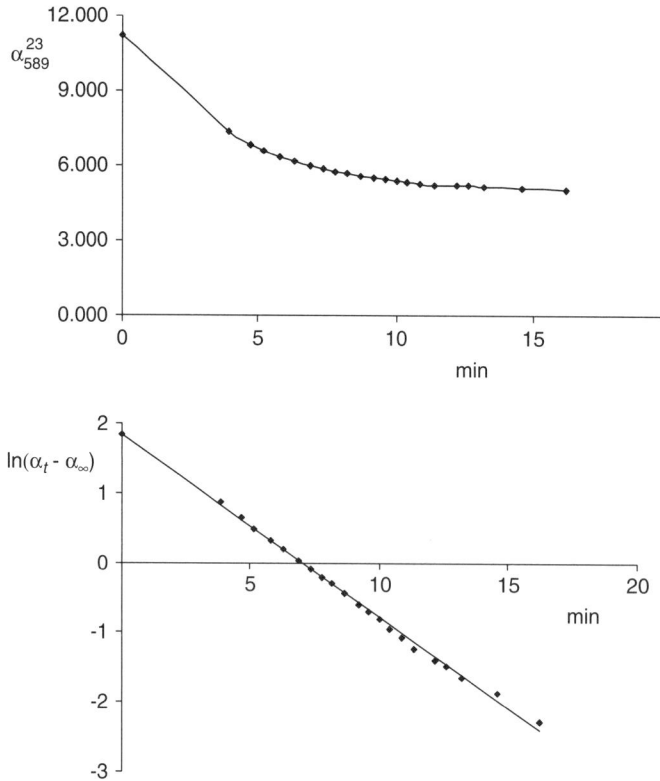

Figure 3.4   Mutarotation of (+)-α-glucose at pH 7.0 in 0.05 M NaH$_2$PO$_4$/Na$_2$HPO$_4$ at 23.0 °C.

An aqueous solution of 0.10 g/mL of (+)-α-D-glucose at neutral pH has an optical rotation of +11.20° which rapidly decreases to the equilibrium value of 4.96°, Figure 3.4. From a plot of $\ln\left(\alpha_t - \alpha_\infty\right)$ versus t, the value of $-(k_1 + k_2)$ can be directly obtained from the slope. The half-life time of mutarotation, $\tau_{1/2}$, can be calculated in accordance with Equation 3.5 as:

$$\tau_{1/2} = \frac{\ln 2}{k_1 + k_2} = \frac{\ln 2}{0.26\,\text{min}^{-1}} = 2.67\,\text{min}$$

The term epimerization was originally coined to describe the change in the absolute configuration at *C*-2 in aldoses. A classical example is the interconversion between glucose and mannose. Today, epimerization generally refers to stereomutation of a single stereocenter in diastereomers that have at least two elements of chirality. Epimerization of natural carbohydrates and cyclic polyols is often achieved with epimerases that catalyze inversion of the absolute configuration at one particular chiral center.[20] Alternatively, stereocontrolled epimerization can be accomplished via acetalization with highly active aldehydes in the presence of a carbodiimide such as DCC, provided that the pyranoside and inositol display a *cis/trans*-sequence of three contiguous hydroxyl groups.[21,22] Treatment of methyl α-L-glucopyranoside with DCC and chloral and subsequent ring flip generate a reactive intermediate that forms a trichloroethylidene acetal through intramolecular nucleophilic substitution involving inversion of the configuration at *C*-4. Reduction of the trichloroethylidene to the corresponding ethylidene acetal facilitates acidic hydrolysis which is followed by basic removal of the carbamoyl protection group to afford methyl α-L-galactopyranoside. Comparison of the $^4C_1$-conformation of the two L-pyranosides reveals regioselective epimeration at *C*-4, Scheme 3.9.

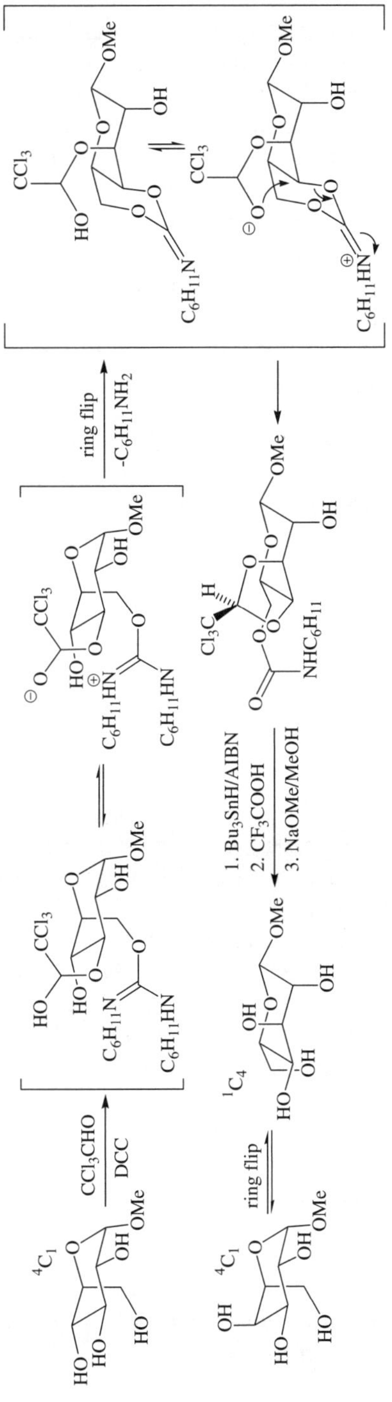

Scheme 3.9 Epimerization of methyl α-L-glucopyranoside.

3.2 STEREOMUTATIONS OF CHIRAL COMPOUNDS: MECHANISMS AND ENERGY BARRIERS

In general, stereomutations of chiral compounds involve either inversion of configuration or interconversion of conformational isomers.[vi] This may require sequential cleavage and formation of one or more covalent bonds. For example, microwave-assisted racemization of (+)-vincadifformine, a naturally occuring indole alkaloid, proceeds via retro Diels–Alder reaction. The achiral intermediate readily undergoes intramolecular [4+2]cycloaddition to regenerate the pentacyclic scaffold which is obtained as a racemate, Scheme 3.10.[23] Epimerization of tilivalline occurs in the presence of a Lewis acid at 50 °C within 24 hours. Reversible ring opening affords a mixture of two diastereomers in a ratio of 87:13 at equilibrium.[24] However, stereomutations do not necessarily require bond cleavage and generation of an achiral intermediate. The dynamic stereochemistry of the tricyclic pyridone shown in Scheme 3.11 provides an interesting case.[25] This compound bears two elements of chirality, and exists as a mixture of four stereoisomers that interconvert via rotation about the chiral 2-tolylpyridone axis and flipping of the seven-membered ring. The free energy barriers to these independent processes are 79.2 and 91.8 kJ/mol, respectively. Axially chiral biaryls and many other atropisomeric compounds racemize via thermally or photochemically induced rotation about a chiral axis, which is discussed in Chapter 3.3.

Numerous racemization and diastereomerization procedures have been reported to date. One can classify the most prominent mechanisms as follows:[26]

1. Thermal reactions
2. Photochemical reactions
3. Acid and base catalysis
4. Radical reactions
5. Oxidation-reduction sequences
6. Enzyme catalysis
7. Nucleophilic substitutions

A mechanistic understanding of these processes is crucial for both prediction and control of the conformational and configurational stability of chiral compounds. As has been pointed out in the preceding chapters, racemization and diastereomerization are undesirable if they reduce the efficiency of asymmetric synthesis or the pharmacological integrity of chiral drugs. However, they can also be advantageous, and have been exploited for the development of molecular propellers and motors or asymmetric strategies based on dynamic kinetic and dynamic thermodynamic resolution. To provide a systematic overview, the wealth of racemization and diastereomerization reactions is discussed below, based on individual functional groups.

3.2.1 Alkanes

Thermal racemization of alkanes devoid of other functional groups is not observed unless very high temperatures are applied. Both photochemically and thermally induced isomerizations generally occur via homolytic bond cleavage and radical formation, which is facilitated in cycloalkanes exhibiting considerable ring strain (see Chapter 3.2.11 for an example of heterolytic cyclopropane ring opening). Racemization and *cis/trans*-isomerization via sequential ring opening and closure of chiral cyclopropanes is usually studied in the gas phase, to avoid side reactions of intermediate diradicals.[27–33] The stereomutations of a series of cycloalkanes have been examined by Baldwin and others using vibrational circular dichroism to accurately monitor the change in the enantiomeric

---

[vi] This is not necessarily the case with catenanes, rotaxanes and other topologically chiral compounds, see Chapter 9.

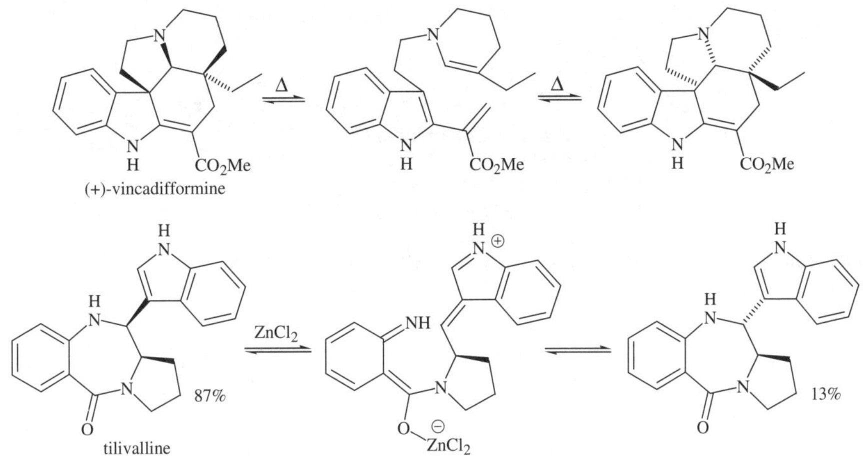

Scheme 3.10 Racemization of vincadifformine (top) and epimerization of tilivalline (bottom).

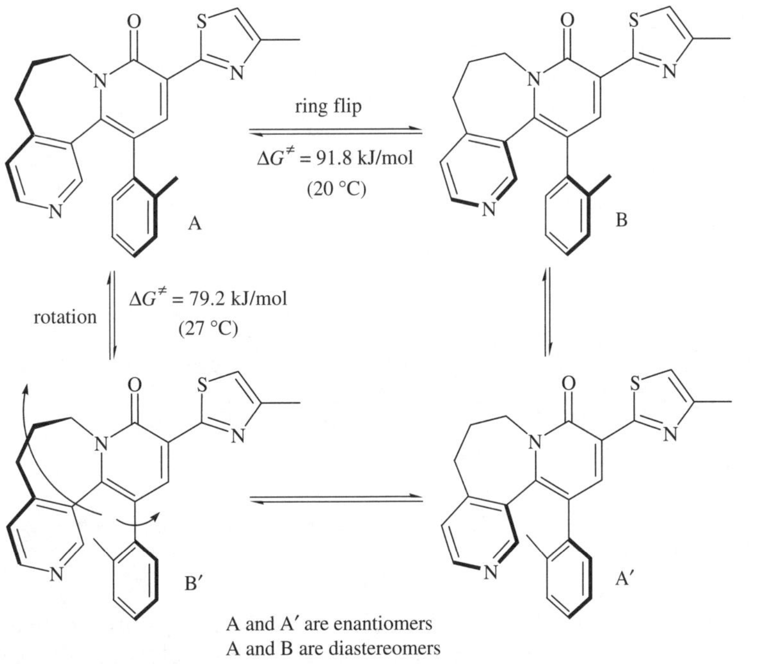

A and A′ are enantiomers
A and B are diastereomers

Scheme 3.11 Ring flipping and atropisomerism of a tricyclic pyridone.

composition of chiral *trans*- and *cis*-isomers. They observed that the rate of enantiomerization of *trans*-1,2-disubstituted cyclopropanes is correlated to the sum of the substituent radical stabilization energies, which indicates that transient 1,3-diradicals are formed. Activation energies range from 145 to 240 kJ/mol, and incorporation of functional groups that stabilize the intermediate diradical increases reaction rates, Scheme 3.12. Racemization and concomitant diastereomerization of chiral cyclobutanes have been observed at temperatures above 400 °C.[34]

1,1,2,2-Tetraarylethanes adopt propeller-shaped ground state conformations resulting in helical chirality even when four equal aryl groups are present.[35] The vicinal methine hydrogens remain on

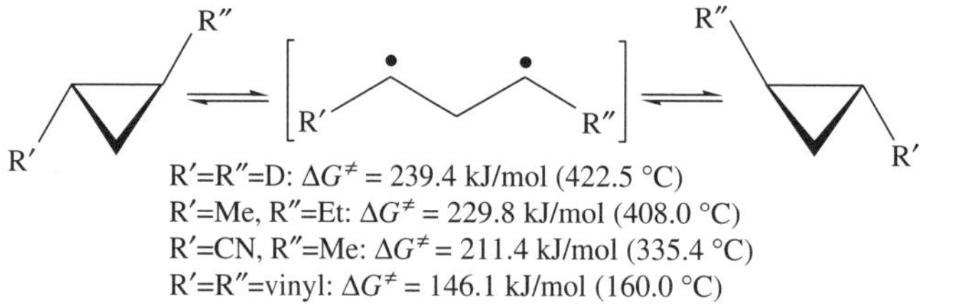

R′=R″=D: $\Delta G^{\neq}$ = 239.4 kJ/mol (422.5 °C)
R′=Me, R″=Et: $\Delta G^{\neq}$ = 229.8 kJ/mol (408.0 °C)
R′=CN, R″=Me: $\Delta G^{\neq}$ = 211.4 kJ/mol (335.4 °C)
R′=R″=vinyl: $\Delta G^{\neq}$ = 146.1 kJ/mol (160.0 °C)

Scheme 3.12 Enantiomerization barriers of 1,2-disubstituted cyclopropanes.

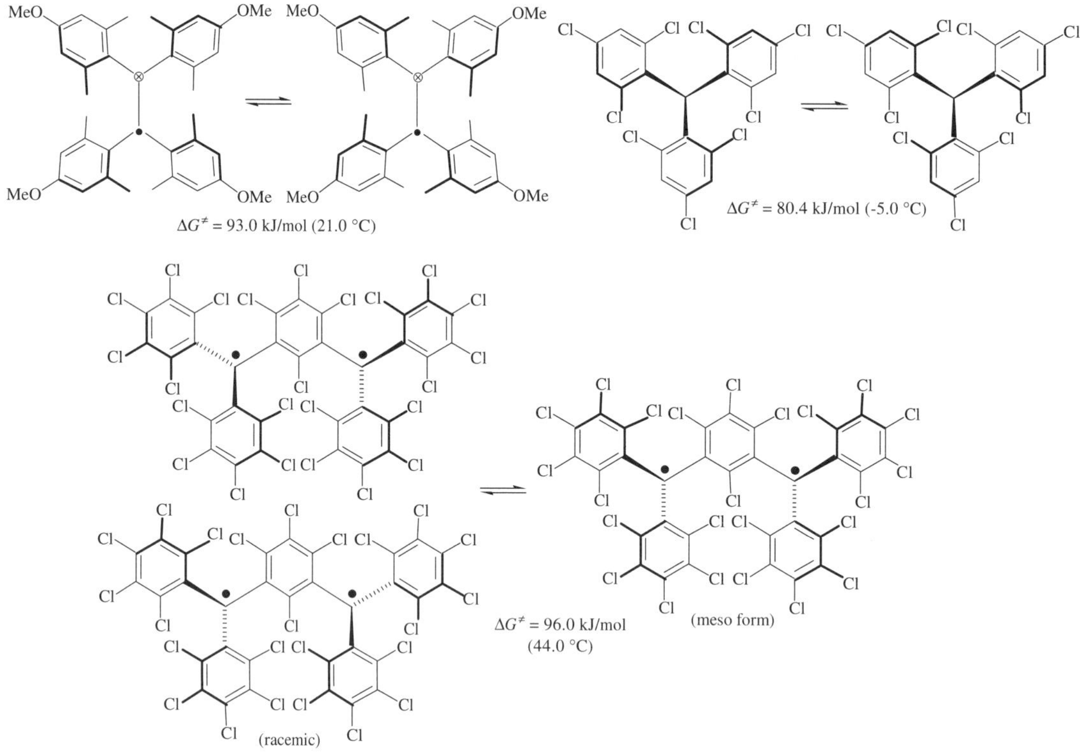

$\Delta G^{\neq}$ = 93.0 kJ/mol (21.0 °C) $\Delta G^{\neq}$ = 80.4 kJ/mol (-5.0 °C)

$\Delta G^{\neq}$ = 96.0 kJ/mol
(44.0 °C)

(racemic)   (meso form)

Scheme 3.13 Stereomutations of helical 1,1,2,2-tetrakis(2,6-dimethyl-4-methoxy)ethane and perchlorinated triarylmethanes.

opposite sides to minimize steric repulsion between the bulky aryl groups. Racemization proceeds via a concerted ring flipping mechanism and does not involve bond cleavage. The geared rotation of the aryl rings in C$_2$-symmetric 1,1,2,2-tetrakis(2,6-dimethyl-4-methoxy)ethane has an energy barrier of 93.0 kJ/mol at 21 °C. Many other triarylmethanes and triarylmethyl radicals show similar stereomutations. For example, tris(2,4,6-trichlorophenyl)methane racemizes with an energy barrier of 80.4 kJ/mol at –5.0 °C, Scheme 3.13.[36] The structures and stereodynamics of other propeller-like molecules are discussed in Chapter 8.

3.2.2 Alkenes and Annulenes

Sterically overcrowded alkenes possess a twisted ground state conformation rather than a planar structure, if severe repulsion between the ethylene substituents renders the latter an energy maximum. By analogy with allenes and biphenyls, selectively substituted nonplanar alkenes can be axially chiral. The barrier to rotation about the stereogenic axis in tetrasubstituted ethylenes is often surprisingly low. This can be a consequence of enhanced ground state energy due to considerable strain and nonbonding interactions between the vinyl substituents, a decreased transition state energy or a combination of both. The structures and interconversion of the stereoisomers of several bistricyclic alkenes have been studied, including disubstituted bifluorenylidenes,[37–40] biphenanthrylidenes,[41–43] biacridanes,[44] bixanthylidenes,[45] bithioxanthylidenes,[46] and bianthrones.[47,48] Rotation about the stereogenic axis can result in enantioconversion and *E/Z*-isomerization, and the corresponding energy barriers vary from 45 to 120 kJ/mol, Figure 3.5.[vii]

Disubstituted bifluorenylidenes favor a twisted ground state geometry reminiscent of biphenyls, Scheme 3.14. In order to optimize both steric repulsion, which is minimized in a perpendicular orientation of the fluorene planes, and π-conjugation, which is at a maximum in a coplanar structure, bifluorenylidenes adopt a torsion angle of approximately 45°. In this conformation substantial π-overlap, which decreases with the cosine of the torsion angle, is maintained while repulsive interactions are effectively reduced. Rotation about the double bond is hindered by two different barriers; one is a consequence of strong steric interactions in a planar structure, and the other stems from loss of stabilizing π-overlap and conjugation at a perpendicular geometry. Rotation about the 0° barrier causes enantiomerization whereas the 90° barrier separates *E/Z*-diastereomers. Investigation of both processes in the 1,1′-disubstituted bifluorenylidene shown in Figure 3.5 revealed that rotational barriers to these stereomutations are quite similar and approximately 84 kJ/mol.

A quite different situation arises with biacridanes, bixanthylidenes and bianthrones. These sterically crowded alkenes can afford a complex mixture of conformers. Usually, so-called wing-shaped folded conformations are more heavily populated than less stable twisted conformations, Scheme 3.15. Isomerization of the folded (*Z*)-conformer *A* via eclipsing of the two smaller

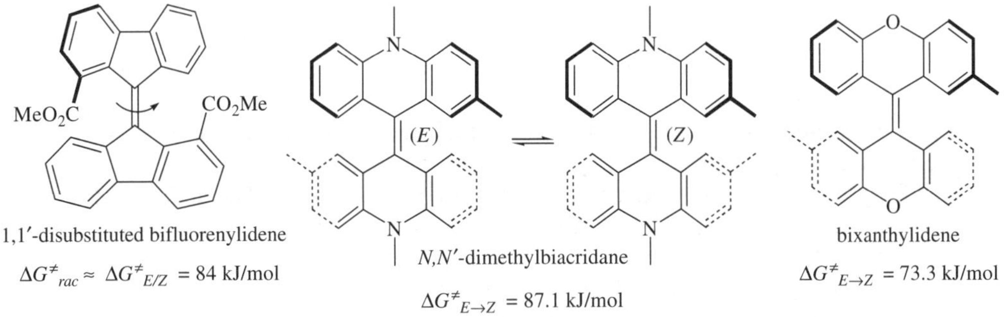

1,1′-disubstituted bifluorenylidene

$\Delta G^{\neq}_{rac} \approx \Delta G^{\neq}_{E/Z} = 84$ kJ/mol

N,N′-dimethylbiacridane

$\Delta G^{\neq}_{E \to Z} = 87.1$ kJ/mol

bixanthylidene

$\Delta G^{\neq}_{E \to Z} = 73.3$ kJ/mol

Figure 3.5 Structures and rotational energy barriers of axially chiral alkenes.

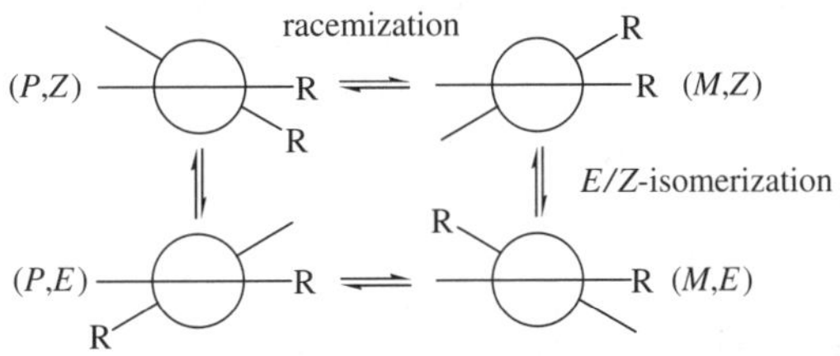

Scheme 3.14 Twisted conformations and rotations of disubstituted bifluorenylidenes.

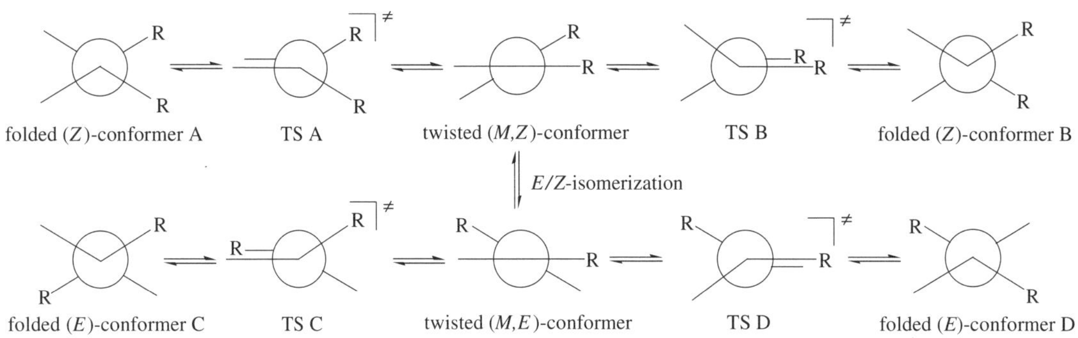

Scheme 3.15 Interconversion pathways of folded alkenes.

substituents (transition state *A*) gives the twisted (*M*,*Z*)-form which may undergo *E*/*Z*-isomerization or formation of conformer *B*, the enantiomer of the original folded isomer *A*. This process occurs via *R*/*R*-eclipsing (transition state *B*). Since *R*/*R*-eclipsing causes more steric repulsion than *H*/*H*-eclipsing, transition state *B* is less stable than transition state *A*.[viii] The corresponding (*E*)-conformer gives rise to similar stereomutations.

Interconversion of the enantiomers of (*E*)- and (*Z*)-1,1′,2,2′,3,3′,4,4′-octahydro-4,4′-biphenanthrylidene exhibiting two helical entities proceeds via consecutive inversion of each tetrahydrophenanthrene ring but without concurrent *E*/*Z*-isomerization.[49,50] Rapid interconversion of the enantiomers of the (*Z*)-isomer is observed at 25 °C while racemization of the less sterically crowded (*E*)-olefin requires temperatures above 55 °C.[51] The Gibbs activation energy for racemization of the (*Z*)-and the (*E*)-olefin is 97.7 and 114.3 kJ/mol, respectively, Scheme 3.16. The racemization processes involve a meso intermediate having either a symmetry plane or an inversion center. The intriguing stereodynamic chemistry and the thermo- and photochromic properties of octahydro-4,4′-biphenanthrylidenes and other overcrowded chiral alkenes have been exploited for the development of molecular switches and motors, see Chapters 8.9 and 8.11.[52,53]

In the case of twisted push-pull or capto-dative ethylenes, the combination of electron-donating and electron-withdrawing groups on opposite sides of the alkene moiety results in stabilization of a zwitterionic resonance structure and delocalization of the π-bond.[54] The reduced double bond character of the chiral axis greatly facilitates rotation, Scheme 3.17.[55,56] In *trans*-cycloalkenes the aliphatic bridge is forced out of the plane of the twisted double bond which breaks the symmetry of the molecule. Accordingly, *trans*-cycloalkenes display planar chirality and racemization can occur by swiveling of the double bond through the saturated ring component. This is indeed a stepwise process, in which the crown conformation of the chiral cycloalkene is first changed to an intermediate chairlike structure that rearranges to the enantiomer of the original crown conformer, Scheme 3.18. The formation of the distorted chair conformation is believed to be the rate-determining step, and the barrier to inversion depends on the ring size and steric interactions between the bridge and additional double bond substituents.[57,58] The enantiomers of (*E*)-cyclooctene can be isolated at room temperature and undergo interconversion at elevated temperatures with an activation barrier of 146.1 kJ/mol at 155 °C.[59] As the ring size increases, there is less steric hindrance to swiveling of the double bond through the bridge. Racemization of (*E*)-cyclononene and (*E*)-cyclodecene is fast at room temperature, and the corresponding energy barriers have been determined as 83.7 and 44.8 kJ/mol, respectively.[60,61] The smallest *trans*-cycloalkene, (*E*)-cycloheptene, has been generated under cryogenic conditions, but it affords severe steric strain and rapidly forms the more stable (*Z*)-isomer at

---

[viii] It should be noted that only one of two racemization courses is shown in Scheme 3.15. Rotation of conformer *A* via a transition state showing *R*/*R*-eclipsing (enantiomeric to transition state *B*) generates the twisted (*P*,*Z*)-isomer which can then generate conformer *B* via a transition state exhibiting *H*/*H*-eclipsing (enantiomeric to transition state *A*).

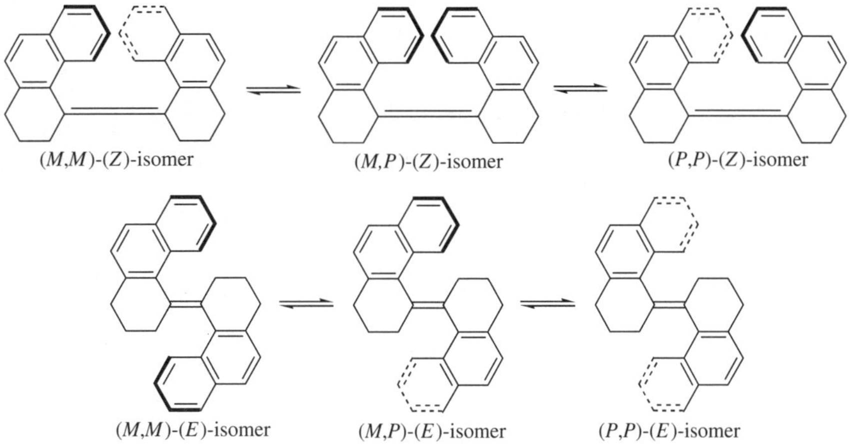

Scheme 3.16 Racemization course of 1,1',2,2',3,3',4,4'-octahydro-4,4'-biphenanthrylidenes.

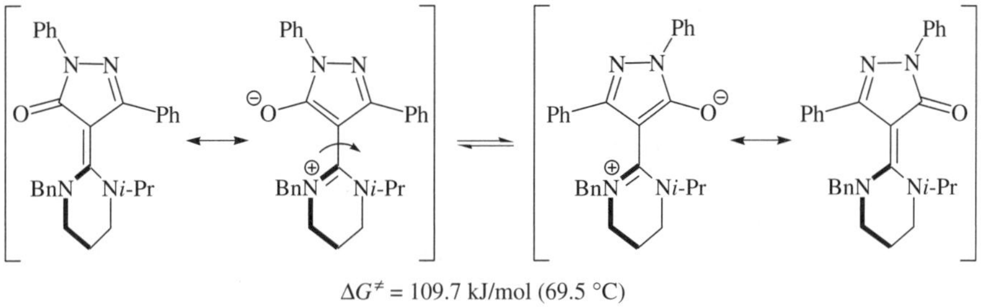

$$\Delta G^{\neq} = 109.7 \text{ kJ/mol } (69.5 \text{ }^\circ\text{C})$$

Scheme 3.17 Racemization of a push-pull ethylene.

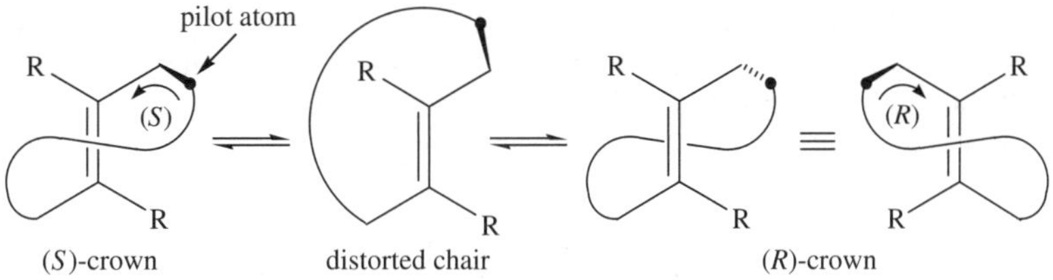

Scheme 3.18 Racemization course of *trans*-cycloalkenes.

room temperature.[62] Incorporation of a larger heteroatom such as silicon into the ring reduces the strain, and the enantiomers of *trans*-silacycloheptene have been isolated by preparative gas chromatography and proved to be thermally stable.[63]

The remarkable stereodynamics of cycloalkenes, *i.e.*, their propensity for *E/Z*-isomerization and enantioconversion, have been exploited for asymmetric synthesis. Inoue and coworkers demonstrated that photoisomerization of cyclooctene in the presence of a chiral sensitizer allows transformation of the achiral (*Z*)-isomer to nonracemic (*E*)-cyclooctene, see Chapter 8.10.[64–69] Bridging of a cycloalkene impedes the swiveling motion and significantly enhances conformational stability. This has been demonstrated with (5*S*)-bicyclo[3.3.1]-1(2)-nonene, a methylene-bridged derivative of (*E*)-cyclooctene, representing an anti-Bredt compound, Figure 3.6.[70] Replacement of the two vinyl hydrogens of *trans*-cycloalkenes by a second bridge generates planar chiral [a.b]betweenanenes. The presence of two rings dramatically increases the racemization barrier, and the enantiomers of [10.10]betweenanene and other homologs can be isolated at room temperature.[71–73]

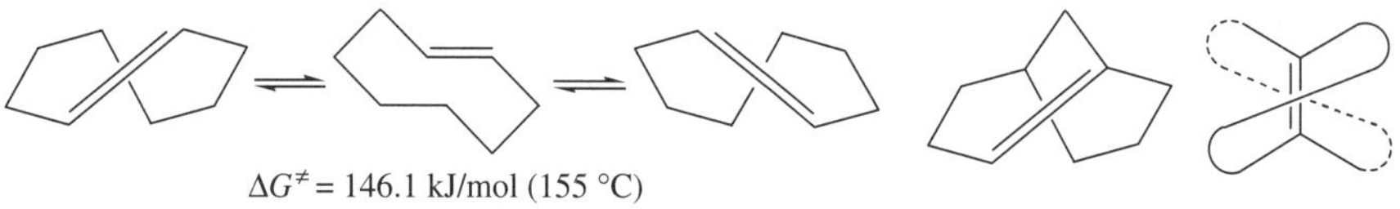

$$\Delta G^{\neq} = 146.1 \text{ kJ/mol } (155 \text{ °C})$$

Figure 3.6   Racemization of (*E*)-cyclooctene (left), and structures of (5*S*)-bicyclo[3.3.1]-1(2)-nonene and an [a.b]betweenanene (right).

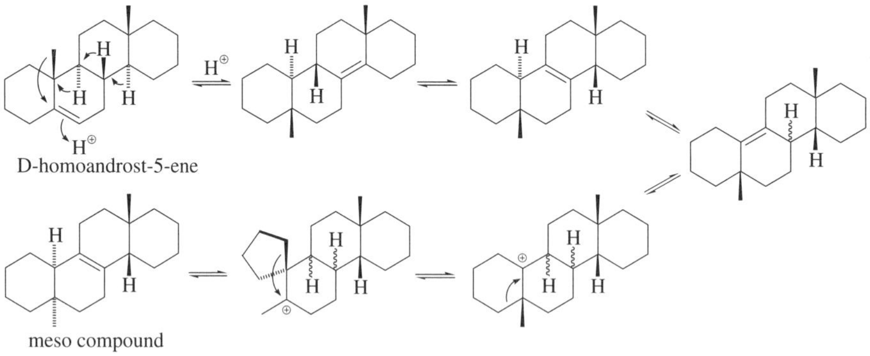

Scheme 3.19   Racemization of D-homoandrost-5-ene.

Backbone rearrangements of olefinic steroids from the cholestane, androstane and pregnane series are known to occur under acidic conditions and usually generate a mixture of constitutional isomers. An interesting exception is the reversible acid-catalyzed rearrangement of D-homo-androst-5-ene.[74,75] The stereochemical outcome has been attributed to the formation of a meso compound which causes complete racemization at thermodynamic equilibrium, Scheme 3.19. It is noteworthy that, unlike the examples discussed above, the chirality of D-homoandrost-5-ene and its isomers does not originate from the presence of an axially chiral double bond. However, the alkene moiety still plays a crucial role in the racemization process.

Some crowded alkenes exhibiting aryl substituents possess helical chirality, reminiscent of propellers, Scheme 3.20. The dynamic stereochemistry of these compounds is due to rotation of the aryl substituents about the stereogenic vinyl–aryl axes, rather than *E/Z*-isomerization of the double bond. Tetramesitylethylene represents a four-bladed molecular propeller that racemizes through correlated rotation of the four aryl rings with an energy barrier of 165.8 kJ/mol.[76] Stereomutations of other molecular propellers can be more complicated and may involve competing processes. For example, (*E*)- and (*Z*)-2-*meta*-methoxymesityl-1,2-dimesitylvinyl acetates undergo two-ring flip enantiomerization and three-ring flip diastereomerization. The three-ring flip has a lower energy barrier than α,β-, α,β'- and β,β'-two-ring rotations. Ring flipping between chiral conformations of triarylvinyl derivatives[77] and cog-wheeling of chiral propeller molecules are discussed in greater detail in Chapter 8.[78–82]

Annulenes are cyclic hydrocarbons with alternating single and double bonds, and are generally classified into aromatic, antiaromatic and nonaromatic compounds. According to Hückel's rule, planar annulenes with ($4N+2$) π-electrons (where *N* is an integer) such as benzene are aromatic. Cyclobutadiene and other annulenes containing $4N$ π-electrons are antiaromatic. The third class of annulenes comprises nonplanar and therefore nonaromatic species that often display interesting stereodynamics. For example, cyclooctatetraene adopts a tub conformation with a ring inversion barrier of 57.3 kJ/mol.[83] Incorporation of substituents into this puckered [8]annulene introduces chirality. The enantiomeric conformations of substituted cyclooctatetraenes can interconvert via

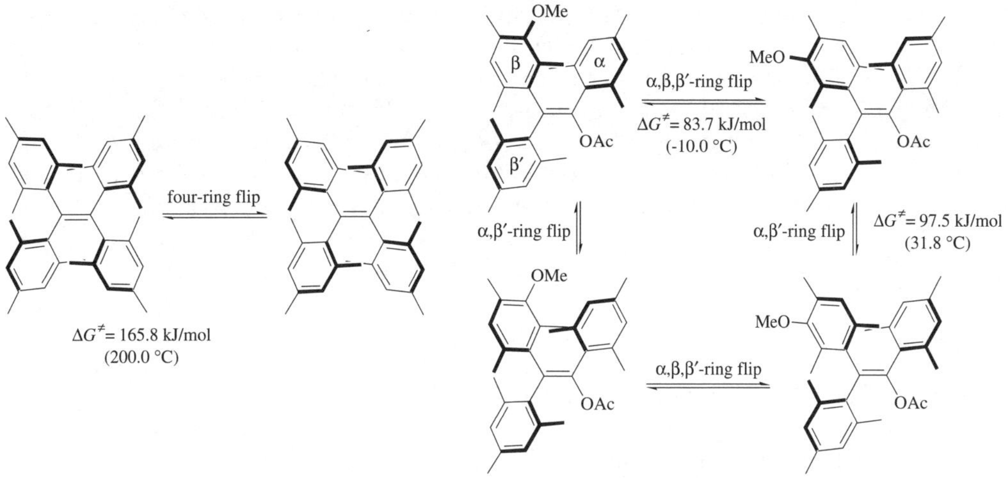

Scheme 3.20 Enantiomerization of tetramesitylethylene (left) and stereomutations of (*E*)-2-*meta*-methoxymesityl-1,2-dimesitylvinyl acetate (right).

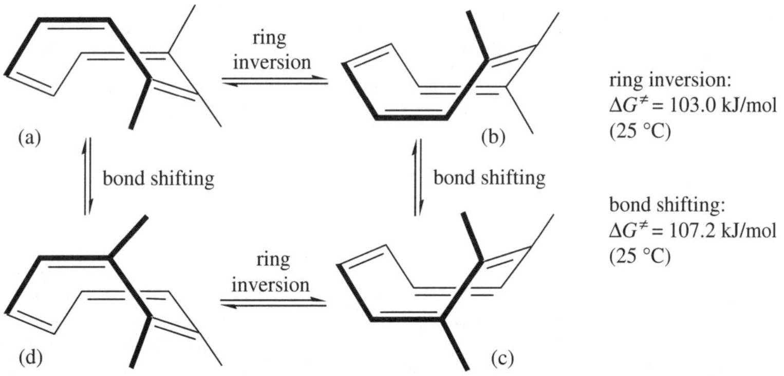

Scheme 3.21 Tub conformation and stereomutations of 1,2,3-trimethylcyclooctatetraene. Note that (a) = (c) and (b) = (d), whereas (a) and (b) are enantiomeric.

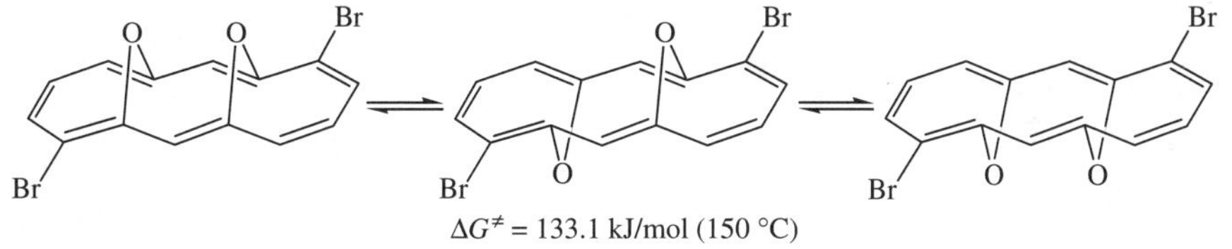

Scheme 3.22 Stepwise interconversion of the isomers of 2,9-dibromo-1,6:8,13-bisoxido[14]annulene.

bond shifting around the whole molecule, resulting in exchange of single and double bonds, or via ring inversion, Scheme 3.21.[84] While both mechanisms might be operative at the same time, the energy barrier to bond shifting usually (but not necessarily) exceeds that for ring inversion.[85–94] Paquette *et al.* found that 1,2,3-trimethylcyclooctatetraene has a barrier to ring inversion and bond shifting of 103.0 and 107.2 kJ/mol, respectively.[95–97] Introduction of another substituent significantly increases the conformational stability. The activation energies for inversion and bond shifting of 1,2,3,4-tetramethylcyclooctatetraene are 133.1 and 134.0 kJ/mol.[98,99]

An interesting situation arises with bridged annulenes that exist as *cis*- and *trans*-isomers.[100,101] Interconversion between the planar and therefore aromatic *cis*-isomers and the puckered nonaromatic *trans*-form of doubly bridged [14]annulenes does not take place in the presence of methylene bridges, but has been observed with a 1,6:8,13-bisoxido[14]annulene, Scheme 3.22.[102] In fact, the enantiomers of *cis*-2,9-dibromo-1,6:8,13-bisoxido[14]annulene interconvert above 120 °C with an activation barrier of 133.1 kJ/mol. The racemization requires consecutive passage of the oxygen bridges through the annulene plane. The *trans*-intermediate possesses an inversion center and is achiral.

3.2.3 Allenes and Cumulenes

Racemization of axially chiral allenes in the absence of a catalyst generally requires high temperatures.[103] Aliphatic allenes such as 1,3-dimethylallene possess rotational energy barriers above 180 kJ/mol, Scheme 3.23. The rotation about the chiral axis in simple allenes is believed to be a two-step process involving a nonlinear allyl diradical intermediate with considerable carbene character.[104] Both conjugation and ring strain in cyclic allenes decrease the barrier to racemization. For example, dimethyl 2,3-pentadienedioate has an interconversion barrier of only 128 kJ/mol, which has been attributed to a resonance-stabilized dipolar structure.[105,106] Allenes are also susceptible to photoracemization at room temperature.[107,108]

Mild racemization of allenes occurs in the presence of stoichiometric amounts of organocuprate and Grignard reagents.[109] Imidozirconium complexes undergo stereospecific [2+2]cycloaddition with chiral allenes to an azazirconacyclobutane intermediate that racemizes via reversible β-hydride

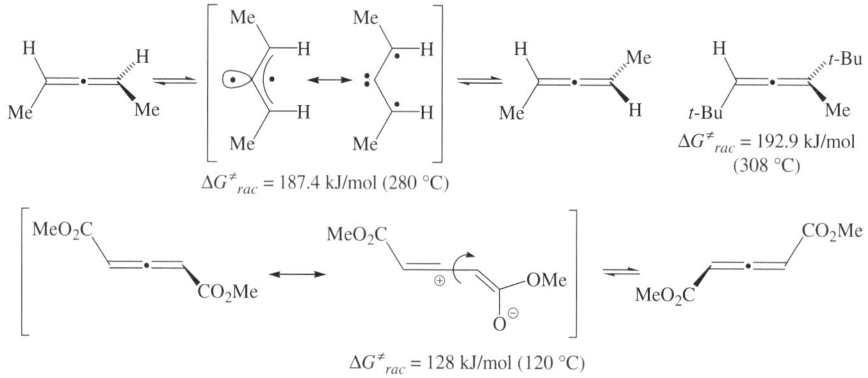

Scheme 3.23   Racemization of allenes.

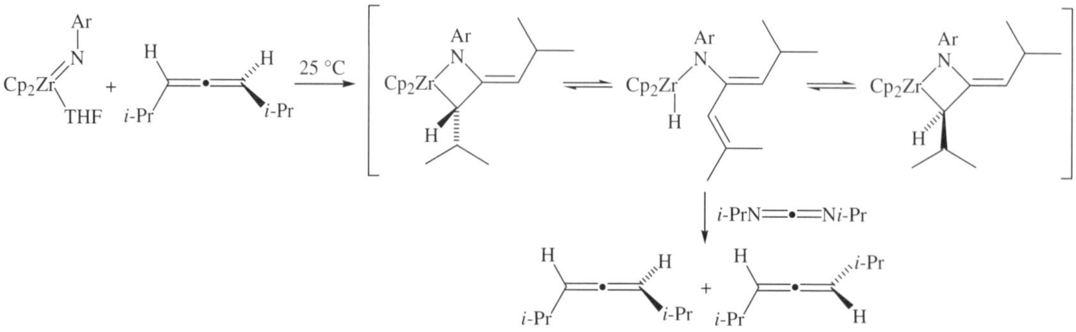

Scheme 3.24   Imidozirconium-promoted racemization of allenes.

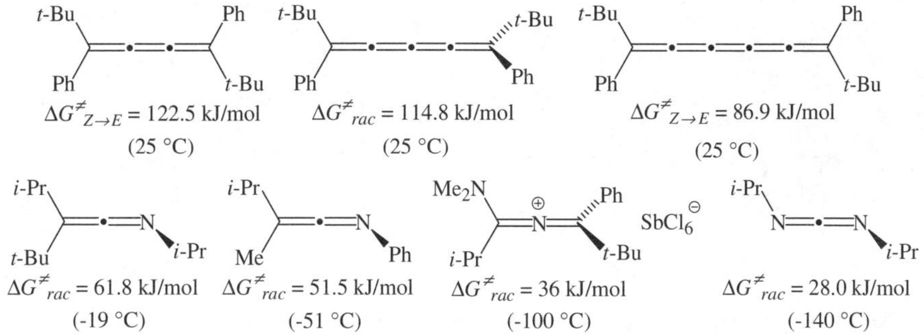

Figure 3.7 Thermal isomerization barriers of cumulenes.

elimination at room temperature.[110,111] Final cleavage of the metallacycle with diisopropylcarbodi-imide thus produces a racemic allene, Scheme 3.24. Synthetically more useful are rapid palladium-catalyzed racemizations suitable to dynamic kinetic resolution (DKR) and dynamic kinetic asymmetric transformation (DYKAT), see Chapters 3.1.1, 7.4.2, and 7.5.[112] Alternatively, deracemization of allenes has been accomplished with $Eu(hfc)_3$ and other chiral lanthanide shift reagents.[113,114]

Properly substituted cumulenes with an even number of unsaturated bonds are chiral due to the orthogonal arrangement of the successive double bonds.[115] In contrast, cumulenes with an odd number of double bonds are planar and therefore achiral. Chiral allenes generally do not show any sign of racemization in the absence of a catalyst such as $Pd(OAc)_2$ at room temperature but higher cumulenes are less stable. As a rule of thumb, the energy barrier to rotation about the stereogenic axis of cumulenes decreases as the number of double bonds increases, Figure 3.7.[116,117] Racemization of other cumulenes including ketene imines,[118,119] 2-azaallenium salts[120] and carbodi-imides[121] is fast and can be observed under cryogenic conditions, but the enantiomers of ketene immonium salts are isolable at 25 °C.[122]

3.2.4 Helicenes and Phenanthrenes

Helicenes consist of *ortho*-condensed aryl rings that form a nonplanar helical structure, Figure 3.8. The racemization barrier of [5]helicene is 100.9 kJ/mol at 57 °C, which corresponds to a half-life of 62.7 minutes. Interestingly, racemization of monoaza- and thiapentahelicenes is significantly faster.[123,124] The enantioconversion of pentahelicenes probably proceeds via stepwise slippage of the two ends through the mean plane of the helicene. The higher homologs exhibit considerable steric hindrance to this process and are much more stable. For example, [6]helicene has a barrier to racemization of 151.5 kJ/mol, and [7]-, [8]- and [9]helicenes possess Gibbs activation energies of 174.5, 177.5 and 182.1 kJ/mol, respectively, at 27 °C.[125–130] Racemization of these helicenes requires substantial molecular deformation towards a C_s-symmetric transition state.[131] Stereomutations of helicenes involve extensive bond bending and torsion of the whole molecule, which explains the modest increase in conformational stability when additional benzene rings are incorporated into heptahelicene. Annulation of seven thiophene rings into a helicene structure provides a chiral oligothiophene that slowly racemizes at 199 °C.[132]

The chirality of helicenes is closely related to that of overcrowded phenanthrenes. In 1947, Newman and Hussey reported the synthesis and partial separation of the enantiomers of (4,5,8-trimethyl-1-phenanthryl)acetic acid which undergoes racemization within a few seconds at room temperature, Scheme 3.25. This finding paved the way to the discovery of helicenes, and has contributed substantially to our current knowledge of molecular helicity.[133–146] The exact barrier to helical inversion of the first chiral phenanthrene was later determined as 78.3 kJ/mol by

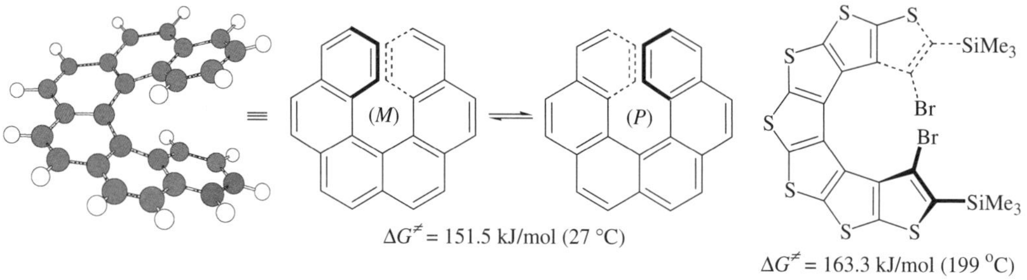

$\Delta G^{\neq} = 151.5$ kJ/mol (27 °C)

$\Delta G^{\neq} = 163.3$ kJ/mol (199 °C)

Figure 3.8   Structure and racemization of hexahelicene (left) and a thiaheptahelicene (right).

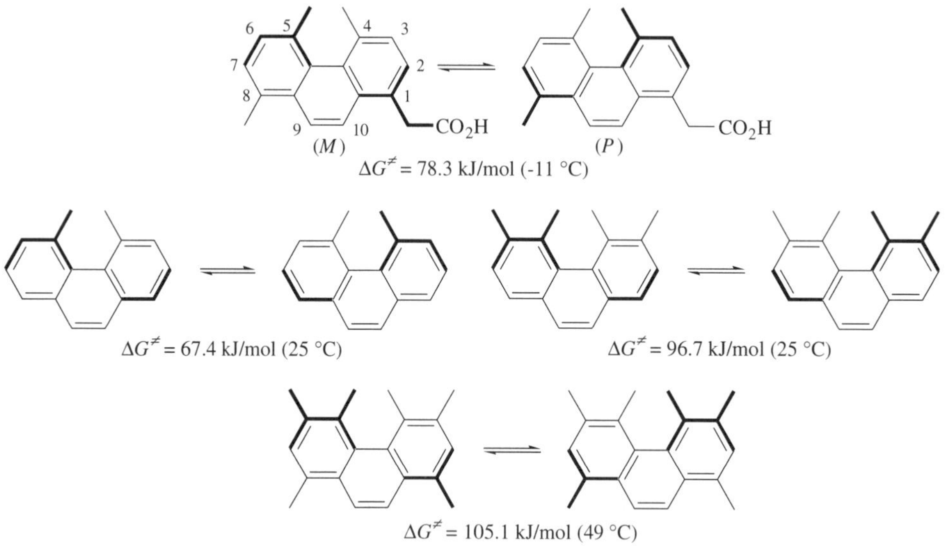

$\Delta G^{\neq} = 78.3$ kJ/mol (-11 °C)

$\Delta G^{\neq} = 67.4$ kJ/mol (25 °C) $\Delta G^{\neq} = 96.7$ kJ/mol (25 °C)

$\Delta G^{\neq} = 105.1$ kJ/mol (49 °C)

Scheme 3.25   Helical inversion of overcrowded phenanthrenes.

Mannschreck and coworkers.[147] The nonplanarity and helicity of 4,5-dimethylphenanthrenes stems from severe nonbonding interactions between the proximate substituents at *C*-4 and *C*-5. The conformational stability is mostly determined by the bulkiness of these substituents which have to pass each other in the transition state.[148] However, comparison of the activation barriers obtained for 4,5-dimethyl- and 3,4,5,6-tetramethylphenanthrene reveals a strong buttressing effect, in other words, the neighboring methyl groups reduce the flexibility of the substituents at positions 4 and 5 and thereby increase steric repulsion in the transition state.[149] Similarly, incorporation of substituents into the positions at *C*-1 and *C*-8 substantially enhances conformational stability, which has been attributed to steric interactions with the adjacent hydrogens at *C*-9 and *C*-10.

3.2.5   Alkyl Halides, Nitriles and Nitro Compounds

Chiral alkyl halides racemize via copper(I)-catalyzed atom transfer reactions[150] or through reversible bimolecular and unimolecular nucleophilic substitutions that are often accelerated by phase-transfer catalysts, Scheme 3.26.[151–155] In particular, the enantiomers of 1-bromo-1-arylalkanes are prone to rapid interconversion at ambient temperatures. Kinetic measurements of the racemization of 1-bromo-1-phenylethane in acetonitrile confirmed that both bimolecular and unimolecular bromide displacements occur concurrently, although the former is usually significantly more rapid

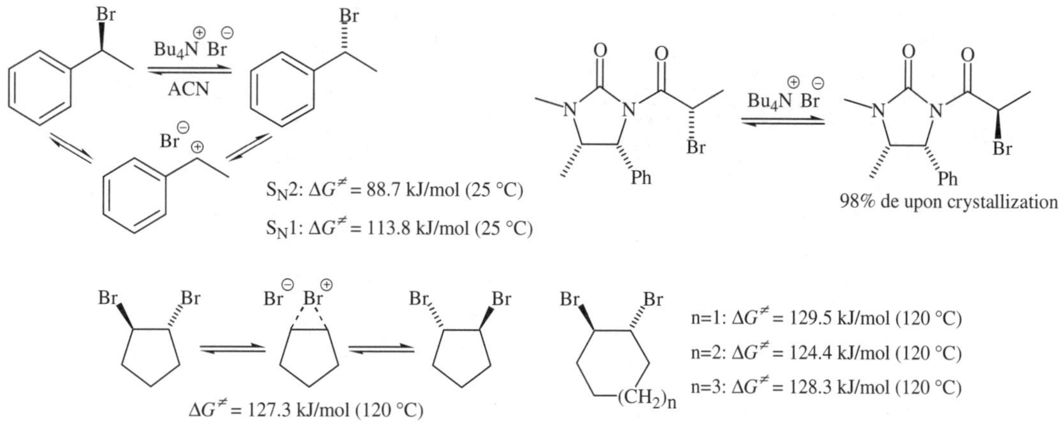

Scheme 3.26 Reversible nucleophilic substitution of 1-bromo-1-phenylethane, CIDR of interconverting diastereomers of an α-bromopropionyl imide and racemization of dihalides.

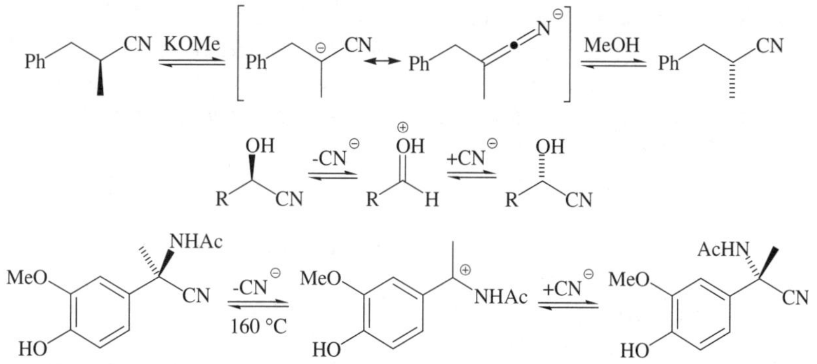

Scheme 3.27 Racemization of 2-methyl-3-phenylpropionitrile (top) and sequential elimination and addition of cyanide observed with cyanohydrins and α-acetamido-α-vanillylpropionitrile (bottom).

in the case of benzyl halides.[156] Stereomutations of configurationally unstable alkyl halides have been exploited for asymmetric synthesis. For example, epimerization of an α-bromopropionyl imide in the presence of tetrabutylammonium bromide, and subsequent crystallization-induced dynamic resolution (CIDR), produces one diastereomer in remarkable 98% de, Scheme 3.26.[157] Acyclic and cyclic *trans*-1,2-dihalides are known to racemize on heating.[158] In general, diaxial halides are more prone to racemization than diequatorial isomers, and reaction rates in the dibromocycloalkane series decrease in the order $C_7 > C_5 > C_8 > C_6$ which resembles the trend of debromination rates. It is therefore assumed that the 1,2-interchange involves an intermediate bromonium/bromide ion pair.

By analogy with aldehydes, ketones and carboxylic acid derivatives, the considerable acidity of α-hydrogens in nitriles ($pK_a \approx 25$) and nitro alkanes ($pK_a \approx 9$) facilitates deprotonation and subsequent racemization.[ix] The rate of racemization depends on substrate structure, base and solvent. Interconversion of the enantiomers of 2-methyl-3-phenylpropionitrile is about 10^9 times faster in DMSO than in methanol, Scheme 3.27.[159–161] Racemization of cyanohydrins and α-amino or α-amidonitriles that are in equilibrium with the corresponding aldehyde, ketone and imine

---

[ix] For comparison, pK_a values of α-hydrogens in ketones, esters and amides are approximately 20, 25 and 28, respectively.

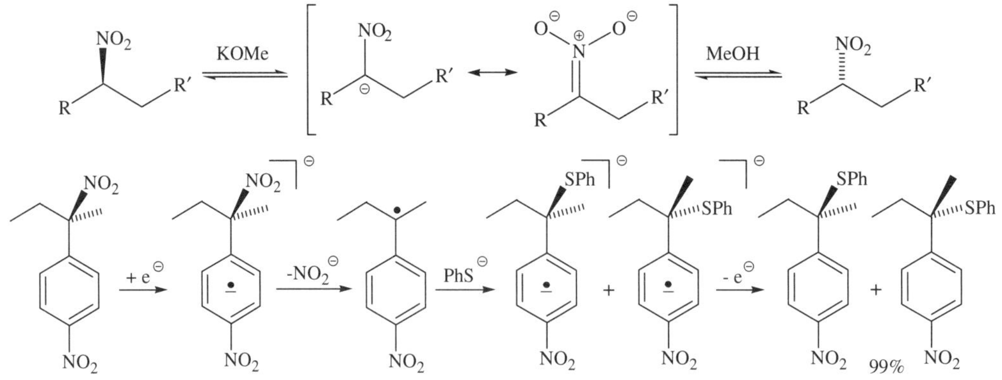

Scheme 3.28   Base-catalyzed racemization (top) and radical-mediated electron-transfer substitution of nitro alkanes (bottom).

precursors does not require deprotonation. Some cyanohydrins rapidly racemize due to sequential elimination and addition of cyanide at room temperature.[162–164] Configurationally unstable chiral amino nitriles have been employed as key intermediates in the asymmetric synthesis of amino acids based on dynamic kinetic resolution and related processes.[165–170] Nonenolizable α-methyl-α-acetamidonitriles including α-acetamido-α-vanillylpropionitrile show reversible elimination of HCN at high temperatures.[171] Racemization via retro aldol reaction of a chiral β-hydroxynitrile has also been reported.[172]

Substitution of the alkyl nitro group of 2-(4-nitrophenyl)-2-nitrobutane by sodium azide, thiophenoxide or other nucleophiles has been achieved via electron-transfer radical chain reactions. The nucleophilic displacement occurs at room temperature, and involves a radical intermediate that either adopts a planar and thus achiral configuration or undergoes rapid pyramidal inversion. As a result, racemic substitution products are obtained, Scheme 3.28.[173]

3.2.6   Amines

Amines with three different substituents are chiral but racemization via pyramidal N-inversion is often too fast for separation of enantiomers at room temperature.[174–176] For instance, the energy barrier to racemization of ethylmethylisopropylamine is 31.4 kJ/mol, Scheme 3.29.[177,178] However, fixation of ethylmethylisopropylamine in the cavity of appropriately sized host molecules and hydrogen bonding or protonation can effectively inhibit inversion and afford configurational stability at room temperature.[179] In general, the barrier to nitrogen inversion decreases as the bulkiness of the substituents increases because of destabilizing steric repulsion in the pyramidal ground state. Inversion at the sp^3-hybridized nitrogen occurs via a planar sp^2-hybridized transition state. An increase in the p-character of the nitrogen bonds therefore impedes N-inversion. Such an increase can be accomplished by introduction of chlorine or other electronegative substituents to the nitrogen atom. The bond angles in N-chloramines are significantly reduced from the near-tetrahedral value of aliphatic amines, and formation of a transition state structure exhibiting 120° bond angles requires a higher activation energy.[180]

Incorporation of the stereogenic nitrogen into three- and four-membered rings of aziridines and azetidines effectively enhances the activation energy for pyramidal inversion as a result of additional ring strain in the planar transition state.[181] The enantiomers of N-chloro- and N-alkoxyaziridines or diaziridines are configurationally stable at 25 °C but nitrogen inversion is observed upon heating.[182–191] The enantiomerization barriers of 1-chloro-2,2-diphenylaziridine and 1,2,3,4-tetramethyldiaziridine, which interconverts via two consecutive nitrogen inversions, are 102.1 kJ/mol

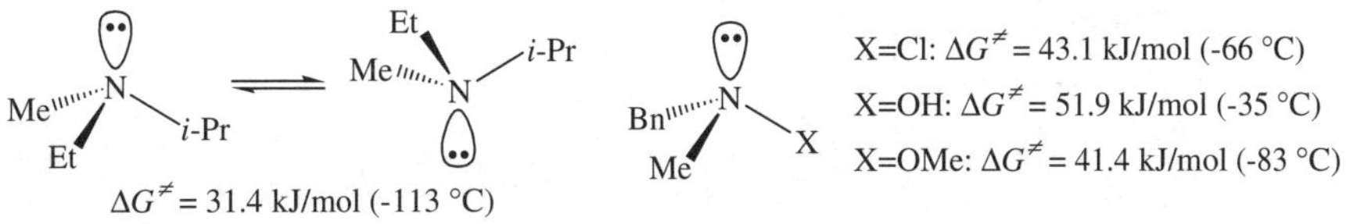

Scheme 3.29 Facile *N*-inversion of acyclic amines.

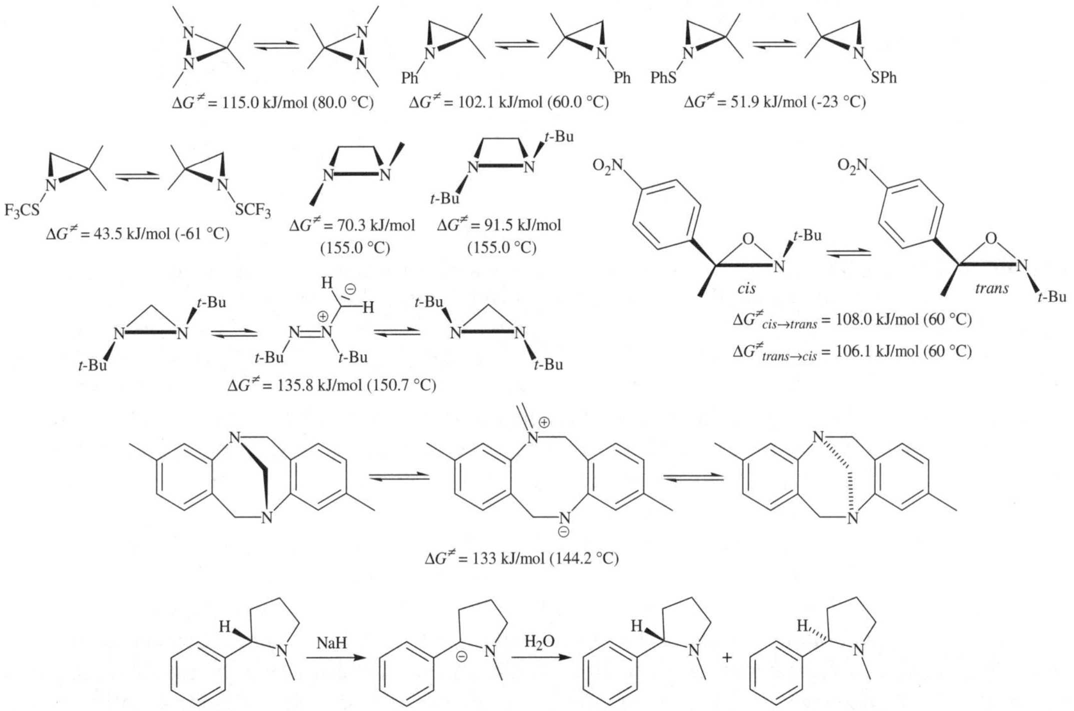

Scheme 3.30 Racemization pathways of amines.

and 115.0 kJ/mol, respectively.[192–199] Oxaziridines exhibit similar configurational stability and show nitrogen inversion at ambient temperatures. The barrier to *cis/trans*-isomerization of 2-*tert*-butyl-3-methyl-3-(4-nitrophenyl)oxaziridine is 108.0 kJ/mol.[200] Interestingly, aryl- and alkyl-sulfenylaziridines readily racemize even under cryogenic conditions because the barrier to *N*-inversion is usually lower than 60 kJ/mol, Scheme 3.30.[201] An increase in ring size facilitates racemization due to reduced strain in the transition state. The barrier to racemization of 1,2-di-*tert*-butyl-1,2-diazetidine is 91.5 kJ/mol at 155 °C which is significantly lower than the inversion barrier of the corresponding diaziridine.[202] Several cyclic amines including 1,3,4-oxadiazolidines and 1,2-di-*tert*-butyldiaziridine have been reported to racemize through a dissociative mechanism and reversible bond breaking.[203] In the case of 1,2-di-*tert*-butyldiaziridine, thermal racemization occurs through reversible conrotatory electrocyclic ring opening with an activation energy of 135.8 kJ/mol, Scheme 3.30.[204] Quaternary ammonium salts and rigid bicyclic amines containing the nitrogen as a bridgehead such as Tröger's base are usually configurationally stable. The two stereogenic nitrogen atoms in Tröger's base can not accomplish pyramidal inversion without bond cleavage. This is observed at high temperatures, probably as a result of retro hetero-Diels–Alder ring opening or formation of a zwitterionic intermediate.[205–207]

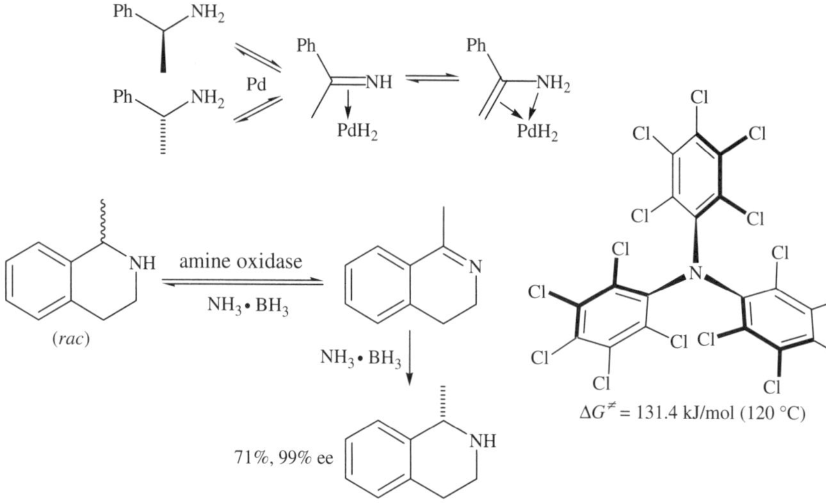

Scheme 3.31 Palladium-catalyzed racemization of α-methylbenzylamine, amine oxidase-catalyzed de-racemization, and the propeller-like structure of perchlorotriphenylamine.

Tröger's base also racemizes in the presence of 1,1′-binaphthyl-2,2′-diyl hydrogenphosphate or other acids, which has been utilized for crystallization-induced asymmetric transformation.[208] Some amines including nicotine racemize in the presence of a strong base, probably via formation of a configurationally labile carbanion intermediate.[209] In accordance with amino acids, formation of a Schiff base provides another venue for racemization of amines, *vide infra*.[210]

Palladium, ruthenium and iridium complexes catalyze racemization of amines through *CH*-bond activation and the formation of achiral enamine and imine complexes.[211–215] Deuterium labeling experiments showed that the mechanism for racemization of α-methylbenzylamine proceeds via coordination to palladium, followed by β-elimination of a PdH species, Scheme 3.31.[216] A similar pathway, *i.e.*, enantioconversion of chiral amines via an achiral imine intermediate, has been exploited for chemoenzymatic deracemization of primary and secondary amines with enantioselective amine oxidases in the presence of a nonselective reducing agent, see Chapter 7.[217,218] An interesting case arises with perchlorotriphenylamine, an example of a three-bladed propeller. The nitrogen atom is chirotopic but nonstereogenic, and the chirality of this molecule originates solely from its propeller-like structure. At 120 °C, racemization of perchlorotriphenylamine occurs due to reversal of the propeller helicity but not because of *N*-inversion. The so-called threshold mechanism presumably involves a two-ring flip with an energy barrier of 131.4 kJ/mol.[219,220] Racemization reactions of molecular propellers and ring-flip mechanisms are discussed in Chapter 8.

3.2.7 Aldehydes, Ketones and Imines

Aldehydes, ketones and imines with a stereogenic tertiary α-carbon undergo base- and acid-catalyzed racemization through keto/enol- or imine/enamine-tautomerization. The base-catalyzed reaction usually follows an S_E1 mechanism and the rate of enolate formation is equal to the rate of racemization, Scheme 3.32. In the presence of acid, a less reactive achiral enol is formed that tautomerizes to either enantiomer. Amfepramone and other drugs exhibiting an enolizable stereocenter are often not stable to racemization under physiological conditions. The study of the stereochemical integrity of such pharmaceuticals has therefore become routine during drug discovery, see Chapter 3.4.[221] Substituent effects on the reaction rate are quite important: 1,2-diphenylpropanone racemizes in ethanolic solution containing sodium ethoxide with a half-life time

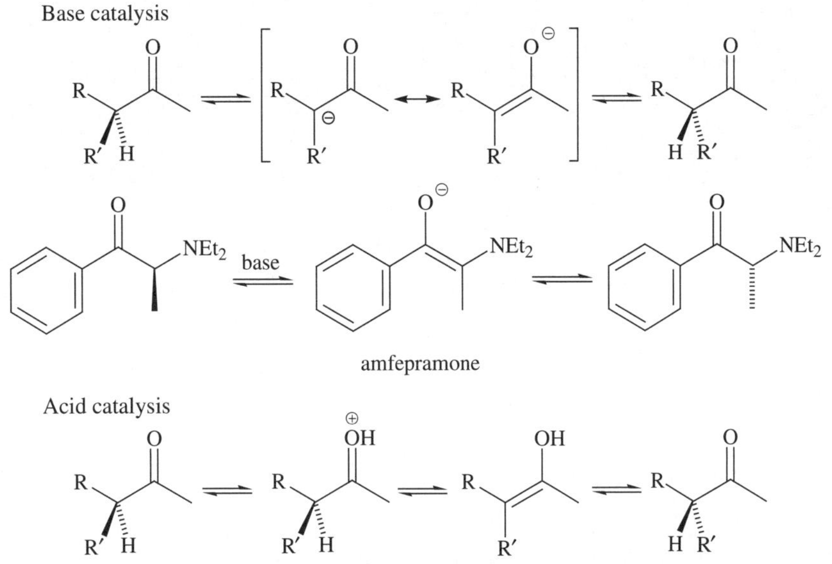

Scheme 3.32 Base- and acid-catalyzed racemization of carbonyl compounds.

of 18.4 minutes while under the same conditions that of 3,3-dimethyl-1,2-diphenylbutanone is 228 hours, Scheme 3.33.[222] Treatment of (−)-menthone with sodium methoxide in methanol causes epimerization to a diastereomeric mixture exhibiting 40% de at equilibrium.[223]

Racemization of enolizable chiral carbonyl compounds can compromise the stereochemical purity of reaction products and requires careful attention during asymmetric synthesis.[224–228] Alternatively, enolization of α-substituted aldehydes[229] and ketones[230–238] has been utilized to enhance stereoselectivity in a number of asymmetric syntheses, for example by base-catalyzed equilibration of diastereomers, Scheme 3.34. Racemization and diastereomerization of aldehydes and ketones often involves enolization but this is not necessarily the case.[239] Ketalization of the monoterpenoid (−)-cryptone with ethylene glycol results in destruction of asymmetry due to double bond migration and formation of an achiral Δ^3-dioxolane intermediate, and racemic cryptone is obtained upon acidic hydrolysis.[240] Chiral α-substituted aldehydes and ketones bearing an electron-withdrawing group in β-position are susceptible to enol formation and racemization under mild conditions. The configurational instability of β-dicarbonyl compounds is crucial to Noyori's (BINAP)RuBr$_2$-catalyzed asymmetric hydrogenation of rapidly interconverting enantiomers of β-keto esters,[241,242] and has been exploited in many other elegant dynamic kinetic resolutions, see Chapter 3.1.2.[243–248] The high acidity of α-hydrogens in 1,3-diketones and their derivatives greatly facilitates enolization even in water,[x] providing an entry to chemoenzymatic DKR of β-diketones,[249] β-keto esters,[250] β-keto nitriles,[251] and β-keto amides.[252]

Imine/enamine-tautomerization of chiral α-substituted imines[253,254] and hydrazones[255] causes racemization or diastereomerization, Scheme 3.35. Although interconversion of the stereoisomers of α-substituted aldehydes and ketones via enolization has been successfully incorporated into many synthetic strategies, it can be advantageous to use the corresponding imines and hydrazones. For example, transformation of (−)-menthone to its tosylhydrazone derivative greatly facilitates acid-catalyzed epimerization and provides a different diastereomeric ratio than equilibration of the original ketone, see Scheme 3.33 for comparison.[256] Condensation of methyl 5-oxo-4-phenylpentanoate and (*R*)-phenylglycinol gives a mixture of diastereomeric imines that readily interconvert via enamine

---

[x] The pK_a of acetylacetone is 13.3 and the equilibrium constant for enolization in water, $K_{enol/keto}$, is 0.23.

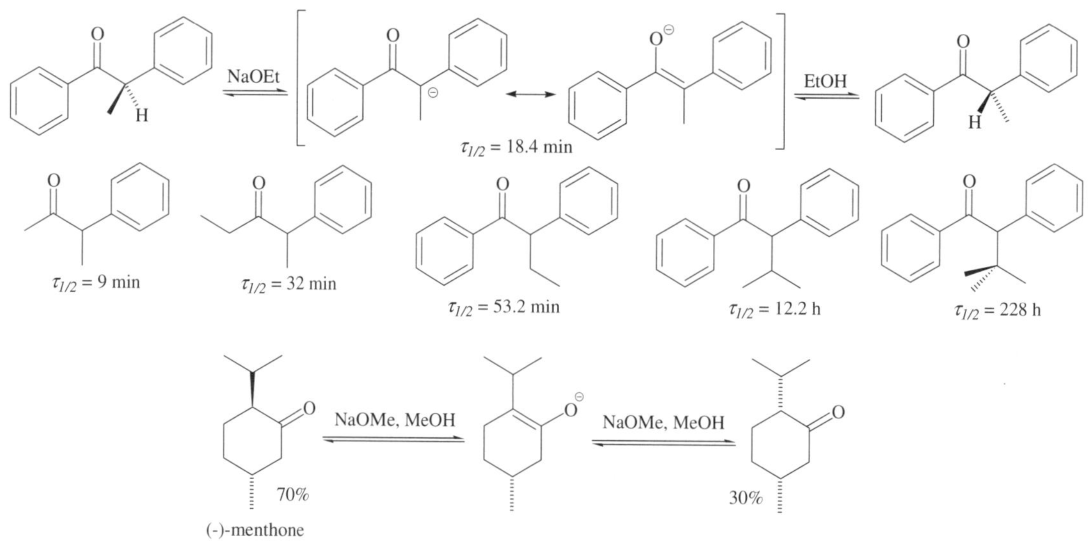

Scheme 3.33 Racemization of ketones via base-catalyzed enolization (top) and epimerization of (–)-menthone (bottom).

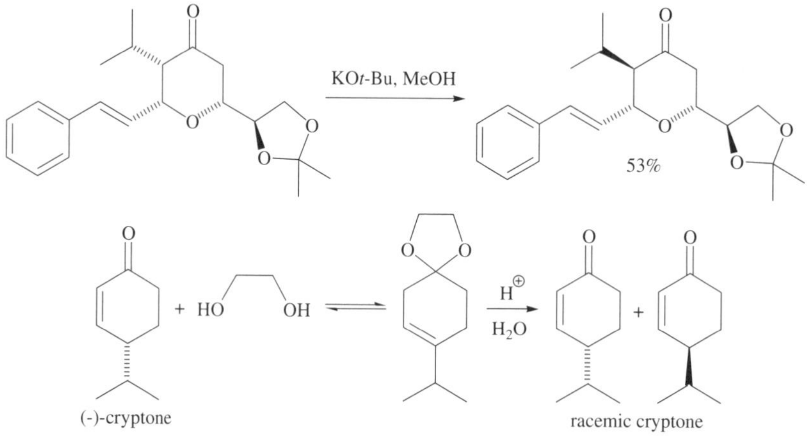

Scheme 3.34 Epimerization of a chiral ketone and racemization of cryptone via a stable achiral intermediate.

formation.[257] Reversible oxazolidine formation followed by irreversible lactamization favors formation of the bicyclic (*R*,*R*,*S*)-isomer which is obtained in 49% yield. The same strategy has been applied in the synthesis of other polysubstituted lactams from δ-keto esters and δ-oxo diesters.[258,259]

3.2.8   Alcohols, Ethers, Acetals, and Ketals

Secondary and tertiary chiral alcohols are configurationally stable under basic conditions but they are prone to acid-catalyzed racemization in aquous solution.[260,xi] In the case of unimolecular dissociation, reversible formation of a planar carbocation intermediate from a chiral alcohol in the presence of a Lewis or Brønsted acid is sometimes accompanied by undesirable rearrangements or elimination.

[xi] Racemization of tertiary alcohols is an S_N1 process while secondary alcohols often prefer the S_N2 mechanism.

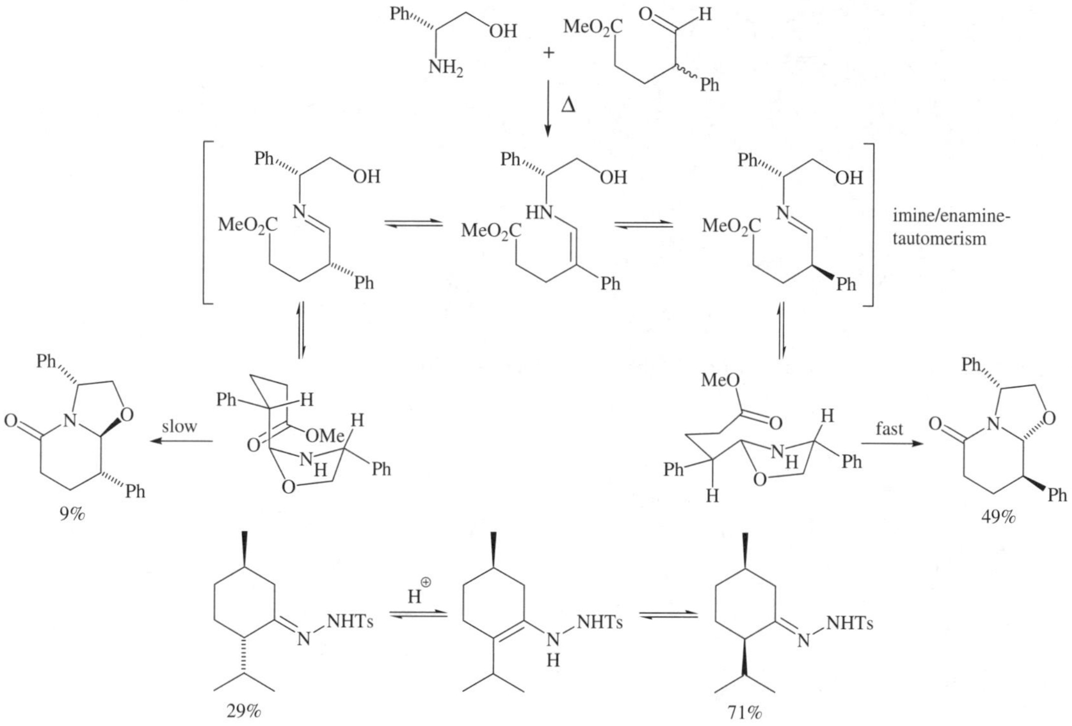

Scheme 3.35 Epimerization of α-substituted imines and hydrazones.

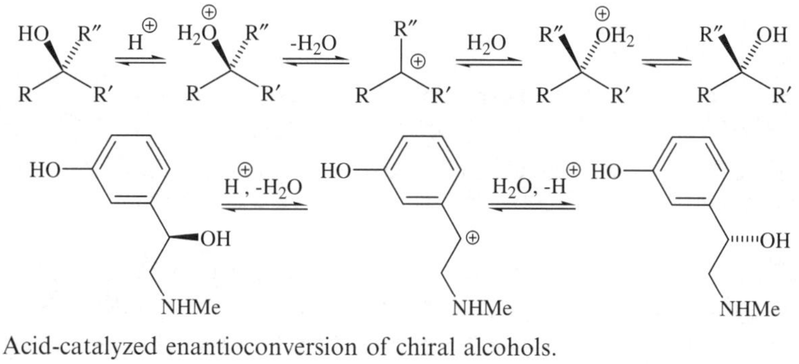

Scheme 3.36 Acid-catalyzed enantioconversion of chiral alcohols.

While formation of elimination products can be avoided at low temperatures, rearrangments are less likely with carbocations derived from chiral benzylic or allylic alcohols due to resonance stabilization, Scheme 3.36.[261] A different mechanism has been proposed for the acid-catalyzed racemization of 1-heteroaryloxy-2,3-propanediols derived from pyridine, thiazole, pyrazine, and pyrimidine.[262] Protonation of the heteroaryl moiety facilitates cyclization which is followed by ring opening to an achiral intermediate that regenerates either enantiomer, Scheme 3.37.

Sequential oxidation and reduction based on reversible ruthenium-catalyzed hydrogen transfer provides a useful method for mild racemization of secondary alcohols, Scheme 3.38.[263–277] This can also be accomplished with rhenium, iridium and rhodium complexes.[278,279] In general, deprotonation of the substrate is followed by formation of a transition metal hydride intermediate and a prochiral ketone that is then nonselectively reduced to the racemic alcohol. Reversible

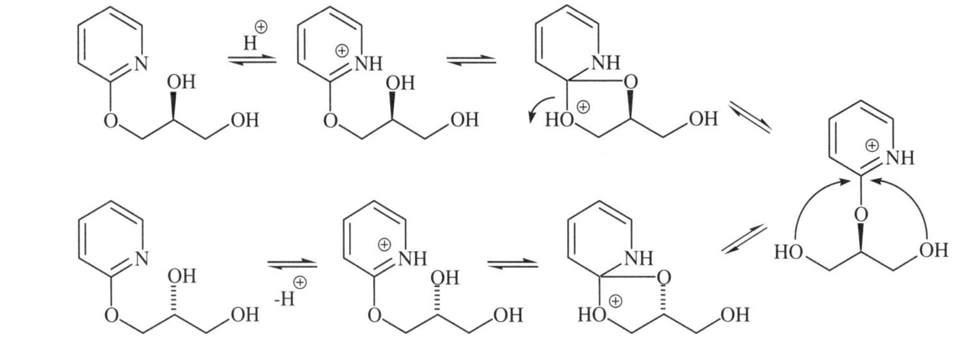

Scheme 3.37   Racemization of 1-(2′-pyridyloxy)-2,3-propanediol.

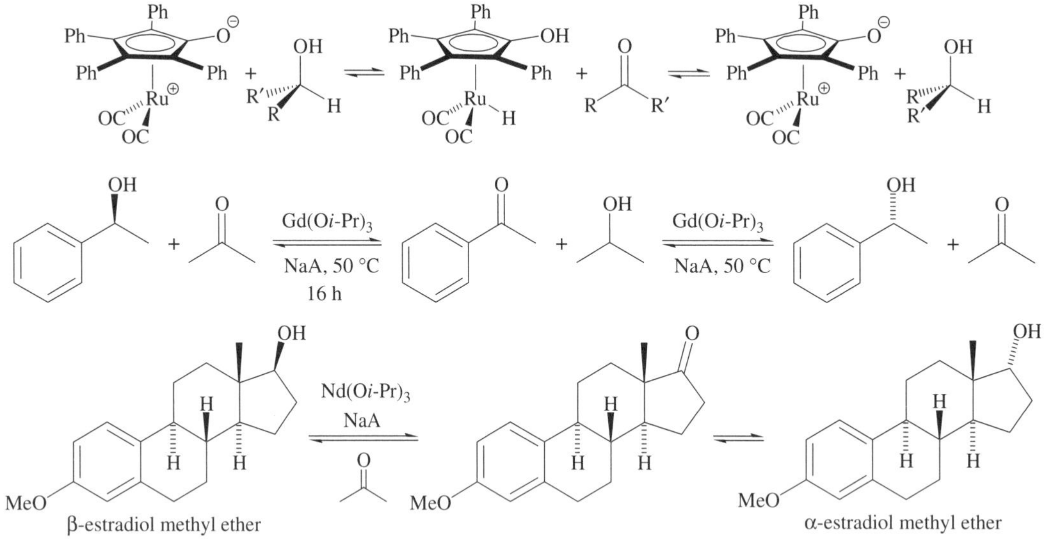

Scheme 3.38   Ruthenium-catalyzed hydrogen transfer with secondary alcohols (top) and reversible Meerwein–Ponndorf–Verley/Oppenauer redox reactions (bottom).

Meerwein–Ponndorf–Verley/Oppenauer redox reactions catalyzed by lanthanide triisopropoxides constitute another viable alternative for racemization of secondary alcohols.[280] Neodymium, gadolinium and samarium isopropoxides allow mild racemization of 1-phenylethanol and have proved more effective than the aluminum salts traditionally used. Acetone is the preferred hydride acceptor and zeolite NaA is added to suppress the competing aldol reaction. The procedure can also be utilized for selective epimerization of steroidal β-estradiol methyl ether.

Reversible crossed aldol reactions that establish a mixture of rapidly equilibrating β-hydroxy aldehyde stereoisomers afford interesting synthetic opportunities. Reyes and Córdova found that the (S)-proline-catalyzed aldol addition of aldehydes to propanal is reversible at room temperature, and can be coupled to a stereoisomer-differentiating reaction between the enantiomeric β-hydroxy-aldehydes and a proline-derived enamine. The reaction sequence is completed by ring closure, providing deoxysugars in excellent ee and de, albeit with low yields, Scheme 3.39.[281] Chiral homoallyl alcohols undergo racemization via reversible aldehyde-mediated allyl transfer in the presence of catalytic amounts of acid. For example, the enantiomeric excess of pure (R)-1-phenylhex-5-en-3-ol decreases to 59% ee within 2 hours upon treatment with 10 mol% of 3-phenylpropanal and p-toluenesulfonic acid at 20 °C.[282]

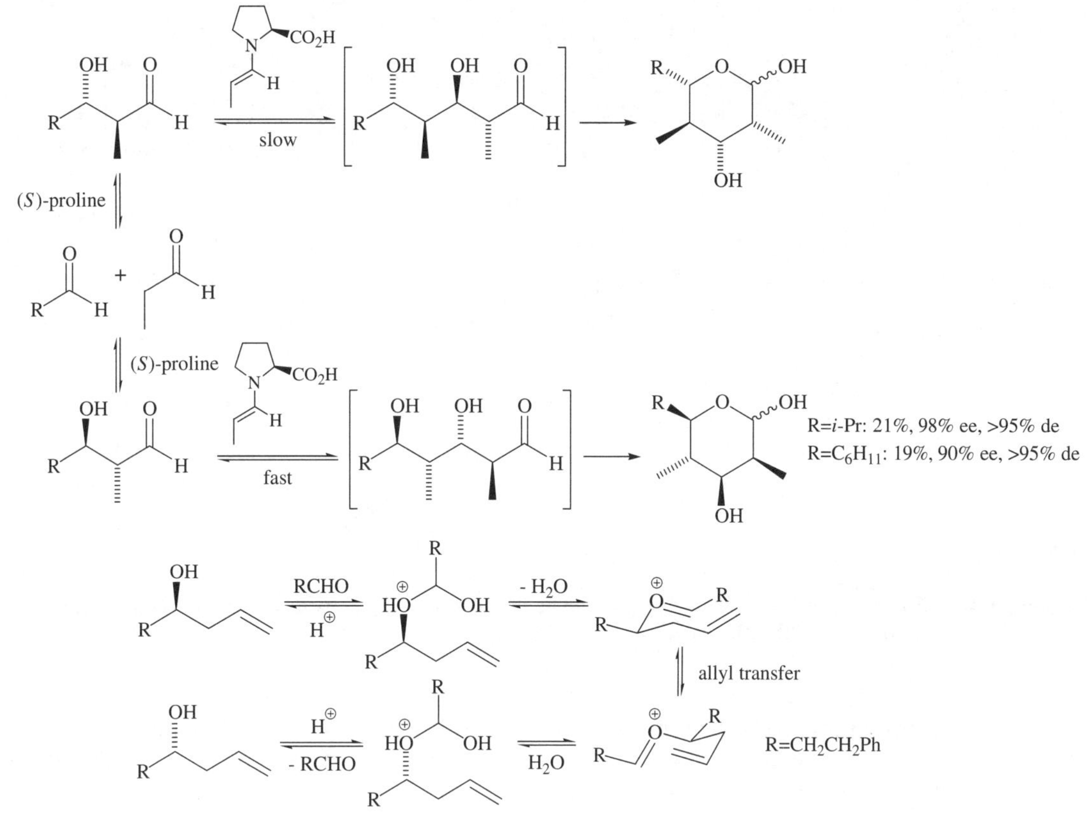

Scheme 3.39 Stereoselective synthesis of deoxysugars based on fast racemization of chiral aldol products (top) and enantioconversion of 1-phenylhex-5-en-3-ol (bottom).

Racemization or diastereomerization of chiral ethers provides access to important building blocks for asymmetric synthesis. An intriguing example is the stereomutation of *C*-galactopyranosides involving a sequence of base-promoted enolization, β-elimination and intramolecular Michael addition, Scheme 3.40. The equilibrium favors formation of the *C*-glycopyranoside having the substituent at *C*-1 in equatorial position.[283,284,xii] Cyanostilbene oxide and other 1,2-diaryloxiranes undergo thermal and photochemical racemization via intermediate carbonyl ylides or diradicals.[285–289] Although metal-catalyzed stereointerconversion of epoxides is often of limited use because epoxides readily rearrange to ketones in the presence of a Lewis acid, a ruthenium-catalyzed *cis/trans*-isomerization procedure has been reported.[290] Jacobsen's group developed a protocol for (salen)Cr(III)N$_3$-catalyzed racemization of epichlorohydrin in the presence of trimethylsilyl cyanide.[291] Alternatively, the enantiomers of epichlorohydrin interconvert via formation of achiral 1,3-dichloro-2-propanol, Scheme 3.40.[292]

Sequential ring opening and cyclization of chiral 2-aryl-2-methyl-2*H*-1-benzopyrans results in racemization. The mechanism entails formation of an achiral intermediate that is probably generated by heterolytic cleavage of the C(*sp³*)-*O* bond.[293] Narwedine possesses a chiral axis in addition to two asymmetric carbon centers and racemizes via ring opening followed by Michael addition in the presence of base. Cyclization of the achiral intermediate can produce either

---

[xii] The diastereomer with an equatorial substituent at *C*-1 is thermodynamically more stable because the anomeric effect is not operative in *C*-glycopyranosides.

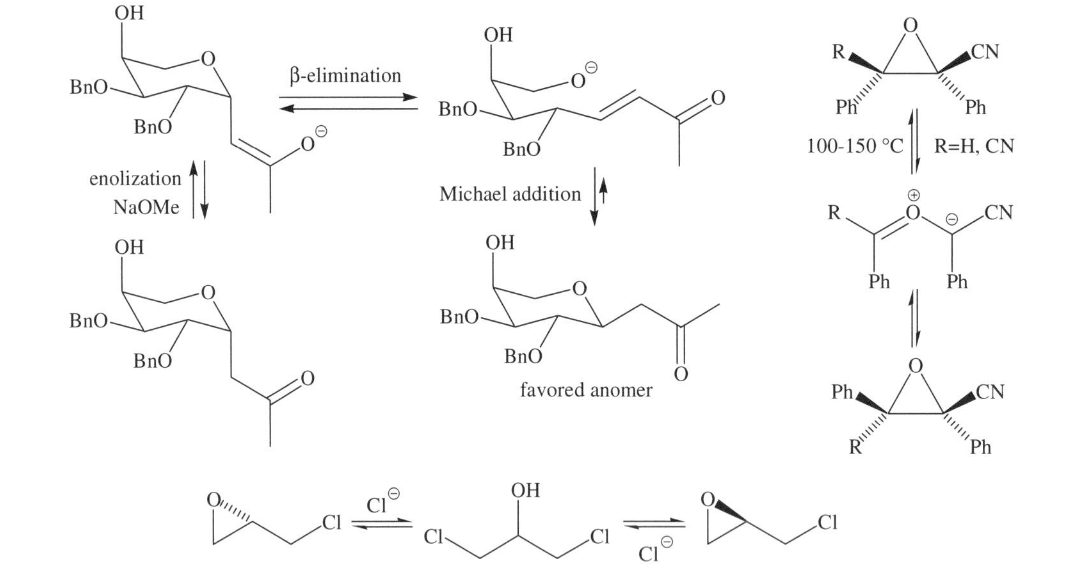

Scheme 3.40   Epimerization of an α-*C*-glycopyranoside (left) and stilbene oxide derivatives (right). Racemization of epichlorohydrin via 1,3-dichloro-2-propanol (bottom).

enantiomer but formation of diastereomers is not observed, Scheme 3.41.[294,xiii] Enantioconversion of homofuran proceeds via an achiral carbonyl ylide-like intermediate generated by a disrotatory electrocyclic reaction upon heating. The ring opening and closure are orbital symmetry-allowed and obey Woodward–Hoffmann rules. The activation energy for enantiomerization of homofuran is 113.8 kJ/mol.[295,296] Similar stereomutations have been observed with homopyrrole, homothiophene and benzoheterocycles, Scheme 3.42.[297]

Chiral hemiacetals, hemiketals, hemithioacetals, and hemiaminals are prone to racemization via reversible ring opening.[298–304] Related epimerization[305,306] and mutarotation processes[307] play a fundamental role in carbohydrate chemistry, see Chapter 3.1.4. Racemization of 5-hydroxyfuranones, which are important chiral building blocks, is fast at room temperature or upon mild heating even in the absence of catalytic amounts of acid or base, Scheme 3.43. The mechanism is based on ring opening and subsequent intramolecular acetalization. The reversibility of hemiacetal formation under aqueous conditions certainly facilitates the development of enzymatic dynamic kinetic resolution processes and several procedures have been reported.[308,309] Thermal racemization due to reversible ring opening is also common with ketals. For example, the oxazolebenzodiazepinone shown in Scheme 3.43 racemizes in refluxing methanol via formation of an achiral quaternary iminium ion.[310] The photochromism and facile reversible ring opening of spiro *O,O*- and *N,O*-ketals at room temperature has paved the way for the development of chiroptical switches.[311–314] The racemization mechanism of 2,2′-spirobichromene[315] and a spiro 2*H*-1-benzopyran-2,2′-indoline[316] is described in Scheme 3.44.

Ring opening and subsequent ring closure of the spiroketal chalcogran, an important beetle aggregation pheromone, occurs at 70–120 °C and causes epimerization, Scheme 3.45. The Gibbs free activation energies for the interconversion of (2*R*,5*R*)- and (2*R*,5*S*)-2-ethyl-1,6-dioxaspiro[4.4]-nonane are 108.0 and 108.5 kJ/mol, respectively.[317] The corresponding low activation enthalpies and the highly negative activation entropies are indicative of a heterolytic dissociative mechanism

---

xiii This is remarkable because stereolabile compounds exhibiting more than one chiral element usually form a mixture of diastereomers.

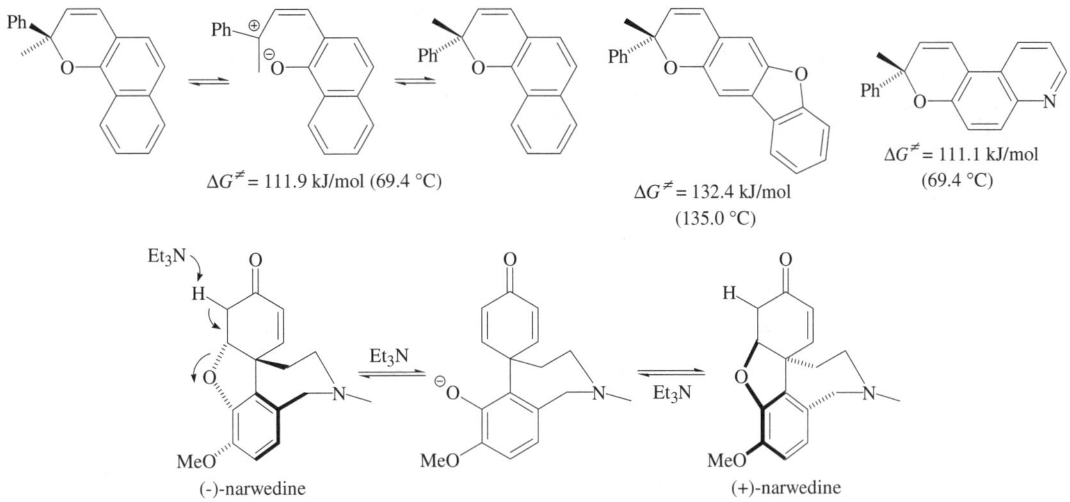

$\Delta G^{\neq} = 111.9$ kJ/mol (69.4 °C)

$\Delta G^{\neq} = 132.4$ kJ/mol (135.0 °C)

$\Delta G^{\neq} = 111.1$ kJ/mol (69.4 °C)

(-)-narwedine

(+)-narwedine

Scheme 3.41 Racemization of cyclic ethers.

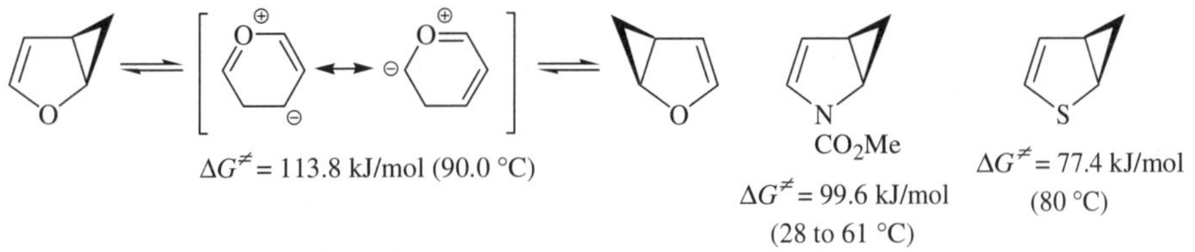

$\Delta G^{\neq} = 113.8$ kJ/mol (90.0 °C)

$\Delta G^{\neq} = 99.6$ kJ/mol (28 to 61 °C)

$\Delta G^{\neq} = 77.4$ kJ/mol (80 °C)

Scheme 3.42 Stereomutations of homofuran and its derivatives.

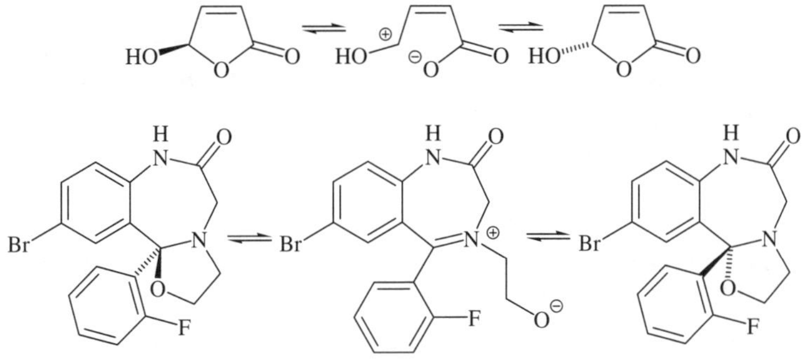

Scheme 3.43 Thermal racemization of 5-hydroxy-5*H*-furan-2-one and an oxazolebenzodiazepinone derivative.

with a zwitterionic enol ether intermediate. Acid-catalyzed ring opening of the spirohydantoin shown in Scheme 3.45 results in complete racemization within two hours upon heating to 90 °C in the presence of 10% HCl. It is assumed that the *N,N*-ketal moiety is temporarily cleaved to form a transient achiral acyliminium ion.[318]

3.2.9 Carboxylic Acids and Derivatives

Carboxylic acids and their derivatives undergo base- and acid-catalyzed enolization. Base-promoted removal of a proton from a stereogenic carbon center usually entails complete racemization

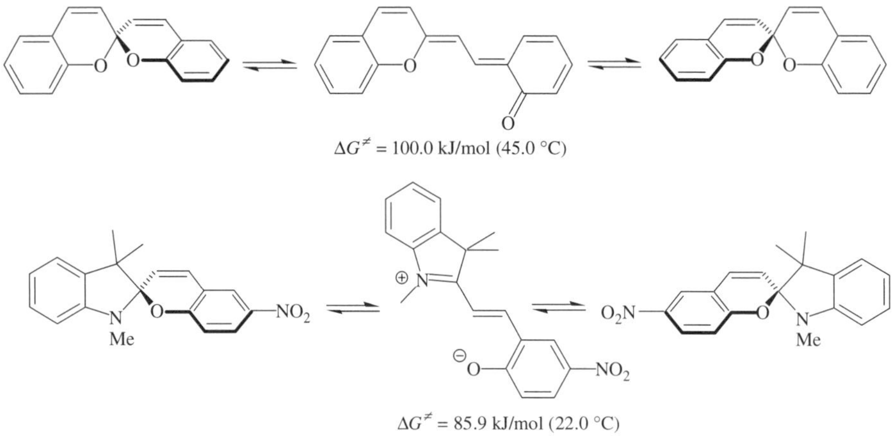

$\Delta G^{\neq} = 100.0$ kJ/mol (45.0 °C)

$\Delta G^{\neq} = 85.9$ kJ/mol (22.0 °C)

Scheme 3.44   Interconversion of the enantiomers of 2,2'-spirobichromene and of a spiro 2*H*-1-benzopyran-2,2'-indoline.

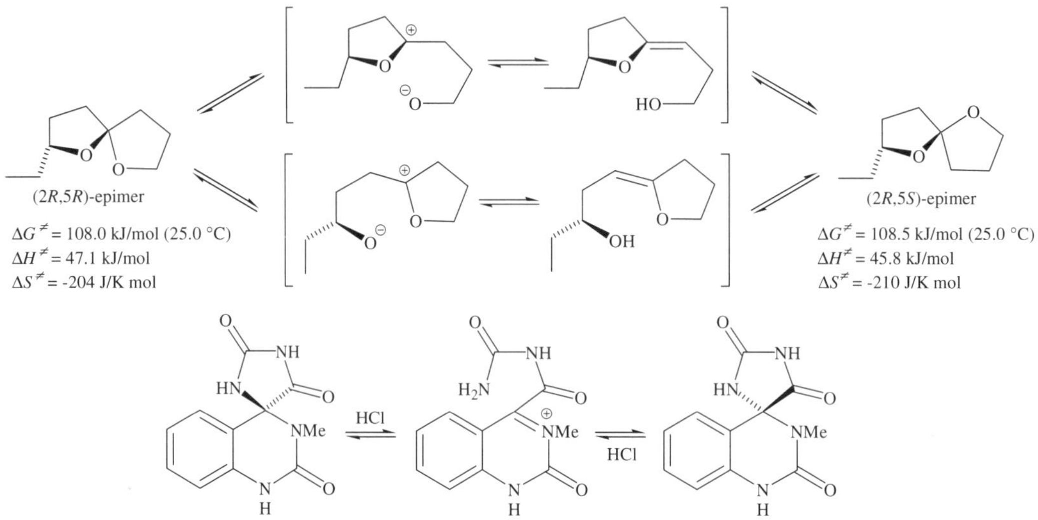

(2*R*,5*R*)-epimer

$\Delta G^{\neq} = 108.0$ kJ/mol (25.0 °C)
$\Delta H^{\neq} = 47.1$ kJ/mol
$\Delta S^{\neq} = -204$ J/K mol

(2*R*,5*S*)-epimer

$\Delta G^{\neq} = 108.5$ kJ/mol (25.0 °C)
$\Delta H^{\neq} = 45.8$ kJ/mol
$\Delta S^{\neq} = -210$ J/K mol

Scheme 3.45   Possible epimerization pathways of chalcogran (top) and racemization of a spirohydantoin (bottom).

due to the inherently low stereochemical integrity of carbanions.[319–323] Nevertheless, carbanions are generally less prone to racemization than carbocations.[324] It has been reported that exchange of the α-proton in *N*-pivaloyl phenylalanine dimethylamide by deuterium proceeds with considerable retention of configuration. Kinetic studies revealed that the isotope exchange process is about 2.4 times faster than the concomitant racemization, Scheme 3.46.[325] The remarkable configurationally stable of some carbanions can be exploited for asymmetric synthesis. For example, nonracemic bromochlorofluoromethane has been prepared by haloform reaction of 1-bromo-1-chloro-1-fluoro-acetone[326,327] and by decarboxylation of the (–)-strychnine salt of bromochlorofluoroacetic acid.[328] Numerous racemization and epimerization procedures based on enolate formation of carboxylic acids,[329,330] esters,[331–340] thioesters,[341,342] and amides[343–348] have been developed and employed in the synthesis of a wide range of chiral compounds. A few examples are depicted in Figure 3.9.

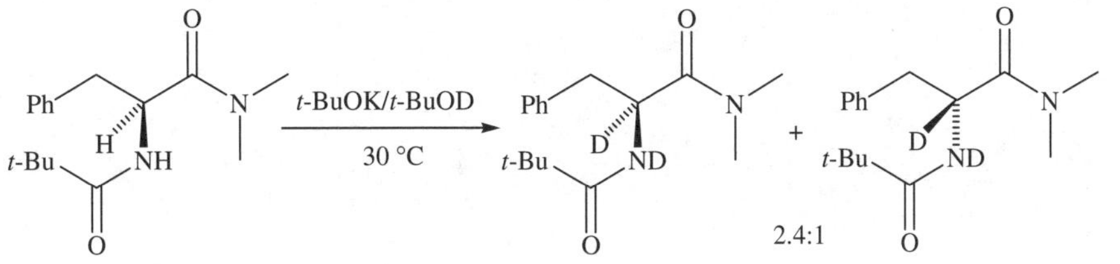

Scheme 3.46 Partial racemization of *N*-pivaloyl phenylalanine dimethylamide during proton/deuterium exchange.

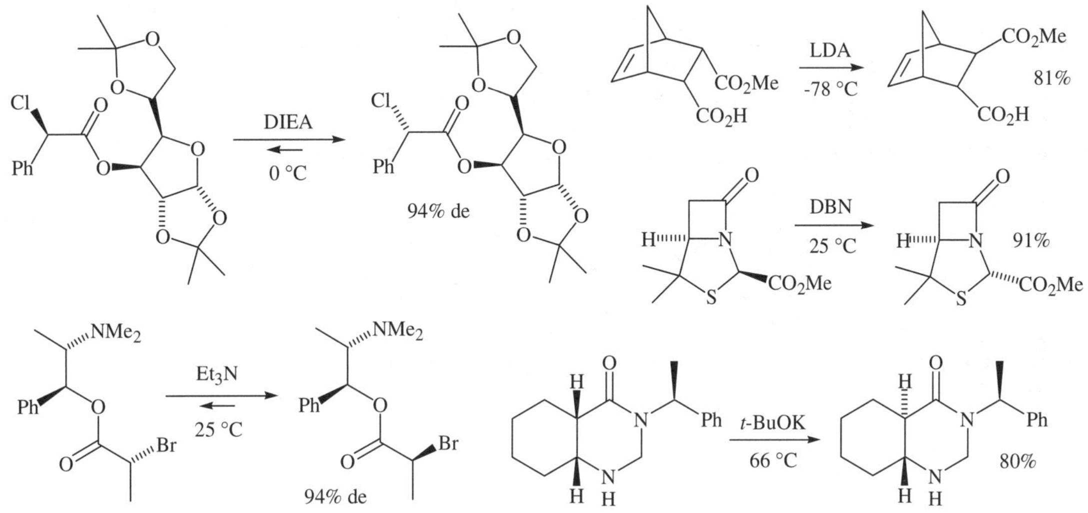

Figure 3.9 Stereoinversions of carboxylic acid derivatives.

Dicarboxylic acid derivatives can racemize through formation of a meso intermediate.[349] This has been observed with 3,3-dimethylcyclopropyl-1,2-dicarboxylic acid monomethyl ester which shows internal displacement of the methoxy group upon heating in the presence of a strong base. Nucleophilic attack by the methoxide at the meso anhydride occurs at either carbonyl group and yields the racemic monomethyl ester, Scheme 3.47.[350] 2-Hydroxy-2-methyl-3-oxobutanoic acid follows an unusual racemization course that is based on a reversible base-catalyzed tertiary ketol rearrangement, Scheme 3.48.[351] The intramolecular migration of the carboxylate group has been verified by ^{13}C NMR experiments with isotopically labeled α-acetolactic acid. The rearrangement of the dianion is degenerate, in that both starting material and product have identical constitution but opposite configuration. The reaction is accompanied by complete racemization of the originally enantiopure acid. Mild enantioconversion of allylic esters can be achieved through formation of achiral π-allylpalladium complexes, Scheme 3.49.[352,353] Fast interconversion of stereolabile π-allyl species generated from γ-acyloxybutenolides and a chiral palladium complex has been exploited by Trost and Toste for the synthesis of γ-alkoxy derivatives based on dynamic kinetic asymmetric transformation, see Chapter 7.5.[354]

Acyl chlorides bearing a stereogenic α-carbon atom readily form achiral ketenes by base-promoted elimination of hydrochloride. The reactive ketene can regenerate racemic starting materials in the absence of an appropriate nucleophile, or alternatively be trapped as an amide or ester with primary and secondary amines or alcohols, Scheme 3.50.[355,356] Ring opening of the acyl chloride of chrysanthemic acid results in the simultaneous loss of both chiral centers.[357]

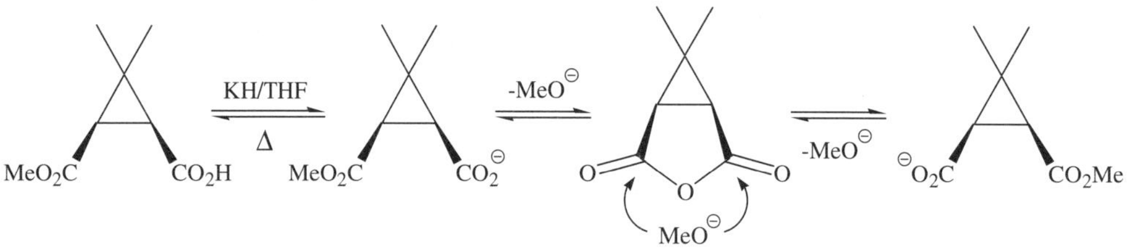

Scheme 3.47   Enantioconversion involving a meso intermediate.

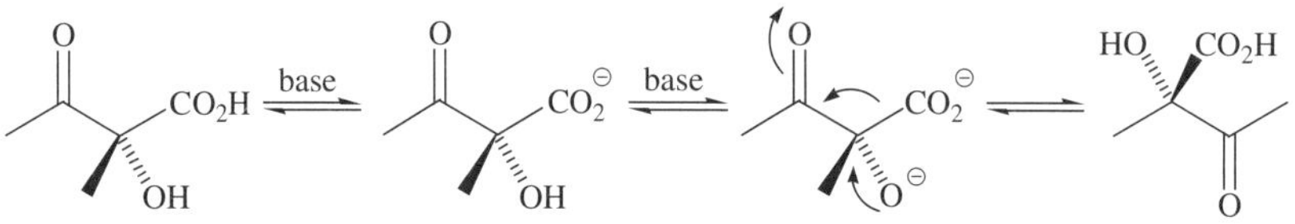

Scheme 3.48   Tertiary ketol rearrangement of 2-hydroxy-2-methyl-3-oxobutanoic acid.

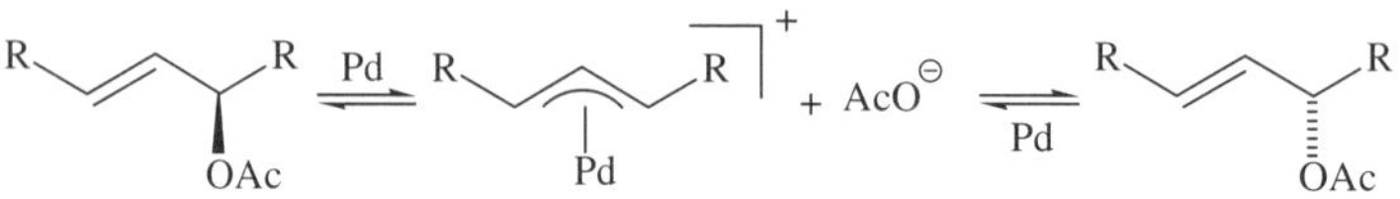

Scheme 3.49   Enantioconversion via reversible formation of a π-allylpalladium complex.

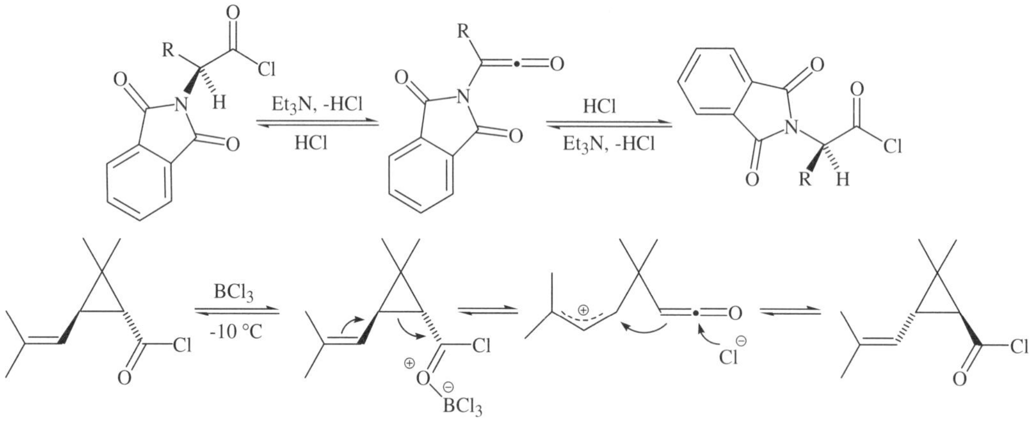

Scheme 3.50   Acid- and base-promoted racemization of acyl chlorides.

The reaction is reversible and occurs under mild conditions in the presence of catalytic amounts of Lewis-acidic boron trichloride. Nucleophilic attack of the chloride ion at the achiral ketene intermediate is followed by ring closure and formation of the racemic *trans*-product in 90% yield.

3.2.10   Amino Acids

Amino acids undergo base- and acid-catalyzed racemization, Scheme 3.51.[358,359] Several factors increase the rate of enantioconversion: the presence of electronegative substituents at the stereogenic carbon atom; resonance stabilization of the intermediate carbanion;[360,361] coordination to

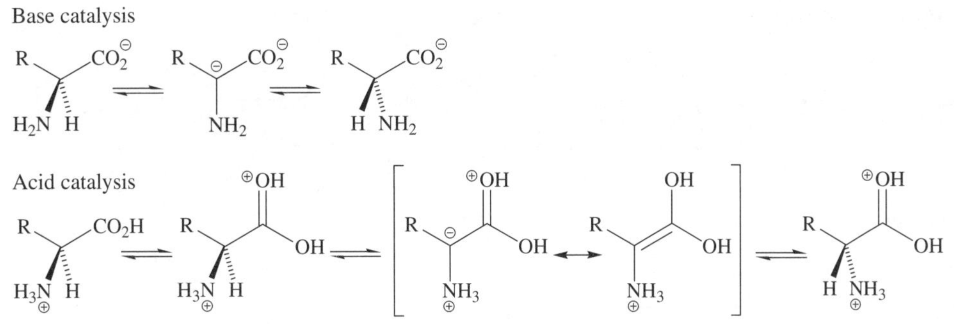

Base catalysis

Acid catalysis

Scheme 3.51   Base- and acid-catalyzed racemization of amino acids.

metal ions;[362] and derivatization of the amino and carboxylic acid groups to amides and esters.[363] This is consistent with the observation that epimerization of peptides is generally faster than racemization of free amino acids. Introduction of protecting groups can reduce the configurational stability of amino acids. In particular, the low stereochemical integrity of *N*-acyl derivatives can be problematic during peptide synthesis,[364,365] or this can alternatively provide an entry to crystallization-induced asymmetric transformation and dynamic kinetic resolution.[366–369] Derivatization of amino acids to hydantoins, oxazolidinediones and succinimides facilitates deprotonation and thus racemization, see Chapter 3.4. Similarly, epimerization of peptides occurs more readily with diketopiperazine derivatives. Racemization pathways involving formation of intermediate azlactones[370–373] and Schiff bases[374–382] that are prone to tautomerization or deprotonation are shown in Scheme 3.52.

Many enzymes utilize aromatic pyridoxal-5′-phosphate (PLP) as a cofactor for Schiff base-mediated racemization of amino acids. For example, alanine racemase catalyzes interconversion of (*S*)- and (*R*)-alanine through a two-base mechanism. In the absence of a substrate in the active site of the enzyme, the aromatic aldehyde is fixed as an aldimine derived from the terminal amino function of Lys39. When this amino acid is replaced by the substrate, PLP is transferred to either (*S*)- or (*R*)-alanine. In the case of (*S*)-alanine, abstraction of the α-proton involves Tyr265. This step is facilitated by formation of a resonance-stabilized quinoid-type structure. Finally, reprotonation generates (*R*)-alanine which is released through replacement with Lys39. As illustrated in Scheme 3.53, Tyr265 is the catalytic base for the interconversion of (*S*)- to (*R*)-alanine, and Lys39 catalyzes the opposite transformation.[383–385] Many enzymes have been isolated and successfully employed in organic synthesis. Although the scope of racemases is often limited by their intrinsic substrate specificity, a wide range of chiral compounds other than amino acids are suitable to enzymatic and microbial racemization or diasteromerization.[386,387] The scope of these asymmetric biotransformations is discussed in Chapter 7.

The stereodynamic properties of amino acids lacking an asymmetric carbon atom are noteworthy, since they can easily be introduced to peptides to alter their 3-dimensional structure and biological activity.[xiv] Achiral amino acids have been incorporated into chiral peptides to stabilize helical structures through asymmetric induction, and to investigate the stereochemistry of peptide folding.[388–391] Even external chiral stimuli can initiate so-called asymmetric noncovalent domino effects and induce helicity in achiral peptide chains.[392,393]

[xiv] Achiral compounds that can adopt chiral conformations can be utilized for amplification and switching of chirality, for example in supramolecular assemblies, see Chapter 3.2.13.

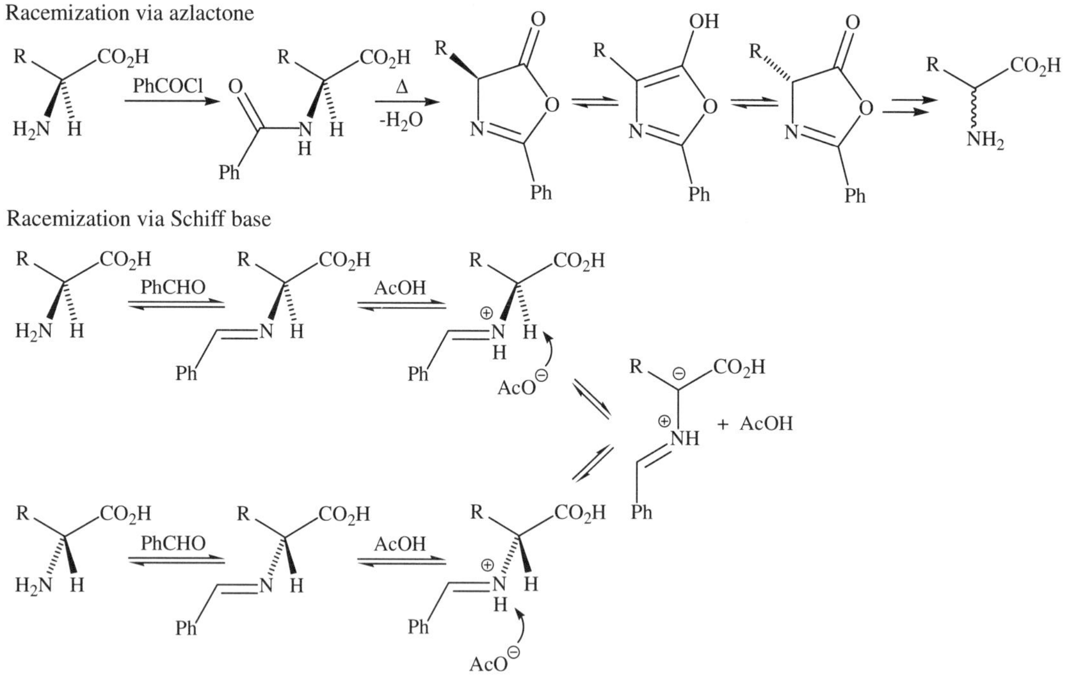

Racemization via azlactone

Racemization via Schiff base

Scheme 3.52 Racemization pathways of amino acids via azlactone (top) and Schiff base intermediates (bottom).

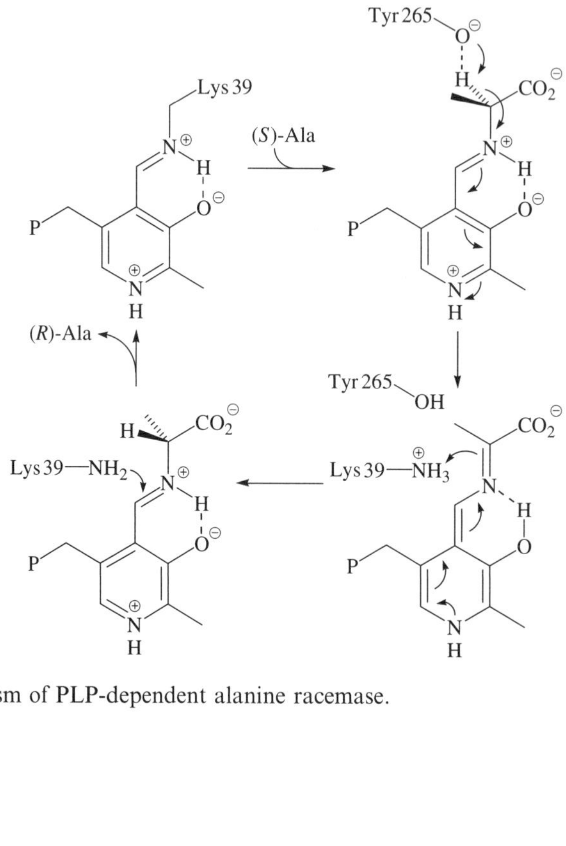

Scheme 3.53 Mechanism of PLP-dependent alanine racemase.

3.2.11 Silicon, Phosphorus and Sulfur Compounds

Chiral compounds bearing a stereogenic third-row element play an important role as reagents, auxiliaries or catalysts in asymmetric synthesis, and are key components of pharmaceuticals such as esomeprazole. The configurational stability of chiral silicon, phosphorus and sulfur centers has therefore received considerable attention. By analogy with compounds possessing an asymmetric carbon center, tetrahedral chiral-at-silicon compounds are generally stable to racemization, and have been successfully employed in stereoselective substituent displacement reactions.[394] For example, treatment of (*S*)-methyl(1-naphthyl)phenylsilyl chloride with butyllithium at –20 °C generates (*R*)-(*n*-butyl)methyl(1-naphthyl)phenylsilane in 72% yield and 92% ee, Scheme 3.54. Silyllithium species are configurationally stable under cryogenic conditions and have been used as key intermediates in the asymmetric synthesis of several chiral silanes.[395–398] However, chiral silyl chlorides are known to undergo facile chloride-induced racemization which has been attributed to the formation of achiral pentacoordinate silicon dichloride derivatives.[399] In the presence of LiCl, 1-chloro-1-phenyl-1,2,3,4-tetrahydro-1-silanaphthalene racemizes at 0 °C within one hour. The racemization process probably entails apical and equatorial attack of the chloride anion, and the corresponding hypervalent silicon species may interconvert via Berry pseudorotation exchanging equatorial and axial positions. Coordination of achiral chelating ligands to silicon can result in the formation of chiral penta- and hexavalent complexes, see Chapter 2.1. Due to facile ligand-site exchange processes, in particular through *Si-N* dissociation and subsequent recombination, these so-called λ⁵- and λ⁶-silicates show rapid racemization in solution.[400,401] Inversion of chirality at the silicon center via an *O,O*-exchange mechanism has also been reported, Scheme 3.54.[402]

Third-row elements have considerably higher pyramidal inversion barriers than amines. Chiral phosphines[403,404] and sulfonium salts[405–407] are configurationally stable at room temperature, whereas acyclic amines readily racemize. Since the bond angle in trivalent phosphorus and sulfur is typically about 100°, a greater amount of distortion is required to reach the planar transition state. The barrier to racemization of *tert*-butylmethylphenylphosphine is 136.9 kJ/mol, Scheme 3.55. Thermodynamically controlled pyramidal inversion of tertiary phosphines can be useful synthetically, even though high temperatures may be required. Epimerization of 1,2-bis(2′-benzylphospholano)benzene proceeds with excellent diastereoselectivity at 190 °C.[408] Incorporation of an acyl substituent significantly reduces configurational stability, and inversion barriers of acyl- and alkoxycarbonylphosphines are generally lower than 100 kJ/mol.[409,410] Pyramidal inversion at

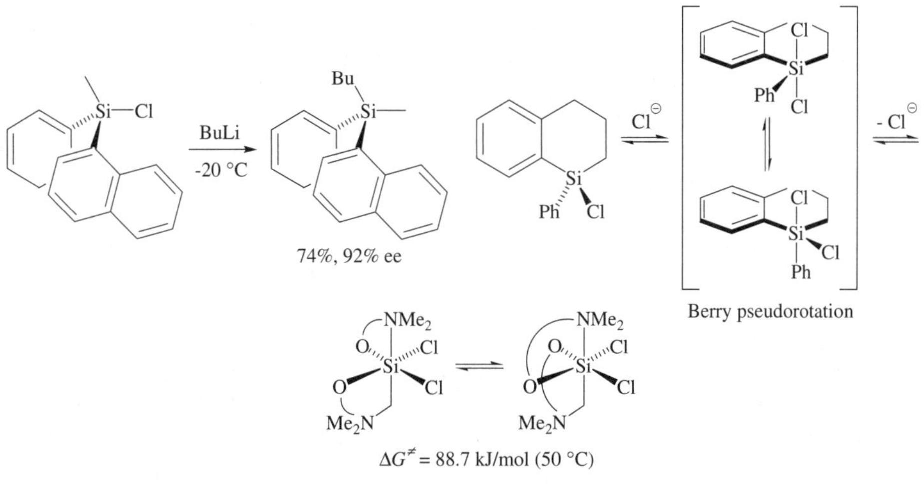

$\Delta G^{\neq} = 88.7$ kJ/mol (50 °C)

Scheme 3.54 Asymmetric transformation and racemization of chlorosilanes (top) and *O,O*-exchange at a λ⁶-silicate (bottom).

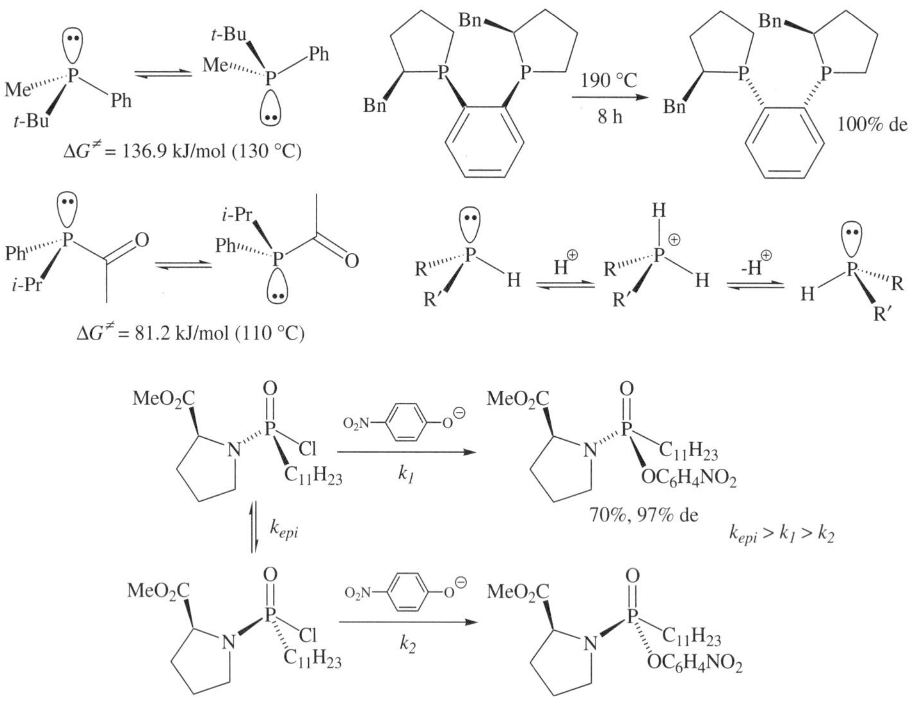

Scheme 3.55 Pyramidal inversion of phosphines, racemization of secondary phosphines via achiral phosphonium ions and DYKAT of a monochlorophosphoramidate.

the stereogenic phosphorus atom is not the only stereomutation pathway of chiral phosphines. Facile acid-catalyzed racemization of secondary phosphines involves formation of an achiral phosphonium ion and often renders isolation of enantiomers difficult.[411,412] Monochlorophosphoramidates and other configurationally unstable phosphorus compounds have been employed in DYKAT and similar stereodynamic synthetic strategies, Scheme 3.55.[413–415]

Isotope experiments proved that chiral phosphates can racemize through a dissociative mechanism with an achiral metaphosphate intermediate.[416] Rotto and Coates uncovered an intriguing stereomutation of a chiral tricyclic phosphate that is configurationally stable in organic solvents but shows thermal diastereomerization and racemization upon heating to 145 °C.[417] This can be explained by formation of a symmetric zwitterion exhibiting both an allylic carbocation and a phosphate anion moiety. Ring closure and allylic rearrangement of the phosphate group generates the other enantiomer through *formal* inversion at all chiral carbon and phosphorus centers, Scheme 3.56.

Most chiral sulfoxides are configurationally stable but undergo pyramidal inversion at high temperatures, Scheme 3.57. This unimolecular process obeys first-order kinetics and is observed with both aliphatic and aromatic sulfoxides at approximately 200 °C. Activation barriers range from 140 to 180 kJ/mol.[418] Interestingly, the barrier to pyramidal inversion of benzothiophene oxides is as low as 100 kJ/mol and enantioconversion can be observed at ambient temperatures.[419] Benzyl *p*-tolyl sulfoxide racemizes approximately 1000 times faster than comparable alkyl aryl sulfoxides. The mechanism for racemization of benzylic sulfoxides involves homolytic cleavage of the carbon-sulfur bond and subsequent recombination of the radical pair.[xv] The energy barrier to

---

[xv] The dissociative mechanism has a positive activation entropy which explains the relatively low energy barrier to racemization of benzyl *p*-tolyl sulfoxide.

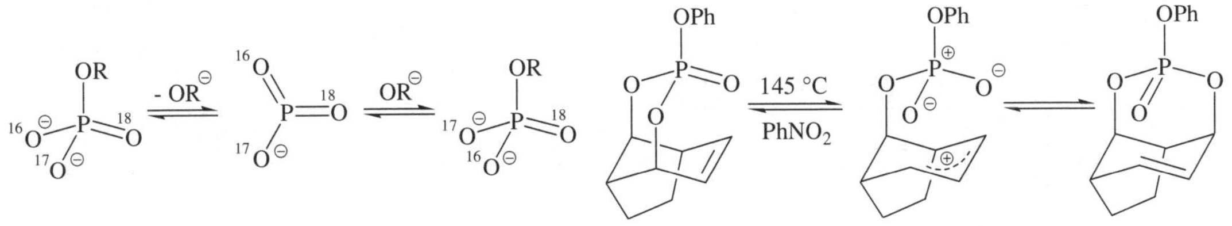

Scheme 3.56 Enantioconversion of isotopically labeled and allylic phosphates.

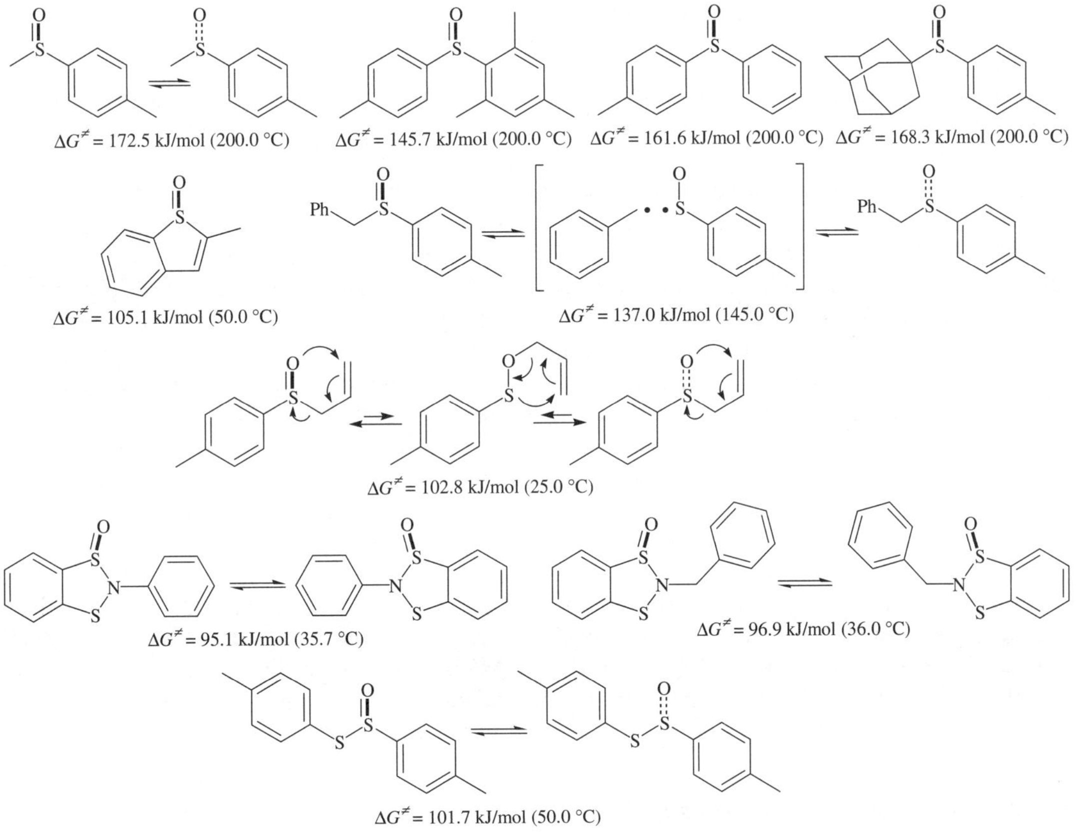

Scheme 3.57 Racemization pathways of sulfoxides and their derivatives.

racemization of benzyl *p*-tolyl sulfoxide is 137.0 kJ/mol.[420] Aryl sulfinamides have also been reported to racemize via a radical mechanism.[421] Although homolytic α-cleavage and recombination of radical pairs is a common photochemical reaction of sulfoxides, photoracemization originates exclusively from direct sulfur inversion, and carbon–sulfur bond scission has been ruled out as an alternative mechanism.[422–427] Enantioconversion of chiral sulfoxides is usually based on homolysis or pyramidal inversion but thermal racemization of allylic and allenic sulfoxides entails a different concerted mechanism.[428,429] Kinetic studies and isotope labeling experiments with allyl *p*-tolyl sulfoxide proved that the reaction proceeds via reversible sigmatropic [2,3]-rearrangement to an achiral allyl sulfenate intermediate with an energy barrier of 102.8 kJ/mol.[430] Sulfoxide derivatives such as *N*-aryl or *N*-benzyl-1,3,2-benzodithiazole-1-oxides[431,432] and aryl arenethiolsulfinates[433] are less configurationally stable than sulfoxides and are prone to sulfur stereoinversion

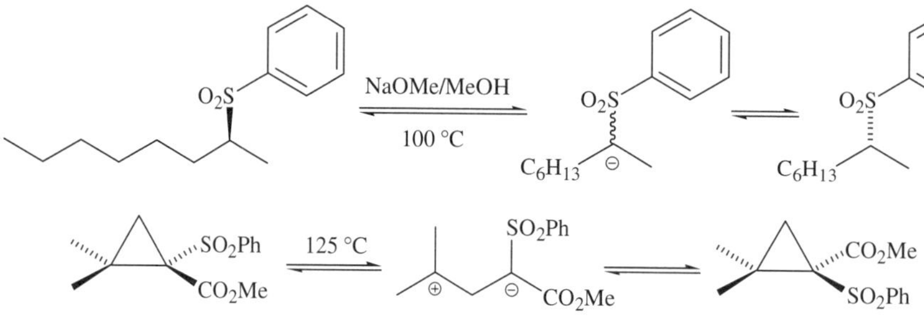

Scheme 3.58 Enantioconversion of chiral sulfones.

under mild conditions. Aryl- and alkylsulfinyl chlorides easily racemize in the presence of amines and this has been utilized for asymmetric synthesis of chiral sulfinate esters and sulfoxides.[434–437]

In contrast to sulfoxides, the sulfur atom in sulfones is not the origin of chirality but it can affect the configurational stability of an adjacent asymmetric carbon center. The racemization course of sulfones usually resembles that of carboxylic acid derivatives and is strikingly different from the stereomutations of sulfoxides discussed above. Since the pK_a value of the α-proton in aliphatic sulfones is about 30, base-catalyzed racemization via an intermediate carbanion is possible at elevated temperatures, Scheme 3.58.[438] Deuterium exchange experiments conducted with aryl 2-octyl sulfones at 100 °C revealed that the rate of isotope exchange is significantly faster than the rate of racemization. This proves a remarkable stereochemical integrity of the intermediate carbanion which is apparently not planar and does not readily undergo pyramidal inversion.[439,440] Cram *et al.* recognized the ability of the sulfone group to stabilize adjacent carbanions, and designed a cyclopropane that favors thermal racemization via heterolytic ring opening and subsequent cyclization over the radical mechanism generally observed with cyclopropanes, see Chapter 3.2.1.[441] The methyl ester of 2,2-dimethyl-1-phenylsulfonylcyclopropane-1-carboxylic acid exhibits carbanion-stabilizing and carbonium ion-stabilizing groups on either side of the strained three-membered ring. The substituents effectively favor formation of an intermediate achiral zwitterion and racemization can be observed in DMSO at 125 °C which is substantially lower than the temperatures usually required for stereomutations of cyclopropanes. Atropisomerization of axially chiral sulfoxides and sulfones is discussed in Chapter 3.3.2.

3.2.12 Organometallic Compounds

The dynamic stereochemistry of chiral transition metal complexes is quite versatile and adds a new dimension to asymmetric synthesis and supramolecular chemistry including stereoselective self-association and chiral amplification.[442,443] During the last 50 years, the ever-increasing impact of organometallic compounds on stereoselective synthesis and chiral recognition has fueled the development of numerous chiral metal complexes. A close look at the configurational stability and reactivity of Grignard and organolithium reagents underscores the fact that synthetic applications of metal complexes are defined by their stereodynamic properties.

The configurational stability of chiral organolithium compounds has been studied by variable-temperature ^1H, ^{13}C, ^6Li, and ^7Li NMR spectroscopy. Stereoinversion at the chiral center requires migration of the lithium atom from one enantiotopic site to the other. This may occur via formation of a solvent-separated ion pair or through a nondissociative process when a proximate Lewis basic site coordinating to the lithium cation is present, Scheme 3.59.[444–451] The configurational stability is often solvent-dependent and varies significantly with the steric bulk and chelating properties of neighboring groups.[452–456] While some organolithium compounds racemize rapidly

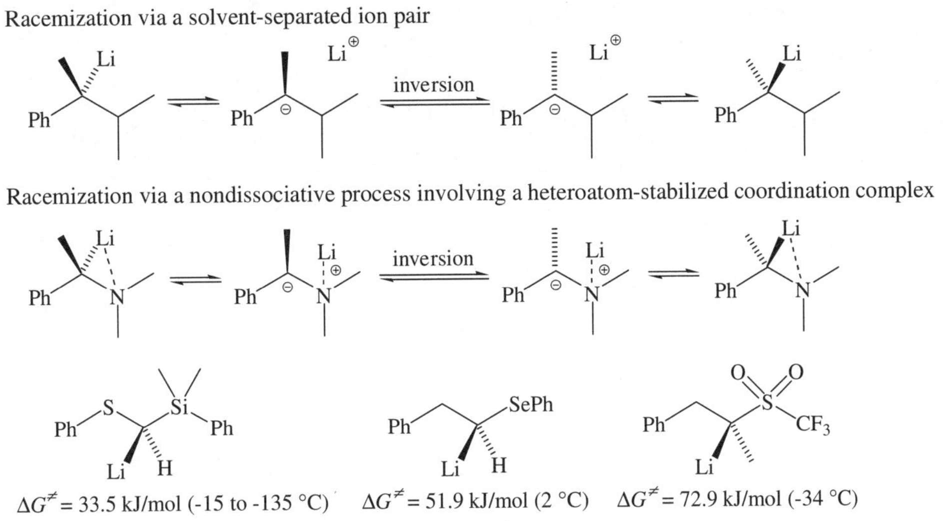

Racemization via a solvent-separated ion pair

Racemization via a nondissociative process involving a heteroatom-stabilized coordination complex

$\Delta G^{\neq} = 33.5$ kJ/mol (-15 to -135 °C) $\Delta G^{\neq} = 51.9$ kJ/mol (2 °C) $\Delta G^{\neq} = 72.9$ kJ/mol (-34 °C)

Scheme 3.59 Racemization pathways of organolithium compounds and energy barriers to racemization of selected examples.

even under cryogenic conditions, others such as α-lithio sulfones can have energy barriers above 70 kJ/mol.[xvi]

For synthetic purposes, it is often not necessary to determine accurately the racemization rate of an organometallic species. If an enantiopure organolithium reagent is consumed by an electrophile before it can undergo measurable racemization, the overall stereoselectivity of the asymmetric reaction is not compromised. It is therefore sufficient to know the *relative* configurational stability with respect to the rate of reaction with a trapping electrophile. This information can be obtained by the Hoffmann test which is based on the principles of KR, DKR and DYKAT, see Chapters 7.4 and 7.5.[457–462] This test requires two experiments that are usually conducted with an excess of an electrophile to guarantee complete conversion: the racemic organolithium species is first treated with a racemic electrophile and then with the enantiomerically pure electrophile. The corresponding energy profiles for an organolithium compound that racemizes rapidly compared to the reaction with the racemic or enantiopure electrophile, is depicted in Figure 3.10. According to the Curtin–Hammett principle, the stereochemical outcome of both scenarios depends solely on the difference in energy of the diastereomeric transition states, $\Delta\Delta G^{\neq}$, and the diastereoselectivities obtained by the two experiments must be the same.[xvii] This is no longer the case when the organolithium compound is configurationally stable or has a racemization half-life time comparable to the rate of reaction with the electrophile, Figure 3.11. Quantitative trapping with a racemic electrophile still affords a mixture of diastereomers that is determined by the energetic difference of the diastereomeric transition states, $\Delta\Delta G^{\neq}$.[xviii] However, when an enantiopure electrophile is used the diastereoselectivity will be different, and the product ratio will be unity if equilibration between the organolithium enantiomers does not occur during the reaction. Based on the Hoffmann test, one can conclude that a chiral organolithium reagent undergoes fast racemization relative to the reaction with an electrophile if both experiments provide the same diastereoselectivity. If the stereochemical outcome is different,

---

[xvi] Usually, chiral nonracemic Grignard and organolithium compounds can not be isolated at room temperature. In the context of asymmetric synthesis, organometallic species are considered configurationally stable if racemization is relatively slow compared to a trapping reaction with an electrophile. This is often the case at very low temperatures.

[xvii] Note that enantiomeric pathways do not alter the diastereoselectivity of the reaction.

[xviii] This is not a consequence of the Curtin–Hammett principle, but is a result of matched and mismatched stereoselection. One would obtain the same ratio of diastereomers with the enantiopure organolithium compound.

a) Reaction of a stereolabile organolithium compound with a racemic electrophile

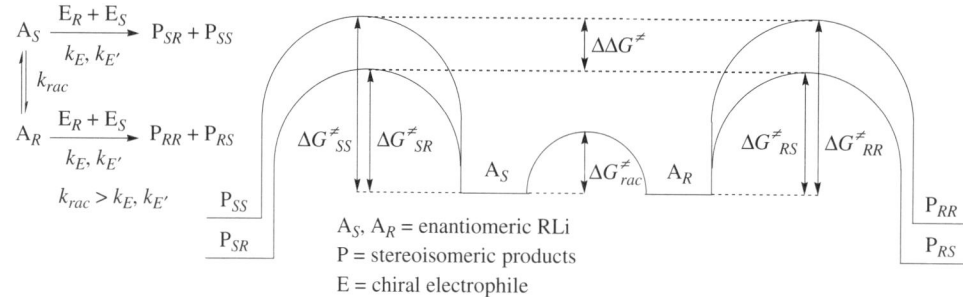

b) Reaction of a stereolabile organolithium compound with an enantiopure electrophile

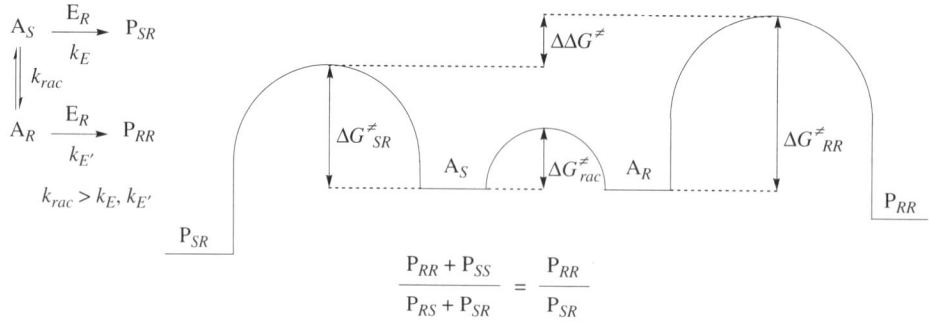

$$\frac{P_{RR} + P_{SS}}{P_{RS} + P_{SR}} = \frac{P_{RR}}{P_{SR}}$$

Figure 3.10   Energy profiles for the reactions between a stereolabile organolithium species and a racemic or enantiopure electrophile.

both racemization and reaction with the electrophile must occur on a similar time scale, or if equal amounts of diastereomers are obtained with an enantiopure electrophile the organolithium species is configurationally stable under the reaction conditions used.

Grignard compounds are certainly among the most widely used organometallic reagents but few asymmetric reactions with nonracemic organomagnesium reagents are known.[463] This is a consequence of the low configurational stability of chiral carbanions under conditions that are normally required to produce Grignard reagents from alkyl halides. Hoffmann's group showed that the preparation of a nonracemic Grignard complex is possible through sulfoxide/magnesium exchange at low temperatures. This approach reduces the risk of racemization which would ultimately compromise the stereoselectivity of a subsequent reaction with an electrophile, Scheme 3.60.[464-466] For example, treatment of $(R_S,1R)$-p-tolyl-1-chloro-2-phenylethyl sulfoxide with ethylmagnesium bromide at $-78\,^{\circ}C$ gives the corresponding (R)-α-chloroalkylmagnesium bromide with retention of configuration. The chiral Grignard reagent can then be trapped with benzaldehyde, providing $(1R,2R)$-1,3-diphenyl-2-chloro-1-hydroxypropane in 77% yield and 93% ee. Even chiral-at-metal magnesium complexes that display two chelating 1,2-dimethoxyethane ligands have been described. The six-coordinate magnesium atom is the only chiral center in these complexes which rapidly racemize in solution, although enantiopure crystals obtained by crystallization-induced asymmetric transformation have been employed in enantioselective additions to aldehydes with some success.[467,468] In contrast to Grignard complexes, nonracemic organolithium compounds have found numerous applications in asymmetric synthesis. The configurational stability of lithiated carbanions can be significantly enhanced by intramolecular coordination of proximate heteroatoms to the lithium atom, see Chapter 6.1. For instance, transmetalation between (R)-1-benzyloxymethoxy-1-tributylstannylpropane and butyllithium at $-78\,^{\circ}C$ followed by methylation with dimethyl sulfate produces (R)-2-benzyloxymethoxybutane in 95% ee, Scheme 3.60.[469]

a) Reaction of a configurationally stable organolithium compound with a racemic electrophile

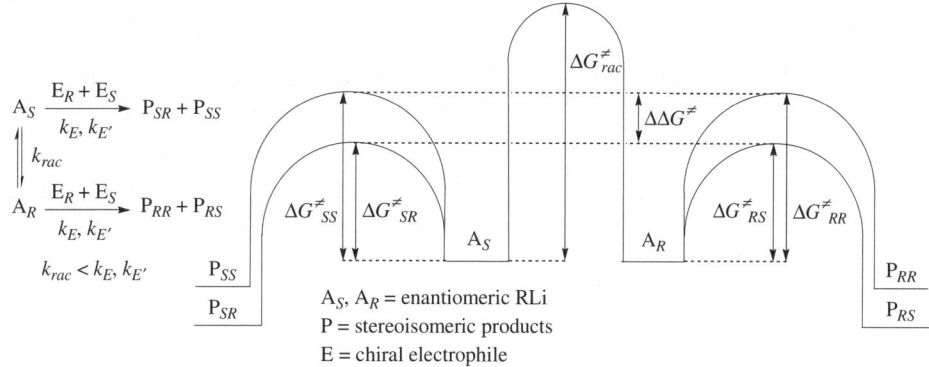

$$A_S, A_R = \text{enantiomeric RLi}$$
$$P = \text{stereoisomeric products}$$
$$E = \text{chiral electrophile}$$

b) Reaction of a configurationally stable organolithium compound with an enantiopure electrophile

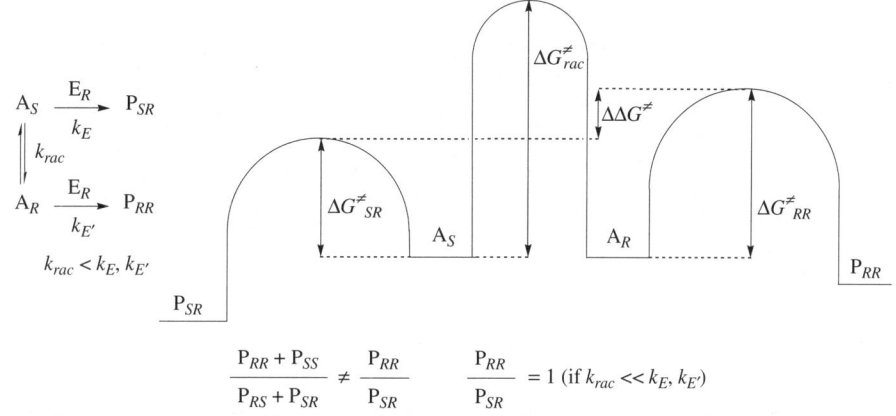

$$\frac{P_{RR} + P_{SS}}{P_{RS} + P_{SR}} \neq \frac{P_{RR}}{P_{SR}} \qquad \frac{P_{RR}}{P_{SR}} = 1 \text{ (if } k_{rac} \ll k_E, k_{E'}\text{)}$$

Figure 3.11 Kinetics of the reactions of a configurationally stable organolithium species with a racemic or enantiopure electrophile.

Scheme 3.60 Asymmetric synthesis with chiral Grignard and organolithium reagents.

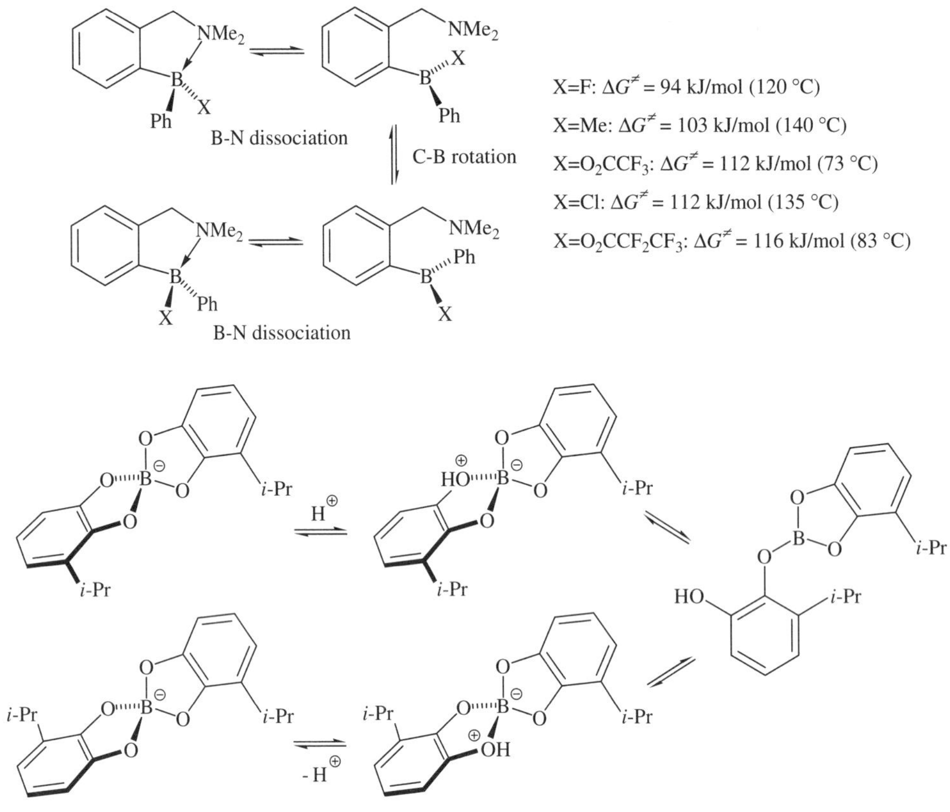
NMe2
B-N dissociation
C-B rotation
Ph
X
X=F: ΔG≠ = 94 kJ/mol (120 °C)
X=Me: ΔG≠ = 103 kJ/mol (140 °C)
X=O2CCF3: ΔG≠ = 112 kJ/mol (73 °C)
X=Cl: ΔG≠ = 112 kJ/mol (135 °C)
X=O2CCF2CF3: ΔG≠ = 116 kJ/mol (83 °C)
i-Pr
H⊕
-H⊕
HO
OH

Scheme 3.61 Racemization of tetrahedral boron complexes.

Chiral-at-boron complexes are usually not configurationally stable in solution due to rapid dissociation of boron–nitrogen and boron–oxygen bonds.[470] The stereodynamics of several four-coordinate oxazaborolidinone-derived boron complexes have been exploited for asymmetric transformation of the second kind.[471–473] Toyota and coworkers demonstrated that substituents have a major influence on the barrier to racemization of amine-borane complexes, Scheme 3.61.[474] The configurational stability of [2-(dimethylaminomethyl)phenyl](pentafluoropropionyloxy)phenylborane was determined as 116 kJ/mol. The remarkable stability to racemization of perfluoroacyl-derived complexes coincides with shorter *B–N* bond lengths and increased thermal stability.[xix] A dissociative acid-catalyzed mechanism has been proposed for the enantioconversion of chiral borates.[475]

Racemization and diastereomerization of kinetically unstable transition metal complexes often involve ligand exchange processes. In the case of $K_3[Yb\{(M)\text{-BINOL}\}_3]$, stepwise replacement of all (*M*)-BINOL ligands upon addition of a molar excess of the (*P*)-enantiomer results in complete chirality inversion because the homochiral complexes are thermodynamically favored over the heterochiral intermediates. Although near-IR CD and NMR studies revealed that the bidentate ligands are rapidly exchanged one-by-one, this is not a trivial process: each replacement requires additional proton transfer steps that may occur via a concerted or sequential mechanism, Scheme 3.62.[476]

[xix] The increased stability to boron–nitrogen bond dissociation is a consequence of the enhanced Lewis acidity of the corresponding trivalent boron complex.

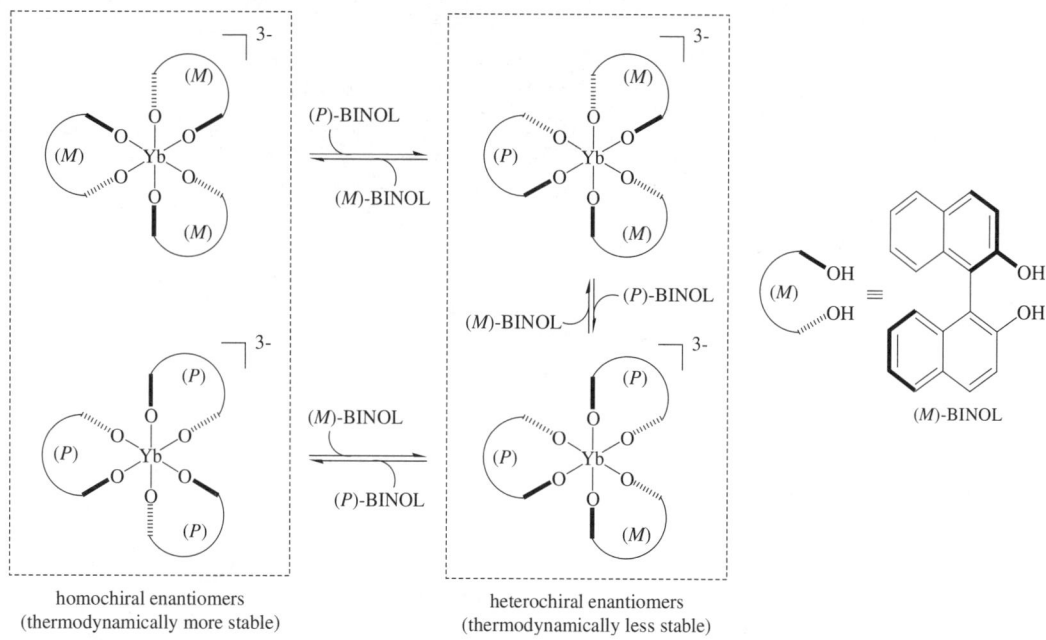

Scheme 3.62 Enantioconversion of [Yb(BINOL)$_3$]$^{3-}$ complexes by ligand exchange.

Most stereomutations of kinetically unstable transition metal complexes occur through a dissociation/association sequence and an intermediate unsaturated 16-electron species that allows rapid reorientation of one or more ligands. Faller postulated two possible mechanisms for the interconversion of chiral ruthenium complexes.[477] The chiral-at-metal complex shown in Scheme 3.63 can racemize either via dissociation and subsequent association of water or by conversion of the chelating *P,O*-donor to a monodentate ligand followed by internal rotation about the *Ru-P* bond and rechelation. In both pathways, formation of an intermediate achiral 16-electron complex yields racemization. Some chiral-at-metal complexes are configurationally stable in solution while others undergo rapid racemization or epimerization if another chiral element is present in the ligand sphere.[478–490] Prediction of the configurational stability of transition metal complexes is difficult and half-lives vary substantially among different metals and ligands, Scheme 3.64.[491–494]

Inversion of the planar chirality of aryl metallocenes can occur via migration of the metal moiety to the opposite arene face. Heating of *syn*-(η-1,2,3,4,5,6-tricarbonyl-2-hydroxymethyl-6-methoxy-2'-methylbiphenyl)chromium in a nonaromatic solvent affords a mixture of two *anti*-diastereomers due to rotation about the chiral axis (atropisomerization) and stereoselective tricarbonylchromium migration to the less sterically hindered arene face. The latter results in inversion of planar chirality and is the predominant reaction. It has been proposed that the migration is assisted by the hydroxymethyl group and that it proceeds in an intermolecular fashion.[495,496] An intramolecular migration of tricarbonylchromium along a naphthohydroquinone skeleton has been incorporated into the design of a reversible chiral switch.[497] The naphthohydroquinone-derived tricarbonyl-chromium complex shown in Scheme 3.65 forms the thermodynamically disfavored (*R*)-isomer by an asymmetric photo-induced metal shift along the aromatic ring. The reaction outcome can be reversed by thermally controlled metal migration to the more stable (*S*)-regioisomeric complex. Both haptotropic migrations proceed without any sign of racemization.

Induction and inversion of chirality play an important role in nature and are known to affect the properties and functions of DNA and proteins. The versatile coordination chemistry and stereo-dynamics of chiral metal complexes can be used to mimic such processes, to devise chiral switches

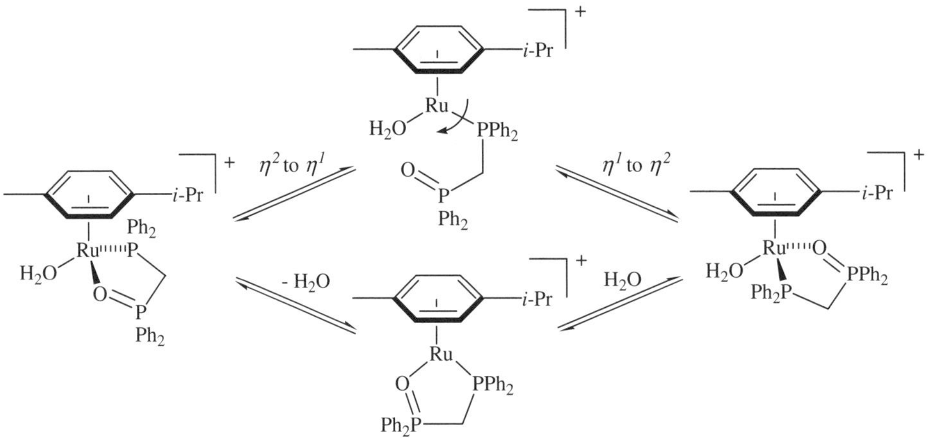

Scheme 3.63   Possible racemization pathways of a chiral ruthenium complex.

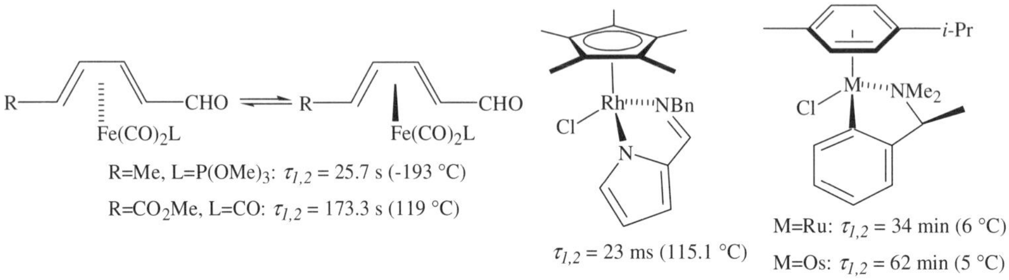

R=Me, L=P(OMe)$_3$: $\tau_{1,2}$ = 25.7 s (-193 °C)
R=CO$_2$Me, L=CO: $\tau_{1,2}$ = 173.3 s (119 °C)

$\tau_{1,2}$ = 23 ms (115.1 °C)

M=Ru: $\tau_{1,2}$ = 34 min (6 °C)
M=Os: $\tau_{1,2}$ = 62 min (5 °C)

Scheme 3.64   Configurational stability of transition metal complexes.

that respond to external stimuli, and to develop supramolecular architectures exhibiting chiral amplification and memory.[498–504] Miyake's group demonstrated that the helicity of kinetically labile octahedral Co(II), Cu(II) and Ni(II) complexes can be reversed upon addition of an achiral stimulus.[505,506] Coordination of a chiral amino acid-derived tetradentate ligand to cobalt(II) affords a thermodynamically favored Λ-isomer with a positive CD signal, Scheme 3.66. However, addition of two equivalents of tetrabutylammonium nitrate completely reverses the CD signal due to displacement of two solvent ligands and cooperative stabilization of the diastereomeric Δ-form.[xx] Because of the kinetic instability of the cobalt(II) complex, the stereochemical response to the presence of nitrate anions resulting in helicity inversion within the metal coordination sphere requires less than a minute.

3.2.13   Supramolecular Structures

The dynamics, cooperative effects and correlated motions of chiral supramolecular structures held together by weak intermolecular interactions such as hydrogen bonding, metal coordination, π–π interactions, Coulomb forces, dipole–dipole interactions, and van der Waals forces have led to the development of complex molecular assemblies that can be used for switching[507–510] and amplification[511–515] of chirality. Lehn's group demonstrated that self-association of achiral compounds adopting a helical conformation can lead to chiral supramolecular aggregates.[516] Intramolecular hydrogen bonding of linear heptamers derived from repeating 2′-pyridyl-2-pyridinecarboxamide

---

[xx] The ratio of the diastereomeric Co(II) complexes is also sensitive to the solvent composition. In this case, dynamic helicity inversion is based on solvato-diastereomerism.

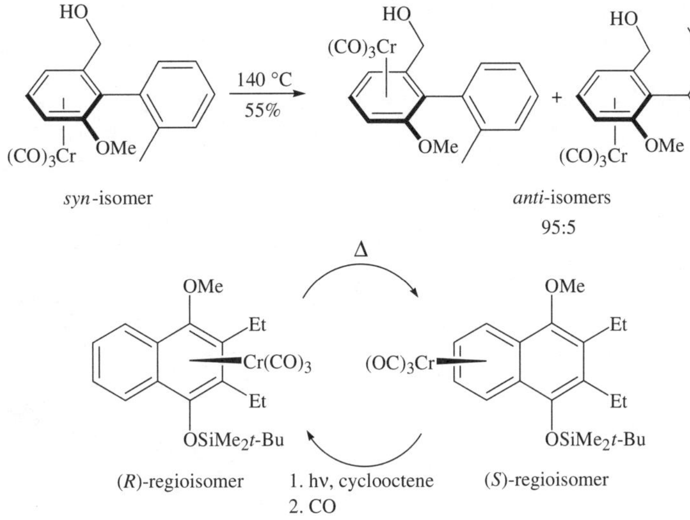

Scheme 3.65 Isomerization of chiral aryl tricarbonylchromium complexes.

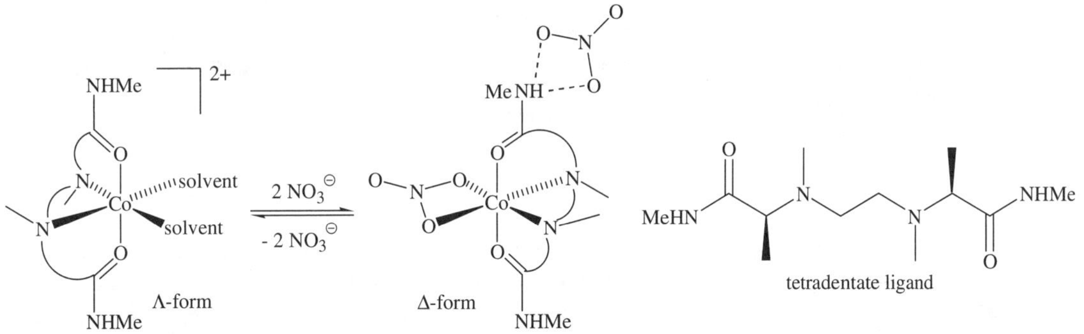

Scheme 3.66 Dynamic helicity inversion of a stereolabile Co(II) complex.

units establishes a racemic mixture of short helical structures consisting of approximately one and a half turns, Figure 3.12. Dimerization of homochiral helices was observed by variable-temperature NMR spectroscopy and has been attributed to spiral sliding of one oligopyridinecarb-oxamide coil into another. This artificial double helix has remarkable stereodynamic features. Partial unwinding due to lateral movements of the individual strands gives rise to translational isomerism which can be monitored at ambient temperatures up to 55 °C. Dissociation of the double-stranded helix requires complete unwinding through translational movement of one coil relative to the other. The dimers are remarkably stable in solution due to cooperative intermolecular π-stacking but dissociation is fast and thermodynamically favored above 105 °C.[xxi]

Chiral inversion and induction processes play an important role in host-guest chemistry, self-assembly and molecular recognition. The fascinating structure and properties of double-stranded helical DNA has inspired Lehn and others to design artificial analogs, in which molecular chirality is transformed into supramolecular helicity.[517–522] For example, charge-assisted hydrogen bonding

---

[xxi] At room temperature, the dimer of the heptamer depicted in Figure 3.12 remains the major species in solution, even at concentrations as low as 300 μM. The dimerization constant in CDCl$_3$ is $6.5 \cdot 10^4$ L mol^{-1}.

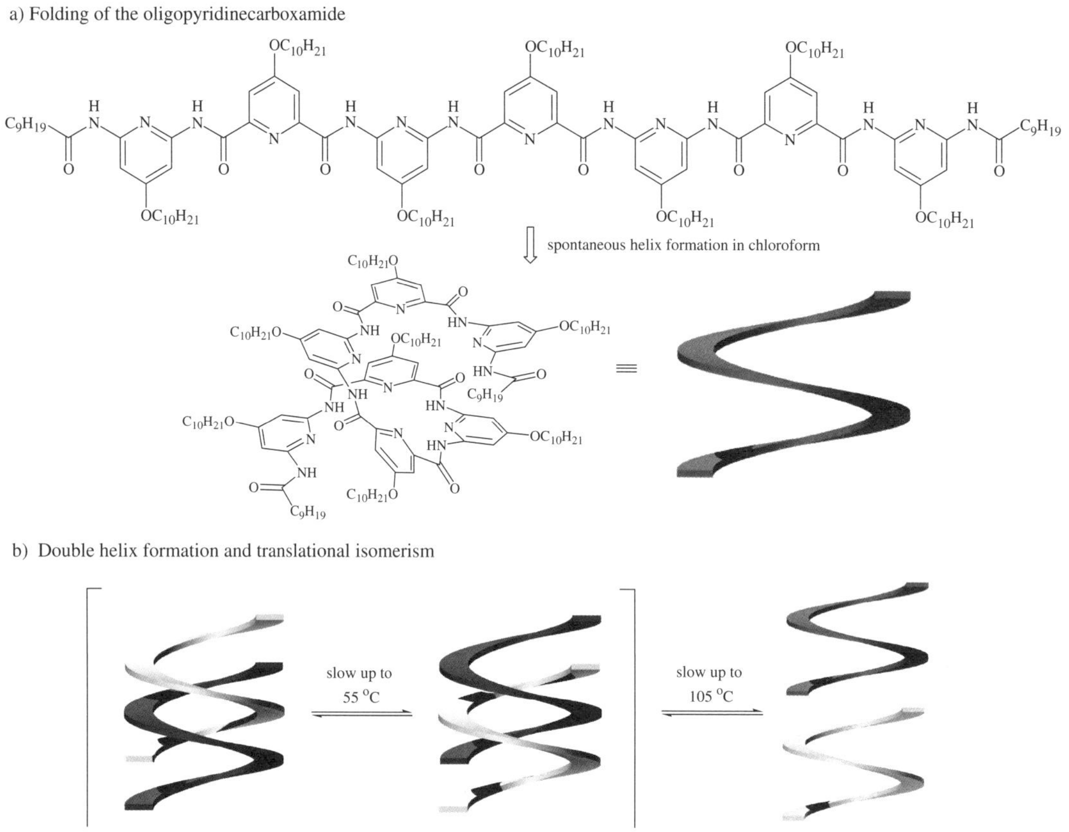

a) Folding of the oligopyridinecarboxamide

⇓ spontaneous helix formation in chloroform

b) Double helix formation and translational isomerism

slow up to 55 °C

slow up to 105 °C

Figure 3.12 Dynamic exchange of artificial single and double helices derived from oligopyridinecarbox-amide strands. [Modified with permission from *Nature* **2000**, *407*, 720-723.]

between two sulfate anions and two enantiomerically pure bisguanidinium strands results in the assembly of a well-defined supramolecular structure. The handedness of the helical arrangement is controlled by the central chirality of the homochiral guanidinium moieties, Figure 3.13.[523] The same principle can be applied to the synthesis of hetero-stranded double helices. Furusho and Yashima reported spontaneous self-assembly of an enantiopure double helix from two complementary *m*-terphenyl strands exhibiting amidine and carboxylate groups, respectively. Cooperative charge-assisted hydrogen bonding between the chiral bisamidine strand and the complementary dicarboxylate strand affords an artificial double helix that is stable in solution and in the solid state. The sense of chirality of the supramolecular structure is determined by the absolute configuration of the amidine units and enantiomeric bisamidine strands have been found to generate assemblies of opposite helicity.[524] Chiral induction is a widespread phenomenon that is not only observed with supramolecular assemblies. Numerous examples of switching and amplification of chirality in oligomers,[525–528] polymers,[529–543] and gels or liquid crystals have been reported.[544–557]

Cyclodextrins consist of six, seven or eight glucose units linked by α-1,4 glycosidic bonds. These naturally occurring macrocycles have been known for a long time and are commonly employed in chromatography, electrophoresis and many other purposes. Cyclodextrins have the shape of a truncated cone that is chiral due to the presence of five stereogenic centers in each glucose unit, Figure 3.14. Inclusion of UV-active achiral molecules into the chiral cavity of cyclodextrins can force the guest into a nonracemic chiral conformation which gives rise to induced circular

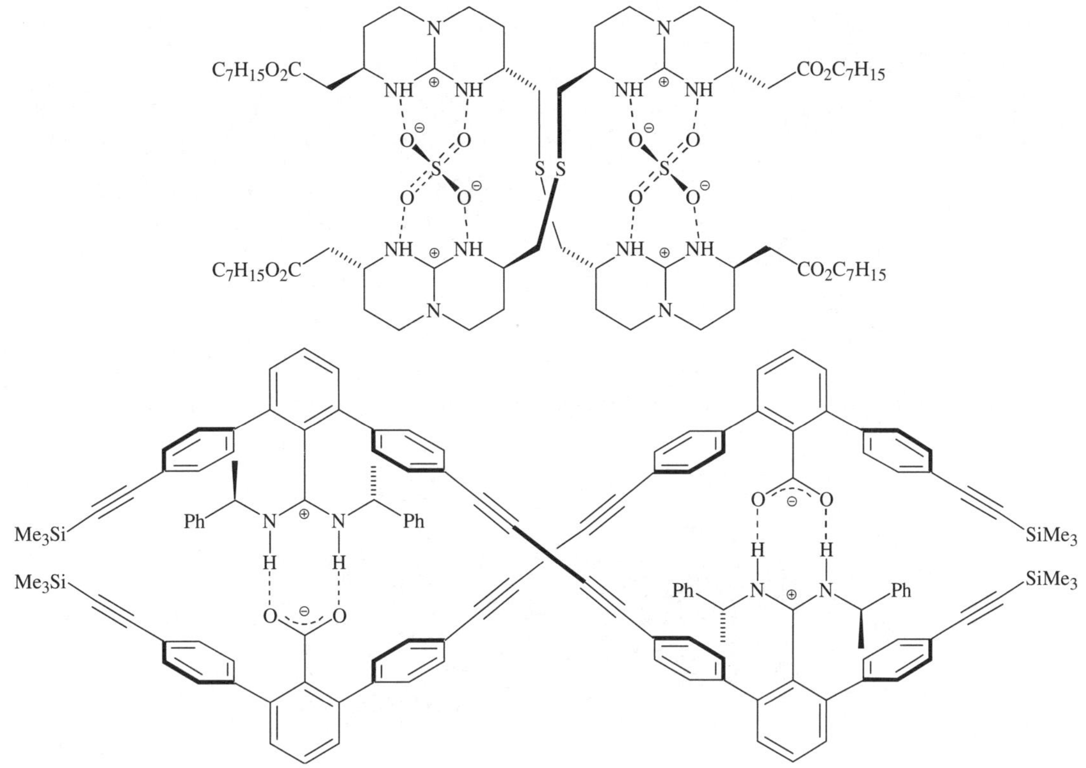

Figure 3.13 Artificial double-stranded helices.

dichroism.[558] Polycyclic aromatic hydrocarbons,[559] benzophenones[560,561] and diarylmethanes[562] have been reported to adopt a helical structure upon complexation by native β- or γ-cyclodextrins. The asymmetric perturbation and relative orientation of aromatic guest molecules in the chiral cavity can be conveniently measured by circular dichroism spectroscopy because native cyclodextrins are CD-silent.

The usefulness of cyclodextrin-derived inclusion complexes has inspired the study of molecular recognition and chiral induction with artificial hosts such as crown ethers, calixarenes and other cavitands.[xxii] Both induction of chirality from a chiral guest to an achiral host, and from a chiral host to a guest molecule that can adopt chiral conformations, have been reported.[563–567] Some macrocycles possess flexible structures that can racemize without bond cleavage. This has been demonstrated with calixarenes derived from selectively substituted arene units. Collet and others found that a partially *O*-methylated calix[4]arene consisting of four 3,4-dimethylphenol units adopts a chiral cone conformation despite the lack of a chiral center, Scheme 3.67.[568,569] Due to intramolecular hydrogen bonding, this calixarene affords only two enantiomeric cone structures while achiral 1,2-alternate and 1,3-alternate conformations exhibiting C_1- and S_4-symmetry, respectively, have not been detected. The enantiomers can be separated by chiral chromatography, but racemize via cone-to-cone ring inversion at ambient temperatures.[570]

[xxii] Cavitands are macrocyclic compounds with an open cavity that can be occupied by small molecules. Calix[n]arenes are a sub-class of cavitands and consist of *n* arene units linked by methylene bridges. Calixarenes exhibit a hydrophobic interior and have found widespread applications in host-guest chemistry.

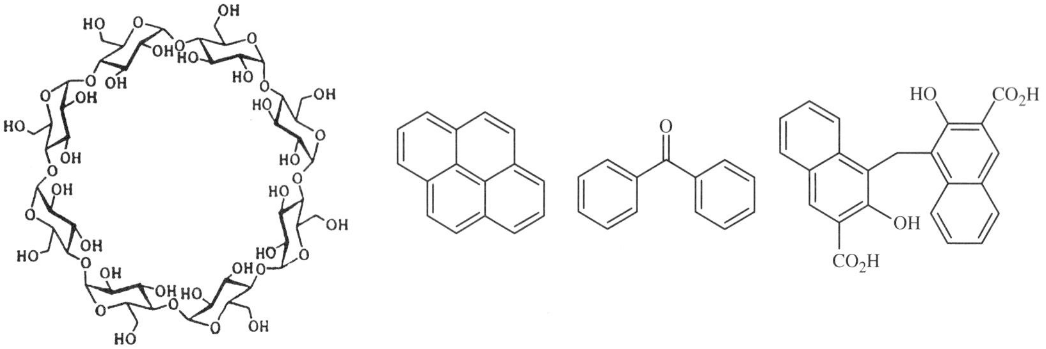

Figure 3.14   Structures of native γ-cyclodextrin and guest molecules that undergo chiral induction upon complexation.

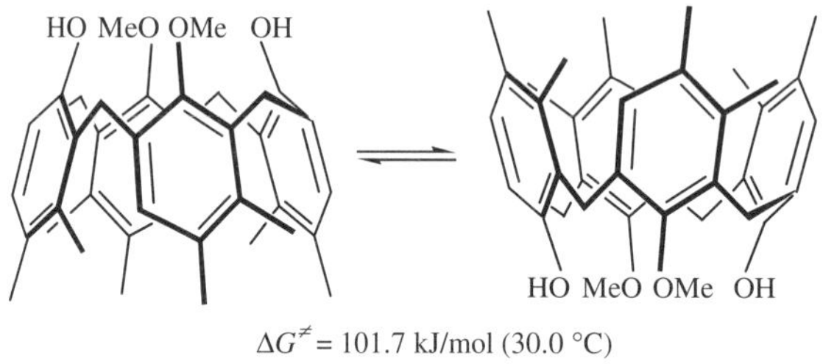

$$\Delta G^{\neq} = 101.7 \text{ kJ/mol} (30.0 \text{ °C})$$

Scheme 3.67   Racemization of a calix[4]arene via cone-to-cone ring inversion.

3.3 ATROPISOMERIZATION

In 1922, Christie and Kenner reported the separation of the enantiomers of 6,6′-dinitrobiphenyl-2,2′-dicarboxylic acid and its 4,4′,6,6′-tetranitrobiphenyl derivative by preferential crystallization of diastereomeric brucine salts, Figure 3.15.[571] But they were unable to resolve the enantiomers of biphenyl-2,2′-dicarboxylic acid. We know today that the latter can not be separated into enantiomers at room temperature because this biphenyl readily racemizes due to facile rotation about the chiral axis. For practical reasons, Kuhn introduced the concept of atropisomerism to describe stereoisomers that result from restricted rotation about a single bond and that can be separated at room temperature.[572] This definition is solely based on the rotational stability of isolable stereoisomers, and suffers from the same drawbacks as the physicochemical classification of configurational and conformational isomers discussed in Chapter 2.2.[573] For example, 2,2′-diiodobiphenyl has a rotational energy barrier of 97.0 kJ/mol at 34 °C and can be separated into enantiomers at room temperature but it slowly racemizes in solution.[574] Since Kuhn's stability criterion is vague, *i.e.*, there is no clearly defined minimum half-life time or free energy barrier to rotation associated with atropisomerism, it remains unclear whether 2,2′-diiodobiphenyl fulfills the requirements of this type of stereoisomerism. A strict interpretation of Kuhn's original definition would include axially chiral biphenyls that are stable to racemization *and* have been separated into stereoisomers, for example 2,2′,3,4,4′,5′,6-heptachlorobiphenyl (PCB 183). On the other hand, the enantiomers of conformationally stable 2,2′,3,3′,4,4′,5,6′-octachlorobiphenyl (PCB 196) would not be considered atropisomeric because they have not been separated to date. Although the concept of atropisomerism is not clearly defined, it is widely used in the literature, irrespective of the

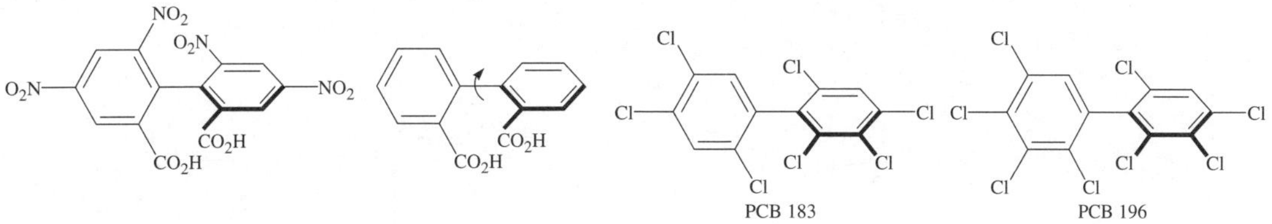

Figure 3.15   Structures of axially chiral biphenyls.

separation criterion, and generously applied to stereoisomers that exhibit hindered rotation about a single bond.[xxiii]

3.3.1   Biaryls, Triaryls and Diarylacetylenes

Since the early findings of Christie and Kenner, the unique relationship between structure and stereodynamics of atropisomeric biaryls and triaryls has received significant attention. Interconversion of the enantiomers of 1,8-bis(2′-phenyl-4′-quinoyl)naphthalene proceeds via a diastereomeric meso intermediate and involves two rotations obeying reversible first-order kinetics, Chapter 3.1.3. In contrast, racemization of chiral biphenyls constitutes a one-step process and is kinetically less complicated which greatly facilitates investigation of stereodynamic properties.[575–578] Wolf and coworkers conducted a systematic study of steric and electronic effects of *ortho*-, *meta*- and *para*-substituents on the rotational energy barrier of axially chiral 2,2′-dialkylbiphenyls utilizing polarimetry and dynamic gas chromatography, Figures 3.16 and 3.19.[579–582]

The conformational stability of unbridged biphenyls is mainly determined by the number and size of *ortho*-substituents that experience strong steric repulsion in the periplanar transition state and thus impede rotation about the pivotal bond.[xxiv] Most tri- and tetra-*ortho*-substituted biphenyls are conformationally stable and can be conveniently resolved into enantiomers at room temperature, unless small fluoro or methoxy *ortho*-substituents are present. With the exception of 2,2′-disubstituted biphenyls bearing bulky isopropyl, phenyl or *tert*-butyl groups, axially chiral biphenyls that have fewer than three *ortho*-substituents are often not stable to racemization at 25 °C. Rotation about the chiral axis of 2,2′-disubstituted biphenyls can in principle occur via two different transition states, in which the substituents pass either one another or a hydrogen atom of the adjacent aryl ring. The latter entails significantly less steric repulsion and is the energetically favored and therefore predominant pathway of atropisomerization, Figure 3.17. A closer look at tri- and tetrasubstituted biphenyls shows that the relative position of *ortho*-substituents is also important. For example, 2-*tert*-butyl-2′-methyl-6-isopropylbiphenyl can access two transition states. Rotation via a transition state in which the 2′-methyl and the 6-isopropyl substituents pass each other will be energetically favored because it avoids the severe steric hindrance that would result from the interaction between the 2′-methyl and the bulky 2-*tert*-butyl group. In the case of 2-*tert*-butyl-2′-methyl-6′-isopropylbiphenyl, enantiomerization requires that either the 2′-methyl or the 6′-isopropyl group passes the 2-*tert*-butyl group of the other phenyl ring. With this biphenyl, the least steric repulsion is encountered in the transition state that has the methyl and the *tert*-butyl group passing each other. A comparison of the accessible transition states of these two constitutional isomers reveals that 2-*tert*-butyl-2′-methyl-6′-isopropylbiphenyl is conformationally more stable.

[xxiii] Atropisomerism is a special case of conformational isomerism. Atropisomers are necessarily conformational isomers but not all conformational isomers fulfill the criteria of atropisomerism, for example the overcrowded alkenes described in Chapter 3.2.2.

[xxiv] The transition state is not perfectly planar and has a slightly twisted geometry, see Figure 3.18.

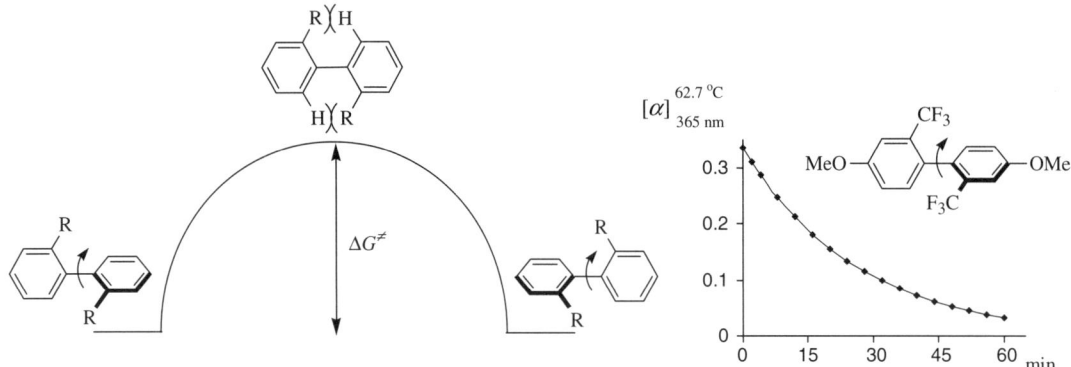

Figure 3.16   Energy profile of the atropisomerization of biphenyls (left) and polarimetric study of the racemization of 4,4′-dimethoxy-2,2′-bis(trifluoromethyl)biphenyl (right).

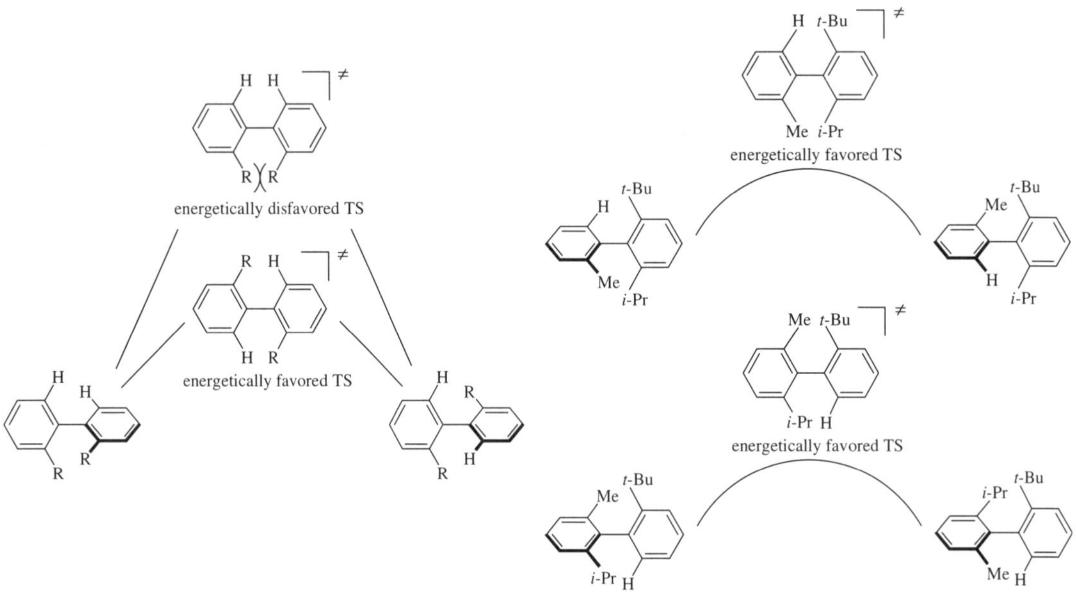

Figure 3.17   Rotation pathways of di- and trisubstituted biphenyls.

The energy barrier to atropisomerization of biphenyls steadily increases with the bulkiness of the substituents due to both enthalpic and negative entropic contributions. The latter originates from the compromised rotational freedom of *ortho*-substituents in the crowded transition state. For example, 2,2′-diisopropylbiphenyl has a Gibbs free energy of activation, $\Delta G^{\neq}$, of 116.6 kJ/mol and an activation entropy, $\Delta S^{\neq}$, of –46 J/K mol at 159.9 °C. Replacement of one isopropyl substituent with a bulky *tert*-butyl group affords 2-*tert*-butyl-2′-isopropylbiphenyl which has a rotational energy barrier of 136.9 kJ/mol at the same temperature. On the other hand, the enantiomers of 2-ethyl-2′-isopropylbiphenyl interconvert rapidly and can not be isolated at room temperature. Substituents in the *meta*-position exhibit a so-called buttressing effect: they reduce the flexibility of the adjacent *ortho*-substituent and therefore enhance steric repulsion during rotation about the chiral axis. The buttressing effect can provide a significant contribution to the overall steric hindrance to rotation. Incorporation of a relatively small ethyl group into the *meta*-position at *C*-3

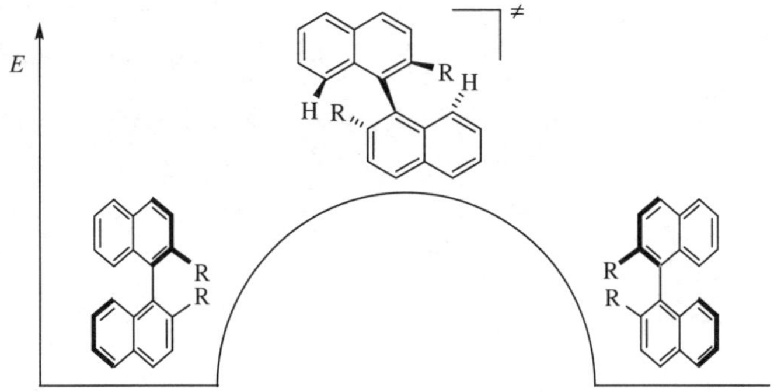

Figure 3.18 Racemization pathway and twisted transition state of 1,1′-binaphthyls.

in 2,2′-bis(trifluoromethyl)biphenyl increases the rotational energy barrier from 114.2 kJ/mol to 126.3 kJ/mol at 128.9 °C. The buttressing effect depends on the size and geometry of the *meta*-group, and on the bulkiness of the adjacent *ortho*-substituent.[xxv] Introduction of ethyl at the opposite *meta*-position at *C*-5 in 2,2′-bis(trifluoromethyl)biphenyl yields a negligible buttressing effect on the adjacent hydrogen attached to *C*-6 and a rotational energy barrier of only 114.5 kJ/mol, Figure 3.19. Quantum mechanical computations of biaryls indicate that rotation about the chiral axis occurs via a transition state that is not perfectly coplanar. As is shown for 1,1′-binaphthyls, the enantio-conversion process involves a transition state with twisted geometry. The distortion of the aryl rings decreases steric hindrance between *ortho*-substituents when the aryl rings go through the same plane in the transition state, Figure 3.18.[583]

Electronic effects of *para*-substituents are less pronounced than steric interactions and alter the rotational energy barrier of 2,2′-disubstituted biphenyls by up to 10%. Changes in the stability of the ground state and of the transition state have to be considered for analysis of substituent effects on the atropisomerization barrier. A comparison of the Gibbs free energy of activation of 2,2′-bis(trifluoromethyl)biphenyl and the corresponding 4,4′-disubstituted analogs reveals that electron-withdrawing groups increase conformational stability, whereas electron-donating groups decrease the rotational barrier, Figure 3.19. The destabilizing effect of methoxy groups and other electron donors has been attributed to enhanced electron density at carbons *C*-1 and *C*-1′. This facilitates out-of-plane bending and thus reduces steric interactions in the crowded transition state, Figure 3.20. Electron acceptors decrease the electron density at the pivotal carbon atoms and impede out-of-plane bending. The presence of these groups therefore increases the barrier to atropisomerization. Kinetic studies of a range of 2,2′-bis(trifluoromethyl)biphenyls, conducted by Wolf *et al.*, established the following order of destabilizing *para*-substituents: $NH_2 >> NHAc > OH > OMe > CH_3 > H \approx Br > NH_3^+ > CF_3 > Cl > NO_2 \approx F$. Interestingly, incorporation of one methoxy and one nitro group into opposite *para*-positions significantly accelerates racemization. The low conformational stability of 4-methoxy-4′-nitro-2,2′-bis(trifluoromethyl)biphenyl has been explained by effective push-pull conjugation which stabilizes the transition state and therefore increases the rate of rotation, Figure 3.20.

The order of destabilizing *para*-substituents established for 2,2′-bis(trifluoromethyl)biphenyls is in agreement with similar trends observed in a wide range of biphenyls. However, electronic contributions can be more complex in biaryls that undergo *CH*/π-interactions, and opposite electronic effects on the conformational stability have been observed with *para*-substituted 2,2′-diisopropylbiphenyls. In this case, electron donors such as amino groups increase the

---

[xxv] The van der Waals radius of a chloro substituent is slightly smaller than that of a methyl group but it provides a stronger buttressing effect because of its spherical geometry.

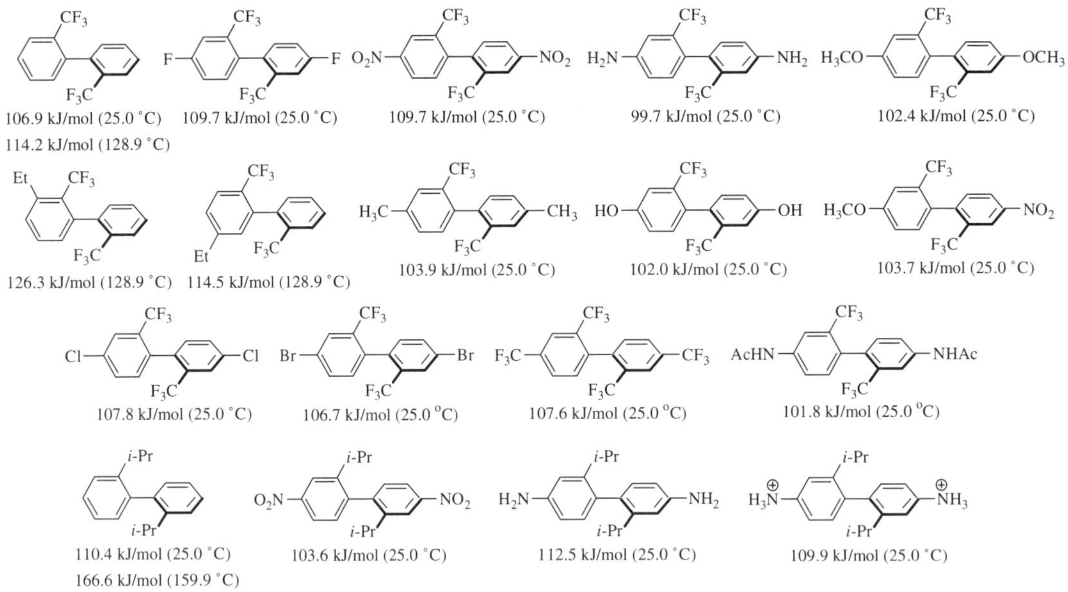

Figure 3.19 Rotational energy barriers of *meta*- and *para*-substituted dialkylbiphenyls.

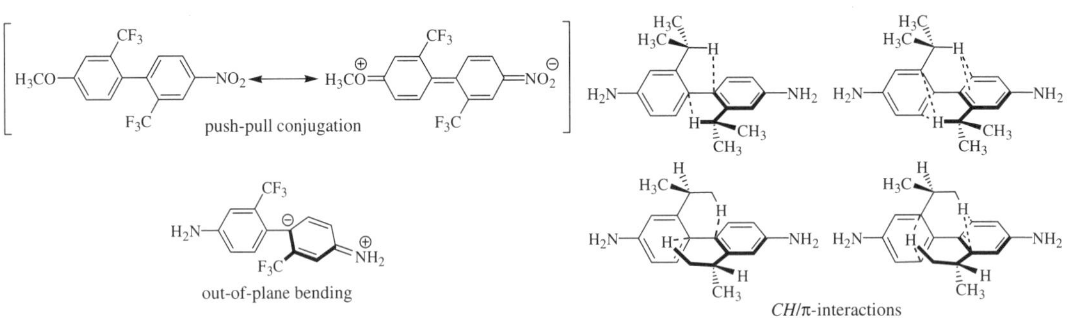

Figure 3.20 Electronic effects of *para*-substituents on the conformational stability of biphenyls and ground state stabilizing intramolecular *CH*/π-interactions in 2,2′-diisopropylbiphenyls.

rotational energy barrier while introduction of electron-withdrawing nitro and ammonium groups facilitates atropisomerization. It is assumed that the electronic effects on the stability of the transition state discussed above are overcompensated by intramolecular *CH*/π-interactions that selectively stabilize the ground state.[584] For geometric reasons, *CH*/π-interactions between soft Lewis-acidic *C–H* bonds of the *ortho*-alkyl groups and the soft Lewis-basic π-system of the adjacent phenyl ring in 2,2′-dialkylbiphenyls are only possible in the ground state but not in the periplanar transition state. Electron-donating groups increase the π-basicity of the aryl moieties and thus enhance intramolecular *CH*/π-interactions. Accordingly, the amino groups in 4,4′-diamino-2,2′-diisopropylbiphenyl strongly increase the stability of the ground state. Electron-withdrawing groups reduce the aryl π-basicity which results in a less stable ground state and consequently in reduced conformational stability, Figure 3.20. Energies related to *CH*/π-bonds are usually less than 10 kJ/mol but these interactions are known to significantly affect the relative stability of conformational isomers of various compounds.[585] For 2,2′-diisopropylbiphenyls, several *CH*/π-interactions involving the methine and methyl protons of both isopropyl groups and the *ipso*- and

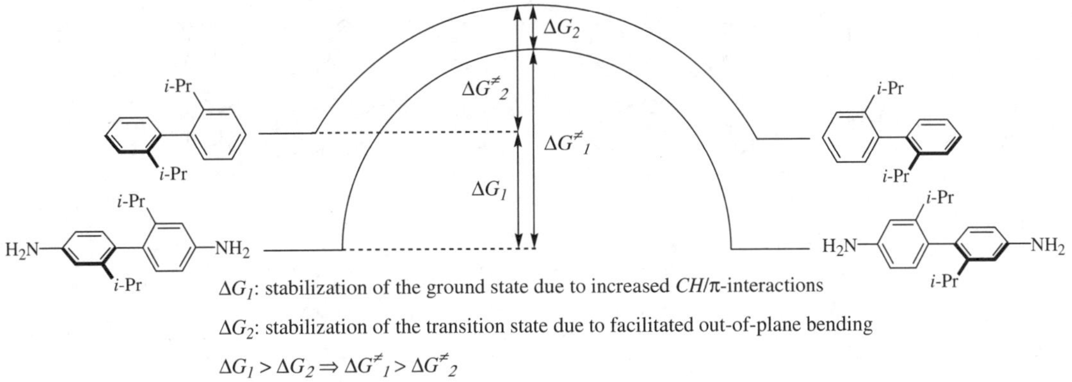

ΔG_1: stabilization of the ground state due to increased CH/π-interactions

ΔG_2: stabilization of the transition state due to facilitated out-of-plane bending

$$\Delta G_1 > \Delta G_2 \Rightarrow \Delta G^{\neq}_1 > \Delta G^{\neq}_2$$

Figure 3.21 Illustration of the energetic effects of electron-donating *para*-substituents on the rotational barrier of 2,2′-diisopropylbiphenyl.

ortho-aryl carbons can be anticipated, Figure 3.20. Consequently, the presence of electron-donating groups leads to substantial stabilization of the ground state, ΔG_1, due to enhanced CH/π-interactions which becomes the predominant effect on the rotational energy barrier as it overcompensates the concurrent stabilization of the transition state via facilitated out-of-plane bending, ΔG_2, Figure 3.21.

The conformational stability of many other biaryls[586–601] and triaryls[602–611] has been investigated, and the rotational energy barriers can usually be explained on the basis of the steric and electronic substituent effects discussed above, Figure 3.22. The *syn/anti*-isomerization of 1,8-bis(2′-phenyl-4′-pyridyl)naphthalene has been studied over a temperature range of more than 100 °C by chromatographic and NMR spectroscopic methods. Rotation of either pyridyl ring of this atropisomer about the pyridyl–naphthalene axis entails interconversion of a meso *syn*- and two chiral *anti*-isomers. The isomerization process involves a sterically crowded T-shaped transition state, in which the edge of the rotating pyridine is directed towards the adjacent pyridyl ring. Because the rotational energy barrier of 1,8-bis(2′-phenyl-4′-pyridyl)naphthalene is only 70.4 kJ/mol, interconversion between the *syn*- and *anti*-diastereomers proceeds rapidly at room temperature. The corresponding activation enthalpy, $\Delta H^{\neq}$, and activation entropy, $\Delta S^{\neq}$, are 57.5 kJ/mol and –43.4 J/K mol, respectively. The considerably negative activation entropy is indicative of a highly ordered transition state which is typical for atropisomerizations of biaryls and triaryls. Interestingly, *N*-oxidation of diheteroaryl-naphthalenes decreases the conformational stability.[xxvi] The reduced rotational energy barrier of diheteroarylnaphthalene *N,N′*-dioxides has been attributed to two synergistic effects on the relative stability of the ground and transition states: repulsive dipole/dipole-interactions between the cofacial rings of 1,8-diheteroarylnaphthalene *N,N′*-dioxides selectively destabilize the ground state and thus decrease the rotational energy barrier (1); and *N*-oxidation increases the electron density at the pivotal carbon atoms, facilitating out-of-plane bending and rotation about the axially chiral naphthyl-heteroaryl bond via a less sterically hindered transition state (2), Scheme 3.68.

The conformational stability of bridged biaryls varies significantly with the ring size. As a rule of thumb, biaryls that possess one bridging atom are not stable to rotation at room temperature even if the remaining two *ortho*-positions are occupied by bulky groups.[612,613] An increase in bridge length enhances the torsion angle between the two aryl rings and raises the energy barrier to racemization. Nevertheless, biaryls with a five- or six-membered bridge may still undergo rotation about the chiral axis, unless bulky *ortho*-substituents are present in the bridged biaryl

---

[xxvi] As discussed above for biaryls, electronic substituent effects on the rotational barrier of triaryls are significantly smaller than steric effects.

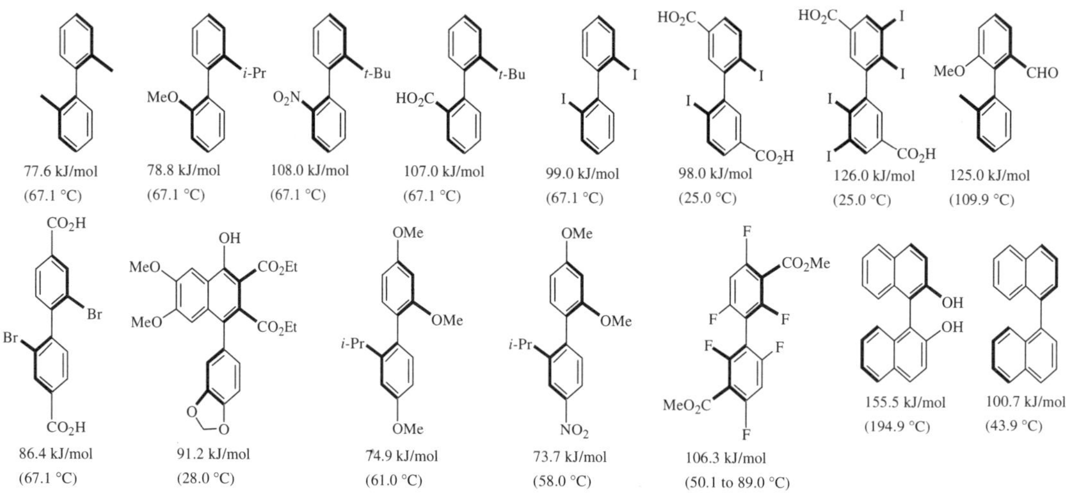

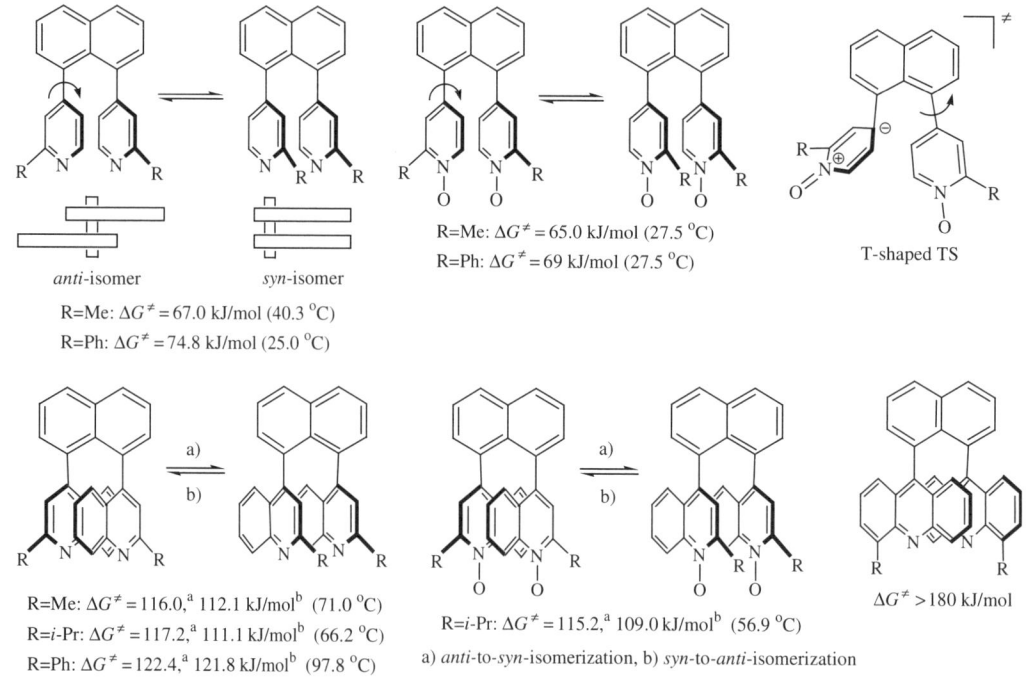

Figure 3.22 Structures and rotational energy barriers of selected biaryls.

Scheme 3.68 Diastereoisomerization of axially chiral 1,8-diheteroarylnaphthalenes.

framework.[614–620] This has been demonstrated with a series of benzonaphthopyranones, Figure 3.23. Biaryls containing seven-membered or larger rings are generally at least as stable as their unbridged analogs.[621–625]

The transient formation of a short bridge can significantly compromise the conformational stability of biaryls.[626,627] According to the discussion above, one would expect that incorporation of four *ortho*-substituents into a biaryl framework establishes considerable conformational

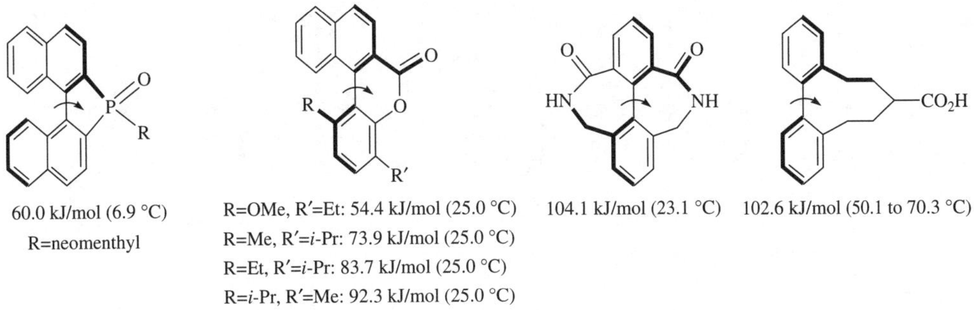

60.0 kJ/mol (6.9 °C)
R=neomenthyl

R=OMe, R′=Et: 54.4 kJ/mol (25.0 °C)
R=Me, R′=*i*-Pr: 73.9 kJ/mol (25.0 °C)
R=Et, R′=*i*-Pr: 83.7 kJ/mol (25.0 °C)
R=*i*-Pr, R′=Me: 92.3 kJ/mol (25.0 °C)

104.1 kJ/mol (23.1 °C)   102.6 kJ/mol (50.1 to 70.3 °C)

Figure 3.23   Rotational energy barriers of bridged biaryls.

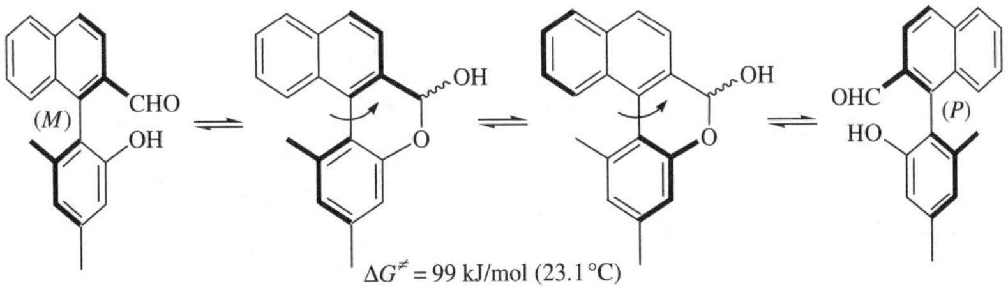

$\Delta G^{\neq} = 99$ kJ/mol (23.1 °C)

Scheme 3.69   Atropisomerization of a 2-hydroxy-2′-formylbiaryl via lactol formation.

stability. However, Bringmann and coworkers reported that an enantiopure sample of the tetra-*ortho*-substituted biaryl depicted in Scheme 3.69 racemizes at room temperature even under neutral conditions due to a surprisingly low rotational energy barrier of 99 kJ/mol. The biaryl has a formyl and a hydroxyl group in opposite aryl rings and forms an intermediate five-membered lactol ring, which reduces both the torsion angle between the two aryl rings and the barrier to rotation about the chiral axis.

Atropisomerization is an intramolecular process that is controlled by steric and electronic substituent effects. In some cases, the barrier to rotation can be altered by external factors.[628] An important example is the acid-catalyzed racemization of BINOL. This biaryl does not show any sign of racemization on heating at 100 °C for 24 hours under neutral conditions but it racemizes at 100 °C in 1.2 *N* hydrochloric acid within this period of time.[629,630] This has been attributed to formation of a protonated nonaromatic intermediate exhibiting a $C(sp^2)$–$C(sp^3)$ bond which reduces steric repulsion during rotation about the pivotal axis, Scheme 3.70. Racemization of BINOL is also catalyzed under basic conditions due to deprotonation and dianion formation.[631] By analogy with the rationalization of electronic effects of electron-donating substituents on the rotational energy barrier of biphenyls, it is believed that the increase in electron density at the central carbon atoms of the naphthoxide moieties facilitates rotation via out-of-plane bending in the transition state.

Atropisomerizations are not necessarily thermal rotation processes and smooth photoracemization of several biaryls has been observed.[632–637] An interesting example is 1,1′-binaphthyl which has a rotational energy barrier of 100.7 kJ/mol at 43.9 °C. LASER photolysis experiments revealed that it undergoes facile racemization in the excited triplet state.[638,639] The diradical character of the photochemically induced triplet state results in increased aryl–aryl bond order and smaller biaryl torsion angle. The higher bond order and the shorter distance between the aryl planes should enhance steric hindrance and thus increase the rotational barrier, but this is apparently overcompensated by the large electron stabilization energy. The rotational energy barrier in the triplet state has been estimated to be as low as 8 kJ/mol, Figure 3.24.

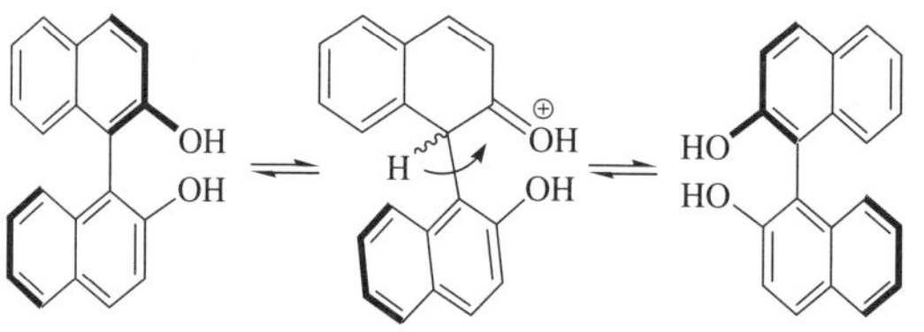

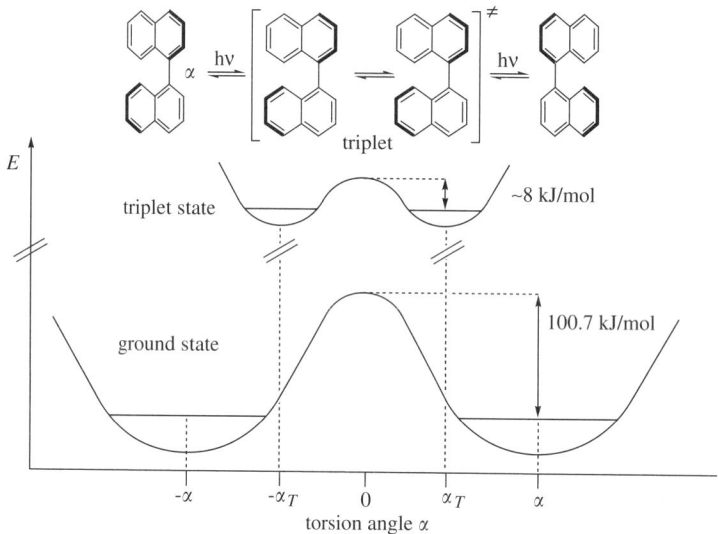

Scheme 3.70   Acid-catalyzed atropisomerization of BINOL.

Figure 3.24   Photoracemization of 1,1′-binaphtyl.

Thermal and photochemical racemization of the vast majority of stereolabile chiral biaryls involves rotation about the aryl–aryl bond, although this is not always the case. The enantiomers of 1-arylpyrimidine-2-thiones and their oxygen analogs are stable to interconversion at room temperature and can be conveniently isolated by chiral chromatography. Atropisomerization of this class of compounds entails ring opening as a consequence of a reversible [3,3]electrocyclic rearrangement and does not require rotation about the chiral axis, Scheme 3.71.[640,641] Decakis(di-chloromethyl)biphenyl gives rise to so-called dynamic molecular gearing, in which the methine protons are wedged between the two chloro substituents of an adjacent dichloromethyl group.[642] The five dichloromethyl groups in each phenyl ring can not rotate individually but do so together in a concerted process. Decasubstituted biphenyls such as decakis(dichloromethyl)biphenyl are chiral due to the directionality of the geared dichloromethyl groups. Racemization does not proceed via rotation about the chiral axis but through concerted rotation of five dichloromethyl groups in one phenyl ring. Reversal of the sense of directionality of the dichloromethyl groups in one aryl moiety generates the other enantiomer, whereas reversal of the directionality of all ten substituents affords topomerization, *i.e.*, degenerate isomerization due to exchange of the positions of identical ligands.

Atropisomeric biphenyls are important building blocks of many biologically active compounds including alkaloids, flavonoids, lignans and peptides.[643–650] The demand for total synthesis and structure elucidation of naturally occurring atropisomers has been fueled by the discovery of important pharmacological properties and applications.[651–653] Natural atropisomers can have distinct physiological activities and often occur in plants or microorganisms in enantiopure form. Bringmann and coworkers investigated the isomerization of macrocyclic bisbiphenyls having

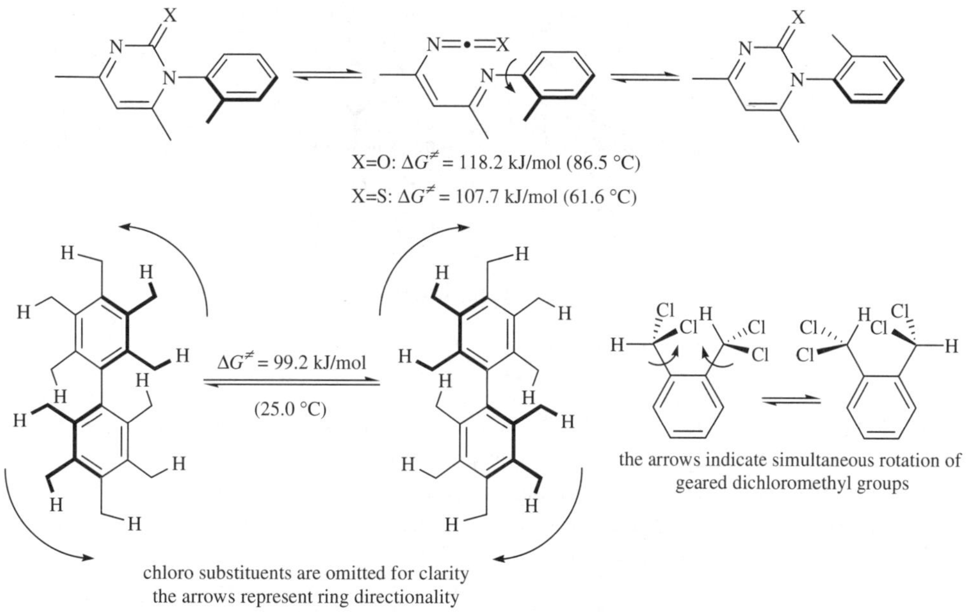

X=O: $\Delta G^{\neq}$ = 118.2 kJ/mol (86.5 °C)

X=S: $\Delta G^{\neq}$ = 107.7 kJ/mol (61.6 °C)

$\Delta G^{\neq}$ = 99.2 kJ/mol

(25.0 °C)

the arrows indicate simultaneous rotation of
geared dichloromethyl groups

chloro substituents are omitted for clarity
the arrows represent ring directionality

Scheme 3.71 Electrocyclic ring opening of 1-(2′-methylphenyl)-4,6-dimethylpyrimidin-2-one (top) and dynamic gearing of decakis(dichloromethyl)biphenyl (bottom).

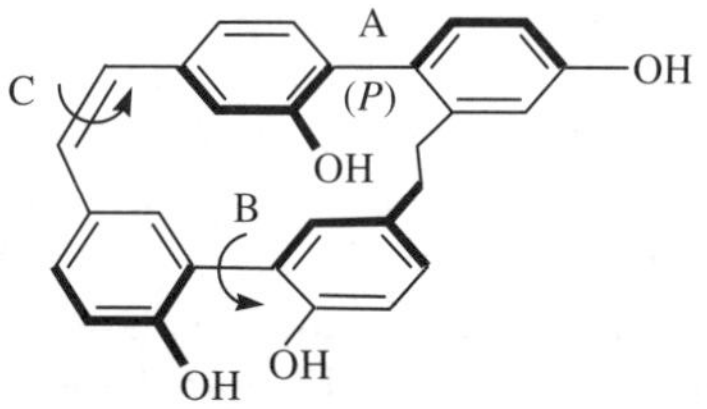

Figure 3.25 Structure of (*P*)-isoplagiochin C.

remarkable antitumor, antibacterial and antimycotic properties, and found that atropisomeric conformations coexist in liverworts.[654] A closer look at the structure of isoplagiochin C reveals the presence of three chiral elements: two axially chiral biphenyl units, *A* and *B*, and one twisted double bond, *C*, Figure 3.25. According to the preceding discussion of the conformational stability of 2,2′-disubstituted biphenyls, rotation about the chiral axes should proceed rapidly at room temperature. Since quantum mechanical calculations suggest fast interconversion between the two helical forms of the chiral *cis*-stilbene unit, one would expect up to eight nonresolvable conformational stereoisomers of isoplagiochin C.[655] Surprisingly, circular dichroism and variable-temperature NMR measurements in conjunction with computational studies revealed that the rigid cyclic structure of this bisbiphenyl affords significant conformational stability of chiral axis *A* while chiral elements *B* and *C* are fluxional at room temperature. The rotational energy barrier of biaryl *A* is 101.6 kJ/mol. Chromatographic and CD spectroscopic analysis proved that isoplagiochin C exists as a separable atropisomeric 85:15 mixture favoring (*P*)-conformation of axis *A* in *Plagiochila deflexa* liverworts.

The rotational isomerism of diarylacetylenes is of interest for the design of turnstiles,[656] gyroscopes[657,658] and similar molecular devices, see also Chapter 8.8. Incorporation of an acetylene unit into a biaryl system extends the axis to approximately 4.0 Å. Because of the elongated

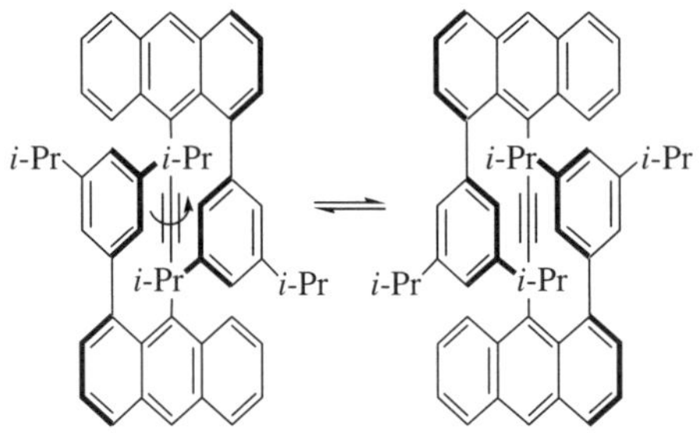

$$\Delta G^{\neq} = 75.3 \text{ kJ/mol} (0 \text{ }^{\circ}\text{C})$$

Scheme 3.72   Rotamers of a chiral diarylacetylene.

separation of the two aryl rings, steric hindrance to rotation is considerably reduced and diaryl-acetylenes are conformationally unstable.[xxvii] The steric hindrance can be increased by introduction of bulky substituents, but isolable atropisomers of this class of compounds are still elusive.[659–662] The congested structure of bis(1-phenyl-9-anthryl)ethynes has an energy barrier to rotation about the chiral axis of 75.3 kJ/mol, Scheme 3.72.[663]

3.3.2   Nonbiaryl Atropisomers

The unique combination of axial chirality, stereodynamics and reactivity of nonbiaryl atropisomers has been exploited for many purposes including dynamic kinetic resolution and dynamic thermo-dynamic resolution, stereochemical relays, and molecular gears.[664] Among the most prominent examples of nonbiaryl atropisomers are aromatic amides, ketones, sulfoxides, and sulfones.

Tertiary aryl amides carrying different aromatic *ortho*-substituents are nonplanar and axially chiral. Rotation about the aryl-carbonyl bond causes enantiomerization or diastereomerization if another chiral element is present in the molecule.[665–671] Even moderate steric repulsion between the aryl ring and the amide group enforces a dihedral angle of approximately 90° on the aryl–carbonyl axis. The rotational barrier is mostly governed by two factors: the bulkiness of the *ortho*-aryl substituents and the size of the amide moiety. A closer look reveals that aryl amides also undergo restricted *C–N* rotation, and correlation of the two isomerization processes can be observed, Figure 3.26.[672–675] Aryl–carbonyl rotation in aromatic amides that display unbranched nitrogen substituents is usually significantly faster than rotation about the amide bond. The former therefore occurs independently and without concomitant *C–N* rotation. Incorporation of branched nitrogen substituents into the amide moiety increases the energy barrier to rotation about the aryl–carbonyl bond and atropisomerization predominantly proceeds in correlation with *C–N* rotation. Finally, introduction of severe steric hindrance to rotation about the aryl-carbonyl bond is observed with 2-substituted 1-naphthamides which do not show any uncorrelated aryl–carbonyl rotation, while some of the less restricted rotation about the amide bond now proceeds freely and at a faster rate. Aryl thioamides have a considerably higher barrier to rotation about the aryl–thiocarbonyl bond than do the corresponding oxygen analogs.[676–679] By analogy with aryl amides, the conformational stability of aryl thioamides increases when sterically demanding groups are introduced. Correlated rotation about the aryl–thiocarbonyl and the thioamide bond has been observed with acrylic thioamides.[680]

The stereodynamics of 1-[*N*-(2,5-*cis*-dimethyl)pyrrolidinylcarbonyl]-2-methylnaphthalene provide a typical example of atropisomerism and gearing of aryl amides, Scheme 3.73.[681,682] This chiral aryl amide exists in the form of two diastereomeric *endo*- and *exo*-conformations that

---

[xxvii] The rotational energy barrier of diphenylacetylene is only 2-3 kJ/mol.

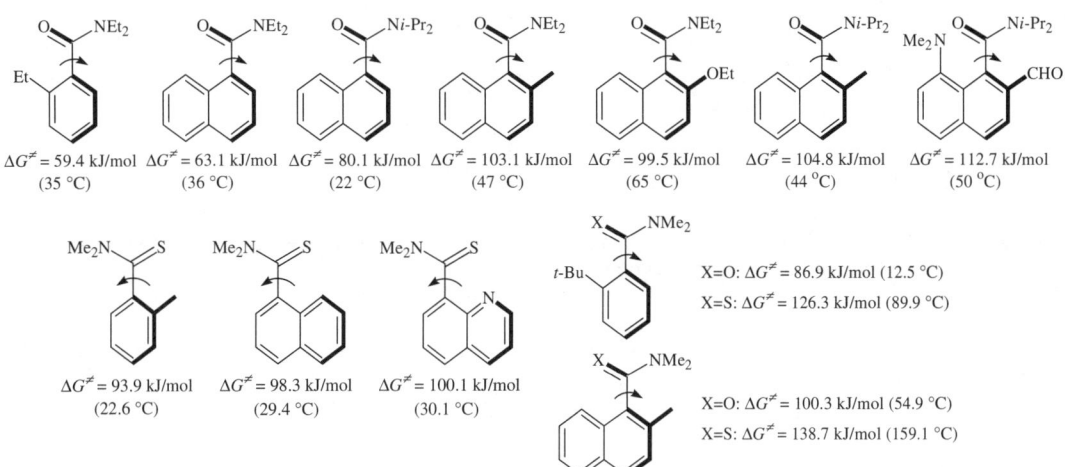

Figure 3.26   Isomerization of tertiary aromatic amides (top) and energy barriers to rotation about the aryl-carbonyl bond of selected examples (bottom).

interconvert via rotation about the aryl–carbonyl axis and the amide bond. Enantiomerization requires completion of both processes, either in concert or sequentially. Since the rate constants for enantiomerization are a magnitude larger than for aryl–carbonyl rotation, at least 90% of the latter must proceed in concert with *C–N* rotation. Rotation about the amide bond is the faster process and also occurs without simultaneous aryl–carbonyl rotation. Comparison of the rates of isomerization proves that sterically hindered 2-substituted 1-naphthamides afford considerable gearing between the two rotational processes but display significantly more slippage (uncorrelated motion) than the triptycyl-derived three-toothed gears discussed in Chapter 8.4.

In contrast to *ortho*-substituted aromatic aldehydes that possess an essentially planar achiral conformation, aryl ketones afford enantiomeric rotamers with almost perfectly orthogonal carbonyl and aryl planes. Rotation about the aryl–carbonyl axis generally proceeds rapidly at room temperature, even in the case of *tert*-butyl 1-(2-methylnaphthyl) ketone, Figure 3.27.[683–687] 1,2-Diacylbenzenes and 1,4-, 1,5- and 1,8-diacylnaphthalenes display two chiral axes, and can in principle exist as a mixture of two chiral *anti*-isomers and one meso *syn*-conformer. However, NMR spectroscopic analysis revealed that 1,2-diacylbenzenes and 1,8-diacylnaphthalenes, with the exception of 1,8-dibenzoylnaphthalenes,[688] predominantly populate the *anti*-conformation.[689–691]

The structure of aryl sulfoxides and aryl sulfones gives rise to similar atropisomerism.[692–694] Rotation about the chiral carbon–sulfur axis of 1-naphthyl sulfones is fast and can be monitored at low temperatures. Naphthyl sulfoxides possess a chiral center at the sulfur atom in addition to

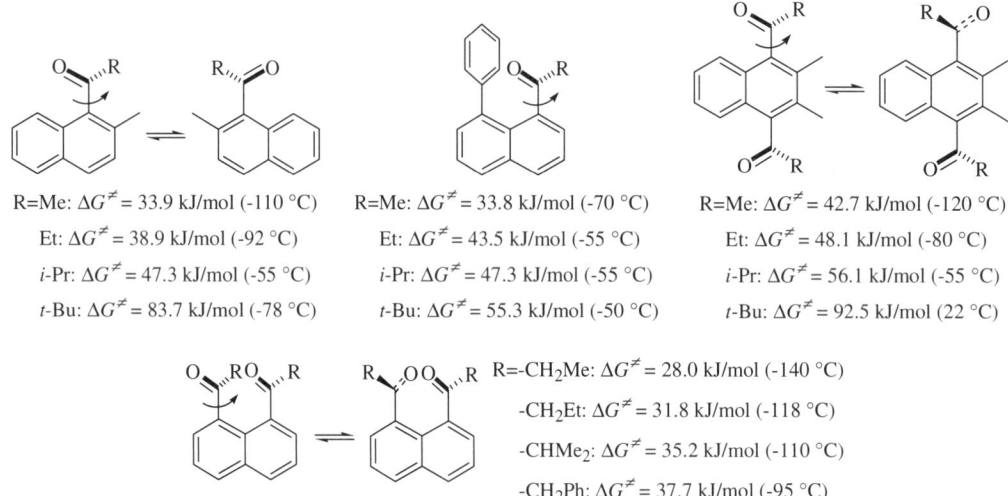

Scheme 3.73   Interconversion of stereoisomers and geared rotation of 1-[*N*-(2,5-*cis*-dimethyl)pyrrolidinyl-carbonyl]-2-methylnaphthalene.

Diastereomerization processes:

$k_{C\text{-}N}$: $3.9 \cdot 10^{-5}$ s^{-1}, $\Delta G^{\neq}_{C\text{-}N}$: 100.8 kJ/mol

$k_{Ar\text{-}CO}$: $0.1 \cdot 10^{-5}$ s^{-1}, $\Delta G^{\neq}_{Ar\text{-}CO}$: 110 kJ/mol

Enantiomerization processes:

k_{exo}: $1.0 \cdot 10^{-5}$ s^{-1}, $\Delta G^{\neq}_{exo}$: 104.3 kJ/mol

k_{endo}: $2.9 \cdot 10^{-5}$ s^{-1}, $\Delta G^{\neq}_{endo}$: 101.6 kJ/mol

R=Me: $\Delta G^{\neq}$ = 33.9 kJ/mol (-110 °C)

Et: $\Delta G^{\neq}$ = 38.9 kJ/mol (-92 °C)

i-Pr: $\Delta G^{\neq}$ = 47.3 kJ/mol (-55 °C)

t-Bu: $\Delta G^{\neq}$ = 83.7 kJ/mol (-78 °C)

R=Me: $\Delta G^{\neq}$ = 33.8 kJ/mol (-70 °C)

Et: $\Delta G^{\neq}$ = 43.5 kJ/mol (-55 °C)

i-Pr: $\Delta G^{\neq}$ = 47.3 kJ/mol (-55 °C)

t-Bu: $\Delta G^{\neq}$ = 55.3 kJ/mol (-50 °C)

R=Me: $\Delta G^{\neq}$ = 42.7 kJ/mol (-120 °C)

Et: $\Delta G^{\neq}$ = 48.1 kJ/mol (-80 °C)

i-Pr: $\Delta G^{\neq}$ = 56.1 kJ/mol (-55 °C)

t-Bu: $\Delta G^{\neq}$ = 92.5 kJ/mol (22 °C)

R=-CH$_2$Me: $\Delta G^{\neq}$ = 28.0 kJ/mol (-140 °C)

-CH$_2$Et: $\Delta G^{\neq}$ = 31.8 kJ/mol (-118 °C)

-CHMe$_2$: $\Delta G^{\neq}$ = 35.2 kJ/mol (-110 °C)

-CH$_2$Ph: $\Delta G^{\neq}$ = 37.7 kJ/mol (-95 °C)

Figure 3.27   Rotational energy barrier of atropisomeric acylarenes.

the chiral carbon–sulfur axis and therefore exist as a pair of two diastereomeric *E/Z*-conformers. The energy barriers to diastereomerization of 1-(alkylsulfinyl)-2-methylnaphthalenes range from 44 to 77 kJ/mol. In general, chiral sulfoxides exhibit high configurational stability at ambient temperature, although racemization via pyramidal sulfur inversion of electron-deficient aryl sulfoxides can occur at 25 °C.[695] Secondary (2-methyl-1-naphthyl)phosphine oxides bearing a configurationally stable chiral phosphorus center undergo diastereomerization due to facile rotation about the chiral aryl–phosphorus axis, Figure 3.28.[696]

Introduction of two different nitrogen substituents to prochiral *ortho*-substituted anilines and 1-aminonaphthalenes provides aryl amines that are axially chiral by virtue of hindered rotation about the aryl–nitrogen bond.[697] In order to minimize steric repulsion between nitrogen

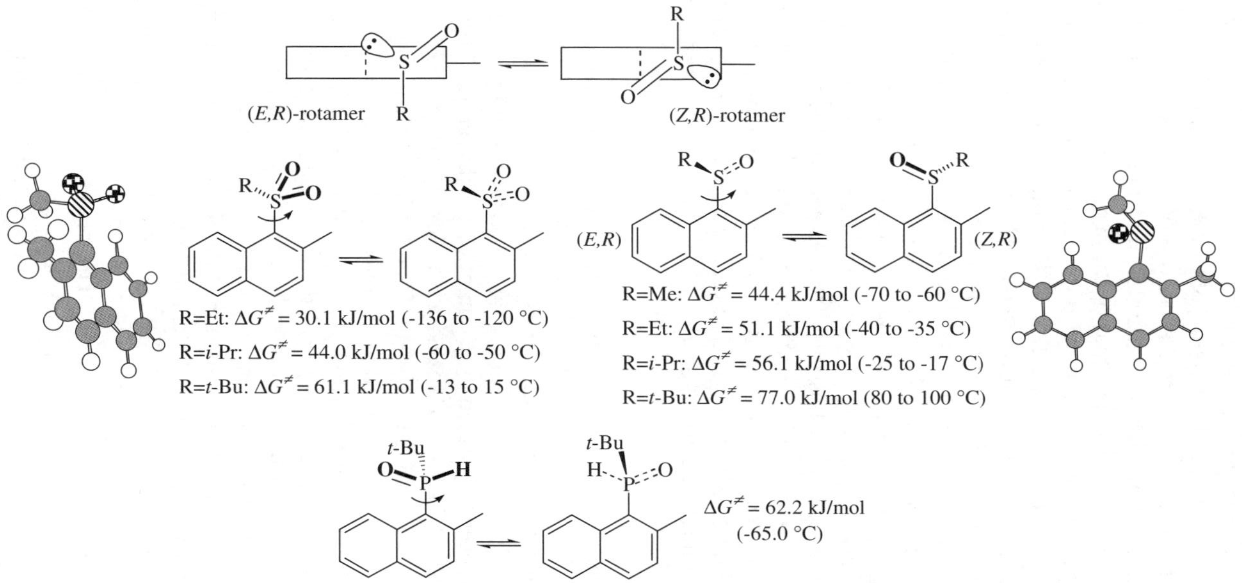

Figure 3.28 Diastereomerization of a 2-methylnaphthyl sulfoxide and rotational energy barriers of naphthyl sulfoxides, sulfones and a secondary phosphine oxide.

substituents and the aryl ring, *N,N*-dialkyl arylamines adopt twisted conformations in the ground state, Figure 3.29. The interconversion of the conformational enantiomers of a series of *N,N*-dialkyl-1-naphthylamines,[698–701] *N*-aryl tetrahydropyrimidines[702] and *N*-aryl-4-pyridones[703] has been studied. Notably, concurrent nitrogen inversion of these compounds has a very low energy barrier and is only observable under cryogenic conditions. This greatly facilitates determination of the free energy of activation for the rotation about the chiral carbon–nitrogen axis. For example, the Gibbs activation energy for nitrogen inversion of *N*-methylaniline is 6.7 kJ/mol whereas the energy barrier to rotation about the aryl–nitrogen bond is 30.3 kJ/mol.[704,705]

The atropisomerism of anilides, imides, barbiturates, and urea derivatives of anilines bearing different *ortho*-substituents is closely related to that of the benzamides and naphthamides discussed above.[706] To minimize steric repulsion, acetanilides populate a conformation in which the aryl ring and the amide group are orthogonal to each other.[707] Rotation about the planar amide function causes *E/Z*-isomerization, whereas rotation about the aryl-nitrogen axis results in interconversion of enantiomeric conformers, Figure 3.30. Substituent and solvent effects on both the ratio of the (*E*)- and (*Z*)-conformers and the energy barrier to *E/Z*-isomerization are difficult to predict. The (*E*)- and (*Z*)-conformers of acetanilides carrying a 2,6-disubstituted phenyl ring can be isolated at room temperature provided at least one of the substituents is a *tert*-butyl group.[708–711] Rotation about the amide bond is usually faster than rotation about the aryl–nitrogen axis, although gearing of the two processes is possible.[712] As expected, the rotation about the axially chiral aryl–nitrogen bond is governed by steric interactions between *ortho*-aryl groups and substituents attached to the nitrogen atom. The atropisomers of amides and lactams derived from either 2,6-disubstituted anilines or 2-*tert*-butyl aniline can be isolated at room temperature.[713–723] In contrast to imides, rhodanines and barbiturates,[724–728] the conformational stability of axially chiral ureas is significantly lower than that of amides, which has been attributed to enhanced ground state strain.[729,730]

Styrene derivatives with bulky vinyl substituents adopt twisted conformations that interconvert via a planar transition state. The presence of different *ortho*-substituents in the aryl ring of a twisted styrene renders the molecule chiral, and enantiomers can be resolved in the case of sufficient steric hindrance to rotation about the aryl–vinyl bond.[731–735] A typical example is β,β-diisopropyl-α,2-dimethylstyrene which has been separated into enantiomers by chiral HPLC, Figure 3.31.[736] Thermal racemization studies at 50–75 °C revealed a rotational barrier, $\Delta G^{\neq}$, of 106.3 kJ/mol and an activation entropy, $\Delta S^{\neq}$, of –80 J/K mol at 66.8 °C. The considerable negative entropic contribution is reminiscent of the racemization of biaryls, and can be attributed to a loss of vibrational and rotational freedom of the vinyl and aryl substituents in the congested transition state. A similar situation arises with *ortho*-substituted benzophenone oximes and benzyl aryl ketoximes.[737,738] The Gibbs activation energy for the enantiomerization of axially chiral *ortho*-halogenated (*Z*)-*O*-methyl benzyl aryloximes increases with the bulkiness of the substituents due to increased steric hindrance in the periplanar transition state. Steric repulsion is significantly reduced in (*E*)-aryloximes which rapidly rotate about the chiral axis.

The nonplanarity of highly substituted 1,3-dienes gives rise to axial chirality and atropisomerism. Nonplanar dienes always belong to the point groups C_1 and C_2 and are inherently chiral, irrespective of the substitution pattern.[739–741,xxviii] The enantiomers of 2,3,4,5-tetrabromo-2,4-hexadiene-1,6-diol are stable to racemization in the solid state but show slow rotation about the chiral axis via an *s-trans* (*transoid*) transition state in solution at room temperature, Figure 3.31.[742] The four bromo substituents impede a coplanar orientation of the double bonds and therefore induce chirality and steric hindrance to racemization. The two hydroxymethyl groups

---

[xxviii] By contrast, a prerequisite for chirality of biphenyls and styrenes is that the aryl rings are not symmetrically substituted, *i.e.*, not all *ortho*- and *meta*-substituents attached to the same aryl group can be identical. For example, 2,2′,6-trimethylbiphenyl has a symmetry plane and is achiral because it bears one symmetrically substituted aryl ring.

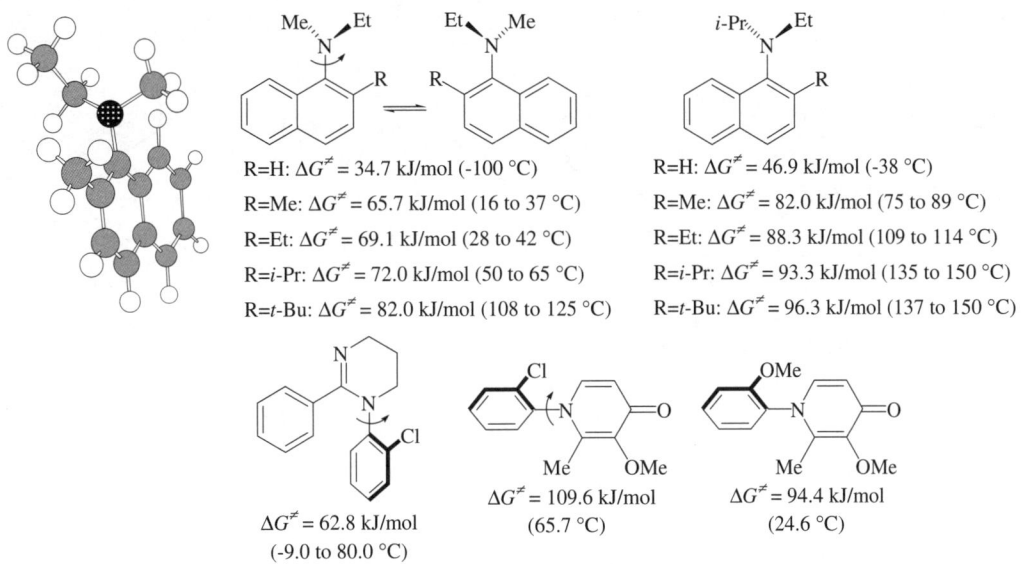

R=H: $\Delta G^{\neq}$ = 34.7 kJ/mol (-100 °C)
R=Me: $\Delta G^{\neq}$ = 65.7 kJ/mol (16 to 37 °C)
R=Et: $\Delta G^{\neq}$ = 69.1 kJ/mol (28 to 42 °C)
R=i-Pr: $\Delta G^{\neq}$ = 72.0 kJ/mol (50 to 65 °C)
R=t-Bu: $\Delta G^{\neq}$ = 82.0 kJ/mol (108 to 125 °C)

R=H: $\Delta G^{\neq}$ = 46.9 kJ/mol (-38 °C)
R=Me: $\Delta G^{\neq}$ = 82.0 kJ/mol (75 to 89 °C)
R=Et: $\Delta G^{\neq}$ = 88.3 kJ/mol (109 to 114 °C)
R=i-Pr: $\Delta G^{\neq}$ = 93.3 kJ/mol (135 to 150 °C)
R=t-Bu: $\Delta G^{\neq}$ = 96.3 kJ/mol (137 to 150 °C)

$\Delta G^{\neq}$ = 62.8 kJ/mol
(-9.0 to 80.0 °C)

$\Delta G^{\neq}$ = 109.6 kJ/mol
(65.7 °C)

$\Delta G^{\neq}$ = 94.4 kJ/mol
(24.6 °C)

Figure 3.29 Favored conformation of *N,N*-dialkyl-1-naphthylamines and rotational energy barriers of atropisomeric aniline derivatives.

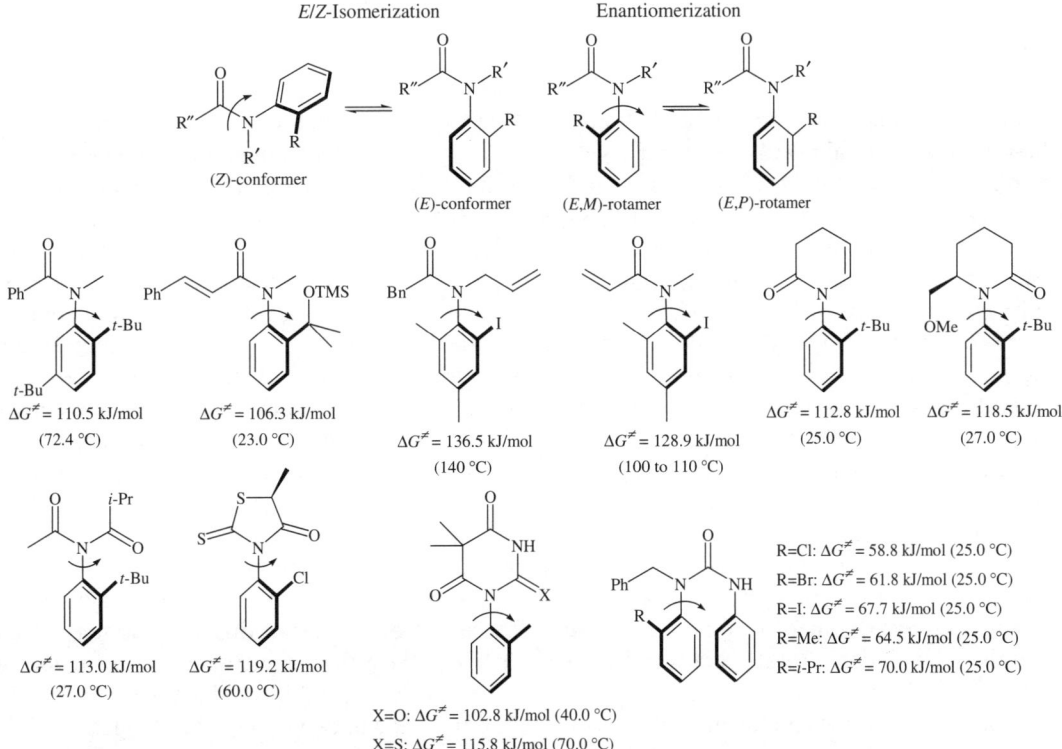

$\Delta G^{\neq}$ = 110.5 kJ/mol
(72.4 °C)

$\Delta G^{\neq}$ = 106.3 kJ/mol
(23.0 °C)

$\Delta G^{\neq}$ = 136.5 kJ/mol
(140 °C)

$\Delta G^{\neq}$ = 128.9 kJ/mol
(100 to 110 °C)

$\Delta G^{\neq}$ = 112.8 kJ/mol
(25.0 °C)

$\Delta G^{\neq}$ = 118.5 kJ/mol
(27.0 °C)

$\Delta G^{\neq}$ = 113.0 kJ/mol
(27.0 °C)

$\Delta G^{\neq}$ = 119.2 kJ/mol
(60.0 °C)

R=Cl: $\Delta G^{\neq}$ = 58.8 kJ/mol (25.0 °C)
R=Br: $\Delta G^{\neq}$ = 61.8 kJ/mol (25.0 °C)
R=I: $\Delta G^{\neq}$ = 67.7 kJ/mol (25.0 °C)
R=Me: $\Delta G^{\neq}$ = 64.5 kJ/mol (25.0 °C)
R=i-Pr: $\Delta G^{\neq}$ = 70.0 kJ/mol (25.0 °C)

X=O: $\Delta G^{\neq}$ = 102.8 kJ/mol (40.0 °C)
X=S: $\Delta G^{\neq}$ = 115.8 kJ/mol (70.0 °C)

Figure 3.30 Stereodynamics of anilides, imides, barbiturates, and ureas.

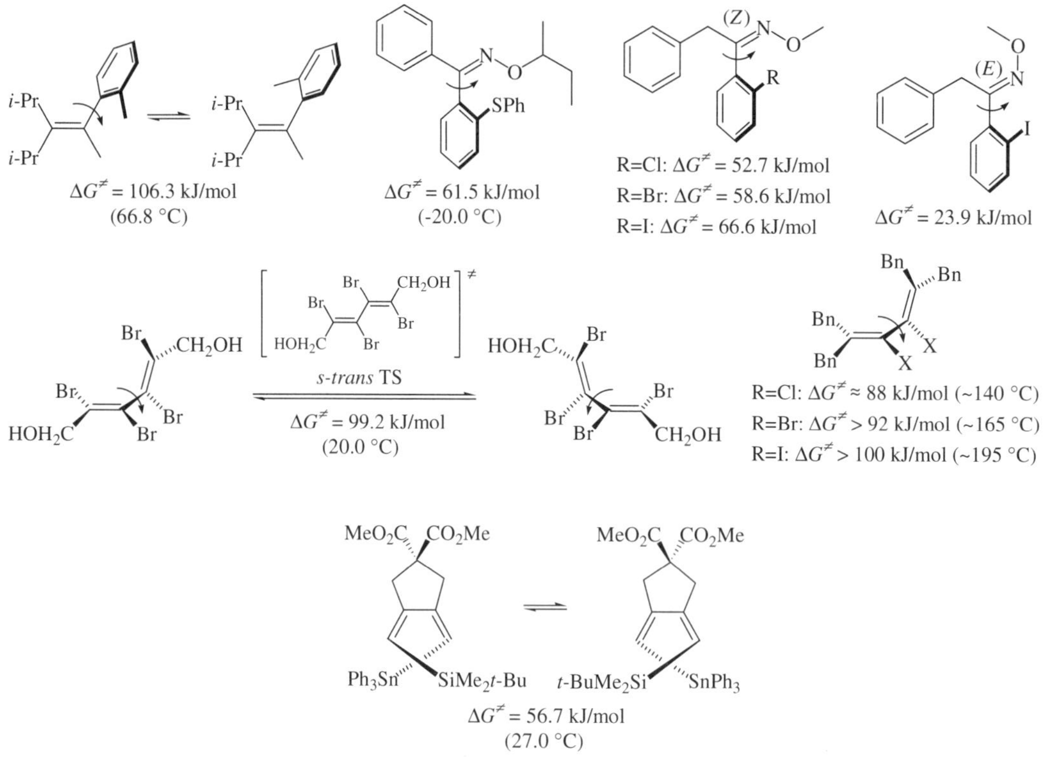

$\Delta G^{\neq}$ = 106.3 kJ/mol
(66.8 °C)

$\Delta G^{\neq}$ = 61.5 kJ/mol
(-20.0 °C)

R=Cl: $\Delta G^{\neq}$ = 52.7 kJ/mol
R=Br: $\Delta G^{\neq}$ = 58.6 kJ/mol
R=I: $\Delta G^{\neq}$ = 66.6 kJ/mol

$\Delta G^{\neq}$ = 23.9 kJ/mol

s-trans TS

$\Delta G^{\neq}$ = 99.2 kJ/mol
(20.0 °C)

R=Cl: $\Delta G^{\neq} \approx$ 88 kJ/mol (~140 °C)
R=Br: $\Delta G^{\neq}$ > 92 kJ/mol (~165 °C)
R=I: $\Delta G^{\neq}$ > 100 kJ/mol (~195 °C)

$\Delta G^{\neq}$ = 56.7 kJ/mol
(27.0 °C)

Figure 3.31 Atropisomeric styrenes, *O*-alkyl aryloximes and 1,3-dienes.

exert a buttressing effect similar to that observed with *meta*-substituents in biaryls and further enhance the rotational energy barrier. Some axially chiral 1,2-dialkylidenecycloalkane derivatives including 1,4-disubstituted (*Z*,*Z*)-1,3-dienes with terminal silyl and stannyl substituents undergo facile enantioconversion.[743–745] The surprisingly low energy barrier to racemization of these compounds is probably a consequence of destabilizing steric repulsion between the bulky substituents in a highly distorted ground state. Stereomutations of *N*-aryl imines are mechanistically more complicated than the rotational racemization pathways of styrenes and similar systems discussed above. By analogy with amines, the atropisomers of *N*-aryl ketimines can undergo nitrogen inversion in addition to aryl–nitrogen rotation, Scheme 3.74.[746] Variable-temperature NMR studies confirmed that the favored pathway for enantiomerization of an *N*-1-naphthyl ketimine derivative bearing two terminal isopropyl groups is rotation about the aryl–nitrogen bond, whereas the enantiomers of the di-*tert*-butyl analog interconvert through nitrogen inversion.[747,748]

Few studies of rotation about the chiral nitrogen–nitrogen axis of C_2-symmetric hydrazines, 3,3'-biquinazoline-4,4'-diones and *N*-nitroso amines have been conducted.[749–751] An interesting example is 3-nitroso-1,5-dimethyl-3-azabicyclo[3.1.0]hexane, a chiral *N*-nitroso amine that can be separated into enantiomers at room temperature. Racemization studies of this atropisomer, which is chiral due to hindered rotation of the nitroso group about the nitrogen–nitrogen axis, revealed an energy barrier of 94.2 kJ/mol, Figure 3.32. Sulfenamides adopt a chiral ground state conformation in which the substituents on the nitrogen and sulfur atoms reside in two perpendicular planes.[752,753] Interconversion of the enantiomers of *N*-(2,4-dinitrobenzenesulfenyl)-2-isopropylacridone via rotation about the *S–N* bond has an activation barrier of 95.0 kJ/mol.

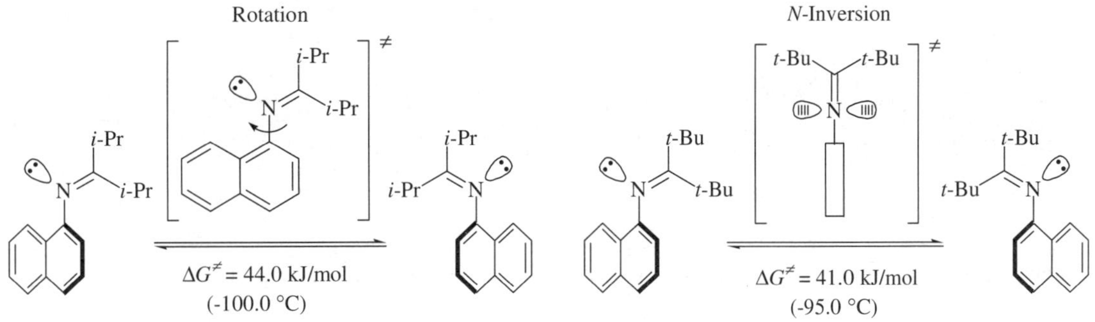

Rotation

N-Inversion

$\Delta G^{\neq} = 44.0$ kJ/mol
(-100.0 °C)

$\Delta G^{\neq} = 41.0$ kJ/mol
(-95.0 °C)

Scheme 3.74 Racemization pathways of *N*-aryl imines.

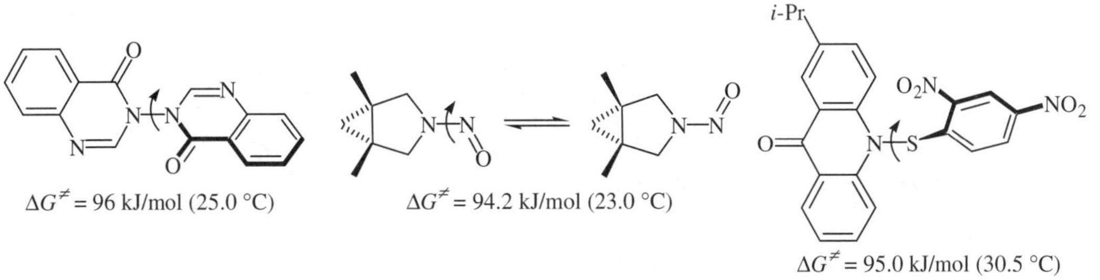

$\Delta G^{\neq} = 96$ kJ/mol (25.0 °C)

$\Delta G^{\neq} = 94.2$ kJ/mol (23.0 °C)

$\Delta G^{\neq} = 95.0$ kJ/mol (30.5 °C)

Figure 3.32 Atropisomerization of 3,3′-biquinazoline-4,4′-dione, 3-nitroso-1,5-dimethyl-3-azabicyclo[3.1.0]hexane and *N*-(2,4-dinitrobenzenesulfenyl)-2-isopropylacridone.

The conformational stability of atropisomers is solely determined by the energetic difference between ground and transition states. According to the analysis of electronic effects of *para*-substituents on the rotational energy barrier of axially chiral biphenyls, a decrease in conformational stability can either result from stabilization of the transition state or destabilization of the ground state, see Chapter 3.3.1. A careful evaluation of the structure and relative stability of ground and transition states is also indispensable for a discussion of the hindered rotation about the $C(sp^2)$–$C(sp^3)$ bond of axially chiral 1-monosubstituted and 1,8-disubstituted naphthalene derivatives.

The barrier to rotation about the aryl–alkyl bond of 1-alkylnaphthalenes devoid of bulky *ortho*-aryl substituents is usually quite low. For example, 1-neopentylnaphthalene exists as a mixture of two relatively unstrained enantiomeric conformers exhibiting the *tert*-butyl group almost exactly perpendicular to the naphthyl ring, Figure 3.33.[754] Interconversion of these two conformers requires that the alkyl group passes at the unsubstituted side at *C*-2 through the naphthalene plane to avoid enhanced steric repulsion with the *peri*-hydrogen at *C*-8. The energy barrier to this enantiomerization process is 21.6 kJ/mol.[755] One would expect that 1,8-dialkylated naphthalenes have significantly higher rotational barriers. Nevertheless, alkyl rotation in 1,8-di-*tert*-butylnaphthalene has an activation energy of only 26.4 kJ/mol.[756] The barrier to the same process in 1,8-bis(trimethylsilyl)naphthalene, and in the corresponding germanium and tin derivatives, decreases further to 25.0, 22.8 and 19.7 kJ/mol, respectively.[757] A closer look reveals that the ground state of 1,8-dialkylnaphthalenes is quite different from that of naphthalene derivatives bearing sp^2-hybridized substituents in the *peri*-positions. Both aryl rings and acyl moieties can adopt relatively stable ground states, in which these planar groups are oriented perpendicular to the naphthalene ring and thus undergo few steric interactions. This is not possible when both *peri*-positions are occupied by sp^3-hybridized substituents. The ground state of 1,8-di-*tert*-butylnaphthalene is considerably

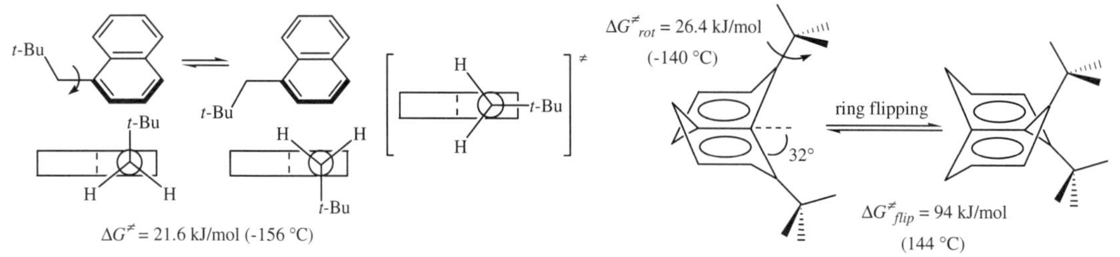

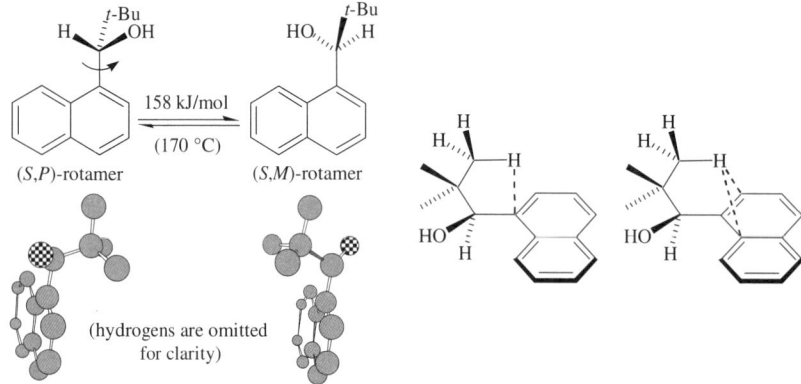

Figure 3.33 Stereomutations of 1-neopentylnaphthalene and 1,8-di-*tert*-butylnaphthalene.

Figure 3.34 Diastereomerization of 1-(1-naphthyl)-2,2-dimethylpropan-1-ol and intramolecular CH/π-interactions. [Reproduced with permission from *Tetrahedron Lett.* **2002**, *43*, 8563-8567.]

destabilized due to strong repulsion between the two bulky tetrahedral groups. As a result of severe molecular distortion, the *tert*-butyl groups are placed on opposite sides of the naphthalene ring and rotation about the $C(sp^2)–C(sp^3)$ bond is fast. The surprisingly low rotational energy barrier of this compound can therefore be attributed to an overcrowded ground state. Interestingly, the distortion renders 1,8-di-*tert*-butylnaphthalene C_2-symmetric and enantiomerization via ring flipping, with an energy barrier of 94 kJ/mol, has been observed by variable-temperature NMR spectroscopy.

Quantum mechanical calculations of the ground state structures of 1-naphthyl-derived carbinols such as 1-(1-naphthyl)-2,2-dimethylpropan-1-ol suggest that these atropisomers populate two conformations in which the bulky *tert*-butyl group is perpendicular to the naphthalene ring.[758] The conformers are stabilized by intramolecular CH/π-interactions between $C–H$ bonds of the *tert*-butyl group and the adjacent aromatic π-system.[759] Notably, all nine hydrogens of the *tert*-butyl group are available to participate in multiple CH/π-interactions involving C_{ipso} and C_{ortho} of the naphthalene ring, Figure 3.34. Rotation about the chiral axis results in interconversion of diastereomeric conformers. To minimize steric hindrance in the transition state, the bulky *tert*-butyl group is likely to pass the *ortho*-hydrogen rather than the *peri*-hydrogen of the naphthalene ring. As a consequence of considerable ground state stabilization due to CH/π-interactions and strong repulsion between the sterically demanding *tert*-butyl group and the naphthalene ring in the transition state, 1-(1-naphthyl)-2,2-dimethylpropan-1-ol is conformationally stable and has a rotational energy barrier of 158 kJ/mol. Similarly, 1-(1-naphthyl)ethylamines possess a chiral center in addition to a stereolabile chiral axis but they exist as a pair of rapidly interconverting diastereoisomers. For example, (*S*)-*N,N*-dibutyl-1-(1-naphthyl)ethylamine affords two diastereomeric rotamers exhibiting the amino group in almost perpendicular orientation to the naphthalene ring, Figure 3.35. Because of steric repulsion between the methyl group attached

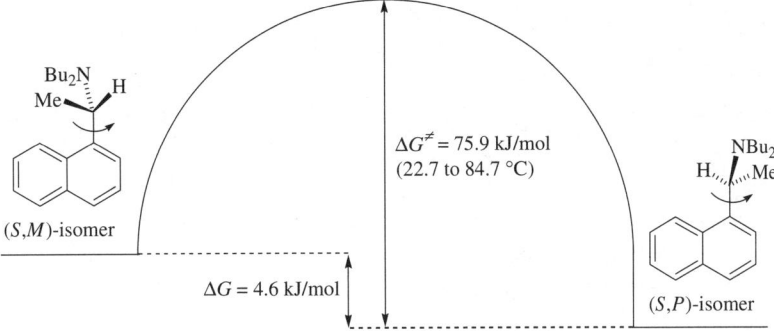

(*S,M*)-isomer

$\Delta G^{\neq} = 75.9$ kJ/mol
(22.7 to 84.7 °C)

(*S,P*)-isomer

$\Delta G = 4.6$ kJ/mol

Figure 3.35 Diasteromerization of *N,N*-dibutyl-1-(1-naphthyl)ethylamine.

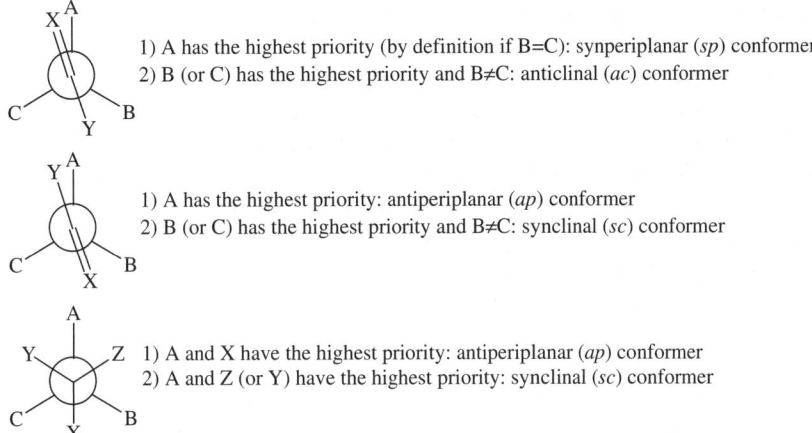

1) A has the highest priority (by definition if B=C): synperiplanar (*sp*) conformer
2) B (or C) has the highest priority and B≠C: anticlinal (*ac*) conformer

1) A has the highest priority: antiperiplanar (*ap*) conformer
2) B (or C) has the highest priority and B≠C: synclinal (*sc*) conformer

1) A and X have the highest priority: antiperiplanar (*ap*) conformer
2) A and Z (or Y) have the highest priority: synclinal (*sc*) conformer

Figure 3.36 Prelog–Klyne designation of atropisomerism about $C(sp^2)$–$C(sp^3)$ and $C(sp^3)$–$C(sp^3)$ bonds.

to the chiral center and the *peri*-hydrogen of the naphthalene ring, the (*S,M*)-rotamer is about 4.6 kJ/mol less stable than the (*S,P*)-diastereoisomer. The energy barrier to conversion of the more stable to the less populated atropisomer has been determined as 75.9 kJ/mol.

The rotation about $C(sp^2)$–$C(sp^3)$ and $C(sp^3)$–$C(sp^3)$ bonds of atropisomeric 9-arylfluorenes and triptycene derivatives has been studied in some depth, and exploited for the development of molecular gears and propellers, see Chapters 8.1 and 8.4.[760–764] The conformational isomerism of these compounds is often described based on the Prelog–Klyne rules, Figure 3.36. 9-Arylfluorenes derived from selectively substituted aryl and fluorenyl moieties possess axial and central chirality, and exist as a mixture of diastereomeric *ac*- and *sc*-rotamers. Rotation about the chiral $C(sp^2)$–$C(sp^3)$ axis results in diasteromerization whereas enantiomerization is not observed due to the configurational stability of the asymmetric carbon atom, Scheme 3.75.[765–768] The rotational energy barrier of 9-arylfluorenes carrying two *ortho*-substituents in the 9-aryl ring is sufficient for isolation of atropisomers but mono-*ortho*-substituted derivatives are usually not conformationally stable at room temperature.

Yamamoto and Oki discovered that sterically hindered dibenzobicyclo[2.2.2]octadienes and 9-substituted triptycenes can be separated into a meso *ap*-isomer and *sc*-enantiomers.[769,770] The dibenzobicyclo[2.2.2]octadiene shown in Scheme 3.76 has a barrier to rotation about the chiral axis of 139.0 kJ/mol.[771–773] At equilibrium, the *sc/ap* ratio is 3.0 which is slightly higher than the statistical ratio of 2. It has been hypothesized that the more flexible methoxycarbonyl group

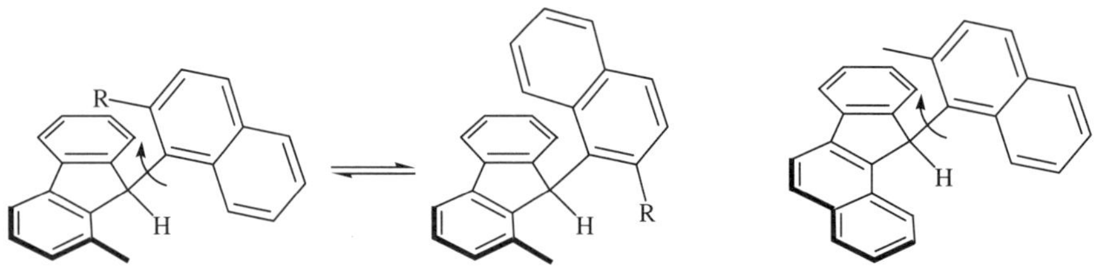

(R)-*ac*-conformer (R)-*sc*-conformer $\Delta G^{\neq}_{ac \to sc} = 137.3$ kJ/mol (170 °C)

R=H: $\Delta G^{\neq}_{ac \to sc} = 89.5$ kJ/mol (160 °C)

R=Me: $\Delta G^{\neq}_{ac \to sc} = 139.3$ kJ/mol (166 °C)

Scheme 3.75   Diastereomerization of 9-(1-naphthyl)fluorenes.

and the proximate benzyl group experience less steric repulsion in the *sc*-conformers than do the fused benzene rings and the benzyl group in the less stable *ap*-atropisomer. Crystallographic analysis and comparison of the conformational stability of a number of 9-*tert*-alkyl triptycenes indicate that incorporation of *peri*-substituents, X, into one benzene moiety causes considerable strain and favors population of the *ap*-isomer.[774] The statistical *ap*/*sc* ratio of 0.5 is only obtained in the presence of *peri*-hydrogens. Introduction of other substituents X disfavors population of the *sc*-rotamers due to enhanced steric repulsion between X and the neighboring benzyl group. The energy barrier to rotation about the pivotal $C(sp^3)$–$C(sp^3)$ bond increases with decreasing size of X, which has been attributed to steric repulsion between the *peri*-substituent and the tetrahedral triptycyl group in the congested ground state. Again, destabilization of the ground state, rather than stabilization of the transition state, is responsible for a decrease in the isomerization barrier. Replacement of the tertiary by a secondary alkyl group significantly reduces the rotational energy barrier of triptycenes, Scheme 3.76.[775–778] Interestingly, the *ap*-isomer of 9-isopropyltriptycenes is usually not observed and the predominant *sc*-enantiomers undergo facile interconversion at ambient temperatures. More examples of atropisomerism involving rotation about $C(sp^2)$–$C(sp^3)$ and $C(sp^3)$–$C(sp^3)$ bonds are presented in the context of molecular propellers, see Chapter 8.

3.3.3   Cyclophanes

Cyclophanes or so-called ansa compounds (Latin *ansa*, handle) were discovered by Lüttringhaus in 1941.[779,780] The chirality of these compounds stems from the fact that the handle, which may contain another aryl unit, resides outside the plane of a selectively substituted aryl ring. Rotation about the axis, defined by the bonds that connect the aryl ring to the bridge, results in enantiomerization. Alternatively, this process can be described as a flip of the handle from one side of the chiral plane to the other. The energy barrier to racemization is generally controlled by the length and rigidity of the handle and the bulkiness of the aryl substituents. However, the size of aryl substituents in monosubstituted paracyclophanes has only a minor effect on the rotational energy barrier, as the handle preferably flips around the unsubstituted and consequently less sterically hindered side of the chiral plane. As a rule of thumb, paracyclophanes with a bridge consisting of fewer than eleven atoms are stable to enantiomerization while chiral [12]paracyclophanes undergo rapid racemization at 25 °C.[781] The enantiomers of [11]paracyclophanes can often be isolated at room temperature but racemization occurs upon heating.[782–785] Monosubstituted diaza-, dioxa- and carbocyclic [11]paracyclophanes have interconversion barriers between 110 and 135 kJ/mol which are pH-dependent in the case of diaza derivatives, Figure 3.37.[786] Typically, 1,11-dioxa[11]paracyclophanes possess a higher steric

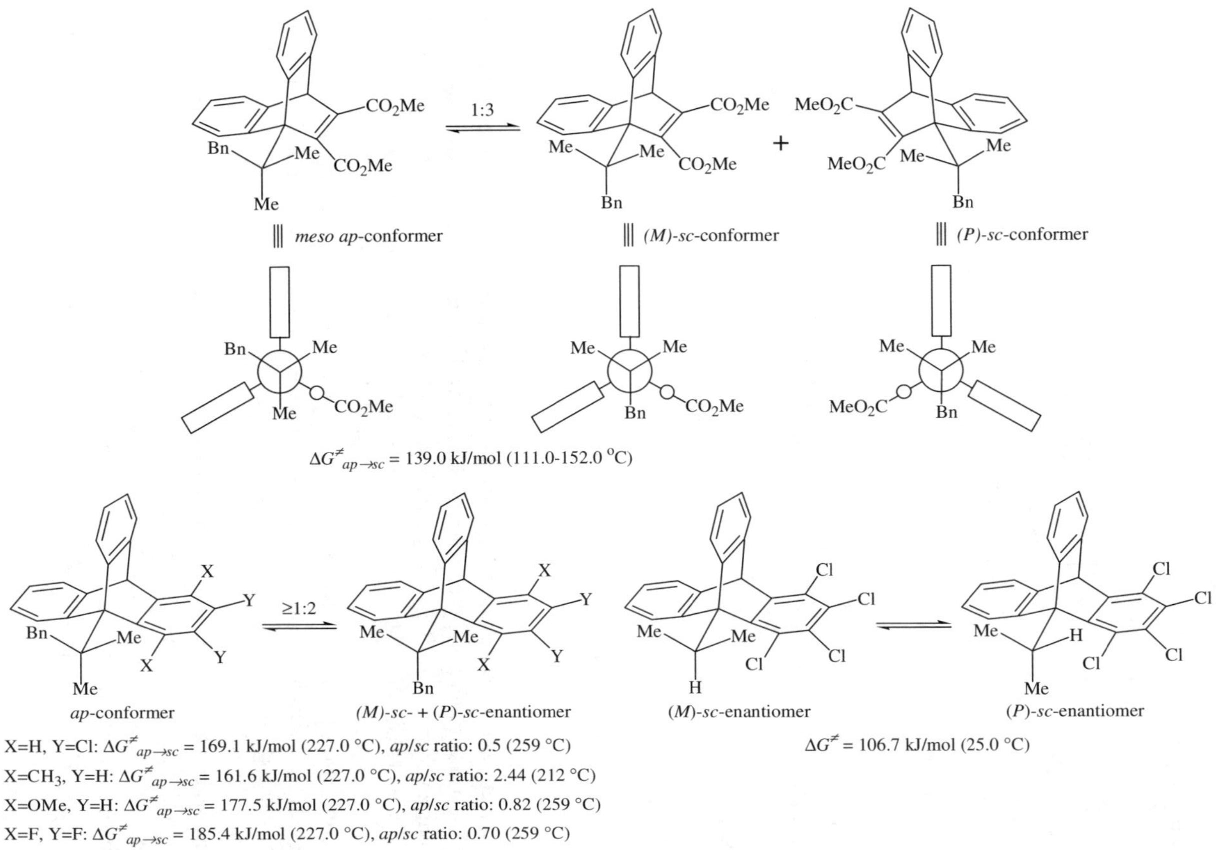

Scheme 3.76 Atropisomerization of *ap*- and *sc*-diastereoisomers of a dibenzobicyclo[2.2.2]octadiene (top) and 9-substituted triptycenes (bottom).

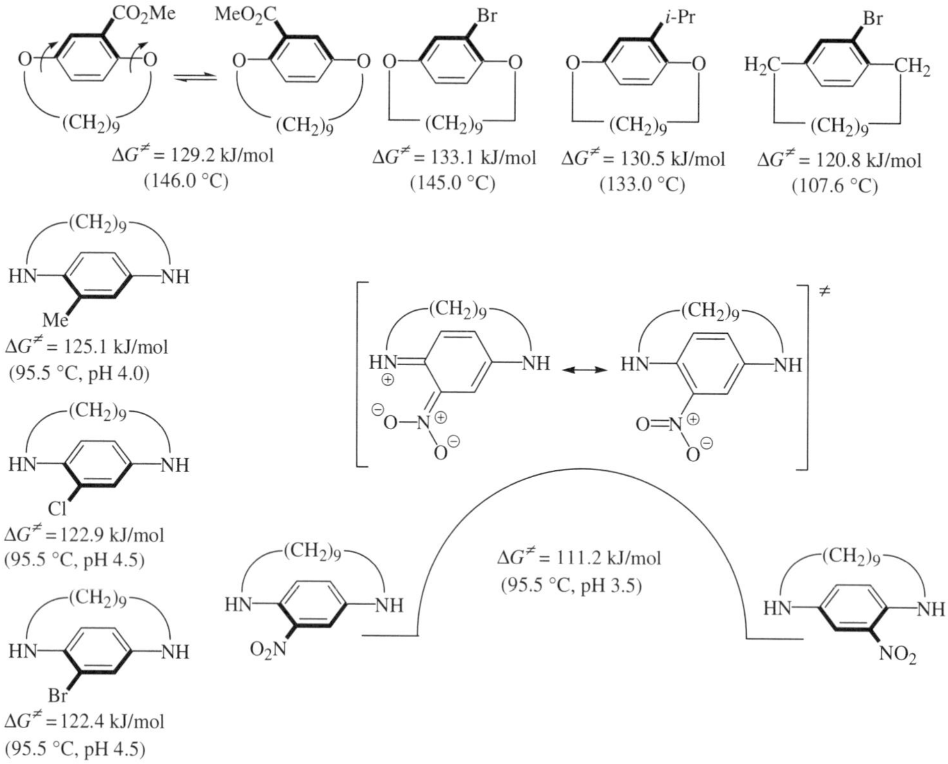

Figure 3.37 Interconversion barriers of [11]paracyclophanes.

hindrance to racemization than their carbocyclic analogs due to the relatively short *C–O* bond distance.[xxix] The rotational energy barrier of 1,11-diaza[11]paracyclophanes is similar to that of carbocyclic paracyclophanes. This is probably a consequence of facile pyramidal nitrogen inversion which increases the flexibility of the handle and its ability to accommodate a planar transition state. Incorporation of strong electron-withdrawing groups into the aryl ring effectively reduces the conformational stability of diazaparacyclophanes. The racemization barrier of 2′-nitro-1,11-diaza[11]paracyclophane is significantly lower than that of other 2′-substituted derivatives. It is assumed that the nitro group stabilizes the planar transition state by push-pull conjugation.

Cyclophanes with a short handle are considerably distorted. The aromatic carbon atoms directly connected to the bridge are bent out of the plane of the arene moiety and afford a benzene boat conformation. The high strain energy, which is about 130 kJ/mol in [2.2]paracyclophanes, facilitates photochemical and thermal racemization via biradial open chain intermediates. For example, 4-carbomethoxy[2.2]paracyclophane racemizes at 200 °C with an activation energy of approximately 160 kJ/mol. This value is slightly lower than the racemization barrier of the corresponding [3.4]paracyclophane derivative.[787–791] In general, chiral metacyclophanes show significantly less steric hindrance to atropisomerization and probably interconvert through a chair inversion process, Scheme 3.77.[792–799]

[xxix] The *C–O* bond length in paracyclophanes is 1.43 Å. A typical *C(sp²)–C(sp³)* bond length is 1.50 Å.

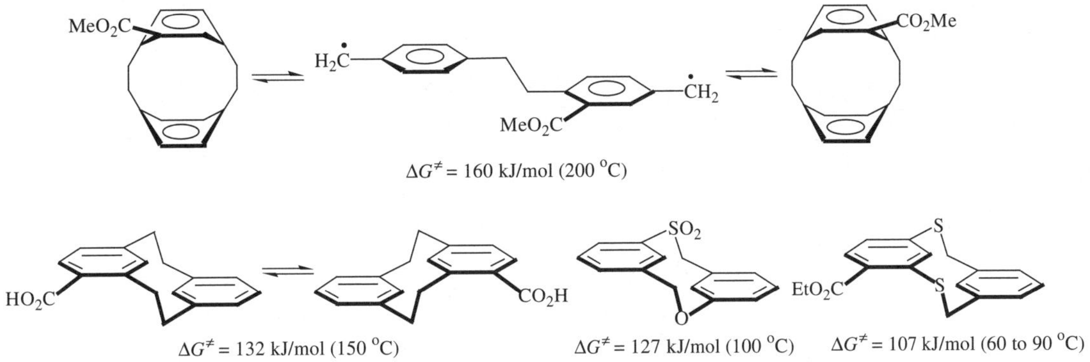

$\Delta G^{\neq} = 160$ kJ/mol (200 °C)

$\Delta G^{\neq} = 132$ kJ/mol (150 °C) $\Delta G^{\neq} = 127$ kJ/mol (100 °C) $\Delta G^{\neq} = 107$ kJ/mol (60 to 90 °C)

Scheme 3.77   Racemization of a [2.2]paracyclophane and chair inversion of metacyclophanes.

3.3.4   Atropisomeric Xenobiotics

The pharmacological and toxicological activities of some pesticides and herbicides are closely related to their chirality. The enantiomers of α-1,2,3,4,5,6-hexachlorocyclohexane (α-HCH), a chiral congener of the pesticide lindane (γ-HCH), exhibit different toxicity and metabolic pathways in biological samples.[800] Analysis of human tissue as well as aquatic samples (salmon, penguin, seal) revealed stereoselective metabolism of chlordane, a widely used pesticide mixture.[801] Apparently the enantiomeric composition of these environmental pollutants changes during biological transformation because uptake, metabolism and excretion proceed enantioselectively. The unique chemical and physical properties of polychlorinated biphenyls (PCBs) have led to the extensive use of these xenobiotics as refrigerants, impregnants for wood and paper, hydraulic fluids, and as plastic and paint additives. In addition, PCBs have been applied as electrical insulators in capacitors and transformers. Since 1929, more than a million tonnes of PCBs have been produced worldwide.[802,xxx] Although their production has been banned by the United Nations treaty on persistent organic pollutants, PCBs have become ubiquitous environmental xenobiotics. Their uptake by invertebrates and fish has introduced PCBs into the food chain. In addition to their carcinogenic potential, PCBs have numerous toxic effects including immunosuppression and thymic atrophy, hyperkeratosis, and hepatotoxicity. The pharmacological and toxicological properties of PCBs depend on their axial chirality as well as on the stability to racemization, Scheme 3.78.[803] Only 78 of the 209 possible PCBs are chiral, and 19 PCBs (45, 84, 88, 91, 95, 131, 132, 135, 136, 139, 144, 149, 171, 174, 175, 176, 183, 196, 197) afford at least three chloro substituents in the *ortho*-position which is a requirement for conformational stability under abiotic conditions, Table 3.3. The energy barrier to racemization of these 19 PCBs is greater than 180 kJ/mol, and most of these so-called PB-type PCBs have been separated into enantiomers by chiral chromatography on selectively substituted cyclodextrins.[804–808]

Pharmacological and toxicological studies of PCBs proved that the biological activities of these xenobiotics vary significantly with the number and position of chloro atoms.[809,810] Based on characteristic structure–activity relationships, one can distinguish between two major classes, namely, 3-methylcholanthrene-type PCBs (MC-type) and phenobarbital-type PCBs (PB-type).[811] MC-type PCBs do not exhibit chloro atoms in *ortho*-position and are not conformationally stable. It is assumed that these PCBs can easily adopt a coplanar structure similar to TCDD (dioxin). MC-type PCBs have been reported to increase benzo[a]pyrene hydroxylase activity and expression of cytochrome P450c and P450d in liver cells. They also cause accumulation of uroporphyrinogen (URO), probably as a consequence of decreasing activity of URO decarboxylase and induction of

---

xxxTechnical mixtures such as Acroclor and Clophen A contain considerable amounts (often 30–60%) of PCBs.

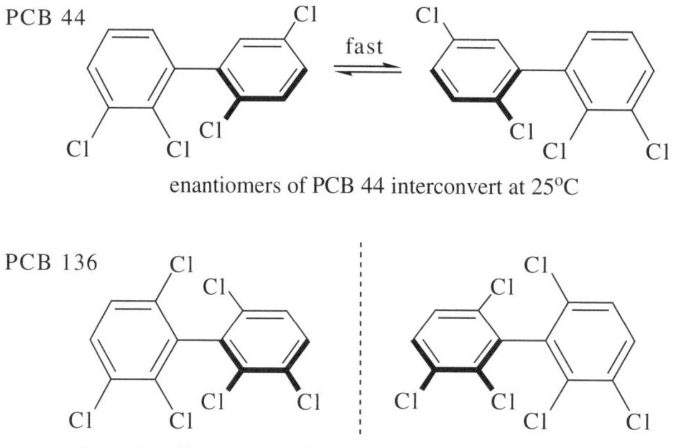

Scheme 3.78 Axially chiral PB-type PCBs 44 and 136.

Table 3.3 Ballschmiter nomenclature for conformationally stable chiral PCBs.

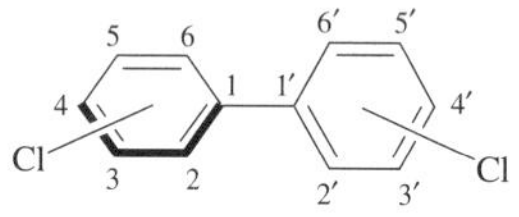

PCB	Substitution pattern
45	2,2′,3,6
84	2,2′,3,3′,6
88	2,2′,3,4,6
91[b]	2,2′,3,4′,6
95[a,b]	2,2′,3,5′,6
131[a]	2,2′,3,3′,4,6
132[a,b]	2,2′,3,3′,4,6′
135[a,b]	2,2′,3,3′,5,6′
136[a,b]	2,2′,3,3′,6,6′
139	2,2′,3,4,4′,6
144[a]	2,2′,3,4,5′,6
149[a,b]	2,2′,3,4′,5′,6
171	2,2′,3,3′,4,4′,6
174[a,b]	2,2′,3,3′,4,5,6′
175	2,2′,3,3′,4,5′,6
176[b]	2,2′,3,3′,4,6,6′
183	2,2′,3,4,4′,5′,6
196[a]	2,2′,3,3′,4,4′,5,6′
197	2,2′,3,3′,4,4′,6,6′

[a] PB-Type PCBs found in butter and human milk.
[b] Enantioselective bioenrichment has been observed.

cytochrome P448.[812–814] Their biological activity seems to be related to the coplanar structure as well as to the number and position of chloro substituents in *meta*- und *para*-position.[815] Phenobarbital-type PCBs have three or four chloro atoms in the *ortho*-positions and are conformationally stable.[816–818] Similarly to α-HCH[819] and metolachlor,[820] the enantiomers of PB-type PCBs show distinctive biological effects and metabolism. They induce cytochrome P450b, P450e, aldrin epoxidase, and aminopyrine *N*-demethylase, and cause proliferation of liver endoplasmic reticulum.

For several reasons it is believed that PB-type PCBs do not undergo racemization under physiological conditions:

- The energy barrier to racemization is mostly determined by *intra*molecular steric interactions, and *inter*molecular interactions with proteins, membranes and solvents are expected to have negligible affects on the conformational stability.
- A PCB racemase, if existing, would have to cleave or elongate the pivotal *C–C* bond between the two chlorinated benzene rings. However, biphenyls are metabolized via hydroxylation or arene oxide formation of the aryl moieties by hepatic oxygenase systems, *i.e.*, enzyme attacks occur remote from the chiral *C–C* bond.[821]
- Enantiomeric enrichment of (+)-PCB 139 in rat liver observed by Püttmann and coworkers is only possible in the absence of an active PCB racemase.

Püttmann studied the biological activity of conformationally stable PB-type PCBs 139 and 197.[822] They found that (+)-PCB 139 induces expression of aminopyrine *N*-demethylase, aldrin epoxidase and cytochrome P450 more effectively than the (–)-enantiomer. Similarly, (+)-PCB 197 induces production of aminopyrine *N*-demethylase more strongly than does (–)-PCB 197. Enantiomeric enrichment of (+)-PCB 139 was discovered in rat liver and attributed to pharmacokinetic effects (absorption, distribution, metabolism or excretion). Rodman *et al.* studied inductive effects of PCBs 88, 139 and 197 using chick embryo hepatocyte cultures to minimize pharmacokinetic effects on cytochrome P450 induction.[823] They reported enhanced induction of cytochrome P450 by (+)-PCB 139. The most striking enantioselective effects are observed with induction of ethoxy-resorufin-*O*-deethylase (EROD). The (+)-enantiomers of PCBs 88, 139 and 197 induce EROD more effectively than the corresponding (–)-enantiomers, and the (+)-enantiomer of PCB 139 exhibits stronger inducing potency for benzphetamin *N*-demethylase (BPDM) than its (–)-enantiomer. By contrast, the (–)-enantiomers of PCBs 88 and 197 are more effective inducers of BPDM than their (+)-enantiomers. The (+)-enantiomer of PCB 139 causes URO accumulation which correlates with substantial increase in EROD activity. The results of this study clearly demonstrate that the chirality of PCBs plays a role during induction of cytochrome P450 and accumulation of URO.

The bio- and geoaccumulation of PCBs originate from their persistence to degradation and high lipophilicity. Some of the PB-type PCBs mentioned are components of Clophen A or Acroclor and have been found in butter and human milk.[824,825] The pronounced enantioselective metabolism in aquatic sediment and biota results in accumulation of nonracemic PCB mixtures in the food chain. To date, enantioenrichment has been reported for PCB 149 in blue mussels,[826] for PCB 132 in human milk,[827] and for PCBs 95, 132, 135, 149, and 176 in dolphins.[828] Enantioenrichment of PCBs 91, 95, 132, 136, 149, 174, and 176 has been observed in river sediments.[829,830] Significant health and environmental risks could be related to enantioselective enrichment of PB-type PCBs in nature.

3.4 PHARMACOLOGICAL AND PHARMACOKINETIC SIGNIFICANCE OF RACEMIZATION

Some chiral pharmaceuticals including thalidomide racemize either within minutes or several hours under physiological conditions (37 °C, pH 7.4). Oxazepam has a racemization half-life of less than five minutes at 37 °C, Scheme 3.79. From a pharmacological and toxicological standpoint this is quite important because stereoisomers can have distinct biological properties, see Chapter 1.[831,xxxi]

[xxxi] Racemization that does not occur within the pharmacological time scale but over a period of months or years after formulation is also important, as it compromises the purity and shelf-life of chiral drugs.

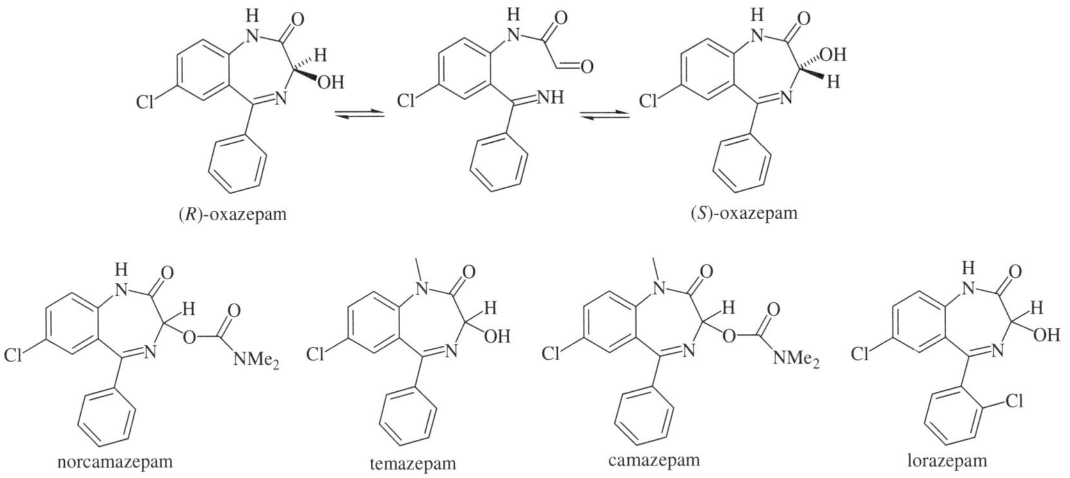

Scheme 3.79   Racemization mechanism of oxazepam and structures of 3-hydroxybenzodiazepinones.

For example, 3-hydroxybenzodiazepinone-derived anti-anxiety agents experience enantioselective metabolism and undergo reversible ring opening via an achiral intermediate.[832] *In vitro* studies with camazepam and other racemic 3-hydroxybenzodiazepinones showed that the (*R*)-enantiomer is selectively metabolized by human liver microsomes.[833] Although the stereochemical integrity of new chiral drugs can often be deduced from previously studied compounds exhibiting similar functionalities, it is indispensable to carefully evaluate the effect of endogenous factors such as pH, ionic strength and the presence of plasma proteins, in particular epimerases and racemases.[834] The ultimate analysis of the stereochemical stability of a chiral drug candidate during preclinical studies requires racemization studies in biological media.[835] Welch *et al.* investigated the configurational stability of a 5-aryl-thiazolidinedione, a peroxisome proliferator-activated receptor (PPAR) agonist, under various conditions.[836] The drug proved to be stereolabile in various solvents irrespective of pH, and rapid racemization in both dog and human plasma was confirmed by enantioselective HPLC-MS analysis. Many pharmaceuticals such as thalidomide, bupivacaine, tolperisone, ritalin, ibuprofen, hyoscyamine, proglumide, scopolamine, tropicamide, and cyclandelate possess a chiral center carrying an acidic proton and are susceptible to base-catalyzed racemization, Scheme 3.80. Chlorthalidone racemizes through acid-catalyzed elimination and base-catalyzed ring opening.[837]

Chiral barbiturates, imidazolines, hydantoins, oxazolines, thiazolines, oxazolidinediones, succinimides, profens, and other widespread pharmacophores racemize due to facile enolization, Figure 3.38.[838–843] The configurational stability of enolizable drugs is often evaluated by proton/deuterium exchange in deuterium oxide. The deuteration process can easily be monitored by ¹H NMR spectroscopy using racemic samples, and in most cases the rate of deuteration is equal to that of racemization.[844] Racemization of drugs can also be studied by HPLC, SFC and GC on chiral stationary phases, electrophoresis with chiral additives, chiroptical methods, and NMR spectroscopy with chiral shift reagents, see Chapter 4.

Nonenzymatic interconversion of (*S*)-amino acids to the biologically rare (*R*)-enantiomer possibly plays an important role in aging. Racemization of (*S*)-aspartic acid proceeds *in vivo* via succinimide formation and is believed to constitute a biological clock for peptide and protein aging, Scheme 3.81.[845,846] Numerous studies have shown that epimerization of dipeptides proceeds faster than racemization of free amino acids due to neighboring effects and formation of diketopiperazines.[847–849] The accumulation of (*R*)-aspartic acid in tissue with no turnover (tooth enamel and bone osteocalcin) allows accurate age determination for forensic purposes.[850] Measurements of the extent of racemization of (*S*)-aspartic acid in urinary molecular fragments

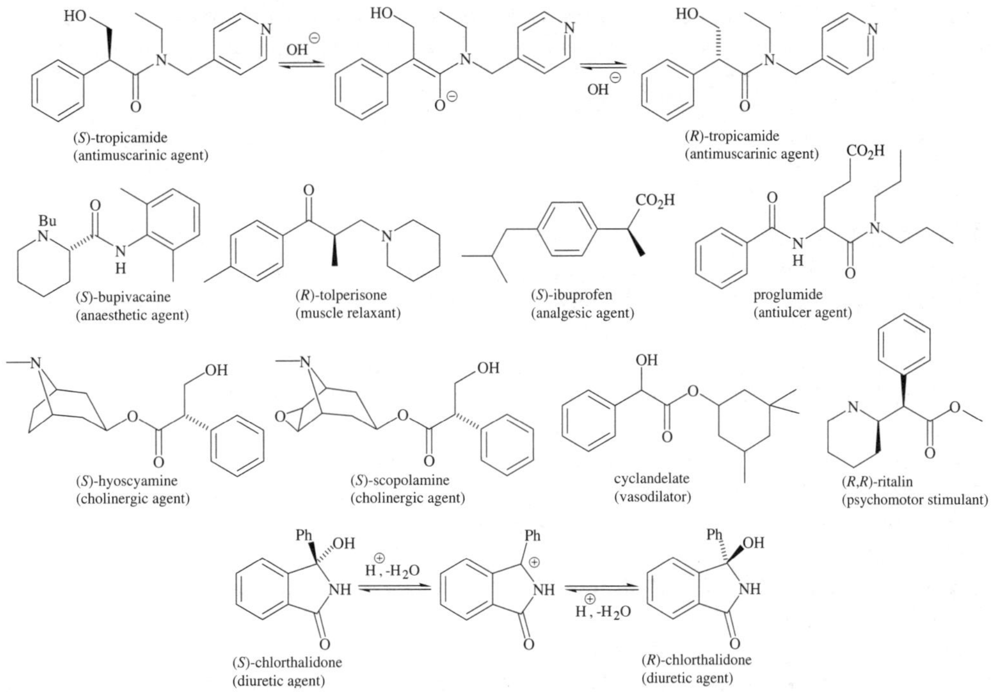

Scheme 3.80   Racemization pathways of chiral drugs.

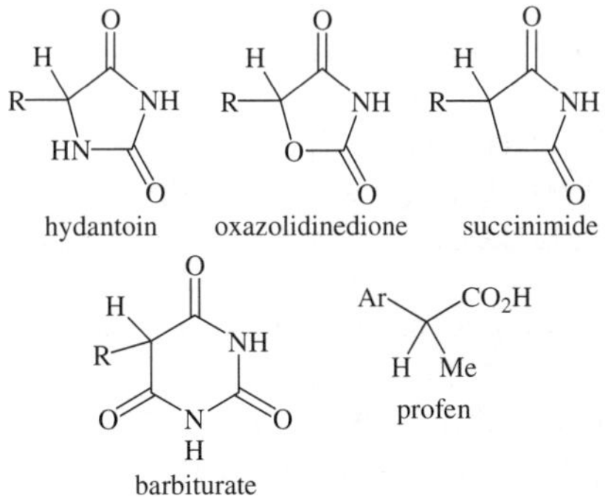

Figure 3.38   Enolizable pharmacophores.

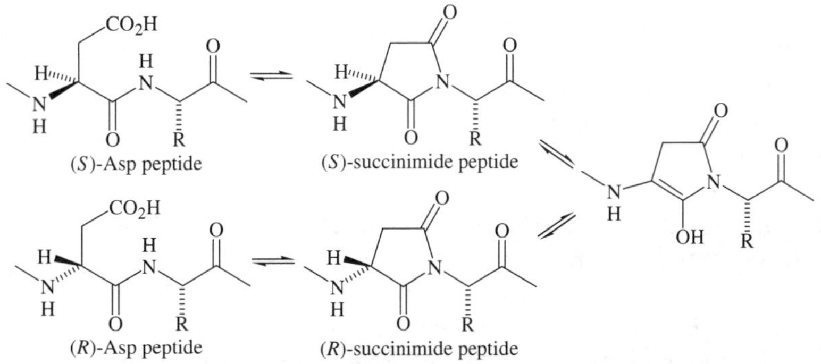

Scheme 3.81   Racemization of (*S*)-aspartic acid via succinimide formation.

containing α1 *C*-terminal telopeptides of type I collagen, a major constituent of extracellular tissue, can be used as a noninvasive tool for estimation of bone turnover and for diagnosis of metabolic bone diseases.

REFERENCES

1. Reist, M.; Testa, B.; Carrupt, P.-A.; Jung, M.; Schurig, V. *Chirality*, **1995**, *7*, 396-400.
2. Hutt, A. J.; Caldwell, J. *J. Pharm. Pharmacol.* **1983**, *35*, 693-704.
3. Mayer, J. M.; Roy-de Vos, M.; Audergon, C.; Testa, B.; Etter, J. C. *Int. J. Tissue React.* **1994**, *16*, 59-72.
4. Campbell, R. D.; Gilow, H. M. *J. Am. Chem. Soc.* **1960**, *82*, 5426-5431.
5. Campbell, R. D.; Gilow, H. M. *J. Am. Chem. Soc.* **1962**, *84*, 1440-1443.
6. Tumambac, G. E.; Francis, C. J.; Wolf, C. *Chirality* **2005**, *17*, 171-176.
7. Horváth, A.; Bäckvall, J.-E. *Chem. Commun.* **2004**, 964-965.
8. Noyori, R.; Ikeda, T.; Ohkuma, T.; Widhalm, M.; Kitamura, M.; Takaya, H.; Akutagawa, S.; Sayo, N.; Saito, T.; Taketomi, T.; Kumobayashi, H. *J. Am. Chem. Soc.* **1989**, *111*, 9134-9135.
9. Kiau, S.; Discordia, R. P.; Madding, G.; Okuniewicz, F. J.; Rosso, V.; Venit, J. J. *J. Org. Chem.* **2004**, *69*, 4256-4261.
10. Tumambac, G. E.; Wolf, C. *J. Org. Chem.* **2005**, *70*, 2930-2938.
11. Ward, D. E.; Sales, M.; Sasmal, P. K. *J. Org. Chem.* **2004**, *69*, 4808-4815.
12. Liu, J.; Swidorski, J. J.; Peters, S. D.; Hsung, R. P. *J. Org. Chem.* **2005**, *70*, 3898-3902.
13. Nelson, T. D.; Meyers, A. I. *Tetrahedron Lett.* **1994**, *35*, 3259-3262.
14. Nelson, T. D.; Meyers, A. I. *J. Org. Chem.* **1994**, *59*, 2577-2580.
15. Evans, D. A.; Wood, M. R.; Trotter, B. W.; Richardson, T. I.; Barrow, J. C.; Katz, J. L. *Angew. Chem., Int. Ed.* **1998**, *37*, 2700-2704.
16. Evans, D. A.; Dinsmore, C. J.; Watson, P. S.; Wood, M. R.; Richardson, T. I.; Trotter, B. W.; Katz, J. L. *Angew. Chem., Int. Ed.* **1998**, *37*, 2704-3708.
17. Boger, D. L.; Miyazaki, S.; Kim, S. H.; Wu, J. H.; Loiseleur, O.; Castle, S. L. *J. Am. Chem. Soc.* **1999**, *121*, 3226-3227.
18. Boger, D. L.; Miyazaki, S.; Kim, S. H.; Wu, J. H.; Castle, S. L.; Loiseleur, O.; Jin, Q. *J. Am. Chem. Soc.* **1999**, *121*, 10004-10011.
19. Armstrong, D. W.; Jin, H. L. *Chirality* **1989**, *1*, 27-37.
20. Samuel, J.; Tanner, M. E. *Nat. Prod. Rep.* **2002**, *19*, 261-277.
21. Frank, M.; Miethchen, R.; Reinke, H. *Eur. J. Org. Chem.* **1999**, 1259-1263.
22. Miethchen, R. *J. Carbohydr. Chem.* **2003**, *22*, 801-825.
23. Takano, S.; Kijima, A.; Sugihara, T.; Satoh, S.; Ogasawara, K. *Chem. Lett.* **1989**, 87-88.
24. Matsumoto, T.; Matsunaga, N.; Kanai, A.; Aoyama, T.; Shioiri, T.; Osawa, E. *Tetrahedron* **1994**, *50*, 9781-9788.
25. Gibson, K. R.; Hitzel, L.; Mortishire-Smith, R. J.; Gerhard, U.; Jelley, R. A.; Reeve, A. J.; Rowley, M.; Nadin, A.; Owens, A. P. *J. Org. Chem.* **2002**, *67*, 9354-9360.
26. Ebbers, E. J.; Ariaans, G. J. A.; Houbiers, J. P. M.; Bruggink, A.; Zwanenburg, B. *Tetrahedron* **1997**, *53*, 9417-9476.
27. Berson, J. A.; Balquist, J. M. *J. Am. Chem. Soc.* **1968**, *90*, 7343-7344.
28. Bergman, R. G.; Carter, W. L. *J. Am. Chem. Soc.* **1969**, *91*, 7411-7425.
29. Okada, K.; Samizo, F.; Oda, M. *J. Chem. Soc., Chem. Commun.* **1986**, 1044-1046.
30. Baldwin, J. E. *J. Chem. Soc., Chem. Commun.* **1988**, 31-32.
31. Cianciosi, S. J.; Ragunathan, N.; Freedman, T. B.; Nafie, L. A.; Baldwin, J. E. *J. Am. Chem. Soc.* **1990**, *112*, 8204-8206.

32. Cianciosi, S. J.; Ragunathan, N.; Freedman, T. B.; Nafie, L. A.; Lewis, D. K.; Glenar, D. A.; Baldwin, J. E. *J. Am. Chem. Soc.* **1991**, *113*, 1864-1866.
33. Freedman, T. B.; Hausch, D. L.; Cianciosi, S. J.; Baldwin, J. E. *Can. J. Chem.* **1998**, *76*, 806-810.
34. Chickos, J. S.; Annamalai, A.; Keiderling, T. A. *J. Am. Chem. Soc.* **1986**, *108*, 4398-4402.
35. Schlögl, K.; Weissensteiner, W.; Widhalm, M. *J. Org. Chem.* **1982**, *47*, 5025-5027.
36. Veciana, J.; Crespo, M. I. *Angew. Chem., Int. Ed. Engl.* **1991**, *30*, 74-77.
37. Gault, I. R.; Ollis, W. D.; Sutherland, I. O. *J. Chem. Soc., Chem. Commun.* **1970**, 269-271.
38. Agranat, I.; Rabinovitz, M.; Weitzen-Dagan, A.; Gosnay, I. *J. Chem. Soc., Chem. Commun.* **1972**, 732-733.
39. Wang, X.-J.; Luh, T. Y. *J. Org. Chem.* **1989**, *54*, 263-265.
40. Yip, Y. C.; Wang, X.-J.; Ng, D. K. P.; Mak, T. C. W.; Chiang, P.; Luh, T.-Y. *J. Org. Chem.* **1990**, *55*, 1881-1889.
41. Harada, N.; Saito, A.; Koumura, N.; Roe, D. C.; Jager, W. F.; Zijlstra, R. W. J.; de Lange, B.; Feringa, B. L. *J. Am. Chem. Soc.* **1997**, *119*, 7249-7255.
42. Zijlstra, R. W. J.; Jager, W. F.; de Lange, B.; van Duijnen, P. T.; Feringa, B. L.; Goto, H.; Saito, A.; Koumura, N.; Harada, N. *J. Org. Chem.* **1999**, *64*, 1667-1674.
43. Kuwahara, S.; Fujita, T.; Harada, N. *Eur. J. Org. Chem.* **2005**, 4544-4556.
44. Agranat, I.; Tapuhi, Y. *J. Am. Chem. Soc.* **1978**, *100*, 5604-5609.
45. Agranat, I.; Tapuhi, Y. *J. Am. Chem. Soc.* **1979**, *101*, 665-671.
46. Feringa, B. L.; Jager, W. F.; de Lange, B. *J. Chem. Soc., Chem. Commun.* **1993**, 288-290.
47. Agranat, I.; Tapuhi, Y. *J. Am. Chem. Soc.* **1979**, *101*, 1941-1948.
48. Agranat, I.; Tapuhi, Y. *J. Am. Chem. Soc.* **1979**, *101*, 1949-1952.
49. Harada, N.; Saito, A.; Koumura, N.; Uda, H.; de Lange, B.; Jager, W. F.; Wynberg, H.; Feringa, B. L. *J. Am. Chem. Soc.* **1997**, *119*, 7241-7248.
50. Harada, N.; Koumura, N.; Feringa, B. L. *J. Am. Chem. Soc.* **1997**, *119*, 7256-7264.
51. Harada, N.; Saito, A.; Koumura, N.; Roe, D. C.; Jager, W. F.; Zijlstra, R. W. J.; de Lange, B.; Feringa, B. L. *J. Am. Chem. Soc.* **1997**, *119*, 7249-7255.
52. Vicario, J.; Meetsma, A.; Feringa, B. L. *Chem. Commun.* **2005**, 5910-5912.
53. Pijper, D.; van Delden, R. A.; Meetsma, A.; Feringa, B. L. *J. Am. Chem. Soc.* **2005**, *127*, 17612-17613.
54. Sandström, J. *Top. Stereochem.* **1983**, *14*, 83-181.
55. Berg, U.; Isaksson, R.; Sandström, J.; Sjöstrand, U.; Eiglsperger, A.; Mannschreck, A. *Tetrahedron Lett.* **1982**, *23*, 4237-4240.
56. Khan, A. Z.-Q.; Sandström, J. *J. Chem. Soc., Perkin Trans. 2* **1994**, 1575-1579.
57. Marshall, J. A.; Konicek, T. R.; Flynn, K. E. *J. Am. Chem. Soc.* **1980**, *102*, 3287-3288.
58. Marshall, J. A.; Audia, V. H.; Jenson, T. M.; Guida, W. C. *Tetrahedron* **1986**, *42*, 1703-1709.
59. Cope, A. C.; Pawson, B. A. *J. Am. Chem. Soc.* **1965**, *87*, 3649-3651.
60. Cope, A. C.; Banholzer, K.; Keller, H.; Pawson, B. A.; Whang, J. J.; Winkler, H. J. S. *J. Am. Chem. Soc.* **1965**, *87*, 3644-3649.
61. Binsch, G.; Roberts, J. D. *J. Am. Chem. Soc.* **1965**, *87*, 5157-5162.
62. Squillacote, M.; Bergmann, A.; De Felippis, J. *Tetrahedron Lett.* **1989**, *30*, 6805-6808.
63. Krebs, A.; Pforr, K.-I.; Raffay, W.; Thölke, B.; König, W. A.; Hardt, I. H.; Boese, R. *Angew. Chem., Int. Ed. Engl.* **1997**, *36*, 159-160.
64. Inoue, Y.; Yamasaki, N.; Yokoyama, T.; Tai, A. *J. Org. Chem.* **1992**, *57*, 1332-1345.
65. Inoue, Y.; Yamasaki, N.; Yokoyama, T.; Tai, A. *J. Org. Chem.* **1993**, *58*, 1011-1018.
66. Inoue, Y.; Dong, F.; Yamamoto, K.; Tong, L.-H.; Tsuneishi, H.; Hakushi, T.; Tai, A. *J. Am. Chem. Soc.* **1995**, *117*, 11033-11034.
67. Inoue, Y.; Ikeda, H.; Kaneda, M.; Sumimura, T.; Everitt, S. R. L.; Wada, T. *J. Am. Chem. Soc.* **2000**, *122*, 406-407.

68. Wada, T.; Shikimi, M.; Inoue, Y.; Lem, G.; Turro, N. J. *Chem. Commun.* **2001**, 1864-1865.

69. Fukuhara, G.; Mori, T.; Wada, T.; Inoue, Y. *Chem. Commun.* **2005**, 4199-4201.

70. Nakazaki, M.; Naemura, K.; Nakahara, S. *J. Org. Chem.* **1979**, *44*, 2438-2441.

71. Marshall, J. A.; Lewellyn, M. E. *J. Am. Chem. Soc.* **1977**, *99*, 3508-3510.

72. Nakazaki, M.; Yamamoto, K.; Yanagi, J. *J. Am. Chem. Soc.* **1979**, *101*, 147-151.

73. Marshall, J. A. *Acc. Chem. Res.* **1980**, *13*, 213-218.

74. Kirk, D. D.; Shaw, P. M. *J. Chem. Soc. D, Chem. Commun.* **1971**, 948-949.

75. Kirk, D. D.; Shaw, P. M. *J. Chem. Soc., Perkin Trans. 1* **1975**, 2284-2294.

76. Gur, E.; Kaida, Y.; Okamoto, Y.; Biali, S.; Rappoport, Z. *J. Org. Chem.* **1992**, *57*, 3689-3693.

77. Rochlin, E.; Rappoport, Z. *J. Org. Chem.* **2003**, *68*, 216-226.

78. Grilli, S.; Lunazzi, L.; Mazzanti, A.; Casarini, D.; Femoni, C. *J. Org. Chem.* **2001**, *66*, 488-495.

79. Grilli, S.; Lunazzi, L.; Mazzanti, A.; Mazzanti, G. *J. Org. Chem.* **2001**, *66*, 748-754.

80. Casarini, D.; Grilli, S.; Lunazzi, L.; Mazzanti, A. *J. Org. Chem.* **2001**, *66*, 2757-2763.

81. Grilli, S.; Lunazzi, L.; Mazzanti, A. *J. Org. Chem.* **2001**, *66*, 4444-4446.

82. Grilli, S.; Lunazzi, L.; Mazzanti, A. *J. Org. Chem.* **2001**, *66*, 5853-5858.

83. Anet, F. A. L. *J. Am. Chem. Soc.* **1962**, *84*, 671-672.

84. Paquette, L. A. *Acc. Chem. Res.* **1993**, *26*, 57-62.

85. Mislow, K.; Perlmutter, H. D. *J. Am. Chem. Soc.* **1962**, *84*, 3591-3592.

86. Gardlik, J. M.; Paquette, L. A.; Gleiter, F. *J. Am. Chem. Soc.* **1979**, *101*, 1617-1620.

87. Paquette, L. A.; Gardlick, J. M. *J. Am. Chem. Soc.* **1980**, *102*, 5033-5035.

88. Paquette, L. A.; Hanzawa, Y.; McCullough, K. J.; Tagle, B.; Swenson, W.; Clardy, J. *J. Am. Chem. Soc.* **1981**, *103*, 2262-2269.

89. Paquette, L. A.; Gardlick, J. M.; McCullough, K. J.; Hanzawa, Y. *J. Am. Chem. Soc.* **1983**, *105*, 7644-7648.

90. Paquette, L. A.; Gardlick, J. M.; McCullough, J. M.; Samodral, R.; DeLucca, G.; Oullette, R. J. *J. Am. Chem. Soc.* **1983**, *105*, 7649-7655.

91. Paquette, L. A.; Wang, T.-H. *J. Am. Chem. Soc.* **1988**, *110*, 8192-8197.

92. Paquette, L. A.; Trova, M. P. *J. Am. Chem. Soc.* **1988**, *110*, 8197-8201.

93. Paquette, L. A.; Trova, M. P.; Luo, J.; Clough, A. E.; Anderson, L. B. *J. Am. Chem. Soc.* **1990**, *112*, 228-239.

94. Paquette, L. A.; Wang, T.-Z.; Luo, J.; Cottrell, C. E.; Clough, A. E.; Anderson, L. B. *J. Am. Chem. Soc.* **1990**, *112*, 239-253.

95. Paquette, L. A.; Gardlick, J. M.; Photis, J. M. *J. Am. Chem. Soc.* **1976**, *98*, 7096-7098.

96. Gardlick, J. M.; Johnson, L. K.; Paquette, L. A.; Solheim, B. A.; Springer, J. P.; Clardy, J. *J. Am. Chem. Soc.* **1979**, *101*, 1615-1617.

97. Paquette, L. A.; Gardlick, J. M. *J. Am. Chem. Soc.* **1980**, *102*, 5016-5025.

98. Paquette, L. A.; Gardlik, J. M.; Johnson, L. K.; McCullough, K. J. *J. Am. Chem. Soc.* **1980**, *102*, 5026-5032.

99. Paquette, L. A.; Hanzawa, Y.; Hefferon, G. J.; Blount, J. F. *J. Org. Chem.* **1982**, *47*, 265-272.

100. Meyer, A.; Schlögl, K.; Lerch, U.; Vogel, E. *Chem. Ber.* **1988**, *121*, 917-922.

101. Meyer, A.; Schlögl, K.; Essert, T.; Jörrens, F.; Klug, W.; Lex, J.; Schmickler, H.; Vogel, E. *Monatsh. Chem.* **1994**, *125*, 783-790.

102. Vogel, E.; Tückmantel, W.; Schlögl, K.; Widhalm, M.; Kraka, E.; Cremer, D. *Tetrahedron Lett.* **1984**, *25*, 4925-4928.

103. Roth, W. R.; Ruf, G.; Ford, P. W. *Chem. Ber.* **1974**, *107*, 48-52.

104. Roth, W. R.; Bastigkeit, T. *Liebigs Ann.* **1996**, 2171-2183.

105. Schurig, V.; Keller, F.; Reich, S.; Fluck, M. *Tetrahedron: Asymm.* **1997**, *8*, 3475-3480.

106. Trapp, V.; Schurig, V. *Chirality* **2002**, *14*, 465-470.
107. Rodriguez, O.; Morrison, H. *J. Chem. Soc. D, Chem. Commun.* **1971**, 679.
108. Stierman, T. J.; Johnson, R. P. *J. Am. Chem. Soc.* **1985**, *107*, 3971-3980.
109. Claesson, A.; Olsson, L.-I. *J. Chem. Soc., Chem. Commun.* **1979**, 524-525.
110. Michael, F. E.; Duncan, A. P.; Sweeney, Z. K.; Bergman, R. G. *J. Am. Chem. Soc.* **2003**, *125*, 7184-7185.
111. Michael, F. E.; Duncan, A. P.; Sweeney, Z. K.; Bergman, R. G. *J. Am. Chem. Soc.* **2005**, *127*, 1752-1764.
112. Trost, B. M.; Fandrick, D. R.; Dinh, D. C. *J. Am. Chem. Soc.* **2005**, *127*, 14186-14187.
113. Naruse, Y.; Watanabe, H.; Inagaki, S. *Tetrahedron: Asymm.* **1992**, *3*, 599-602.
114. Naruse, Y.; Watanabe, H.; Ishiyama, Y.; Yoshida, T. *J. Org. Chem.* **1997**, *62*, 3862-3866.
115. Kuhn, R.; Fischer, H.; Fischer, H. *Chem. Ber.* **1964**, *97*, 1760-1766.
116. Bertsch, K.; Jochims, J. C. *Tetrahedron Lett.* **1977**, *18*, 4379-4382.
117. Bertsch, K.; Rahman, M. A.; Jochims, J. C. *Chem. Ber.* **1979**, *112*, 567-576.
118. Lambrecht, J.; Gambke, B.; von Seyerl, J.; Huttner, G.; Kollmannsberger-von Nell, G.; Herzberger, S.; Jochims, J. C. *Chem. Ber.* **1981**, *114*, 3751-3771.
119. Jochims, J. C.; Lambrecht, J.; Burkert, U.; Huttner, G. *Tetrahedron Lett.* **1984**, *40*, 893-903.
120. Jochims, J. C.; Abu-El-Halawa, R.; Jibril, I.; Huttner, G. *Chem. Ber.* **1984**, *117*, 1900-1912.
121. Anet, F. A. L.; Jochims, J. C.; Bradley, C. H. *J. Am. Chem. Soc.* **1970**, *92*, 2557-2558.
122. Lambrecht, J.; Zsolnai, L.; Huttner, G.; Jochims, J. C. *Chem. Ber.* **1982**, *115*, 172-184.
123. Yamada, K.; Nakagawa, H.; Kawazura, H. *Bull. Chem. Soc. Jpn.* **1986**, *59*, 2429-2432.
124. Lebon, F.; Longhi, G.; Gangemi, F.; Abbate, S.; Priess, J.; Juza, M.; Bazzini, C.; Caronna, T.; Mele, A. *J. Phys. Chem A* **2004**, *108*, 11752-11761.
125. Goedicke, C.; Stegemeyer, H. *Tetrahedron Lett.* **1970**, *11*, 937-940.
126. Martin, R. H.; Marchant, M.-J. *Tetrahedron Lett.* **1972**, *13*, 3707-3708.
127. Martin, R. H.; Marchant, M.-J. *Tetrahedron* **1974**, *30*, 347-349.
128. Martin, R. H. *Angew. Chem., Int. Ed. Engl.* **1974**, *13*, 649-660.
129. Borkent, J. H.; Laarhoven, W. H. *Tetrahedron* **1978**, *34*, 2565-2567.
130. Martin, R. H.; Libert, V. *J. Chem. Research (S)* **1980**, 130-131.
131. Janke, R. H.; Haufe, G.; Würthwein, E.-U.; Borkent, J. H. *J. Am. Chem. Soc.* **1996**, *118*, 6031-6035.
132. Rajca, A.; Miyasaka, M.; Pink, M.; Wang, H.; Rajca, S. *J. Am. Chem. Soc.* **2004**, *126*, 15211-14222.
133. Newman, M. S.; Hussey, A. S. *J. Am. Chem. Soc.* **1947**, *69*, 3023-3027.
134. Hall, D. M.; Turner, E. E.; Hamlett, K. E. *J. Chem. Soc.* **1955**, 1242-1251.
135. Newman, M. S.; Wise, R. M. *J. Am. Chem. Soc.* **1956**, *78*, 450-454.
136. Hall, D. M. *J. Chem. Soc.* **1956**, 3674.
137. Newman, M. S.; Lednicer, D. *J. Am. Chem. Soc.* **1956**, *78*, 4765-4770.
138. Crawford, M.; Mackinnon, R. A. M.; Supanekar, V. R. *J. Chem. Soc.* **1959**, 2807-2812.
139. Mislow, K.; Graeve, R.; Gordon, A. J.; Wahl Jr., G. H. *J. Am. Chem. Soc.* **1963**, *85*, 1199-1200.
140. Newman, M. S.; Mentzer, R. G.; Slomp, G. *J. Am. Chem. Soc.* **1963**, *85*, 4018-4020.
141. Mislow, K.; Graeve, R.; Gordon, A. J.; Wahl Jr., G. H. *J. Am. Chem. Soc.* **1964**, *86*, 1733-1741.
142. Laarhoven, W. H.; Peters, W. H. M.; Tinnemans, A. H. A. *Tetrahedron* **1978**, *34*, 769-777.
143. Darnow, J. N.; Armstrong, R. N. *J. Am. Chem. Soc.* **1990**, *112*, 6725-6726.
144. Fritsch, R.; Hartmann, E.; Andert, D.; Mannschreck, A. *Chem. Ber.* **1992**, *125*, 849-855.
145. Gao, J. P.; Meng, X. S.; Bender, T. P.; MacKinnon, S.; Grand, V.; Wang, Z. Y. *Chem. Commun.* **1999**, 1281-1282.

146. Kiefl, C.; Zinner, H.; Cuyegkeng, M. A.; Eiglsperger, A. *Tetrahedron: Asymm.* **2000**, *11*, 3503-3513.

147. Mannschreck, A.; Gmahl, E.; Burgemeister, T.; Kastner, F.; Sinnwell, V. *Angew. Chem., Int. Ed. Engl.* **1988**, *27*, 270-271.

148. Scherübl, H.; Fritsche, U.; Mannschreck, A. *Chem. Ber.* **1984**, *117*, 336-343.

149. Armstrong, R. N.; Ammon, H. L.; Darnow, J. N. *J. Am. Chem. Soc.* **1987**, *109*, 2077-2082.

150. Matyjaszewski, K.; Paik, H.-J.; Shipp, D. A.; Isobe, Y.; Okamoto, Y. *Macomolecules* **2001**, *34*, 3127-3129.

151. Greuter, H.; Dingwall, J.; Martin, P.; Bellus, D. *Helv. Chim. Acta* **1981**, *64*, 2812-2820.

152. Koh, K.; Durst, T. *J. Org. Chem.* **1994**, *59*, 4683-4686.

153. O'Meara, J. A.; Jung, M.; Durst, T. *Tetrahedron Lett.* **1995**, *36*, 2559-2562.

154. Scott, J. M. W.; Summers, D. *Can. J. Chem.* **1998**, *76*, 643-648.

155. Ben, R. N.; Durst, T. *J. Org. Chem.* **1999**, *64*, 7700-7706.

156. Stein, A. R. *Can. J. Chem.* **1994**, *72*, 1789-1796.

157. Caddick, S.; Jenkins, K. *Tetrahedron Lett.* **1996**, *37*, 1301-1304.

158. Bellucci, G.; Marsili, A.; Mastrorilli, E.; Morelli, I.; Scartoni, V. *J. Chem. Soc., Perkin Trans. 2* **1974**, 201-204.

159. Cram, D. J.; Rickborn, B.; Kingsbury, C. A.; Haberfield, P. *J. Am. Chem. Soc.* **1961**, *83*, 3678-3687.

160. Wong, S. M.; Fischer, H. P.; Cram, D. J. *J. Am. Chem. Soc.* **1971**, *93*, 2235-2243.

161. Bergman, N.-Å.; Källson, I. *Acta Chem. Scand.* **1976**, *A 30*, 411-417.

162. Toda, F.; Tanaka, K. *Chem. Lett.* **1983**, 661-664.

163. Inagaki, M.; Hiratake, J.; Nishioka, T.; Oda, J. *J. Am. Chem. Soc.* **1991**, *113*, 9360-9361.

164. Paizs, C.; Tähtinen, P.; Lundell, K.; Poppe, L.; Irimie, F.-D.; Kanerva, L. T. *Tetrahedron: Asymm.* **2003**, *14*, 1895-1904.

165. Hassan, N. A.; Bayer, E.; Jochims, J. C. *J. Chem. Soc., Perkin Trans. 1* **1998**, 3747-3757.

166. Lopez-Serrano, P.; Jongejan, J. A.; van Rantwijk, F.; Sheldon, R. A. *Tetrahedron: Asymm.* **2001**, *12*, 219-228.

167. Boesten, W. H. J.; Seerden, J.-P. G.; de Lange, B.; Dielemans, H. J. A.; Elsenberg, H. L. M.; Kaptein, B.; Moody, H. M.; Kellogg, R. M.; Broxterman, Q. B. *Org. Lett.* **2001**, *3*, 1121-1124.

168. Sakurai, R.; Suzuki, S.; Hashimoto, J.; Baba, M.; Itoh, O.; Uchida, A.; Hattori, T.; Miyano, S.; Yamaura, M. *Org. Lett.* **2004**, *6*, 2241-2244.

169. Chaplin, J. A.; Levin, M. D.; Morgan, B.; Farid, N.; Li, J.; Zhu, Z.; McQuaid, J.; Nicholson, L. W.; Rand, C. A.; Burk, M. J. *Tetrahedron: Asymm.* **2004**, *15*, 2793-2796.

170. Sakurai, R.; Itoh, O.; Uchida, A.; Hattori, T.; Miyano, S.; Yamaura, M. *Tetrahedron* **2004**, *60*, 10553-10557.

171. Firestone, R. A.; Reinhold, D. F.; Gaines, W. A.; Chemerda, J. M.; Sletzinger, M. *J. Org. Chem.* **1968**, *33*, 1213-1218.

172. Kimura, T.; Yamamoto, N.; Suzuki, Y.; Kawano, K.; Norimine, Y.; Ito, K.; Nagato, S.; Iimura, Y.; Yonaga, M. *J. Org. Chem.* **2002**, *67*, 6228-6231.

173. Kornblum, N.; Wade, P. A. *J. Org. Chem.* **1987**, *52*, 5301-5305.

174. Danehey Jr., C. T.; Grady, G. L.; Bonneau, P. R.; Bushweller, C. H. *J. Am. Chem. Soc.* **1988**, *110*, 7269-7279.

175. Brown, J. H.; Bushweller, C. H. *J. Phys. Chem.* **1994**, *98*, 11411-11419.

176. Brown, J. H.; Bushweller, C. H. *J. Org. Chem.* **2001**, *66*, 903-909.

177. Lunazzi, L.; Macciantelli, D.; Grossi, L. *Tetrahedron* **1983**, *39*, 305-308.

178. Bushweller, C. H.; Wang, C. Y.; Reny, J.; Lorandos, M. Z. *J. Am. Chem. Soc.* **1977**, *99*, 3938-3941.

179. Wash, P. L.; Renslo, A. R.; Rebek Jr., J. *Angew. Chem., Int. Ed.* **2001**, *40*, 1221-1222.
180. Griffith, D. L.; Olson, B. L.; Roberts, J. D. *J. Am. Chem. Soc.* **1971**, *93*, 1648-1649.
181. Andose, J. D.; Lehn, J. M.; Mislow, K.; Wagner, J. *J. Am. Chem. Soc.* **1970**, *92*, 4050-4056.
182. Brois, S. J. *J. Am. Chem. Soc.* **1968**, *90*, 506-508.
183. Brois, S. J. *J. Am. Chem. Soc.* **1968**, *90*, 508-509.
184. Lehn, J. M.; Wagner, J. *Chem. Commun. (London)* **1968**, 148-150.
185. Felix, D.; Eschenmoser, A. *Angew. Chem., Int. Ed. Engl.* **1968**, *7*, 224-225.
186. Brois, S. J. *J. Am. Chem. Soc.* **1970**, *92*, 1079-1080.
187. Kostyanovsky, R. G.; Rudchenko, V. F.; Dyachenko, O. A.; Chervin, I. I.; Zolotoi, A. B.; Atovmyan, L. O. *Tetrahedron* **1979**, *35*, 213-224.
188. Kostyanovsky, R. G.; Shustov, G. V.; Zaichenko, N. L. *Tetrahedron* **1982**, *38*, 949-960.
189. Rudchenko, V. F.; Dyachenko, O. A.; Zolotoi, A. B.; Atovmyan, L. O.; Chervin, I. I.; Kostyanovsky, R. G. *Tetrahedron* **1982**, *38*, 961-975.
190. Shustov, G. V.; Kadorkina, G. K.; Kostyanovsky, R. G.; Rauk, A. *J. Am. Chem. Soc.* **1988**, *110*, 1719-1724.
191. Shustov, G. V.; Varlamov, S. V.; Chervin, I. I.; Aliev, A. E.; Kostyanovsky, R. G.; Kim, R. G.; Rauk, A. *J. Am. Chem. Soc.* **1989**, *111*, 4210-4215.
192. Forni, A.; Moretti, I.; Prosyanik, A. V.; Torre, G. *J. Chem. Soc., Chem. Commun.* **1981**, 588-590.
193. Bürkle, W.; Karfunkel, H.; Schurig, V. *J. Chromatogr.* **1984**, *288*, 1-14.
194. Forni, A.; Moretti, I.; Torre, G.; Brückner, S.; Malpezzi, L.; Di Silvestro, G. *J. Chem. Soc., Perkin Trans. 2* **1984**, 791-797.
195. Spada, G. P.; Tampieri, A.; Gottarelli, G.; Moretti, I.; Torre, G. *J. Chem. Soc., Perkin Trans. 2* **1989**, 513-516.
196. Schurig, V.; Leyrer, U. *Tetrahedron: Asymm.* **1990**, *1*, 865-868.
197. Jung, M. Schurig, V. *J. Am. Chem. Soc.* **1992**, *114*, 529-534.
198. Reich, S.; Trapp, O.; Schurig, V. *J. Chromatogr. A* **2000**, *892*, 487-498.
199. Trapp, O.; Schurig, V. *Chirality* **2002**, *14*, 465-470.
200. Bjorgo, J.; Boyd, D. R.; Campbell, R. M.; Thompson, N. J.; Jennings, W. B. *J. Chem. Soc., Perkin Trans. 2* **1976**, 606-609.
201. Kost, D.; Raban, M. *J. Am. Chem. Soc.* **1976**, *98*, 8333-8338.
202. Hall, J. H.; Bigard, W. S. *J. Org. Chem.* **1978**, *43*, 2785-2788.
203. Kostyanovsky, R. G.; Kadorkina, G. K.; Kostyanovsky, V. R.; Schurig, V.; Trapp, O. *Angew. Chem., Int. Ed.* **2000**, *39*, 2938-2940.
204. Trapp, O.; Schurig, V.; Kostyanovsky, R. G. *Chem. Eur. J.* **2004**, *10*, 951-957.
205. Prelog, V.; Wieland, P. *Helv. Chim. Acta.* **1944**, *27*, 1127-1134.
206. Greenberg, A.; Molinaro, N.; Lang, M. *J. Org. Chem.* **1984**, *49*, 1127-1130.
207. Trapp, O.; Schurig, V. *J. Am. Chem. Soc.* **2000**, *122*, 1424-1430.
208. Wilen, S. H.; Qi, J. Z. *J. Org. Chem.* **1991**, *56*, 485-487.
209. Bowman, E. R.; McKennis Jr., H.; Martin, B. R. *Synth. Commun.* **1982**, *12*, 871-879.
210. Reider, P. J.; Davis, P.; Hughes, D. L.; Grabowski, E. J. J. *J. Org. Chem.* **1987**, *52*, 957-958.
211. Reetz, M. T.; Schimossek, K. *Chimia* **1996**, *50*, 668-669.
212. Pàmies, O.; Éll, A. H.; Samec, J. S. M.; Hermanns, N.; Bäckvall, J.-E. *Tetrahedron Lett.* **2002**, *43*, 4699-4702.
213. Dorta, R.; Broggini, D.; Kissner, R.; Togni, A. *Chem. Eur. J.* **2004**, *10*, 4546-4555.
214. Parvulescu, A.; de Vos, D.; Jacobs, P. *Chem. Commun.* **2005**, 5307-5309.
215. Paetzold, J.; Bäckvall, J.-E. *J. Am. Chem. Soc.* **2005**, *127*, 17620-17621.
216. Murahashi, S.-I.; Yoshimura, N.; Tsumiyama, T.; Kojima, T. *J. Am. Chem. Soc.* **1983**, *105*, 5002-5011.

217. Alexeeva, M.; Enright, A.; Dawson, M. J.; Mahmoudian, N. J.; Turner, N. J. *Angew. Chem., Int. Ed.* **2002**, *41*, 3177-3180.

218. Carr, R.; Alexeeva, M.; Dawson, M. J.; Gotor-Fernandez, V.; Humphrey, C. E.; Turner, N. J. *ChemBioChem* **2005**, *6*, 637-639.

219. Hayes, K. S.; Nagumo, M.; Blount, J. F.; Mislow, K. *J. Am. Chem. Soc.* **1980**, *102*, 2773-2776.

220. Okamoto, Y.; Yashima, E.; Hatada, K.; Mislow, K. *J. Org. Chem.* **1984**, *49*, 557-558.

221. Mey, B.; Paulus, H.; Lamparter, E.; Blaschke, G. *Chirality* **1998**, *10*, 307-315.

222. Mills, A. K.; Smith, A. E. W. *Helv. Chim. Acta* **1960**, *43*, 1915-1927.

223. Read, J.; Robertson, G. J.; Cook, A. M. R. *J. Chem. Soc.* **1927**, 1276-1284.

224. Rodriguez, J. B. *Tetrahedron* **1999**, *55*, 2157-2170.

225. Yoshikawa, N.; Yamada, Y. M. A.; Das, J.; Sasai, H.; Shibasaki, M. *J. Am. Chem. Soc.* **1999**, *121*, 4168-4178.

226. Myers, A. G.; Kung, D. W.; Zhong, B.; Movassaghi, M.; Kwon, S. *J. Am. Chem. Soc.* **1999**, *121*, 8401-8402.

227. Kelly, C. L.; Lawrie, K. W. M.; Morgan, P.; Willis, C. L. *Tetrahedron Lett.* **2000**, *41*, 8001-8005.

228. Risgaard, T.; Gothelf, K. V.; Jørgensen, K. A. *Org. Biomol. Chem.* **2003**, *1*, 153-156.

229. Rein, T.; Kreuder, R.; von Zezschwitz, P.; Wulff, C.; Reiser, O. *Angew. Chem., Int. Ed. Engl.* **1995**, *34*, 1023-1025.

230. Murata, K.; Okano, K.; Miyagi, M.; Iwane, H.; Noyori, R.; Ikariya, T. *Org. Lett.* **1999**, *1*, 1119-1121.

231. Roy, O.; Diekmann, M.; Riahi, A.; Henin, F.; Muzart, J. *Chem. Commun.* **2001**, 533-534.

232. Jurkauskas, V.; Buchwald, S. L. *J. Am. Chem. Soc.* **2002**, *124*, 2892-2893.

233. Urbaneja, L. M.; Alexakis, A.; Krause, N. *Tetrahedron Lett.* **2002**, *43*, 7887-7890.

234. Heckrodt, T. J.; Mulzer, J. *J. Am. Chem. Soc.* **2003**, *125*, 4680-4681.

235. Nakamura, S.; Nakayama, J.-I.; Toru, T. *J. Org. Chem.* **2003**, *68*, 5766-5768.

236. Ward, D. E.; Jheengut, V.; Akinnusi, O. T. *Org. Lett.* **2005**, *7*, 1181-1184.

237. Ruijter, E.; Schültingkemper, H.; Wessjohann, L. A. *J. Org. Chem.* **2005**, *70*, 2820-2823.

238. Ödman, P.; Wessjohann, L. A.; Bornscheuer, U. T. *J. Org. Chem.* **2005**, *70*, 9551-9555.

239. Uncuta, C.; Caraman, G. B.; Tanase, C. I.; Bartha, E.; Kravtsov, C. C.; Simonov, Y. A.; Lipkowski, J.; Vanthuyne, N.; Roussel, C. *Chirality* **2005**, *17*, 63-72.

240. Soffer, M. D.; Günay, G. E. *Tetrahedron Lett.* **1965**, *6*, 1355-1358.

241. Noyori, R.; Ikeda, T.; Ohkuma, T.; Widhalm, M.; Kitamura, M.; Takaya, H.; Akutagawa, S.; Sayo, N.; Saito, T.; Taketomi, T.; Kumobayashi, H. *J. Am. Chem. Soc.* **1989**, *111*, 9134-9135.

242. Kitamura, M. Tokunaga, M.; Noyori, R. *J. Am. Chem. Soc.* **1993**, *115*, 144-152.

243. Noyori, R.; Ohkuma, T. *Angew. Chem., Int. Ed.* **2001**, *40*, 40-73.

244. Eustache, F.; Dalko, P. I.; Cossy, J. *Org. Lett.* **2002**, *4*, 1263-1265.

245. Labeeuw, O.; Phansavath, P.; Genêt, J.-P. *Tetrahedron: Asymm.* **2004**, *15*, 1899-1908.

246. Mordant, C.; Dünkelmann, P.; Ratovelomanana-Vidal, V.; Genêt, J.-P. *Eur. J. Org. Chem.* **2004**, 3017-3026.

247. Xu, K.; Lalic, G.; Sheehan, S. M.; Shair, M. D. *Angew. Chem., Int. Ed.* **2005**, *44*, 2259-2261.

248. Makino, K.; Hiroki, Y.; Hamada, Y. *J. Am. Chem. Soc.* **2005**, *127*, 5784-5785.

249. Ji, A.; Wolberg, M; Hummel, W.; Wandrey, C.; Müller, M. *Chem. Commun.* **2001**, 57-58.

250. Danchet, S.; Bigot, C.; Buisson, D. Azerad, R. *Tetrahedron: Asymm.* **1997**, *8*, 1735-1739.

251. Dehli, J. R.; Gotor, V. *J. Org. Chem.* **2002**, *67*, 6816-6819.

252. Quiros, M.; Rebolledo, F.; Gotor, V. *Tetrahedron: Asymm.* **1999**, *10*, 473-486.

253. Polyak, F.; Lubell, W. D. *J. Org. Chem.* **1998**, *63*, 5937-5949.

254. Kosmrlj, J.; Weigel, L. O.; Evans, D. A.; Downey, C. W.; Wu, J. *J. Am. Chem. Soc.* **2003**, *125*, 3208-3209.

255. Enders, D.; Maaßen, R.; Runsink, J. *Tetrahedron: Asymm.* **1998**, *9*, 2155-2180.

256. Garner, C. M.; Mossman, B. C.; Prince, M. E. *Tetrahedron Lett.* **1993**, *34*, 4263-4276.

257. Amat, M.; Canto, M.; Llor, N.; Escolano, C.; Molins, E.; Espinosa, E.; Bosch, J. *J. Org. Chem.* **2002**, *67*, 5343-5351.

258. Amat, M.; Canto, M.; Llor, N.; Ponzo, V.; Perez, M.; Bosch, J. *Angew. Chem., Int. Ed.* **2002**, *41*, 335-338.

259. Amat, M.; Bassas, O.; Pericas, M. A.; Pasto, M.; Bosch, J. *Chem. Commun.* **2005**, 1327-1329.

260. Troiani, A.; Gasparrini, F.; Grandinetti, F.; Speranza, M. *J. Am. Chem. Soc.* **1997**, *119*, 4525-4534.

261. Kisbye, J.; Madsen, N. B. *Arch. Pharm. Chem. Sci. Ed.* **1977**, *5*, 97-104.

262. Barlow, J. J.; Block, M. H.; Hudson, J. A.; Leach, A.; Longridge, J. L.; Main, B. G.; Nicholson, S. *J. Org. Chem.* **1992**, *57*, 5158-5162.

263. Koh, J. H.; Jeong, H. M.; Park, J. *Tetrahedron Lett.* **1998**, *39*, 5545-5548.

264. Persson, B. A.; Larsson, A. L. E.; Le Ray, M.; Bäckvall, J.-E. *J. Am. Chem. Soc.* **1999**, *121*, 1645-1650.

265. Koh, J. H.; Jung, H. M.; Kim, M.-J.; Park, J. *Tetrahedron Lett.* **1999**, *40*, 6281-6284.

266. Lee, D.; Huh, E. A.; Kim, M.-J.; Jung, H. M.; Koh, J. H.; Park, J. *Org. Lett.* **2000**, *2*, 2377-2379.

267. Huerta, F. F.; Bäckvall, J.-E. *Org. Lett.* **2001**, *3*, 1209-1212.

268. Kim, M.-J.; Choi, Y. K.; Choi, M. Y.; Kim, M. J.; Park, J. *J. Org. Chem.* **2001**, *66*, 4736-4738.

269. Pàmies, O.; Bäckvall, J.-E. *Chem. Eur. J.* **2001**, *7*, 5052-5058.

270. Dijksman, A.; Elzinga, J. M.; Li, Y.-X.; Arends, I. W. C. E.; Sheldon, R. A. *Tetrahedron: Asymm.* **2002**, *13*, 879-884.

271. Pàmies, O.; Bäckvall, J.-E. *J. Org. Chem.* **2002**, *67*, 1261-1265.

272. Choi, J. H.; Kim, Y. H.; Nam, S. H.; Shin, S. T.; Kim, M.-J.; Park, J. *Angew. Chem., Int. Ed.* **2002**, *41*, 2373-2376.

273. Pàmies, O.; Bäckvall, J.-E. *J. Org. Chem.* **2002**, *67*, 9006-9010.

274. Ito, M.; Osaku, A.; Kitahara, S.; Hirakawa, M.; Ikariya, T. *Tetrahedron Lett.* **2003**, *44*, 7521-7523.

275. Martin-Matute, B. Bäckvall, J.-E. *J. Org. Chem.* **2004**, *69*, 9191-9195.

276. Riermeier, T. H.; Gross, P.; Monsees, A.; Hoff, M.; Trauthwein, H. *Tetrahedron Lett.* **2005**, *46*, 3403-3406.

277. Yamaguchi, K.; Koike, T.; Kotani, M.; Matsushita, M.; Shinachi, S.; Mizuno, N. *Chem. Eur. J.* **2005**, *11*, 6574-6582.

278. Saura-Llamas, I.; Gladysz, J. A. *J. Am. Chem. Soc.* **1992**, *114*, 2136-2144.

279. Dinh, P. M.; Howarth, J. A.; Hudnott, A. R.; Williams, J. M. J. *Tetrahedron Lett.* **1996**, *37*, 7623-7626.

280. Klomp, D.; Djanashvili, K.; Cianfanelli Svennum, N.; Chantapariyavat, N.; Wong, C.-S.; Vilela, F.; Maschmeyer, T.; Peters, J. A.; Hanefeld, U. *Org. Biomol. Chem.* **2005**, *3*, 483-489.

281. Reyes, E.; Córdova, A. *Tetrahedron Lett.* **2005**, *46*, 6605-6609.

282. Hussain, I.; Komasaka, T.; Ohga, M.; Nokami, J. *Synlett* **2002**, 640-642.

283. Wang, Z.; Shao, H.; Lacroix, E.; Wu, S.-H.; Jennings, H. J.; Zou, W. *J. Org. Chem.* **2003**, *68*, 8097-8105.

284. Shao, H.; Wang, Z.; Lacroix, E.; Wu, S.-H.; Jennings, H. J.; Zou, W. *J. Am. Chem. Soc.* **2002**, *124*, 2130-2131.

285. Griffin, G. W. *Angew. Chem., Int. Ed. Engl.* **1971**, *10*, 537-547.

286. Huisgen, R. *Angew. Chem., Int. Ed. Engl.* **1977**, *16*, 572-585.

287. Rau, H. *Chem. Rev.* **1983**, *83*, 535-547.

288. Manring, L. E.; Peters, K. S. *J. Am. Chem. Soc.* **1984**, *106*, 8077-8079.

289. Inoue, Y.; Yamasaki, N.; Shimoyama, H.; Tai, A. *J. Org. Chem.* **1993**, *58*, 1785-1793.

290. Lo, C.-Y.; Pal, S.; Odedra, A.; Liu, R.-S. *Tetrahedron Lett.* **2003**, *44*, 3143-3146.

291. Schaus, S. E.; Jacobsen, E. N. *Tetrahedron Lett.* **1996**, *37*, 7937-7940.

292. Spelberg, J. H. L.; Tang, L.; Kellog, R. M.; Janssen, D. B. *Tetrahedron: Asymm.* **2004**, *15*, 1095-1102.

293. Harié, G.; Samat, A.; Gugliemetti, R.; van Parys, I.; Saeyens, W.; de Keukeleire, D.; Lorenz, K.; Mannschreck, A. *Helv. Chim. Acta* **1997**, *80*, 1122-1132.

294. Shieh, W.-C.; Carlson, J. A. *J. Org. Chem.* **1994**, *59*, 5463-5465.

295. Klärner, F.-G.; Schröer, D. *Angew. Chem., Int. Ed. Engl.* **1987**, *26*, 1294-1295.

296. Schurig, V.; Jung, M.; Schleimer, M.; Klärner, F.-G. *Chem. Ber.* **1992**, *125*, 1301-1303.

297. Klärner, F.-G.; Kleine, A. E.; Oebels, D.; Scheidt, F. *Tetrahedron: Asymm.* **1993**, *4*, 479-490.

298. Brand, S.; Jones, M. F.; Rayner, C. M. *Tetrahedron Lett.* **1995**, *36*, 8493-8496.

299. Yamada, S.; Noguchi, E. *Tetrahedron Lett.* **2001**, *42*, 3621-3624.

300. Sharfuddin, M.; Narumi, A.; Iwai, Y.; Miyazawa, K.; Yamada, S.; Kakuchi, T.; Kaga, H. *Tetrahedron: Asymm.* **2003**, *14*, 1581-1585.

301. Amat, M.; Escolano, C.; Lozano, O.; Llor, N.; Bosch, J. *Org. Lett.* **2003**, *5*, 3139-3142.

302. Harada, M.; Nakai, T.; Tomooka, K. *Synlett* **2004**, 365-367.

303. Li, L.; Thompson, R.; Sowa Jr.; J. R.; Clausen, A.; Dowling, T. *J. Chromatogr. A* **2004**, *1043*, 171-175.

304. Szatmari, I.; Martinek, T. A.; Lazar, L.; Koch, A.; Kleinpeter, E.; Neuvonen, K.; Fülöp, F. *J. Org. Chem.* **2004**, *69*, 3645-3653.

305. Durrwachter, J. R.; Wong, C.-H. *J. Org. Chem.* **1988**, *53*, 4175-4181.

306. Miethchen, R.; Sowa, C.; Frank, M.; Michalik, M.; Reinke, H. *Carbohydr. Res.* **2002**, *337*, 1-9.

307. Ryu, K.-S.; Kim, C.; Park, C.; Choi, B.-S. *J. Am. Chem. Soc.* **2004**, *126*, 9180-9181.

308. Van der Deen, H.; Cuiper, A. D.; Hof, R. P.; van Oeveren, A.; Feringa, B. L.; Kellogg, R. M. *J. Am. Chem. Soc.* **1996**, *118*, 3801-3803.

309. Thuring, J. W. J. F.; Klunder, A. J. H.; Nefkens, G. H. L.; Wegman, M. A.; Zwanenburg, B. *Tetrahedron Lett.* **1996**, *37*, 4759-4760.

310. Okada, Y.; Takebayashi, T.; Hashimoto, M.; Kasuga, S.; Sata, S.; Tamura, C. *J. Chem. Soc., Chem. Commun.* **1983**, 784-785.

311. Eggers, L.; Buss, V. *Angew. Chem., Int. Ed. Engl.* **1997**, *36*, 881-883.

312. Hobley, J.; Malatesta, V.; Millini, R.; Montanari, L.; Parker Jr., W. O. N. *Phys. Chem. Chem. Phys.* **1999**, *1*, 3259-3267.

313. Loncar-Tomaskovic, L.; Lorenz, K.; Hergold-Brundic, A.; Mrvos-Sermek, D.; Nagl, A.; Mintas, M.; Mannschreck, A. *Chirality* **1999**, *11*, 363-372.

314. Sakata, T.; Yan, Y.; Marriott, G. *J. Org. Chem.* **2005**, *70*, 2009-2013.

315. Stephan, B.; Zinner, H.; Kastner, F.; Mannschreck, A. *Chimia* **1990**, *44*, 336-338.

316. Kießwetter, R.; Pustet, N.; Brandl, F.; Mannschreck, A. *Tetrahedron: Asymm.* **1999**, *10*, 4677-4687.

317. Trapp, O.; Schurig, V. *Chem. Eur. J.* **2001**, *7*, 1495-1502.

318. Yamagishi, M.; Yamada, Y.; Ozaki, K.-I.; Da-Te, T.; Okamura, K.; Suzuki, M.; Matsumoto, K. *J. Org. Chem.* **1992**, *57*, 1568-1571.

319. Ashley, W. C.; Shriner, R. L. *J. Am. Chem. Soc.* **1932**, *54*, 4410-4414.

320. Bonner, W.; Hurd, C. D. *J. Am. Chem. Soc.* **1951**, *73*, 4290-4294.

321. Brown, E.; Daugan, A. *Tetrahedron Lett.* **1985**, *26*, 3997-3998.

322. Wirz, B.; Spur, P. *Tetrahedron: Asymm.* **1995**, *6*, 669-670.
323. Chung, J. Y. L.; Hughes, D. L.; Zhao, D.; Song, Z.; Mathre, D. J.; Ho, G.-J.; McNamara, J. M.; Douglas, A. W.; Reamer, R. A.; Tsay, F.-R.; Varsolona, R.; McCauley, J.; Grabowski, E. J. J.; Reider, P. J. *J. Org. Chem.* **1996**, *61*, 215-222.
324. Henderson, J. W. *Chem. Soc. Rev.* **1973**, *2*, 397-413.
325. Guthrie, R. D.; Nicolas, E. C. *J. Am. Chem. Soc.* **1981**, *103*, 4637-4638.
326. Hargreaves, M. K.; Modarai, B. *J. Chem. Soc. C* **1971**, 1013-1015.
327. Wilen, S. H.; Bunding, K. A.; Kascheres, C. M.; Wieder, M. J. *J. Am. Chem. Soc.* **1985**, *107*, 6997-6998.
328. Doyle, T. R.; Vogl, O. *J. Am. Chem. Soc.* **1989**, *111*, 8510-8511.
329. Kiau, S.; Discordia, R. P.; Madding, G.; Okuniewicz, F. J.; Rosso, V.; Venit, J. V. *J. Org. Chem.* **2004**, *69*, 4256-4261.
330. Xu, H.-W.; Wang, Q.-W.; Zhu, J.; Deng, J.-G.; Cun, L.-F.; Cui, X.; Wu, J.; Xu, X.-L.; Wu, Y.-L. *Org. Biomol. Chem.* **2005**, *3*, 4227-4232.
331. Lombart, H.-G.; Lubell, W. D. *J. Org. Chem.* **1996**, *61*, 9437-9446.
332. Camps, P.; Pérez, F.; Soldevilla, N. *Tetrahedron: Asymm.* **1997**, *8*, 1877-1894.
333. O'Meara, J. A.; Gardee, N.; Jung, M.; Ben, R. N.; Durst, T. *J. Org. Chem.* **1998**, *63*, 3117-3119.
334. Maison, W.; Kosten, M.; Charpy, A.; Kintscher-Langenhagen, J.; Schlemminger, I.; Lützen, A.; Westerhoff, O.; Martens, J. *Eur. J. Org. Chem.* **1999**, 2433-2441.
335. Lee, S.-K.; Nam, J.; Park, Y. S. *Synlett* **2002**, 790-792.
336. Nam, J.; Lee, S.-K.; Kim, K. Y.; Park, Y. S. *Tetrahedron Lett.* **2002**, *43*, 8253-8255.
337. Tanyeli, C.; Özçubukçu, S. *Tetrahedron: Asymm.* **2003**, *14*, 1167-1170.
338. Nam, J.; Lee, S.-K.; Park, Y. S. *Tetrahedron* **2003**, *59*, 2397-2401.
339. Ammazzalorso, A.; Amoroso, R.; Bettoni, G.; De Filippis, B.; Giampietro, L.; Maccallini, C.; Tricca, M. L. *ARKIVOC* **2004**, *V*, 375-381.
340. Kim, H. J.; Shin, E.-K.; Chang, J.-Y.; Kim, Y.; Park, Y. S. *Tetrahedron Lett.* **2005**, *46*, 4115-4117.
341. Clericuzio, M.; Degani, I.; Dughera, S.; Fochi, R. *Tetrahedron: Asymm.* **2003**, *14*, 119-125.
342. Kato, D.-I.; Miyamoto, K.; Ohta, H. *Tetrahedron: Asymm.* **2004**, *15*, 2965-2973.
343. Van Maanen, H. L.; Jastrzebski, J. T. B. H.; Verweij, J.; Kieboom, A. P. G.; Spek, A. L.; van Koten, G. *Tetrahedron: Asymm.* **1993**, *4*, 1441-1444.
344. Kubota, H.; Kubo, A.; Takahashi, M.; Shimizu, R.; Da-Te, T.; Okamura, K.; Nunami, K.-I. *J. Org. Chem.* **1995**, *60*, 6776-6784.
345. Hanessian, S.; Andreotti, D.; Gomtsyan, A. *J. Am. Chem. Soc.* **1995**, *117*, 10393-10394.
346. Priego, J.; Flores, P.; Ortiz-Nava, C.; Escalante, J. *Tetrahedron: Asymm.* **2004**, *15*, 3545-3549.
347. Mitchell, D.; Hay, L. A.; Koenig, T. M.; McDaniel, S.; Nissen, J. S.; Audia, J. E. *Tetrahedron: Asymm.* **2005**, *16*, 3814-3819.
348. McNeil, A. J.; Collum, D. B. *J. Am. Chem. Soc.* **2005**, *127*, 5655-5661.
349. Ohtani, M.; Matsuura, T.; Watanabe, F.; Narisada, M. *J. Org. Chem.* **1991**, *56*, 2122-2127.
350. De Vos, M.-J.; Krief, A. *J. Am. Chem. Soc.* **1982**, *104*, 4282-4283.
351. Crout, D. H. G.; Hedgecock, C. J. R. *J. Chem. Soc., Perkin Trans 1* **1979**, 1982-1989.
352. Granberg, K. L.; Bäckvall, J.-E. *J. Am. Chem. Soc.* **1992**, *114*, 6858-6863.
353. Lüssem, B. J.; Gais, H.-J. *J. Org. Chem.* **2004**, *69*, 4041-4052.
354. Trost, B. M.; Toste, F. D. *J. Am. Chem. Soc.* **1999**, *121*, 3543-3544.
355. Pracejus, G. *Liebigs Ann. Chem.* **1957**, *622*, 10-22.
356. Camps, P.; Gimenez, S. *Tetrahedron: Asymm.* **1996**, *7*, 1227-1234.
357. Suzukamo, G.; Fukao, M.; Nagase, T. *Chem. Lett.* **1984**, 1799-1802.
358. Baum, R.; Smith, G. G. *J. Am. Chem. Soc.* **1986**, *108*, 7325-7327.

359. Shiraiwa, T.; Furukawa, T.; Tsuchida, T.; Sakata, S.; Sunami, M.; Kurokawa, H. *Bull. Chem. Soc. Jpn.* **1991**, *64*, 3729-3731.

360. Smith, G. G.; Sivakua, T. *J. Org. Chem.* **1983**, *48*, 627-634.

361. Van Maanen, H. L.; Kleijn, H.; Jastrzebski, J. T. B. H.; Verweji, J.; Kieboom, A. P. G.; van Koten, G. *J. Org. Chem.* **1995**, *60*, 4331-4338.

362. Smith, G. G.; Khatib, A.; Reddy; G. S. *J. Am. Chem. Soc.* **1983**, *105*, 293-295.

363. Shiraiwa, T.; Sakata, S.; Natsuyama, H.; Fujishima, K.; Miyazaki, H.; Kubo, S.; Nitta, T.; Kurokawa, H. *Bull. Chem. Soc. Jpn.* **1992**, *65*, 965-970.

364. Schön, I.; Szirtes, T.; Rill, A.; Balogh, G.; Vadasz, Z.; Seprödi, J.; Teplan, I.; Chino, N.; Kumogaye, K. Y.; Sakakibara, S. *J. Chem. Soc., Perkin Trans. 1* **1991**, 3213-3223.

365. Tedeschi, T.; Corradini, R.; Marchelli, R.; Pushl, A.; Nielsen, P. E. *Tetrahedron: Asymm.* **2002**, *13*, 1629-1636.

366. Hongo, C.; Yamada, S.; Chibata, I. *Bull. Chem. Soc. Jpn.* **1981**, *54*, 3286-3290.

367. Calmes, M.; Glot, C.; Michel, T.; Rolland, M.; Martinez, J. *Tetrahedron: Asymm.* **2000**, *11*, 737-741.

368. Camps, P.; Perez, F.; Soldevilla, N.; Borrego, M. A. *Tetrahedron: Asymm.* **1999**, *10*, 493-509.

369. Hang, J.; Li, H.; Deng, L. *Org. Lett.* **2002**, *4*, 3321-3324.

370. Fryzuk, M. D.; Bosnich, B. *J. Am. Chem. Soc.* **1977**, *99*, 6262-6267.

371. Liang, J.; Ruble, J. C.; Fu, G. C. *J. Org. Chem.* **1998**, *63*, 3154-3155.

372. Berkessel, A.; Cleemann, F.; Mukherjee, S.; Müller, T. N.; Lex, J. *Angew. Chem., Int. Ed.* **2005**, *44*, 807-811.

373. Berkessel, A.; Mukherjee, S.; Cleemann, F.; Müller, T. N.; Lex, J. *Chem. Commun.* **2005**, 1898-1900.

374. Clark, J. C.; Phillipps, G. H.; Steer, M. R. *J. Chem. Soc., Perkin Trans. 1* **1976**, 475-81.

375. Yamada, S.; Hongo, C.; Yoshioka, R.; Chibata, I. *J. Org. Chem.* **1983**, *48*, 843-846.

376. Hongo, C.; Yoshioka, R.; Tohyama, M.; Yamada, S.; Chibata, I. *Bull. Chem. Soc. Jpn.* **1983**, *56*, 3744-3747.

377. Hongo, C.; Tohyama, M.; Yoshioka, R.; Yamada, S.; Chibata, I. *Bull. Chem. Soc. Jpn.* **1985**, *58*, 433-436.

378. Tabushi, I.; Kuroda, Y.; Yamada, M. *Tetrahedron Lett.* **1987**, *28*, 5695-5698.

379. Yoshioka, R.; Tohyama, M.; Ohtsuki, O.; Yamada, S.; Chibata, I. *Bull. Chem. Soc. Jpn.* **1987**, *60*, 4321-4323.

380. Shiraiwa, T.; Sjinjo, K.; Kurokawa, H.; Chibata, I. *Bull. Chem. Soc. Jpn.* **1991**, *64*, 3251-3255.

381. Solladie-Cavallo, A.; Sedy, O.; Salisova, M.; Schmitt, M. *Eur. J. Org. Chem.* **2002**, 3042-3049.

382. Liljeblad, A.; Kiviniemi, A.; Kanerva, L. T. *Tetrahedron* **2004**, *60*, 671-677.

383. Walsh, C. T. *J. Biol. Chem.* **1989**, *264*, 2393-2396.

384. Sawada, S.; Tanaka, Y.; Hayashi, S.; Ryu, M.; Hasegawa, T.; Yukio, Y.; Esaki, N.; Soda, K.; Takahashi, S. *Biosci. Biotechnol. Biochem.* **1994**, *58*, 807-811.

385. Shaw, J. P.; Petsko, G. P.; Ringe, D. *Biochemistry* **1997**, *36*, 1329-1342.

386. Schnell, B.; Faber, K.; Kroutil, W. *Adv. Synth. Catal.* **2003**, *345*, 653-666.

387. Kato, D.-I.; Miyamoto, K.; Ohta, H. *Tetrahedron: Asymm.* **2004**, *15*, 2965-2973.

388. Inai, Y.; Kurokawa, Y.; Ida, A.; Hirabayashi, T. *Bull. Chem. Soc. Jpn.* **1999**, *72*, 55-61.

389. Ramesh, K.; Balaram, P. *Bioorg. Med. Chem.* **1999**, *7*, 105-117.

390. Heinonen, P.; Virta, P.; Lonnberg, H. *Tetrahedron* **1999**, *55*, 7613-7624.

391. Abele, S.; Seebach, D. *Eur. J. Org. Chem.* **2000**, *1*, 1-15.

392. Inai, Y.; Tagawa, K.; Takasu, A.; Hirabayashi, T.; Oshikawa, T.; Yamashita, M. *J. Am. Chem. Soc.* **2000**, *122*, 11731-11732.

393. Inai, Y.; Ousaka, N.; Okabe, T. *J. Am. Chem. Soc.* **2003**, *125*, 8151-8162.

394. Klebe, J. F.; Finkbeiner, H. *J. Am. Chem. Soc.* **1968**, *90*, 7255-7261.

395. Lambert, J. B.; Urdaneta,-Perez, M. *J. Am. Chem. Soc.* **1978**, *100*, 157-162.

396. Omote, M.; Tokita, T.; Shimizu, Y.; Imae, I.; Shirakawa, E.; Kawakami, Y. *J. Organomet. Chem.* **2000**, *611*, 20-25.

397. Flock, M.; Marschner, C. *Chem. Eur. J.* **2002**, *8*, 1024-1030.

398. Suzuki, K.; Kawakami, Y.; Velmurugan, D.; Yamane, T. *J. Org. Chem.* **2004**, *69*, 5383-5389.

399. Oestreich, M.; Auer, G.; Keller, M. *Eur. J. Org. Chem.* **2005**, 184-195.

400. Girshberg, O.; Kalikhman, I.; Stalke, D.; Walfort, B.; Kost, D. *J. Mol. Struct.* **2003**, *661-662*, 259-264.

401. Tacke, R.; Bertermann, R.; Burschka, C.; Dragota, S.; Penka, M.; Richter, I. *J. Am. Chem. Soc.* **2004**, *126*, 14493-14505.

402. Kalikhman, I.; Girshberg, O.; Lameyer, L.; Stalke, D.; Kost, D. *Organometallics* **2000**, *19*, 1927-1934.

403. Baechler, R. D.; Mislow, K. *J. Am. Chem. Soc.* **1970**, *92*, 3090-3093.

404. Baechler, R. D.; Mislow, K. *J. Am. Chem. Soc.* **1971**, *93*, 773-774.

405. Scartazzini, R.; Mislow, K. *Tetrahedron Lett.* **1967**, *8*, 2719-2722.

406. Darwish, D.; Hui, S. H.; Tomilson, R. *J. Am. Chem. Soc.* **1968**, *90*, 5631-5632.

407. Andersen, K. K.; Cinquini, M.; Papanikolaou, N. E. *J. Org. Chem.* **1970**, *35*, 706-710.

408. Hoge, G. *J. Am. Chem. Soc.* **2004**, *126*, 9920-9921.

409. Egan, W.; Mislow, K. *J. Am. Chem. Soc.* **1971**, *93*, 1805-1806.

410. Vedejs, E.; Donde, Y. *J. Am. Chem. Soc.* **1997**, *119*, 9293-9294.

411. Bader, A.; Pabel, M.; Wild, S. B. *J. Chem. Soc., Chem. Commun.* **1994**, 1405-1406.

412. Albert, J.; Cadena, J. M.; Granell, J.; Muller, G.; Panyella, D.; Sanudo, C. *Eur. J. Inorg. Chem.* **2000**, 1283-1286.

413. Cavalier, J.-F.; Fotiadu, F.; Verger, R.; Buono, G. *Synlett* **1998**, 73-75.

414. Wolfe, B.; Livinghouse, T. *J. Am. Chem. Soc.* **1998**, *120*, 5116-5117.

415. Hayakawa, Y.; Hyodo, M.; Kimura, K.; Kataoka, M. *Chem. Commun.* **2003**, 1704-1705.

416. Friedman, J. M.; Freeman, S.; Knowles, J. R. *J. Am. Chem. Soc.* **1988**, *110*, 1268-1275.

417. Rotto, N. T.; Coates, R. M. *J. Am. Chem. Soc.* **1989**, *111*, 8941-8943.

418. Rayner, D. R.; Gordon, A. J.; Mislow, K. *J. Am. Chem. Soc.* **1968**, *90*, 4854-4860.

419. Boyd, D. R.; Sharma, N. D.; Haughey, S. A.; Malone, J. F.; McMurray, B. T.; Sheldrake, G. N.; Allen, C. C. R.; Dalton, H. *Chem. Commun.* **1996**, 2363-2364.

420. Miller, E. G.; Rayner, D. R.; Thomas, H. T.; Mislow, K. *J. Am. Chem. Soc.* **1968**, *90*, 4861-4868.

421. Booms, R. E.; Cram, D. J. *J. Am. Chem. Soc.* **1972**, *94*, 5438-5446.

422. Mislow, K.; Axelrod, M.; Rayner, D. R.; Gotthardt, H.; Coyne, L. M.; Hammond, G. S. *J. Am. Chem. Soc.* **1967**, *87*, 4958-4959.

423. Hammond, G. S.; Gotthardt, H.; Coyne, L. M.; Axelrod, M.; Rayner, D. R.; Mislow, K. *J. Am. Chem. Soc.* **1967**, *87*, 4959-4960.

424. Cooke, R. S.; Hammond, G. S. *J. Am. Chem. Soc.* **1970**, *92*, 2739-2745.

425. Guo, Y.; Jenks, W. S. *J. Org. Chem.* **1997**, *62*, 857-864.

426. Lee, W.; Jenks, W. S. *J. Org. Chem.* **2001**, *66*, 474-480.

427. Vos, B. W.; Jenks, W. S. *J. Am. Chem. Soc.* **2002**, *124*, 2544-2547.

428. Cinquini, M.; Colonna, S. *J. Chem. Soc., Chem. Commun.* **1975**, 256-257.

429. Cinquini, M.; Colonna, S.; Cozzi, F.; Stirling, C. J. M. *J. Chem. Soc., Perkin Trans. 1* **1976**, 2061-2067.

430. Bickart, P.; Carson, F. W.; Jacobus, J.; Miller, E. G.; Mislow, K. *J. Am. Chem. Soc.* **1968**, *90*, 4869-4876.

431. Oxelbark, J.; Allenmark, S. *J. Org. Chem.* **1999**, *64*, 1483-1486.

432. Oxelbark, J.; Allenmark, S. *J. Chem. Soc., Perkin Trans. 2* **1999**, 1587-1589.

433. Koch, P.; Fava, A. *J. Am. Chem. Soc.* **1968**, *90*, 3867-3868.

434. Khiar, N.; Alcudia, F.; Espartero, J.-L.; Rodriguez, L.; Fernandez, I. *J. Am. Chem. Soc.* **2000**, *122*, 7598-7599.

435. Khiar, N.; Araujo, C. S.; Alcudia, F.; Fernandez, I. *J. Org. Chem.* **2002**, *67*, 345-356.

436. Evans, J. W.; Fierman, M. B.; Miller, S. J.; Ellman, J. A. *J. Am. Chem. Soc.* **2004**, *126*, 8134-8135.

437. Shibata, N.; Matsunaga, M.; Nakagawa, M.; Fukuzumi, T.; Nakamura, S.; Toru, T. *J. Am. Chem. Soc.* **2005**, *127*, 1374-1375.

438. Goering, H. L.; Towns, D. L.; Dittmar, B. *J. Org. Chem.* **1962**, *27*, 736-739.

439. Cram, D. J.; Nielsen, W. D.; Rickborn, B. *J. Am. Chem. Soc.* **1960**, *82*, 6415-6416.

440. Corey, E. J.; Kaiser, E. T. *J. Am. Chem. Soc.* **1961**, *83*, 490-491.

441. Cram, D. J.; Ratajczak, A. *J. Am. Chem. Soc.* **1968**, *90*, 2198-2200.

442. Knof, U.; von Zelewsky, A.; *Angew. Chem., Int. Ed.* **1999**, *38*, 302-322.

443. Knight, P. D.; Scott, P. *Coord. Chem. Rev.* **2003**, *242*, 125-143.

444. Boche, G.; Marsch, M.; Harbach, J.; Harms, K.; Ledig, B.; Schubert, F.; Lohrenz, J. C. W.; Ahlbrecht, H. *Chem. Ber.* **1993**, *126*, 1887-1894.

445. Ruhland, T.; Dress, R. W; Hoffmann, R. W. *Angew. Chem., Int. Ed. Engl.* **1993**, *32*, 1467-1468.

446. Reich, H. J.; Dykstra, R. R. *Angew. Chem., Int. Ed. Engl.* **1993**, *32*, 1469-1470.

447. Fraenkel, G.; Cabral, J. A. *J. Am. Chem. Soc.* **1993**, *115*, 1551-1557.

448. Reich, H. J.; Dykstra, R. R. *J. Am. Chem. Soc.* **1993**, *115*, 7041-7042.

449. Ahlbrecht, H.; Harbach, J.; Hoffmann, R. W.; Ruhland, T. *Liebigs Ann.* **1995**, 211-216.

450. Fraenkel, G.; Martin, K. V. *J. Am. Chem. Soc.* **1995**, *117*, 10336-10344.

451. Basu, A.; Thayumanavan, S. *Angew. Chem., Int. Ed.* **2002**, *41*, 716-738.

452. Gais, H.-J.; Hellmann, G.; Günther, H.; Lopez, F.; Lindner, H. J.; Braun, S. *Angew. Chem., Int. Ed. Engl.* **1989**, *28*, 1025-1028.

453. Gais, H.-J.; Hellmann, G. *J. Am. Chem. Soc.* **1992**, *114*, 4439-4440.

454. Ruhland, T.; Dress, R.; Hoffmann, R. W. *Angew. Chem., Int. Ed. Engl.* **1993**, *32*, 1467-1468.

455. Hoffmann, R. W.; Dress, R. K.; Ruhland, T.; Wenzel, A. *Chem. Ber.* **1995**, *128*, 861-870.

456. Reich, H. J.; Kulicke, K. J. *J. Am. Chem. Soc.* **1995**, *117*, 6621-6622.

457. Hirsch, R.; Hoffmann, R. W. *Chem. Ber.* **1992**, *125*, 975-982.

458. Klute, W.; Dress, R.; Hoffmann, R. W. *J. Chem. Soc., Chem. Commun.* **1993**, 1409-1411.

459. Hoffmann, R. W.; Julius, M.; Chemla, F.; Ruhland, T.; Frenzen, G. *Tetrahedron* **1994**, *50*, 6049-6060.

460. Basu, A.; Gallagher, D. J.; Beak, P. *J. Org. Chem.* **1996**, *61*, 5718-5719.

461. Behrens, K.; Fröhlich, R.; Meyer, O.; Hoppe, D. *Eur. J. Org. Chem.* **1998**, 2397-2403.

462. Weisenburger, G. A.; Faibish, N. C.; Pippel, D. J.; Beak, P. *J. Am. Chem. Soc.* **1999**, *121*, 9522-9530.

463. McGarvey, G. J.; Kimura, M. *J. Org. Chem.* **1982**, *47*, 5422-5424.

464. Hoffmann, R. W.; Nell, P. G.; Leo, R.; Harms, K. *Chem. Eur. J.* **2000**, *6*, 3359-3365.

465. Hoffmann, R. W.; Hölzer, B. *Chem. Commun.* **2001**, 491-492.

466. Hölzer, B.; Hoffmann, R. W. *Chem. Commun.* **2003**, 732-733.

467. Vestergren, M.; Eriksson, J.; Hakansson, M. *J. Organomet. Chem.* **2003**, *681*, 215-224.

468. Vestergren, M.; Eriksson, J.; Hakansson, M. *Chem. Eur. J.* **2003**, *9*, 4678-4686.

469. Still, W. C.; Sreekumar, C. *J. Am. Chem. Soc.* **1980**, *102*, 1201-1202.

470. Györi, B.; Emri, J. *J. Organomet. Chem.* **1982**, *238*, 159-170.

471. Vedejs, E.; Fields, S. C.; Lin, S.; Schrimpf, M. R. *J. Org. Chem.* **1995**, *60*, 3028-3034.

472. Vedejs, E.; Fields, S. C.; Hayashi, R.; Hitchcock, S. R.; Powell, D. R.; Schrimpf, M. R. *J. Am. Chem. Soc.* **1999**, *121*, 2460-2470.

473. Vedejs, E.; Chapman, R. W.; Müller, M.; Powell, D. R. *J. Am. Chem. Soc.* **2000**, *122*, 3047-3052.

474. Toyota, S.; Ito, F.; Nitta, N.; Hakamata, T. *Bull. Chem. Soc. Jpn.* **2004**, *77*, 2081-2088.
475. Green, S.; Nelson, A.; Warriner, S.; Whittaker, B. *Chem. Commun.* **2000**, 4403-4408.
476. Di Bari, L.; Lelli, M.; Salvadori, P. *Chem. Eur. J.* **2003**, *10*, 4594-4598.
477. Faller, J. W.; Patel, B. P.; Albrizzio, M. A.; Curtis, M. *Organometallics* **1999**, *18*, 3096-3104.
478. Kawaguchi, H.; Fukaki, H.; Ama, T.; Yasui, T.; Okamoto, K.-I.; Hidaka, J. *Bull. Chem. Soc. Jpn.* **1988**, *61*, 2359-2364.
479. Fernandez, J. M.; Gladysz, J. A. *Organometallics* **1989**, *8*, 207-219.
480. Metcalf, D. H.; Snyder, S. W.; Demas, J. N.; Richardson, F. S. *J. Am. Chem. Soc.* **1990**, *112*, 469-479.
481. Metcalf, D. H.; Snyder, S. W.; Demas, J. N.; Richardson, F. S. *J. Phys. Chem.* **1990**, *94*, 7143-7153.
482. Mendez, N. Q.; Mayne, C. L.; Gladysz, J. A. *Angew. Chem., Int. Ed. Engl.* **1990**, *29*, 1475-1476.
483. Faller, J. W.; Ma, Y. *Organometallics* **1992**, *11*, 2726-2729.
484. Peng, T.-S.; Gladysz, J. A. *J. Am. Chem. Soc.* **1992**, *114*, 4174-4181.
485. Carmona, D.; Lahoz, F. J.; Oro, L. A.; Lamata, M. P.; Viguri, F.; Jose, E. S. *Organometallics* **1996**, *15*, 2961-2966.
486. Dewey, M. A.; Stark, G. A.; Gladysz, J. A. *Organometallics* **1996**, *15*, 4798-4807.
487. Brunner, H.; Klankermayer, J.; Zabel, M. *Eur. J. Inorg. Chem.* **2002**, 2494-2501.
488. Carmona, D.; Lahoz, F. J.; Elipe, S. Oro, L. A.; Lamata, M. P.; Viguri, F.; Sanchez, F.; Martinez, S.; Cativiela, C.; Lopez-Ram de Viu, M. P. *Organometallics* **2002**, *21*, 5100-5114.
489. Brunner, H.; Zwack, T.; Zabel, M.; Beck, W.; Böhm, A. *Organometallics* **2003**, *22*, 1741-1750.
490. Hamelin, O.; Pecaut, J.; Fontecave, M. *Chem. Eur. J.* **2004**, *10*, 2548-2554.
491. Howell, J. A. S.; Squibb, A. D.; Bell, A. G.; McArdle, P.; Cunningham, D.; Goldschmidt, Z.; Gottlieb, H. E.; Hezroni-Langerman, D. *Organometallics* **1994**, *13*, 4336-4351.
492. Brunner, H.; Köllnberger, A.; Burgemeister, T.; Zabel, M. *Polyhedron* **2000**, *19*, 1519-1526.
493. Brunner, H.; Köllnberger, A.; Zabel, M. *Polyhedron* **2003**, *22*, 2639-2646.
494. Brunner, H.; Zwack, T; Zabel, M. *Polyhedron* **2003**, *22*, 861-865.
495. Kamikawa, K.; Sakamoto, T.; Uemura, M. *Synlett* **2003**, 516-518.
496. Kamikawa, K.; Sakamoto, T.; Tanaka, Y.; Uemura, M. *J. Org. Chem.* **2003**, *68*, 9356-9363.
497. Jahr, H. C.; Nieger, M.; Dötz, K. H. *Chem. Commun.* **2003**, 2866-2867.
498. Zahn, S.; Canary, J. W. *Science* **2000**, *288*, 1404-1407.
499. Lauceri, R.; Raudino, A.; Scolaro, L. M.; Micali, N.; Purrello, R. *J. Am. Chem. Soc.* **2002**, *124*, 894-895.
500. Mazet, C.; Gade, L. H. *Chem. Eur. J.* **2002**, *8*, 4308-4318.
501. Purrello, R. *Nature Mater.* **2003**, *2*, 216-217.
502. Ziegler, M.; Davis, A. V.; Johnson, D. W.; Raymond, K. N. *Angew. Chem., Int. Ed.* **2003**, *42*, 665-668.
503. Tashiro, R.; Sugiyama, H. *J. Am. Chem. Soc.* **2005**, *127*, 2094-2097.
504. Lauceri, R.; Purrello, R. *Supramol. Chem.* **2005**, *17*, 61-66.
505. Miyake, H.; Yoshida, K.; Sugimoto, H.; Tsukube, H. *J. Am. Chem. Soc.* **2004**, *126*, 6524-6525.
506. Miyake, H.; Sugimoto, H.; Tamiaki, H.; Tsukube, H. *Chem. Commun.* **2005**, 4219-4293.
507. Goto, H.; Yashima, E. *J. Am. Chem. Soc.* **2002**, *124*, 7943-7949.
508. Hofacker, A.; Parquette, J. R. *Angew. Chem., Int. Ed.* **2005**, *44*, 1053-1057.
509. Maurizot, V.; Dolain, C.; Huc, I. *Eur. J. Org. Chem.* **2005**, 1293-1301.
510. De Jong, J. J. D.; Lucas, L. N.; Kellogg, R. M.; van Esch, J. H.; Feringa, B. L. *Nature* **2004**, *304*, 278-281.
511. Ribó, J. M.; Crusats, J.; Sagués, F.; Claret, J.; Rubires, R. *Science* **2001**, *292*, 2063-2066.
512. Feringa, B. L. *Science* **2001**, *292*, 2021-2022.

513. Prins, L. J.; Timmermann, P.; Reinhoudt, D. N. *J. Am. Chem. Soc.* **2001**, *123*, 10153-10163.
514. Prins, L. J.; Verhage, J. J.; de Jong, F.; Timmermann, P.; Reinhoudt, D. N. *Chem. Eur. J.* **2002**, *8*, 2302-2313.
515. Ishi-I, T.; Crego-Calama, M.; Timmermann, P.; Reinhoudt, D. N.; Shinkai, S. *J. Am. Chem. Soc.* **2002**, *124*, 14631-14641.
516. Berl, V.; Huc, I.; Khoury, R. G.; Krische, M. J.; Lehn, J.-M. *Nature* **2000**, *407*, 720-723.
517. Lehn, J.-M.; Rigault, A.; Siegel, J.; Harrowfield, J.; Chevrier, B.; Moras, D. *Proc. Natl. Acad. Sci.* **1987**, *84*, 2565-2569.
518. Bell, T. W.; Jousselin, H. *Nature* **1994**, *367*, 441-444.
519. Piguet, C.; Bernardinelli, G.; Hopfgartner, G. *Chem. Rev.* **1997**, *97*, 2005-2062.
520. Tanaka, K.; Tengeiji, A.; Kato, T.; Toyoma, N.; Shionoya, M. *Science* **2003**, *299*, 1212-1213.
521. Huc, I. *Eur. J. Org. Chem.* **2004**, 17-29.
522. Albrecht, M. *Angew. Chem., Int. Ed.* **2005**, *44*, 6448-6451.
523. Sanchez-Quesada, J.; Seel, C.; Prados, P.; de Mendoza, J. *J. Am. Chem. Soc.* **1996**, *118*, 277-278.
524. Tanaka, Y.; Katagiri, H.; Furusho, Y.; Yashima, E. *Angew. Chem., Int. Ed.* **2005**, *44*, 3867-3870.
525. Inai, Y.; Tagawa, K.; Takasu, A.; Hirabayashi, T.; Oshikawa, T.; Yamashita, M. *J. Am. Chem. Soc.* **2000**, *122*, 11731-11732.
526. Inai, Y.; Ousaka, N.; Okabe, T. *J. Am. Chem. Soc.* **2003**, *125*, 8151-8162.
527. Tsubaki, K.; Miura, M.; Morikawa, H.; Tanaka, H.; Kawabata, T.; Furuta, T.; Tanaka, K.; Fuji, K. *J. Am. Chem. Soc.* **2003**, *125*, 16200-16201.
528. Tanaka, M.; Anan, K.; Demizu, Y.; Kurihara, M.; Doi. M.; Suemune, H. *J. Am. Chem. Soc.* **2005**, *127*, 11570-11571.
529. Williams, D. J.; Colquhoun, H. M.; O'Mahoney, C. A. *J. Chem. Soc., Chem. Commun.* **1994**, 1643-1644.
530. Porsch, M.; Sigl-Seifert, G.; Daub, J. *Adv. Mater.* **1997**, *9*, 635-639.
531. Yashima, E.; Maeda, K.; Okamoto, Y. *Nature* **1999**, *399*, 449-451.
532. Green, M. M.; Park, J.-W.; Sato, T.; Teramoto, A.; Lifson, S.; Selinger, R. L. B.; Selinger, J. V. *Angew. Chem., Int. Ed.* **1999**, *38*, 3138-3154.
533. Fujiki, M. *J. Organomet. Chem.* **2003**, *685*, 15-34.
534. Yu, Z.; Wan, X.; Zhang, H.; Chen, X.; Zhou, Q. *Chem. Commun.* **2003**, 974-975.
535. Yashima, E.; Maeda, K.; Nishimura, T. *Chem. Eur. J.* **2004**, *10*, 42-51.
536. Maeda, K.; Morino, K.; Okamoto, Y.; Sato, T.; Yashima, E. *J. Am. Chem. Soc.* **2004**, *126*, 4329-4342.
537. Morino, K.; Watase, N.; Maeda, K.; Yashima, E. *Chem. Eur. J.* **2004**, *10*, 4703-4707.
538. Wilson, A. J.; Masuda, M.; Sijbesma, R. P.; Meijer, E. W. *Angew. Chem., Int. Ed.* **2005**, *44*, 2275-2279.
539. Onouchi, H.; Miyagawa, T.; Furuko, A.; Maeda, K.; Yashima, E. *J. Am. Chem. Soc.* **2005**, *127*, 2960-2965.
540. Li, C.; Numata, M.; Bae, A.-H.; Sakurai, K.; Shinkai, S. *J. Am. Chem. Soc.* **2005**, *127*, 4548-4549.
541. Jana, S.; Sherrington, D. C. *Angew. Chem., Int. Ed.* **2005**, *44*, 4804-4808.
542. Miyagawa, T.; Furuko, A. Maeda, K.; Katagiri, H.; Furusho, Y.; Yashima, E. *J. Am. Chem. Soc.* **2005**, *127*, 5018-5019.
543. Abe, H.; Masuda, N.; Waki, M.; Inouye, M. *J. Am. Chem. Soc.* **2005**, *127*, 16189-16196.
544. Lemieux, R. P.; Schuster, G. B. *J. Org. Chem.* **1993**, *58*, 100-110.
545. Suarez, M.; Schuster, G. B. *J. Am. Chem. Soc.* **1995**, *117*, 6732-6738.
546. Janicki, S. Z.; Schuster, G. B. *J. Am. Chem. Soc.* **1995**, *117*, 8524-8527.

547. Feringa, B. L.; Huck, N. P. M.; van Doren, H. A. *J. Am. Chem. Soc.* **1995**, *117*, 9929-9930.

548. Feringa, B. L.; van Delden, R. A.; Koumura, N.; Geertsema, E. M. *Chem. Rev.* **2000**, *100*, 1789-1816.

549. Iftime, G.; Lagugné Labarthet, F.; Natansohn, A.; Rochon, P. *J. Am. Chem. Soc.* **2000**, *122*, 12646-12650.

550. Ruslim, C.; Ichimura, K. *Adv. Mater.* **2001**, *13*, 37-40.

551. Van Delden, R. A.; Koumura, N.; Harada, N.; Feringa, B. L. *Proc. Natl. Acad. Sci.* **2002**, *99*, 4945-4949.

552. Pieraccini, S.; Masiero, S.; Piero Spada, G.; Gottarelli, G. *Chem. Commun.* **2003**, 598-599.

553. Van Delden, R. A.; Mecca, T.; Rosini, C.; Feringa, B. L. *Chem. Eur. J.* **2004**, *10*, 61-70.

554. George, S. J.; Ajayaghosh, A.; Jonkheijm, P.; Schenning, A. P. H. J.; Meijer, E. W. *Angew. Chem., Int. Ed.* **2004**, *43*, 3422-3425.

555. Pieraccini, S.; Gottarelli, G.; Labruto, R.; Masiero, S.; Pandoli, O.; Piero Spada, G. *Chem. Eur. J.* **2004**, *10*, 5632-5639.

556. Holzwarth, R.; Bartsch, R.; Cherkaoui, Z.; Solladié, G. *Eur. J. Org. Chem.* **2005**, 3536-3541.

557. Eelkema, R.; Feringa, B. L. *J. Am. Chem. Soc.* **2005**, *127*, 13480-13481.

558. Zhdanov, Y. A.; Alekseev, Y. E.; Kompantseva, E.; Vergeichik, E. N. *Russ. Chem. Rev.* **1992**, *61*, 1025-1042.

559. Kobayashi, N.; Saito, R.; Hino, H.; Hino, Y.; Ueno, A.; Osa, T. *J. Chem. Soc., Perkin Trans 2* **1983**, 1031-1035.

560. Takenaka, S.; Matsuura, N.; Tokura, N. *Tetrahedron Lett.* **1974**, *15*, 2325-2328.

561. Matsuura, N.; Takenaka, S.; Tokura, N. *J. Chem. Soc., Perkin Trans 2* **1977**, 1419-1421.

562. Kano, K.; Tatsumi, M.; Hashimoto, S. *J. Org. Chem.* **1991**, *56*, 6579-6585.

563. Kikuchi, Y.; Kobayashi, K.; Aoyama, Y. *J. Am. Chem. Soc.* **1992**, *114*, 1351-1358.

564. Morozumi, T.; Shinkai, S. *J. Chem. Soc., Chem. Commun.* **1994**, 1219-1220.

565. Rivera, J. M.; Craig, S. L.; Martin, T.; Rebek Jr., J. *Angew. Chem., Int. Ed.* **2000**, *39*, 2130-2132.

566. Trembleau, L.; Rebek Jr., J. *Science* **2003**, *301*, 1219-1220.

567. Allenmark, S. *Chirality* **2003**, *15*, 409-422.

568. Collet, A.; Gabard, J. *J. Org. Chem.* **1980**, *45*, 5400-5401.

569. Biali, S. E.; Böhmer, V.; Cohen, S.; Ferguson, G.; Grüttner, C.; Grynszpan, F.; Paulus, E. F.; Thondorf, I.; Vogt, W. *J. Am. Chem. Soc.* **1996**, *118*, 12938-12949.

570. Kusano, T.; Tabatabai, M.; Okamoto, Y.; Böhmer, V. *J. Am. Chem. Soc.* **1999**, *121*, 3789-3790.

571. Christie, G. H.; Kenner, J. *J. Chem. Soc.* **1922**, *121*, 614-620.

572. Kuhn, R. In: Freudenberg, K. (ed.) *Stereochemie*, Deuticke, Leipzig, 1933, pp. 803, 810.

573. Ernst, L. *CHIUZ* **1983**, *17*, 21-30.

574. Ling. C. K.; Harris, M. M. *J. Chem. Soc.* **1964**, 1825-1835.

575. Hall, D. M.; Harris, M. M. *J. Chem. Soc.* **1960**, 490-494.

576. Melander, L.; Carter, R. E. *J. Am. Chem. Soc.* **1964**, *86*, 295-296.

577. Ling, C. K.; Harris, M. M. *J. Chem. Soc.* **1964**, 1825-1835.

578. Mislow, K.; Gust, D. *J. Am. Chem. Soc.* **1973**, *95*, 1535-1547.

579. Wolf, C.; König, W. A.; Roussel, C. *Liebigs Ann.* **1995**, 781-786.

580. Wolf, C.; Hochmuth, D. H.; König, W. A.; Roussel, C. *Liebigs Ann.* **1996**, 357-363.

581. Weseloh, G.; Wolf, C.; König, W. A. *Angew. Chem., Int. Ed. Engl.* **1995**, *34*, 1635-1636.

582. Weseloh, G.; Wolf, C.; König, W. A. *Chirality* **1996**, *8*, 441-445.

583. Meca, L.; Reha, D.; Havlas, Z. *J. Org. Chem.* **2003**, *68*, 5677-5680.

584. Nishio, M.; Hirota, M. *Tetrahedron* **1989**, *45*, 7201-7245.

585. Nakai, Y.; Yamamoto, G.; Oki, M. *Chem. Lett.* **1987**, 89-92.

586. Rieger, M.; Westheimer, F. H. *J. Am. Chem. Soc.* **1950**, *72*, 19-28.

587. Oki, M.; Yamamoto, G. *Bull. Chem. Soc. Jpn.* **1971**, *44*, 266-270.

588. Bott, G.; Field, L. D.; Sternhell, S. *J. Am. Chem. Soc.* **1980**, *102*, 5618-5626.

589. Hasaka, N.; Okigawa, M.; Kouno, I.; Kawano, N. *Bull. Chem. Soc. Jpn.* **1982**, *55*, 3828-3830.

590. Meyers, A. I.; Himmelsbach, R. J. *J. Am. Chem. Soc.* **1985**, *107*, 682-685.

591. Wolf, C.; König, W. A.; Roussel, C. *Chirality* **1995**, *7*, 610-611.

592. Schurig, V.; Glausch, A.; Fluck, M. *Tetrahedron: Asymm.* **1995**, *6*, 2161-2164.

593. Charlton, J. L.; Oleschuk, C. J.; Chee, G.-L. *J. Org. Chem.* **1996**, *61*, 3452-3457.

594. Biedermann, P. U.; Schurig, V.; Agranat, I. *Chirality* **1997**, *9*, 350-353.

595. Wolf, C.; Pirkle, W. H.; Welch, C. J.; Hochmuth, D. H.; König, W. A.; Chee, G.-L.; Charlton, J. L. *J. Org. Chem.* **1997**, *62*, 5208-5210.

596. Baker, R. W.; Brkic, Z.; Sargent, M. V.; Skelton, B. W.; White, A. H. *Aust. J. Chem.* **2000**, *53*, 925-938.

597. Chow, H.-F.; Wan, C.-W. *J. Org. Chem.* **2001**, *66*, 5042-5047.

598. Spivey, A. C.; Charbonneau, P.; Fekner, T.; Hochmuth, D. H.; Maddaford, A.; Malardier-Jugroot, C.; Redgrave, A. J.; Whitehead, M. A. *J. Org. Chem.* **2001**, *66*, 7394-7401.

599. Tochtermann, W.; Kuckling, D.; Meints, C.; Kraus, J.; Bringmann, G. *Tetrahedron* **2003**, *59*, 7791-7801.

600. Leroux, F.; Maurin, M.; Nicod, N.; Scopelliti, R. *Tetrahedron Lett.* **2004**, *45*, 1899-1902.

601. Bringmann, G. Gulder, T. A. M.; Maksimenka, K.; Kuckling, D.; Tochtermann, W. *Tetrahedron* **2005**, *61*, 7241-7246.

602. Knauf, A. E.; Shildneck, P. R; Adams, R. *J. Am. Chem. Soc.* **1934**, *56*, 2109-2111.

603. Marriott, P. J.; Lai, Y.-H. *J. Chromatogr.* **1988**, *447*, 29-41.

604. Maier, N. M.; Zoltewicz, J. A. *Tetrahedron* **1997**, *53*, 465-468.

605. Dell'Erba, C.; Gasparrini, F.; Grilli, S.; Lunazzi, L.; Mazzanti, A.; Novi, M.; Pierini, M.; Tavani, C.; Villani, C. *J. Org. Chem.* **2002**, *67*, 1663-1668.

606. Grilli, S.; Lunazzi, L.; Mazzanti, A.; Pinamonti, M. *Tetrahedron* **2004**, *60*, 4451-4458.

607. Wolf, C.; Ghebremariam, T. *Tetrahedron: Asymm.* **2002**, *13*, 1153-1156.

608. Wolf, C.; Tumambac, G. E. *J. Phys. Chem.* **2003**, *107*, 815-817.

609. Wolf, C.; Mei, X. *J. Am. Chem. Soc.* **2003**, *125*, 10651-10658.

610. Tumambac, G. E.; Wolf, C. *J. Org. Chem.* **2004**, *69*, 2048-2055.

611. Tumambac, G. E.; Wolf, C. *J. Org. Chem.* **2005**, *70*, 2930-2938.

612. Watson, A. A.; Willis, A. C.; Wild, S. B. *J. Organomet. Chem.* **1993**, *445*, 71-78.

613. Gladiali, S.; Dore, A.; Fabbri, D.; De Lucchi, O.; Valle, G. *J. Org. Chem.* **1994**, *59*, 6363-6371.

614. Bringmann, G.; Hartung, T.; Göbel, L.; Schupp, O.; Ewers, C. L. J.; Schöner, B.; Zagst, R.; Peters, K.; von Schnering, H. G.; Burschka, C. *Liebigs Ann.* **1992**, 225-232.

615. Fritsch, R.; Hartmann, E.; Brandl, G.; Mannschreck, A. *Tetrahedron: Asymm.* **1993**, *4*, 433-455.

616. Bringmann, G.; Keller, P. A.; Rölfing, K. *Synlett* **1994**, 423-424.

617. Bringmann, G.; Busse, H.; Dauer, U.; Güssregen, S.; Stahl, M. *Tetrahedron* **1995**, *51*, 3149-3158.

618. Bringmann, G.; Breuning, M.; Endress, H.; Vitt, D.; Peters, K.; Peters, E.-M. *Tetrahedron* **1998**, *54*, 10677-10690.

619. Ohmori, K.; Kitamura, M.; Suzuki, K. *Angew. Chem., Int. Ed.* **1999**, *38*, 1226-1229.

620. Bringmann, G.; Heubes, M.; Breuning, M.; Göbel, L.; Ochse, M.; Schöner, B.; Schupp, O. *J. Org. Chem.* **2000**, *65*, 722-728.

621. Adams, R.; Kornblum, N. *J. Am. Chem. Soc.* **1941**, *63*, 188-200.

622. Mislow, K.; Hyden, S.; Schaefer, H. *J. Am. Chem. Soc.* **1962**, *84*, 1449-1455.

623. Tichy, M.; Ridvan, L.; Holy, P.; Zavada, J.; Cisarova, I.; Podlaha, J. *Tetrahedron: Asymm.* **1998**, *9*, 227-234.

624. Superchi, S.; Casarini, D.; Laurita, A.; Bavoso, A.; Rosini, C. *Angew. Chem., Int. Ed.* **2001**, *40*, 451-454.

625. Hatsuda, M.; Hiramatsu, H.; Yamada, S.-I.; Shimizu, T.; Seki, M. *J. Org. Chem.* **2001**, *66*, 4437-4439.

626. Bringmann, G.; Hartung, T. *Liebigs Ann.* **1994**, 313-316.

627. Bringmann, G.; Vitt, D.; Kraus, J.; Breuning, M. *Tetrahedron* **1998**, *54*, 10691-10698.

628. Edwards, D. J.; Pritchard, R. G.; Wallace, T. W. *Tetrahedron Lett.* **2003**, *44*, 4665-4668.

629. Kyba, E. P.; Gokel, G. W.; de Jong, F.; Koga, K.; Sousa, L. R.; Siegel, M. G.; Kaplan, L.; Sogah, G. D. Y.; Cram, D. J. *J. Org. Chem.* **1977**, *42*, 4173-4184.

630. Shimada, T.; Kina, A.; Hayashi, T. *J. Org. Chem.* **2003**, *68*, 6329-6337.

631. Boyd, M. R.; Hallock, Y. F.; Cardellina II, J. H.; Manfredi, K. P.; Blunt, J. W.; McMahon, J. B.; Buckheit Jr., R. W.; Bringmann, G.; Schäffer, M.; Cragg, G. M.; Thomas, D. W.; Jato, J. G.; Lough, A. *J. Med. Chem.* **1994**, *37*, 1740-1745.

632. Mislow, K.; Gordon, A. J. *J. Am. Chem. Soc.* **1963**, *85*, 3521-3521.

633. Zimmermann, H. E.; Crumrine, D. S. *J. Am. Chem. Soc.* **1972**, *94*, 498-506.

634. Zhang, M.; Schuster, G. B. *J. Phys. Chem.* **1992**, *96*, 3063-3067.

635. Cavazza, M.; Zandomeneghi, M.; Ouchi, A.; Koga, Y. *J. Am. Chem. Soc.* **1996**, *118*, 9990-9991.

636. Burnham, K. S.; Schuster, G. B. *J. Am. Chem. Soc.* **1998**, *120*, 12619-12625.

637. Hattori, T.; Shimazumi, Y.; Goto, H.; Yamabe, O.; Morohashi, N.; Kawai, W.; Miyano, S. *J. Org. Chem.* **2003**, *68*, 2099-2108.

638. Irie, M.; Yoshida, K.; Hayashi, K. *J. Phys. Chem.* **1977**, *81*, 969-972.

639. Irie, M.; Yorozu, T.; Yoshida, K.; Hayashi, K. *J. Phys. Chem.* **1977**, *81*, 973-976.

640. Roussel, C.; Adjimi, M.; Chemlal, A.; Djafri, A. *J. Org. Chem.* **1988**, *53*, 5076-5080.

641. Newell, L. M.; Sekhar, V. C.; DeVries, K. M.; Staigers, T. L.; Finneman. J. I. *J. Chem. Soc., Perkin Trans. 2* **2001**, 961-963.

642. Biali, S. E.; Kahr, B.; Okamoto, Y.; Aburatani, R.; Mislow, K. *J. Am. Chem. Soc.* **1988**, *110*, 1917-1922.

643. Schmidt, O. T.; Grünewald, H. H. *Liebigs Ann.* **1957**, 183-188.

644. Khan, N. U.; Ilyas, M.; Rahman, W.; Mashima, T.; Okigawa, M.; Kawano, N. *Tetrahedron* **1972**, *28*, 5689-5695.

645. Kupchan, S. M.; Britton, R. W.; Ziegler, M. F.; Gilmore, C. J.; Restivo, R. J.; Bryan, R. F. *J. Am. Chem. Soc.* **1973**, *95*, 1335-1336.

646. Govindachari, T. R.; Nagarajan, K.; Parthasarathy, P. C.; Rajagopalan, T. G.; Deasi, H. K.; Kartha, G.; Chen, S. L.; Naganishi, K. *J. Chem. Soc., Perkin Trans. 1* **1974**, 1413-1417.

647. Begley, M. J.; Campbell, V.M.; Crombie, L.; Tuck, B.; Whiting, D. A. *J. Chem. Soc. C* **1971**, 3634-3642.

648. Raistrick, H.; Stickings, C. E.; Thomas, R. *J. Biochem.* **1953**, *55*, 421-433.

649. Coombe, R. G.; Jacobs, J. J.; Watson, T. R. *Aust. J. Chem.* **1970**, *23*, 2343-2351.

650. Bracher, F.; Eisenreich, W. J.; Mühlbacher, J.; Dreyer, M.; Bringmann, G. *J. Org. Chem.* **2004**, *69*, 8602-8608.

651. Tomioka, K.; Ishiguro, T.; Koga, K. *Tetrahedron Lett.* **1980**, *21*, 2973-2976.

652. Meyers, A. I.; Flisak, J. R.; Aitken, R. A. *J. Am. Chem. Soc.* **1987**, *109*, 5446-5452.

653. Tucci, F. C.; Hu, T.; Mesleh, M. F.; Bokser, A.; Allsopp, E.; Gross, T. D.; Guo, Z.; Zhu, Y.-F.; Struthers, R. S.; Ling, N.; Chen, C. *Chirality* **2005**, *17*, 559-564.

654. Bringmann, G.; Mühlbacher, J.; Reichert, M.; Dreyer, M.; Kolz, J.; Speicher, A. *J. Am. Chem. Soc.* **2004**, *126*, 9283-9290.

655. Monaco, R. R.; Gardiner, W. C.; Kirschner, S. *Int. J. Quantum Chem.* **1999**, *71*, 57-62.

656. Bedard, T. C.; Moore, J. S. *J. Am. Chem. Soc.* **1995**, *117*, 10662-10671.

657. Dominguez, Z.; Dang, H.; Strouse, M. J.; Garcia-Garibay, M. A. *J. Am. Chem. Soc.* **2002**, *124*, 2398-2399.

658. Dominguez, Z.; Dang, H.; Strouse, M. J.; Garcia-Garibay, M. A. *J. Am. Chem. Soc.* **2002**, *124*, 7719-7727.

659. Koo Tze Mew, P.; Vögtle, F. *Angew. Chem., Int. Ed. Engl.* **1979**, *18*, 159-161.

660. Toyota, S.; Yamamori, T.; Asakura, M.; Oki, M. *Bull. Chem. Soc. Jpn.* **2000**, *73*, 205-213.

661. Toyota, S.; Iida, T.; Kunizane, C.; Tanifuji, N.; Yoshida, Y. *Org. Biomol. Chem.* **2003**, *1*, 2298-2302.

662. Miljanic, O. S.; Han, S.; Holmes, D.; Schaller, G. R.; Vollhardt, K. P. C. *Chem. Commun.* **2005**, 2606-2608.

663. Toyota, S.; Makino, T. *Tetrahedron Lett.* **2003**, *44*, 7775-7778.

664. Clayden, J. *Chem. Commun.* **2004**, 127-135.

665. Adams, R.; Gordon, J. R. *J. Am. Chem. Soc.* **1950**, *72*, 2454-2457.

666. Adams, R.; Gordon, J. R. *J. Am. Chem. Soc.* **1950**, *72*, 2458-2460.

667. Mannschreck, A.; Zinner, H.; Pustet, N. *Chimia* **1989**, *43*, 165-166.

668. Gasparrini, F.; Misiti, D.; Pierini, M.; Villani, C. *Tetrahedron: Asymm.* **1997**, *8*, 2069-2073.

669. Ahmed, A.; Bragg, R. A.; Clayden, J.; Wah Lai, L.; McCarthy, C.; Pink, J. H.; Westlund, N.; Yasin, S. A. *Tetrahedron* **1998**, *54*, 13277-13294.

670. Clayden, J.; Johnson, P.; Pink, J. H.; Helliwell, M. *J. Org. Chem.* **2000**, *65*, 7033-7040.

671. Rios, R.; Jimeno, C.; Carroll, P. J.; Walsh, P. J. *J. Am. Chem. Soc.* **2002**, *124*, 10272-10273.

672. Clayden, J.; McCarthy, C.; Helliwell, M. *Chem. Commun.* **1999**, 2059-2060.

673. Clayden, J.; Westlund, N.; Wilson, F. X. *Tetrahedron Lett.* **1999**, *40*, 7883-7887.

674. Bragg, R. A.; Clayden, J. *Org. Lett.* **2000**, *2*, 3351-3354.

675. Bragg, R. A.; Clayden, J.; Morris, G. A.; Pink, J. H. *Chem. Eur. J.* **2002**, *8*, 1279-1289.

676. Fraser, R. R.; Taymaz, K. *Tetrahedron Lett.* **1976**, *50*, 4573-4576.

677. Berg, U.; Sandstrom, J.; Jennings, W. B.; Randall, D. *J. Chem. Soc., Perkin Trans. 2* **1980**, 949-956.

678. Eiglsperger, A.; Kastner, F.; Mannschreck, A. *J. Mol. Structure* **1985**, *126*, 421-432.

679. Cuyegkeng, M. A.; Mannschreck, A. *Chem. Ber.* **1987**, *120*, 803-809.

680. Kuttenberger, M.; Frieser, M.; Hofweber, M.; Mannschreck, A. *Tetrahedron: Asymm.* **1998**, *9*, 3629-3645.

681. Johnston, E. R.; Fortt, R.; Barborak, J. C. *Magn. Reson. Chem.* **2000**, *38*, 932-936.

682. Clayden, J.; Pink, J. H. *Angew. Chem., Int. Ed.* **1998**, *37*, 1937-1938.

683. Casarini, D.; Lunazzi, L.; Pasquali, F.; Gasparrini, F.; Villani, C. *J. Am. Chem. Soc.* **1992**, *114*, 6521-6527.

684. Kiefl, C.; Mannschreck, A. *Synthesis* **1995**, 1033-1037.

685. Lunazzi, L.; Mazzanti, A.; Alvarez, A. M. *J. Org. Chem.* **2000**, *65*, 3200-3206.

686. Kiefl, C. *Eur. J. Org. Chem.* **2000**, 3279-3286.

687. Leardini, R.; Lunazzi, L.; Mazzanti, A.; Nanni, D. *J. Org. Chem.* **2001**, *66*, 7879-7882.

688. Staab, H. A.; Chi, C.-S.; Dabrowski, J. *Tetrahedron* **1982**, *38*, 3499-3505.

689. Casarini, D.; Lunazzi, L. *J. Org. Chem.* **1994**, *59*, 4637-4641.

690. Casarini, D.; Lunazzi, L.; Mazzanti, A. *J. Org. Chem.* **1997**, *62*, 7592-7596.

691. Casarini, D.; Lunazzi, L.; Mazzanti, A. *J. Org. Chem.* **1998**, *63*, 4991-4995.

692. Casarini, D.; Foresti, E.; Gasparrini, F.; Lunazzi, L.; Macciantelli, D.; Misiti, D.; Villani, C. *J. Org. Chem.* **1993**, *58*, 5674-5682.

693. Casarini, D.; Lunazzi, L.; Gasparrini, F.; Villani, C.; Cirilli, M.; Gavuzzo, E. *J. Org. Chem.* **1995**, *60*, 97-102.

694. Casarini, D.; Lunazzi, L. *J. Org. Chem.* **1995**, *60*, 5515-5519.

695. Yuste, F.; Ortiz, B.; Pérez, J. I.; Rodríguez-Hernández, A.; Sánchez-Obregón, R.; Walls, F.; García Ruano, J. L. *Tetrahedron* **2002**, *58*, 2613-2620.

696. Gasparrini, F.; Lunazzi, L.; Mazzanti, A.; Pierini, M.; Pietrusiewicz, K. M.; Villani, C. *J. Am. Chem Soc.* **2000**, *122*, 4776-4780.

697. Mino, T.; Tanaka, Y.; Yabusaki, T.; Okumura, D.; Sakamoto, M.; Fujita, T. *Tetrahedron: Asymm.* **2003**, *14*, 2503-2506.

698. Casarini, D.; Lunazzi, L.; Placucci, G. *J. Org. Chem.* **1987**, *52*, 4721-4726.

699. Casarini, D.; Lunazzi, L. *J. Org. Chem.* **1988**, *53*, 182-185.

700. Casarini, D.; Foresti, E.; Lunazzi, L.; Macciantelli, D. *J. Am. Chem. Soc.* **1988**, *110*, 4527-4532.

701. Davalli, S.; Lunazzi, L.; Macciantelli, D. *J. Org. Chem.* **1991**, *56*, 1739-1747.

702. Garcia, M. B.; Grilli, S.; Lunazzi, L.; Mazzanti, A.; Orelli, L. R. *J. Org. Chem.* **2001**, *66*, 6679-6684.

703. Mintas, M.; Orhanović, Z.; Jakopčić, K.; Koller, H.; Stühler, G.; Mannschreck, A. *Tetrahedron* **1985**, *41*, 229-233.

704. Lunazzi, L.; Magagnoli, C.; Guerra, M.; Macciantelli, D. *Tetrahedron Lett.* **1979**, *20*, 3031-3032.

705. Casarini, D.; Davalli, S.; Lunazzi, L. *J. Org. Chem.* **1989**, *54*, 4616-4619.

706. Avalos, M.; Babiano, R.; Cintas. P.; Higes, F. J.; Jimenez, J. L.; Palacios, J. C.; Silvero, G.; Valencia, C. *Tetrahedron* **1999**, *55*, 4401-4426.

707. Curran, D. P.; Qi, H.; Geib, S. J.; DeMello, N. C. *J. Am. Chem. Soc.* **1994**, *116*, 3131-3132.

708. Kessler, H.; Rieker, A. *Liebigs Ann. Chem.* **1967**, 57-68.

709. Stewart, W. H.; Siddall III, T. H. *Chem. Rev.* **1970**, *70*, 517-551.

710. Chupp, J. P.; Olin, J. F. *J. Org. Chem.* **1967**, *32*, 2297-2303.

711. Oki, M. *Top. Stereochem.* **1984**, *14*, 10-19.

712. Curran, D. P.; Hale, G. R.; Geib, S. J.; Balog, A.; Cass, Q. B.; Degani, A. L. G.; Hernandes, M. Z.; Freitas, L. C. G. *Tetrahedron: Asymm.* **1997**, *8*, 3955-3975.

713. Shvo, Y.; Taylor, E. C.; Mislow, K.; Raban, M. *J. Am. Chem. Soc.* **1967**, *89*, 4910-4917.

714. Siddall III, T. H.; Stewart, W. E. *J. Phys. Chem.* **1969**, *73*, 40-45.

715. Mintas, M.; Mihaljevic, V.; Koller, H.; Schuster, D.; Mannschreck, A. *J. Chem. Soc., Perkin Trans 2* **1990**, 619-624.

716. Saito, K.; Yamamoto, M.; Yamada, K. *Tetrahedron* **1993**, *49*, 4549-4558.

717. Kondo, K.; Fujita, H.; Suzuki, T.; Murakami, Y. *Tetrahedron Lett.* **1999**, *40*, 5577-5580.

718. Curran, D. P.; Liu, W.; Chen, C. H.-T. *J. Am. Chem. Soc.* **1999**, *121*, 11012-11013.

719. Kitagawa, O.; Fujita, M.; Kohriyama, M.; Hasegawa, H.; Taguchi, T. *Tetrahedron Lett.* **2000**, *41*, 8539-8544.

720. Jog, P. V.; Brown, R. E.; Bates, D. K. *J. Org. Chem.* **2003**, *68*, 8240-8243.

721. Curran, D. P.; Chen, C. H.-T.; Geib, S. J.; Lapierre, A. J. B. *Tetrahedron* **2004**, *60*, 4413-4424.

722. Bennett, D. J.; Blake, A. J.; Cooke, P. A.; Gofrey, C. R. A.; Pickering, P. L.; Simpkins, N. S.; Walker, M. D.; Wilson, C. *Tetrahedron* **2004**, *60*, 4491-4451.

723. Petit, M.; Lapierre, A. J. B.; Curran, D. P. *J. Am. Chem. Soc.* **2005**, *127*, 14994-14995.

724. Dogan, I.; Pustet, N.; Mannschreck, A. *J. Chem. Soc., Perkin Trans 2* **1993**, 1557-1560.

725. Kondo, K.; Iida, T.; Fujita, H.; Suzuki, T.; Yamaguchi, K.; Murakami, Y. *Tetrahedron* **2000**, *56*, 8883-8891.

726. Chen, Y.; Smith, M. D.; Shimizu, K. D. *Tetrahedron Lett.* **2001**, *42*, 7185-7187.

727. Oguz, S. F.; Dogan, I. *Tetrahedron: Asymm.* **2003**, *14*, 1857-1864.

728. Aydeniz, Y.; Oğuz, F.; Yaman, A.; Konuklar, A. S.; Doğan, I.; Aviyente, V.; Klein, R. A. *Org. Biomol. Chem.* **2004**, *2*, 2426-2436.

729. Haushalter, K. A.; Lau, J.; Roberts, J. D. *J. Am. Chem. Soc.* **1996**, *118*, 8891-8896.

730. Adler, T.; Bonjoch, J.; Clayden, J.; Font-Bardia, M.; Pickworth, M.; Solaus, X.; Sole, D.; Vallverdu, L. *Org. Biomol. Chem.* **2005**, *3*, 3173-3183.

731. Adams, R.; Miller, M. W. *J. Am. Chem. Soc.* **1940**, *62*, 53-56.

732. Adams, R.; Anderson, A. W.; Miller, M. W. *J. Am. Chem. Soc.* **1941**, *63*, 1589-1593.

733. Adams, R.; Mecorney, J. W. *J. Am. Chem. Soc.* **1945**, *67*, 798-802.

734. Anderson, J. E.; Hazlehurst, C. J. *J. Chem. Soc., Chem. Commun.* **1980**, 1188-1189.

735. Baker, R. W.; Taylor, J. A. *Tetrahedron Lett.* **2000**, *41*, 4471-4473.

736. Pettersson, I.; Berg, U. *J. Chem. Research (S)* **1984**, 208-209.

737. Leardini, R.; Lunazzi, L.; Mazzanti, A.; McNab, H.; Nanni, D. *Eur. J. Org. Chem.* **2000**, 3439-3446.

738. Gasparrini, F.; Grilli, S.; Leardini, R.; Lunazzi, L.; Mazzanti, A.; Nanni, D.; Pierini, M.; Pinamonti, M. *J. Org. Chem.* **2002**, *67*, 3089-3095.

739. Köbrich, G.; Mannschreck, A.; Misra, R. A.; Rissmann, G.; Rösner, M.; Zündorf, W *Chem. Ber.* **1972**, *105*, 3794-3806.

740. Mannschreck, A.; Jonas, V.; Bödecker, H.-O.; Elbe, H.-L.; Köbrich, G. *Tetrahedron Lett.* **1974**, *25*, 2153-2156.

741. Becher, G.; Mannschreck, A. *Chem. Ber.* **1981**, *114*, 2365-2368.

742. Rösner, M.; Köbrich, G. *Angew. Chem., Int. Ed. Engl.* **1974**, *13*, 741-742.

743. Pasto, D.; J.; Borchardt, J. K. *J. Am. Chem. Soc.* **1974**, *96*, 6220-6221.

744. Jelinski, L. W.; Kiefer, E. F. *J. Am. Chem. Soc.* **1976**, *98*, 281-282.

745. Warren, S.; Chow, A.; Fraenkel, G.; RajanBabu, T. V. *J. Am. Chem. Soc.* **2003**, *125*, 15402-15410.

746. Kessler, H.; Leibfritz, D. *Chem. Ber.* **1971**, *104*, 2143-2157.

747. Casarini, D.; Lunazzi, L.; Macciantelli, D. *J. Chem. Soc., Perkin Trans. 2* **1992**, 1363-1370.

748. Guerra, A.; Lunazzi, L. *J. Org. Chem.* **1995**, *60*, 7959-7965.

749. Coogan, M. P.; Hibbs, D. E.; Smart, E. *Chem. Commun.* **1999**, 1991-1992.

750. Coogan, M. P.; Passey, S. C. *J. Chem. Soc., Perkin Trans. 2* **2000**, 2060-2066.

751. Olszewska, T.; Milewska, M. J.; Gdaniec, M.; Maluszyńska, H.; Poloński, T. *J. Org. Chem.* **2001**, *66*, 501-506.

752. Raban, M.; Martin, V. A.; Craine, L. *J. Org. Chem.* **1990**, *55*, 4311-4316.

753. Blanca, M. B.-D.; Yamamoto, C.; Okamoto, Y.; Biali, S. E.; Kost, D. *J. Org. Chem.* **2000**, *65*, 8613-8620.

754. Anderson, J. E.; Barkel, D. J. D. *J. Chem. Soc., Perkin Trans. 2* **1984**, 1053-1057.

755. Anderson, J. E.; Franck, R. W. *J. Chem. Soc., Perkin Trans. 2* **1984**, 1581-1582.

756. Anderson, J. E.; Franck, R. W.; Mandella, W. L. *J. Am. Chem. Soc.* **1972**, *94*, 4608-4614.

757. Anet, F. A. L.; Donovan, D.; Sjostrand, U.; Cozzi, F.; Mislow, K. *J. Am. Chem. Soc.* **1980**, *102*, 1748-1749.

758. Wolf, C.; Pranatharthiharan, L.; Ramagosa, R. B. *Tetrahedron Lett.* **2002**, *43*, 8563-8567.

759. Nishio, M.; Hirota, M. *Tetrahedron* **1989**, *45*, 7201-7245.

760. Rieker, A.; Kessler, H. *Tetrahedron Lett.* **1969**, *10*, 1227-1230.

761. Siddall, T. H.; Stewart, W. E. *J. Org. Chem.* **1969**, *34*, 233-237.

762. Nakamura, M.; Oki, M. *Tetrahedron Lett.* **1974**, *13*, 505-508.

763. Nakamura, M.; Oki, M. *Bull. Chem. Soc. Jpn.* **1980**, *53*, 2977-2980.

764. Mori, T.; Nakamura, N.; Oki, M. *Bull. Chem. Soc. Jpn.* **1981**, *54*, 1199-1202.

765. Lomas, J. S.; Dubois, J.-E. *J. Org. Chem.* **1976**, *41*, 3033-3034.

766. Ford, W. T.; Thompson, T. B.; Snoble, K. A. J.; Timko, J. M. *J. Am. Chem. Soc.* **1975**, *97*, 95-101.

767. Nakamura, M.; Nakamura, N.; Oki, M. *Bull. Chem. Soc. Jpn.* **1977**, *50*, 2986-2990.

768. Nakamura, M.; Oki, M. *Bull. Chem. Soc. Jpn.* **1980**, *53*, 3248-3251.

769. Oki, M.; Suda, M. *Bull. Chem. Soc. Jpn.* **1971**, *44*, 1876-1880.

770. Yamamoto, G.; Suzuki, M. Oki, M. *Angew. Chem., Int. Ed. Engl.* **1981**, *20*, 607-608.

771. Yamamoto, G.; Oki, M. *Chem. Lett.* **1972**, 45-48.

772. Yamamoto, G.; Oki, M. *Bull. Chem. Soc. Jpn.* **1975**, *48*, 2592-2596.

773. Yamamoto, G.; Oki, M. *Bull. Chem. Soc. Jpn.* **1975**, *48*, 3686-3690.

774. Mikami, M.; Toriumi, K.; Kondo, M.; Saito, Y. *Acta Cryst.* **1975**, *B31*, 2474-2478.

775. Suzuki, F.; Oki, M.; Nakanishi, H. *Bull. Chem. Soc. Jpn.* **1974**, *47*, 3114-3120.

776. Nakanishi, H.; Kitagawa, Y.; Yamamoto, O.; Oki, M. *Org. Magn. Reson.* **1977**, *9*, 118-120.

777. Suzuki, M.; Yamamoto, G.; Kikuchi, H.; Oki, M. *Bull. Chem. Soc. Jpn.* **1981**, *54*, 2383-2386.

778. Kikuchi, H.; Hatakeyama, S.; Yamamoto, G.; Oki, M. *Bull. Chem. Soc. Jpn.* **1981**, *54*, 3832-3836.

779. Lüttringhaus, A.; Grahlheer, H. *Justus Liebigs Ann.* **1941**, 67-98.

780. Lüttringhaus, A.; Grahlheer, H. *Justus Liebigs Ann.* **1947**, 112-120.

781. Schlögl, K. *Top. Curr. Chem.* **1984**, *125*, 1-48.

782. Vögtle, F.; Neumann, P. *Tetrahedron Lett.* **1970**, *11*, 115-118.

783. Hochmuth, D. H.; König, W. A. *Liebigs Ann.* **1996**, 947-951.

784. Grimme, S.; Harren, J.; Sobanski, A.; Vögtle, F. *Eur. J. Org. Chem.* **1998**, 1491-1509.

785. Hochmuth, D. H.; König, W. A. *Tetrahedron: Asymm.* **1999**, *10*, 1089-1097.

786. Scharwächter, K. P.; Hochmuth, D. H.; Dittmann, H.; König, W. A. *Chirality* **2001**, *13*, 679-690.

787. Reich, H. J.; Cram, D. J. *J. Am. Chem. Soc.* **1967**, *89*, 3078-3080.

788. Reich, H. J.; Cram, D. J. *J. Am. Chem. Soc.* **1969**, *91*, 3517-3626.

789. Delton, M. H.; Cram, D. J. *J. Am. Chem. Soc.* **1970**, *92*, 7623-7625.

790. Delton, M. H.; Gilman, R. E.; Cram, D. J. *J. Am. Chem. Soc.* **1971**, *93*, 2329-2330.

791. Delton, M. H.; Gilman, R. E.; Cram, D. J. *J. Am. Chem. Soc.* **1972**, *94*, 2478-2482.

792. Glotzmann, C.; Langer, E.; Lehner, H.; Schlögl, K. *Monatsh. Chem.* **1974**, *105*, 907-916.

793. Krois, D.; Lehner, H. *Tetrahedron* **1982**, *38*, 3319-3324.

794. Wittek, M.; Vögtle, F.; Stühler, G.; Mannschreck, A.; Lang, B.; Irngartner, H. *Chem. Ber.* **1983**, *116*, 207-214.

795. Vögtle, F.; Meurer, K.; Mannschreck, A.; Stühler, G.; Puff, H.; Roloff, A.; Sievers, R. *Chem. Ber.* **1983**, *116*, 2630-2640.

796. Vögtle, F.; Mittelbach, K.; Struck, J.; Nieger, M. *J. Chem. Soc., Chem. Commun.* **1989**, 65-67.

797. Billen, S.; Vögtle, F. *Chem. Ber.* **1989**, *122*, 1113-1117.

798. Sako, K.; Shinmyozu, T.; Takemura, H.; Suenaga, M.; Inazu, T. *J. Org. Chem.* **1992**, *57*, 6536-6541.

799. Barrett, S.; Bartlett, S.; Bolt, A.; Ironmonger, A.; Joce, C.; Nelson, A.; Woodhall, T. *Chem. Eur. J.* **2005**, *11*, 6277-6285.

800. Kallenborn, R.; Hühnerfuss, H.; König, W. A. *Angew. Chem., Int. Ed. Engl.* **1991**, *30*, 320-321.

801. Buser, H.-R.; Müller, M. D. *Anal. Chem.* **1992**, *64*, 3168-3175.

802. Ballschmiter, K.; Zell, M. *Fresenius Z. Anal. Chem.* **1980**, *302*, 20-31.

803. Pham-Tuan, H.; Larsson, C.; Hoffmann, F.; Bergman, A.; Fröba, M.; Hühnerfuss, H. *Chirality* **2005**, *17*, 266-280.

804. Schurig, V.; Glausch, A.; Fluck, M. *Tetrahedron: Asymm.* **1995**, *6*, 2161-2164.

805. König, W. A.; Gehrcke, B.; Runge, T.; Wolf, C. *J. High Resolut. Chromatogr.* **1993**, *16*, 376-377.

806. Vetter, W.; Klobes, U.; Luckas, B.; Hottinger, G. *J. Chromatogr. A* **1997**, *769*, 247-252.

807. Glausch, A.; Schurig, V. *Naturwissenschaften* **1993**, *80*, 468-469.

808. Glausch, A.; Nicholson, G. J.; Fluck, M.; Schurig, V. *J. High Resolut. Chromatogr.* **1994**, *17*, 347-349.

809. De Jongh, J.; Nieboer, R.; Schröders, I.; Seinen, W.; van den Berg, M. *Arch. Toxicol.* **1993**, *67*, 598-604.

810. Kross, G.; Meyer-Rogge, D.; Seubert, S.; Seubert, A.; Losekam, M. *Arch. Toxicol.* **1993**, *67*, 651-654.

811. Parkinson, A.; Safe, S. H.; Robertson, L. W.; Thomas, P. E.; Ryan, D. E.; Reik, L. M.; Levin, W. *J. Biol. Chem.* **1983**, *258*, 5967-5976.

812. Sinclair, P. R.; Bement, W. J.; Bonkovsky, H. L.; Sinclair, J. F. *Biochem. J.* **1984**, *222*, 737-748.

813. Sinclair, P. R.; Bement, W. J.; Bonkovsky, H. L.; Lambrecht, R. W.; Frezza, J. E.; Sinclair, J. F.; Uroquhart, A. J.; Elder, G. H. *Biochem. J.* **1986**, *237*, 63-71.

814. Lambrecht, R. W.; Sinclair, P. R.; Bement, W. J.; Sinclair, J. F.; Carpenter, H. M.; Buhler, D. R.; Uroquhart, A. J.; Elder, G. H. *Biochem. J.* **1988**, *253*, 131-138.

815. Sassa, S.; Sugita, O.; Ohnuma, N.; Imajo, S.; Okumura, T.; Noguchi, T.; Kappas, A. *Biochem. J.* **1986**, *235*, 291-296.

816. White, J.; Adams, R. *J. Am. Chem. Soc.* **1932**, *54*, 2104-2108.

817. Mannschreck, A.; Pustet, N.; Robertson, L. W.; Oesch, F.; Püttmann, M. *Liebigs Ann.* **1985**, 2101-2103.

818. Tang, T.-H.; Nowakowska, M.; Guillet, L. E.; Csizmadia, I. G. *J. Mol. Struct. (Theochem.)* **1991**, *232*, 133-146.

819. König, W. A.; Icheln, D.; Runge, T.; Pfaffenberger, B.; Ludwig, P.; Hühnerfuss, H. *J. High Resolut. Chromatogr.* **1991**, *14*, 530-536.

820. Moser, H.; Rihs, G.; Sauter, H. *Z. Naturforsch.* **1982**, *37b*, 451-462.

821. Ariyoshi, N.; Yoshimura, H.; Oguri, K. *Biol. Pharm. Bull.* **1993**, *16*, 852-857.

822. Püttmann, M.; Mannschreck, A.; Oesch, F.; Robertson, L. *Biochem. Pharmacol.* **1989**, *38*, 1345.

823. Rodman, L. E.; Shedlofsky, S. I.; Mannschreck, A. Püttmann, M.; Swim, A. T.; Robertson, L. W. *Biochem. Pharmacol.* **1991**, *41*, 915-922.

824. Böhm, V.; Schulte, E.; Thier, H.-P. *Z. Lebensm. Unters. Forsch.* **1993**, *196*, 435-440.

825. Schulte, E.; Malisch, R. *Fresenius Z. Anal. Chem.* **1984**, *319*, 54-59.

826. Hardt, I. H.; Wolf, C.; Gehrcke, B.; Hochmuth, D. H.; Pfaffenberger, B.; Hühnerfuss, H.; König, W. A. *J. High Resolut. Chromatogr.* **1994**, *17*, 859-864.

827. Glausch, A.; Hahn, J.; Schurig, V. *Chemosphere* **1995**, *30*, 2079-2085.

828. Reich, S.; Jimenez, B.; Marsili, L.; Hernandez, L. M.; Schurig, V.; Gonzalez, M. J. *Environ. Sci. Technol.* **1999**, *33*, 1787-1793.

829. Wong, C. S.; Garrison, A. W.; Foreman, W. T. *Environ. Sci. Technol.* **2001**, *35*, 33-39.

830. Wong, C. S.; Garrison, A. W.; Smith, P. D.; Foreman, W. T. *Environ. Sci. Technol.* **2001**, *35*, 2448-2454.

831. Caldwell, J. *J. Chromatogr. A* **1995**, *694*, 39-48.

832. Yang, S. K. *Chirality* **1999**, *11*, 179-186.

833. Lu, X.-L.; Yang, S. K. *J. Chromatogr. A* **1994**, *666*, 249-257.

834. Reist, M.; Testa, B.; Carrupt, P.-A. *Enantiomer* **1997**, *2*, 147-155.

835. Schoetz, G.; Trapp, O.; Schurig, V. *Anal. Chem.* **2000**, *72*, 2758-2764.

836. Welch, C. J.; Kress, M. H.; Beconi, M.; Mathre, D. J. *Chirality* **2003**, *15*, 143-147.

837. Lamparter E.; Blaschke, G.; Schluter, J. *Chirality* **1993**, *5*, 105-111.

838. Bovarnick, M.; Clarke, H. T. *J. Am. Chem. Soc.* **1938**, *60*, 2426-2430.

839. Yonetani, K.; Hirotsu, Y.; Shiba, T. *Bull. Chem. Soc. Jpn.* **1975**, *48*, 3302-3305.

840. Shibata, S.; Matsushita, H.; Kato, K.; Noguchi, M.; Saburi, M.; Yoshikawa, S. *Bull. Chem. Soc. Jpn.* **1979**, *52*, 2938-2941.

841. Lazarus, R. A. *J. Org. Chem.* **1990**, *55*, 4755-4757.

842. Wipf, P.; Fritch, P. C. *Tetrahedron Lett.* **1994**, *35*, 5397-5400.

843. Cabrera, K.; Jung, M.; Fluck, M.; Schurig, V. *J. Chromatogr. A* **1996**, *731*, 315-321.

844. Reist, M.; Carrupt, P.-A.; Testa, B.; Lehmann, S.; Hansen, J. J. *Helv. Chim. Acta* **1996**, *79*, 767-778.

845. Cloos, P. A. C.; Fledelius, C. *Biochem. J.* **2000**, *345*, 473-480.

846. Gineyts, E.; Cloos, P. A. C.; Borel, O.; Grimaud, L.; Delmas, P. D.; Garnero, P. *Biochem. J.* **2000**, *345*, 480-485.

847. Smith, G. G.; de Sol, B. S. *Science* **1980**, *207*, 765-767.

848. Steinberg, S. M.; Bada, J. L. *Science* **1981**, *213*, 544-545.

849. Smith, G. G.; Baum, R. *J. Org. Chem.* **1987**, *52*, 2248-2255.

850. Ritz, S.; Turzynski, A.; Schütz, H. W. *Forensic Sci. Int.* **1994**, *69*, 149-159.

CHAPTER 4

Analytical Methods

The study of racemization and diastereomerization reactions of chiral compounds is a fundamental aspect of dynamic stereochemistry. The determination of the relative thermodynamic stability of stereoisomers and kinetic studies that are aimed at a mechanistic understanding of enantiomer and diastereomer interconversion play a crucial role in the development of asymmetric reactions and molecular gears, switches, motors and other stereodynamic devices. Rate constants and activation parameters ($\Delta G^{\neq}$, $\Delta H^{\neq}$, $\Delta S^{\neq}$) are routinely obtained by spectroscopic, chiroptical and chromatographic techniques to provide today's chemist with invaluable information about the configurational and conformational stability of chiral compounds. In the case of an irreversible first-order reaction, an activation energy of 100 kJ/mol corresponds to a reaction rate of $1.89 \cdot 10^{-5} \, \text{s}^{-1}$ at 25 °C. In general:

$$k = 2.084 \cdot 10^{10} \, T \, e^{-\Delta G^{\neq}/8.314 \, T} \tag{4.1}$$

where T is the temperature in Kelvin and $\Delta G^{\neq}$ is the activation energy in J/mol.

Fast dynamic molecular processes can be investigated by femtosecond or picosecond pump-probe methods,[1] optical calorimetry measurements,[2] stopped-flow methods, and flash photolysis, in conjunction with time-resolved spectroscopy or spectrometry. If the reaction time-scale permits, routine methods can be used, such as microwave, fluorescence, infrared, electron spin resonance (ESR), and nuclear magnetic resonance (NMR) spectroscopy.[3,4] The interconversion of stereoisomers exhibiting activation barriers between 20 and 100 kJ/mol is usually monitored by variable-temperature (dynamic) NMR spectroscopy, whereas chiroptical techniques (polarimetry, circular dichroism and optical rotary dispersion) and chromatographic methods are often employed in kinetic studies of more stable compounds. The choice of a suitable technique for the study of an enantiomerization or diastereomerization reaction mainly depends on the reaction rate, the availability of isolated enantiomers or diastereomers, and the capacity of the method to differentiate between interconverting stereoisomers. In contrast to chiroptical methods, variable-temperature NMR spectroscopy, dynamic chromatography and chromatographic stopped-flow techniques render the isolation of pure enantiomers or diastereoisomers unnecessary.

Stereoisomers with an interconversion barrier above 100 kJ/mol can often be isolated by chromatography or crystallization at room temperature, which greatly facilitates stereodynamic analysis. For example, Tumambac and Wolf were able to separate the meso *syn-* and enantiomeric *anti-*isomers of axially chiral 1,8-bis(2′-methyl-4′-quinolyl)naphthalene by semipreparative chiral chromatography.[5,6] Slow atropisomerization of the enantiopure *anti-*isomer was observed upon heating to 71.0 °C and monitored by chiral HPLC at room temperature, Figure 4.1. Integration and quantification, using

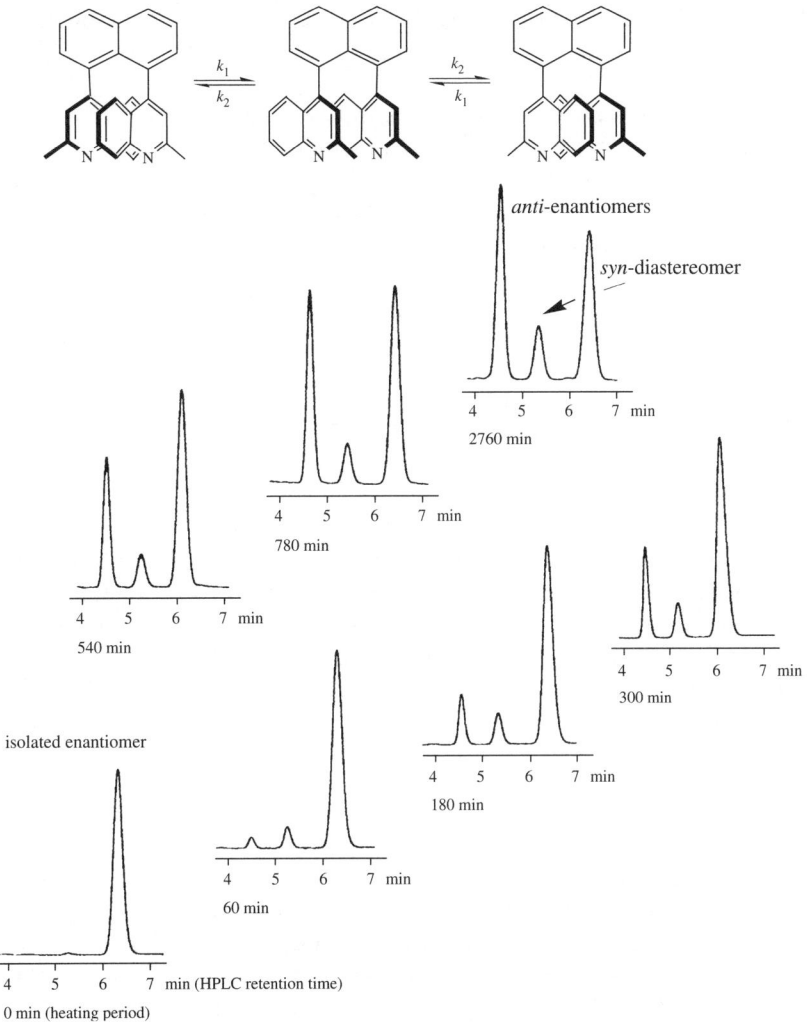

Figure 4.1 Chromatographic analysis of the atropisomerization of 1,8-bis(2′-methyl-4′-quinolyl)-naphthalene at 71 °C.[ii] [Reproduced with permission from *J. Org. Chem.* **2004**, *69*, 2048-2055.]

individual UV response factors for each conformer, allowed accurate analysis of the isomer composition at various reaction times. The concentration of the *syn*-isomer increased to 7.5% within the first three hours and then continued to change very slowly to its final value as the rate of conversion of the *syn*-isomer to the *anti*-isomers increased significantly relative to its rate of formation. The fast initial increase in the concentration of the intermediate *syn*-isomer followed by a slow change to the final steady state value is typical for consecutive reversible reactions. After 46 hours, an equilibrium consisting of 44.4% of each *anti*-conformer and 11.2% of the meso *syn*-isomer was obtained, which corresponds to a difference in Gibbs free energy, ΔG, of 3.4 kJ/mol according to the Boltzmann equation:[i]

$$\frac{n_{syn}}{2\, n_{anti}} = \mathrm{e}^{-\Delta G/RT} \tag{4.2}$$

[i] The factor of 2 accounts for the coexistence of two enantiomeric *anti*-conformations.
[ii] For stereoselective analysis, small aliquots of the reaction mixture (71 °C) were rapidly cooled to room temperature and the three stereoisomers were separated by HPLC on a Chiralpak AD column.

where R is the universal gas constant ($8.3144\,\mathrm{J\,K^{-1}\,mol^{-1}}$); T is the temperature in Kelvin; n_{syn} and n_{anti} describe the population of the *syn-* and *anti-*conformation; and $\Delta G = G_{syn} - G_{anti}$.

With knowledge of the *syn/anti* ratio at equilibrium, the rate constants for the reversible isomerization reaction of 1,8-bis(2′-methyl-4′-quinolyl)naphthalene, k_1 and k_2, were determined as $3.45 \cdot 10^{-5}\,\mathrm{s^{-1}}$ for the *anti*-to-*syn*- and as $1.38 \cdot 10^{-4}\,\mathrm{s^{-1}}$ for the *syn*-to-*anti*-atropisomerization. Using the Eyring equation 4.3, the free Gibbs activation energy, $\Delta G^{\neq}$, was calculated as 116.0 for the conversion of the *anti*- to the *syn*-isomer and 112.1 kJ/mol for the conversion of the *syn*- to the *anti*-isomer.

$$k = \kappa \frac{\mathrm{k_B}\,T}{\mathrm{h}} \mathrm{e}^{-\Delta G^{\neq}/\mathrm{R}T} \qquad (4.3)$$

where k is the rate constant at the temperature T in Kelvin; h is Planck's constant ($6.6261 \cdot 10^{-34}\,\mathrm{J\,s}$); $\mathrm{k_B}$ is the Boltzmann constant ($1.3807 \cdot 10^{-23}\,\mathrm{J\,K^{-1}}$); $\Delta G^{\neq}$ is the Gibbs activation energy in $\mathrm{J\,mol^{-1}}$; R is the universal gas constant ($8.3144\,\mathrm{J\,K^{-1}\,mol^{-1}}$); and κ is the transmission coefficient.

Chromatography is not always suitable for preparative separation of stereoisomers or for the study of racemization and distereomerization reactions of chiral compounds. In these cases, isolation of stereoisomers by crystallization and NMR spectroscopic analysis of the isomerization course can provide a viable alternative. In contrast to the 1,8-diheteroarylnaphthalene discussed above, the conformers of 1,8-bis(2′-isopropyl-4′-quinolyl)naphthalene N,N'-dioxide can not be resolved by chiral chromatography, but the *syn*- and *anti*-diastereoisomers are separable by crystallization. The diastereoisomers show distinguishable ^{1}H NMR spectra and the interconversion process can be directly monitored by NMR spectroscopy in deuterated DMSO at 56.9 °C. Using reversible first-order reaction kinetics that account for interconversion of stereoisomers via two consecutive reaction steps, as described in Chapter 3.1.3, the energy barrier to *anti*-to-*syn*-atropisomerization was determined as 115.2 kJ/mol, and that to *syn*-to-*anti*-atropisomerization was 109.0 kJ/mol. At equilibrium, the diastereomers afford an *anti/syn* ratio of 9.6:1, which corresponds to a difference in Gibbs free energy, ΔG, of 4.9 kJ/mol, Figure 4.2.

The stereodynamic analysis of 1,8-diheteroarylnaphthalenes discussed above was accomplished with chromatographic and spectroscopic methods that were well suited to monitoring racemization and diastereomerization of isolated isomers. Similar approaches have been employed in kinetic studies of many other compounds, for instance decasubstituted biphenyls exhibiting dynamic molecular gearing and crowded helical alkenes that behave like molecular propellers, Scheme 4.1. Mislow and coworkers separated semipreparative amounts of the enantiomers of decakis(dichloromethyl)biphenyl by HPLC on cellulose tris(3,5-dimethylphenylcarbamate).[7] The same HPLC method was then used to monitor the decreasing enantiomeric excess as a function of time, and the activation barrier to racemization was determined as 99.2 kJ/mol. Interestingly, interconversion of the enantiomers does not proceed via rotation about the chiral biphenyl axis but through concerted internal rotation of five dichloromethyl groups in one phenyl moiety.[iii] Biali, Rappoport and others resolved the enantiomers of tetramesitylethylene, an example of a tetraarylvinyl propeller, by HPLC on poly(triphenylmethyl)methacrylate.[8] Helical inversion of this molecular propeller involves correlated rotation of the four aryl rings via conrotatory ring flipping with an

---

[iii] Reversal of the sense of directionality of the dichloromethyl groups in one aryl ring results in enantiomerization of decakis(dichloromethyl)biphenyl, whereas reversal of the directionality of all ten substituents affords topomerization (degenerate isomerization) due to exchange of the positions of identical ligands.

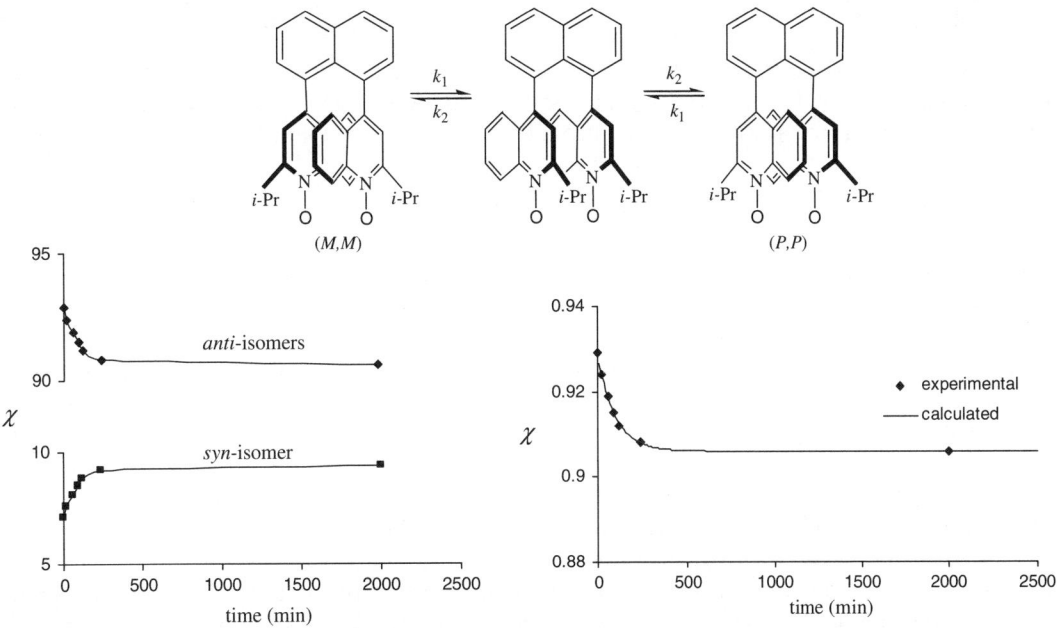

Figure 4.2 Plot of the mole fraction, χ, of the *syn*- and *anti*-isomers of 1,8-bis(2′-isopropyl-4′-quinolyl)naphthalene N,N'-dioxide *versus* time (left), and curve fit using the mathematical treatment for reversible first-order interconversion of axially chiral triaryls discussed in Chapter 3.1.3 (right). [Reproduced with permission from *J. Org. Chem.* **2005**, *70*, 2930-2938.]

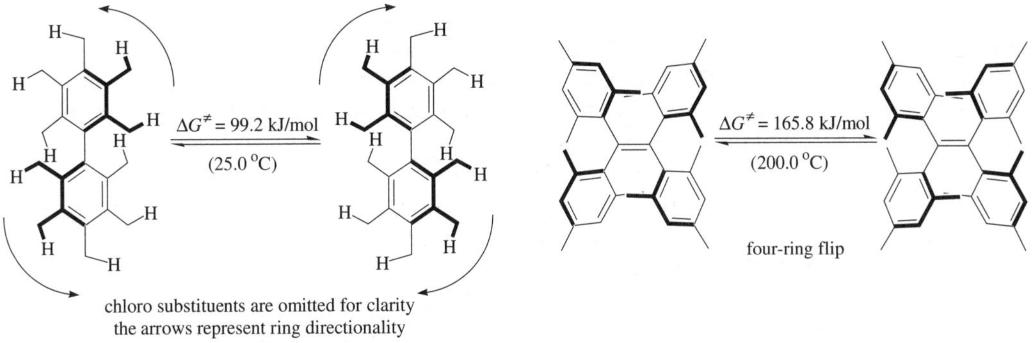

Scheme 4.1 Enantioconversion of decakis(dichloromethyl)biphenyl and tetramesitylethylene.

activation energy of 165.8 kJ/mol. This value was obtained by external heating of an isolated enantiomer and chiral HPLC analysis of small aliquots of the reaction mixture at room temperature.

In general, it is desirable to avoid time-consuming preparative separation steps prior to kinetic analysis. This is often possible when NMR spectroscopy, dynamic chromatography and chromatographic stopped-flow methods can be used. Nevertheless, the stereodynamics of a broad variety of chiral compounds have been studied by thermal interconversion of isolated stereoisomers in combination with chiroptical measurements.

4.1 CHIROPTICAL METHODS

Polarimetry has traditionally been used for analysis of the enantiomeric purity of chiral compounds with known specific rotation based on Equation 4.4, but NMR spectroscopy and chiral chromatography are preferred nowadays since they provide more accurate results.

$$[\alpha]_\lambda^T = \frac{[\alpha]}{c\,l} \tag{4.4}$$

where $[\alpha]_\lambda^T$ is the specific rotation and $[\alpha]$ is the measured optical rotation in degrees; l is the length of the cuvette in dm; and c is the sample concentration in g/mL.

The optical rotation of a chiral compound depends on the wavelength of the linearly polarized light and the solvent, temperature, and sample concentration. The presence of small amounts of (chiral) impurities can significantly affect the accuracy of polarimetric measurements, in particular when compounds with small rotation angles are analyzed. Horeau and others have pointed out that, in the case of associating chiral analytes such as carboxylic acids, determination of the optical rotation for stereochemical analysis can be misleading.[9,10] The optical rotation of a chiral compound does not always increase linearly with the enantiomeric purity and is therefore not necessarily representative of the actual enantiomeric composition. It is important to remember that there is no simple relationship between the sign of the optical rotation of a chiral compound and its absolute configuration. Accordingly, polarimetry is generally not used for determination of absolute configurations unless a reference is available. A noteworthy exception is the diastereoselective esterification of chiral secondary alcohols with an excess of racemic 2-phenylbutyric anhydride (or 2-phenylbutyric chloride), followed by polarimetric analysis of the free carboxylic acid obtained after hydrolysis of residual anhydride (or acyl chloride). According to Horeau's rule, the sign of optical rotation of the remaining 2-phenylbutyric acid, which is enantiomerically enriched due to kinetic resolution, allows deduction of the absolute configuration of the secondary alcohol. This concept has even been applied to alcohols that are chiral only by virtue of isotopic substitution (RCHDOH).[11–13] Other chiroptical techniques, including optical rotary dispersion (ORD), circular dichroism (CD), vibrational ORD and CD, and circular polarization of emission (CPE), afford more structural information about chiral compounds than polarimetric measurements. Some of these techniques have been used for determination of enantiomeric composition as well as for conformational and configurational analysis based on the octant rule and other semiempirical methods.[14,15] In particular, CD[16–20] and ORD[21,22] spectroscopy have become invaluable tools for elucidation of the absolute configuration of chiral compounds, although crystallography remains the ultimate method of choice.

In addition to structural analysis, chiroptical methods have found widespread use in kinetic studies of racemization and diastereomerization reactions. As discussed in Chapter 3.1.4, polarimetric analysis of the decrease in the optical rotation of an enantiopure or enantiomerically enriched sample provides convenient access to racemization and diastereomerization rate constants, half-life times and the corresponding activation parameters. A major drawback of polarimetry, which can be applied in the study of stereomutations with energy barriers between 80 and 180 kJ/mol, is that at least semipreparative amounts ($>10\,\mathrm{mg}$) of purified stereoisomers are required. Small amounts of chiral impurities must be carefully removed prior to analysis in order to exclude interference with chiroptical measurements.

Mannschreck and coworkers utilized preparative HPLC on cellulose- and polymethacrylate-derived chiral stationary phases for separation of the enantiomers of spiro 2*H*-1-benzopyran-2,2′-indolines,[23] 2-aryl-2-methyl-2*H*-1-benzopyrans,[24] sterically crowded tetrasubstituted phenanthrenes,[25] *N*-aryl-4-pyridones,[26] twisted push-pull ethylenes,[27] and atropisomeric *N*,*N*-dimethyl thiobenzamides[28] to investigate the mechanism and barrier to thermal interconversion by

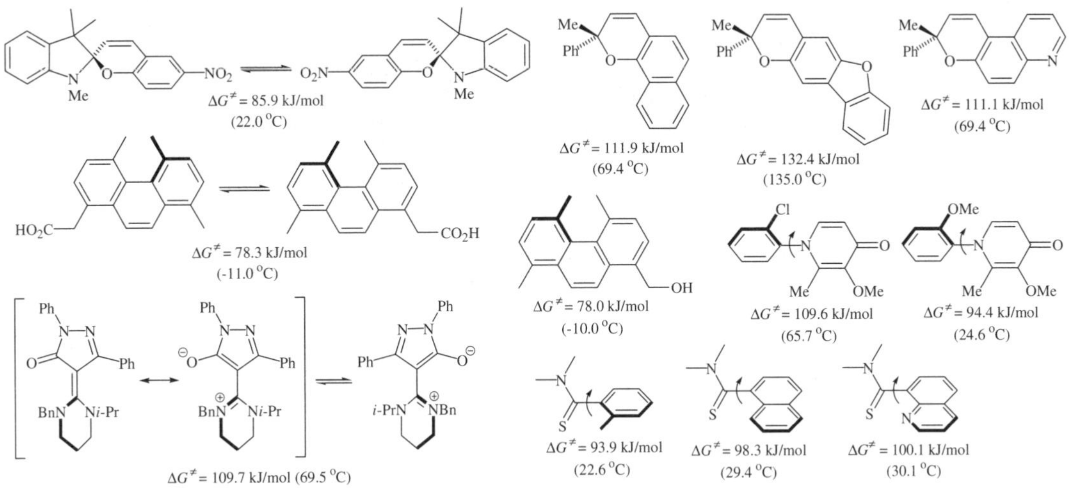

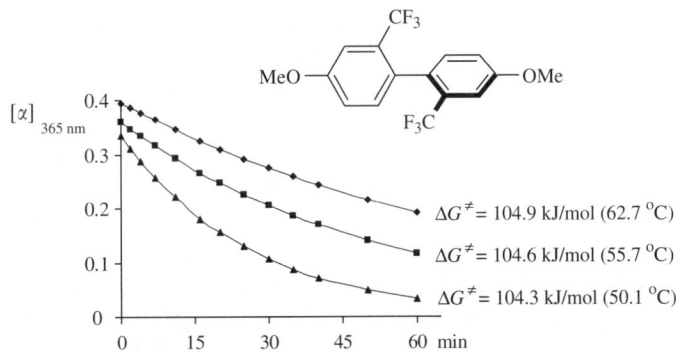

Figure 4.3 Racemization barriers of selected chiral compounds determined by polarimetric analysis of chromatographically enriched or isolated enantiomers.

Figure 4.4 Polarimetric studies of the racemization of enantiomerically enriched 4,4'-dimethoxy-2,2'-bis(trifluoromethyl)biphenyl at various temperatures.

polarimetry, Figure 4.3.[29] Similarly, Wolf *et al.* separated or enriched the enantiomers of axially chiral biphenyls by HPLC on microcrystalline triacetyl cellulose to determine the rotational barrier by monitoring the decay of the optical rotation at various temperatures, Figure 4.4.[30]

Stereochemical analysis of sterically crowded triarylvinyl acetates and thioxanthene-derived alkenes by NMR and CD spectroscopy, combined with polarimetric racemization studies, has provided insights into ring flipping mechanisms of molecular propellers and paved the way for the development of chiroptical switches.[31–34] Rappoport *et al.* were able to deconvolute the energy barriers to two-ring flip enantiomerization and three-ring flip diastereomerization of (*E*)- and (*Z*)-2-*meta*-methoxymesityl-1,2-dimesitylvinyl acetates, based on DNMR and polarimetric studies. They found that the three-ring flip affords a lower energy barrier than α,β-, α,β'-, and β,β'-two-ring circuits in which the nonflipping ring passes through the double bond plane, Scheme 4.2. Polarimetry has also been used for stereodynamic analysis of atropisomeric amides,[35] chiral tricyclic phosphates,[36] biquinazolines exhibiting hindered rotation about a chiral *N*–*N* bond,[37] and biologically relevant compounds such as adrenaline.[38]

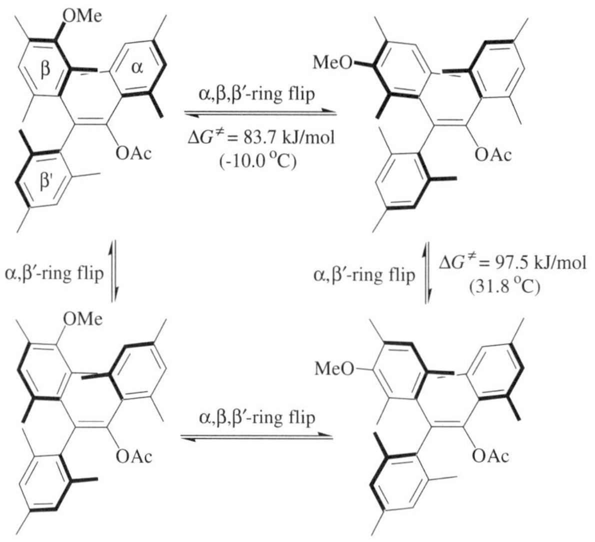

Scheme 4.2   Rotational energy barriers for the α,β,β'-ring flip diastereomerization and α,β'-ring flip enantiomerization of (E)-2-*meta*-methoxymesityl-1,2-dimesitylvinyl acetate.

Alternatively, racemization can be monitored by CD spectroscopy. Harada's group isolated the enantiomers of the (E)- and (Z)-isomers of a crowded octahydrobiphenanthrylidene by HPLC on (+)-poly(triphenylmethylmethacrylate). Thermal racemization studies revealed that enantioconversion of the (Z)-isomer occurs at room temperature, while stereomutation of the (E)-olefin requires temperatures above 55 °C.[39] The barrier to enantioconversion of (Z)-1,1',2,2',3,3',4,4'-octahydro-4,4'-biphenanthrylidene was determined as 97.7 kJ/mol at 24.9 °C, Figure 4.5.[iv]

Resolution of the enantiomers of *N*-nitroso piperidines via crystallization of diastereomeric TADDOL salts enabled Poloński and coworkers to apply CD spectroscopy in racemization studies of these atropisomers which are chiral due to hindered rotation of the nitroso group about the *N–N* bond, Figure 4.6.[40] In addition to racemization studies of polysubstituted helical phenanthrenes,[41] propeller-shaped tetraarylethanes,[42] axially chiral biphenyls,[43] and azapentahelicenes,[44] circular dichroism has also been used to monitor imidozirconium-promoted enantioconversion of allenes[45] and stereomutations of supramolecular assemblies.[46] Gas phase racemization and epimerization kinetics of cyclopropanes, chiral only by virtue of isotope substitution, have been examined by vibrational CD spectroscopy which extends the application scope of chiroptical methods to compounds devoid of a chromophore.[47,48]

The examples of stereochemical and stereodynamic analysis given above underscore the usefulness of chiroptical measurements. However, the need for semipreparative isolation of stereoisomers prior to kinetic analysis is a major drawback of these methods. To overcome this problem, Metcalf and Richardson introduced time-resolved chiroptical luminescence spectroscopy as an elegant alternative. Excitation of a racemic mixture of chiral lanthanide complexes having stereolabile three-bladed propeller-like structures by a pulse of circularly polarized light (CPL) generates a nonracemic excited state population. The detection of emission of circularly polarized light by enantiomerically enriched excited species allows determination of racemization kinetics based on time-resolved circularly polarized luminescence (TR-CPL). Although TR-CPL provides an additional means for kinetic studies with racemic samples, this method is only applicable to luminescent samples such as trisdipicolinate europium complexes. Time-resolved CPL measures molecular

---

[iv] Chiroptical and NMR measurements proved that enantioconversion of neither alkene coincides with *cis*/*trans*-isomerization.

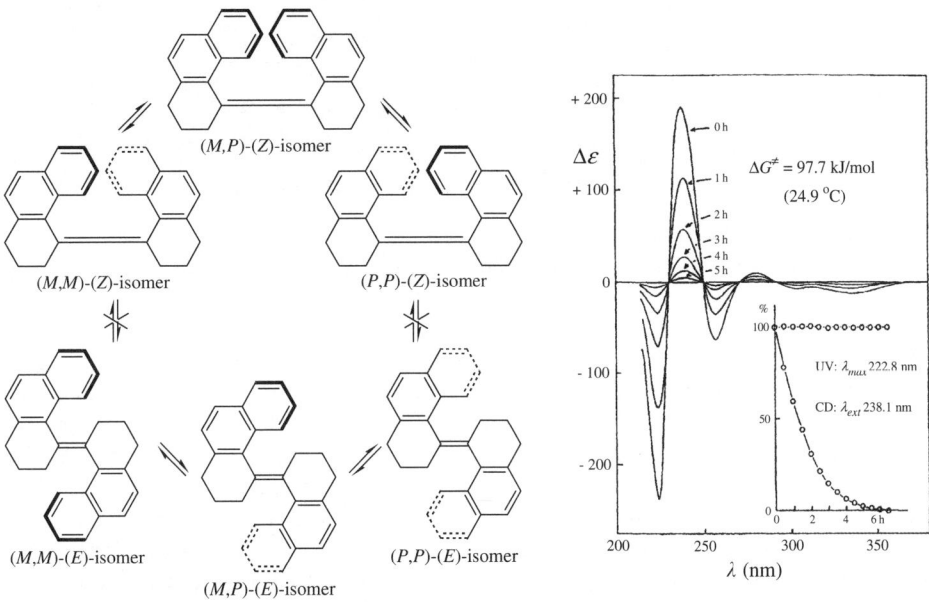

Figure 4.5 Racemization pathways of 1,1′,2,2′,3,3′,4,4′-octahydro-4,4′-biphenanthrylidenes and decrease in the CD signal of the (*M,M*)-(*Z*)-isomer at 24.9 °C. [Reproduced with permission from *J. Am. Chem. Soc.* **1997**, *119*, 7249-7255.]

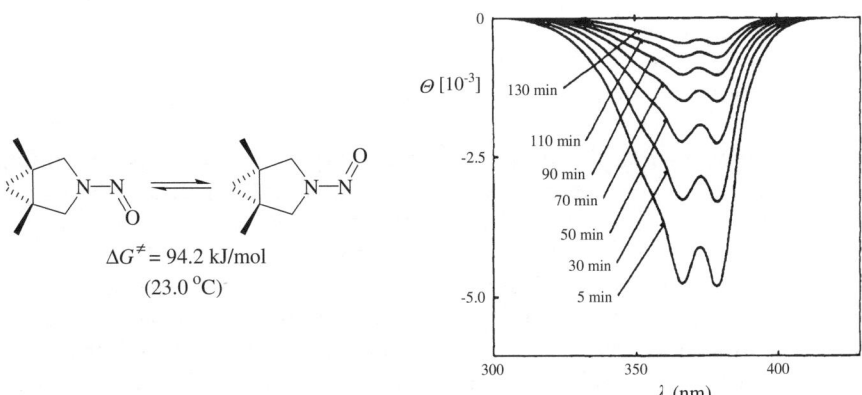

Figure 4.6 Racemization and CD signal decay of 3-nitroso-1,5-dimethyl-3-azabicyclo[3.1.0]hexane. [Reproduced with permission from *J. Org. Chem.* **2001**, *66*, 501-506.]

dynamics of excited states, but it does not afford information about interconversion of species in the ground state.[49–52]

4.2 NMR SPECTROSCOPY

The determination of the composition of enantiomeric or diastereomeric mixtures is routinely performed by NMR spectroscopy. In addition, NMR spectroscopy is invaluable for conformational and configurational analysis of chiral compounds in solution. Although ^{1}H and ^{13}C are the

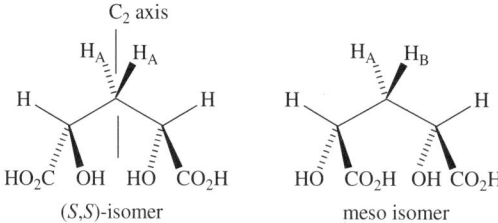

Figure 4.7   Chemical (non)equivalence of methylene protons in 2,4-dihydroxyglutaric acids.

most relevant nuclei, ^{15}N, ^{19}F, ^{29}Si, ^{31}P, and [33]S NMR measurements are also quite useful.[v] The elucidation of the three-dimensional structure or stereoisomeric composition of chiral molecules by NMR spectroscopy requires assessment of homotopic and heterotopic moieties. Homotopic and enantiotopic groups have isochronous NMR signals, but diastereotopic nuclei are chemically nonequivalent and usually exhibit different (anisochronous) chemical shifts. Chemical nonequivalence can originate from the presence of a proximate stereogenic element. However, nuclei remain equivalent when they are exchangeable by a symmetry operation that applies to the whole molecule. For example, the methylene protons of (*S*,*S*)-2,4-dihydroxyglutaric acid can be exchanged by rotation about a C_2-axis, Figure 4.7. The protons are therefore chemically equivalent and part of an A_2B system, whereas the meso isomer carries diastereotopic methylene protons that establish an *ABC* system. Conformational analysis of chiral compounds generally entails correlation of chemical shifts and coupling constants in conjunction with nuclear Overhauser effect spectroscopy (NOESY) and other NMR experiments. The absolute configuration of chiral compounds can be deduced in many cases from empirical chemical shift correlation of covalent or noncovalent adducts. The development of rationally designed chiral derivatizing agents (CDAs),[53–61] chiral solvating agents (CSAs),[62–67] and chiral lanthanide shift reagents (CSRs)[68–70] has extended the application spectrum of structural NMR analysis to a wide range of compounds, Figure 4.8.[71,72]

Diastereoisomers have different chemical and physical properties and often afford distinguishable NMR spectra, which greatly simplifies analysis of diastereomeric mixtures and diastereomerization processes. Since NMR spectra of enantiomers are identical, stereodifferentiation requires formation of diastereomers using a CDA,[73–80] CSA,[81–86] or paramagnetic CSR[87–90] such as lanthanide (Eu, Pr, Yb) chelate complexes, Figure 4.9. The determination of the enantiomeric composition of a sample requires the use of a 100% enantiopure CDA and quantitative conversion of enantiomers to diastereoisomers in order to avoid chiral discrimination effects that can originate from kinetic resolution. This problem can be avoided by formation of noncovalent adducts with a CSA or CSR which, even if not enantiopure, give accurate results if signal integration is not compromised due to reduced signal resolution. The formation of diastereomeric complexes from enantiomeric analytes and a CSA often gives rise to anisochronous NMR shifts. Since CSA/solute association is an equilibrium process, one observes averaged NMR signals for each diastereomeric complex and its corresponding free enantiomer. The observed difference in chemical shifts upon formation of diastereomeric adducts is usually quite small and diminishes with increasing solvent polarity. This can be compensated with a stronger magnetic field, by reducing the temperature if the solubility of the analyte is not compromised, or by addition of an achiral lanthanide shift reagent. The usefulness of CSRs for enantiomeric analysis is limited to compounds that form a coordination complex with Lewis-acidic lanthanides, and frequently suffers from severe peak broadening due to fast chemical

---

[v]The usefulness of nuclei with spin quantum numbers greater than $I = 1/2$ such as [33]S is somewhat limited because the quadrupole moment of these nuclei can cause significant line broadening and thus compromise resolution.

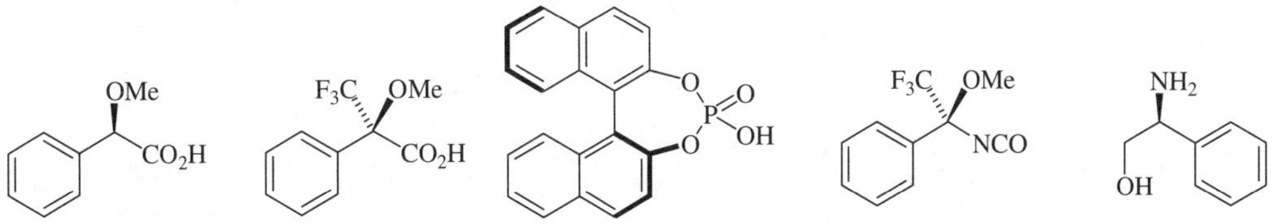

(R)-O-methylmandelic acid   MTPA (Mosher's reagent)   (P)-binaphthyl-2,2′-diyl-hydrogen phosphate   (S)-α-metyhoxy-α-trifluoro-methylbenzyl isocyanate   (S)-phenylglycinol

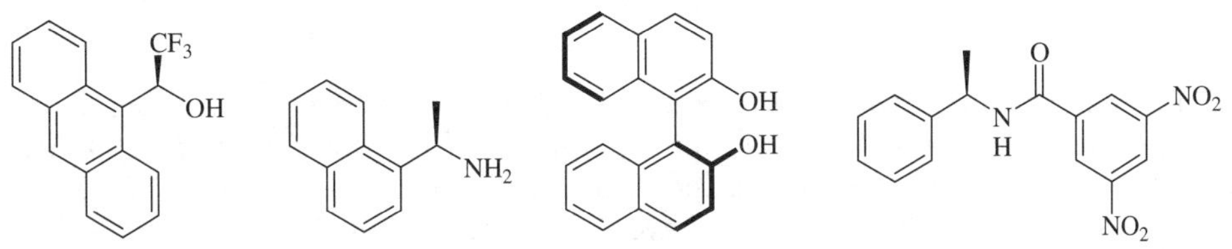

(S)-1-(9′-anthryl)-2,2,2-trifluoroethanol   (R)-1-(1′-naphthyl)-ethylamine   (P)-BINOL   (R)-N-(3,5-dinitrobenzoyl)-α-methylbenzylamine

Figure 4.8 Chiral derivatizing agents (top) and chiral solvating agents (bottom) commonly employed in stereochemical NMR analysis.

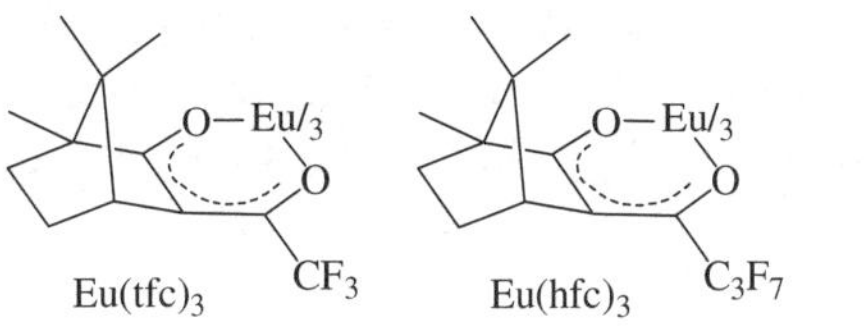

$Eu(tfc)_3$ $Eu(hfc)_3$

Figure 4.9 Structures of chiral NMR lanthanide shift reagents tris(3-trifluoroacetylcamphorato)europium, $Eu(tfc)_3$, and the heptafluorobutyryl analog, $Eu(hfc)_3$.

exchange processes and short longitudinal relaxation times. A major shortcoming of NMR spectroscopy is its inherently low sensitivity, which makes it difficult to detect and quantify impurities below 2%.

Isomerization reactions that involve proton exchange, for example keto/enol-tautomerization, can often be conveniently monitored by ^{1}H NMR spectroscopy in deuterated protic solvents.[91–93] However, this method provides proton/deuterium exchange rates that sometimes deviate from the corresponding rate of racemization. Soda and coworkers examined the enzyme-catalyzed racemization of α-amino-ε-caprolactam in the presence of a pyridoxal-P racemase in deuterium oxide at p^2H 8.8.[94] Subsequent replacement of the α-proton attached to the chiral center of α-amino-ε-caprolactam by deuterium results in the disappearance of the corresponding ^{1}H NMR signal. The isotope exchange was monitored by integration of the proton NMR signal and the proton/deuterium exchange rates of the (R)- and (S)-enantiomer were determined as 170 and 98 min^{-1}, respectively. Polarimetric analysis using isolated enantiomers of α-amino-ε-caprolactam under the same conditions showed that the actual rates of racemization are substantially higher, *i.e.*, 220 and 170 min^{-1}. Comparison of the individual exchange rates obtained by NMR spectroscopy with racemization rates determined by polarimetry thus reveals a pronounced primary isotope effect.

Dynamic processes exhibiting an energy barrier between 20 and 100 kJ/mol can often be studied by analysis of temperature-dependent NMR spectra of the interconverting species.[95] In an achiral environment, the NMR spectra of enantiomers are indistinguishable and this is usually also observed with mixtures of enantiomers. In rare cases, however, pronounced diastereomeric solute/solute-interactions in nonracemic mixtures give rise to distinctive NMR spectra.[96,97] The study of

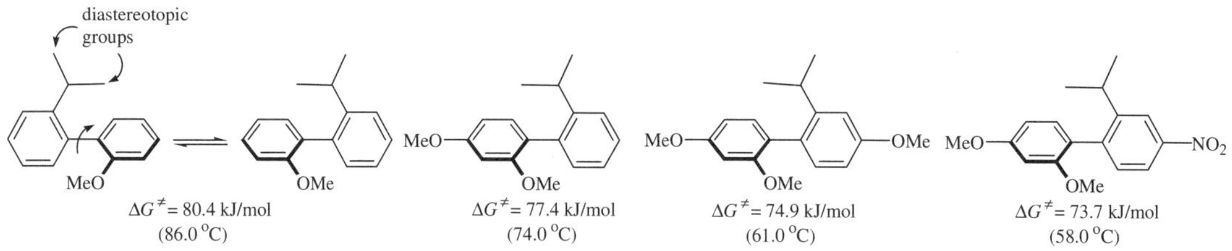

Figure 4.10   Enantioconversion of 2-isopropyl-2′-methoxybiphenyls studied by DNMR.

enantiomerization reactions requires that enantiotopic atoms or groups of the interconverting species are rendered diastereotopic and chemically nonequivalent. This can be accomplished with a nonracemic CSR or CSA.[vi] Diastereomerization reactions can often be monitored in the absence of any additives, although achiral lanthanide shift reagents have been used to increase anisochrony for deconvolution of diastereotopic signals. The study of enantiomerization reactions is greatly facilitated when interconverting species possess diastereotopic nuclei even in the absence of a chiral additive. For example, the two substituents Y in $-CXY_2$, such as the two methyl moieties in an isopropyl group, become diastereotopic and therefore magnetically nonequivalent if the molecule contains a chiral element. This has been observed by Oki and Yamamoto with 2-isopropyl-2′-methoxybiphenyl derivatives.[98] Examination of the ^{1}H NMR spectra of 2-isopropyl-2′-methoxybiphenyl and its derivatives showed that the presence of the chiral axis renders the methyl protons of the isopropyl group diastereotopic. At increased temperature, rotation about the chiral biphenyl axis is relatively fast compared to the NMR time scale and one observes coalescence of the diastereotopic methyl signals. Analysis of the temperature-dependent NMR spectra of 2-isopropyl-2′-methoxybiphenyl gave an activation energy of 80.4 kJ/mol at 86 °C, Figure 4.10. Introduction of methoxy and nitro groups into the *para*-positions of 2-isopropyl-2′-methoxybiphenyl reduces the rotational energy barrier, which can be explained by facilitated out-of-plane bending and push-pull conjugation in the coplanar transition state, see Chapter 3.3.1.

The interconversion of stereoisomers exhibiting diastereotopic nuclei can be monitored by variable-temperature NMR spectroscopy (dynamic NMR) when the reaction is slow on the NMR time scale. This is the case when:

$$k \ll \pi \frac{\Delta v}{\sqrt{2}} \tag{4.5}$$

where k is the interconversion rate constant and Δv is the NMR shift separation of the signals at low temperatures at which exchange does not occur.

A particularly simple situation arises when two nuclei (*AB* case) provide sharp signals with equal intensity and undergo chemical exchange resulting in line broadening and coalescence upon heating, Figure 4.11. In such a case, one can obtain well resolved signals at low temperatures where the two nuclei do not show measurable exchange. As the temperature is increased, the exchange rate becomes fast relative to the NMR time scale and only one averaged species is observed due to signal coalescence. In the absence of any coupling, the first-order interconversion rate constant can be calculated as:

$$k_{T_c} = \pi \frac{\Delta v}{\sqrt{2}} \tag{4.6}$$

[vi] It is important to realize that both CSR and CSA can alter the actual reaction rate and the thermodynamic equilibrium of the interconverting species.

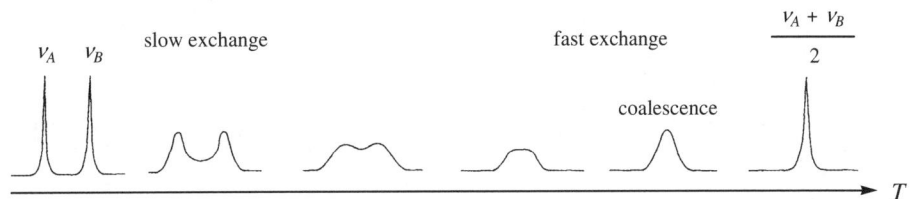

Figure 4.11 Variable-temperature NMR spectra of an uncoupled dynamic *AB* system.

where k_{T_c} is the rate constant at the coalescence temperature T_c, and Δv is the difference in the chemical shifts of the two signals without exchange (at low temperatures, $T \ll T_c$).

Incorporation of Equation 4.3 provides the free energy of activation:

$$\Delta G^{\neq} = 19.1 \cdot 10^{-3} \, T_c \, (9.97 + \lg T_c - \lg |v_A - v_B|) \tag{4.7}$$

where $\Delta G^{\neq}$ is the free energy of activation in kJ/mol; T_c is the coalescence temperature in Kelvin; and v_A and v_B are the chemical shifts of nuclei *A* and *B* in Hz.

If coupling between the exchanging nuclei occurs, the rate constant can be estimated using the following equation:

$$k_{T_c} \approx 2.22 \, \sqrt{(\Delta v^2 + 6 \, J_{AB}^2)} \tag{4.8}$$

where J_{AB} is the coupling constant between nuclei *A* and *B* in Hz.

Variable-temperature NMR spectroscopy was employed by Wolf and Ghebremariam in the study of the diastereomerization of axially chiral 1,8-dipyridylnaphthalenes which exist as almost equimolar mixtures of *syn*- and *anti*-isomers.[99] The sharp proton NMR signals of the naphthalene ring undergo only minor changes within the temperature range studied. In contrast, the diastereotopic methyl proton signals of 1,8-bis(2′-methyl-4′-pyridyl)naphthalene and its *N,N′*-dioxide are well resolved at −17.8 °C but show broadening at increasing temperature and coalescence at 40.3 °C and 11.2 °C, respectively, Figure 4.12. Using Equations 4.6 and 4.7, the corresponding rotational energy barriers were calculated as 67 kJ/mol and 64 kJ/mol. It is noteworthy that the pyridyl signals of the *N,N′*-dioxide are also well resolved and coalesce at 27.5 °C, which corresponds to an energy barrier of 65 kJ/mol.

According to Equation 4.6, k_{T_c} depends on the difference in the chemical shifts of the exchanging signals, Δv, which increases with the NMR measurement frequency and the applied magnetic field. Both k_{T_c} and T_c therefore increase with the magnetic field strength. As described above, a coalescence phenomenon is associated with a specific set of signals, and a compound can have more than one pair of exchanging nuclei with individual coalescence temperatures. If coalescence events are a consequence of the same stereomutation, Equations 4.6 to 4.8 must yield the same activation energy within experimental error.[vii] Although dynamic processes are usually studied by ¹H NMR spectroscopy, DNMR experiments with other nuclei including ¹⁹F, ³¹P, and ¹³C provide valuable alternatives. The inherently larger NMR scale of these nuclei provides superior resolution of chemical shifts and facilitates quantitative analysis of rapidly exchanging signals. The accuracy of the coalescence method is often limited to approximately ±0.5 to 1.0 kJ/mol due to difficulties in determination of the exact coalescence temperature.

[vii] This has been verified by analysis of the exchanging pyridyl and methyl protons, respectively, of 1,8-bis(2′-methyl-4′-pyridyl)naphthalene *N,N′*-dioxide shown in Figure 4.12. The difference in the rotational energy barrier of 1 kJ/mol is within the limited accuracy of the coalescence DNMR method.

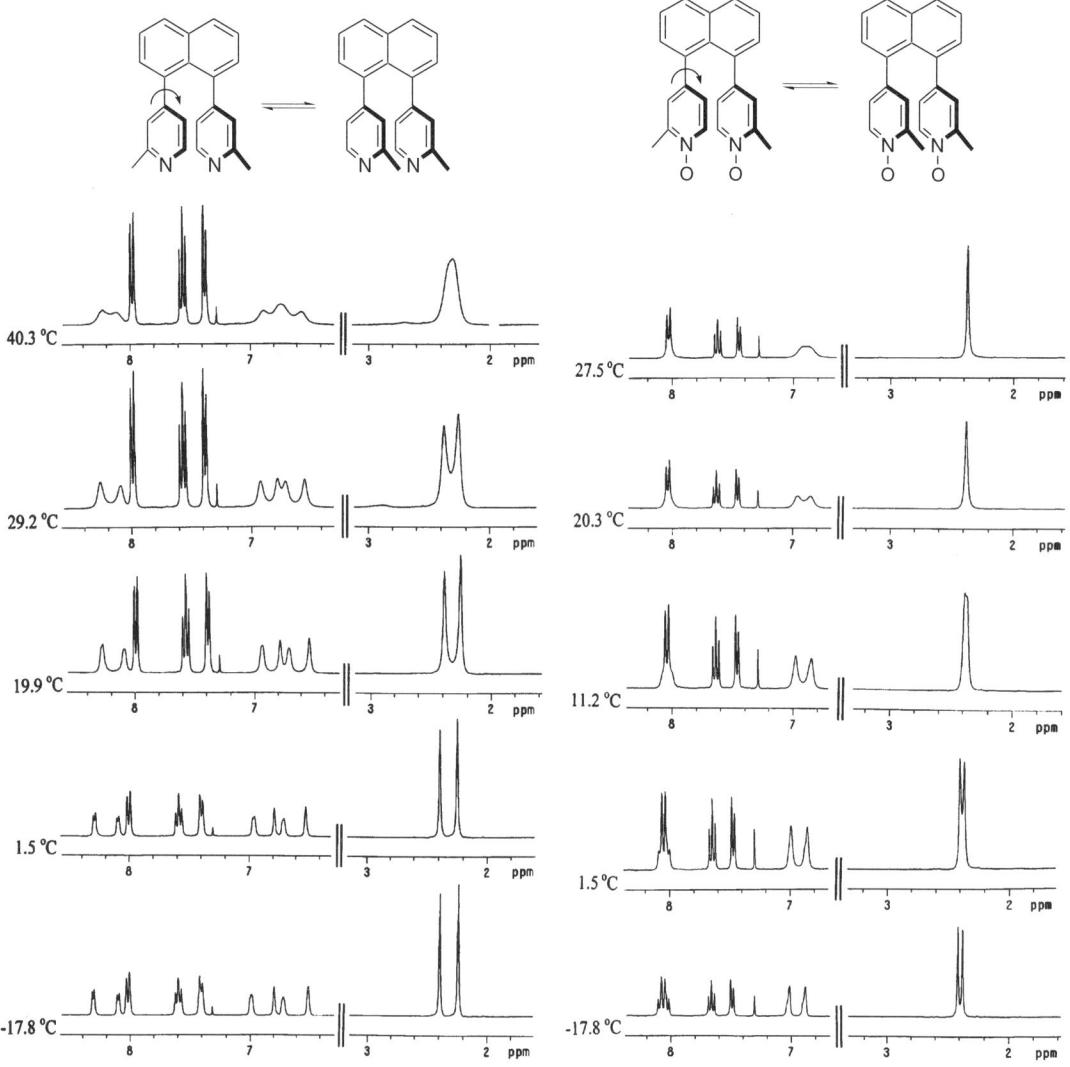

Figure 4.12   Variable-temperature NMR spectra of 1,8-bis(2′-methyl-4′-pyridyl)naphthalene (left) and the
N,N′-dioxide derivative (right). [Reproduced with permission from *Tetrahedron: Asymm.*
2002, *13*, 1153-1156.]

The simplicity of the coalescence method has certainly contributed to its widespread use.
However, this method is not applicable to interconversions of species with different thermodynamic
stability, nor to compounds that possess complicate NMR spectra with multiple coupling patterns.
In these cases, accurate results can be obtained by NMR line shape analysis. The atropisomerization of 1-(1-naphthyl)ethylamines, which exist as a pair of diastereoisomeric racemates bearing a
chiral center and a stereolabile chiral axis, provides a good example, Figure 4.13.[100] According to
PM3 calculations, *N,N*-dibutyl-1-(1-naphthyl)ethylamine affords two diastereomeric rotamers
having the amine moiety in almost perpendicular orientation to the naphthalene ring. Because of
increasing steric repulsion between the methyl group attached to the chiral center and the *peri*-
hydrogen at *C*-8 in the naphthalene ring, the (*S*)-(*M*)-rotamer is considered to be less stable than
the (*S*)-(*P*)-diastereoisomer. The diastereomers have different NMR spectra and integration of well

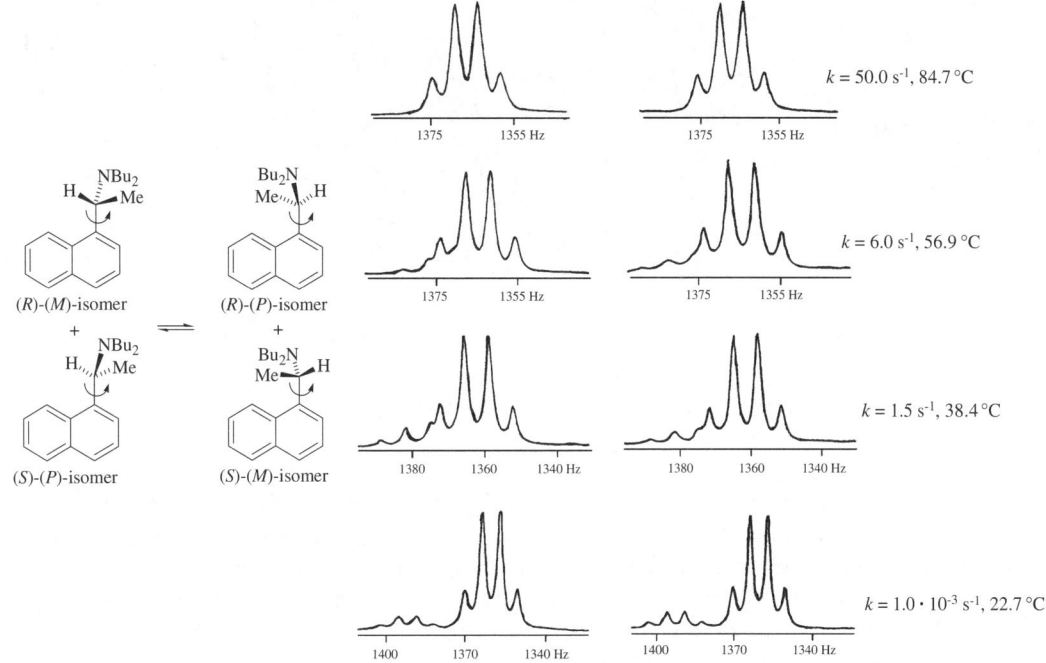

Figure 4.13 Experimental ^{1}H NMR spectra of the methine protons of *N,N*-dibutyl-1-(1-naphthyl)ethyl-amine (left), and computer simulations and corresponding rate constants for the interconversion of the more stable to the less stable rotamer (right). [Reproduced with permission from *Tetrahedron Lett.* **2002**, *43*, 8563-8567.]

resolved signals gave an atropisomeric ratio of 6.5:1, which corresponds to a difference in the Gibbs free energy of 4.6 kJ/mol based on the Boltzmann equation. Screening of various solvents revealed that the diastereotopic quartet signals of the methine protons attached to the chiral center of *N,N*-dibutyl-1-(1-naphthyl)ethylamine are resolved in *ortho*-xylene-d$_{10}$, and variable-temperature NMR spectroscopy yielded coalescence at 84.7 °C.[viii] Individual rate constants were obtained by line shape analysis of experimentally obtained ^{1}H NMR spectra and the rotational energy barrier for the isomerization of the more stable to the less stable atropisomer was calculated as 75.9 ($\pm$0.2) kJ/mol using Equation 4.3.

Variable-temperature NMR spectroscopy has been applied extensively in kinetic studies of stereodynamic processes with energy barriers between 20 and 130 kJ/mol. The energy range covered by DNMR is limited by the NMR time scale, signal resolution and the practical NMR temperature range (-150 °C to $+160$ °C) which ultimately depends on the melting and boiling points of available deuterated solvents. Since DNMR can be used to study stereolabile compounds that rapidly interconvert at room temperature, it has become a powerful tool for analysis of single bond rotations. Similarly to the NMR study of 2-isopropyl-2′-methoxybiphenyls mentioned above, atropisomerization barriers of many other biaryls have been determined by the coalescence method and line shape analysis. Charlton and coworkers investigated hindered rotation about the chiral axis of arylnaphthalene lignans having half-lives of less than 10 minutes. They observed coalescence of diastereotopic protons of justicidin A and other natural products and determined energy barriers to rotation about the chiral biaryl axis, ranging from 70 to 90 kJ/mol, Figure 4.14.[101] Lunazzi and others examined stereomutations of diastereomeric diarylbenzenes

---

[viii] The coalescence method and Equation 4.6 can and not be used for kinetic analysis of this atropisomerization reaction because the rotamers exist in unequal amounts and afford two distinct interconversion rates.

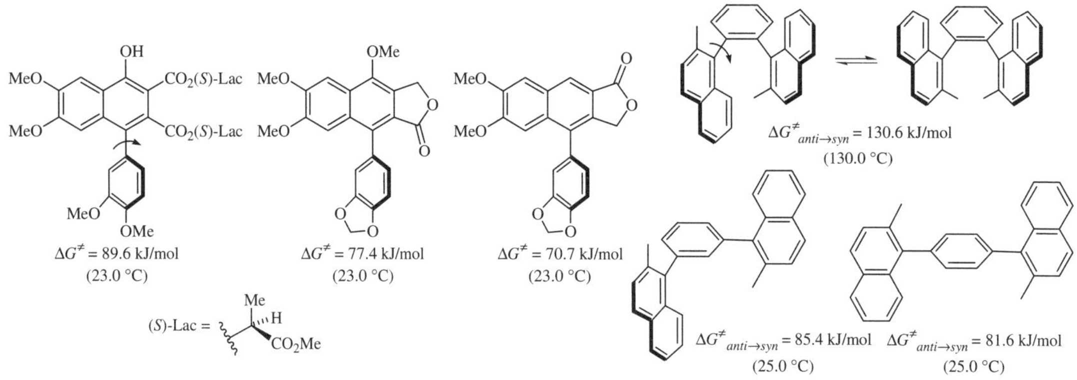

Figure 4.14   Rotational energy barrier of diastereomeric lignan biaryls and triaryls determined by line shape
analysis.

such as 1,2-bis(2′-methyl-1′-naphthyl)benzene.[102,103] By analogy with 1,8-diarylnaphthalenes, these
triaryls exist as a mixture of meso *syn-* and chiral *anti*-isomers. An exception is 1,4-bis(2′-methyl-1′-
naphthyl)benzene, which is achiral in both conformations due to the presence of a symmetry plane
in the *syn-* and an inversion center in the *anti*-rotamer. Line shape analysis of experimentally
obtained NMR spectra revealed that the energy barriers for the *anti*-to-*syn*-isomerization of these
atropisomers span a range of 80 to more than 130 kJ/mol.

Alkyl aryl ketones afford a mixture of enantiomeric conformations exhibiting almost perfectly
orthogonal carbonyl and aryl planes. Gasparrini, Lunazzi and others showed that the existence
of twisted conformations and rotation about the aryl–carbonyl bond can be detected by ^{1}H and
^{13}C NMR spectroscopy when prochiral probes such as an isopropyl group are present, or if
necessary with the help of chiral solvating agents, Scheme 4.3.[104–106] Activation energies obtained by
simulation of the exchange of different proton and carbon nuclei were in excellent agreement and
solvent effects on the carbonyl rotation proved to be negligible. Interconversion of diastereoisomeric
diacyl arenes and atropisomerization of bridged biaryl lactones have also been explored by
DNMR.[107–110] The dynamic stereochemistry of many other nonbiaryl atropisomers has been
studied by variable-temperature NMR spectroscopy. An interesting situation arises with naphthylamines that adopt twisted conformations in the ground state to minimize steric repulsion between
nitrogen substituents and the aryl ring. The atropisomerism of a series of hindered *N,N*-dialkyl-1-
naphthylamines has been examined by Lunazzi and coworkers using line shape analysis of NMR
spectra recorded between 16 and 165 °C, Figure 4.15.[111,ix] Similar structures and conformational
equilibria have been observed with aryl sulfoxides and aryl sulfones.[112,113] Rotation about the chiral
carbon–sulfur axis of 1-naphthyl sulfones is evident from the exchange of the diastereotopic signals
of the geminal hydrogens in 1-(ethanesulfonyl)-2-methylnaphthalene and of the geminal methyl
groups in 1-(isopropylsulfonyl)-2-methylnaphthalene. The absence of a diastereotopic probe in the
tert-butyl sulfone derivative requires the use of a chiral solvating agent such as (*S*)-2,2,2-trifluoro-
1-(9′-anthryl)ethanol for NMR analysis. By contrast, alkyl aryl sulfoxides possess a chiral center at
the sulfur atom in addition to the chiral *C–S* axis, and therefore exist as a pair of diastereomeric
E/Z-conformers that can easily be distinguished by NMR spectroscopy in the absence of a chiral
additive. Diastereomerization of several 1-(alkylsulfinyl)-2-methylnaphthalenes has been examined
between −70 and +100 °C, Figure 4.15.

[ix] Concurrent nitrogen inversion of these compounds has a very low energy barrier and is only observable under cryogenic
conditions. The two processes therefore do not interfere, which facilitates determination of the free energy of activation for
the rotation about the chiral carbon–nitrogen axis.

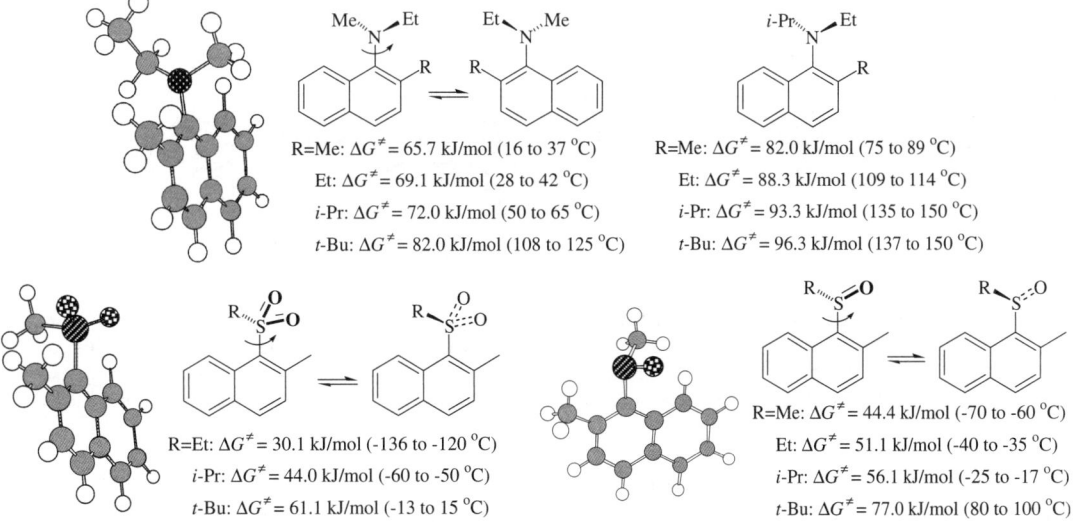

Scheme 4.3 Rotational energy barrier of atropisomeric acyl arenes and biaryl lactones.

R=Me: $\Delta G^{\neq}$ = 33.9 kJ/mol (−110 °C)
Et: $\Delta G^{\neq}$ = 38.9 kJ/mol (−92 °C)
i-Pr: $\Delta G^{\neq}$ = 47.3 kJ/mol (−55 °C)
t-Bu: $\Delta G^{\neq}$ = 83.7 kJ/mol (−78 °C)

R=Me: $\Delta G^{\neq}$ = 33.8 kJ/mol (−70 °C)
Et: $\Delta G^{\neq}$ = 43.5 kJ/mol (−55 °C)
i-Pr: $\Delta G^{\neq}$ = 47.3 kJ/mol (−55 °C)
t-Bu: $\Delta G^{\neq}$ = 55.3 kJ/mol (−50 °C)

R=Me: $\Delta G^{\neq}$ = 42.7 kJ/mol (−120 °C)
Et: $\Delta G^{\neq}$ = 48.1 kJ/mol (−80 °C)
i-Pr: $\Delta G^{\neq}$ = 56.1 kJ/mol (−55 °C)
t-Bu: $\Delta G^{\neq}$ = 92.5 kJ/mol (22 °C)

R=CH$_2$Me: $\Delta G^{\neq}$ = 28.0 kJ/mol (−140 °C)
CH$_2$Et: $\Delta G^{\neq}$ = 31.8 kJ/mol (−118 °C)
CHMe$_2$: $\Delta G^{\neq}$ = 35.2 kJ/mol (−110 °C)
CH$_2$Ph: $\Delta G^{\neq}$ = 37.7 kJ/mol (−95 °C)

R=Me, R′=*i*-Pr: $\Delta G^{\neq}$ = 73.9 kJ/mol (25 °C)
R=Et, R′=*i*-Pr: $\Delta G^{\neq}$ = 83.0 kJ/mol (25 °C)
R=OMe, R′=Et: $\Delta G^{\neq}$ = 54.4 kJ/mol (25 °C)
R=Me, R′=Et: $\Delta G^{\neq}$ = 72.1 kJ/mol (25 °C)

R=Me: $\Delta G^{\neq}$ = 65.7 kJ/mol (16 to 37 °C)
Et: $\Delta G^{\neq}$ = 69.1 kJ/mol (28 to 42 °C)
i-Pr: $\Delta G^{\neq}$ = 72.0 kJ/mol (50 to 65 °C)
t-Bu: $\Delta G^{\neq}$ = 82.0 kJ/mol (108 to 125 °C)

R=Me: $\Delta G^{\neq}$ = 82.0 kJ/mol (75 to 89 °C)
Et: $\Delta G^{\neq}$ = 88.3 kJ/mol (109 to 114 °C)
i-Pr: $\Delta G^{\neq}$ = 93.3 kJ/mol (135 to 150 °C)
t-Bu: $\Delta G^{\neq}$ = 96.3 kJ/mol (137 to 150 °C)

R=Et: $\Delta G^{\neq}$ = 30.1 kJ/mol (-136 to -120 °C)
i-Pr: $\Delta G^{\neq}$ = 44.0 kJ/mol (-60 to -50 °C)
t-Bu: $\Delta G^{\neq}$ = 61.1 kJ/mol (-13 to 15 °C)

R=Me: $\Delta G^{\neq}$ = 44.4 kJ/mol (-70 to -60 °C)
Et: $\Delta G^{\neq}$ = 51.1 kJ/mol (-40 to -35 °C)
i-Pr: $\Delta G^{\neq}$ = 56.1 kJ/mol (-25 to -17 °C)
t-Bu: $\Delta G^{\neq}$ = 77.0 kJ/mol (80 to 100 °C)

Figure 4.15 Atropisomerism, PM3 computed ground states, and rotational energy barriers of selected naphthylamines, sulfones and sulfoxides.

Line shape analysis of temperature-dependent NMR spectra has been applied to numerous atropisomers including bis(1-phenyl-9-anthryl)ethynes,[114] 3-(*ortho*-aryl)-5-methylrhodanines,[115] *ortho*-substituted (*Z*)-benzophenone oximes and benzyl aryl oximes,[116,117] *N*-aryl tetrahydropyrimidines,[118] *N*-naphthyl imines,[119] highly substituted 1,3-dienes,[120] pyridine-derived adamantanes,[121] and secondary

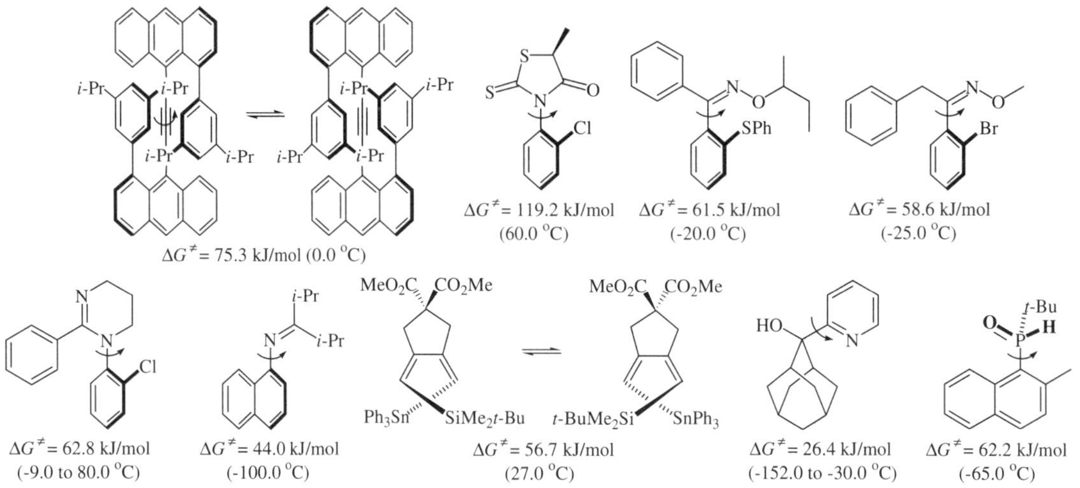

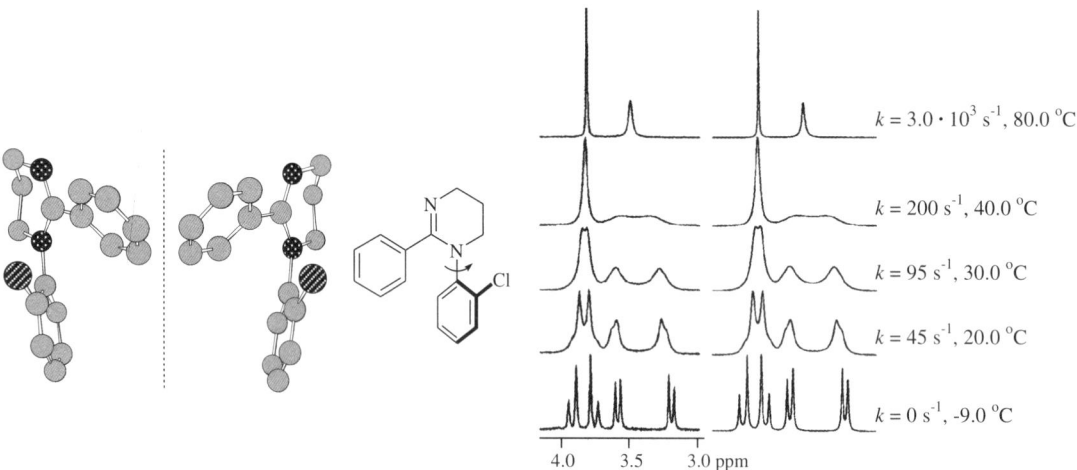

Figure 4.16   Stereomutations studied by DNMR spectroscopy.

Figure 4.17   Computed ground states of the interconverting enantiomers of 1-(2′-chlorophenyl)-2-phenyl-1,4,5,6-tetrahydropyrimidine (left), and experimental and simulated NMR signals of the tetrahydropyrimidine methylene protons at various temperatures (right). Hydrogens are omitted for clarity. [Reproduced with permission from *J. Org. Chem.* **2001**, *66*, 6679-6684.]

phosphine oxides, Figures 4.16 and 4.17.[122] Deconvolution of concurrent atropisomerization and ring flipping of tricyclic pyridones derived from a nonplanar seven-membered ring has been accomplished by a combination of HPLC racemization studies of isolated enantiomers and two-dimensional NMR analysis (EXSY) of relatively fast diastereomerization resulting from rotation about the chiral axis, Scheme 4.4.[123] Dynamic EXSY has also been used by Paquette *et al.* to determine the rates of ring inversion and bond shifting in chiral 1,3-bridged cyclooctatetraenes.[124] Other stereodynamic processes that have been investigated by DNMR spectroscopy include enantiomerization of chiral cyclohexene oxide conformers in solution and in the solid state,[125,126] electrocyclic rearrangements of homofuran and its derivatives,[127] interconversion of helical conformations of overcrowded naphthalenes,[128] ring flipping between chiral conformations of

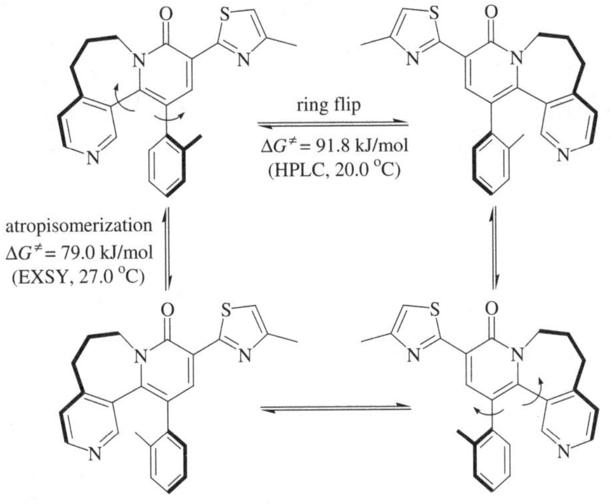

Scheme 4.4 Stereodynamics of a tricyclic pyridone.

triarylvinyl derivatives,[129] cog-wheeling of propeller molecules,[130–134] helical inversion of dinuclear bis(oxazolinyl)pyrrole-derived palladium complexes,[135] epimerization of a uranyl-salophen complex,[136] and stereomutations of supramolecular assemblies such as helical inversion of donor-acceptor catenanes.[137]

4.3 DYNAMIC CHROMATOGRAPHY

Chromatography (GC, HPLC, SFC, SubFC, CEC, MEKC, and TLC) and capillary zone electrophoresis (CZE) provide a powerful means for the determination of the enantiomeric and diastereomeric purity of chiral compounds.[138] Chiral chromatography is also very well suited for preparative isolation of stereoisomers. Repetitive batch and simulated moving bed chromatography (SMB) are commonly used for HPLC, SubFC and SFC separation of stereoisomers on the semipreparative (mg to g amounts) or industrial scale (kg quantities).[139,140] Prior to the development of chiral stationary phases, chromatographic analysis of an enantiomeric mixture was accomplished by formation of diastereoisomers with an enantiopure chiral derivatizing agent (CDA) and subsequent separation on an achiral column, Figure 4.18. A shortcoming of this approach is that detection of diastereomers by UV spectroscopy or other methods necessitates cumbersome correction of integration results with individual response factors in order to obtain accurate ee values.

The choice of a suitable CDA depends on the structure and functional groups of the analyte. Basic requirements for common GC/FID analysis, *e.g.*, thermal stability and volatility, or sufficient solubility for SFC, SubFC, CEC, TLC, MEKC, and HPLC separation, also need to be addressed. As is discussed in Chapter 4.2, stereochemical analysis with chiral derivatizing agents is generally not very attractive because it entails an additional synthetic step and has other severe drawbacks. The preparation of separable diastereomers for analysis of the enantiomeric composition of a chiral sample is not a trivial task; enantiomer discrimination due to incomplete derivatization or partial isolation of the diastereomeric mixture gives false results. Similarly, enantiomeric impurities in a CDA alter the diastereomeric product composition and lead to incorrect calculation of the ee. This can be avoided by direct enantioseparation in the presence of chiral mobile phase additives that form transient diastereomeric complexes during chromatographic separation. But the scope of this

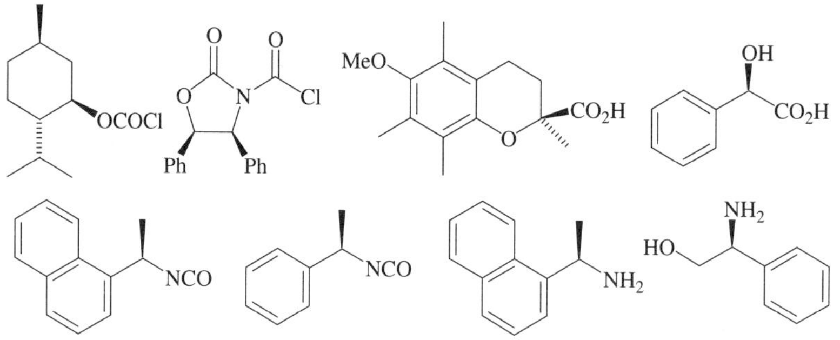

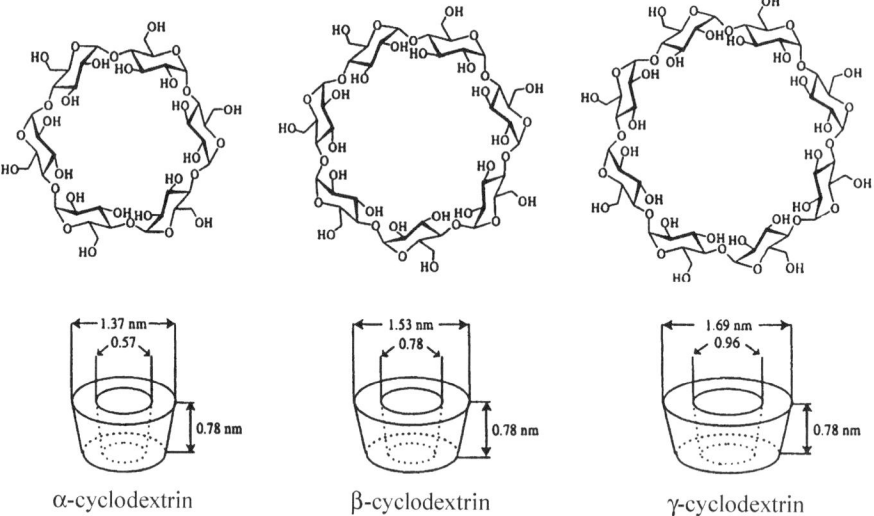

Figure 4.18   Chiral derivatizing agents used for chromatographic analysis of enantiomeric mixtures.

α-cyclodextrin   β-cyclodextrin   γ-cyclodextrin

Figure 4.19   Structures and dimensions of native cyclodextrins that can be selectively derivatized in positions 2, 3 and 6 of each glucose unit.

approach is limited due to the low stability and insufficient resolution of noncovalent adducts.[141,142] The shortcomings of the methods described above have been overcome by the introduction of chiral stationary phases (CSPs). To date, a broad variety of CSPs has been developed for direct separation of enantiomers. Enantioselective chromatography greatly facilitates quantification because enantiomers have the same individual response factors (in an achiral environment) and the ratio can readily be obtained by peak area integration.[x] Numerous CSPs derived from α-, β- and γ-cyclodextrins, short peptides, and chiral metal chelates have been introduced to gas chromatography (GC), Figures 4.19 and 4.20.[143–146] In particular, GC on selectively acylated, alkylated or silylated cyclodextrins and metal camphorate complexes has been established by König, Schurig and others.[147–151] Water-soluble cyclodextrin derivatives are frequently employed

---

[x] For convenience, the term chiral chromatography has been coined to refer to separation of enantiomers on a CSP and is commonly used instead of the more accurate term enantioselective chromatography. Chromatography is of course enantioselective and is not inherently chiral. A similar situation arises with the term asymmetric synthesis which is widely used to describe enantioselective and diastereoselective synthesis.

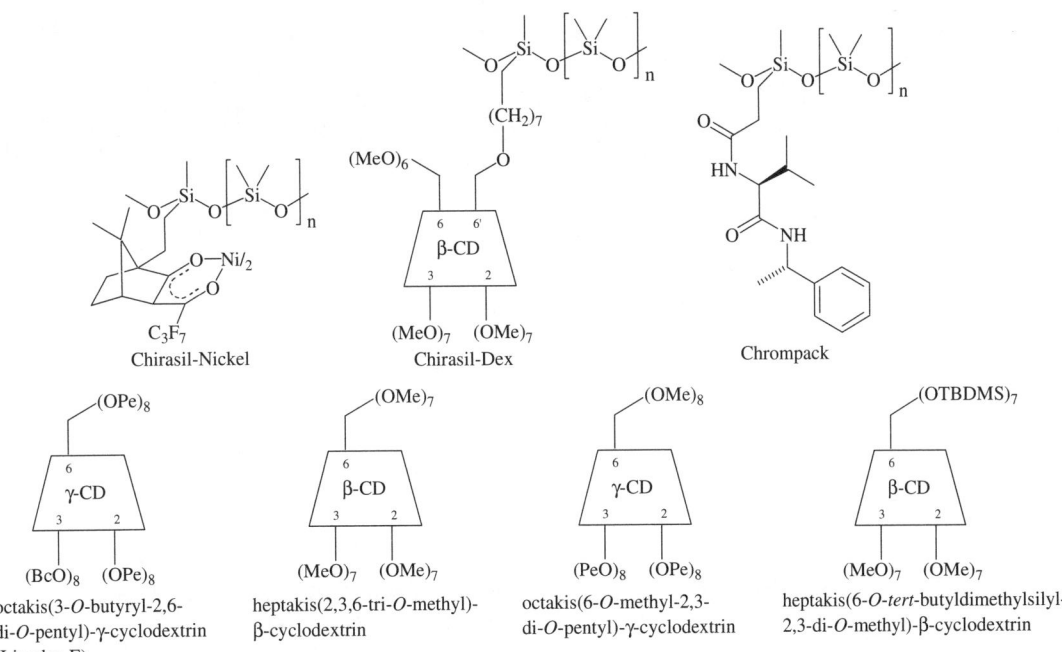

Chirasil-Nickel

Chirasil-Dex

Chrompack

octakis(3-*O*-butyryl-2,6-
di-*O*-pentyl)-γ-cyclodextrin
(Lipodex E)

heptakis(2,3,6-tri-*O*-methyl)-
β-cyclodextrin

octakis(6-*O*-methyl-2,3-
di-*O*-pentyl)-γ-cyclodextrin

heptakis(6-*O*-*tert*-butyldimethylsilyl-
2,3-di-*O*-methyl)-β-cyclodextrin

Figure 4.20 Structures of CSPs commonly used for GC enantioseparations.

as chiral additives in enantioselective CZE, or in combination with surfactants in micellar electrokinetic chromatography (MEKC).[152–156]

Cellulose and amylose derivatives, including microcrystalline cellulose triacetate, Chiralcel OD and Chiralpak AD, and so-called brush-type CSPs developed by Pirkle and coworkers, such as Whelk-O 1, *N*-(3,5-dinitrobenzoyl)phenylglycine and other amino acid derivatives, allow convenient enantioseparation of a broad range of chiral compounds by HPLC,[157–164] SFC,[165–170] SubFC,[171–174] and CEC,[175–180] Figure 4.21. Importantly, rationally designed CSPs exhibiting a reasonably well understood chiral recognition mechanism can be used to elucidate the absolute configuration of chiral compounds based on predictable elution orders.[181,182] Gas chromatography (GC), high performance liquid chromatography (HPLC) and supercritical fluid chromatography (SFC) are the most widely used techniques for separation of enantiomers on both analytical and preparative scales. Direct chromatographic resolution of enantiomers is based on the formation of transient diastereomeric complexes between enantiomeric solutes and a chiral stationary phase. The separation is thus a result of different partitioning of the enantiomers between an achiral mobile phase and a CSP. Alternatively, enantioseparation can be achieved by MEKC with sodium cholate or other chiral mobile phase additives. This technique is based on selective distribution of solutes between an electroosmotically driven mobile phase and chiral micelles. In MEKC, chromatographic separation results from enantioselective partitioning and different electrophoretic mobilities of the free solutes and their micellar complexes.

By analogy with variable-temperature NMR spectroscopy, peak coalescence phenomena can be observed in chromatography as a consequence of simultaneous resolution and on-column interconversion. A successful chromatographic separation of stereoisomers affords two distinct peaks. However, stereolabile compounds can undergo interconversion during the chromatographic process. Competition between separation and isomerization gives rise to elution profiles showing a plateau between the peaks. Since plateau formation is due to on-column isomerization, both the reaction rate and the height of the plateau increase with temperature. When the interconversion is

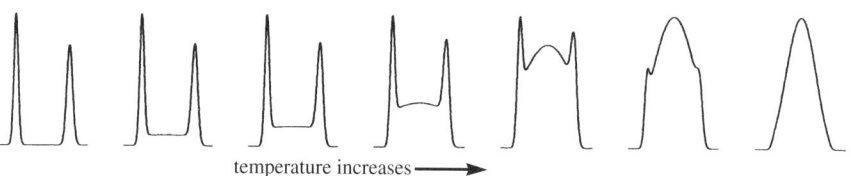

Figure 4.21 Structures of CSPs commonly employed in enantioselective HPLC, SFC and SubFC.

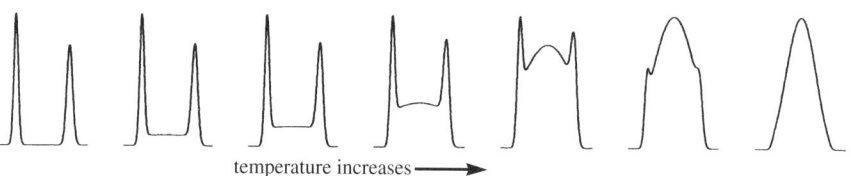

Figure 4.22 Temperature-dependent plateaus and peak coalescence due to simultaneous chromatographic resolution and on-column isomerization.

much faster than the chromatographic separation one observes peak coalescence, Figure 4.22. Computer simulation of chromatographic elution profiles provides individual rate constants and activation parameters of the isomerization.[183]

Based on the rate of enantioconversion of a stereolabile compound one can expect three different chromatographic scenarios: (a) two baseline-separated peaks are obtained if enantiomerization is slow compared to the chromatographic time scale; (b) competition between chromatographic resolution and enantiomerization generates a temperature-dependent plateau between the two peaks; and (c) only one peak is observed when enantiomerization is fast relative to the time required for the separation. These three cases are observed when a mixture of 2,2'-diisopropyl-, 2-*tert*-butyl-2'-isopropyl-, and 2,2'-di-*tert*-butylbiphenyl is analyzed by chiral GC at 165 °C, Figure 4.23.[184] Because of the high conformational stability of 2,2'-di-*tert*-butylbiphenyl, both enantiomers are separated and show no sign of interconversion. As is discussed in Chapter 3.3.1, replacement of the sterically demanding *tert*-butyl moiety by the smaller isopropyl group decreases the rotational energy barrier, and 2-*tert*-butyl-2'-isopropylbiphenyl undergoes significant isomerization in under 10 minutes at 165 °C. Finally, 2,2'-diisopropylbiphenyl, which can be separated into enantiomers on cyclodextrin-derived CSPs at lower temperatures, is not stable to enantiomerization at 165 °C and only one racemic peak is observed.

Chromatography is often described as a discontinuous process using a theoretical plate model in which the column is segmented into *N* theoretical plates. Each plate affords selective distribution of the analytes between the mobile and the stationary phase. Upon completion of the partitioning

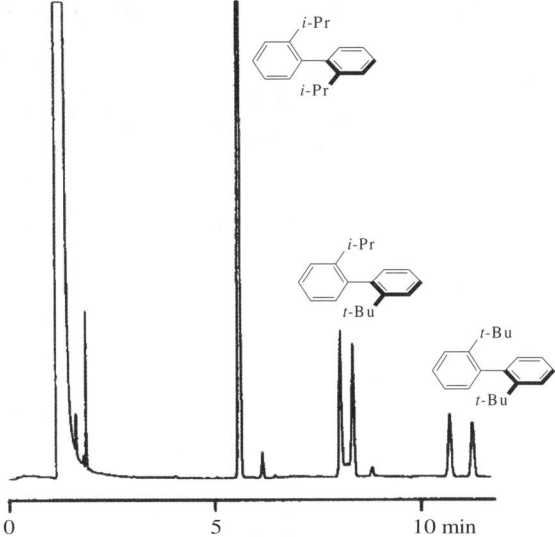

Figure 4.23 GC resolution of 2,2′-disubstituted biphenyls at 165 °C on heptakis(6-*O*-*tert*-butyldimethylsilyl-2,3-di-*O*-methyl)-β-cyclodextrin. [Reproduced with permission from *Liebigs Ann.* **1995**, 781-786.]

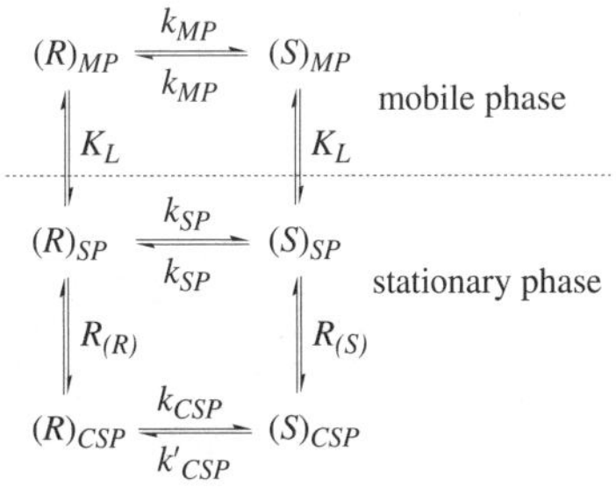

$$
\begin{array}{ccc}
(R)_{MP} & \underset{k_{MP}}{\overset{k_{MP}}{\rightleftharpoons}} & (S)_{MP} \\
\Big\updownarrow K_L & & \Big\updownarrow K_L \qquad \text{mobile phase} \\
(R)_{SP} & \underset{k_{SP}}{\overset{k_{SP}}{\rightleftharpoons}} & (S)_{SP} \\
\Big\updownarrow R_{(R)} & & \Big\updownarrow R_{(S)} \qquad \text{stationary phase} \\
(R)_{CSP} & \underset{k'_{CSP}}{\overset{k_{CSP}}{\rightleftharpoons}} & (S)_{CSP}
\end{array}
$$

Scheme 4.5 Equilibria between enantiomers of a chiral compound in a theoretical plate.

process in each plate, the mobile phase and its components are shifted to the next theoretical plate to participate in another separation step, and so on. To account for isomerization that is competing with the chromatographic separation processes, each theoretical plate is considered a chemical reactor, Scheme 4.5.[185–188] Schurig and coworkers were first to incorporate both resolution and enantiomerization of chiral compounds into a computer program based on the theoretical plate model.[189] Their seminal work was followed by the development of user-friendly software packages that simplify simulation of enantioconversion and partitioning equilibria, including achiral and chiral contributions to chromatographic retention. The partition coefficient, K_L, describes the distribution of the enantiomers between the mobile phase, MP, and achiral sites in the stationary phase, SP. The retention increment, R, originates from selective partitioning of the enantiomers between the achiral and the chiral sites within the stationary phase. Enantioconversion occurs in the gas phase and in the achiral and chiral part of the stationary phase, which is described by four different rate constants, k_{MP}, k_{SP}, k_{CSP}, and k'_{CSP}. The simulation of elution profiles requires determination of the number of theoretical plates, N, from the retention time of a peak, t_R, and the peak width at half height, $b_{0.5}$, as well as the enantioselectivity factor, α, which can be obtained from the void volume, t_0, and the retention times of the enantiomers, t_1 and t_2, according to Equations 4.9 and 4.10:

$$N = 5.54\left(\frac{t_R}{b_{0.5}}\right)^2 \tag{4.9}$$

$$\alpha = \frac{t_1 - t_0}{t_2 - t_0} \text{ where } t_1 > t_2 \tag{4.10}$$

Iterative optimization of enantiomerization rates provides simulated elution profiles that are superimposable with experimentally obtained chromatograms. Wolf and coworkers observed plateau formation during GC enantioseparation of 2,2'-bis(trifluoromethyl)biphenyl on octa-kis(6-*O*-methyl-2,3-di-*O*-pentyl)-γ-cyclodextrin at different temperatures, Figure 4.24.[190] Repetitive computer simulation of the experimentally obtained chromatograms with varying rates of enantiomerization furnished identical elution profiles, and a rotational energy barrier of 109.7 (±0.2) kJ/mol. These results were confirmed by polarimetric studies of enantioenriched samples, proving compatibility of the two techniques. Dynamic chromatography allows determination of reaction rates and activation parameters with high precision. A comparison of computer simulations obtained by variation of the energy barrier to enantiomerization demonstrates that an energetic difference of less than 0.2 kJ/mol can easily be visualized, Figure 4.25.

Since all parameters required for the computer simulation are readily obtained from the chromatogram, dynamic chromatography has become a widely used technique. In particular, DGC and DHPLC have been exploited for the determination of enantiomerization barriers of a broad variety of chiral compounds and constitute a powerful alternative to NMR and chiroptical

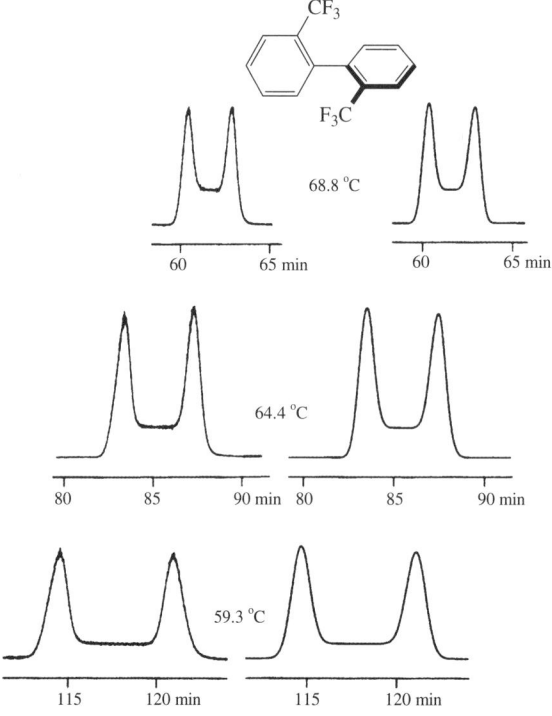

Figure 4.24 Comparison of experimentally obtained (left) and simulated chromatograms (right) of 2,2'-bis(trifluoromethyl)biphenyl. [Reproduced with permission from *Liebigs Ann.* **1996**, 357-363.]

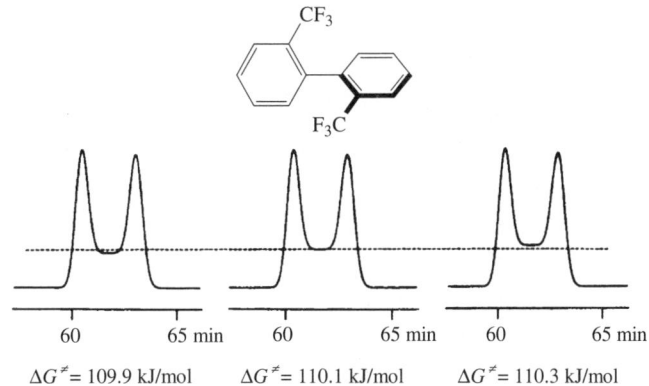

Figure 4.25 Comparison of elution profiles computed with different rate constants and energy barriers.

techniques.[191–193] A major advantage of dynamic chromatography over chiroptical methods is that preparative separation of enantiomers is not necessary. Furthermore, only minute amounts of a racemic sample are used, and chiral or achiral impurities do not interfere with the measurements as they are usually removed during the chromatographic process. On the other hand, a CSP capable of separating the enantiomers of interest at various temperatures remains an indispensable prerequisite. It is important to realize that on-column enantiomerization proceeds partly in a chiral environment which can alter activation parameters. Because of the inherently different time scales and operating temperatures, stereomutations spanning a wide range of activation energies can be investigated with dynamic HPLC, GC, SFC, SubFC, and MEKC, *vide infra*.[194]

4.3.1 Dynamic High Performance Liquid Chromatography

In 1982, Horváth and coworkers observed *cis/trans*-isomerization of small peptides using reversed phase HPLC. They found that proline-derived peptides undergo rotation about the amide bond during chromatographic separation under neutral and acidic conditions at ambient temperatures. Investigating peptide isomerization by chromatographic means, Horváth's team was first to employ HPLC in kinetic studies of interconverting stereoisomers. They were able to determine the two rate constants for the reversible first-order interconversion of *cis*- and *trans*-diastereoisomers of (*S*)-alanine-(*S*)-proline and (*S*)-valine-(*S*)-proline dipeptides.[195–197] Mannschreck's group then extended dynamic HPLC to enantiomerization reactions. Using various chiral stationary phases in conjunction with both photometric and polarimetric detection, they discovered that the enantiomers of 2,2′-spirobichromenes interconvert during the chromatographic process due to reversible thermal ring opening.[198] Simulation of the elution profiles obtained by HPLC on tribenzoyl cellulose revealed an energy barrier to ring opening of 100.0 kJ/mol at 45.0 °C, Figure 4.29. This was confirmed by polarimetric studies of the racemization of enantiomerically enriched samples of 2,2′-spirobichromene under the same conditions. Mannschreck's group also examined the enantiomerization of axially chiral 1-dimethylamino-8-dimethylcarbamoylnaphthalene and helical 1,3,7-trimethylbenzo[c]phenanthrene by HPLC on triacetyl cellulose.[199,200] Veciana and Crespo exploited DHPLC for stereodynamic analysis of tris(2,4,6-trichlorophenyl)methane and obtained an enantiomerization barrier of 80.4 kJ/mol at −5.0 °C using a (+)-poly(triphenylmethylmethacrylate) column, Figure 4.29.[201] Cabrera *et al.* introduced DHPLC on ChiraDex to enantiomerization studies of chiral pharmaceutical drugs such as chlorthalidone, oxazepam and prominal, and found that the stereochemical stability and pharmacokinetic integrity of these chiral drugs strongly depend on the solvent and pH, see also Chapter 3.4.[202,203] Other interesting

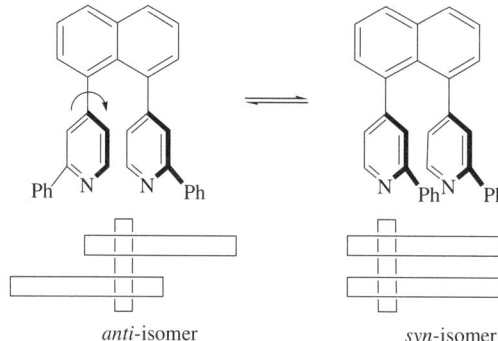

anti-isomer   *syn*-isomer

Scheme 4.6 Isomerization of the conformational isomers of 1,8-bis(2′-phenyl-4′-pyridyl)naphthalene having either parallel (*syn*-isomer) or antiparallel (*anti*-isomer) 2-phenylpyridyl rings.

DHPLC studies have been performed with stereolabile trityloxymethyl butyrolactol[204] and with axially chiral biaryl derivatives of 4-(dimethylamino)pyridine in order to identify conformationally stable catalyst candidates for kinetic resolution of secondary alcohols.[205]

Wolf and Tumambac explored the isomerization of 1,8-bis(2′-phenyl-4′-pyridyl)naphthalene over a temperature range of more than 100 °C using dynamic HPLC and variable-temperature NMR spectroscopy. Rotation of either pyridyl ring about the pyridyl–naphthalene bond results in interconversion of a meso *syn*- and two axially chiral *anti*-isomers, Scheme 4.6.[206] The reaction involves a sterically crowded T-shaped transition state in which the edge of the rotating pyridyl ring is directed towards the adjacent pyridyl moiety. Dynamic NMR studies showed that diastereomerization between the almost equienergetic rotamers of 1,8-bis(2′-phenyl-4′-pyridyl)naphthalene proceeds rapidly at room temperature and has an energy barrier to rotation about the chiral naphthyl–pyridyl axis of 73 kJ/mol at 40.3 °C. Chromatographic baseline separation of the *syn*- and *anti*-isomers of this atropisomer was achieved on an achiral HPLC column at −70.0 °C. Plateau formation occurred when the column temperature was above −65.0 °C, and peak coalescence was observed at −39.0 °C as a result of rapid diastereomerization relative to the HPLC time scale, Figure 4.26. Computer simulation of the experimentally obtained chromatograms allowed accurate determination of the activation energy over a wide temperature range, Figure 4.27 and Table 4.1. A comparison of DHPLC and NMR data reveals that the conformational stability of 1,8-bis(2′-phenyl-4′-pyridyl)naphthalene increases steadily with temperature, which can be attributed to a congested transition state exhibiting a considerably negative entropy of activation. The Eyring plot for the isomerization of this atropisomer provides an activation enthalpy, $\Delta H^{\neq}$, and an activation entropy, $\Delta S^{\neq}$, of 57.5 kJ/mol and −43.4 J/K mol, respectively. The plot demonstrates that DHPLC and DNMR data are in excellent agreement and that they can be combined for stereodynamic analysis over a temperature range spanning more than 100 °C, Figure 4.28.

Trapp and coworkers examined the reversible enantiomerization of a chiral tetrabenzoxazine resorcarene and observed plateau formation by HPLC on Chiralpak AD. This macrocycle displays a labile hemiaminal group that is readily cleaved under slightly acidic conditions. The enantiomers interconvert by consecutive ring opening of each oxazine ring via an iminium intermediate and ring closure involving the opposite hydroxyl group of the resorcinol unit. Computer simulation of DHPLC elution profiles obtained at various temperatures gave a free activation energy, $\Delta G^{\neq}$, of 92 kJ/mol, and the activation parameters, $\Delta H^{\neq}$ and $\Delta S^{\neq}$, were determined as 53 kJ/mol and −131 J/K mol, respectively. The low activation enthalpy and the highly negative activation entropy are in agreement with the postulated dissociative enantiomerization mechanism, Scheme 4.7.[207]

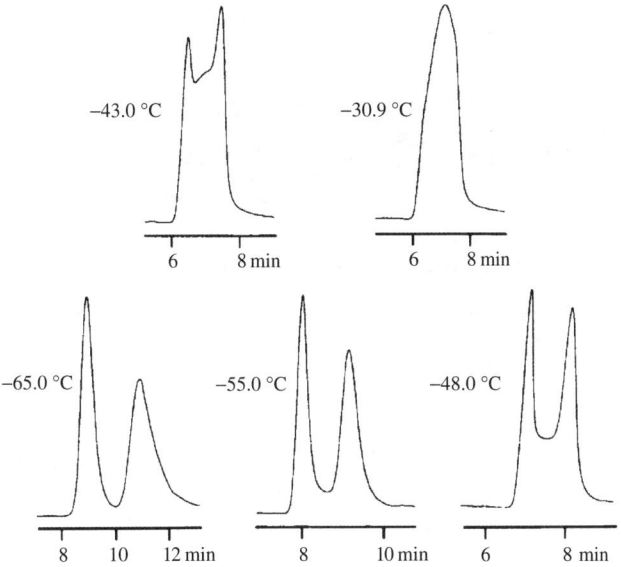

Figure 4.26 Cryogenic HPLC separation of the *syn*- and *anti*-isomers of 1,8-bis(2'-phenyl-4'-pyridyl)naphthalene and plateau formation at various temperatures. [Reproduced with permission from *J. Phys. Chem.* **2003**, *107*, 815-817.]

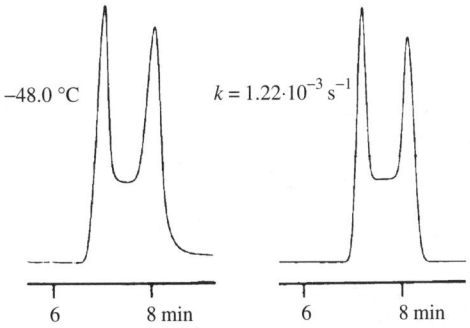

Figure 4.27 Experimental (left) and simulated (right) elution profile of 1,8-bis(2'-phenyl-4'-pyridyl)naphthalene. [Reproduced with permission from *J. Phys. Chem.* **2003**, *107*, 815-817.]

Table 4.1 Rotational energy barrier of 1,8-bis(2'-phenyl-4'-pyridyl)naphthalene obtained by DHPLC and DNMR.

Method	T [°C]	ΔG$^{\neq}$ [kJ/mol]	k [s^{-1}]
DHPLC	−65.0	67.8	$8.3 \cdot 10^{-5}$
DHPLC	−55.0	68.3	$4.0 \cdot 10^{-4}$
DHPLC	−48.0	68.5	$1.22 \cdot 10^{-3}$
DHPLC	−33.0	68.7	$2.50 \cdot 10^{-3}$
DNMR	40.3	73.0	9.03

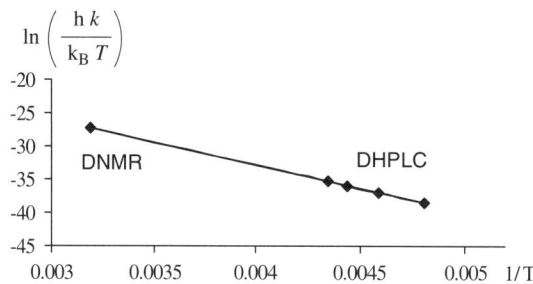

Figure 4.28 Eyring plot of DNMR and DHPLC data. [Reproduced with permission from *J. Phys. Chem.* **2003**, *107*, 815-817.]

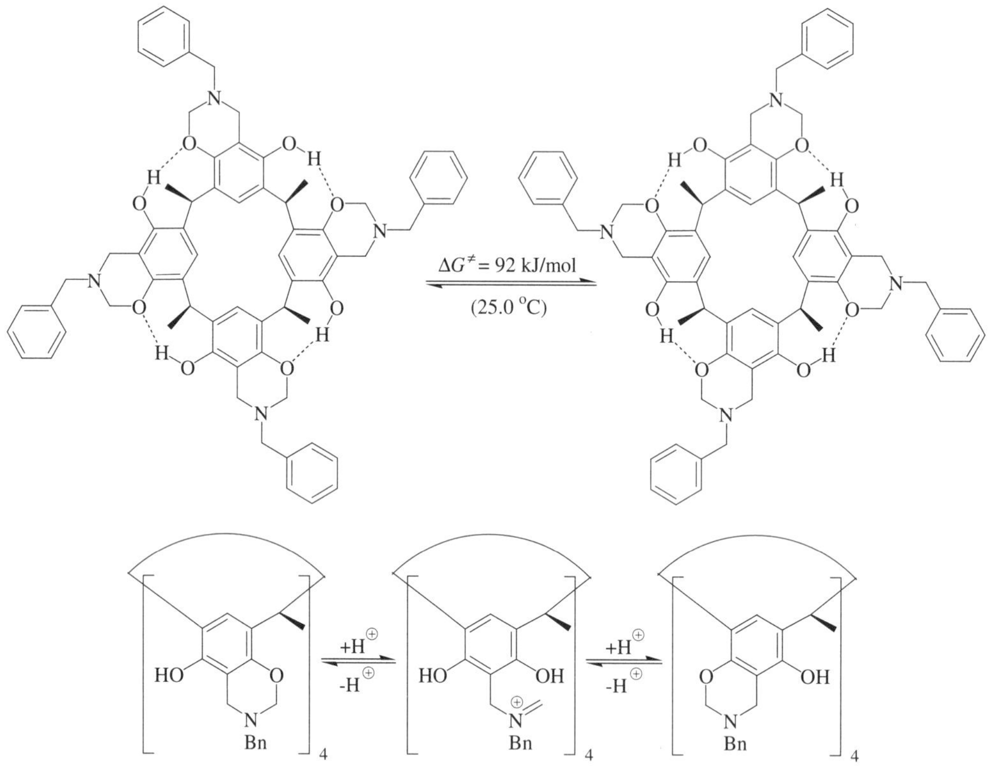

Scheme 4.7 Enantiomerization of a chiral tetrabenzoxazine resorcarene derivative and mechanism of reversible oxazine ring opening.

Villani *et al.* investigated the stereodynamics of numerous axially chiral 2-methyl- and 2-ethoxy-1-naphthyl carboxamides, sulfones and sulfoxides by DHPLC using a (3*R*,4*S*)-Whelk-O 1 CSP.[208,209] Comparison of the rotational energy barriers determined by computer simulation of elution profiles obtained by DHPLC with results from off-column racemization experiments revealed that the CSP increases the energy barrier of these atropisomers by 1-4 kJ/mol. Similarly, Oxelbark and Allenmark determined the configurational stability of *N*-aryl- and *N*-benzyl-1,3,2-benzodithiazole-1-oxides by both DHPLC and off-column racemization monitored by CD spectroscopy.[210,211] Again, the CSP was found to increase the barrier to enantiomerization by

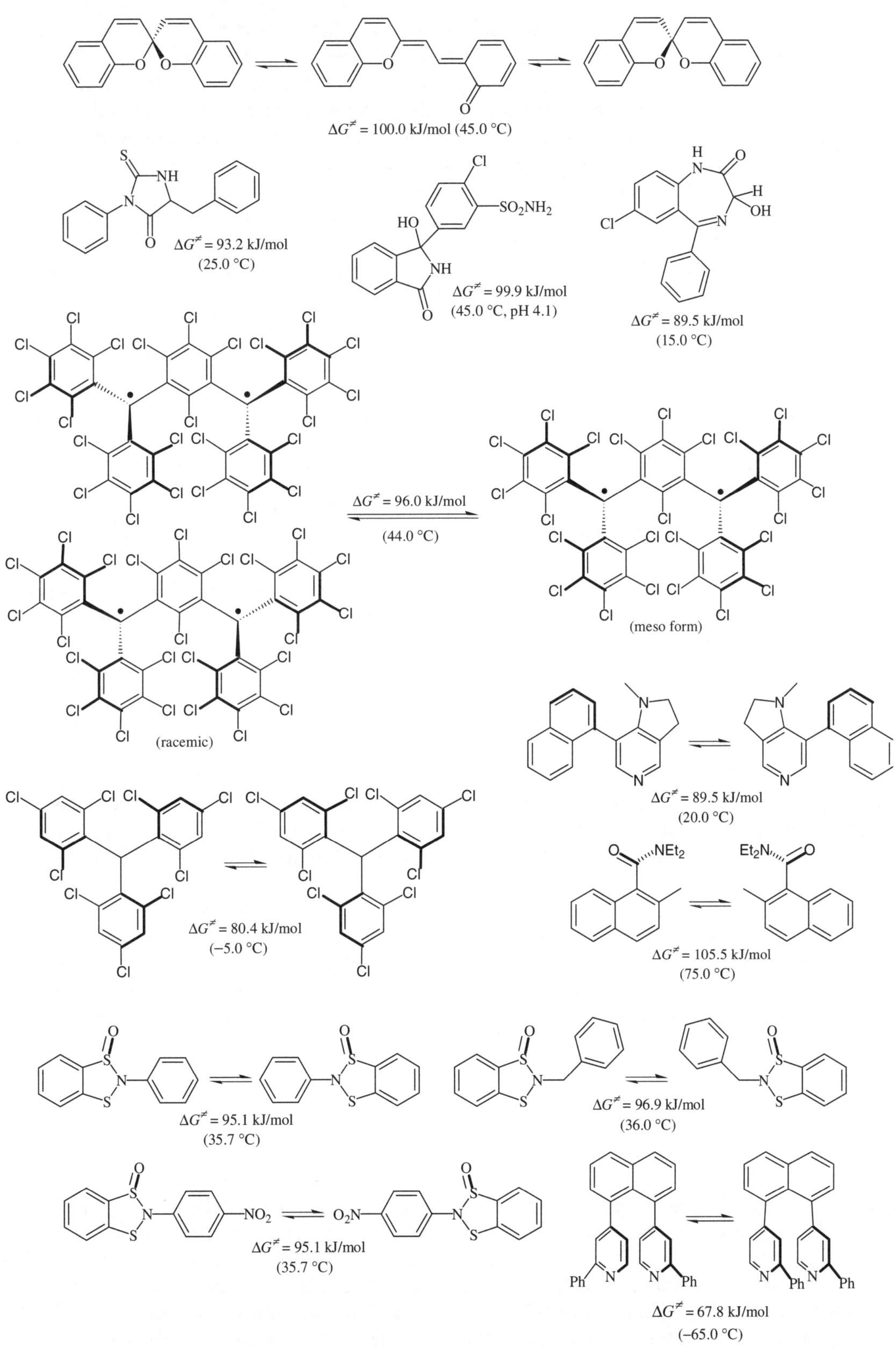

$\Delta G^{\neq} = 100.0$ kJ/mol (45.0 °C)

$\Delta G^{\neq} = 93.2$ kJ/mol
(25.0 °C)

$\Delta G^{\neq} = 99.9$ kJ/mol
(45.0 °C, pH 4.1)

$\Delta G^{\neq} = 89.5$ kJ/mol
(15.0 °C)

$\Delta G^{\neq} = 96.0$ kJ/mol
(44.0 °C)

(meso form)

(racemic)

$\Delta G^{\neq} = 89.5$ kJ/mol
(20.0 °C)

$\Delta G^{\neq} = 105.5$ kJ/mol
(75.0 °C)

$\Delta G^{\neq} = 80.4$ kJ/mol
(−5.0 °C)

$\Delta G^{\neq} = 95.1$ kJ/mol
(35.7 °C)

$\Delta G^{\neq} = 96.9$ kJ/mol
(36.0 °C)

$\Delta G^{\neq} = 95.1$ kJ/mol
(35.7 °C)

$\Delta G^{\neq} = 67.8$ kJ/mol
(−65.0 °C)

Figure 4.29 Isomerization reactions studied by DHPLC.

approximately 2 kJ/mol. For example, the enantiomerization barrier of 2-benzyl-1,3,2-benzodithiazole-1-oxide was determined as 96.9 kJ/mol at 36.0 °C by DHPLC on a (3*R*,4*S*)-Whelk-O 1 column, but an energy barrier of 94.7 kJ/mol was obtained in the absence of the chiral selector. The increase in the stereochemical stability of these compounds is commonly attributed to stabilization of the ground state of the interconverting stereoisomers during complexation by the CSP. This is by no means a general trend. Other chiral stationary phases such as poly(triphenylmethylmethacrylate) and triacetyl cellulose have been reported to exhibit small destabilizing effects on the conformational or configurational stability of chiral compounds. Comparison of isomerization barriers obtained by dynamic chromatography and variable-temperature NMR spectroscopy or polarimetric studies usually shows excellent agreement between the methods.[212] To date, DHPLC and computer simulation has been applied in kinetic studies of a variety of enantiomerization and diastereomerization reactions with Gibbs activation energies ranging from 60 to 120 kJ/mol, Figure 4.29.

4.3.2 Dynamic Gas Chromatography

At the same time, and independent of the DHPLC studies reported by Horváth in 1982, Schurig and coworkers observed plateau formation due to on-column enantiomerization of 1-chloro-2,2-dimethylaziridine and 1,6-dioxaspiro[4.4]nonane during complexation gas chromatography using bis[(1*R*)-3-heptafluorobutyrylcamphorate]nickel(II) dissolved in squalane as chiral stationary phase.[213] Having developed the theoretical background for dynamic chromatography and the necessary tools for peak shape analysis, they found that the enantiomerization barrier of 1-chloro-2,2-dimethylaziridine is 104.9 kJ/mol at 60.0 °C.[214] Klärner and Schröer determined the enantiomerization barrier of homofuran as 113.8 kJ/mol at 90.0 °C and concluded that the reaction proceeds via a carbonyl ylide-like intermediate that is formed by orbital symmetry-allowed disrotatory electrocyclic ring opening obeying Woodward-Hoffmann rules. Importantly, enantiomerization barriers of homofuran obtained by DGC and polarimetry are in excellent agreement, see Figure 4.31.[215,216] Dynamic gas chromatography has also been applied to *syn/anti*-isomerizations of axially chiral phenanthrenes and anthracenes.[217] For instance, peak shape analysis of the plateau formed by on-column interconversion of the *anti*-isomers and the meso *syn*-form of 9,10-bis(2-methylphenyl)phenanthrene at 240.0 °C gave an energy barrier of 132 kJ/mol. Schurig's group employed Chirasil–Dex (permethylated β-cyclodextrin immobilized on dimethylpolysiloxane) and Chirasil–Nickel (polysiloxane-anchored bis[(1*R*)-3-heptafluorobutyrylcamphorate]nickel(II)) in DGC analysis of several diaziridines and allenes.[218–220] Computer simulation of elution profiles revealed an enantiomerization barrier for 1,2,3,4-tetramethyldiaziridine, which interconverts via two consecutive nitrogen inversions, of 115.0 kJ/mol at 80.0 °C. Chiral stationary phases commonly used in DGC can have measurable effects on isomerization rates, confirming similar observations with DHPLC measurements. For example, Chirasil–Dex accelerates the rate of enantiomerization of 1,2,3,4-tetramethyldiaziridine and reduces the activation energy by 1–3 kJ/mol. A more pronounced effect was uncovered with axially chiral dimethyl 2,3-pentadienedioate. This allene has an interconversion barrier of 128 kJ/mol at 120 °C according to stopped-flow multidimensional gas chromatographic analysis conducted in the absence of a CSP, but DGC measurements using Chirasil–Nickel and Chirasil–Dex furnished energy barriers of 117.5 and 114.9 kJ/mol, respectively, at the same temperature, Figure 4.31. Investigating the isomerization of spiroketal chalcogran, an important beetle aggregation pheromone, Trapp and Schurig extended DGC to epimerization reactions. Because of the inherently different ground state energy of epimers, individual energy barriers for the forward and backward reaction had to be considered. Computer simulation of elution profiles yielded individual Gibbs free activation energies of 108.0 and 108.5 kJ/mol, respectively, for the interconversion of (2*R*,5*R*)- and (2*R*,5*S*)-2-ethyl-1,6-dioxaspiro[4.4]nonane.[221] The corresponding low activation enthalpies and the highly negative activation entropies are indicative of a dissociative heterolytic mechanism proceeding via a zwitterionic enol ether intermediate, Scheme 4.8.

Wolf and coworkers studied the rotational energy barrier of several axially chiral biphenyls by DGC on selectively modified cyclodextrins, Figure 4.30.[222] Energy barriers determined by DGC were in excellent agreement with results obtained by polarimetric analysis of enantioenriched

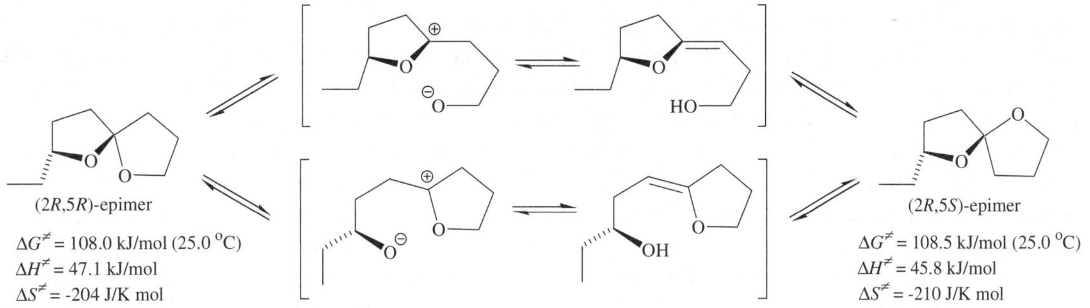

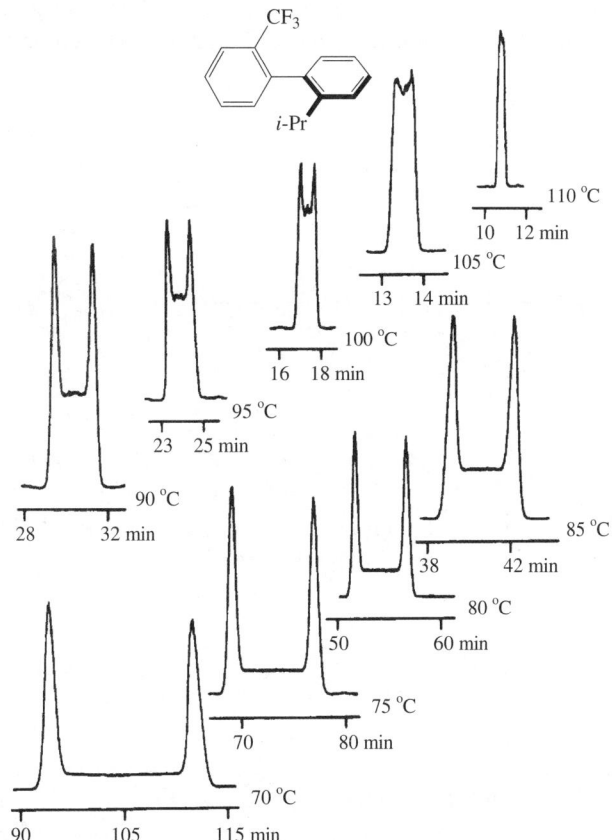

(2R,5R)-epimer

$\Delta G^{\neq}$ = 108.0 kJ/mol (25.0 °C)
$\Delta H^{\neq}$ = 47.1 kJ/mol
$\Delta S^{\neq}$ = -204 J/K mol

(2R,5S)-epimer

$\Delta G^{\neq}$ = 108.5 kJ/mol (25.0 °C)
$\Delta H^{\neq}$ = 45.8 kJ/mol
$\Delta S^{\neq}$ = -210 J/K mol

Scheme 4.8 Dissociative epimerization mechanism of chalcogran.

CF$_3$

i-Pr

110 °C
10 12 min

105 °C
13 14 min

100 °C
16 18 min

95 °C
23 25 min

90 °C
28 32 min

85 °C
38 42 min

80 °C
50 60 min

75 °C
70 80 min

70 °C
90 105 115 min

Figure 4.30 DGC elution profiles of 2-isopropyl-2'-trifluoromethylbiphenyl using octakis(6-*O*-methyl-2,3-di-*O*-pentyl)-γ-cyclodextrin as CSP. [Reproduced with permission from *Liebigs Ann.* **1995**, 781-786.]

samples, but small effects of CSPs on the biaryl atropisomerization barrier were reported. Simulation of experimentally recorded elution profiles of 2,2′-diisopropylbiphenyl using heptakis(6-*O*-*tert*-butyldimethylsilyl-2,3-di-*O*-pentyl)-β-cyclodextrin as CSP gave an energy barrier of 112.4 (±0.2) kJ/mol (78.5 to 83.5 °C), whereas a value of 114.6 (±0.2) kJ/mol (78.4 to 88.1 °C) was obtained with octakis(6-*O*-methyl-2,3-di-*O*-pentyl)-γ-cyclodextrin. Polarimetric studies of enantiomerically enriched 2,2′-diisopropylbiphenyl, isolated by HPLC on microcrystalline triacetyl cellulose, were conducted in the same temperature range and gave an energy barrier of 112.8 (±0.1) kJ/mol. Since additional polarimetric studies in various solvents including ethanol and hexanes showed that solvent effects on the conformational stability of 2,2′-diisopropylbiphenyl are negligible, the difference in the energy barriers calculated by DGC on cyclodextrins can be attributed to small, albeit unpredictable, stabilization or destabilization effects of the CSPs used.

Hochmuth and König used DGC to study the enantiomerization of substituted [10]-, [11]- and [12]paracyclophanes exhibiting a chiral plane.[223,224] They reported baseline separations of the enantiomers of various [10]paracyclophanes but did not observe any sign of on-column interconversion, even at 170 °C, due to the high conformational stability of these atropisomers. In contrast, [12]paracyclophanes undergo fast interconversion compared to the GC time scale and enantiomers can not be resolved above 40 °C.[xi] As expected, enantiomerization of substituted dioxa- and carbocyclic [11]paracyclophanes is suitable for DGC analysis and formation of a temperature-dependent plateau during separation on cyclodextrin-derived CSPs can be observed. The stereo-dynamics of several [11]paracyclophanes with enantiomerization barriers between 115 and 135 kJ/mol have been analyzed by DGC, Figure 4.31. Since its introduction by Schurig in 1982, DGC has been applied in kinetic studies of a variety of enantiomerization and diastereomerization reactions. As a

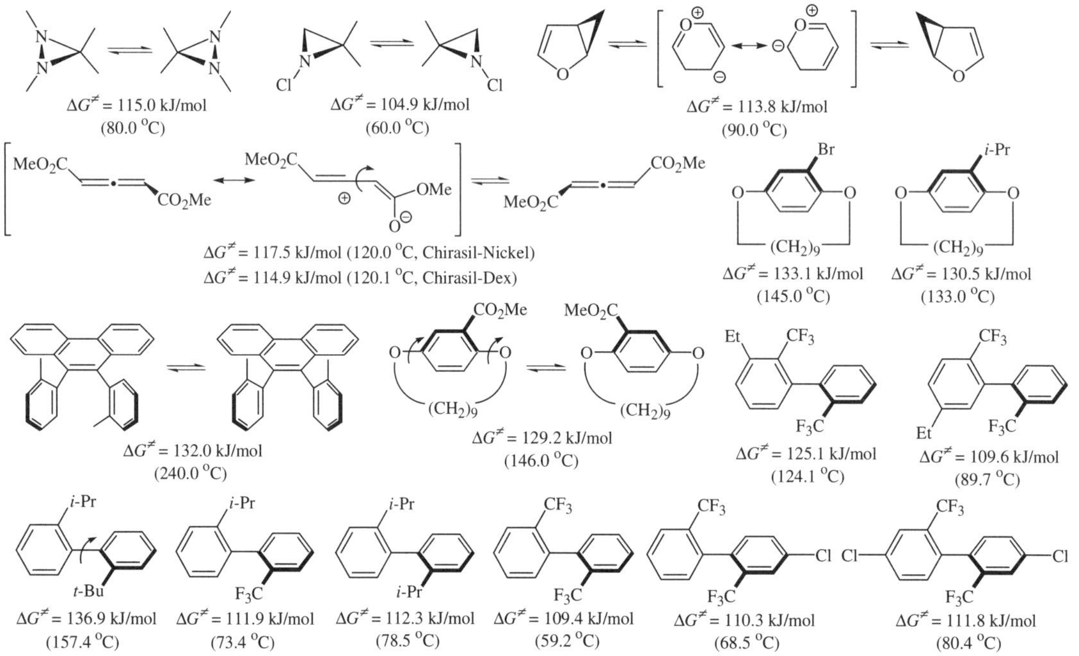

Figure 4.31   Stereomutations studied by DGC.

[xi] In principle, the stereodynamics of [12]paracyclophanes can be studied at lower temperatures, but this lies outside the scope of DGC due to insufficient volatility of the analytes below 40 °C.

rule of thumb, stereomutations with Gibbs free activation energies, $\Delta G^{\neq}$, between 100 and 150 kJ/mol can be examined by this method.

4.3.3 Dynamic Supercritical Fluid Chromatography and Electrokinetic Chromatography

The principles of dynamic chromatography and computer simulation have been applied in several separation techniques. Although competition between isomerization and separation of chiral compounds has mainly been observed by HPLC and GC, other pressure-driven methods such as super- and subcritical fluid chromatography and electroosmotically driven separations further extend the usefulness of dynamic chromatography. The application spectrum of DHPLC and DGC is confined by the chromatographic time scale, solubility and thermal stability of analytes, and the temperature range inherent to these techniques. Because of decreasing analyte solubility and increasing viscosity of the mobile phase causing high back pressures at low temperature, HPLC is commonly performed between 0 and 120 °C, although some brush-type CSPs are compatible with operating temperatures as low as -50 °C. In contrast, GC requires sufficient volatility and thermal stability of the analyte, and chiral separations are usually conducted between 40 and 180 °C. A further extension of this temperature range is not possible because of the thermal instability of CSPs above 180 °C. The intrinsically high diffusivity and low viscosity of subcritical fluids provide a significant advantage of subcritical fluid chromatography (SubFC) over HPLC when cryogenic conditions are required. Wolf and coworkers were first to use dynamic subcritical fluid chromato-graphy (DSubFC) for the study of rapidly interconverting stereoisomers.[225] Enantioseparation of stereolabile axially chiral arylnaphthalene lignans under cryogenic conditions was accomplished on a polyWhelk-O CSP. Simulation of the temperature-dependent elution profiles gave rotational energy barriers ranging from 75.0 to 92.0 (± 0.2) kJ/mol. Sub- and supercritical fluid chromato-graphy significantly expand the application spectrum of DGC and DHPLC, Figure 4.32. Because SubFC and SFC cover the combined temperature range of HPLC and GC, these techniques can be applied to isomerization processes having energy barriers between 60 and 150 kJ/mol.

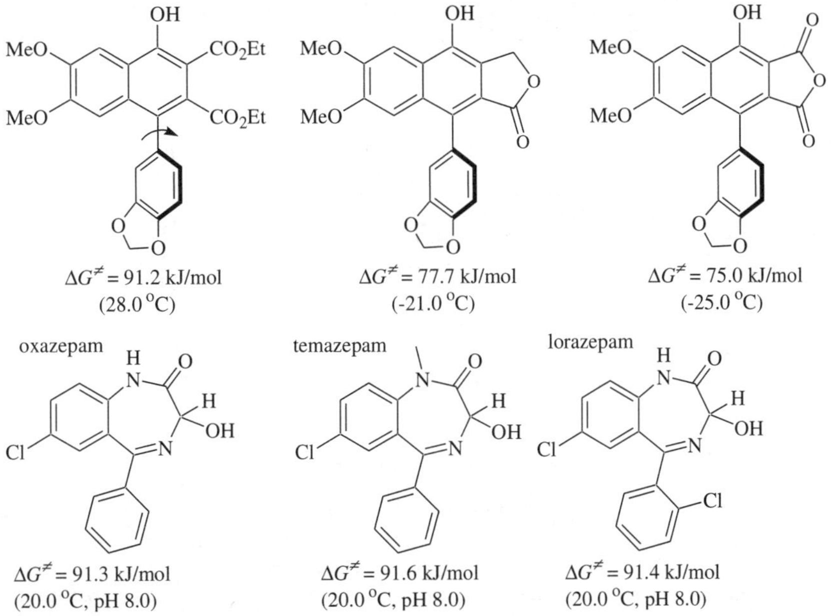

Figure 4.32 Enantiomerization of chiral compounds studied by DSubFC (top) and DMEKC (bottom).

Schurig *et al.* investigated the racemization of chiral benzodiazepine drugs such as oxazepam, temazepam and lorazepam using dynamic micellar electrokinetic chromatography (DMEKC) and sodium cholate as micelle-forming surfactant.[226] Dynamic MEKC is applicable to reactions exhibiting energy barriers between 80 and 120 kJ/mol. Since separations can be performed with aqueous buffers and at ambient temperatures, MEKC provides an important opportunity to study isomerization processes under biologically relevant conditions.

4.4 CHROMATOGRAPHIC AND ELECTROPHORETIC STOPPED-FLOW ANALYSIS

Chiral chromatography is invaluable for analytical and preparative separation of stereoisomers and for kinetic analysis of stereomutations via computer simulation of temperature-dependent elution profiles. The development of stopped-flow procedures suitable to isomerization studies of chiral compounds by Weseloh, Wolf and König has introduced another powerful methodology.[227] They discovered that the enantiomers of 4,4'-diammonium-2,2'-diisopropylbiphenyl can be separated with high selectivity and efficiency by capillary zone electrophoresis (CZE) when heptakis(2,3,6-tri-*O*-methyl)-β-cyclodextrin is added to the electrophoretic buffer, Figure 4.33. A racemic sample of 4,4'-diammonium-2,2'-diisopropylbiphenyl was injected electrokinetically into the CZE capillary and the enantiomers were separated at 25 °C. The voltage was switched off at exactly the middle of the electrophoretic separation process and the temperature was increased to 82.6 °C to allow interconversion of the atropisomers. At this point, the enantiomers were already well separated and located at a similar distance either in front of or behind the midpoint of the capillary. After a certain time, the on-column interconversion was stopped and the buffer was cooled to the original

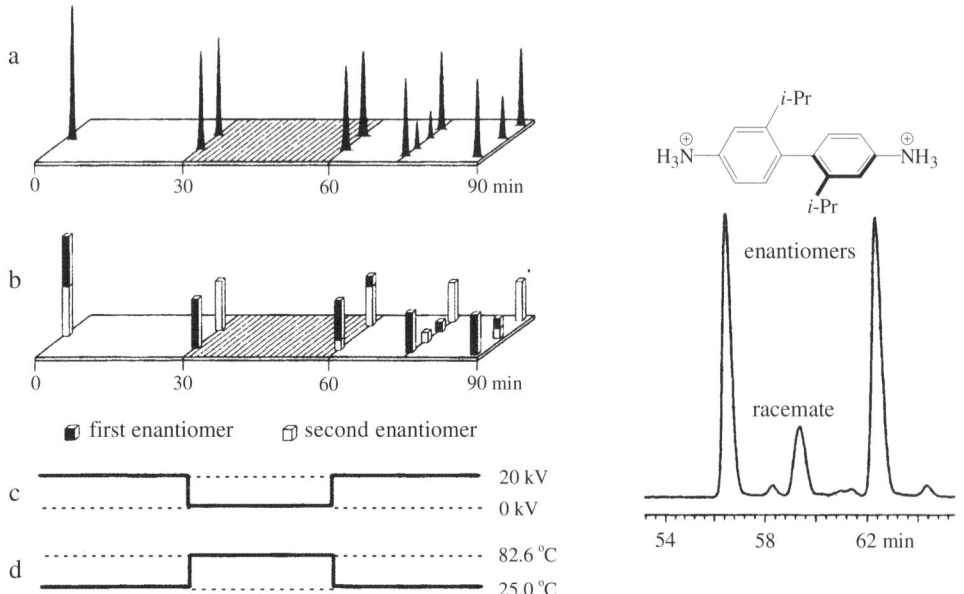

Figure 4.33 Illustration of the electrophoretic separation and interconversion of the enantiomers of 4,4'-diammonium-2,2'-diisopropylbiphenyl: (a) on-column enantiomerization experiment and (b) enantiomeric composition of each peak. Enantioseparation proceeds in sections *A* while enantiomerization takes place in section *B*. The applied voltage and temperature are illustrated at the bottom (c and d) and a typical electropherogram is shown at the right. [Reproduced with permission from *Angew. Chem., Int. Ed. Engl.* **1995**, *34*, 1635-1636.]

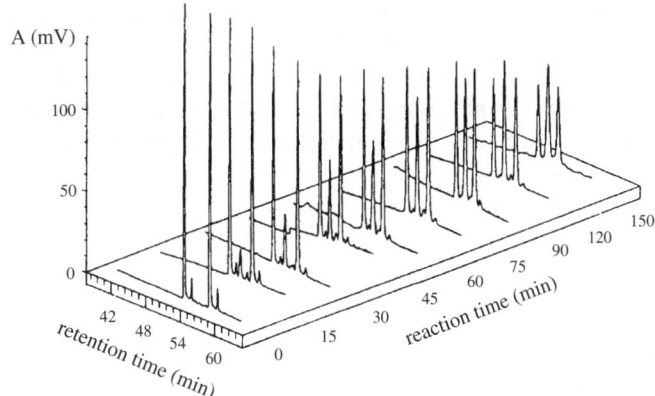

Figure 4.34   Electropherograms obtained by variation of the on-column enantiomerization time. The central peak corresponds to the amount of racemate formed. [Reproduced with permission from *Angew. Chem., Int. Ed. Engl.* **1995**, *34*, 1635-1636.]

temperature. The electrophoretic process was then resumed by switching on the applied voltage of 20 kV and the separation continued through the second part of the capillary. The stopped-flow experiment illustrated in Figure 4.33 affords an electropherogram with three peaks. In contrast to a typical CZE enantioseparation, one observes an additional racemic peak between the signals that correspond to the enantiomeric portions that do not undergo on-column enantiomerization during the allotted reaction time, Figure 4.34. The rotational energy barrier of 4,4'-diammonium-2,2'-diisopropylbiphenyl was calculated as 115.2 kJ/mol, based on integrated peak areas, the applied temperature and the period of heating. An attractive feature of this method is that it requires only minute amounts of racemate but eliminates the need for computer simulation, which is indispensable for kinetic analysis based on dynamic chromatography. Since electrophoretic stopped-flow analysis is usually conducted with aqueous buffers, isomerization reactions can be investigated at temperatures up to 95 °C and energy barriers to isomerization ranging from 100 to 130 kJ/mol have been determined with high accuracy (±0.2 kJ/mol).

The preceding discussion of DHPLC and DGC reveals that the presence of a chiral selector can cause a measurable increase or decrease in enantiomerization barriers due to ground state stabilization or catalysis of the interconversion process. The stopped-flow method was therefore modified to allow kinetic analysis in the absence of a chiral additive and to exclude stabilizing or destabilizing interactions between the analyte and the chiral cyclodextrin host.[228] Through careful rinsing procedures and control of the electroosmotic flow, the CZE capillary was segmented into three different buffer zones. Following the stopped-flow procedure described above, the enantiomers of 4,4'-diammonium-2,2'-bis(trifluoromethyl)biphenyl were separated in the first and third buffer zone containing the cyclodextrin selector, and the enantiomerization experiment was conducted in the middle section which did not contain any chiral additive. It is noteworthy that heating of only one part of the capillary provides a means for selective racemization of one enantiomer. Segmentation of the capillary into different buffer and heating zones thus enables one to determine selectively the rotational energy barrier of either enantiomer in the presence and in the absence of a chiral additive, Figure 4.35. The investigation of the stereodynamics of 4,4'-diammonium-2,2'-bis(trifluoromethyl)biphenyl in a 30 mM phosphate buffer at pH 2.4 by this modified stopped-flow procedure gave a rotational energy barrier of 106.9 kJ/mol at 71.0 °C in the absence of a cyclodextrin additive, which is in excellent agreement with polarimetric studies of an enantioenriched sample under similar conditions. The stopped-flow technique is a highly versatile analytical tool for investigation of stereomutations in the presence and absence of a chiral additive and can be utilized for studying stereoselective interactions between stereolabile compounds and cyclodextrin hosts or other chiral selectors. By analogy with dynamic chromatography, stopped-flow CZE confirmed that cyclodextrins can alter the rate of biaryl atropisomerization. Analysis of the conformational stability of 4,4′-diammonium-2,2′-bis(trifluoromethyl)biphenyl in the presence of permethylated β-cyclodextrin at 71 °C revealed a rotational energy barrier of 105.2 kJ/mol, while addition of hydroxypropyl-β-cyclodextrin increased the conformational stability to 108.3 kJ/mol.

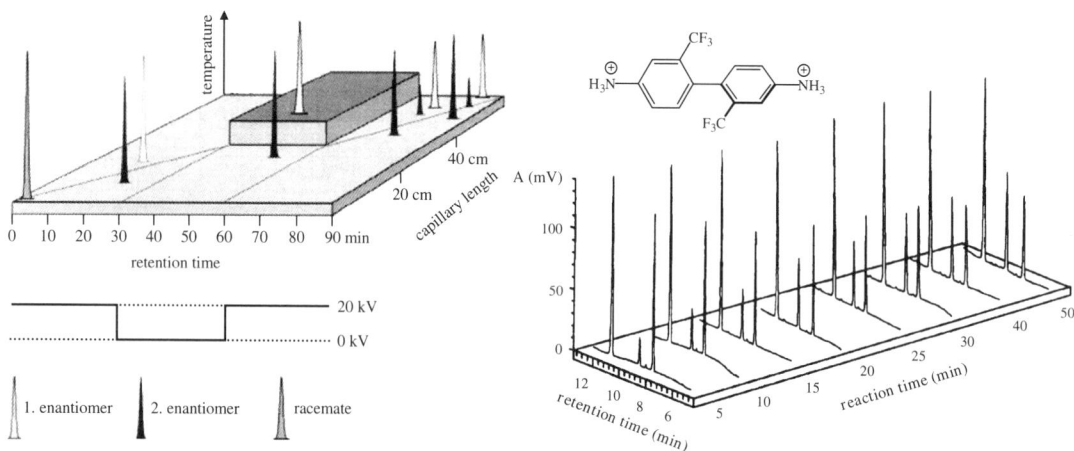

Figure 4.35　Illustration of CZE enantioseparation and on-column racemization of one enantiomer of 4,4′-diammonium-2,2′-bis(trifluoromethyl)biphenyl (left), and electropherograms obtained after different times of racemization (right). The emerging peak in the middle corresponds to the amount of enantiomer formed during selective on-column heating of the less retained enantiomer. [Reproduced with permission from *Chirality* **1996**, *8*, 441–445 and *Angew. Chem., Int. Ed. Engl.* **1995**, *34*, 1635-1636.]

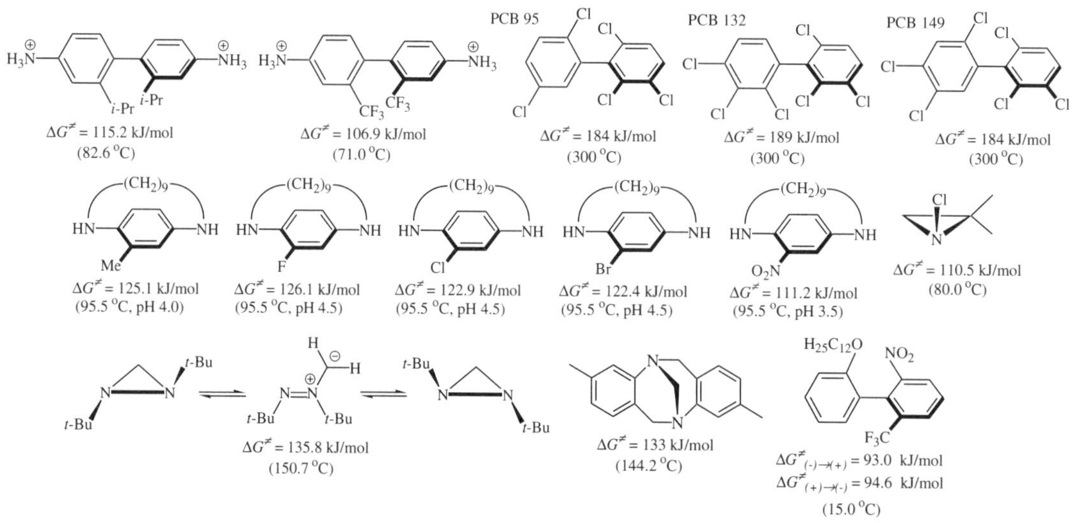

Figure 4.36　Enantiomerization of chiral compounds studied by chromatographic and electrophoretic stopped-flow analysis.

Table 4.2 Comparison of methods frequently used for investigation of isomerization reactions of chiral compounds.

Method	Polarimetry	DNMR	DGC	DSubFC/ DSFC	DHPLC	DMEKC	Stopped-flow analysis
Principle	Isomerization of one stereoisomer is monitored by polarimetry	Coalescence of NMR signals is analyzed by complete line shape analysis (CLA)	Simultaneous chromatographic separation of stereoisomers and on-column interconversion results in plateau formation; simulation of experimentally obtained elution profiles provides rate constants				Chromatographic or electrophoretic separation and intermediate on-column isomerization
Energy range (kJ/mol)	80–180	20–130	100–150	60–150	60–120	80–120	80–150
Requirements	Semipreparative isolation or enrichment of stereoisomers	Sufficient resolution of diastereotopic signals for accurate integration	Chromatographic conditions that allow baseline separation of stereoisomers				Chromatographic or electrophoretic conditions for baseline separation of enantiomers; control of diffusion during heating
Chiral impurities	Interfere	Might not interfere if signals are resolved and interactions with the analyte are negligible	Impurities are usually separated during chromatography and do not affect the enantiomerization or diastereomerization process				No interference if resolved
Sample amount	>10 mg	1–10 mg	<10 µg				<10 µg
Accuracy (kJ/mol)	±0.5	±0.2 (CLA) ±0.5–1.0 (coalescence method)	±0.5 (interactions between the analyte and the CSP or chiral additive can increase or decrease the energy barrier)				±0.2–1.0
Precision (kJ/mol)	±0.2	±0.2	±0.2				±0.2–1.0

Stopped-flow electrophoresis has also been used to explore the stereodynamics of other chiral compounds including 1,11-diaza[11]paracyclophanes, Figure 4.36.[229] Schurig applied the concept of stopped-flow analysis to gas chromatography. The rotational barrier of hexachlorobiphenyl PCB 132 was determined in the presence of a cyclodextrin-derived chiral stationary phase to be approximately 184 ($\pm$2) kJ/mol.[230] In general, stopped-flow multi-dimensional gas chromatography is suitable for investigation of isomerization reactions with activation energies between 70 and 200 kJ/mol.[231,232] This technique has been used for analytical separation of the enantiomers of 1-chloro-2,2-dimethylaziridine on a cyclodextrin column and subsequent selective introduction of one enantiomer to a second achiral column. This enantiomer was then subjected to racemization at high temperature and the partially racemized sample was subsequently separated into enantiomers on a third chiral column at lower temperatures. Integration of peak areas allowed determination of the barrier to *N*-inversion as 110.5 ($\pm$0.5) kJ/mol based on the known racemization temperature and the time that the analyte spent in the achiral reactor column. Similarly, the enantiomerization barrier of 1,2-di-*tert*-butyldiaziridine, which probably undergoes reversible conrotatory electrocyclic ring opening, was determined as 135.8 ($\pm$0.2) kJ/mol at 150.7 °C, Figure 4.36.[233] Stopped-flow multi-dimensional gas chromatography has also been used to examine the conformational stability of PCBs 95, 132 and 136, and Tröger's base, a widely used chiral solvating agent. This chiral amine has two asymmetric bridgehead nitrogen atoms that are stable to pyramidal inversion at room temperature. However, stopped-flow multi-dimensional gas chromatography proved that the enantiomers of Tröger's base interconvert at 144 °C with an activation energy of 133 ($\pm$1.5) kJ/mol.[234] Lindner *et al.* extended the concept of stopped-flow chromatographic analysis to HPLC and capillary electrochromatography (CEC) to study the rotational energy barrier of 2'-dodecyloxy-6-nitrobiphenyl-2-carboxylic acid.[235] Using a quinine-derived chiral stationary phase they determined the activation barrier to enantiomerization of the levorotatory and dextrorotatory enantiomer to be 93.0 and 94.6 kJ/mol, respectively. The difference in the rotational energy barriers of the enantiomers of 2'-dodecyloxy-6-nitrobiphenyl-2-carboxylic acid in the presence of the quinine CSP was exploited for on-column deracemization based on asymmetric transformation of the second kind, see Chapter 7.3.2.

4.5 COMPARISON OF ANALYTICAL METHODS

Polarimetry, variable-temperature NMR spectroscopy, dynamic chromatography, and stopped-flow chromatography and electrophoresis are widely used techniques for stereodynamic analysis of chiral compounds. The majority of enantiomerization and diastereomerization reactions discussed in Chapter 3 has been examined using these methods. The principles and scope of their application are summarized in Table 4.2.

REFERENCES

1. Zijlstra, R. W. J.; van Duijnen, P. T.; Feringa, B. L.; Steffen, T.; Duppen, K.; Wiersma, D. A. *J. Phys. Chem. A* **1997**, *101*, 9828-9836.
2. Ma, J.; Dutt, G. B.; Waldeck, D. H.; Zimmt, M. B. *J. Am. Chem. Soc.* **1994**, *116*, 10619-10629.
3. Kessler, H. *Angew. Chem., Int. Ed. Engl.* **1970**, *3*, 219-235.
4. Grevels, F.-W.; Jacke, J.; Klotzbücher, W. E.; Krüger, C.; Seevogel, K.; Tsay, Y.-H. *Angew. Chem., Int. Ed. Engl.* **1987**, *26*, 885-887.
5. Tumambac, G. E.; Wolf, C. *J. Org. Chem.* **2004**, *69*, 2048-2055.
6. Tumambac, G. E.; Wolf, C. *J. Org. Chem.* **2005**, *70*, 2930-2938.
7. Biali, S. E.; Kahr, B.; Okamoto, Y.; Aburatani, R.; Mislow, K. *J. Am. Chem. Soc.* **1988**, *110*, 1917-1922.

8. Gur, E.; Kaida, Y.; Okamoto, Y.; Biali, S.; Rappoport, Z. *J. Org. Chem.* **1992**, *57*, 3689-3693.
9. Horeau, A. *Tetrahedron Lett.* **1969**, *10*, 3121-3124.
10. Horeau, A.; Guetté, J.-P. *Tetrahedron* **1974**, *30*, 1923-1931.
11. Horeau, A.; Nouaille, A. *Tetrahedron Lett.* **1966**, *7*, 3953-3959.
12. Brewster, J. H. *Assignment of Stereochemical Configuration by Chemical Methods*, In: Bentley, K. W.; Kirby, G. W. (Eds.) *Elucidation of Organic Structures by Physical and Chemical Methods*, Vol. IV, Part III, 2nd ed., Wiley-Interscience, New York, 1972, pp. 1-249.
13. Horeau, A. *Determination of the Configuration of Secondary Alcohols by Partial Resolution*, In Kagan, H. B. (Ed.) *Stereochemistry, Fundamentals and Methods*, Vol. 3, Georg Thieme, Stuttgart, 1977.
14. Eliel, E. L.; Wilen, S. H. *Stereochemistry of Organic Compounds*, Wiley & Sons, New York, 1994, pp. 991-1118.
15. Sandström, J. *Chirality* **1995**, *7*, 181-193.
16. Freedman, T. B.; Cao, X.; Dukor, R. K.; Nafie, L. A. *Chirality* **2003**, *15*, 743-758.
17. Polavarapu, P. L.; He, J. *Anal. Chem.* **2004**, *76*, 61A-67A.
18. Superchi, S.; Giorgio, E.; Rosini, C. *Chirality* **2004**, *16*, 422-451.
19. Watanabe, M.; Suzuki, H.; Tanaka, Y.; Ishida, T.; Oshikawa, T.; Tori, A. *J. Org. Chem.* **2004**, *69*, 7794-7801.
20. Vandyck, K.; Matthys, B.; van der Eycken, J. *Tetrahedron Lett.* **2005**, *46*, 75-78.
21. Specht, K. M.; Nam, J.; Ho, D. M.; Berova, N.; Kondru, R. K.; Beratan, D. N.; Wipf, P.; Pascal Jr., R. A.; Kahne, D. *J. Am. Chem. Soc.* **2001**, *123*, 8961-8966.
22. Giorgio, E.; Viglione, R. G.; Zanasi, R.; Rosini, C. *J. Am. Chem. Soc.* **2004**, *126*, 12968-12976.
23. Kießwetter, R.; Pustet, N.; Brandl, F.; Mannschreck, A. *Tetrahedron: Asymm.* **1999**, *10*, 4677-4687.
24. Harié, G.; Samat, A.; Gugliemetti, R.; van Parys, I.; Saeyens, W.; de Keukeleire, D.; Lorenz, K.; Mannschreck, A. *Helv. Chim. Acta* **1997**, *80*, 1122-1132.
25. Mannschreck, A.; Gmahl, E.; Burgemeister, T.; Kastner, F.; Sinnwell, V. *Angew. Chem., Int. Ed. Engl.* **1988**, *27*, 270-271.
26. Mintas, M.; Orhanović, Z.; Jakopčić, K.; Koller, H.; Stühler, G.; Mannschreck, A. *Tetrahedron* **1985**, *41*, 229-233.
27. Berg, U.; Isaksson, R.; Sandström, J.; Sjöstrand, U.; Eiglsperger, A.; Mannschreck, A. *Tetrahedron Lett.* **1982**, *23*, 4237-4240.
28. Eiglsperger, A.; Kastner, F.; Mannschreck, A. *J. Mol. Structure* **1985**, *126*, 421-432.
29. Mannschreck, A.; Andert, D.; Eiglsperger, A.; Gmahl, E.; Buchner, H. *Chromatographia* **1988**, *25*, 182-188.
30. Wolf, C.; König, W. A.; Roussel, C. *Liebigs Ann.* **1995**, 781-786.
31. Biali, S. E.; Rappoport, Z.; Mannschreck, A.; Pustet, N. *Angew. Chem., Int. Ed. Engl.* **1989**, *28*, 199-201.
32. Rochlin, E.; Rappoport, Z.; Kastner, F.; Pustet, N.; Mannschreck, A. *J. Org. Chem.* **1999**, *64*, 8840-8845.
33. Feringa, B. L.; Jager, W. F.; de Lange, B. *Tetrahedron Lett.* **1992**, *33*, 2887-2890.
34. Feringa, B. L.; Jager, W. F.; de Lange, B. *J. Chem. Soc., Chem. Commun.* **1993**, 288-290.
35. Clayden, J.; Johnson, P.; Pink, J. H.; Helliwell, M. *J. Org. Chem.* **2000**, *65*, 7033-7040.
36. Rotto, N. T.; Coates, R. M. *J. Am. Chem. Soc.* **1989**, *111*, 8941-8943.
37. Coogan, M. P.; Passey, S. C. *J. Chem. Soc., Perkin Trans. 2* **2000**, 2060-2066.
38. Alibrandi, G.; Coppolino, S.; D'Aliberti, S.; Ficarra, P.; Micali, N.; Villari, A. *J. Pharm. Biomed. Anal.* **2002**, *29*, 1025-1029.
39. Harada, N.; Saito, A.; Koumura, N.; Roe, D. C.; Jager, W. F.; Zijlstra, R. W. J.; de Lange, B.; Feringa, B. L. *J. Am. Chem. Soc.* **1997**, *119*, 7249-7255.

40. Olszewska, T.; Milewska, M. J.; Gdaniec, M.; Maluszyńska, H.; Poloński, T. *J. Org. Chem.* **2001**, *66*, 501-506.

41. Armstrong, R. N.; Ammon, H. L.; Darnow, J. N. *J. Am. Chem. Soc.* **1987**, *109*, 2077-2082.

42. Schlögl, K.; Weissensteiner, W.; Widhalm, M. *J. Org. Chem.* **1982**, *47*, 5025-5027.

43. Ceccacci, F.; Mancini, G.; Mencarelli, P.; Villani, C. *Tetrahedron Asymm.* **2003**, *14*, 3117-3122.

44. Lebon, F.; Longhi, G.; Gangemi, F.; Abbate, S.; Priess, J.; Juza, M.; Bazzini, C.; Caronna, T.; Mele, A. *J. Phys. Chem. A* **2004**, *108*, 11752-11761.

45. Michael, F. E.; Duncan, A. P.; Sweeney, Z. K.; Bergman, R. G. *J. Am. Chem. Soc.* **2003**, *125*, 7184-7185.

46. Ishi-I, T.; Crego-Calama, M.; Timmermann, P.; Reinhoudt, D. N.; Shinkai, S. *J. Am. Chem. Soc.* **2002**, *124*, 14631-14641.

47. Freedman, T. B.; Hausch, D. L.; Cianciosi, S. J.; Baldwin, J. E. *Can. J. Chem.* **1998**, *76*, 806-810.

48. Cianciosi, S. J.; Ragunathan, N.; Freedman, T. B.; Nafie, L. A.; Lewis, D. K.; Glenar, D. A.; Baldwin, J. E. *J. Am. Chem. Soc.* **1991**, *113*, 1864-1866.

49. Metcalf, D. H.; Snyder, S. W.; Demas, J. N.; Richardson, F. S. *J. Am. Chem. Soc.* **1990**, *112*, 469-482.

50. Metcalf, D. H.; Snyder, S. W.; Demas, J. N.; Richardson, F. S. *J. Am. Chem. Soc.* **1990**, *112*, 5681-5697.

51. Metcalf, D. H.; Snyder, S. W.; Demas, J. N.; Richardson, F. S. *J. Phys. Chem.* **1990**, *94*, 7143-7153.

52. Glover-Fischer, D. P.; Metcalf, D. H.; Hopkins, T. A.; Pugh, V. J.; Chisdes, S. J.; Kankare, J.; Richardson, F. S. *Inorg. Chem.* **1998**, *37*, 3026-3033.

53. Seco, J. M.; Quinoa, E.; Riguera, R. *J. Org. Chem.* **1999**, *64*, 4669-4675.

54. Ferreiro, M. J.; Latypov, S. K.; Quinoa, E.; Riguera, R. *J. Org. Chem.* **2000**, *65*, 2658-2666.

55. Takahashi, T.; Fukuishima, A.; Tanaka, Y.; Takeuchi, Y.; Kabuto, K.; Kabuto, C. *Chem. Commun.* **2000**, 787-788.

56. Weibel, D. B.; Walker, T. R.; Schroeder, F. C.; Meinwald, J. *Org. Lett.* **2000**, *2*, 2381-2383.

57. Bravo, J.; Cativiela, C.; Chaves, J. E.; Navarro, R.; Urriolabeitia, E. P. *Inorg. Chem.* **2003**, *42*, 1006-1013.

58. Porto, S.; Duran, J.; Seco, J. M.; Quinoa, E.; Riguera, R. *Org. Lett.* **2003**, *5*, 2979-2982.

59. Sureshan, K. M.; Miyasou, T.; Miyamori, S.; Watanabe, Y. *Tetrahedron: Asymm.* **2004**, 15, 3357-3364.

60. Kasai, Y.; Taji, H.; Fujita, T.; Yamamoto, Y.; Akagi, M.; Sugio, A.; Kuwahara, S.; Watanabe, M.; Harada, N.; Ichikawa, A.; Schurig, V. *Chirality* **2004**, *16*, 569-585.

61. Takeuchi, Y.; Fujisawa, H.; Noyori, R. *Org. Lett.* **2004**, *6*, 4607-4610.

62. Kobayashi, Y.; Hayashi, N.; Tan, C.-H.; Kishi, Y. *Org. Lett.* **2001**, *3*, 2245-2248.

63. Hayashi, N.; Kobayashi, Y.; Kishi, Y. *Org. Lett.* **2001**, *3*, 2249-2252.

64. Kobayashi, Y.; Hayashi, N.; Kishi, Y. *Org. Lett.* **2001**, *3*, 2253-2255.

65. Kobayashi, Y.; Hayashi, N.; Kishi, Y. *Org. Lett.* **2002**, *4*, 411-414.

66. Kobayashi, Y.; Hayashi, N.; Kishi, Y. *Tetrahedron Lett.* **2003**, *44*, 7489-7491.

67. Pazos, Y.; Leiro, V.; Seco, J. M.; Quinoa, E.; Riguera, R. *Tetrahedron: Asymm.* **2004**, *15*, 1825-1829.

68. Inamoto, A.; Ogasawara, K.; Omata, K.; Kabuto, K.; Sasaki, Y. *Org. Lett.* **2000**, *2*, 3543-3545.

69. Omata, K.; Aoyagi, S.; Kabuto, K. *Tetrahedron: Asymm.* **2004**, *15*, 2351-2356.

70. Ghosh, I.; Zeng, H.; Kishi, Y. *Org. Lett.* **2004**, *6*, 4715-4718.

71. Wenzel, T. J.; Wilcox, J. D. *Chirality* **2003**, *15*, 256-270.

72. Seco, J. M.; Quinoa, E.; Riguera, R. *Chem. Rev.* **2004**, *104*, 17-117.

73. Jacobus, J.; Raban, M.; Mislow, K. *J. Org. Chem.* **1968**, *33*, 1142-1145.

74. Dale, J. A.; Dull, D. L.; Mosher, H. S. *J. Org. Chem.* **1969**, *34*, 2543-2549.
75. Anderson, R. C.; Shapiro, M. J. *J. Org. Chem.* **1984**, *49*, 1304-1305.
76. Takeuchi, Y.; Itoh, N.; Koizumi, T. *J. Chem. Soc., Chem. Commun.* **1992**, 1514-1515.
77. Uccello-Barretta, G.; Bernardini, R.; Lazzaroni, R.; Salvadori, P. *J. Organomet. Chem.* **2000**, *598*, 174-178.
78. Alexakis, A.; Chauvin, A.-S. *Tetrahedron: Asymm.* **2001**, *12*, 1411-1415.
79. Alexakis, A.; Chauvin, A.-S. *Tetrahedron: Asymm.* **2001**, *12*, 4245-4248.
80. Blazewska, K.; Gajda, T. *Tetrahedron: Asymm.* **2002**, *13*, 671-674.
81. Pirkle, W. H. *J. Am. Chem. Soc.* **1966**, *88*, 1837.
82. Burlingame, T. G.; Pirkle, W. H. *J. Am. Chem. Soc.* **1966**, *88*, 5150-5155.
83. Pirkle, W. H.; Beare, S. D. *J. Am. Chem. Soc.* **1969**, *91*, 4294.
84. Deshmukh, M.; Dunach, E.; Juge, S.; Kagan, H. B. *Tetrahedron Lett.* **1984**, *25*, 3467-3470.
85. Parker, D. *Chem. Rev.* **1991**, *91*, 1441-1457.
86. Wenzel, T. J.; Amonoo, E. P.; Shariff, S. S.; Aniagyei, S. E. *Tetrahedron: Asymm.* **2003**, *14*, 3099-3104.
87. Fraser, R. R.; Petit, M. A.; Saunders, J. K. *J. Chem. Soc., Chem. Commun.* **1971**, 1450-1451.
88. Goering, H. L.; Eikenberry, J. N.; Koermer, G. S. *J. Am. Chem. Soc.* **1971**, *93*, 5913-5914.
89. Goering, H. L.; Eikenberry, J. N.; Koermer, G. S.; Lattimer, C. J. *J. Am. Chem. Soc.* **1974**, *96*, 1493-1501.
90. Yeh, H. J. C.; Balani, S. K.; Yagi, H.; Greene, R. M. E.; Sharma, N. D.; Boyd, D. R.; Jerina, D. M. *J. Org. Chem.* **1986**, *51*, 5439-5443.
91. Vagg, R. S.; Williams, P. A. *Inorg. Chem.* **1983**, *22*, 355-357.
92. Rhys-Williams, W.; McCarthy, F.; Baker, J.; Hung, Y.-F.; Thomason, M. J.; Lloyd, A. W.; Hanlon, G. W. *Enzyme Microb. Technol.* **1998**, *22*, 281-287.
93. Hobley, J.; Malatesta, V.; Millini, R.; Montanari, L.; Parker Jr., W. O. N. *Phys. Chem. Chem. Phys.* **1999**, *1*, 3259-3267.
94. Ahmed, S. A.; Esaki, N.; Tanaka, H.; Soda, K. *Biochemistry* **1986**, *25*, 385-388.
95. Gasparrini, F.; Lunazzi, L.; Misiti, D.; Villani, C. *Acc. Chem. Res.* **1995**, *28*, 163-170.
96. Luchinat, C.; Roelens, S. *J. Am. Chem. Soc.* **1986**, *108*, 4873-4878.
97. Parker, D. *Chem. Rev.* **1991**, *91*, 1441-1457.
98. Oki, M.; Yamamoto, G. *Bull. Chem. Soc. Jpn.* **1971**, *44*, 266-270.
99. Wolf, C.; Ghebremariam, T. *Tetrahedron: Asymm.* **2002**, *13*, 1153-1156.
100. Wolf, C.; Pranatharthiharan, L.; Ramagosa, R. B. *Tetrahedron Lett.* **2002**, *43*, 8563-8567.
101. Charlton, J. L.; Oleschuk, C. J.; Chee, G.-L. *J. Org. Chem.* **1996**, *61*, 3452-3457.
102. Dell'Erba, C.; Gasparrini, F.; Grilli, S.; Lunazzi, L.; Mazzanti, A.; Novi, M.; Pierini, M.; Tavani, C.; Villani, C. *J. Org. Chem.* **2002**, *67*, 1663-1668.
103. Grilli, S.; Lunazzi, L.; Mazzanti, A.; Pinamonti, M. *Tetrahedron* **2004**, *60*, 4451-4458.
104. Casarini, D.; Lunazzi, L.; Pasquali, F.; Gasparrini, F.; Villani, C. *J. Am. Chem. Soc.* **1992**, *114*, 6521-6527.
105. Lunazzi, L.; Mazzanti, A.; Alvarez, A. M. *J. Org. Chem.* **2000**, *65*, 3200-3206.
106. Leardini, R.; Lunazzi, L.; Mazzanti, A.; Nanni, D. *J. Org. Chem.* **2001**, *66*, 7879-7882.
107. Casarini, D.; Lunazzi, L. *J. Org. Chem.* **1994**, *59*, 4637-4641.
108. Casarini, D.; Lunazzi, L.; Mazzanti, A. *J. Org. Chem.* **1997**, *62*, 7592-7596.
109. Casarini, D.; Lunazzi, L.; Mazzanti, A. *J. Org. Chem.* **1998**, *63*, 4991-4995.
110. Bringmann, G.; Heubes, M.; Breuning, M.; Göbel, L.; Ochse, M.; Schöner, B.; Schupp, O. *J. Org. Chem.* **2000**, *65*, 722-728.
111. Davalli, S.; Lunazzi, L.; Macciantelli, D. *J. Org. Chem.* **1991**, *56*, 1739-1747.
112. Casarini, D.; Foresti, E.; Gasparrini, F.; Lunazzi, L.; Macciantelli, D.; Misiti, D.; Villani, C. *J. Org. Chem.* **1993**, *58*, 5674-5682.
113. Casarini, D.; Lunazzi, L. *J. Org. Chem.* **1995**, *60*, 5515-5519.

114. Toyota, S.; Makino, T. *Tetrahedron Lett.* **2003**, *44*, 7775-7778.

115. Aydeniz, Y.; Oğuz, F.; Yaman, A.; Konuklar, A. S.; Doğan, I.; Aviyente, V.; Klein, R. A. *Org. Biomol. Chem.* **2004**, *2*, 2426-2436.

116. Leardini, R.; Lunazzi, L.; Mazzanti, A.; McNab, H.; Nanni, D. *Eur. J. Org. Chem.* **2000**, 3439-3446.

117. Gasparrini, F.; Grilli, S.; Leardini, R.; Lunazzi, L.; Mazzanti, A.; Nanni, D.; Pierini, M.; Pinamonti, M. *J. Org. Chem.* **2002**, *67*, 3089-3095.

118. Garcia, M. B.; Grilli, S.; Lunazzi, L.; Mazzanti, A.; Orelli, L. R. *J. Org. Chem.* **2001**, *66*, 6679-6684.

119. Guerra, A.; Lunazzi, L. *J. Org. Chem.* **1995**, *60*, 7959-7965.

120. Warren, S.; Chow, A.; Fraenkel, G.; RajanBabu, T. V. *J. Am. Chem. Soc.* **2003**, *125*, 15402-15410.

121. Casarini, D.; Coluccini, C.; Lunazzi, L.; Mazzanti, A.; Rompietti, R. *J. Org. Chem.* **2004**, *69*, 5746-5748.

122. Gasparrini, F.; Lunazzi, L.; Mazzanti, A.; Pierini, M.; Pietrusiewicz, K. M.; Villani, C. *J. Am. Chem. Soc.* **2000**, *122*, 4776-4780.

123. Gibson, K. R.; Hitzel, L.; Mortishire-Smith, R. J.; Gerhard, U.; Jelley, R. A.; Reeve, A. J.; Rowley, M.; Nadin, A.; Owens, A. P. *J. Org. Chem.* **2002**, *67*, 9354-9360.

124. Paquette, L. A.; Wang, T.-Z.; Luo, J.; Cottrell, C. E.; Clough, A. E.; Anderson, L. B. *J. Am. Chem. Soc.* **1990**, *112*, 239-253.

125. Pawar, D.; Noe, E. A. *J. Am. Chem. Soc.* **1998**, *120*, 1485-1488.

126. Casarini, D.; Lunazzi, L.; Mazzanti, A.; Simon, G. *J. Org. Chem.* **2000**, *65*, 3207-3208.

127. Klaerner, F.-G.; Kleine, A. E.; Oebels, D.; Scheidt, F. *Tetrahedron: Asymm.* **1993**, *4*, 479-90.

128. Simaan, S.; Marks, V.; Gottlieb, H. E.; Stanger, A.; Biali, S. E. *J. Org. Chem.* **2003**, *68*, 637-640.

129. Rochlin, E.; Rappoport, Z. *J. Org. Chem.* **2003**, *68*, 216-226.

130. Grilli, S.; Lunazzi, L.; Mazzanti, A.; Casarini, D.; Femoni, C. *J. Org. Chem.* **2001**, *66*, 488-495.

131. Grilli, S.; Lunazzi, L.; Mazzanti, A.; Mazzanti, G. *J. Org. Chem.* **2001**, *66*, 748-754.

132. Casarini, D.; Grilli, S.; Lunazzi, L.; Mazzanti, A. *J. Org. Chem.* **2001**, *66*, 2757-2763.

133. Grilli, S.; Lunazzi, L.; Mazzanti, A. *J. Org. Chem.* **2001**, *66*, 4444-4446.

134. Grilli, S.; Lunazzi, L.; Mazzanti, A. *J. Org. Chem.* **2001**, *66*, 5853-5858.

135. Mazet, C.; Gade, L. H. *Chem. Eur. J.* **2002**, *8*, 4308-4318.

136. Cort, A. D.; Mandolini, L.; Pasquini, C.; Schiaffino, L. *Org. Lett.* **2004**, *6*, 1697-1700.

137. Vignon, S. A.; Wong, J.; Tseng, H.-R.; Stoddart, J. F. *Org. Lett.* **2004**, *6*, 1095-1098.

138. Ward, T. J. *Anal. Chem.* **2002**, *74*, 2863-2872.

139. Terfloth, G. *LC-GC* **1999**, *17*, 400-405.

140. Zenoni, G.; Pedeferri, M.; Mazzotti, M.; Morbidelli, M. *J. Chromatogr. A* **2000**, *888*, 73-83.

141. Pirkle, W. H.; Sikkenga, D. L. *J. Chromatogr.* **1976**, *123*, 400-404.

142. Pettersson, C.; Schill, G. *J. Chromatogr.* **1981**, *204*, 179-183.

143. König, W. A. *The Practice of Enantiomer Separation by Capillary Gas Chromatography*, Hüthig, Heidelberg, 1987.

144. Frank, H.; Nicholson, G. J.; Bayer, E. *J. Chromatogr. Sci.* **1977**, *15*, 174-176.

145. Gil-Av, E.; Freibush, B.; Charles-Sigler, R. *Tetrahedron Lett.* **1966**, *7*, 1009-1015.

146. Schurig, V. *J. Chromatogr.* **1988**, *441*, 135-153.

147. Schurig, V.; Novotny, H. P. *Angew. Chem., Int. Ed. Engl.* **1990**, *29*, 939-957.

148. König, W. A.; Krebber, R.; Mischnick, P. *J. High Resolut. Chromatogr.* **1989**, *12*, 732-738.

149. Schurig, V. *J. Chromatogr. A* **1994**, *666*, 111-129.

150. König, W. A. *J. High Resolut. Chromatogr.* **1993**, *16*, 569-586.

151. Schurig, V. *J. Chromatogr. A* **2002**, *965*, 315-356.

152. Gassman, E.; Kuo, J. E.; Zare, R. N. *Science* **1985**, *230*, 813-814.
153. Engelhardt, H.; Beck, W.; Kohr, J.; Schmitt, T. *Angew. Chem., Int. Ed. Engl.* **1993**, *32*, 629-649.
154. Rogan, M. M.; Altria, K. D.; Goodall, D. M. *Chirality* **1994**, *6*, 25-40.
155. Fanali, S. *J. Chromatogr. A* **1997**, *792*, 227-268.
156. Scriba, G. K. E. *Electrophoresis* **2003**, *24*, 2409-2421.
157. Pirkle, W. H.; Pochapsky, T. C.; Mahler, G. S.; Corey, D. E.; Reno, D. S.; Alessi, D. M. *J. Org. Chem.* **1986**, *51*, 4991-5000.
158. Pirkle, W. H.; Welch, C. J.; Lamm, B. *J. Org. Chem.* **1992**, *57*, 3854-3860.
159. Okamoto, Y.; Yashima, E. *Angew. Chem., Int. Ed.* **1998**, *37*, 1021-1043.
160. Pirkle, W. H.; Liu, Y. *J. Org. Chem.* **1994**, *59*, 6911-6916.
161. Pirkle, W. H.; Gan, K. Z. *Tetrahedron: Asymm.* **1997**, *8*, 811-814.
162. Subramanian, G. *A Practical Approach to Chiral Separations by Liquid Chromatography*, Verlag Chemie, Weinheim, 1994.
163. Wolf, C.; Pirkle, W. H. *J. Chromatogr. A* **1998**, *799*, 177-184.
164. Welch, C. J. *J. Chromatogr. A* **1994**, *666*, 3-26.
165. Petersson, P.; Markides, K. E. *J. Chromatogr. A* **1994**, *666*, 381-394.
166. Terfloth, G. J.; Pirkle, W. H.; Lynam, K. G.; Nicolas, E. C. *J. Chromatogr. A* **1995**, *705*, 185-194.
167. Wolf, C.; Pirkle, W. H. *LC-GC* **1997**, *15*, 352-363.
168. Wolf, C.; Pirkle, W. H. *Enantiomer Separation by Sub- and Supercritical Fluid Chromatography on Rationally Designed Chiral Stationary Phases*, In: Anton, K.; Berger, C. (Eds.) *Supercritical Fluid Chromatography with Packed Columns*, Marcel Dekker, New York, 1997, pp. 251-271.
169. Smith, R. M. *Chiral Chromatography Using Sub- and Supercritical Fluids*, In: Anton, K.; Berger, C. (Eds.) *Supercritical Fluid Chromatography with Packed Columns*, Marcel Dekker, New York, 1997, pp. 223-249.
170. Liu, Y.; Lantz, A. W.; Armstrong, D. W. *J. Liq. Chromatogr. Related Technol.* **2004**, *27*, 1121-1178.
171. Pirkle, W. H.; Brice, L. J.; Terfloth, G. J. *J. Chromatogr. A* **1996**, *753*, 109-119.
172. Medvedovici, A.; Sandra, P.; Toribio, L.; David, F. *J. Chromatogr. A* **1997**, *785*, 159-171.
173. Wolf, C.; Pirkle, W. H. *J. Chromatogr. A* **1997**, *785*, 173-178.
174. Terfloth, G. *J. Chromatogr. A* **2001**, *906*, 301-307.
175. Wolf, C.; Spence, P. L.; Pirkle, W. H.; Derrico, E. M.; Cavender, D. M.; Rozing, G. P. *J. Chromatogr. A* **1997**, *782*, 175-179.
176. Wolf, C.; Spence, P. L.; Pirkle, W. H.; Cavender, D. M.; Derrico, E. M. *Electrophoresis* **2000**, *21*, 917-924.
177. Kang, J.; Wistuba, D.; Schurig, V. *Electrophoresis* **2002**, *23*, 4005-4021.
178. Laemmerhofer, M.; Tobler, E.; Zarbl, E.; Lindner, W.; Svec, F.; Frechet, J. M. J. *Electrophoresis* **2003**, *24*, 2986-2999.
179. Wistuba, D.; Schurig, V. *J. Chromatogr. A* **2000**, *875*, 255-276.
180. Guebitz, G.; Schmid, M. G. *Electrophoresis* **2004**, *25*, 3981-3996.
181. Wolf, C.; Pranatharthiharan, L.; Volpe, E. C. *J. Org. Chem.* **2003**, *68*, 3287-3290.
182. Roussel, C.; Del Rio, A.; Pierrot-Sanders, J.; Piras, P.; Vanthuyne, N. *J. Chromatogr. A* **2004**, *1037*, 311-328.
183. Wolf, C. *Chem. Rev.* **2005**, *34*, 595-608.
184. König, W. A.; Gehrcke, B.; Runge, T.; Wolf, C. *J. High Resolut. Chromatogr.* **1993**, *16*, 376-378.
185. Kallen, J.; Heilbronner, E. *Helv. Chim. Acta* **1960**, *43*, 489-500.
186. Bassett, D. W.; Habgood, H. W. *J. Phys. Chem.* **1960**, *64*, 769-773.

187. Langer, S. H.; Yurchak, J. Y.; Patton, J. E. *Ind. Eng. Chem.* **1969**, *61*, 11-21.

188. Kramer, R. *J. Chromatogr.* **1975**, *107*, 241-252.

189. Bürkle, W.; Karfunkel, H.; Schurig, V. *J. Chromatogr.* **1984**, *288*, 1-14.

190. Wolf, C.; Hochmuth, D. H.; König, W. A.; Roussel, C. *Liebigs Ann.* **1996**, 357-363.

191. Keller, R. A.; Giddings, J. C. *J. Chromatogr.* **1960**, *3*, 205-220.

192. Jung, M. *QCPE Bull.* **1992**, *3*, 12.

193. Trapp, O.; Schurig, V. *Comput. Chem.* **2001**, *25*, 187-195.

194. Trapp, O.; Schoetz, G.; Schurig, V. *Chirality* **2001**, *13*, 403-414.

195. Melander, W. R.; Lin, H.-J.; Jacobsen, J.; Horváth, C. *J. Chromatogr.* **1982**, *234*, 269-276.

196. Melander, W. R.; Lin, H.-J.; Jacobsen, J.; Horváth, C. *J. Phys. Chem.* **1984**, *88*, 4527-4536.

197. Jacobsen, J.; Melander, W.; Vaisnis, G.; Horváth, C. *J. Phys. Chem.* **1984**, *88*, 4536-4542.

198. Stephan, B.; Zinner, H.; Kastner, F.; Mannschreck, A. *Chimia* **1990**, *44*, 336-338.

199. Mannschreck, A.; Zinner, H.; Pustet, N. *Chimia* **1989**, *43*, 165-166.

200. Mannschreck, A.; Kießl, L. *Chromatographia* **1989**, *28*, 263-266.

201. Veciana, J.; Crespo, M. I. *Angew. Chem., Int. Ed. Engl.* **1991**, *30*, 74-77.

202. Cabrera, K.; Lubda, D. *J. Chromatogr. A* **1994**, *666*, 433-438.

203. Cabrera, K.; Jung, M.; Fluck, M.; Schurig, V. *J. Chromatogr. A* **1996**, *731*, 315-321.

204. Li, L.; Thompson, R.; Sowa Jr., J. R.; Clausen, A.; Dowling, T. *J. Chromatogr. A* **2004**, *1043*, 171-175.

205. Spivey, A. C.; Charbonneau, P.; Fekner, T.; Hochmuth, D. H.; Maddaford, A.; Malardier-Jugroot, C.; Redgrave, A. J.; Whitehead, M. A. *J. Org. Chem.* **2001**, *66*, 7394-7401.

206. Wolf, C.; Tumambac, G. E. *J. Phys. Chem.* **2003**, *107*, 815-817.

207. Trapp, O.; Caccamese, S.; Schmidt, C.; Böhmer, V.; Schurig, V. *Tetrahedron: Asymm.* **2001**, *12*, 1395-1398.

208. Villani, C.; Pirkle, W. H. *Tetrahedron: Asymm.* **1995**, *6*, 27-30.

209. Gasparrini, F.; Misiti, D.; Pierini, M.; Villani, C. *Tetrahedron: Asymm.* **1997**, *8*, 2069-2073.

210. Oxelbark, J.; Allenmark, S. *J. Org. Chem.* **1999**, *64*, 1483-1486.

211. Oxelbark, J.; Allenmark, S. *J. Chem. Soc., Perkin Trans. 2* **1999**, 1587-1589.

212. Gasparrini, F.; Lunazzi, L.; Misiti, D.; Villani, C. *Acc. Chem. Res.* **1995**, *28*, 163-170.

213. Schurig, V.; Bürkle, W. *J. Am. Chem. Soc.* **1982**, *104*, 7573-7580.

214. Bürkle, W.; Karfunkel, H.; Schurig, V. *J. Chromatogr.* **1984**, *288*, 1-14.

215. Klärner, F.-G.; Schröer, D. *Angew. Chem., Int. Ed. Engl.* **1987**, *26*, 1294-1295.

216. Schurig, V.; Jung, M.; Schleimer, M.; Klärner, F.-G. *Chem. Ber.* **1992**, *125*, 1301-1303.

217. Marriott, P. J.; Lai, Y.-H. *J. Chromatogr.* **1988**, *447*, 29-41.

218. Jung, M.; Schurig, V. *J. Am. Chem. Soc.* **1992**, *114*, 529-534.

219. Schurig, V.; Keller, F.; Reich, S.; Fluck, M. *Tetrahedron: Asymm.* **1997**, *8*, 3475-3480.

220. Trapp, O.; Schurig, V. *Chirality* **2002**, *14*, 465-470.

221. Trapp, O.; Schurig, V. *Chem. Eur. J.* **2001**, *7*, 1495-1502.

222. Wolf, C.; Hochmuth, D. H.; König, W. A.; Roussel, C. *Liebigs Ann.* **1996**, 357-363.

223. Hochmuth, D. H.; König, W. A. *Liebigs Ann.* **1996**, 947-951.

224. Hochmuth, D. H.; König, W. A. *Tetrahedron: Asymm.* **1999**, *10*, 1089-1097.

225. Wolf, C.; Pirkle, W. H.; Welch, C. J.; Hochmuth, D. H.; König, W. A.; Chee, G.-L.; Charlton, J. L. *J. Org. Chem.* **1997**, *62*, 5208-5210.

226. Schoetz, G.; Trapp, O.; Schurig, V. *Anal. Chem.* **2000**, *72*, 2758-2764.

227. Weseloh, G.; Wolf C.; König, W. A. *Angew. Chem., Int. Ed. Engl.* **1995**, *34*, 1635-1636.

228. Weseloh, G.; Wolf, C.; König, W. A. *Chirality* **1996**, *8*, 441-445.

229. Scharwächter, K. P.; Hochmuth, D. H.; Dittmann, H.; König, W. A. *Chirality* **2001**, *13*, 679-690.

230. Schurig, V.; Glausch, A.; Fluck, M. *Tetrahedron: Asymm.* **1995**, *6*, 2161-2164.
231. Schurig, V.; Reich, S. *Chirality* **1998**, *10*, 316-320.
232. Reich, S.; Trapp, O.; Schurig, V. *J. Chromatogr. A* **2000**, *892*, 487-498.
233. Trapp, O.; Schurig, V.; Kostyanovsky, R. G. *Chem. Eur. J.* **2004**, *10*, 951-957.
234. Trapp, O.; Schurig, V. *J. Am. Chem. Soc.* **2000**, *122*, 1424-1430.
235. Tobler, E.; Lämmerhofer, M.; Mancini, G.; Lindner, W. *Chirality* **2001**, *13*, 641-647.

Principles of Asymmetric Synthesis

Many pharmaceuticals, agrochemicals, flavors, fragrances, nutrients, and other biologically active compounds are chiral, and the majority of today's top-selling drugs is formulated and marketed in enantiopure form. The omnipresence and significance of chirality has led to an ever-increasing industrial and academic demand for enantiopure compounds and methods to make them. This can in principle be accomplished by formation of racemic products and subsequent preparative enantioseparation or through asymmetric synthesis. The former approach is generally less attractive because it requires an additional separation step and is limited to a maximum yield of 50% unless the undesired enantiomer can be recycled and employed in an enantioconvergent process. Asymmetric synthesis is aimed at selective formation of a single stereoisomer and therefore affords superior atom economy. Numerous synthetic methods utilizing a wealth of chiral catalysts, auxiliaries and reagents have been developed to date.[1-4] Asymmetric catalysis is particularly attractive because it provides efficient access to enantiopure compounds.[5,6] An asymmetric reaction can be defined as follows:

> An asymmetric reaction selectively introduces one or more than one element of chirality to a substrate, in such a way that stereoisomers are formed in unequal amounts.

5.1 CLASSIFICATION OF ASYMMETRIC REACTIONS

Asymmetric synthesis requires the use of a chiral catalyst, auxiliary or reagent that either recognizes heterotopic atoms, groups and faces of a substrate, or distinguishes between competing stereoisomers. While *asymmetric synthesis* entails selective introduction and manipulation of elements of chirality, the more broadly defined term *stereoselective synthesis* includes formation of achiral diastereomers such as (Z)- and (E)-alkenes. It is helpful to differentiate between enantioselective and diastereoselective processes:[7]

A. Enantioselective synthesis:

1. Enantiotopos-differentiating reactions
2. Enantioface-differentiating reactions
3. Enantiomer-differentiating reactions

B. Diastereoselective synthesis:

1. Diastereotopos-differentiating reactions
2. Diastereoface-differentiating reactions
3. Diastereomer-differentiating reactions

Selective modification or replacement of heterotopic atoms or groups can introduce a range of new stereogenic elements to a substrate, for instance a chiral center, a chiral axis or a 1,2-disubstituted double bond, Scheme 5.1. An enantiotopos-selective (diastereotopos-selective) reaction generates unequal amounts of enantiomers (diastereomers). Typical examples of heterotopos-differentiating reactions are (−)-sparteine-mediated deprotonation of achiral carbamates and desymmetrization of prochiral biaryls, see Chapters 6.1 and 6.2.4. Compounds with heterotopic faces, for example aldehydes and ketones with different substituents attached to the carbonyl group, can either undergo enantioface-selective (if no chiral element is present) or diastereoface-selective (if the substrate is chiral) reactions. Common examples are chiral Lewis acid-catalyzed Diels–Alder reactions, aldol-type reactions and Grignard additions to chiral aldehydes and ketones, see Chapter 5.3. Stereoisomer differentiation occurs when enantiomers or diastereomers undergo the same

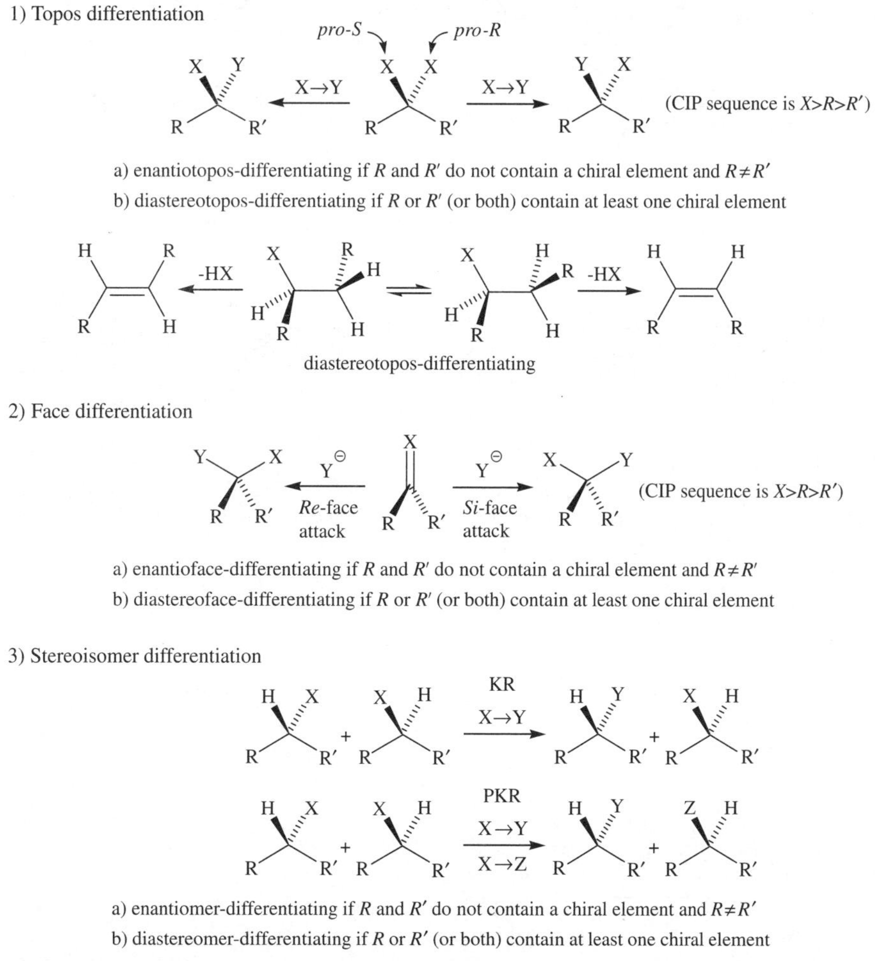

Scheme 5.1 Categories of stereoselective reactions.

reaction with different rates or when they are selectively transformed to nonisomeric products by different processes. This is observed during kinetic resolution (KR), parallel kinetic resolution (PKR), dynamic kinetic resolution (DKR) and dynamic kinetic asymmetric transformation (DYKAT). The principles and scope of these methods are discussed in Chapter 7.

Organic reactions are also commonly classified into stereoselective and stereospecific processes, Scheme 5.2. By convention, the term stereospecific is solely reserved for reactions that produce different stereoisomers from stereoisomeric starting materials under identical conditions. It follows from this definition and the discussion above that stereospecificity is a sufficient condition for stereoselectivity but that a stereoselective reaction is not necessarily stereospecific. The family of stereospecific reactions includes *trans*-addition of bromine to (*E*)- and (*Z*)-alkenes, cheletropic *syn*-addition of singlet carbenes to alkenes, electrocyclic reactions such as disrotatory ring closures, and the sigmatropic Claisen rearrangement of the *cis*- and *trans*-isomers of (4*S*)-vinyloxypent-2-enes. These reactions have in common that stereoisomeric substrates are transformed to stereoisomeric products. For example, bromination of (*Z*)-butene gives racemic 2,3-dibromobutane but (*E*)-butene is converted to the corresponding meso isomer under the same reaction conditions.

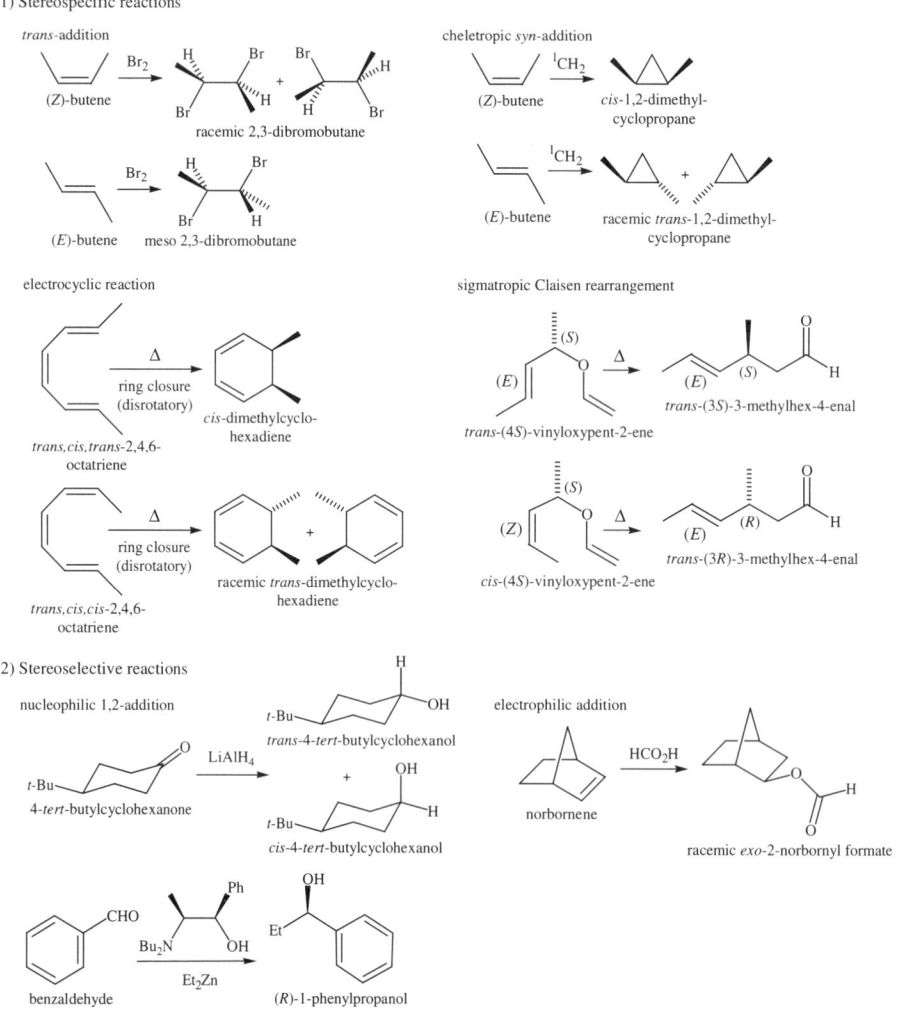

Scheme 5.2   Examples of stereospecific and stereoselective reactions.

In a stereoselective reaction, a single reactant may be transformed to several stereoisomers but one is obtained predominantly or even exclusively. This is observed during stereoselective addition of formic acid to norbornene, which yields diastereomerically pure *exo*-2-norbornyl formate; diastereoselective reduction of 4-*tert*-butylcyclohexanone with lithium aluminum hydride, providing *trans*-4-*tert*-butylcyclohexanol in 80% de; and enantioselective alkylation of benzaldehyde with organozinc reagents in the presence of catalytic amounts of (1R,2S)-N,N-dibutylnorephedrine. Reactions that proceed with perfect stereoselectivity are often inaccurately described in the literature as stereospecific. Electrophilic addition of formic acid to norbornene yields exclusively the racemic *exo*-ester and no *endo*-product is formed. However, this does not constitute a stereospecific process but is an example of a highly diastereoselective reaction.

5.2   KINETIC AND THERMODYNAMIC CONTROL

A basic concept of asymmetric synthesis is to effectively embed a substrate into the chiral environment of a catalyst, auxiliary or reagent. In all cases, chiral induction is accomplished either under kinetic or thermodynamic reaction control. In a thermodynamically controlled reaction, product formation is reversible and the stereochemical outcome is governed by the relative stability of the diastereomeric products, Figure 5.1. For example, asymmetric transformation of compound X to diastereomers Y and Z produces a mixture with the more stable species Y as the major isomer. When products and starting materials are in equilibrium, the product ratio is solely determined by the difference in the free energy, ΔG, of the diastereomers Y and Z at a given temperature, as defined by Equation 5.1. The energies of the transition states leading to the formation of Y and Z are irrelevant and do not affect the diastereoselectivity of the reaction. Enantioselective synthesis under thermodynamic control is not possible because enantiomers are isoenergetic and a racemate must be obtained at equilibrium unless a chiral solvent or resolving agent is used.[i]

$$\frac{[Y]}{[Z]} = K_{eq} = e^{-\Delta G/RT} \tag{5.1}$$

where $\Delta G = G_Y - G_Z$

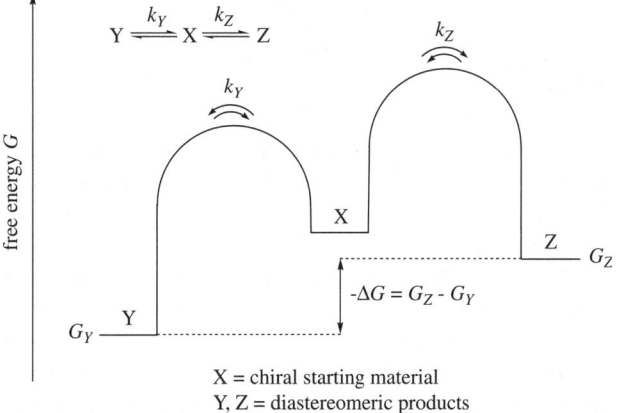

Figure 5.1   Energy profile for a thermodynamically controlled asymmetric reaction.

---

[i]In fact, this constitutes a diastereoselective process. The presence of a chiral additive or solvent generates diastereomeric adducts that can have different thermodynamic stability and may therefore be formed in unequal amounts.

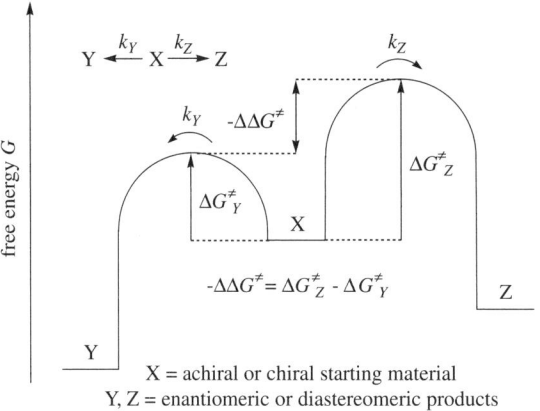

Figure 5.2   Energy profile for a kinetically controlled asymmetric reaction.

When a reaction is carried out under kinetic control the products can not equilibrate and the stereoselective outcome solely depends on the rates of product formation, r_Y and r_Z, Figure 5.2. In this case, X is converted to enantiomeric or diastereomeric products Y and Z under irreversible conditions, and the product ratio is determined by the difference in the free energy of the corresponding transition states of Y and Z, $\Delta\Delta G^{\neq}$, at a given temperature as defined in Equation 5.4. Because conversion of X to Y has a lower free energy of activation, $\Delta G^{\neq}_Y$, than the competing formation of Z, Y will be the predominant product. Under kinetic control, the relative stability of the products Y and Z is irrelevant and does not affect the stereoselectivity of the reaction. Enantioselective synthesis is feasible when nonequienergetic diastereomeric transition states can be established under irreversible conditions.

$$r_Y = [X]k_Y \qquad (5.2)$$

where $k_Y = \kappa \frac{k_B T}{h} e^{-\Delta G^{\neq}_Y / RT}$

$$r_Z = [X]k_Z \qquad (5.3)$$

where $k_Z = \kappa \frac{k_B T}{h} e^{-\Delta G^{\neq}_Z / RT}$

$$\frac{r_Y}{r_Z} = \frac{[Y]}{[Z]} = \frac{k_Y}{k_Z} = e^{-\Delta\Delta G^{\neq} / RT} \qquad (5.4)$$

where $\Delta\Delta G^{\neq} = \Delta G^{\neq}_Y - \Delta G^{\neq}_Z$

Equations 5.1 and 5.4 describe the exponential dependence of the product ratio Y/Z on the difference in the free energy of the products, ΔG, or of the corresponding transition states, $\Delta\Delta G^{\neq}$, at a given temperature. The stereoselectivity of an asymmetric reaction improves as the difference in the relative stability of the products or transition states increases. This is illustrated in Figure 5.3 and Table 5.1. At 0 °C, a product ratio of 9 : 1 (80% ee or de) can be achieved when ΔG or $\Delta\Delta G^{\neq}$ is about 5 kJ/mol. A further increase in ΔG or $\Delta\Delta G^{\neq}$ to approximately 6.7 kJ/mol enhances the enantiomeric or diastereomeric excess to 90%. The same stereoselectivity can be obtained with energy differences of 4.8 and 7.3 kJ/mol at −78 °C and 25 °C, respectively. It is important to realize that these values are fairly small and similar to the energy of a weak hydrogen bond or the energetic differences of conformational isomers.[ii] Because of the exponential relationship between stereoselectivity and the difference in the free energy or free activation energy, the curves shown in Figure 5.3 start with a

---

[ii] For example, the energetic difference between the axial and equatorial conformation of methylcyclohexane is approximately 7 kJ/mol.

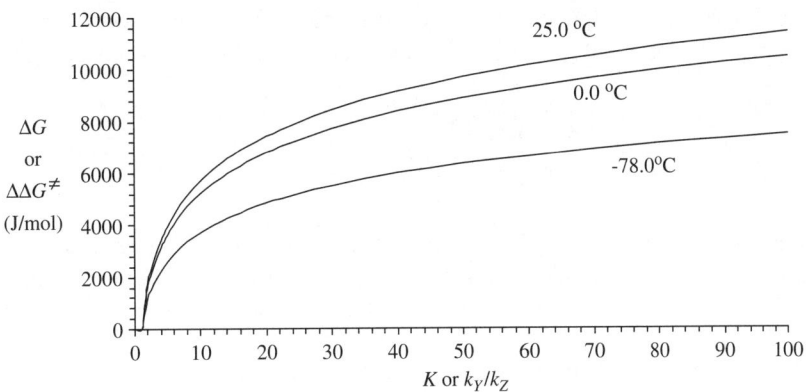

Figure 5.3 Dependence of stereoselectivity on the difference in the free energy of the products, ΔG, or of the corresponding transition states, $\Delta\Delta G^{\neq}$, at frequently used temperatures.

Table 5.1 Dependence of the product ratio of Y and Z and the corresponding ee or de on the difference in the free energy of the products, ΔG, and of the corresponding transition states, $\Delta\Delta G^{\neq}$.

K or k_Y/k_Z	ee or de (%)	ΔG or $\Delta\Delta G^{\neq}$ [J/mol][a]		
		$-78.0\,^{\circ}C$	$0.0\,^{\circ}C$	$25.0\,^{\circ}C$
3	50	1782	2495	2723
7	75	3157	4419	4824
9	80	3565	4990	5447
19	90	4777	6687	7299
24	92	5156	7217	7878
49	96	6314	8838	9647
99	98	7455	10435	11390

[a] $1.0\,\mathrm{J/mol} = 0.239\,\mathrm{cal/mol}$

steep slope that gradually decreases until an essentially straight line is obtained at Y/Z ratios above 20. A change in stereoselectivity from 3 : 1 to 9 : 1 requires doubling of ΔG (thermodynamic control) or $\Delta\Delta G^{\neq}$ (kinetic control) from 2.5 to 5 kJ/mol at 0 °C, Table 5.1. In contrast, small energetic changes have a profound effect on stereoselectivity in the flat part of the curve. A further increase of ΔG or $\Delta\Delta G^{\neq}$ by another 50% to 7.5 kJ/mol drastically improves selectivity to more than 27 : 1, which is close to 93% ee or de. Since the flat part of the curve is wider at -78 °C than it is at 0 °C or 25 °C, small energetic differences generally have a stronger impact on stereoselectivity at lower temperatures.

In general, stereoselectivity increases at lower temperatures as described above. But this is not necessarily the case because both ΔG and $\Delta\Delta G^{\neq}$ depend on the absolute temperature according to the Gibbs–Helmholtz equation:

$$\Delta G = \Delta H - T\Delta S \qquad (5.5)$$

Combination of the Gibbs–Helmholtz relationship and Equation 5.1 gives:

$$\frac{[Y]}{[Z]} = K_{eq} = (e^{-\Delta H/RT})(e^{\Delta S/R}) \qquad (5.6)$$

where $\Delta H = H_Y - H_Z$ and $\Delta S = S_Y - S_Z$

Equation 5.6 describes enthalpic and entropic contributions in a thermodynamically controlled reaction. In the case of a kinetically controlled reaction, combination of Equations 5.4 and 5.5 gives:

$$\frac{[Y]}{[Z]} = \frac{k_Y}{k_Z} = (e^{-\Delta\Delta H^{\neq}/RT})(e^{-\Delta\Delta S^{\neq}/R}) \qquad (5.7)$$

where $\Delta\Delta H^{\neq} = \Delta H^{\neq}_Y - \Delta H^{\neq}_Z$ and $\Delta\Delta S^{\neq} = \Delta S^{\neq}_Y - \Delta S^{\neq}_Z$

A closer look at Equations 5.6 and 5.7 reveals that a negative enthalpy term favors formation of Y over Z and stereoselectivity should therefore increase at lower temperatures. This is true for the majority of reactions which are enthalpy controlled because the enthalpic term outweighs entropic contributions. When the entropic term predominates, the reaction is entropy controlled and one can imagine two scenarios.[8–13] If both enthalpic and entropic changes are in favor of formation of the same stereoisomer, a decrease in temperature will improve selectivity although the effect is diminished since only the relatively small enthalpic term increases.[iii] If enthalpy and entropy favor formation of opposite stereoisomers, a decrease in temperature will be detrimental to asymmetric induction. In this case, the enthalpic contribution favoring the minor stereoisomer will increase relative to the entropic term favoring the major isomer. As a result, the difference in ΔG or $\Delta\Delta G^{\neq}$ and therefore the stereoselectivity of such an entropically controlled reaction must decrease.

The preceding discussion of thermodynamically and kinetically controlled asymmetric reactions is based on the comparison of just two competing reaction pathways leading either to enantiomeric or diastereomeric products. In many cases, however, more than two reaction courses exhibiting individual transition states and different asymmetric induction are operative. This does not affect the stereochemical outcome of thermodynamically controlled reactions if the individual pathways afford the same products. But the stereoselectivity of kinetically controlled reactions can be significantly compromised because the success of asymmetric induction under irreversible reaction conditions depends on both the relative stability and the number of coexisting diastereomeric transition states.

5.3 ASYMMETRIC INDUCTION

As discussed above, the stereoselectivity of an asymmetric reaction is determined by (a) the relative rate of competing reaction pathways if it is conducted under kinetic control, or (b) the relative stability of diastereoisomeric products if the reaction is thermodynamically controlled. Much effort has been devoted to the development of synthetic strategies that generate a single or at least a predominant well-defined transition state in order to achieve effective and predictable asymmetric induction. This often implies the use of a chiral catalyst or auxiliary that is designed to control the number of possible orientations and conformations of a substrate in the activated complex. The success of these approaches and the usefulness of an asymmetric reaction can be evaluated based on the following criteria:[14]

1. The reaction must be highly stereoselective and provide high yields.
2. The chiral catalyst or auxiliary must be recoverable without racemization.
3. The chiral catalyst or auxiliary must be inexpensive and readily available in both enantiopure forms.
4. The reaction must be applicable to a wide range of substrates.

These requirements can not be generally applied to assess the usefulness of every asymmetric reaction and have to be carefully evaluated case by case. Chiral catalysts are usually employed in enantiopure form but this is not always necessary. An important example is the enantioselective

---

[iii] Since the reactions are entropy controlled, the second term of Equations 5.6 and 5.7 outweighs the temperature-dependent enthalpic term which in this case has a relatively small effect on the product ratio.

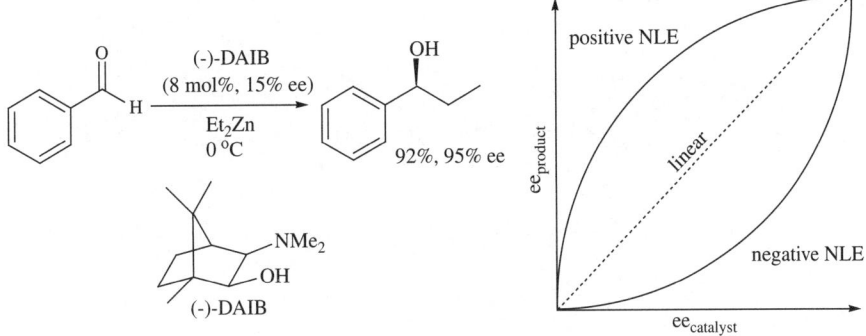

Figure 5.4 Asymmetric alkylation of benzaldehyde catalyzed by (−)-DAIB (15% ee), and possible relationships between enantioselectivity and ee of a catalyst.

alkylation of aldehydes with dialkylzinc reagents which is catalyzed by a wide range of readily available β-amino alcohols.[15–24] Because of the significance of chiral secondary alcohols in asymmetric synthesis, the ease of operation, and the observation of intriguing nonlinear relationships between the ee of the product and the enantiopurity of the catalyst used, this reaction has become one of the most intensively studied asymmetric *C–C* bond formations.[25–29] Noyori *et al.* observed that alkylation of benzaldehyde with diethylzinc in the presence of (−)-3-*exo*-(dimethylamino)isoborneol, (−)-DAIB, affords (*S*)-1-phenylpropanol in 92% yield and 95% ee even when the enantiopurity of the catalyst is very low, Figure 5.4. The reaction shows a positive deviation from the usually observed linear relationship between enantioselectivity and enantiomeric purity of the catalyst. Because of this positive nonlinear effect (NLE) a highly enantioenriched or enantiopure catalyst is not required to produce the alcohol in excellent yield and ee.

Both positive and negative nonlinear effects have been encountered in a variety of asymmetric reactions[30–43] and used as powerful mechanistic probes.[44–48] The origin of the (+)-NLE in the DAIB-catalyzed enantioselective alkylation of aldehydes has been attributed to the formation of homochiral and heterochiral complexes with different thermodynamic stability, Scheme 5.3. Initial reaction between DAIB (which is actually a precatalyst) and one equivalent of the organozinc reagent generates a catalytically active tricoordinate alkylzinc complex that undergoes reversible dimerization to inactive species consisting of a four-membered μ-oxo ring. In the presence of both enantiomers of DAIB, this process can afford homochiral *syn*-dimers bearing both alkyl groups on the same side, or a heterochiral *anti*-dimer with both alkyl groups on opposite sides of the four-membered ring. Due to less steric repulsion between the alkyl moieties, the *anti*-dimer is approximately 12 kJ/mol more stable than the *syn*-isomers.[49,50] The equilibrium between the monomeric and dimeric species therefore strongly favors formation of the heterochiral complex which serves as a trap for the minor enantiomer. When (−)-DAIB exhibiting 15% ee is employed in the reaction between benzaldehyde and diethylzinc, the minor (+)-enantiomer is almost quantitatively consumed in the catalytically inactive dimeric (−)-(+)-DAIB complex. Consequently, the reaction is mainly catalyzed by highly enantioenriched (−)-DAIB which is easily generated from the less stable homochiral (−)-(−)-DAIB complex. The term reservoir effect has been coined to illustrate that the minor enantiomer of the catalyst is trapped while the excess of the major enantiomer remains catalytically active.[51–53]

Even more impressive results are observed when positive nonlinear effects occur in autocatalytic asymmetric reactions.[54–63] Such a process is the enantioselective alkylation of heteroaryl carbaldehydes with dialkylzinc reagents. In asymmetric autocatalysis the chiral product accelerates its own stereoselective formation, which means that the structure of the catalyst and the product is the same. Soai and coworkers demonstrated that the reaction between a 2-alkynylpyrimidine-5-carbaldehyde and diisopropylzinc in the presence of catalytic amounts of the corresponding

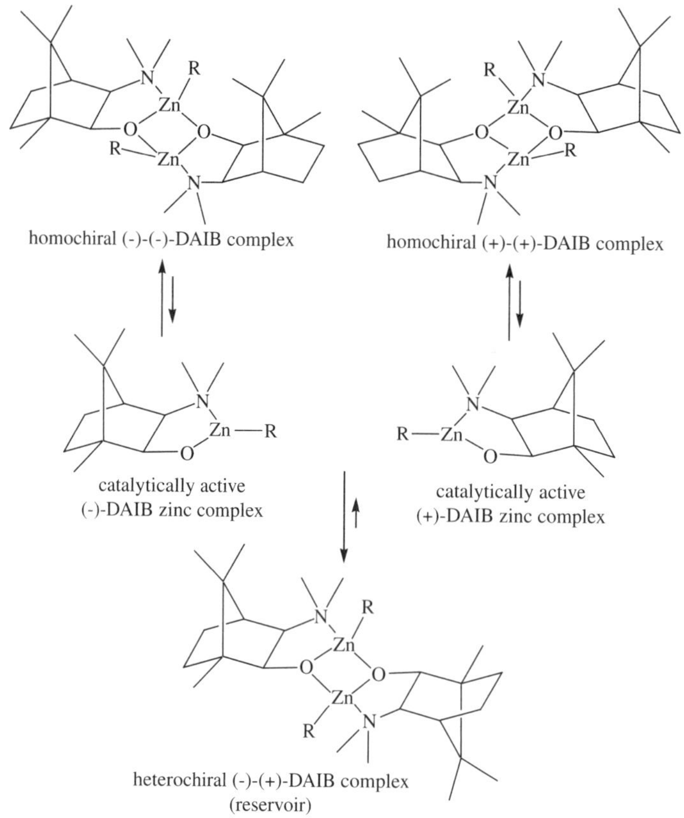

Scheme 5.3 Equilibrium between monomeric DAIB-derived alkylzinc complexes and homo- and hetero-chiral dimers.

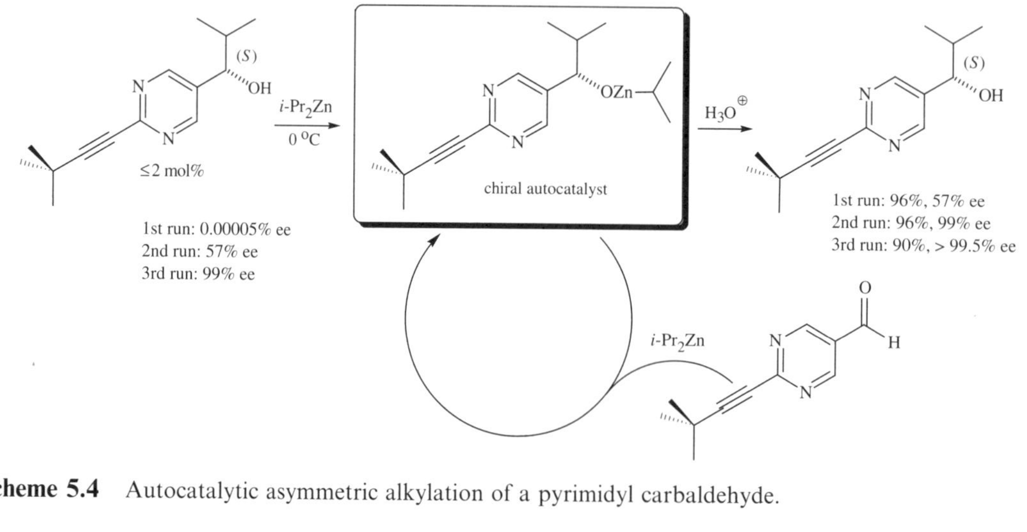

Scheme 5.4 Autocatalytic asymmetric alkylation of a pyrimidyl carbaldehyde.

(*S*)-1(-2-alkynyl-5-pyrimidyl)-2-methylpropanol exhibiting only $5 \cdot 10^{-5}\%$ ee generates the (*S*)-pyrimidyl alcohol in 57% ee, Scheme 5.4.[64] The active catalyst is initially formed by reaction of the alcohol and the organozinc reagent. The resulting zinc alkoxide then catalyzes the asymmetric alkylation of the aldehyde, resulting in remarkably stereoselective replication which is a consequence of a positive nonlinear effect based on the principles of Kagan's reservoir model.[65-68] Reiteration of this process using the (*S*)-enantiomer with 57% ee as the catalyst provides another batch with 99% ee. The enantiomeric excess of the (*S*)-pyrimidyl alcohol can be further enhanced to above 99.5% in a third run. This example demonstrates that a small imbalance of enantiomers can be effectively transformed to a highly enantioenriched mixture after just a few consecutive runs without an additional source of chirality. This so-called amplification of chirality can be observed in asymmetric autocatalytic processes that involve a positive nonlinear effect and may explain the origin of biological homochirality such as the dominance of (*S*)-amino acids in nature.

5.3.1 Control of Molecular Orientation and Conformation

A common feature of successful chiral catalysts is a rigid C_2-symmetric structure which reduces the number of possible catalyst–substrate arrangements and isomeric transition states, Figure 5.5.[69] Competing reaction pathways are therefore eliminated and, when compared with a chiral catalyst devoid of a C_2-axis, superior stereoselectivities and yields are often obtained.[iv] Mechanistic studies and prediction of the sense of asymmetric induction are generally simplified because fewer structures of coexisting transition states have to be considered. The introduction of the C_2-symmetric ligand DIOP to asymmetric catalysis by Kagan in 1971 has prompted the development of other equally successful ligand designs.[70,71] In particular, chiral catalysts derived from rigid C_2-symmetric ligands including binaphthols such as BINOL and BINAP,[72-75] bisoxazolines,[76-81] semicorrins,[82] salen,[83,84] DUPHOS,[85] and TADDOL[86] have proved to be exceptionally versatile and useful in a wide range of asymmetric reactions. For example, bisoxazoline-derived chiral Lewis acids (CLAs) have been very successfully applied in catalytic asymmetric Diels–Alder and ene reactions,[87-93] Mukaiyama aldol reactions,[94,95] cyclopropanations,[96-100] and aziridinations.[101-103] Coordination of a prochiral bidentate substrate to the metal center of a square-planar bisoxazoline-derived copper(II) complex can occur with two different orientations. However, the C_2-symmetry of the catalyst renders the two possible CLA-dienophile arrangements identical, Scheme 5.5. Formation of a complex between the Lewis acid and the *N*-acryloyl oxazolidinone substrate reduces the energy of the LUMO of the dienophile and activates it for Diels–Alder reaction with cyclopentadiene. Due to the C_2-symmetry of the (*S,S*)-bisoxazoline ligand, the *Si*-face of the dienophile is shielded while the *Re*-face is accessible independent of the substrate orientation in the activated complex. This effectively limits the number of possible transition states and therefore enhances enantiofacial control. The *endo*-approach of the diene to the well-defined copper coordination complex provides the (*S*)-cycloadduct.

In order to establish a well-defined reaction course one has to control the relative orientation of the reactants in the transition state, and particularly in the case of acyclic substrates the equilibrium between conformational isomers. The inherent fluxionality of acyclic compounds can often be controlled through chelation of activated intermediates. Again, this can be illustrated with a chiral Lewis acid-catalyzed Diels–Alder reaction using Evans' *N*-acryloyl oxazolidinone or other bidentate substrates, Scheme 5.6.[104-106] As discussed above, coordination of an *N*-acryloyl oxazolidinone to a bisoxazoline-derived Lewis acid generates a well-defined arrangement of the activated dienophile within the C_2-symmetric environment of the catalyst. Coordination of the bidentate substrate also impedes rotation about the acryloyl imide bond. Importantly, the rotational freedom of the substrate is further reduced by $A^{1,3}$ strain between the enone moiety and the oxazolidinone ring which forces the molecule into an *s-cis* conformation. Both chelation and steric repulsion lock the activated dienophile into a single geometry which gives rise to a highly stereoselective Diels–Alder reaction with cyclopentadiene, thus providing the corresponding bicyclic adduct in 86% yield

---

[iv] The presence of a symmetry axis (C_2 or C_3) is a common feature of many powerful chiral catalysts but this is by no means a strict requirement for effective asymmetric induction.

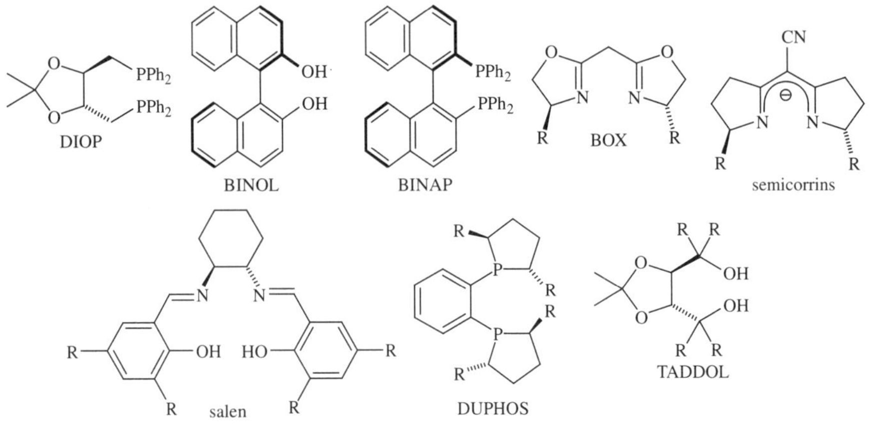

Figure 5.5 Structures of C$_2$-symmetric ligands.

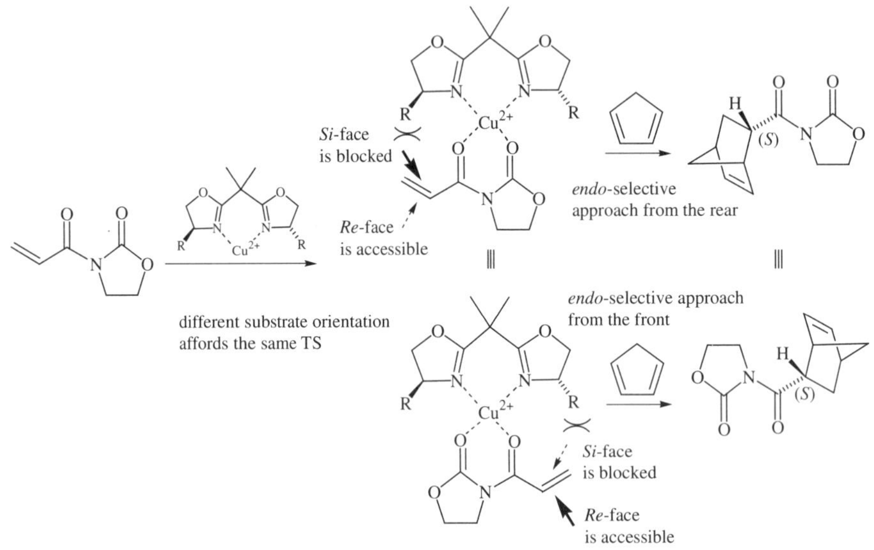

Scheme 5.5 Enantiofacial control with a C$_2$-symmetric catalyst. The two possible catalyst-substrate arrangements are identical and lead to the same transition state and product.

and more than 98% ee. This concept is not limited to oxazolidinones and has been applied to a number of other reactions including nucleophilic Michael additions and radical reactions.[107–118] Another example is the CLA-catalyzed hetero-Diels–Alder reaction between α-keto esters and Danishefsky's diene.[119–124] Coordination of methyl pyruvate to an (*S*,*S*)-bisoxazoline-derived copper(II) complex effectively reduces the conformational freedom of the activated dienophile which then undergoes cycloaddition with the diene to give the corresponding (*S*)-dihydropyranone in 83% yield and 91% ee.

Chelation also plays an important role in diastereoselective reactions. The stereochemical outcome of Grignard additions to chiral substrates that possess a stereocenter bearing a hetero-atom in α- or β-position to the carbonyl function is often rationalized using Cram's chelation model.[125] Cram and others realized that α-alkoxy ketones and Grignard reagents form a cyclic structure with the alkoxy group synperiplanar to the carbonyl function, Scheme 5.7. The restricted conformational freedom results in a highly diastereoselective transition state in which the

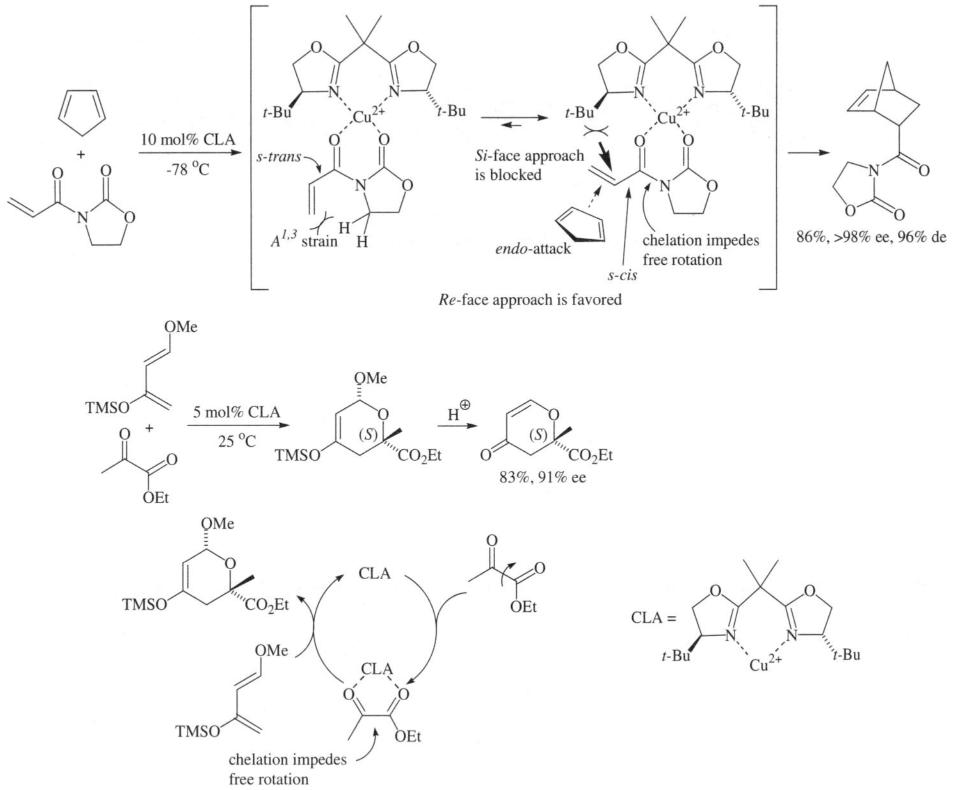

Scheme 5.6   Examples of catalytic enantioselective Diels–Alder reactions using a chiral Lewis acid derived from copper(II) triflate and a bisoxazoline ligand.

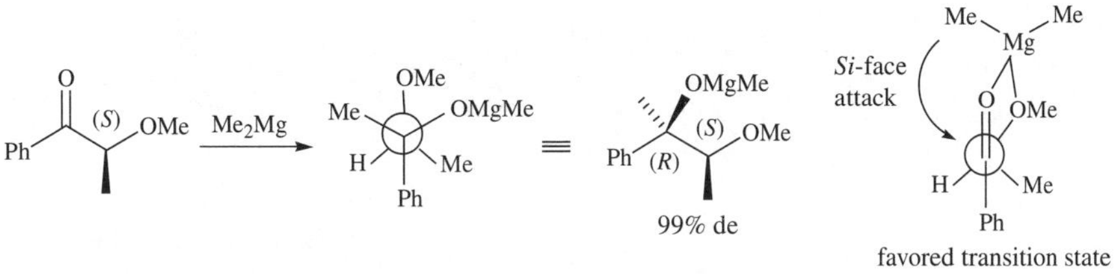

Scheme 5.7   Rationalization of the Grignard reaction of (*S*)-2-methoxy-1-phenylpropanone using Cram's chelation model.

nucleophile is delivered from the less sterically hindered side displaying the smaller ligand *S*. Attack from the opposite face is disfavored due to steric repulsion by the larger ligand *L*. For example, formation of an intermediate dimethylmagnesium complex of (*S*)-2-methoxy-1-phenylpropanone locks the substrate into a single conformer and thus restricts the number of possible transition states. Comparison of the diastereotopic faces of the five-membered chelate explains why the nucleophilic attack on the carbonyl group occurs predominantly from the *Si*-face, producing (2*R*,3*S*)-3-methoxy-2-phenylbutan-2-ol in excellent diastereoselective excess.[126,v] Similarly, restriction of the conformational flexibility of α-hydroxy ketones during complexation to zinc

---

v The mechanism of the Grignard addition to α- or β-alkoxy aldehydes, ketones and their derivatives is further complicated by Schlenk equilibria, competition between acyclic and cyclic reaction pathways, and stereoelectronic and solvation effects. However, Cram's model correctly predicts the stereochemical outcome of many π-facial diastereoselective nucleophilic additions, see Chapter 2.2.

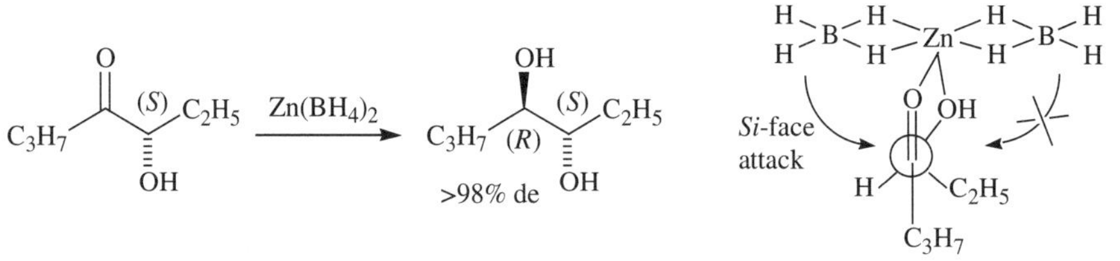

Scheme 5.8 Stereoselective reduction of (*S*)-4-hydroxyheptan-3-one with zinc borohydride.

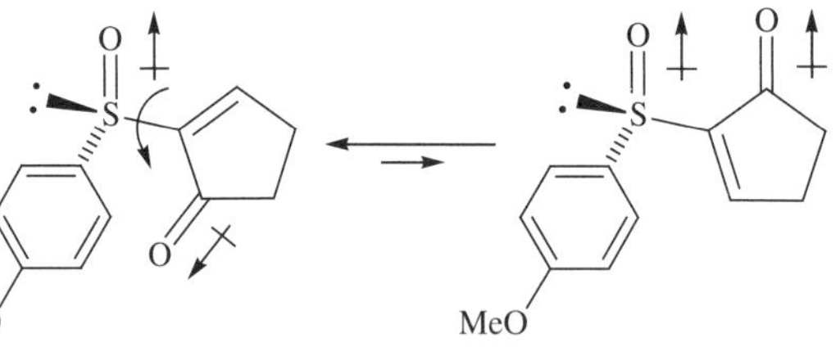

Scheme 5.9 Conformational isomers of (*S*)-2-(4-methoxyphenylsulfinyl)-2-cyclopentenone.

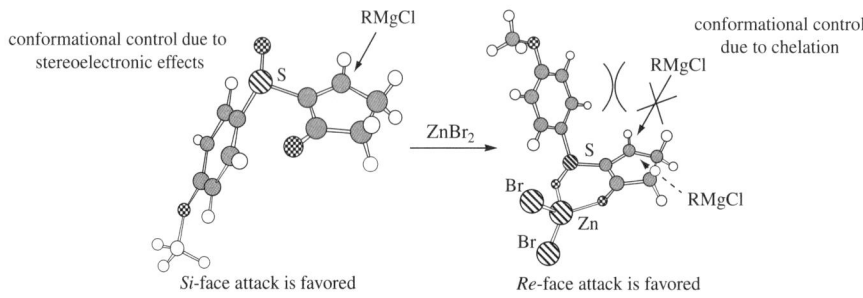

Scheme 5.10 Chelation-controlled diastereoselective Michael addition of Grignard reagents to (*S*)-2-(4-methoxyphenylsulfinyl)-2-cyclopentenone.

borohydride affords diols with high diastereoselectivity due to chelation, Scheme 5.8. Formation of a zinc complex locks (S)-4-hydroxyheptan-3-one into a single conformation that undergoes selective intramolecular reduction at the less sterically hindered *Si*-face of the carbonyl group, yielding the erythro diol in more than 98% de.[127]

Asymmetric reactions may become more versatile, but also more complicated, if stereoelectronic effects are operative in addition to chelation and steric interactions. Posner *et al.* utilized chelation control to switch the asymmetric induction in the diastereofacial Michael addition of Grignard reagents to chiral 2-(4-anisylsulfinyl)-2-cycloalkenones.[128] In the absence of a chelating metal ion, 2-(arylsulfinyl)-2-cycloalkenones prefer a conformation with the ketone and the sulfoxide groups pointing in opposite directions to minimize destabilizing dipole–dipole interactions, Scheme 5.9. Because of this stereoelectronic effect the bulky aryl group in (S_S)-2-(4-methoxyphenylsulfinyl)-2-cyclopentenone preferentially occupies the *Re*-face of the Michael acceptor, favoring nucleophilic attack from the *Si*-face and formation of (S)-3-alkylcyclopentanone derivatives, Scheme 5.10.[129] Chelation with a Lewis acid such as zinc dibromide has two important affects on the Grignard reaction: it increases the reactivity of the Michael acceptor for an electrophilic attack and it generates a rigid conformationally stable coordination complex exhibiting parallel ketone and sulfoxide groups. As a result, the chiral auxiliary blocks the *Si*-face and the Grignard addition occurs from the *Re*-face, producing the (S_S,R)-diastereoisomer. Stepwise Zn(II)-promoted Michael addition and reductive cleavage of the chiral auxiliary therefore furnishes 3-substituted

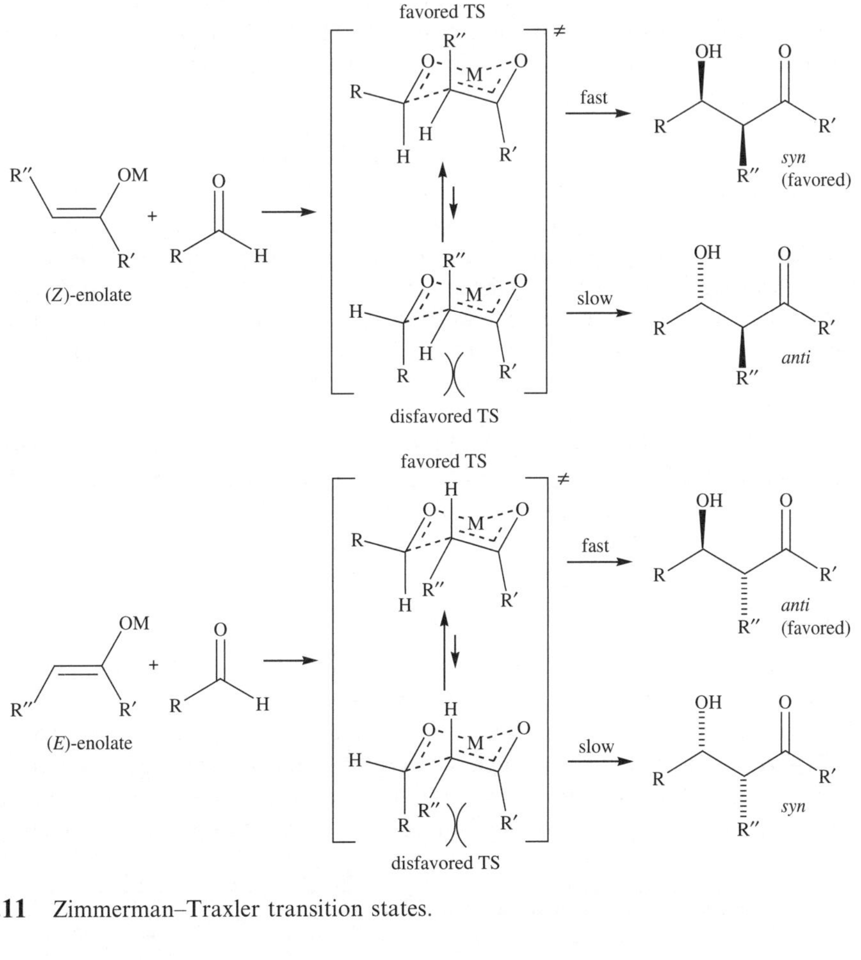

Scheme 5.11 Zimmerman–Traxler transition states.

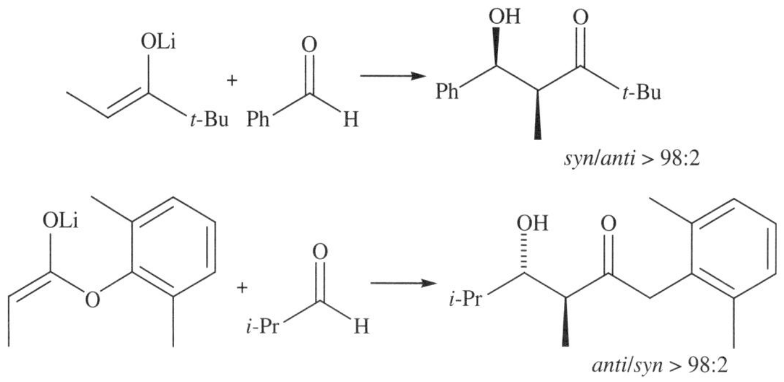

Scheme 5.12   Diastereoselective aldol reactions.

(*R*)-cyclopentanones. For example, ZnBr$_2$-mediated Michael addition of neopentylmagnesium bromide to (*S*$_S$)-2-(4-methoxyphenylsulfinyl)-2-cyclopentenone and subsequent reductive removal of the arylsulfinyl auxiliary gives (*R*)-3-neopentylcyclopentanone in 69% yield and 87% ee.

The selectivity of crossed aldol reactions is governed by chelation as well as by steric and stereoelectronic effects, in particular when chiral substrates[130–146] or chiral Lewis acids[147–158] are employed. Because of its important role in polyketide synthesis the stereochemical course of the aldol reaction has been studied extensively. A closer look at a diastereoselective aldol reaction between an achiral aldehyde and an achiral enolate reveals how the relative orientation of the substrates is controlled by metal coordination, Scheme 5.11. Both reactants are prochiral and have two heterotopic faces that can be attacked from either the *Si*- or the *Re*-face, thus generating aldol products with two chiral centers. According to the Zimmerman–Traxler model, coordination of both the (*Z*)-enolate and the aldehyde to a metal center leads to two diastereomeric chairlike transition states with different substrate orientations.[159] This simple but powerful model is applicable to kinetically controlled aldol-type reactions and assumes that the relative stability of the two possible transition states is determined by repulsive steric interactions between the aldehyde substituent *R* and the α-substituent *R′* in the (*Z*)-enolate. Destabilizing 1,3-diaxial repulsion between these substituents only occurs in the transition state that leads to the *anti*-aldol product. Accordingly, reaction via the more stable transition state with the aldehyde substituent in equatorial position proceeds more rapidly and preferential formation of the *syn*-aldol product can be expected. The same considerations explain why the (*E*)-enolate predominantly reacts to the *anti*-aldol product. The Zimmerman–Traxler model has been confirmed experimentally by Heathcock and coworkers.[160,161] They observed that the lithium (*Z*)-enolate prepared from *tert*-butyl ethyl ketone and benzaldehyde affords the *syn*-product in more than 96% de. In contrast, aldol reaction between the lithium (*E*)-enolate derived from 2,6-dimethylphenyl propionate and 2-methyl-propanal favors formation of the *anti*-β-hydroxy ester, Scheme 5.12.

5.3.2   Single and Double Stereodifferentiation

The stereochemical outcome of the majority of the enantioselective and diastereoselective reactions described above is controlled by a single chiral element present in either the substrate or the catalyst.[vi] In these cases, selective formation of one stereoisomer over another is a consequence of so-called single asymmetric induction or single stereodifferentiation. Many asymmetric reactions of chiral substrates with achiral or chiral reagents are highly diastereoselective due to effective substrate-based stereocontrol, *i.e.*, the stereochemical course and the sense of asymmetric induction

---

[vi] This is not the case in the crossed aldol reactions shown in Schemes 5.11 and 5.12.

is overwhelmingly determined by the chiral element in the substrate. A more complicated scenario arises when two or more elements of chirality influence stereoselectivity, for example in a kinetically controlled aldol reaction between a chiral aldehyde and a chiral (Z)-enolate, Scheme 5.13.[162] In such a double stereodifferentiation, the sense and efficiency of asymmetric induction depend on the absolute configuration of the two chiral centers, and their proximity and relative arrangement in the diastereomeric transition states. Inspection of the single asymmetric induction of both lithium (2S)-(Z)-2-phenyl-2-trimethylsiloxypent-3-en-3-olate and (S)-2-cyclohexylpropanal in the reaction with an achiral reactant reveals moderate stereofacial selectivity and ineffective substrate-based stereocontrol. The chiral lithium enolate favors attack at the *Re*-face of benzaldehyde and generates the corresponding *syn*-diastereomer in 56% de. The outcome of the reaction between an achiral lithium enolate and (S)-2-cyclohexylpropanal proves that the latter prefers attack at the *Re*-face because the *syn*-aldol product is formed in 46% de. Comparison of the individual hetero-facial preferences of the chiral (S)-(Z)-enolate and (S)-2-cyclohexylpropanal suggests that a combination of these will result in a synergistic effect and improved diastereoselectivity. In fact, aldol reaction between the two produces the *syn*-isomer in 78% de. This combination is an example of a double stereodifferentiation involving a so-called matched pair. In the reaction of a matched pair the stereochemical biases of the chiral reactants are alike and both prefer formation of the same stereoisomer. In contrast, the reaction of the (S)-aldehyde with the (R)-(Z)-enolate constitutes a double stereodifferentiation with a mismatched pair and ultimately affords low diastereoselectivity.

Double stereodifferentiation plays a fundamental role in dynamic kinetic resolution, dynamic kinetic asymmetric transformation and other synthetic strategies and has been applied as a mechanistic probe in a range of asymmetric reactions.[163–174] The effect of double stereodifferen-tiation on the CLA-catalyzed Diels–Alder reaction introduced above becomes evident when both a chiral bisoxazoline-derived copper(II) complex and an acryloyl imide template bearing a chiral auxiliary are employed, Scheme 5.14.[175] Diastereoselective cycloaddition of cyclopentadiene and (4R)-3-acryloyl-4-benzyl-oxazolidin-2-one in the presence of an (S,S)-bisoxazoline-derived cop-per(II) catalyst gives the *endo*-product in excellent yield and de. Apparently, the (S,S)-catalyst and the (R)-configured chiral auxiliary constitute a matched pair because both shield the *Si*-face at the α-carbon of the dienophile. In contrast, the (S)-benzyloxazolidinone-derived template shields the *Re*-face of the substrate. This has important consequences: the diastereoselectivity decreases to only 36%; and the reaction proceeds much more slowly and low yields of the cycloadduct are obtained because the approach of the diene from both the *Re*- and the *Si*-face is sterically hindered. The results demonstrate that the concept of double stereodifferentiation is well suited to obtaining mechanistic insights into asymmetric reactions. It is known that copper(II) complexes often accommodate four to six ligands forming various coordination geometries which ultimately affects the sense of asymmetric induction. Crystallographic analysis and the results of double stereodiffer-entiation described above are in agreement with the assumption that the Diels–Alder reaction is catalyzed by a square planar copper complex, as shown in Scheme 5.14.

The effect of double stereodifferentiation on both reactivity and selectivity can be understood from the energy profiles of kinetically controlled reactions involving a matched and a mismatched pair, respectively, Figure 5.6. Under irreversible conditions, a diastereoselective reaction of a chiral substrate X and an achiral reagent favors formation of Y over Z if the pathway leading to the former has a lower transition state energy. The reaction is controlled by the stereoselective bias of the substrate. Such a single asymmetric induction becomes more complicated when a chiral catalyst, auxiliary or reagent C is employed. In the case of a matched pair, the individual stereogenic elements of X and C favor formation of the same stereoisomer by further lowering the activation energy for the path towards Y and increasing the activation energy for the reaction towards Z. Consequently, the difference in the transition state energies is higher and double stereodifferentiation increases selectivity *and* reaction rate. In the case of a mismatched pair the

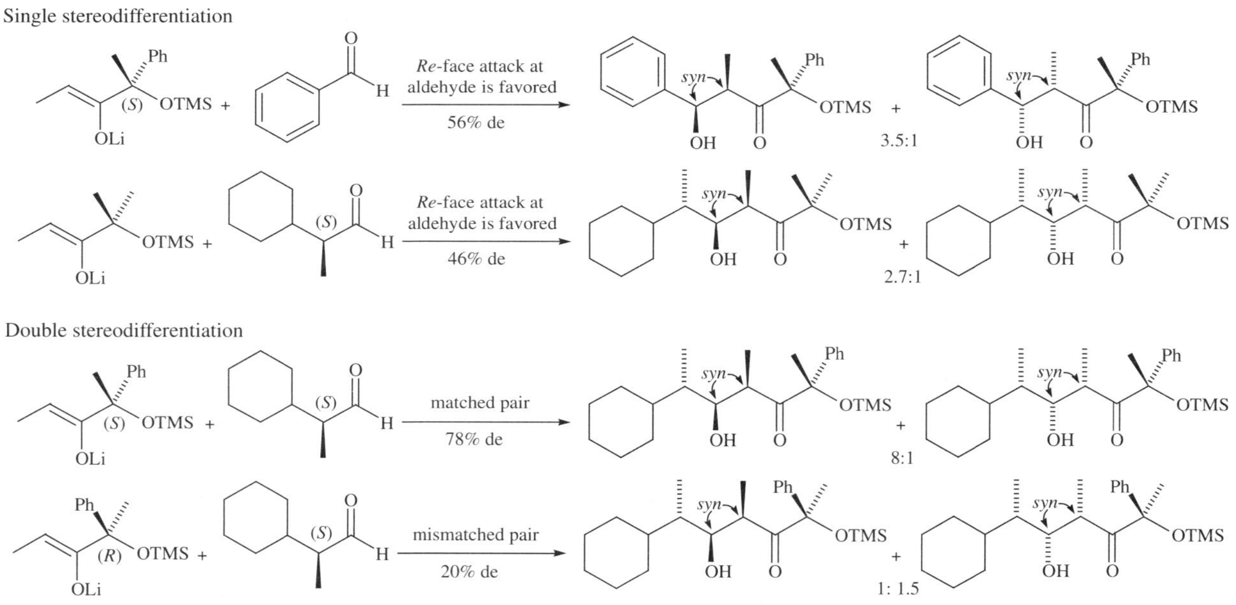

Scheme 5.13 Single and double stereodifferentiation in the aldol reaction of (*Z*)-enolates and aldehydes.

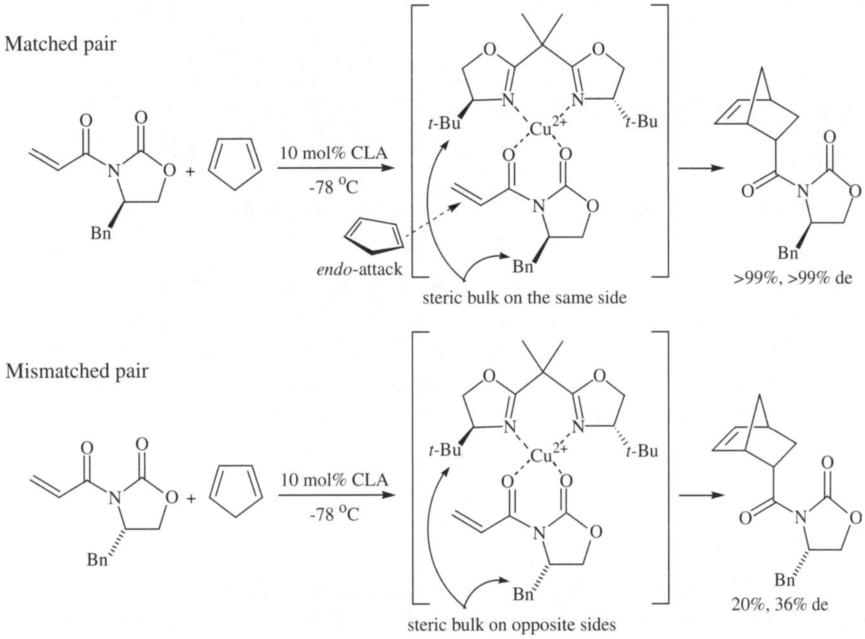

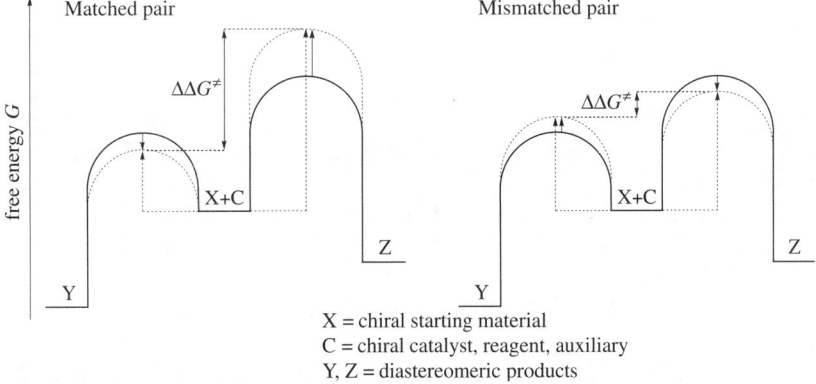

Scheme 5.14 Double stereodifferentiation in a Diels–Alder reaction using an (*S*,*S*)-bisoxazoline copper(II) complex and an acryloyl imide template bearing a chiral auxiliary.

Figure 5.6 Energy diagram for matched and mismatched pairs. Solid lines refer to single asymmetric induction and dashed lines to double stereodifferentiation.

stereoselective biases of *X* and *C* are opposite and the difference in the free energy of the transition states is reduced. This combination therefore results in a less stereoselective reaction.

As has been pointed out above, a useful asymmetric reaction must afford excellent yields and stereoselectivities for a wide range of substrates. In the case of diastereoselective reactions of chiral compounds showing insufficient substrate-based stereocontrol, it is advantageous if the stereochemical outcome of a reaction can be overwhelmingly controlled by a chiral catalyst or reagent. Ideally, the sense of asymmetric induction is predictable and independent of any stereogenic element present in the substrate and is solely determined by the catalyst or reagent used. The Sharpless asymmetric epoxidation is a very good example of this concept which is often referred to

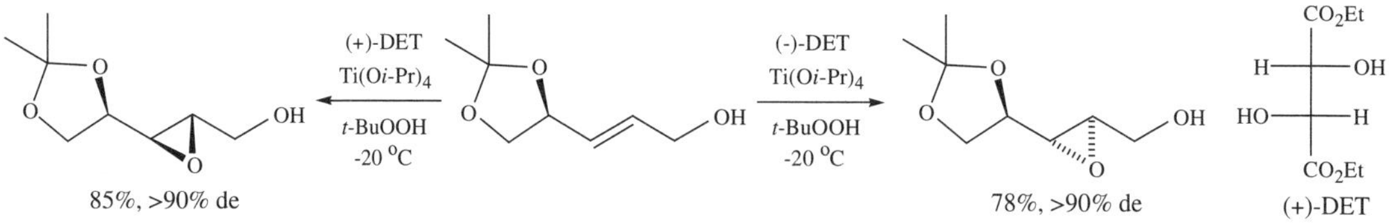

Scheme 5.15 Sharpless asymmetric epoxidation of a chiral allylic alcohol.

as reagent-based stereocontrol. Oxidation of a chiral acetonide-derived (*E*)-allyl alcohol with *tert*-butyl hydroperoxide in the presence of titanium tetraisopropoxide and either (+)- or (−)-diethyl tartrate (DET) provides the corresponding epoxides in 78-85% yield and more than 90% de.[176] The stereochemical outcome and yield of this reaction are clearly controlled by the catalyst and are barely affected by the chiral center of the substrate, Scheme 5.15.

REFERENCES

1. Helmchen, G.; Hoffmann, R. W.; Mulzer, J.; Schaumann, E. (Eds.) *Stereoselective Synthesis*, In: *Methods of Organic Chemistry*, Houben-Weyl, Vol. 21a-21f, 4th ed., Thieme, Stuttgart, 1995.
2. Gawley, R. E.; Aubé, J. *Principles of Asymmetric Synthesis*, Tetrahedron Organic Chemistry Series, Elsevier, New York, 1996.
3. Ho, T.-L. *Stereoselectivity in Synthesis*, Wiley-VCH, New York, 1999.
4. Lin, G.-Q.; Li, Y.-M.; Chan, A. S. C. *Principles and Applications of Asymmetric Synthesis*, Wiley-VCH, New York, 2001.
5. Jonathan, M. J. W. *Catalysis in Asymmetric Synthesis*, Sheffield Academic Press, Sheffield, 1999.
6. Ojima, I. *Catalytic Asymmetric Synthesis*, 2nd ed., Wiley-VCH, New York, 2000.
7. Eliel, E. L.; Wilen, S. H. *Stereochemistry of Organic Compounds*, Wiley, New York, 1994, pp. 835-845.
8. Cainelli, G.; Giacomini, D.; Galletti, P. *Eur. J. Org. Chem.* **1999**, 61-65.
9. Inoue, Y.; Ikeda, H.; Kaneda, M.; Sumimura, T.; Everitt, S. R. L.; Wada, T. *J. Am. Chem. Soc.* **2000**, *122*, 406-407.
10. Sugimura, T.; Tei, T.; Mori, A.; Okuyama, T.; Tai, A. *J. Am. Chem. Soc.* **2000**, *122*, 2128-2129.
11. Sugimura, T.; Hagiya, K.; Sato, Y.; Tei, T.; Tai, A.; Okuyama, T. *Org. Lett.* **2001**, *3*, 37-40.
12. Inoue, Y.; Sugahara, N.; Wada, T. *Pure Appl. Chem.* **2001**, *73*, 475-480.
13. Tei, T.; Sato, Y.; Hagiya, K.; Tai, A.; Okuyama, T.; Sugimura, T. *J. Org. Chem.* **2002**, *67*, 6593-6598.
14. Eliel, E. L. *Tetrahedron Lett.* **1974**, *30*, 1503-1513.
15. Reetz, M. T.; Kükenhöhner, T.; Weinig, P. *Tetrahedron Lett.* **1986**, *27*, 5711-5714.
16. Kitamura, M.; Suga, S.; Kawai, K.; Noyori, R. *J. Am. Chem. Soc.* **1986**, *108*, 6071-6072.
17. Soai, K.; Ookawa, A.; Kaba, T.; Ogawa, K. *J. Am. Chem. Soc.* **1987**, *109*, 7111-7115.
18. Soai, K.; Yokohama, S.; Ebihara, K.; Hayasaka, T. *J. Chem. Soc., Chem. Commun.* **1987**, 1690-1691.
19. Kitamura, M.; Okada, S.; Suga, S.; Noyori, R. *J. Am. Chem. Soc.* **1989**, *111*, 4028-4036.
20. Liu, G.; Ellman, J. A. *J. Org. Chem.* **1995**, *60*, 7712-7713.
21. Solà, L.; Reddy, K. S.; Vidal-Ferran, A.; Moyano, A.; Pericàs, M. A.; Riera, A.; Alvarez-Larena, A.; Piniella, J.-F. *J. Org. Chem.* **1998**, *63*, 7078-7082.

22. Liu, D. X.; Zhang, L. C.; Wang, Q.; Da, C. S.; Xin, Z. Q.; Wang, R.; Choi, M. C. K.; Chan, A. S. C. *Org. Lett.* **2001**, *3*, 2733-2735.

23. Sato, I.; Urabe, H.; Ishii, S.; Tanji, S.; Soai, K. *Org. Lett.* **2001**, *3*, 3851-3854.

24. Priego, J.; Mancheno, O. G.; Cabrera, S.; Carretero, J. C. *J. Org. Chem.* **2002**, *67*, 1346-1353.

25. Noyori, R.; Kitamura, M. *Angew. Chem., Int. Ed. Engl.* **1991**, *30*, 49-69.

26. Soai, K.; Niwa, S. *Chem. Rev.* **1992**, *92*, 833-856.

27. Vidal-Ferran, A.; Moyano, A.; Pericas, M. A.; Riera, A. *Tetrahedron Lett.* **1997**, *38*, 8773-8776.

28. Goldfuss, B.; Houk, K. N. *J. Org. Chem.* **1998**, *63*, 8998-9006.

29. Wolf, C.; Francis, C. J.; Hawes, P. A.; Shah, M. *Tetrahedron: Asymm.* **2002**, *13*, 1733-1741.

30. Puchot, C.; Samuel, O.; Dunach, E.; Zhao, S.; Agami, C.; Kagan, H. B. *J. Am. Chem. Soc.* **1986**, *108*, 2353-2357.

31. Mikami, K.; Motoyama, Y.; Terada, M. *J. Am. Chem. Soc.* **1994**, *116*, 2812-2820.

32. Aggarwal, V. K.; Mereu, A.; Tarver, G. J.; McCague, R. *J. Org. Chem.* **1998**, *63*, 7183-7189.

33. Evans, D. A.; Kozlowski, M. C.; Murry, J. A.; Burgey, C. S.; Campos, K. R.; Connell, B. T.; Staples, R. J. *J. Am. Chem. Soc.* **1999**, *121*, 669-685.

34. Furuno, H.; Hanamoto, T.; Sugimoto, Y.; Inanaga, J. *Org. Lett.* **2000**, *2*, 49-52.

35. Hanessian, S.; Pham, V. *Org. Lett.* **2000**, *2*, 2975-2978.

36. Chen, Y. K.; Costa, A. M.; Walsh, P. J. *J. Am. Chem. Soc.* **2001**, *123*, 5378-5379.

37. Yuan, Y.; Li, X.; Sun, J.; Ding, K. *J. Am. Chem. Soc.* **2002**, *124*, 14866-14867.

38. Melchiorre, P.; Jørgensen, K. A. *J. Org. Chem.* **2003**, *68*, 4151-4157.

39. Fu, Y.; Guo, X.-X.; Zhu, S.-F.; Hu, A.-G.; Xie, J.-H.; Zhou, Q.-L. *J. Org. Chem.* **2004**, *69*, 4648-4655.

40. Mikami, K.; Matsumoto, Y. *Tetrahedron* **2004**, *60*, 7715-7719.

41. Qin, Y.-C.; Liu, L.; Pu, L. *Org. Lett.* **2005**, *7*, 2381-2383.

42. Portada, T.; Roje, M.; Hamersak, Z.; Zinic, M. *Tetrahedron Lett.* **2005**, *46*, 5957-5959.

43. Franzen, J.; Marigo, M.; Fielenbach, D.; Wabnitz, T. C.; Kjrsgaard, A.; Jørgensen, K. A. *J. Am. Chem. Soc.* **2005**, *127*, 18296-18304.

44. Blackmond, D. G. *J. Am. Chem. Soc.* **1998**, *120*, 13349-13353.

45. Blackmond, D. G. *Acc. Chem. Res.* **2000**, *33*, 402-411.

46. Blackmond, D. G. *J. Am. Chem. Soc.* **2001**, *123*, 545-553.

47. Buono, F.; Walsh, P. J.; Blackmond, D. G. *J. Am. Chem. Soc.* **2002**, *124*, 13652-13653.

48. Reetz, M. T.; Meiswinkel, A.; Mehler, G.; Angermund, K.; Graf, M.; Thiel, W.; Mynott, R.; Blackmond, D. G. *J. Am. Chem. Soc.* **2005**, *127*, 10305-10313.

49. Yamakawa, M.; Noyori, R. *J. Am. Chem. Soc.* **1995**, *117*, 6327-6335.

50. Kitamura, M.; Suga, S.; Oka, H.; Noyori, R. *J. Am. Chem. Soc.* **1998**, *120*, 9800-9809.

51. Avalos, M.; Babiano, R.; Cintas, P.; Jimenez, J. L.; Palacios, J. C. *Tetrahedron: Asymm.* **1997**, *8*, 2997-3017.

52. Girard, C.; Kagan, H. B. *Angew. Chem., Int. Ed.* **1998**, *37*, 2922-2959.

53. Kagan, H. B. *Adv. Synth. Catal.* **2001**, *343*, 227-233.

54. Soai, K.; Shibata, T.; Morioka, H.; Choji, K. *Nature* **1995**, *378*, 767-768.

55. Bolm, C.; Bienewald, F.; Seger, A. *Angew. Chem., Int. Ed. Engl.* **1996**, *35*, 1657-1659.

56. Soai, K.; Shibata, T.; Sato, I. *Acc. Chem. Res.* **2000**, *33*, 382-390.

57. Todd, M. H. *Chem. Soc. Rev.* **2002**, *31*, 211-222.

58. Sato, I.; Ohno, A.; Aoyama, Y.; Kasahara, T.; Soai, K. *Org. Biomol. Chem.* **2003**, *1*, 244-246.

59. Soai, K.; Shibata, T.; Sato, I. *Bull. Chem. Soc. Jpn.* **2004**, *77*, 1063-1073.

60. Sato, I.; Sugie, R.; Matsueda, Y.; Furumura, Y.; Soai, K. *Angew. Chem., Int. Ed.* **2004**, *43*, 4490-4492.

61. Kawasaki, T.; Jo, K.; Igarashi, H.; Sato, I.; Nagano, M.; Koshima, H.; Soai, K. *Angew. Chem., Int. Ed.* **2005**, *44*, 2774-2777.

62. Podlech, J.; Gehring, T. *Angew. Chem., Int. Ed.* **2005**, *44*, 5776-5777.

63. Lutz, F.; Igarashi, T.; Kawasaki, T.; Soai, K. *J. Am. Chem. Soc.* **2005**, *127*, 12206-12207.

64. Sato, I.; Urabe, H.; Ishiguro, S.; Shibata, T.; Soai, K. *Angew. Chem., Int. Ed.* **2003**, *42*, 315-317.

65. Guillaneux, D.; Zhao, S. H.; Samuel, O.; Rainford, D.; Kagam, H. B. *J. Am. Chem. Soc.* **1994**, *116*, 9430-9439.

66. Gridnev, I. D.; Serafimov, J. M.; Quiney, H.; Brown, J. M. *Org. Biomol. Chem.* **2003**, *1*, 3811-3819.

67. Buono, F. G.; Iwamura, H.; Blackmond, D. G. *Angew. Chem., Int. Ed.* **2004**, *43*, 2099-2103.

68. Gridnev, I. D. ; Serafimov, J. M.; Brown, J. M. *Angew. Chem., Int. Ed.* **2004**, *43*, 4884-4887.

69. Yoon, T. P.; Jacobsen, E. N. *Science* **2003**, *299*, 1691-1693.

70. Dang, T. P.; Kagan, H. B. *J. Chem. Soc. D, Chem. Commun.* **1971**, 481.

71. Kagan, H. B.; Dang, T. P. *J. Am. Chem. Soc.* **1972**, *94*, 6429-6433.

72. Noyori, R. *Adv. Synth. Catal.* **2003**, *345*, 15-32.

73. Chen, Y.; Yekta, S.; Yudin, A. K. *Chem, Rev.* **2003**, *103*, 3155-3212.

74. Gao, G.; Xie, R.-G.; Pu, L. *Proc. Nat. Acad. Sci.* **2004**, *101*, 5417-5420.

75. Shibasaki, M.; Matsunaga, S. *Chem. Soc. Rev.* **2006**, *35*, 269-279.

76. Lowenthal, R. E.; Abiko, A.; Masamune, S. *Tetrahedron Lett.* **1990**, *31*, 6005-6008.

77. Hall, J.; Lehn, J.-M.; DeCian, A.; Fischer, J. *Helv. Chim. Acta* **1991**, *74*, 1-6.

78. Müller, D.; Umbricht, G.; Weber, B.; Pfaltz, A. *Helv. Chim. Acta* **1991**, *74*, 232-240.

79. Evans, D. A.; Woerpel, K. A.; Hinman, M. M.; Faul, M. M. *J. Am. Chem. Soc.* **1991**, *113*, 726-728.

80. Corey, E. J.; Imai, N.; Zhang, H.-Y. *J. Am. Chem. Soc.* **1991**, *113*, 728-729.

81. Helmchen, G.; Krotz, A.; Ganz K. T.; Hansen, D. *Synlett* **1991**, 257-259.

82. Pfaltz, A. *Acc. Chem. Res.* **1993**, *26*, 339-345.

83. Zhang, W.; Loebach, J. L.; Wilson, S. R.; Jacobsen, E. N. *J. Am. Chem. Soc.* **1990**, *112*, 2801-2803.

84. Jacobsen, E. N.; Zhang, W.; Muci, A. R.; Ecker, J. R.; Deng, L. *J. Am. Chem. Soc.* **1991**, *113*, 7063-7064.

85. Burk, M. J.; Gross, M. F.; Harper, G. P.; Kalberg, C. S.; Lee, J. R.; Martinez, J. P. *Pure Appl. Chem.* **1996**, *68*, 37-44.

86. Seebach, D.; Beck, A. K.; Heckel, A. *Angew. Chem., Int. Ed. Engl.* **2001**, *40*, 92-138.

87. Evans, D. A.; Burgey, C. S.; Paras, N. A.; Vojkovsky, T.; Tregay, S. W. *J. Am. Chem. Soc.* **1998**, *120*, 5824-5825.

88. Evans, D. A.; Johnson, J. S.; Burgey, C. S.; Campos, K. R. *Tetrahedron Lett.* **1999**, *40*, 2879-2882.

89. Evans, D. A.; Barnes, D. M.; Johnson, J. S.; Lectka, T.; Matt, P. V.; Miller, S. J.; Murry, J. A.; Norcross, R. D.; Shaughnessy, E. A.; Campos, K. J. *J. Am. Chem. Soc.* **1999**, *121*, 7582-7594.

90. Sibi, M. P.; Venkatraman, L.; Liu, M.; Jasperse, C. P. *J. Am. Chem. Soc.* **2001**, *123*, 8444-8445.

91. Bayer, A.; Gautun, O. R. *Tetrahedron: Asymm.* **2001**, *12*, 2937-2939.

92. Rechavi, D.; Lemaire, M. *Chem. Rev.* **2002**, *102*, 3467-3493.

93. Aburel, P. S.; Zhuang, W.; Hazell, R. G.; Jørgensen, K. A. *Org. Biomol. Chem.* **2005**, *3*, 2344-2349.

94. Matsunaga, H.; Yamada, Y.; Tsukasa, I. *Tetrahedron: Asymm.* **1999**, *10*, 3095-3098.

95. Benaglia, M.; Cinquini, M.; Cozzi, F.; Celentano, G. *Org. Biomol. Chem.* **2004**, *2*, 3401-3407.

96. Bedekar, A. V.; Koroleva, E. B.; Andersson, P. G. *J. Org. Chem.* **1997**, *62*, 2518-2526.

97. Clariana, J.; Comelles, J.; Moreno-Manas, M.; Vallribera, A. *Tetrahedron: Asymm.* **2002**, *13*, 1551-1554.

98. Schinnerl, M.; Bohm, C.; Seitz, M.; Reiser, O. *Tetrahedron: Asymm.* **2003**, *14*, 765-771.

99. Charette, A. B.; Janes, M. K.; Lebel, H. *Tetrahedron: Asymm.* **2003**, *14*, 867-872.

100. Itagaki, M.; Masumoto, K.; Yamamoto, Y. *J. Org. Chem.* **2005**, *70*, 3292-3295.

101. Evans, D. A.; Faul, M. M.; Bilodeau, M. T.; Anderson, B. A.; Barnes, D. M. *J. Am. Chem. Soc.* **1993**, *115*, 5328-5329.

102. Juhl, K.; Hazell, R. G.; Jørgensen, K. A. *J. Chem. Soc., Perkin Trans. 1* **1999**, 2293-2297.

103. Krumper, J. R.; Gerisch, M.; Suh, J. M.; Bergman, R. G.; Tilley, T. D. *J. Org. Chem.* **2003**, *68*, 9705-9710.

104. Corey, E. J.; Ishihara, K. *Tetrahedron Lett.* **1992**, *33*, 6807-6810.

105. Evans, D. A.; Miller, S. J.; Lectka, T. *J. Am. Chem. Soc.* **1993**, *115*, 6460-6461.

106. Johnson, J. S.; Evans, D. A. *Acc. Chem. Res.* **2000**, *33*, 325-335.

107. Wu, J. H.; Radinov, R.; Porter, N. A. *J. Am. Chem. Soc.* **1995**, *117*, 11029-11030.

108. Sibi, M. P.; Ji, J.; Wu, J. H.; Gurtler, S.; Porter, N. A. *J. Am. Chem. Soc.* **1996**, *118*, 9200-9201.

109. Sibi, M. P.; Shay, J. J.; Ji, J. *Tetrahedron Lett.* **1997**, *38*, 5955-5958.

110. Sibi, M. P.; Liu, M. *Enantiomer* **1999**, *4*, 575-590.

111. Sibi, M. P.; Porter, N. A. *Acc. Chem. Res.* **1999**, *32*, 163-171.

112. Iserloh, U.; Curran, D. P.; Kanemasa, S. *Tetrahedron: Asymm.* **1999**, *10*, 2417-2428.

113. Sibi, M. P.; Manyem, S. *Tetrahedron* **2000**, *56*, 8033-8061.

114. Feng, H.; Kavrakova, I. K.; Pratt, D. A.; Tellinghuisen, J.; Porter, N. A. *J. Org. Chem.* **2000**, *67*, 6050-6054.

115. Sibi, M. P.; Manyem, S. *Org. Lett.* **2002**, *4*, 2929-2932.

116. Sibi, M. P.; Zimmerman, J.; Rheault, T. *Angew. Chem., Int. Ed.* **2003**, *42*, 4521-4523.

117. Friestad, G. K.; Shen, Y.; Ruggles, E. L. *Angew. Chem., Int. Ed.* **2003**, *42*, 5061-5063.

118. Sibi, M. P.; He, L. *Org. Lett.* **2004**, *6*, 1749-1752.

119. Yao, S.; Johannsen, M.; Audrain, H.; Hazell, R. G.; Jørgensen, K. A. *J. Am. Chem. Soc.* **1998**, *120*, 8599-8605.

120. Yao, S.; Roberson, F. R.; Hazell, R. G.; Jørgensen, K. A. *J. Org. Chem.* **1999**, *64*, 6677-6687.

121. Evans, D. A.; Johnson, J. S.; Olhava, E. J. *J. Am. Chem. Soc.* **2000**, *122*, 1635-1649.

122. Jørgensen, K. A. *Angew. Chem., Int. Ed.* **2000**, *39*, 3558-3588.

123. Joly, G. D.; Jacobsen, E. N. *Org. Lett.* **2002**, *4*, 1795-1798.

124. Wolf, C.; Fadul, Z.; Hawes, P. A.; Volpe, E. C. *Tetrahedron: Asymm.* **2004**, *15*, 1987-1993.

125. Cram, D. J.; Kopecky, K. R. *J. Am. Chem. Soc.* **1959**, *81*, 2748-2755.

126. Chen, X.; Hortelano, E. R.; Eliel, E. L.; Frye, S. V. *J. Am. Chem. Soc.* **1992**, *114*, 1778-1784.

127. Nakata, T.; Tanaka, T. Oishi, T. *Tetrahedron Lett.* **1983**, *24*, 2653-2656.

128. Posner, G. H.; Frye, L. L.; Hulce, M. *Tetrahedron* **1984**, *40*, 1401-1407.

129. Posner, G. H.; Mallamo, J. P. K.; Hulce, M.; Frye, L. L. *J. Am. Chem. Soc.* **1982**, *104*, 4180-4185.

130. Heathcock, C. H. *Science* **1981**, *214*, 395-400.

131. Evans, D. A.; Bartroli, J.; Shih, T. L. *J. Am. Chem. Soc.* **1981**, *103*, 2127-2129.

132. Evans, D. A.; McGee, L. R. *J. Am. Chem. Soc.* **1981**, *103*, 2876-2878.

133. Evans, D. A.; Nelson, J. V.; Taber, T. R. *Top. Stereochem.* **1982**, *13*, 1-115.

134. Helmchen, G.; Leikauf, U.; Taufer, Knoepfel, I. *Angew. Chem., Int. Ed. Engl.* **1985**, *24*, 874-875.

135. Masamune, S.; Kim, B. M.; Petersen, J. S.; Sato, T.; Veenstra, S. J. *J. Am. Chem. Soc.* **1985**, *107*, 4549-4551.

136. Masamune, S.; Sato, T.; Kim, B. M.; Wollmann, T. A. *J. Am. Chem. Soc.* **1986**, *108*, 8279-8281.

137. Braun, M. *Angew. Chem., Int. Ed. Engl.* **1987**, *26*, 24-37.

138. Hoffmann, R. W. *Angew. Chem., Int. Ed. Engl.* **1987**, *26*, 503-517.

139. Paterson, I.; Lister, M. A. *Tetrahedron Lett.* **1988**, *29*, 585-588.

140. Oppolzer, W.; Starkemann, C.; Rodriguez, I.; Bernardinelli, G. *Tetrahedron Lett.* **1991**, *32*, 61-64.

141. Draanen, N. A. V.; Arseniyades, S.; Crimmins, M. T.; Heathcock, C. H. *J. Org. Chem.* **1991**, *56*, 2499-2506.

142. Oppolzer, W.; Starkemann, C. *Tetrahedron Lett.* **1992**, *33*, 2439-2442.

143. Gennari, C.; Hewkin, C. T.; Molinari, F.; Bernardi, A.; Comotti, A.; Goodman, J. M.; Paterson, I. *J. Org. Chem.* **1992**, *57*, 5173-5177.

144. Paterson, I.; Tillyer, R. D. *J. Org. Chem.* **1993**, *58*, 4182-4184.

145. Saito, S.; Hatanaka, K.; Kano, T.; Yamamoto, H. *Angew. Chem., Int. Ed.* **1998**, 37, 3378-3381.

146. Itoh, Y.; Yamanaka, M.; Mikami, K. *J. Am. Chem. Soc.* **2004**, *126*, 13174-13175.

147. Furuta, K.; Maruyama, T.; Yamamoto, H. *J. Am. Chem. Soc.* **1991**, *113*, 1041-1042.

148. Parmee, E. R.; Tempkin, O.; Masamune, S.; Abiko, A. *J. Am. Chem. Soc.* **1991**, *113*, 9365-9366.

149. Kiyooka, S.-I.; Kaneko, Y.; Kume, K.-I. *Tetrahedron Lett.* **1992**, *33*, 4927-4930.

150. Corey, E. J.; Cywin, C. L.; Roper, T. D. *Tetrahedron Lett.* **1992**, *33*, 6907-6910.

151. Evans, D. A.; Kozlowski, M. C.; Burgey, C. S.; MacMillan, D. W. C. *J. Am. Chem. Soc.* **1997**, *119*, 7893-7894.

152. Kobayashi, S.; Horibe, M. *Chem. Eur. J.* **1997**, *3*, 1472-1481.

153. Evans, D. A.; Kozlowski, M. C.; Murry, J. A.; Burgey, C. S.; Campos, K. R.; Connell, B. T.; Staples, R. J. *J. Am. Chem. Soc.* **1999**, *121*, 669-685.

154. Yoshikawa, N.; Yamada, Y. M. A.; Das, J.; Sasai, H.; Shibasaki, M. *J. Am. Chem. Soc.* **1999**, *121*, 4168-4178.

155. Mahrwald, R. *Chem. Rev.* **1999**, *99*, 1095-1120.

156. Evans, D. A.; Allison, B. D.; Yang, M. G.; Masse, C. E. *J. Am. Chem. Soc.* **2001**, *123*, 10840-10852.

157. Gathergood, N.; Juhl, K.; Poulsen, T. B.; Thordrup, K.; Jørgensen, K. A. *Org. Biomol. Chem.* **2004**, *2*, 1077-1085.

158. Liu, S.-Y.; Hills, I. D.; Fu, G. C. *J. Am. Chem. Soc.* **2005**, *127*, 15352-15353.

159. Zimmerman, H. E.; Traxler, M. D. *J. Am. Chem. Soc.* **1957**, *79*, 1920-1923.

160. Heathcock, C. H.; Buse, C. T.; Kleschick, W. A.; Pirrung, M. C.; Sohn, J. E.; Lampe, J. *J. Org. Chem.* **1980**, *45*, 1066-1081.

161. Heathcock, C. H.; Pirrung, M. C.; Montgomery, S. H.; Lampe, J. *Tetrahedron* **1981**, *37*, 4087-4095.

162. Masamune, S.; Choy, W.; Petersen, J. S.; Sita, L. R. *Angew. Chem., Int. Ed. Engl.* **1985**, *24*, 1-30.

163. Yamamoto, Y.; Komatsu, T.; Maruyama, K. *J. Chem. Soc., Chem. Commun.* **1985**, 814-816.

164. Molander, G. A.; Haar Jr., J. P. *J. Am. Chem. Soc.* **1991**, *113*, 3608-3610.

165. Hattori, K.; Miyata, M.; Yamamoto, H. *J. Am. Chem. Soc.* **1993**, *115*, 151-152.

166. Evans, D. A.; Yang, M. G.; Dart, M. J.; Duffy, J. L.; Kim, A. S. *J. Am. Chem. Soc.* **1995**, *117*, 9598-9599.

167. Denmark, S. E.; Fujimori, S. *Org. Lett.* **2002**, *4*, 3473-3476.

168. Metallinos, C.; Snieckus, V. *Org. Lett.* **2002**, *4*, 1935-1938.

169. Suginome, M.; Ohmura, T.; Miyake, Y.; Mitani, S.; Ito, Y.; Murakami, M. *J. Am. Chem. Soc.* **2003**, *125*, 11174-11175.

170. Davies, S. G.; Hermann, G. J.; Sweet, M. J.; Smith, A. D. *Chem. Commun.* **2004**, 1128-1129.

171. Harb, W.; Ruiz-Lopez, M. F.; Coutrot, F.; Grison, C.; Coutrot, P. *J. Am. Chem. Soc.* **2004**, *126*, 6996-7008.
172. Krishna, P. R.; Sachwani, R.; Kannan, V. *Chem. Commun.* **2004**, 2580-2581.
173. Doyle, M. P.; Morgan, J. P.; Fettinger, J. C.; Zavalij, P. Y.; Colyer, J. T.; Timmons, D. J.; Carducci, M. D. *J. Org. Chem.* **2005**, *70*, 5291-5301.
174. Ma, M.; Peng, L.; Li, C.; Zhang, X.; Wang, J. *J. Am. Chem. Soc.* **2005**, *127*, 15016-15017.
175. Evans, D. A.; Rovis, T.; Johnson, J. S. *Pure Appl. Chem.* **1999**, *71*, 1407-1415.
176. Katsuki, T.; Lee, A. W. M.; Ma, P.; Martin, V. S.; Masamune, S.; Sharpless, K. B.; Tuddenham, D.; Walker, F. J. *J. Org. Chem.* **1982**, *47*, 1373-1378.

Asymmetric Synthesis with Stereodynamic Compounds: Introduction, Conversion and Transfer of Chirality

Asymmetric synthesis of complex chiral molecules, in particular natural products, generally entails incorporation of several chiral elements in addition to strategic carbon–carbon bond formation. Once molecular chirality is established it might be necessary to further manipulate it, for example by intramolecular chirality transfer or interconversion of elements of chirality. Methods that allow selective translocation of a chiral element along a carbon framework or change of a chiral center to a chiral axis with complete conservation of chirality provide invaluable access to complex structures.[i] A profound understanding of the stability of a chiral target compound and starting materials to racemization and diastereomerization is indispensable for planning an efficient synthetic route. Configurationally and conformationally unstable chiral compounds constitute both a challenge and an opportunity at the same time. The determination of the energy barrier to isomerization of a stereolabile substrate, reagent or product is the first step to stereodynamic control and successful multiple-step synthesis. It is crucial to identify and to carefully select reaction conditions that rule out undesirable racemization or diastereomerization because these processes can significantly compromise the efficacy of asymmetric synthesis. The formation of chiral carbanionic species, for instance Grignard reagents prepared from nonracemic alkyl halides, is often accompanied by rapid inversion of configuration and loss of stereochemical purity. Nevertheless, a variety of powerful asymmetric methods using chiral organolithium compounds that are configurationally stable under cryogenic conditions has been developed. The control of the stereodynamics of axially chiral compounds plays a key role in atroposelective biaryl synthesis. The challenge here is to implement both axial chirality and sufficient conformational stability to avoid atropoisomerization of the product. However, interconversion of stereoisomers can also be advantageous. Coupling of racemization or diastereomerization with stereoselective reactions has become a powerful strategy, and many synthetic methods exploit conformationally or configurationally unstable compounds. Examples include asymmetric catalysis with chiral relays or stereolabile chiral ligands, as well as dynamic kinetic resolution (DKR), dynamic kinetic asymmetric transformation (DYKAT) and dynamic thermodynamic resolution (DTR).

---

[i] The term conservation of chirality refers to a process in which the original enantiomeric and diastereomeric purity of a chiral compound is not compromised.

6.1 ASYMMETRIC SYNTHESIS WITH CHIRAL ORGANOLITHIUM REAGENTS

Probably the first asymmetric reaction involving an enantiomerically enriched chiral organolithium compound was reported by Letsinger in 1950.[1] He observed that lithiation of (–)-2-iodooctane followed by trapping of the 2-octyllithium intermediate with carbon dioxide at –70 °C yields nonracemic (–)-2-methyloctanoic acid. Although the reaction is accompanied by considerable racemization, the metal/halogen exchange generates a chiral organolithium species exhibiting remarkable configurational stability under cryogenic conditions, Scheme 6.1.[2,3] Since then, chiral organolithium reagents have found numerous applications in asymmetric synthesis. This is a consequence of (a) the ease of preparation of nonracemic chiral organolithium compounds,[ii] (b) the availability of a range of functional groups that stabilize the absolute configuration at the lithiated carbon atom, thus effectively preventing racemization or diastereomerization, and (c) the usefulness of organolithium reagents for carbon–carbon bond formation. It should be noted, however, that the configurational stability of chiral organolithium compounds is not necessarily a prerequisite for asymmetric synthesis.[4] Powerful methods such as dynamic thermodynamic resolution (DTR) utilize configurationally unstable organolithium compounds, see Chapter 7.6.

Nonracemic organolithium compounds can be prepared by deprotonation of carbamates and esters derived from a chiral alcohol, thiol or amine,[5] transmetalation with isolable chiral stannanes,[6,7] and enantiotopic deprotonation of prochiral carbamates with organolithium reagents in the presence of an enantiopure amine such as (–)-sparteine, Scheme 6.2.[8] The naturally occurring alkaloid (–)-sparteine is certainly the most popular choice for enantiotopic lithiation. Since sparteine is only available in one enantiomeric form, several (+)-sparteine surrogates have been developed to provide additional means for the synthesis of chiral carbamoyl-stabilized carbanions.[9–16] In the context of asymmetric synthesis, a chiral organometallic species is considered configurationally stable if inversion of configuration is relatively slow compared to reaction with an electrophile, see also Chapter 3.2.12. Chiral organolithium compounds are usually generated under cryogenic conditions and are instantly trapped by transmetalation or carbon–carbon bond formation, but they can not be isolated or stored at room temperature. To prevent inversion of configuration even at very low temperatures, lithiated chiral carbanions are generally stabilized by incorporation of proximate heteroatoms that coordinate to the lithium cation and thus significantly reduce the rate of inversion.[iii] Nonracemic organolithium intermediates are frequently subjected to metal exchange with borane, aluminum, tin, titanium or copper reagents prior to asymmetric carbon–carbon bond formation.[17–21] Conversion of organolithium intermediates to other organometallic compounds has several advantages. In particular, stannylation has become a routine procedure for the generation of isolable derivatives which allows one to study the stereochemical integrity of chiral organolithium compounds. Formation of isolable organostannanes also opens a venue for enantioselective HPLC purification in cases of unsatisfactory

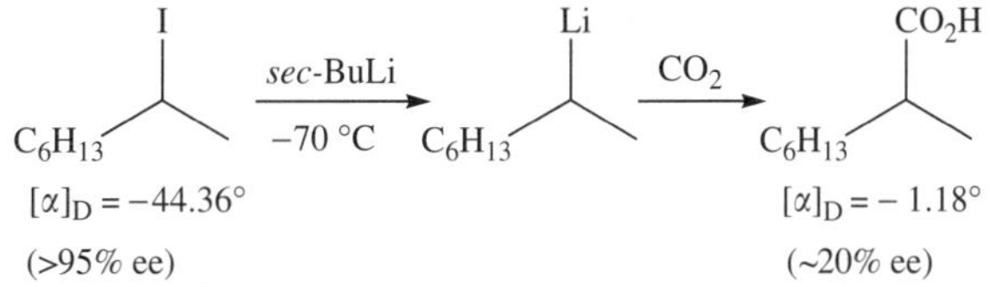

Scheme 6.1 First asymmetric synthesis with a chiral organolithium reagent.

[ii] Common methods for asymmetric synthesis of chiral organolithium compounds are based on lithiation of readily available carbamates derived from enantiopure chiral secondary alcohols or (–)-sparteine-assisted enantiotopic deprotonation of prochiral carbamates and esters.

[iii] Nonfunctionalized aliphatic organolithium compounds are generally of limited use due to their inherently low configurational stability which decreases even further in the presence of polar organic solvents.

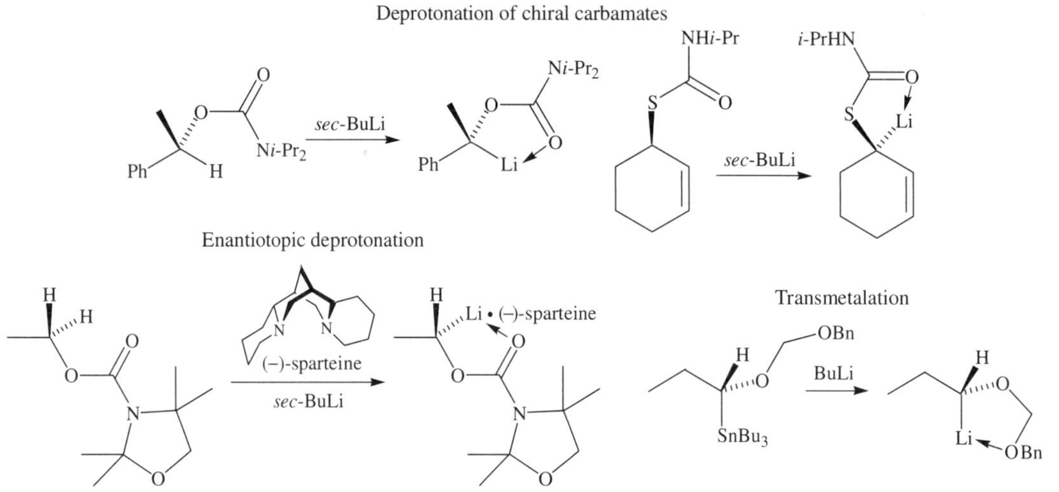

Deprotonation of chiral carbamates

Enantiotopic deprotonation

Transmetalation

Scheme 6.2   Generation of nonracemic organolithium compounds.

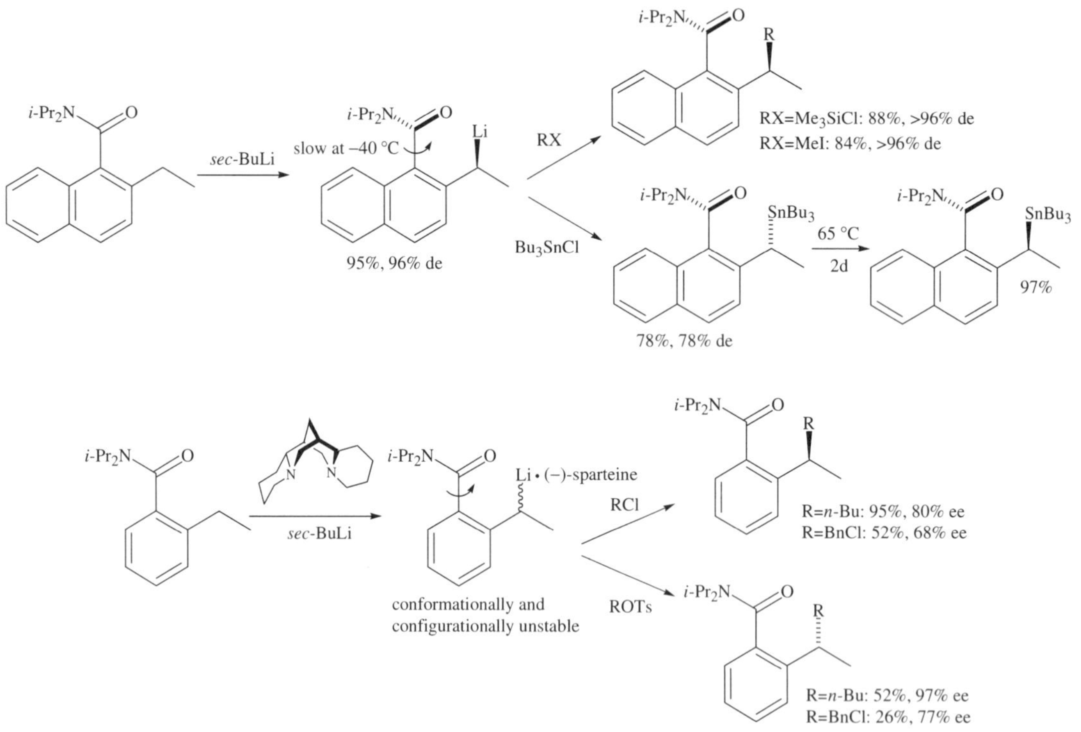

Scheme 6.3   Asymmetric alkylation of laterally lithiated *N,N*-diisopropyl-2-ethylnaphthamide (top) and
N,N-diisopropyl-2-ethylbenzamide (bottom).

enantiopurity of organolithium precursors.[22] Finally, chiral stannanes are often more suitable for
asymmetric reactions that require temperatures at which organolithium compounds undergo racemization or diastereomerization as a result of uncontrolled inversion of configuration.

Inversion of configuration of chiral benzylic organolithium compounds is usually fast.[23–25]
Hoppe, Beak and Clayden have shown that this process can be controlled through intramolecular

lithium coordination with proximate carbamoyl[26,27] or arylamide groups present in anilides or naphthamides.[28–32] Since naphthamides such as *N,N*-diisopropyl-2-ethylnaphthamide exist in the form of two atropisomers due to restricted rotation about the aryl–amide bond, lateral lithiation with *sec*-butyllithium affords a mixture of *syn-* and *anti*-diastereoisomers. Clayden's group found that formation of the *syn*-diastereoisomer is favored and proved that it is configurationally stable up to –40 °C. Interestingly, alkylation and silylation occur with retention of configuration while stannylation gives inversion of configuration, Scheme 6.3. Heating of a solution of the stannylated *anti*-product to 65 °C for two days results in almost complete atropisomerization to the thermo-dynamically more stable *syn*-diastereoisomer that has the tributylstannyl moiety and the two isopropyl groups of the tertiary amide on opposite sides. But not all laterally lithiated aryl amides are configurationally stable under cryogenic conditions.[33] Lithiation of *N,N*-diisopropyl-2-ethyl-benzamide in the presence of (–)-sparteine at –78 °C gives rapidly interconverting diastereomers that nevertheless undergo stereoselective alkylation. Interestingly, the sense of chiral induction is reversed when alkyl halides and tosylates are used. Mechanistic studies including Hoffmann's test revealed that the selectivity is not due to diastereotopic deprotonation and formation of non-equilibrating diastereomeric (–)-sparteine-organolithium adducts. In fact, stereoselectivity is estab-lished in the second step and is a consequence of diastereomeric transition states of two competing alkylation pathways, which constitutes a dynamic kinetic asymmetric transformation, see Chapter 7.5.

6.1.1 α-Alkoxy- and α-Amino-substituted Organolithium Compounds

The vast majority of configurationally stable organolithium compounds possess α-alkoxy or α-amino substituents.[34,iv] Chiral α-alkoxy-substituted carbanions are usually derived from ethers, carbamates and esters. In 1980, Still and Sreekumar discovered that lithiation of an enantiopure α-alkoxy stannane through metal exchange with butyllithium under cryogenic conditions and subsequent methylation with dimethyl sulfate proceeds with no sign of racemization, Scheme 6.4.[35] Diastereoselective addition of α-lithiated ethers or Grignard analogs to aldehydes at –65 °C provides diols with remarkable *syn*-selectivity.[36,v] The configurational stability of the organo-lithium intermediate is a consequence of both α-heteroatom substitution and intramolecular

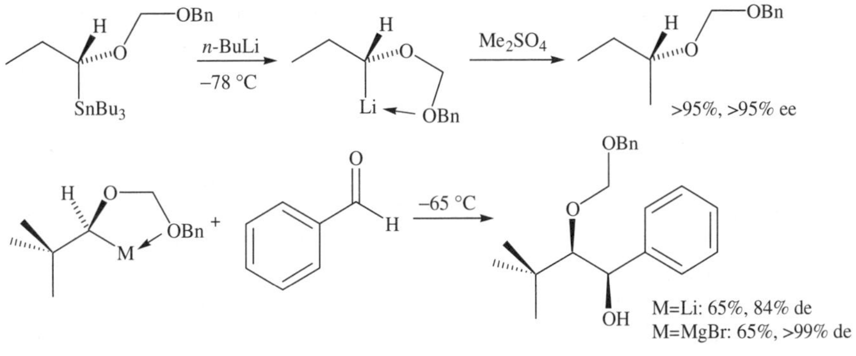

Scheme 6.4 Lithiation of an α-alkoxy stannane followed by methylation with retention of configuration (top) and diastereoselective alkylation of benzaldehyde using configurationally stable organo-lithium and Grignard reagents (bottom).

[iv] For convenience, a chiral organolithium compound is considered configurationally stable if racemization is relatively slow compared to reaction with an electrophile. With organolithium compounds this situation is often realized under cryogenic conditions while racemization would be relatively fast at room temperature. In this regard, configurational stability is not an inherent property of the compound but is dependent on reaction conditions.

[v] Examples of asymmetric reactions with chiral Grignard reagents are given in Chapter 3.2.12.

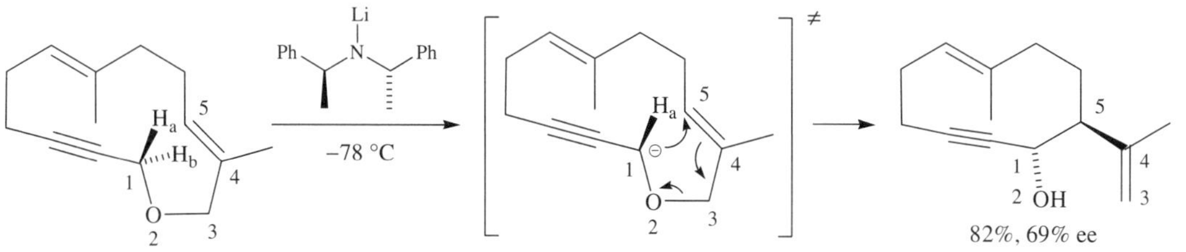

Scheme 6.5 Enantiotopic deprotonation of a macrocyclic propargylic allyl ether and subsequent [2,3]-Wittig ring contraction.

chelation. While incorporation of a heteroatom is known to increase pyramidal inversion barriers of carbanions, chelation stabilizes the position of lithium and thus impedes racemization.

Lithiation of achiral ethers can initiate sigmatropic [2,3]-Wittig rearrangements of configurationally stable carbanions.[37–43] An intriguing example is the enantioselective synthesis of macrocyclic alcohols based on [2,3]-Wittig ring contraction of allylic ethers, Scheme 6.5.[44,45] Enantiotopic deprotonation of an achiral 13-membered cyclic propargylic allyl ether with lithio bis[(S)-1-phenylethyl]amide generates a configurationally stable chiral α-alkoxy lithium intermediate that spontaneously rearranges to a 10-membered propargylic alcohol.[vi] Rearrangements of deprotonated epoxides and aziridines are also known. Enantioselective (–)-sparteine-assisted lithiation of meso epoxides and aziridines constitutes an important desymmetrization strategy. The resulting chiral oxiranyl or aziridinyl anions can either be trapped by electrophiles or, in the absence of the latter, rearrange to acyclic unsaturated diols, amino alcohols and amines.[46–49]

Carbamates derived from primary or secondary alcohols are of great synthetic value. Hoppe and others have exploited a range of sterically hindered carbamates derived from primary alkyl, allyl and alkynyl alcohols for enantiotopic deprotonation in the presence of (–)-sparteine or another chiral base and subsequent reaction with an electrophile. The latter reaction often proceeds with retention of configuration, and deprotonation of primary alkyl carbamates with *sec*-butyllithium/(–)-sparteine generally favors abstraction of the *pro-S* proton, Scheme 6.6.[50–63] The (–)-sparteine-controlled lithiation/substitution reaction sequence with achiral carbamates is quite versatile and has also been used for the synthesis of natural products, for example (R)-pantolactone. Reactions of chiral organolithium compounds do not necessarily proceed with retention of configuration and the stereochemical outcome often depends on the nature of the electrophile used. Trapping of the chiral lithium carbanion pair can involve both inversion and retention of configuration, which is a characteristic of asymmetric synthesis with organolithium compounds. Hoppe's group has extended this methodology to 1-alkenyl carbamates.[64,65] In this case, enantiotopic abstraction of a γ-proton generates an enantiomerically enriched allylic lithium intermediate that favors stereospecific *anti*-S$_E$′ reaction with silyl and stannyl chlorides but shows *syn*-S$_E$′ displacement with a range of carbonyl electrophiles.

Lithiation of carbamates derived from nonracemic chiral secondary alcohols generates organolithium intermediates that often have remarkable configurational stability due to dipole stabilization and intramolecular chelation of the lithium counterion by the carbamoyl group, as indicated in Scheme 6.2.[66] This provides a convenient synthetic entry to tertiary alcohols because chiral secondary alcohols are readily available in enantiopure form, for example by asymmetric alkylation of aldehydes with organozinc reagents in the presence of catalytic amounts of a chiral amino alcohol. Lithiation of Cb-protected (R)-1-phenylethanol at –78 °C produces a configurationally stable carbanion that undergoes electrophile-dependent stereodivergent substitution, Scheme 6.7.[67–69]

[vi]The stereochemical course of sigmatropic [2,3]-Wittig rearrangements is discussed in greater detail in Chapter 6.4.2.

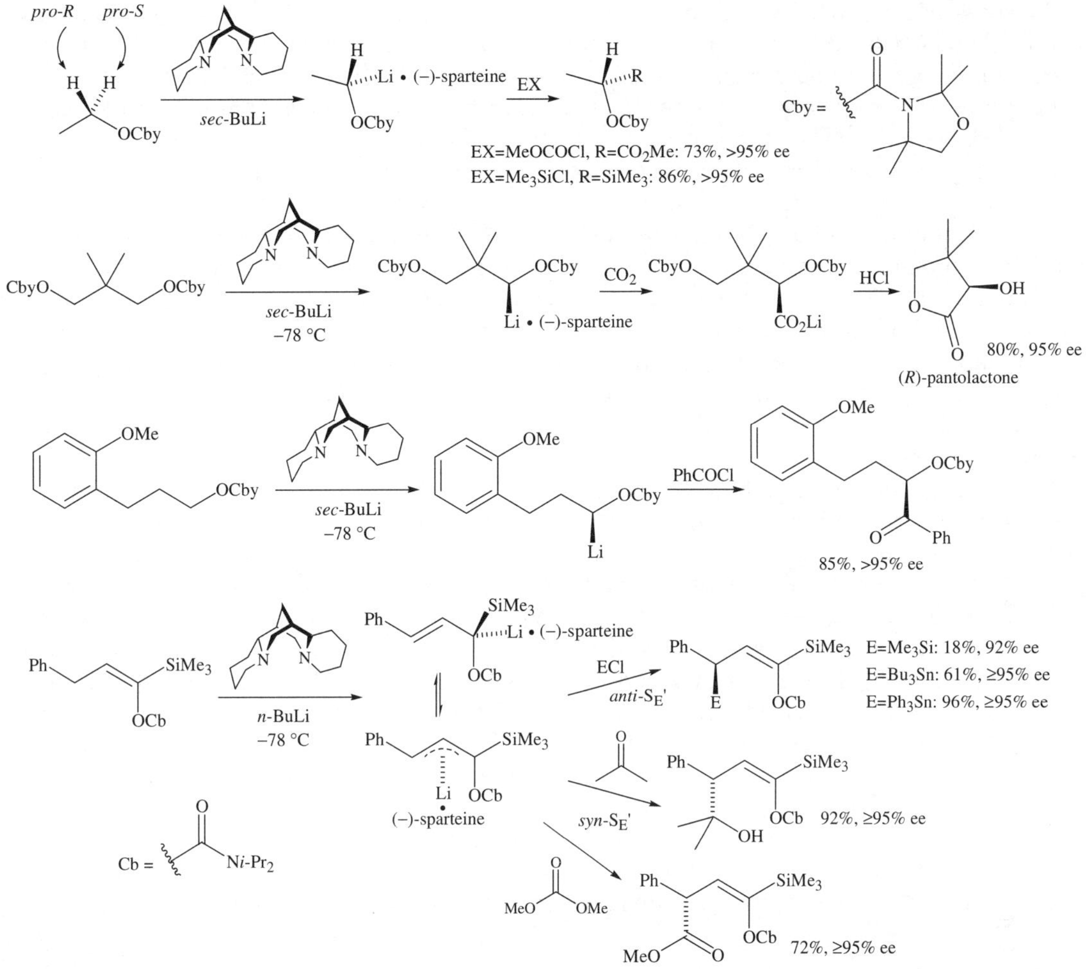

Scheme 6.6 Examples of (–)-sparteine-controlled enantiotopic deprotonation of achiral carbamates and reaction with various electrophiles.

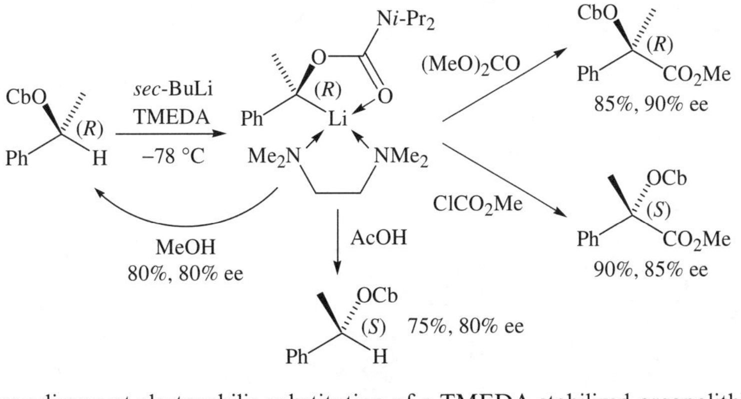

Scheme 6.7 Stereodivergent electrophilic substitution of a TMEDA-stabilized organolithium species.

Protonation of the TMEDA-stabilized organolithium species proceeds with inversion of configuration yielding Cb-protected (*S*)-1-phenylethanol when acetic acid is used, whereas methanol regenerates the original (*R*)-carbamate. Similarly, reaction with dimethyl carbonate and methyl chloroformate gives opposite enantiomers of the α-carbamoyl methyl ester in high yields and up to 90% ee.

In general, α-lithiated carbamates display higher configurational stability than esters.[70–72] In particular, benzyllithium derivatives often racemize considerably prior to reaction with an electrophile. Lithiation of chiral 2,4,6-triisopropylbenzoate derivatives of indanol and tetralol and immediate trapping with deuterated methanol gives the corresponding deuterated esters in high yields but in only 70–78% ee, Scheme 6.8. The stereochemical outcome of reactions of α-lithiated esters is often difficult to predict and can be reversed by subtle changes in the substrate structure, *e.g.*, lithiation and subsequent stannylation of the 2,4,6-triisopropylbenzoyl-derived indanol ester proceeds with net retention but the corresponding tetralol ester gives net inversion of configuration. The propensity for racemization and the unpredictable stereoselectivity limit the synthetic value of α-lithiated esters.

In contrast, α-amino organolithium compounds have become invaluable reagents for asymmetric synthesis. The preparation of α-lithio amines often entails tin/lithium exchange or (−)-sparteine-controlled enantiotopic deprotonation of carbamates.[73–78] Gawley and Zhang discovered the remarkable configurational integrity of cyclic α-lithio amines.[79–81] 2-Lithio-*N*-methylpiperidine and pyrrolidine are stable to racemization for up to 45 minutes at −40 °C in the presence of TMEDA. This configurational stability has been attributed to bridging of lithium across the carbon–nitrogen bond, which is in agreement with the observation of ^{15}N–^{6}Li coupling in the NMR spectrum. The bridging holds the lithium more strongly on one side of the heterocycle which ultimately impedes inversion of configuration. Transmetalation of *N*-methyl-2-(tributylstannyl)piperidine or its pyrrolidine analog with *n*-butyllithium at −80 °C, and trapping of the intermediate lithium adducts with carbon dioxide or dimethyl carbonate, produces the corresponding amino acid and amino ester with excellent enantioselectivity, Scheme 6.9. Cyclic α-lithio amines are versatile nucleophiles for stereoselective carbon–carbon bond formation. Usually, reaction with carbonyl electrophiles and stannanes proceeds with retention of configuration, and inversion of configuration

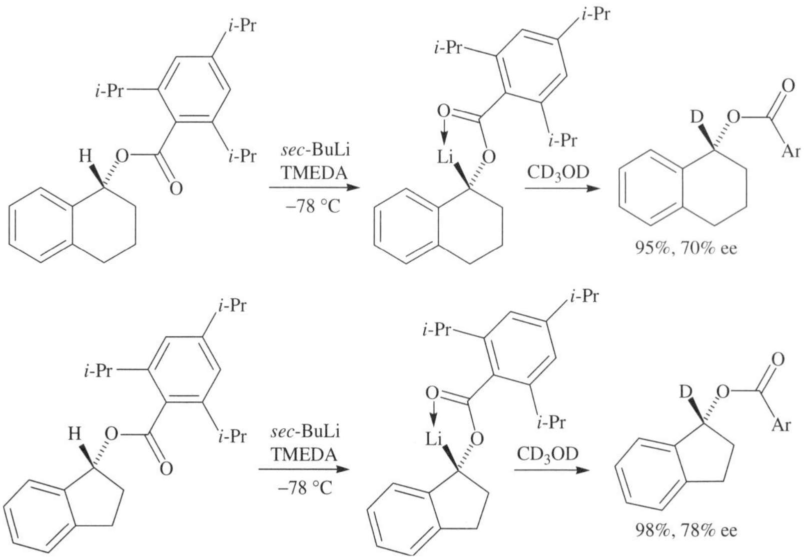

Scheme 6.8 Lithiation and deuteration of 2,4,6-triisopropylbenzoyl derivatives of indanol and tetralol.

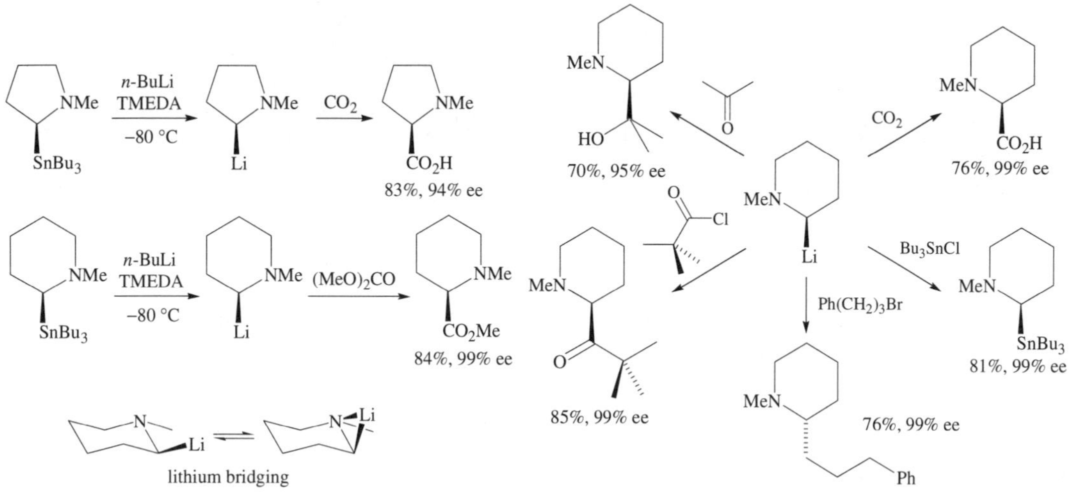

Scheme 6.9 Asymmetric synthesis with 2-lithio-*N*-methyl derivatives of pyrrolidine and piperidine and configurational stabilization due to lithium bridging across the carbon–nitrogen bond.

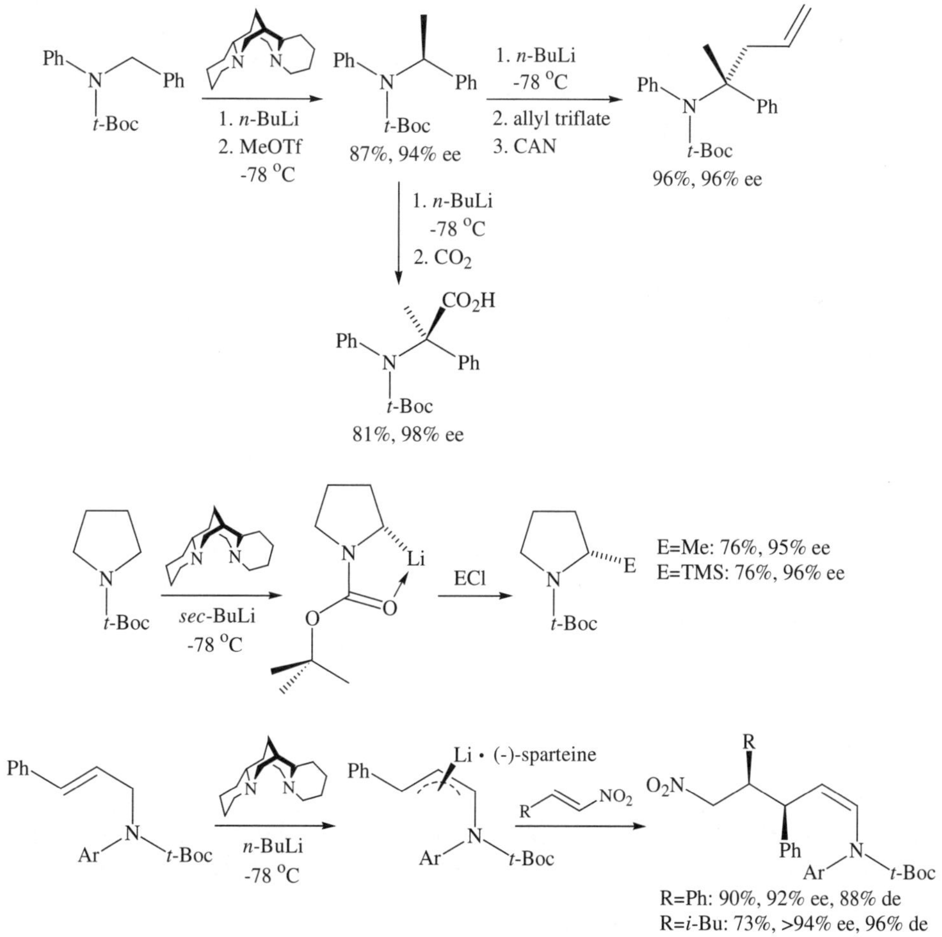

Scheme 6.10 Enantioselective deprotonation of acyclic and cyclic carbamates followed by substitution or conjugate addition.

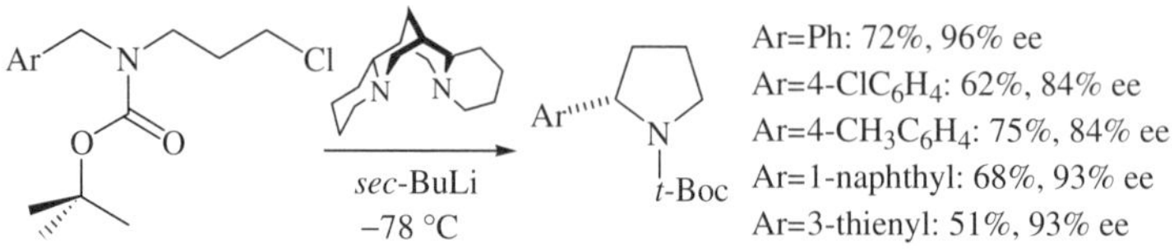

Ar=Ph: 72%, 96% ee
Ar=4-ClC$_6$H$_4$: 62%, 84% ee
Ar=4-CH$_3$C$_6$H$_4$: 75%, 84% ee
Ar=1-naphthyl: 68%, 93% ee
Ar=3-thienyl: 51%, 93% ee

Scheme 6.11 Asymmetric lithiation and cyclization of *t*-Boc-protected benzylic 3-chloropropylamines.

predominates when akyl halides are used.[82] Enantiotopic deprotonation of achiral carbamates with organolithium reagents in the presence of (–)-sparteine and subsequent trapping of the dipole-stabilized α-lithium intermediate gives access to a range of chiral amines. Many examples exploiting both cyclic and acyclic carbamates have been reported by Beak and others, Scheme 6.10.[83–97] Mechanistic studies revealed that when reactions are carried out under cryogenic conditions asymmetric induction takes place during the (–)-sparteine-controlled deprotonation step rather than during the following substitution step. In other words, the reaction sequence involves enantioselective formation of a configurationally stable organolithium species when temperatures as low as –78 °C are applied.[vii] However, both dynamic kinetic and dynamic thermodynamic resolution pathways may be operative at higher temperatures or when configurationally less stable α-lithio carbamates are used.[98,99] As has been mentioned above for α-alkoxy and α-amino organolithium derivatives, lithiated carbamates can react with electrophiles through both inversion and retention of configuration. For example, lithiation of *t*-Boc-protected (*S*)-*N*-phenyl-α-methylbenzylamine and subsequent reaction with allyl triflate results in net retention of configuration but net inversion of configuration is observed when carbon dioxide is used as electrophile. Alternatively, configurationally stable organolithium nucleophiles can be employed in stereoselective conjugate additions to nitroalkenes or other suitable Michael acceptors.[100–103] An enantioselective synthetic route to substituted pyrrolidines based on asymmetric deprotonation of acyclic *t*-Boc-protected benzylic 3-chloropropylamines has been described.[104] At –78 °C, intramolecular substitution of the configurationally stable organolithium intermediate occurs significantly more quickly than racemization and a range of 2-arylsubstituted pyrrolidines is obtained in up to 96% ee, Scheme 6.11.

6.1.2 Sulfur-, Phosphorus- and Halogen-stabilized Organolithium Compounds

In general, α-lithio sulfides, selenides and halides exhibit lower configurational stability than the α-amino and α-alkoxy analogs discussed in the preceding chapter.[105] Because of the practicality of the concept of *Umpolung* (German, reversal of polarity) which involves transformation of electrophilic aldehydes to dithianes or dithiolanes and subsequent deprotonation with organolithium reagents to generate nucleophilic species for carbon–carbon bond formation, α-lithiated compounds have received considerable attention. Stereolabile α-lithio sulfides have been employed in dynamic kinetic and dynamic thermodynamic resolution.[106–109] Similarly, α-sulfonyl carbanions[110–112] and α-lithiated selenides[113,114] are prone to racemization but at very low temperatures these compounds can be sufficiently configurationally stable for stereocontrolled substitution. Many diastereoselective reactions of sulfur-stabilized organolithium intermediates such as chiral α-lithio sulfoxides are known, albeit few enantioselective variations with configurationally stable α-lithiated sulfides have been reported.[115–118] A noteworthy example is Hoffmann's retrocarbolithiation of a chiral (*E*)-1-arylthio-2-(methylthiopropenyl)cyclopropane derivative. Lithiation of the enantiomerically enriched methyl selenoether generates a sulfur-stabilized lithium species with a

---

[vii] This is not always the case. Lithiation of *N,N*-diisopropyl-2-ethylbenzamide at –78 °C produces stereolabile (–)-sparteine adducts that undergo dynamic kinetic asymmetric transformation, see Scheme 6.3.

racemization half-life time of 90 minutes at −78 °C, which corresponds to an inversion barrier of 55.7 kJ/mol.[119] The α-lithio sulfide is configurationally stable at −108 °C and can be trapped with methyl iodide. Both the intramolecular ring opening and the methylation step proceed with retention of configuration, Scheme 6.12.

Following the successful examples reported for α-lithiated chiral carbamates, Hoppe and coworkers explored the usefulness of chiral thiocarbamates, Scheme 6.13.[120–123] Deprotonation of a sterically hindered (S)-1-phenylethanethiol-derived carbamate, which is readily prepared by Mitsunobu reaction from (R)-1-phenylethanol and thioacetic acid, gives a TMEDA-stabilized α-lithio sulfide species that shows no sign of racemization at 0 °C for up to 10 minutes and proves to be configurationally stable at −78 °C. With the exception of anhydrides and protic reagents such as methanol and acetic acid, reaction with typical electrophiles occurs with inversion of configuration.

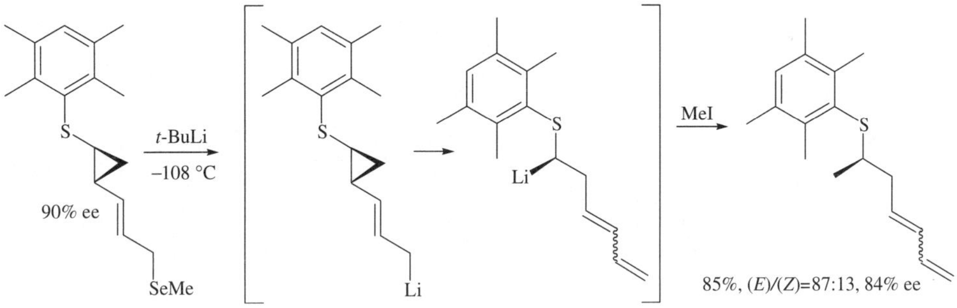

Scheme 6.12   Asymmetric formation of an α-lithio sulfide through cyclopropane ring opening and subsequent methylation.

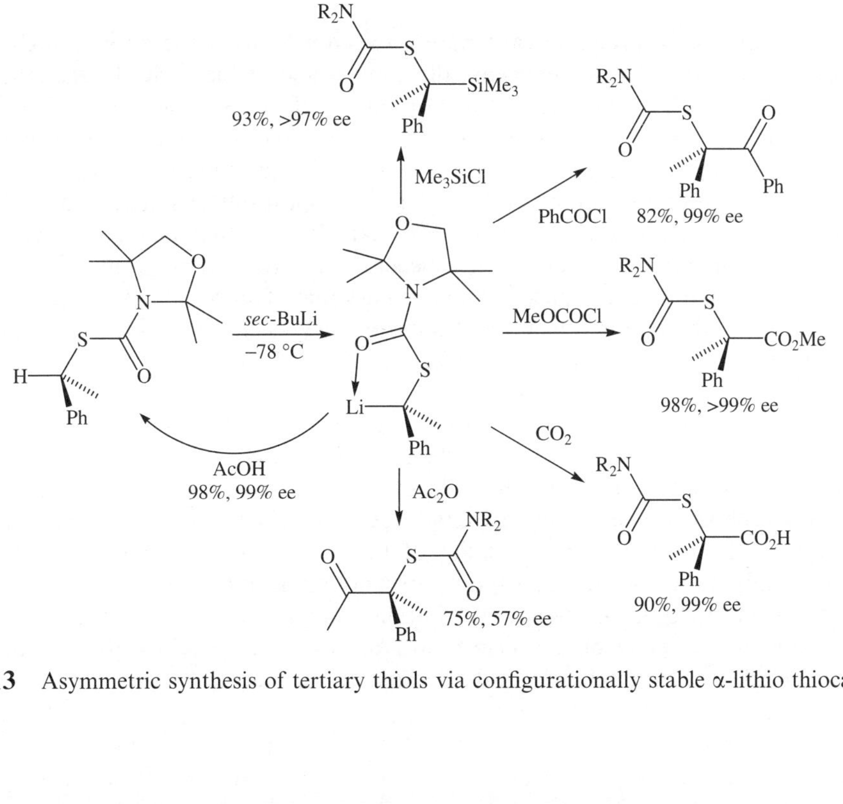

Scheme 6.13   Asymmetric synthesis of tertiary thiols via configurationally stable α-lithio thiocarbamates.

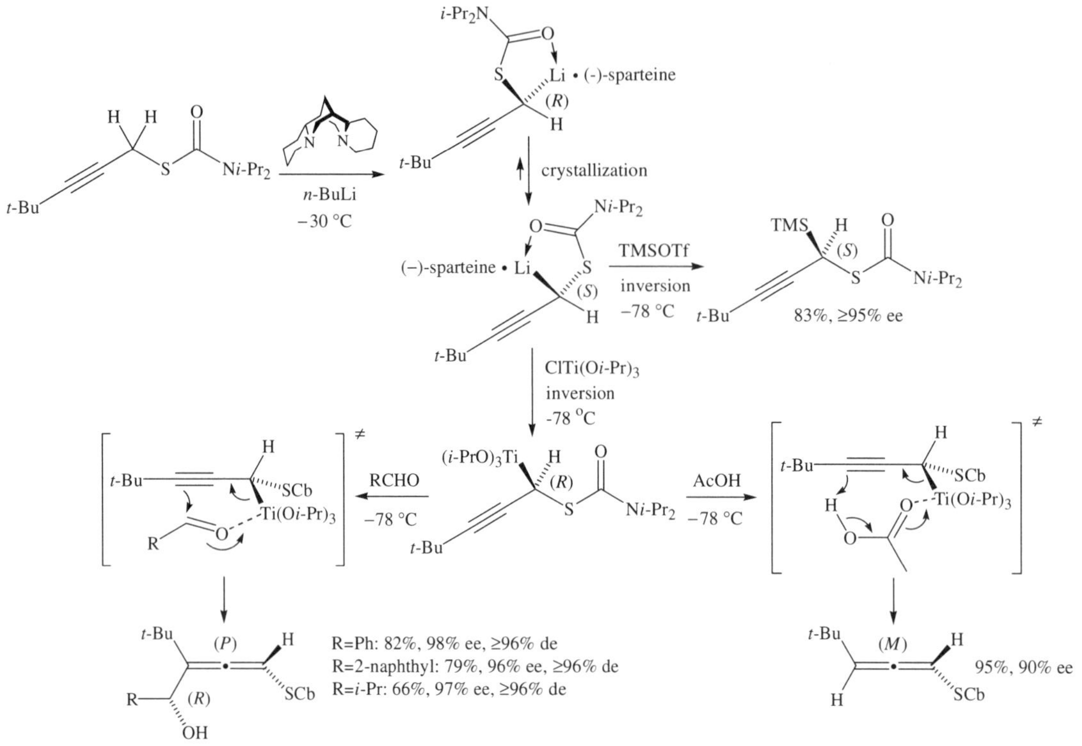

Scheme 6.14 Asymmetric transformation of the second kind and enantioselective reactions of an α-carbamoylthio-substituted propargyllithium-(−)-sparteine complex.

Enantiotopic lithiation of prochiral carbamates does not always proceed with high stereoselectivity. In some cases, however, stereochemically pure (−)-sparteine-derived organolithium complexes can be obtained by asymmetric transformation of the second kind. Investigating the lithiation of a prochiral alkynyl thiocarbamate, Hoppe's group observed that the propargyl/lithium ion pair undergoes rapid pyramidal inversion at −30 °C and subsequent precipitation of one diastereomer from solution in pentane, Scheme 6.14.[124] As a result of simultaneous isomerization and diastereoselective crystallization, the mixture can be quantitatively converted to a single stereoisomer. The synthetic utility of the stereochemically pure (*S*)-α-carbamoylthio-substituted propargyllithium-(−)-sparteine complex is quite remarkable. Trapping with trimethylsilyl triflate and transmetalation with titanium(IV) chlorotriisopropoxide occurs with inversion of configuration. The propargyltitanium intermediate can also be transformed to axially chiral 1-thio-substituted allenes with acetic acid or aldehydes.

Phosphorus- and halogen-stabilized chiral organolithium compounds have rarely been employed in asymmetric synthesis. Deprotonation of diethyl phosphates derived from chiral secondary benzylic alcohols with butyllithium or LDA affords dipole-stabilized chiral carbanions that rearrange to α-hydroxyphosphonates having up to 98% ee.[125,126] The excellent conservation of chirality indicates considerable configurational stability of the intermediate lithium ion pair at −78 °C, Scheme 6.15. Configurationally stable α-bromo organolithium intermediates have been prepared from chiral stannanes by transmetalation with butyllithium at −110 °C. The lithiated alkyl halides can be trapped with acetone prior to racemization and the corresponding alkoxide intermediates readily form epoxides via S_Ni replacement of the adjacent bromide.[127]

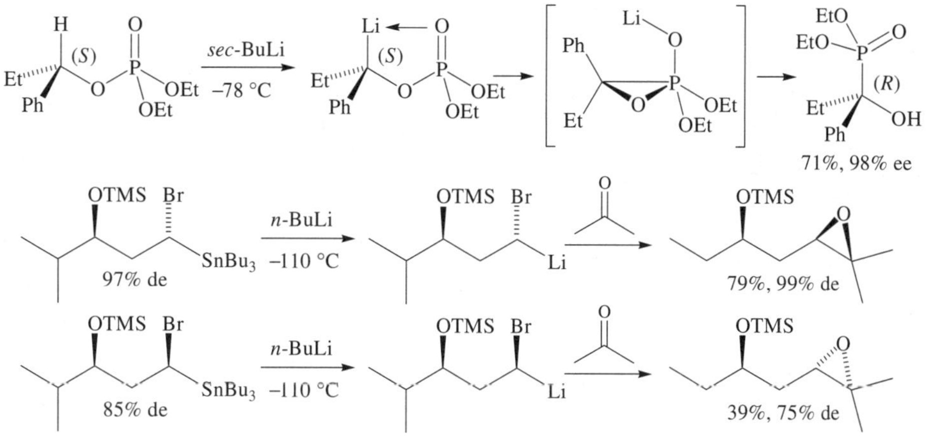

Scheme 6.15 Phosphate/phosphonate rearrangement via diethoxyphosphoryloxy-derived benzylic carbanions (top) and asymmetric epoxide formation with α-halo organolithium intermediates (bottom).

6.2 ATROPOSELECTIVE SYNTHESIS OF AXIALLY CHIRAL BIARYLS

Axially chiral biaryls display a unique stereochemical motif that is present in a variety of natural compounds and pharmaceutical drugs, Figure 6.1. Important examples are the antibiotic glycopeptide van-comycin,[128,129] the cytotoxic tubulin-binding lignan steganacin,[130] the anti-HIV agent michellamine B,[131] knipholone,[132] which has been found in varying and highly characteristic enantiomeric compositions in nature, neurotrophic mastigophorene A,[133] and murrastifoline-F, which exhibits an axially chiral *C–N* bond.[134,135] Some of these natural products possess other chiral elements in addition to the chiral axis but the latter is usually crucial to their biological activity. Many powerful asymmetric catalysts and auxiliaries possess a chiral axis, Figure 6.2.[136,137] In particular, BINOL,[138–141] BINAP,[142–146] their derivatives,[147–154] binaphthyl-based quaternary ammonium phase-transfer catalysts,[155–159] 2,2′-bipyridine *N*-monoxides,[160–163] and 2,2′-bipyridine *N,N′*-dioxides[164–172] afford excellent enantioselectivity in a wide range of asymmetric reactions. Highly congested atropisomeric 1,8-diquinolyl- and 1,8-diacridylnaphthalene-derived fluorosen-sors providing a C2-symmetric cleft for enantioselective recognition of chiral carboxylic acids and amino acids have also been reported.[173–175]

Atroposelective synthesis of axially chiral biaryls has attracted considerable attention due to the increasingly recognized importance of this class of compounds.[176] The total synthesis of complex chiral biaryls such as streptonigrin and other antibiotic natural products constitutes a remarkable challenge, as one must achieve both effective asymmetric induction and control of atropisomeri-zation of multi-functional intermediates and products.[177] The unique dynamic stereochemistry of biaryls has been exploited in a number of ways to gain synthetic access to this class of compounds. A prominent example is the preparation of the axially chiral aglycon of vancomycin via diastereo-selective *C–C* bond formation and atroposelective equilibration, see Chapter 7.2.[178,179] The most popular asymmetric strategies towards axially chiral biaryls can be classified as follows:

1. Intramolecular aryl–aryl coupling
2. Intermolecular aryl–aryl coupling
3. Atroposelective ring construction
4. Desymmetrization of conformationally stable achiral biaryls
5. Stereodynamic methods such as atroposelective cleavage of conformationally fluxional biaryl lactones
6. Interconversion of central to axial chirality which is discussed separately in Chapter 6.4.5

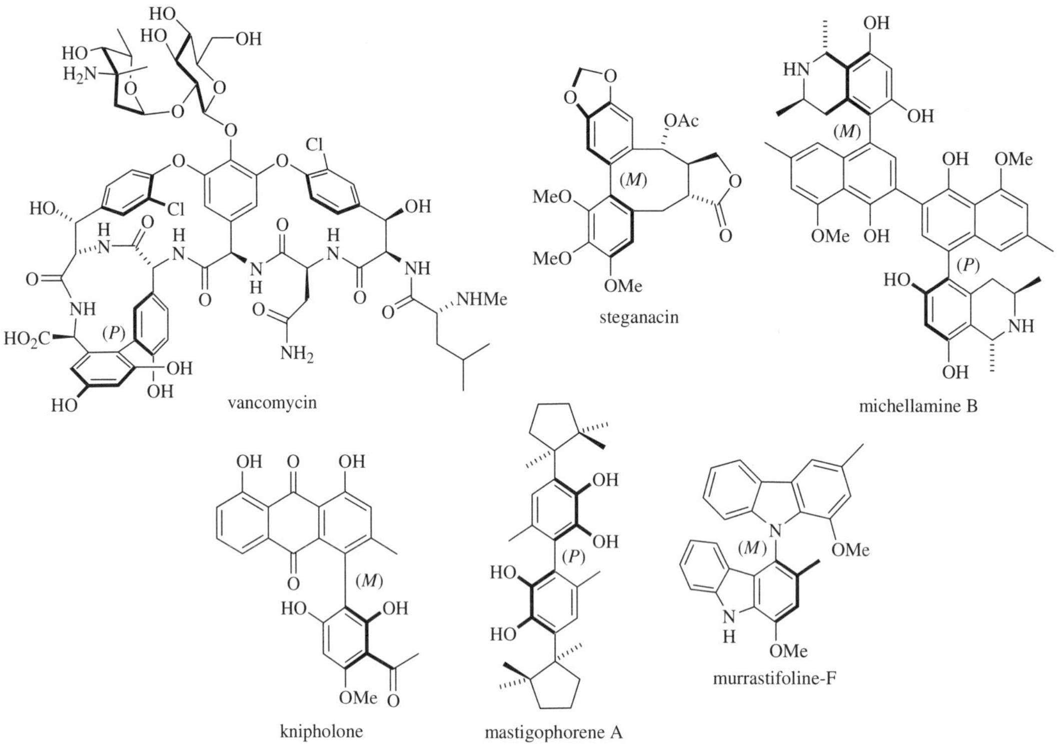

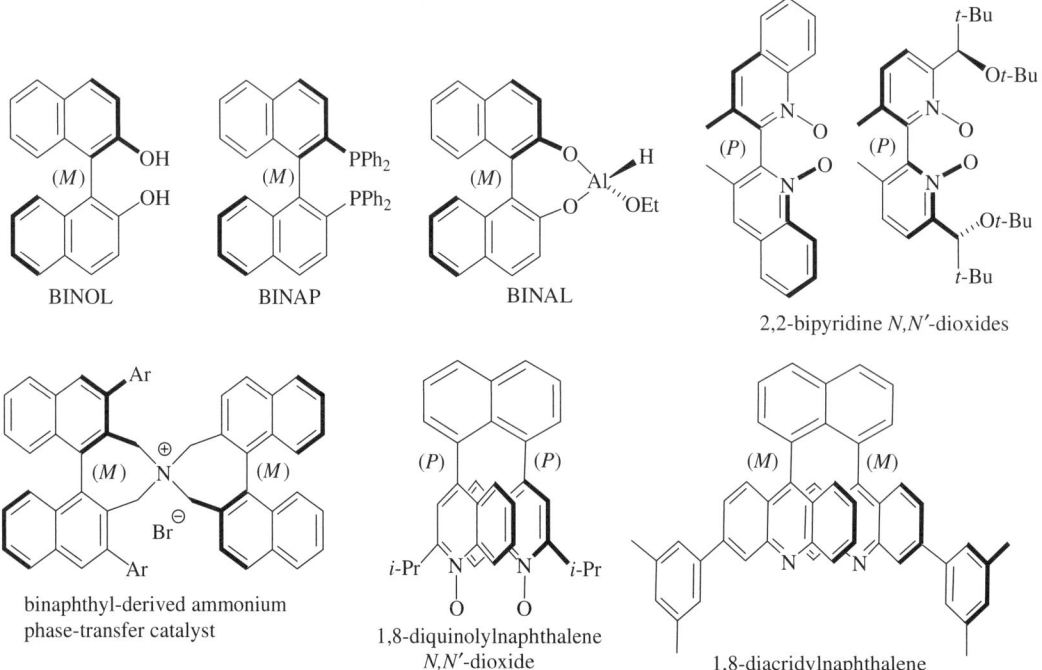

Figure 6.1   Structures of axially chiral natural products.

Figure 6.2   Structures of axially chiral catalysts, reagents and sensors.

6.2.1 Intramolecular Atroposelective Biaryl Synthesis

Atroposelective aryl–aryl bond formation requires construction of a chiral axis under conditions that do not compromise the stereochemical purity of the target compound. In other words, the stereoselective homo- or cross-coupling step must occur at temperatures at which the biaryl is stable to rotation about the incipient chiral axis. As is discussed in Chapter 3.3.1, the conformational stability of biaryls mainly depends on the presence of bulky *ortho*-substituents, and *meta*-substituents further enhance steric hindrance to rotation through the so-called buttressing effect. In contrast to these predominant steric interactions, electronic effects of *para*-substituents on the rotational energy barrier of biaryls play a minor role.[180,181] Incorporation of *ortho*-substituents into the coupling components is necessary to avoid atropisomerization of the biaryl product but this causes significant steric repulsion during *C–C* bond formation. It is therefore difficult to find reaction conditions that allow formation of conformationally stable biaryls in reasonable yield, while avoiding concomitant rotation about the chiral axis which would ultimately diminish atroposelectivity. This challenge is evident in Miyano's asymmetric biaryl synthesis based on intramolecular Ullmann coupling of (*M*)-BINOL-tethered aryl halides, Scheme 6.16.[182–184] Enantiopure 1,1′-bi(2-naphthol)-bridged diesters of 2-halobenzoic acid derivatives were prepared and employed in Cu-mediated homocoupling to atropisomeric diphenic acids. In this approach, the sense of axial chirality in the C_2-symmetric BINOL template determines the relative orientation of the two aryl halides and thus governs the chiral induction during Ullmann reaction. The (*M*)-BINOL-directed asymmetric coupling produces (*M*)-diphenic acid-derived 12-membered ring structures in excellent de. However, highly congested 2,2′,6,6′-tetra-*ortho*-substituted biaryls bearing fused benzene rings are obtained in only moderate yields. Basic hydrolysis of the diesters gives the free (*M*)-diphenic acids without loss of enantiomeric purity and the BINOL template can be recovered.

Exploiting Miyano's pioneering work, Lipshutz *et al.* utilized oxidative coupling of intermediate diarylcyanocuprates for template-mediated diastereoselective biaryl synthesis, Scheme 6.17.[185,186] Anchoring of two 1-bromo-2-naphthol units on a C_2-symmetric tartaric acid-derived (*R,R*)-diol bridge followed by treatment with butyllithium, copper cyanide, and exposure to oxygen allows mild formation of the (*P*)-binaphthyl atropisomer. Removal of the tether is accomplished by double benzylic oxidation with NBS and basic hydrolysis. The advantage of this procedure is that it avoids harsh Ullmann reaction conditions, albeit several steps are required to attach and cleave the chiral

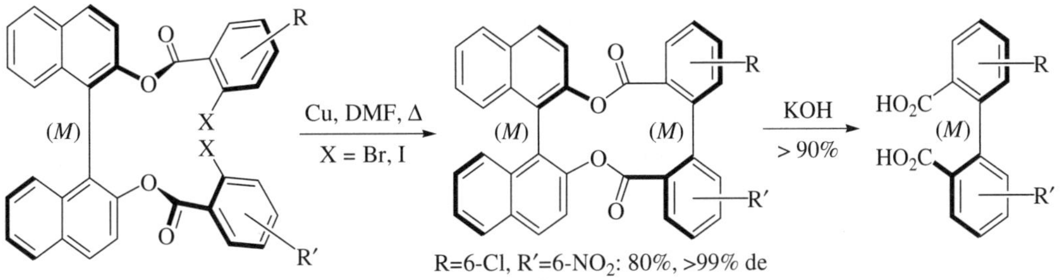

R=6-Cl, R′=6-NO$_2$: 80%, >99% de
R=R′=6-NO$_2$: 70%, 85% de
R=R′=5,6-benzo: 39%, >99% de
R=5,6-benzo, R′=6-Cl: 33%, >99% de
R=5,6-benzo, R′=6-NO$_2$: 44%, >99% de

Scheme 6.16 Atroposelective intramolecular Ullmann coupling of aryl halides using a rigid (*M*)-BINOL template.

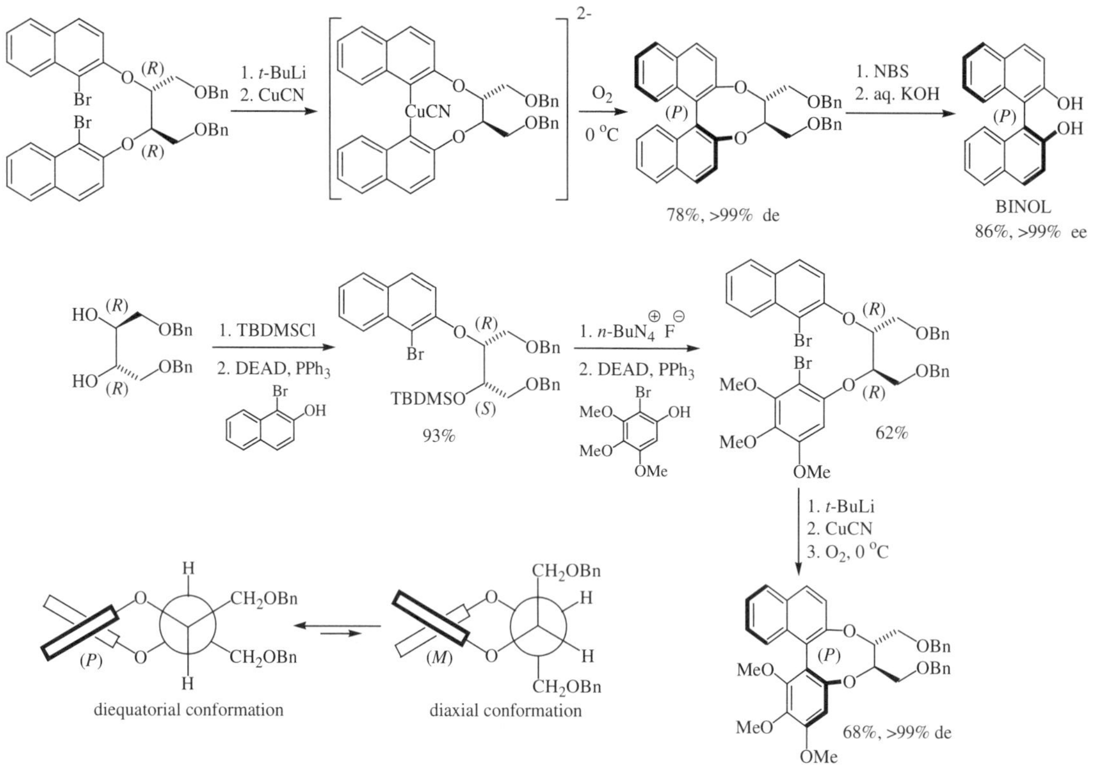

Scheme 6.17 Atroposelective biaryl synthesis based on oxidative coupling of intermediate arylcyanocuprates.

template. Stepwise introduction of 1-bromo-2-naphthol and 2-bromo-3,4,5-trimethoxyphenol to the diol tether via two sequential Mitsunobu reactions and intermediate protection of one alcohol group with *tert*-butyldimethylsilyl chloride has been used to extend this method to asymmetric biaryls consisting of differently substituted aryl moieties. The diastereoselective formation of the (*P*)-biaryl atropisomer can be attributed to preferential population of a diequatorial gauche conformation of the intermediate diarylcyanocuprate.[187] Bridging of 2-naphthols and 2-naphthylamines by a chiral tether and oxidative coupling with Mn(VI) or Cu(II) salts provides access to cyclo-BINOL[188,189] and cyclo-NOBIN[190] derivatives, Scheme 6.18. The atroposelective biaryl synthesis of 7,7'-tethered BINOL and NOBIN, which were originally introduced by Noyori[191] and Kocovsky,[192] is noteworthy because it facilitates immobilization of these chiral ligands on polystyrene and other polymers for heterogeneous asymmetric catalysis.[193]

Atroposelective intramolecular aryl–aryl coupling mediated by chiral tethers has been adapted by other groups[194–198] and applied to the synthesis of complex natural products[199] including *O*-permethyltellimagrandin II,[200] kotanin[201] and rhazilinam.[202] Genêt and Chan have utilized the same strategy as a convenient entry to axially chiral diphosphine ligands, Scheme 6.19.[203–205] In this approach, sequential attachment of 1-iodo-2-naphthol and 2-bromo-3,4,5-trimethylphenol to (*R,R*)-pentanediol via Mitsunobu reaction with diisopropyl azodicarboxylate is followed by oxidative cross-coupling using Lipshutz' procedure. The (*R,R*)-pentanediol-assisted *C–C* bond formation yields a diastereomerically pure (*M*)-biaryl which can be converted to the corresponding (*M*)-diphosphine in three steps.[viii]

[viii] This diphosphine is a useful ligand for asymmetric catalysis and affords excellent results in rhodium-catalyzed hydrogenation of β-keto esters and 1,4-addition of boronic acids to α,β-unsaturated ketones.

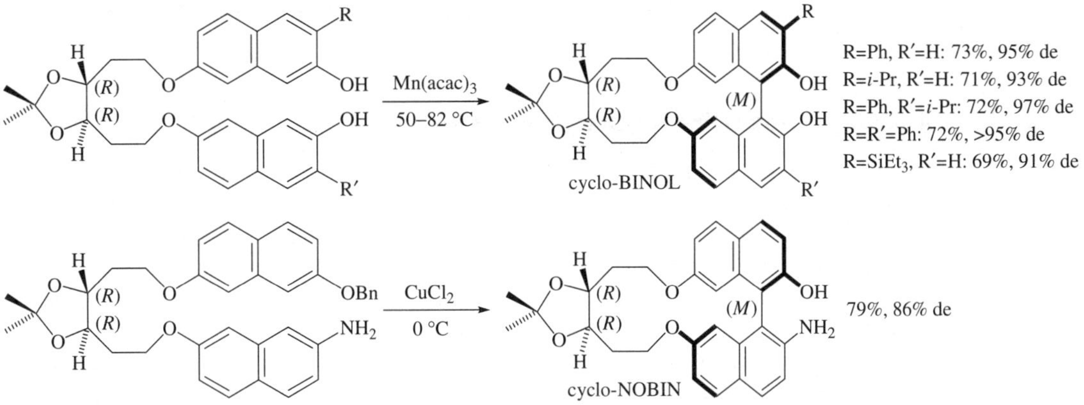

Scheme 6.18 Diastereoselective synthesis of cyclo-BINOL and cyclo-NOBIN derivatives.

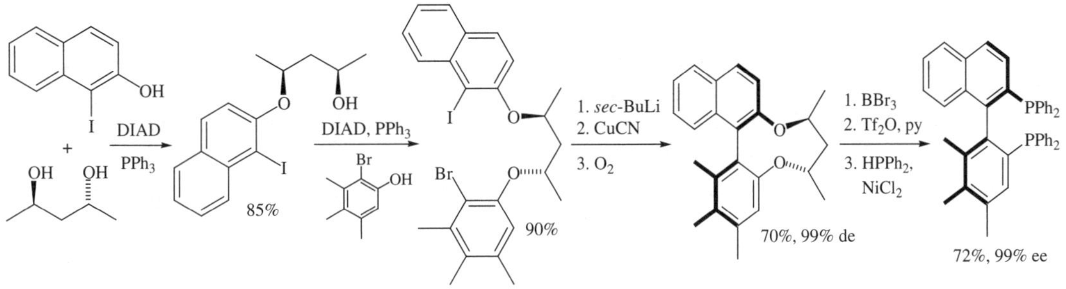

Scheme 6.19 Atroposelective synthesis of an axially chiral diphosphine ligand.

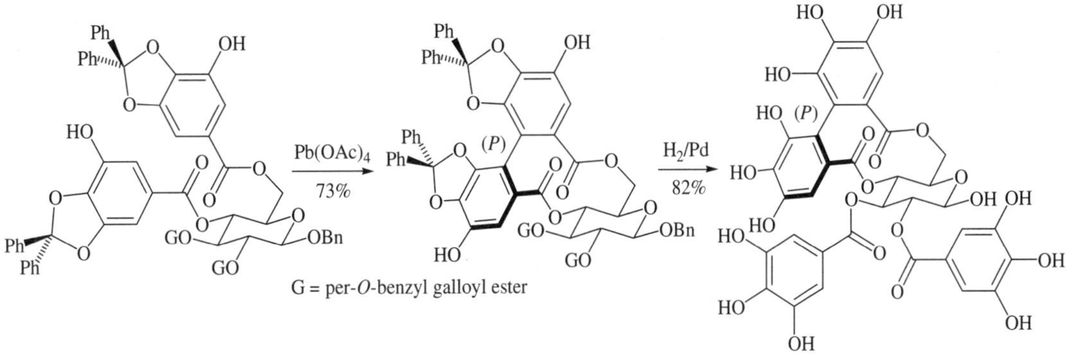

Scheme 6.20 Synthesis of tellimagrandin I by biomimetic atroposelective intramolecular coupling of glucose-bridged galloyl esters.

   Some natural products including ellagitannins possess axially chiral biaryl units that are bridged by a glucose core. It has therefore been postulated that the carbohydrate tether could be the source of asymmetric induction during atroposelective biosynthesis of ellagitannins. This hypothesis inspired the development of powerful biomimetic methods.[206–208] Feldman's group showed that tellimagrandin I, an ellagitannin exhibiting a (*P*)-hexahydroxydiphenic acid core, can be prepared in diastereomerically pure form by oxidative coupling of glucopyranose-bridged 3,4,5-trihydroxybenzoate esters, so-called galloyl esters. This confirmed that the glucose tether does indeed provide effective stereocontrol during atropisomer formation, Scheme 6.20.[209] Biomimetic oxidative biaryl coupling has also been utilized in the total synthesis of vancomycin,[210,211] diazonamide A[212] and stegane lignans.[213–215]

6.2.2 Intermolecular Atroposelective Biaryl Synthesis

Synthetic strategies that are based on *inter*molecular atroposelective biaryl formation are generally more attractive than the *intra*molecular methods discussed above. Diastereoselective intermolecular coupling is applicable to a broader variety of biaryl precursors, because only one aromatic ring has to be modified with a chiral auxiliary and incorporation of a chiral bridge is not necessary. One of the most successful strategies for diastereoselective biaryl synthesis has been introduced by Meyers and coworkers. They observed that the presence of an oxazoline group promotes nucleophilic aromatic substitution of adjacent methoxy and fluoro substituents by alkyl and aryl Grignard reagents.[216] They extended the scope of this nucleophilic displacement reaction to atropodiastereoselective biaryl formation simply by using chiral aryl oxazolines which can be prepared from (*S*)-valinol or (*S*)-*tert*-leucinol and 2,3-dimethoxybenzoic acid. In particular, aryl Grignard reagents with a methoxy group in *ortho*-position yield axially chiral biaryls in remarkable diastereomeric excess.[217–220] The reaction course and sense of asymmetric induction is governed by magnesium chelation and metal coordination of methoxy substituents from both the aryl oxazoline and the Grignard reagent, Scheme 6.21. The aryl Grignard, which is generated *in situ*, can approach the (*S*)-valinol-derived aryl oxazoline from two sides to form a magnesium chelation complex. Since the face opposite the bulky isopropyl group is more accessible, the aryl Grignard undergoes nucleophilic attack at the *ortho*-methoxy group of the aryl oxazoline from this side to form an aza-enolate intermediate. The oxazoline auxiliary apparently fulfills two functions: control of the stereochemical course during carbon–carbon bond formation, and activation of the arene for nucleophilic attack by resonance stabilization of the intermediate negative charge. Although the aryl ring can rotate about the new bond, complexation of the *ortho*-methoxy group to the magnesium ion favors an arrangement that affords the (*P*)-biaryl after elimination of methoxymagnesium bromide. Removal of the chiral auxiliary by acidic hydrolysis and subsequent modification of the carboxylic acid function proceeds without concomitant racemization if the biphenyl

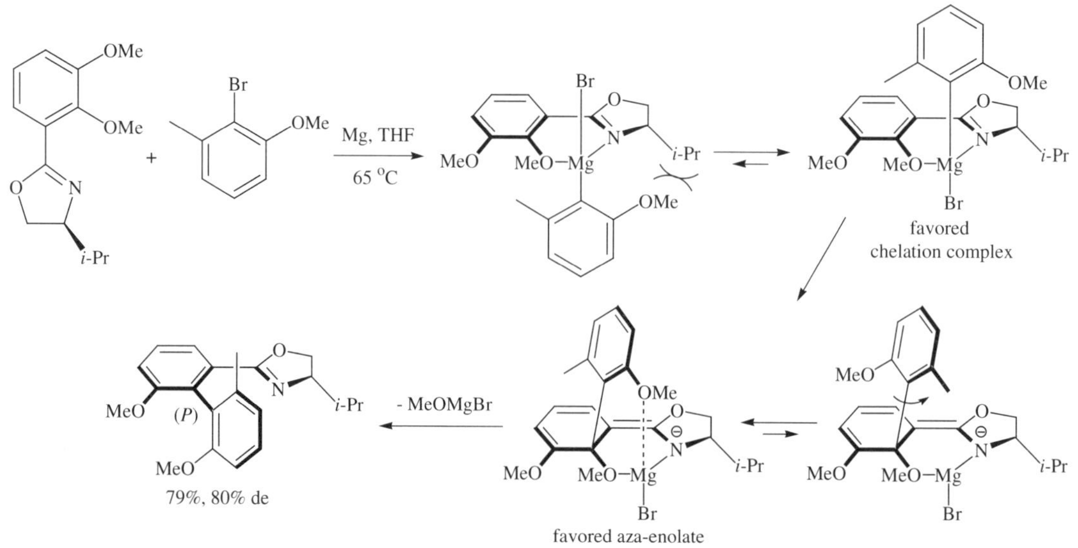

Scheme 6.21   Meyers' atropodiastereoselective biaryl formation with chiral aryl oxazolines.

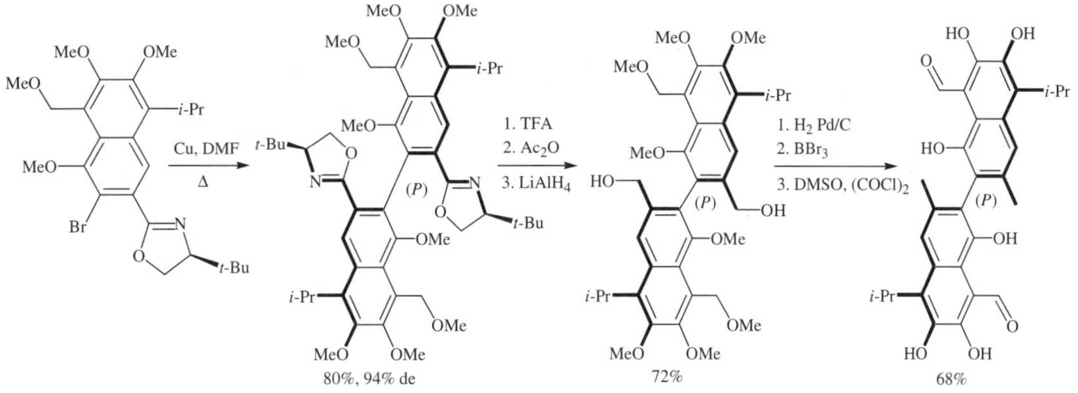

Scheme 6.22 Total synthesis of (+)-gossypol.

has four *ortho*-substituents. A drawback of Meyers' approach is that it requires the presence of methoxy groups in the *ortho*-positions of both coupling reagents. However, oxazoline-mediated cross-coupling provides convenient access to axially chiral biaryls consisting of two different aryl rings, and a variety of natural products including (−)-steganone,[221] (−)-*O*-methylancistrocladine[222] and (−)-schizandrin[223] has been prepared using this method. Aryl oxazolines have also been employed in Ullmann coupling reactions.[224–227] The first total synthesis of (+)-gossypol was accomplished by diastereoselective copper-mediated coupling of (*S*)-*tert*-leucinol-derived bromo(oxazolinyl)arenes.[228,229] The asymmetric coupling step gives the sterically crowded C_2-symmetric biaryl in 80% yield and 94% de. Hydrolysis of the chiral auxiliary and reduction converts the oxazoline groups to methyl substituents and (*P*)-(+)-gossypol is finally obtained after Lewis acid-catalyzed ether cleavage and Swern oxidation, Scheme 6.22.

Following Meyers' seminal work, Lipshutz and others further investigated the scope of atroposelective synthesis with biaryl precursors bearing chiral auxiliaries.[230–233] The korupensamine A skeleton, an important component of the antiviral alkaloid michellamine B, has been prepared by Suzuki coupling of a selectively substituted 1-naphthylboronate ester and a chiral iodotetrahydroisoquinoline carrying a chelating diphenylphosphanyl moiety, Scheme 6.23. Employment of the (*S*)-tetrahydroisoquinoline-derived aryl iodide in the palladium-catalyzed cross-coupling reaction furnishes the desired diastereomerically pure (*P*)-atropisomer. The stereochemical outcome has been attributed to an intermediate Pd(II) species, in which the chiral auxiliary and the dppf ligand coordinate to the transition metal in such a way that the bulky 1-naphthyl moiety is forced above the plane of the aromatic portion of the tetrahydroisoquinoline ring. The remote chiral center of the latter thus controls the relative orientation of the coupling partners prior to generation of the chiral axis during reductive elimination.

A remaining shortcoming of atropodiastereoselective synthesis with chiral aryl precursors is that the covalently attached auxiliary often has to undergo extensive modification after the homo- or cross-coupling step in order to establish the desired biaryl substitution pattern, unless the chiral auxiliary is directly incorporated into the leaving group and is thus removed during nucleophilic aromatic substitution.[234–237] Planar chiral tricarbonyl(arene)chromium complexes provide an elegant solution to this problem, because the chiral auxiliary can easily be removed by oxidative photocleavage. The usefulness of (η^6-arene)chromium complexes for atroposelective synthesis of a range of biaryl compounds including natural products has been demonstrated by Uemura, Scheme 6.24.[238–241] Aryl chromium complexes are suitable for biaryl synthesis via palladium-catalyzed Suzuki cross-coupling and nucleophilic aromatic substitution. In both cases, the electron-withdrawing (η^6-arene)chromium core fulfills two functions: it is the source of chiral induction during atroposelective *C–C*

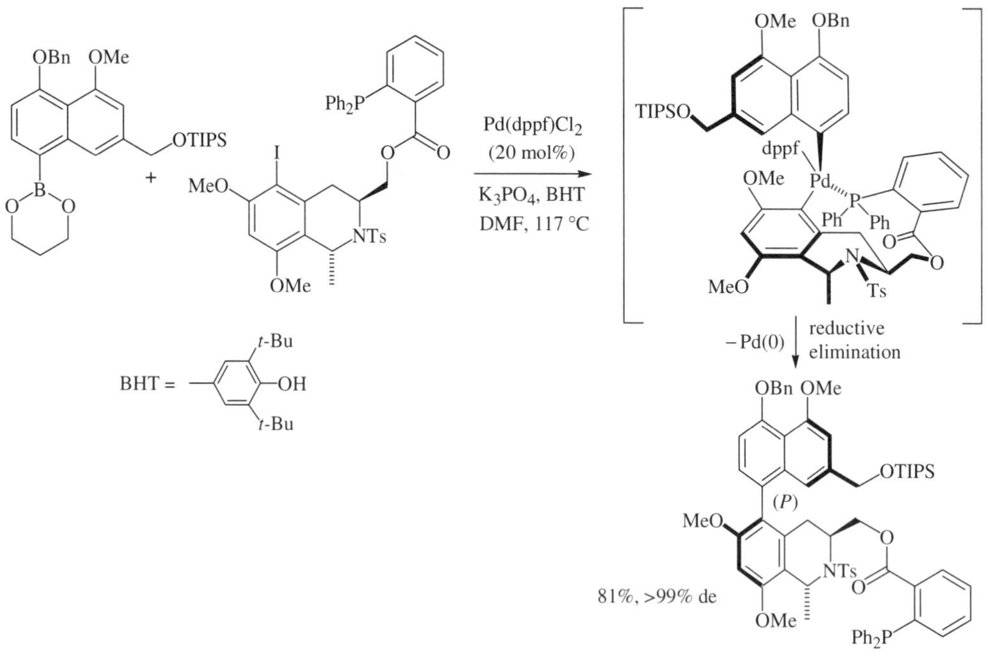

Scheme 6.23 Atroposelective synthesis of the (*P*)-korupensamine A skeleton.

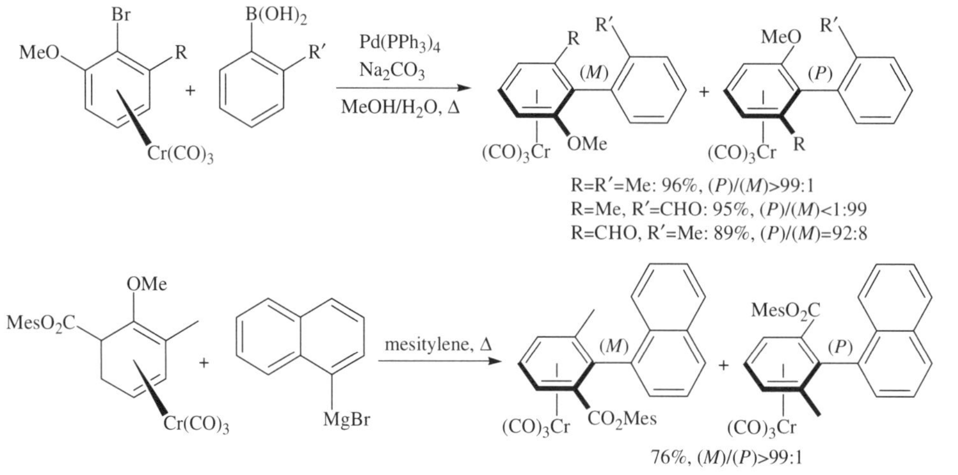

R=R′=Me: 96%, (*P*)/(*M*)>99:1
R=Me, R′=CHO: 95%, (*P*)/(*M*)<1:99
R=CHO, R′=Me: 89%, (*P*)/(*M*)=92:8

76%, (*M*)/(*P*)>99:1

Scheme 6.24 Atroposelective biaryl synthesis with (η^6-arene)chromium complexes.

bond formation, and it activates one of the reaction partners. Coordination of an aryl halide to the tricarbonylchromium fragment facilitates the rate-limiting oxidative addition to Pd(0) which is followed by transmetalation with an aryl boronic acid and subsequent reductive elimination to complete the catalytic cycle.[242–246] Alternatively, formation of an electron-deficient (*ortho*-methoxybenzoate)chromium complex accelerates the nucleophilic displacement of the methoxy group by an aryl Grignard reagent.[247] Unfortunately, both yield and diastereoselectivity decrease when bulky substituents are present and preparation of sterically crowded tetra-*ortho*-substituted biaryls is usually not feasible.

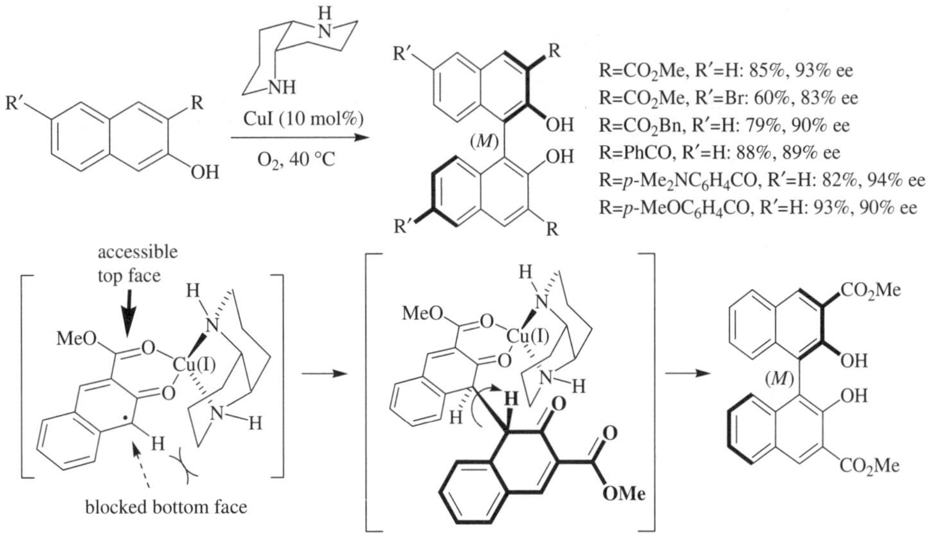

R=CO$_2$Me, R'=H: 85%, 93% ee
R=CO$_2$Me, R'=Br: 60%, 83% ee
R=CO$_2$Bn, R'=H: 79%, 90% ee
R=PhCO, R'=H: 88%, 89% ee
R=p-Me$_2$NC$_6$H$_4$CO, R'=H: 82%, 94% ee
R=p-MeOC$_6$H$_4$CO, R'=H: 93%, 90% ee

Scheme 6.25 Enantioselective synthesis of 3,3'-disubstituted BINOL derivatives via oxidative coupling of naphthols in the presence of an (*S*,*S*)-1,5-diazadecalin-derived copper catalyst.

Intermolecular biaryl synthesis has also been realized by coupling of achiral aryl units with enantioselective catalysts. For example, oxidative dimerization of 3-substituted 2-naphthols using an (*S*,*S*)-1,5-diazadecalin-derived copper catalyst in the presence of oxygen has been reported by Kozlowski's group.[248–250] The procedure allows preparation of 3,3'-disubstituted BINOL derivatives in up to 94% ee, Scheme 6.25. A catalytic cycle involving formation of an intermediate tetrahedral copper(I) complex exhibiting a carbon-centered radical has been proposed.[251] Accordingly, the chelating (*S*,*S*)-diamine blocks the bottom face of the bidentate naphthol derivative while the second substrate can easily approach from the top face. Dimerization is then followed by tautomerization and concomitant clockwise rotation about the incipient chiral axis. Since rotation past the chelating carbonyl moieties rather than the *peri*-hydrogens is favored, the (*M*)-atropisomer is produced in high enantiopurity. The development of atroposelective access to BINOL derivatives is important because of the outstanding results obtained with this class of catalysts. A range of selectively substituted BINOLs has been prepared by enantioselective transition metal-catalyzed oxidative homocoupling of naphthols in the presence of (–)-sparteine,[252,253] salen,[254] and proline- or binaphthyl-derived diamines.[255–257]

Atropoenantioselective cross-coupling has been achieved with aryl Grignard reagents[258,259] (Kumada coupling) and arylboronic acids[260] (Suzuki coupling) but examples with organozinc compounds (Negishi coupling), silanes (Hiyama coupling) and stannanes (Stille coupling) are still elusive. Hayashi and Ito showed that cross-coupling of (2-alkyl-1-naphthyl)magnesium bromides and 1-bromonapthalenes, in the presence of a chiral nickel catalyst formed *in situ* from (*S*)-1-[(*R*)-2-(diphenylphosphino)ferrocenyl]ethylmethyl ether and nickel bromide, provides practical access to 1,1'-binaphthyls in up to 95% ee, Scheme 6.26.[261,262] Suzuki coupling with chiral catalysts prepared from palladium and ferrocenylphosphines[263–265] or BINAP[266–268] tolerates a wide range of functional groups and is probably the most widely used method. Impressive results were obtained by Buchwald and coworkers who discovered that 1-bromo-2-naphthylphosphonates and *ortho*-alkylated phenylboronic acids can be coupled with excellent enantioselectivity when minute amounts of palladium and an axially chiral electron-rich aminophosphine ligand are employed, Scheme 6.26.[269] Transition metal-free atropoenantioselective coupling of aryl lead compounds and *ortho*-lithiated phenols in the presence of brucine has also been reported.[270,271]

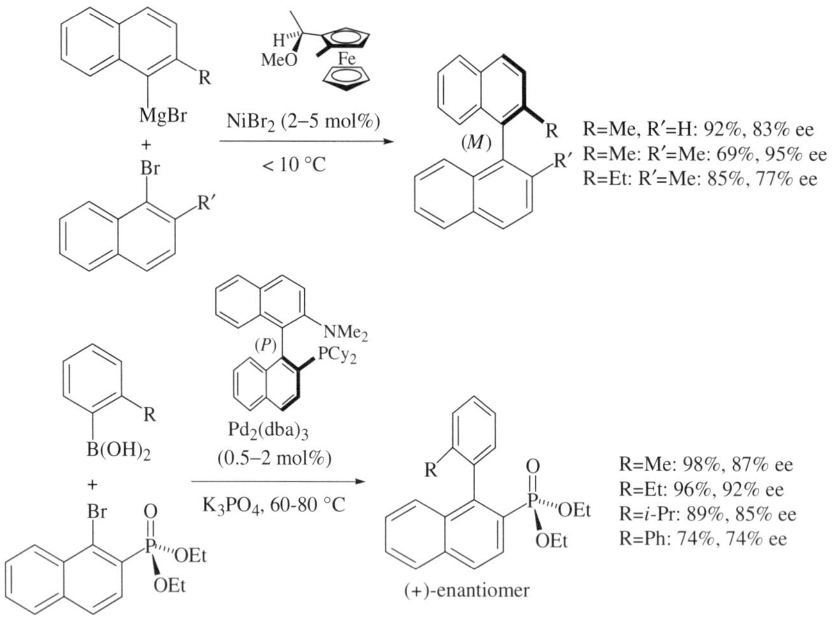

Scheme 6.26 Atropoenantioselective Kumada (top) and Suzuki coupling (bottom).

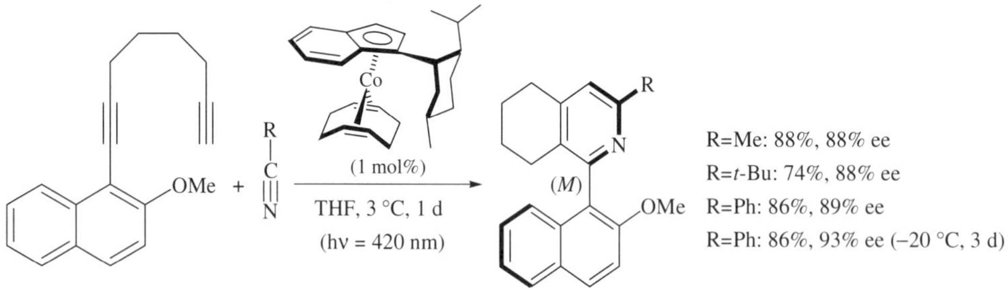

Scheme 6.27 Synthesis of 1-aryl-5,6,7,8-tetrahydroisoquinolines by Co(I)-catalyzed atroposelective [2+2+2]cycloaddition.

6.2.3 Atroposelective Ring Construction

Atroposelective [2+2+2]cycloaddition of alkynes and nitriles can accomplish both arene construction and incorporation of a chiral axis in a single step, thus providing a highly effective alternative to aryl–aryl coupling methods. The use of a photochemically activated chiral Co(I)-complex in the cycloaddition of 2-methoxy-1-(1,7-octadiynyl)naphthalene and nitriles allows asymmetric synthesis of 1-aryl-5,6,7,8-tetrahydroisoquinolines, Scheme 6.27.[272] This cyclotrimerization proceeds with excellent regioselectivity and formation of constitutional isomers is not observed. Cobalt(I)-catalyzed asymmetric ring construction is fairly slow when the temperature is decreased to −20 °C but the enantioselectivity improves under cryogenic conditions. Transition metal-catalyzed cyclotrimerization has been applied in the synthesis of other biaryls including polyaryls with more than one chiral axis.[273,274] Shibata *et al.* demonstrated that cycloaddition of α,ω-binaphthyldiynes and 1,4-dimethoxy-but-2-yne or other symmetric alkynes in the presence of an (*S,S*)-MeDUPHOS-derived iridium catalyst affords axially chiral teraryls. Ring construction and incorporation of two chiral axes occur with excellent stereoselectivity and formation of the meso isomer is not observed,

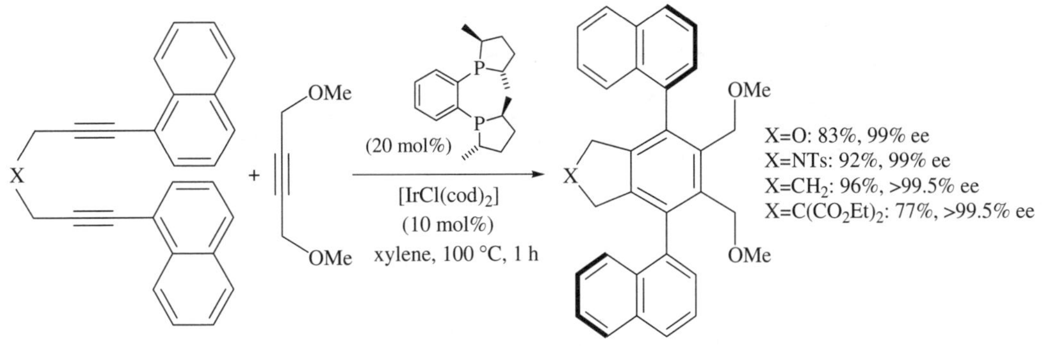

Scheme 6.28 Atroposelective formation of teraryls using an (*S*,*S*)-MeDUPHOS-derived iridium catalyst.

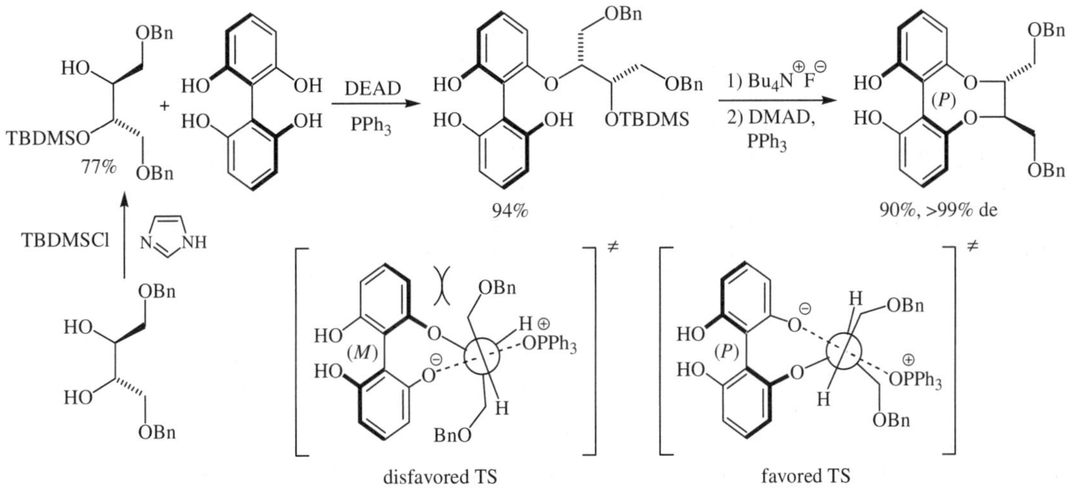

Scheme 6.29 Desymmetrization of 2,2′,6,6′-biphenyltetrol with (2*S*,3*S*)-1,4-di-*O*-benzyl-L-threitol.

Scheme 6.28.[275] Diastereoselective biaryl synthesis based on benzannulation of chiral chromium carbene complexes and 1,3-butadiynes has also been accomplished.[276–278]

6.2.4 Desymmetrization of Conformationally Stable Prochiral Biaryls

Topos-selective modification of conformationally stable achiral biaryls is an attractive alternative to atroposelective aryl–aryl bond formation. Desymmetrization of a prochiral biaryl by enantio-topos- or diastereotopos-selective transformation of one of two identical aryl substituents allows asymmetric introduction of axial chirality. An intrinsic advantage of this strategy is that a wide range of effective methods including Suzuki, Stille, Hiyama, Kumada, and Negishi cross-coupling can be used for preparation of an achiral biaryl framework in high yields prior to desymmetrization, and both procedures can be optimized independently.

Harada's group proved the feasibility of diastereotopos-differentiation of the *ortho*-substituents in 2,2′,6,6′-biphenyltetrol during acetalization with menthone[279,280] and etherification with either (2*S*,3*S*)-1,4-di-*O*-benzyl-L-threitol[281] or the bis(mesylate) of (*S*)-1,2-propanediol,[282] Scheme 6.29. Intermediate protection of one alcohol group of 1,4-di-*O*-benzyl-L-threitol with *tert*-butyldimethylsilyl chloride furnishes a monosilyl ether that undergoes Mitsunobu reaction with the achiral tetrol in the presence of diethyl azodicarboxylate. Deprotection of the silyl ether

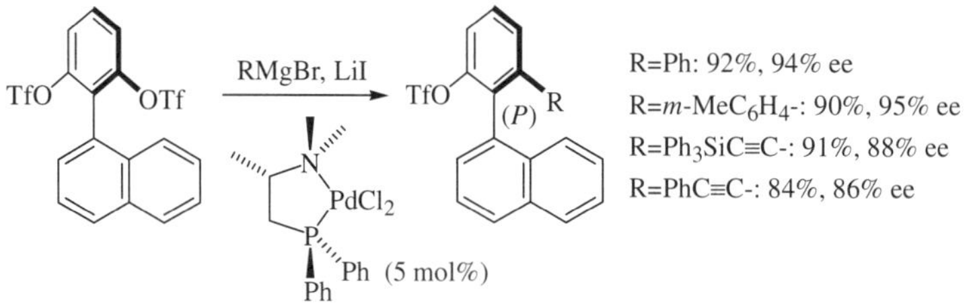

Scheme 6.30 Enantiotopos-selective transformation of a prochiral biaryl ditriflate.

followed by a second intramolecular Mitsunobu coupling using dimethyl azodicarboxylate then yields the diastereomerically pure C_2-symmetric (*P*)-biphenyl-2,2'-diol. It is assumed that the stereochemical outcome of the desymmetrization is determined during the Mitsunobu cyclization step. The S_N2-type transition state generating the (*M*)-atropisomer is disfavored due to destabilizing steric interactions between one phenyl ring and one benzyloxymethyl group of the chiral auxiliary.

An enantiotopos-selective monoarylation and alkynylation process with prochiral biaryl di-triflates and Grignard reagents has been developed by Hayashi.[283–285] The palladium-catalyzed desymmetrization proceeds with excellent enantioselectivity in the presence of an (*S*)-β-(dimethyl-amino)alkyldiphenylphosphine ligand, Scheme 6.30. The remaining triflate group can be further derivatized by another cross-coupling reaction, which underscores the usefulness of this method. Few other atropoenantiotopos-differentiating procedures have been reported to date and the synthetic potential of this approach remains to be fully explored.[286,287]

6.2.5 Asymmetric Transformation of Stereodynamic Biaryls

The challenges encountered with atroposelective biaryl synthesis have led to the development of alternative strategies that disconnect the aryl–aryl coupling step from the actual incorporation of axially chiral bias. This can be accomplished if a stereolabile biaryl scaffold is prepared first and then stereoselectively transformed to a conformationally stable atropisomer in a separate step.[ix] One advantage of this approach is that the synthesis of stereolabile biaryls possessing less steric hindrance than conformationally stable analogs usually proceeds with better yields. In addition, the stereoselective conversion of a fluxional biaryl to a conformationally stable atropisomer can be optimized separately.

The rotational energy barrier of biaryls can be increased by introduction of a rigid nonplanar bridge, and incorporation of two *ortho*-substituents of a stereolabile biaryl into such a chiral linkage has been successfully exploited for atropodiastereoselective synthesis.[288,289] A remarkable example of diastereoselective construction of a conformationally stable 5,7-fused bicyclic lactam from a stereolabile 2,2'-di-*ortho*-substituted biaryl precursor was reported by Levacher *et al.*, Scheme 6.31.[290] The nonisolable enantiomers of ethyl 2-acetylbiphenyl-2'-carboxylate undergo stereoselective oxazolidine formation with (*R*)-phenylglycinol. Lactamization of the rapidly inter-converting diastereomers then occurs preferentially from the side opposite to the phenyl and methyl groups of the oxazolidine unit via counterclockwise rotation of the aryl ester moiety about the chiral axis. Apparently, the methyl group of the *N,O*-ketal sterically hinders cyclization via clockwise rotation and thus controls the stereochemical outcome of the biaryl bridging. The rigid (*M*)-atropisomer is obtained in 95% yield and more than 95% de after refluxing ethyl 2-acetylbi-phenyl-2'-carboxylate and the chiral auxiliary in toluene for six days. A similar

---

[ix] This is often accomplished under kinetic reaction control. Alternatively, the ratio of atropodiastereoisomers that are susceptible to rotation about the chiral axis at elevated temperatures can be manipulated according to the principles of asymmetric transformation of the first kind (thermodynamic control).

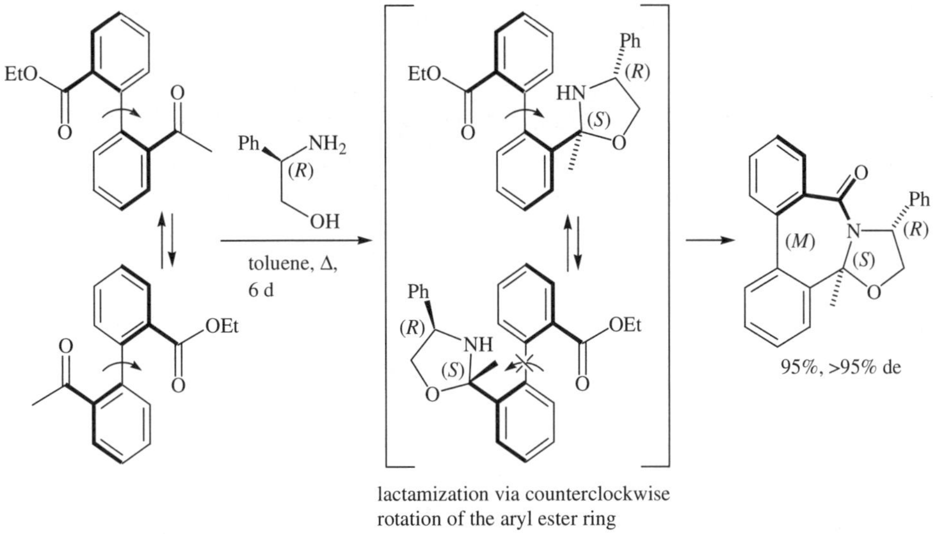

lactamization via counterclockwise
rotation of the aryl ester ring

Scheme 6.31 Atropodiastereoselective formation of a bridged biaryl using a chiral oxazolidine auxiliary.

atropodiastereoselective condensation of 2-formylbiphenyl-2′-carboxylic acid and (S)-valinol has been reported by Wallace.[291]

An intriguing asymmetric route towards conformationally stable 2,2′-bipyridine N,N′-dioxides from fluxional 2,2′-bipyridine precursors has been developed by Hayashi, Scheme 6.32.[292] Dynamic kinetic resolution of 3,3′-bis(hydroxymethyl)-6,6′-disubstituted-2,2′-bipyridines with (M)-2,2′-bis(chlorocarbonyl)-1,1′-binaphthalene affords the heterochiral diesters in good yield and up to 92% de. The bridged macrocycles are stable to diastereomerization at room temperature (activation energies range from 102 to 104 kJ/mol) which permits chromatographic isolation of homo- and heterochiral products. Subsequent oxidation with *m*-chloroperbenzoic acid and saponification yields enantiopure (P)-2,2′-bipyridine N,N′-dioxides. Alternatively, refluxing of the (M,P)-macrocycles in toluene for two days or treatment with trifluoroacetic acid results in asymmetric transformation of the first kind, generating the thermodynamically favored homochiral (M,M)-diester in diastereomerically pure form.[x,xi]

By analogy with the stabilization of interconverting axially chiral ligands by incorporation of a rigid bridge, coordination of a bidentate biaryl to a metal center can significantly enhance the rotational energy barrier. For example, 2,2′-bis(diarylphosphino)biphenyls such as BIPHEP and DM-BIPHEP undergo rapid rotation about the chiral axis at room temperature. However, the conformational stability of these diphosphines increases upon transition metal complexation. The addition of enantiopure diamines to ruthenium and rhodium complexes of BIPHEP and DM-BIPHEP has been reported to give atropodiastereoisomers that are stable to isomerization at room temperature. Heating of an initially equimolar mixture of diastereomeric ruthenium complexes of DM-BIPHEP and (S,S)-N,N′-dimethyl-1,2-diamino-1,2-diphenylethane (DM-DPEN) results in equilibration towards the thermodynamically favored (P)-atropisomer, Scheme 6.33.[293] According to NMR studies, the diastereomerically pure (S,S,P)-ruthenium complex is obtained by atropisomerization at 50 °C within one hour. This concept has been used by Mikami, Noyori, Gagné and others for deracemation of axially chiral 2,2′-bis(diarylphosphino)biphenyl-derived transition metal complexes in order to generate new enantioselective hydrogenation catalysts, see Chapter 6.6.2.

---

[x] This method provides enantiodivergent access to enantiopure (P)- and (M)-2,2′-bipyridine N,N′-dioxides which have been successfully employed as chiral Lewis base catalysts in the asymmetric allylation of aldehydes with allyltrichlorosilanes.
[xi] More examples of thermodynamically controlled atropodiastereoselective biaryl synthesis are given in Chapter 7.2.

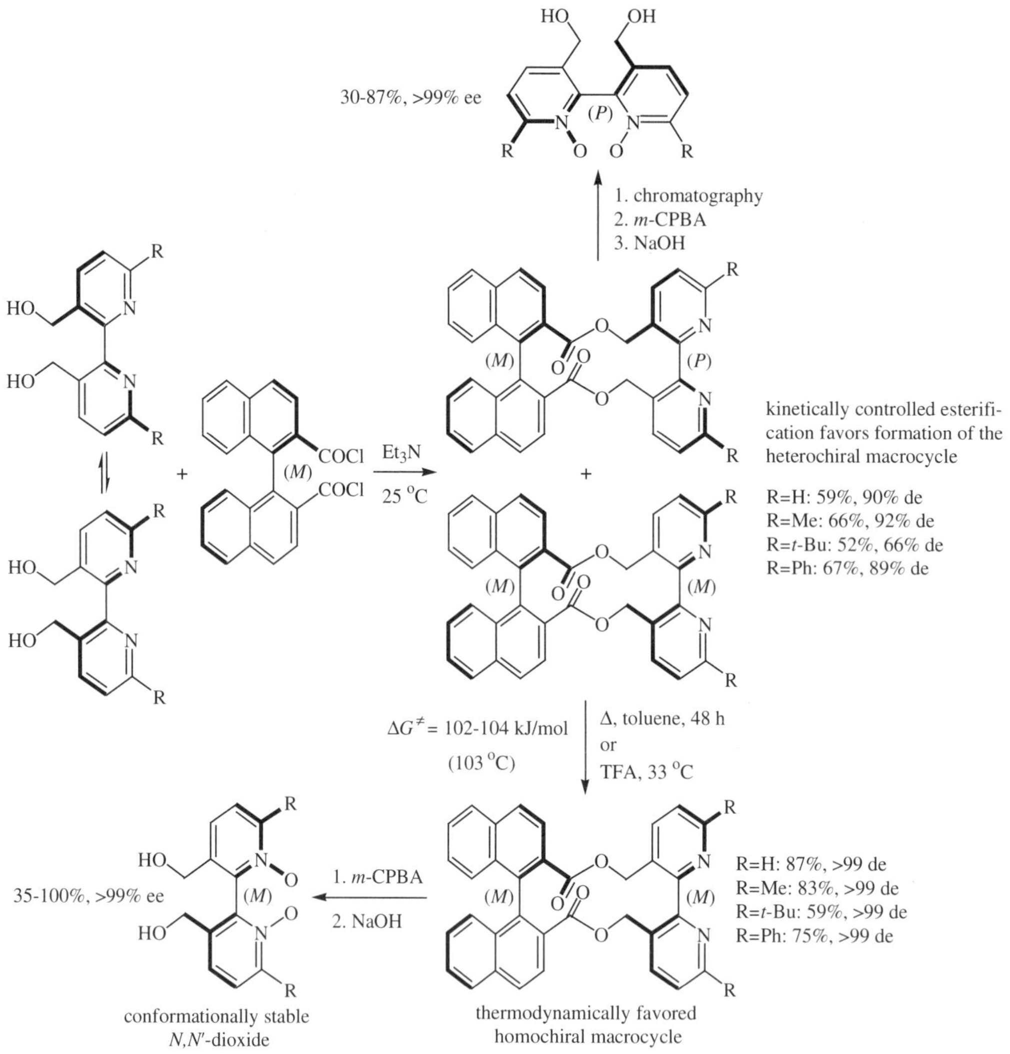

Scheme 6.32   Synthesis of enantiopure (*M*)-2,2′-bipyridine *N,N′*-dioxides from racemic stereolabile bipyridine precursors.

The conformational stability of 2,2′-bridged biaryls depends strongly on the ring size and bulkiness of the *ortho*-substituents. The majority of six-membered lactone-bridged biaryls shows rapid enantiomerization at room temperature. Bringmann *et al.* recognized the unique stereodynamics of these compounds and their general susceptibility to nucleophilic ring opening. They developed a broadly applicable atroposelective synthetic strategy based on dynamic kinetic resolution (DKR) of stereolabile lactone-bridged biaryls, Scheme 6.34.[294–296] In accordance with some of the methods discussed above, this so-called lactone concept disconnects the aryl–aryl coupling step from the actual incorporation of chiral bias which is accomplished separately by stereoselective transformation of interconverting biaryl enantiomers to a conformationally stable atropisomer. In the first step, esterification of a 2-bromobenzoic acid and a phenol derivative establishes a rigid scaffold that preorganizes the two aryl rings. Nonstereoselective cross-coupling with a palladium(0) complex then generates a stereolabile racemic biaryl lactone. The planar lactone bridge fulfills two purposes: it promotes intramolecular *C–C* coupling, and more

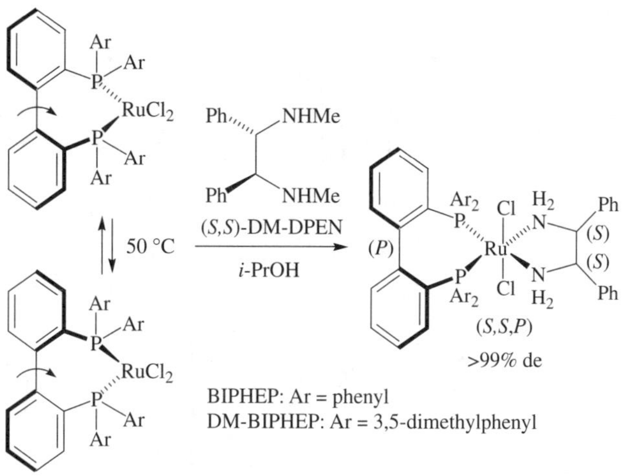

Scheme 6.33 Atropodiastereomerization of 2,2′-bis(diarylphosphino)biphenyl-derived ruthenium complexes in the presence of DM-DPEN.

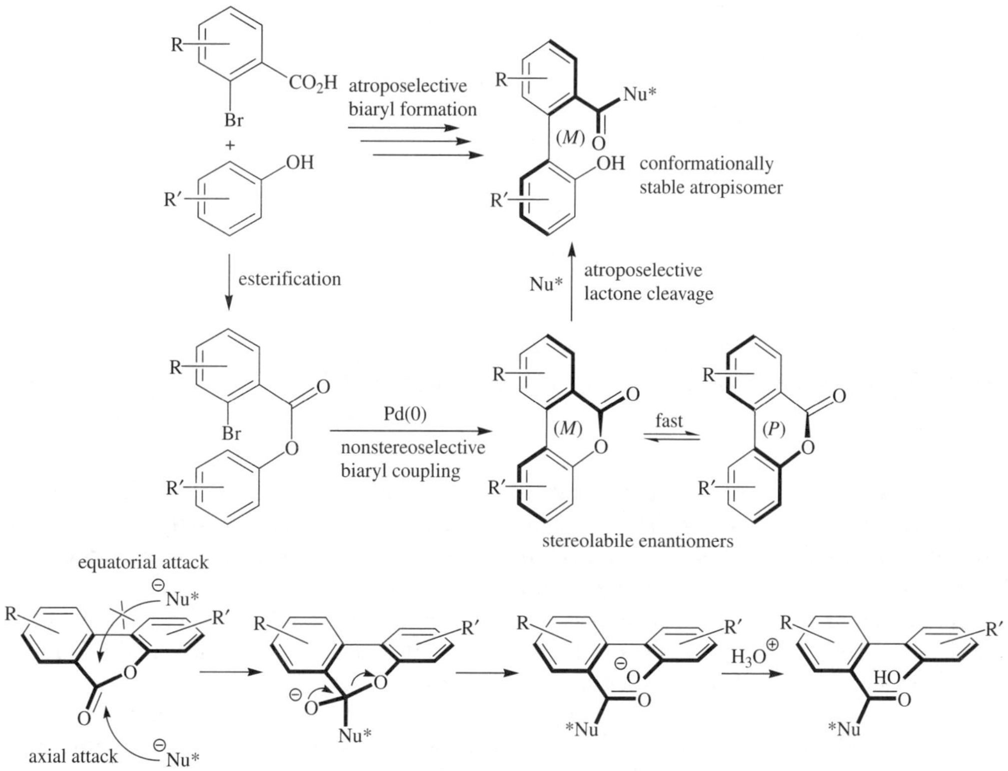

Scheme 6.34 Atroposelective biaryl synthesis based on the lactone concept (top) and axial attack of a chiral nucleophile at the lactone-bridged (*M*)-biaryl (bottom).

importantly it dramatically reduces the rotational barrier of the biaryl axis. In the final step, atroposelective lactone cleavage with a chiral nucleophile or with an achiral reagent in the presence of a chiral catalyst results in DKR and preferential formation of one conformationally stable biaryl atropisomer. According to quantum mechanical calculations, an axial approach of the nucleophile

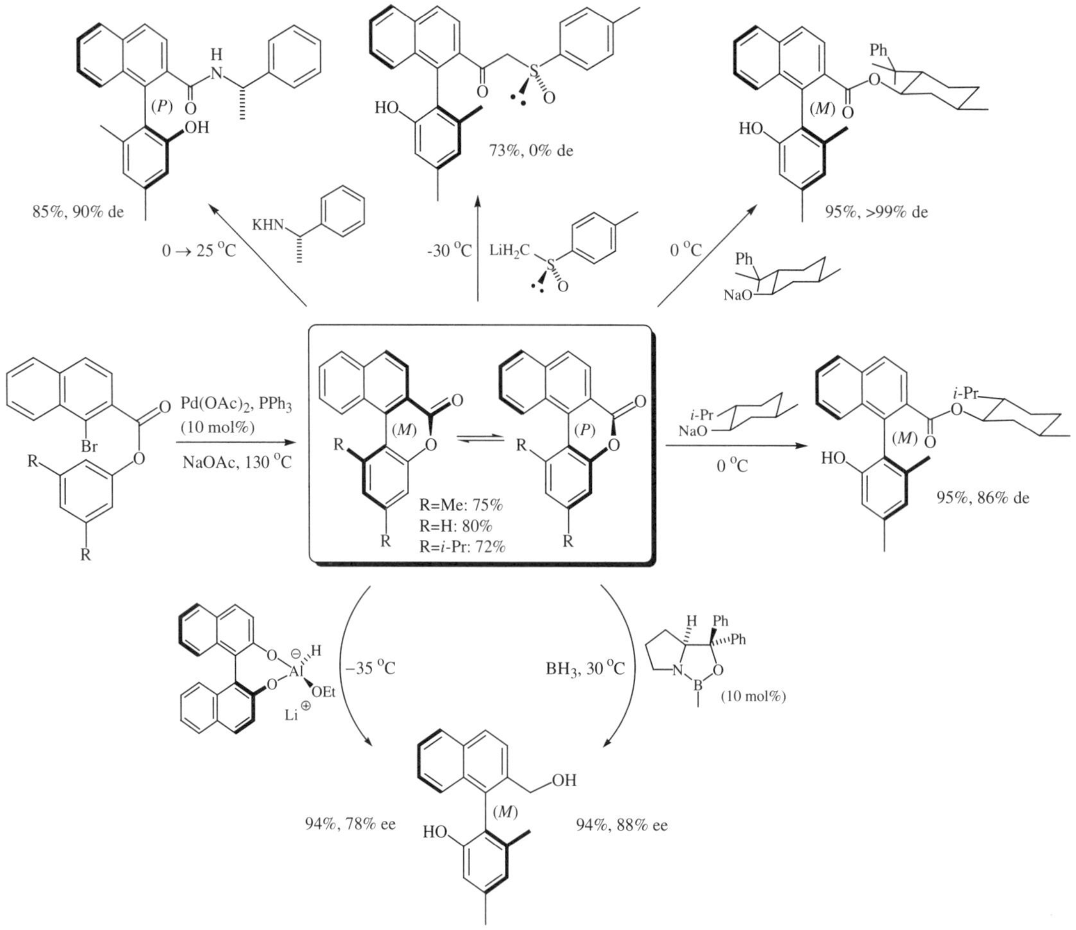

Scheme 6.35   Atroposelective ring opening of a lactone-bridged biaryl.

to the lactone unit is favored over an equatorial attack.[297,298] Selective addition of the nucleophile to the (*M*)-biaryl lactone produces a tetrahedral intermediate that forms the open (*M*)-biaryl phenoxide and the corresponding phenol after aqueous work-up.[xii] It is crucial that the nucleophilic ring opening proceeds with high stereoselectivity and incorporates sufficient steric bulk into the *ortho*-positions to establish conformational stability.

The scope of the lactone concept is demonstrated in Scheme 6.35.[299–304] Palladium-catalyzed intramolecular cross-coupling of esters prepared from 1-bromo-2-napthoic acid and 3,5-dimethyl-phenol or other phenols affords stereolabile 2,2′-bridged biaryl lactones.[305,306] Atroposelective lactone opening with potassium (*S*)-1-phenylethylamide gives the (*P*)-biaryl amide in 85% yield and 90% de. Employment of (*R*)-menthoxide and (*R*)-8-phenylmenthyloxide in the same enantio-mer-differentiating reaction favors formation of (*M*)-biaryls. Unfortunately, treatment of the lactone with a lithium sulfinyl carbanion generated from methyl 4-tolyl sulfoxide provides equal amounts of both diastereomers of the β-keto sulfoxide-derived biaryl. This has been attributed to fast equilibration of the 2-keto-2′-hydroxy-6′-methylbiaryl product rather than to lack of atropo-selectivity during lactone cleavage. Enantioselective reduction with either stoichiometric amounts

---

[xii] The stereochemical outcome is controlled by the chiral nucleophile or catalyst used. In the example shown in Scheme 6.34 the chiral nucleophile reacts more rapidly with the bridged (*M*)-biaryl.

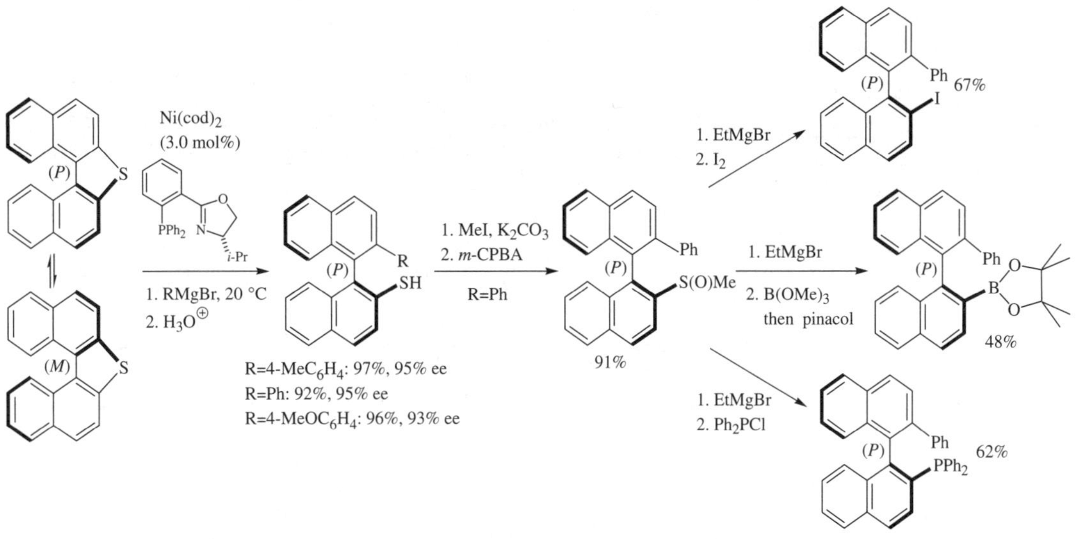

Scheme 6.36 Dynamic kinetic resolution of dinaphthothiophene with aryl Grignard reagents using a chiral oxazolidinylphenylphosphine-derived nickel catalyst.

of (*P*)-BINAL or borane in the presence of catalytic amounts of an (*S*)-2-methyl-CBS-oxazaborolidine produces the (*M*)-2-hydroxy-2'-hydroxymethylbiaryl in 94% yield and 78–88% ee. Because of the synthetic versatility and the mild conditions required for the atroposelective ring opening step, the lactone concept has found numerous applications. It has been applied in the synthesis of complex axially chiral biaryls including (–)-steganone,[307] (+)-isoschizandrin,[308] mastigophorene A and B,[309] knipholone,[310] korupensamine A,[311] dioncophylline C,[312] and other targets.[313–315] Atropodiastereoselective synthesis exploiting the chiral plane of (η^6-arene)ruthenium complexes of biaryl lactones for intramolecular stereocontrol during nucleophilic ring cleavage is a useful alternative.[316,317]

Shortening the length of a biaryl bridge considerably decreases conformational stability, and axially chiral biaryls having a five-membered bridge rapidly racemize at room temperature. Variable-temperature NMR analysis revealed that the rotational energy barrier of dinaphthothiophenes is less than 50 kJ/mol.[318] Applying the lactone concept to sulfur-bridged biaryls, Hayashi *et al.* introduced atropoenantioselective DKR of dinaphthothiophene and related dibenzothiophenes using Grignard reagents in the presence of a chiral nickel catalyst, Scheme 6.36. This procedure affords conformationally stable biaryl thiophenols in excellent yield and ee. Importantly, the mercapto group can easily be converted to the corresponding sulfoxide and then to a range of other synthetically versatile functionalities such as aryl iodides, boronates and phosphines.[319]

6.3 NONBIARYL ATROPISOMERS

Biaryls such as BINOL and BINAP are certainly the most widely recognized atropisomers but other axially chiral compounds that do not possess a biaryl framework have received increasing attention.[320] In particular, aromatic amides, imides and anilides have emerged as versatile chiral auxiliaries[321–343] and ligands for asymmetric catalysis.[344–347] Atropisomeric nonbiaryls have also been incorporated into molecular gears[348–350] and used as stereochemical relays, Figure 6.3.[351–357]

Some of the synthetic strategies described in the preceding sections have been applied to axially chiral nonbiaryls.[358–363] Uemura's group discovered that atroposelective synthesis of nonbiaryls is

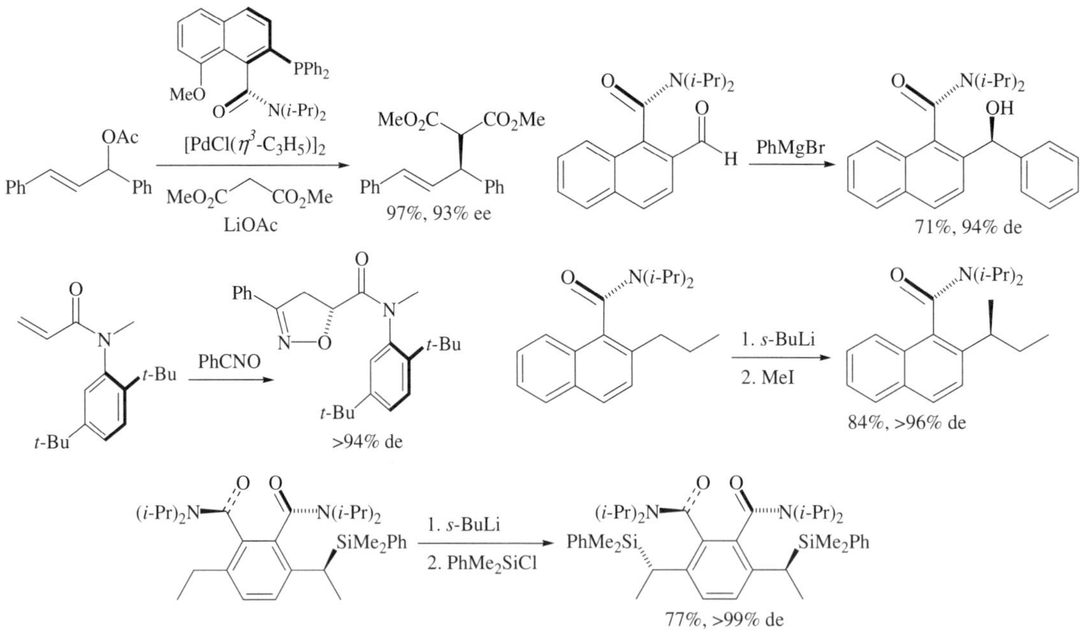

Figure 6.3   Examples of asymmetric synthesis utilizing nonbiaryl atropisomers as auxiliaries, catalyst ligands and relays.

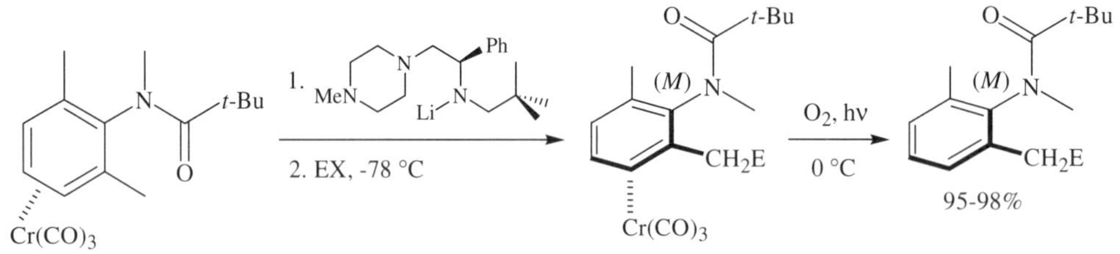

EX=MeI: 90%, 95% ee
EX=BnBr: 90%, 97% ee
EX=PhC≡CCH₂Br: 84%, 96% ee

Scheme 6.37   Desymmetrization of a prochiral anilide chromium complex.

feasible with planar chiral (η^6-arene)chromium complexes.[364,365] Selective desymmetrization of prochiral chromium complexes of aryl amides and anilides is also possible.[366,367] For example, enantiotopic deprotonation of tricarbonyl(*N*-methyl-*N*-pivaloyl-2,6-dimethylaniline)chromium with a chiral lithium amide and subsequent electrophilic trapping yields axially chiral anilides. The chromium moiety can be removed by mild oxidative photocleavage without concomitant racemization, Scheme 6.37. Taguchi *et al.* developed a catalytic asymmetric *N*-arylation procedure for the synthesis of atropisomeric anilides. A palladium-catalyzed nitrogen–carbon coupling step converts a stereolabile axially chiral 2-*tert*-butylanilide into the corresponding *N*-aryl atropisomer which apparently does not racemize under the reaction conditions. This dynamic kinetic resolution is applicable to inter- and intramolecular asymmetric C–N bond formation, Scheme 6.38.[368] Walsh *et al.* exploited a proline-catalyzed asymmetric aldol reaction for DKR of axially 2-formylbenzamides and 2-formylnaphthamides, described in detail in Chapter 7.4.2.[369]

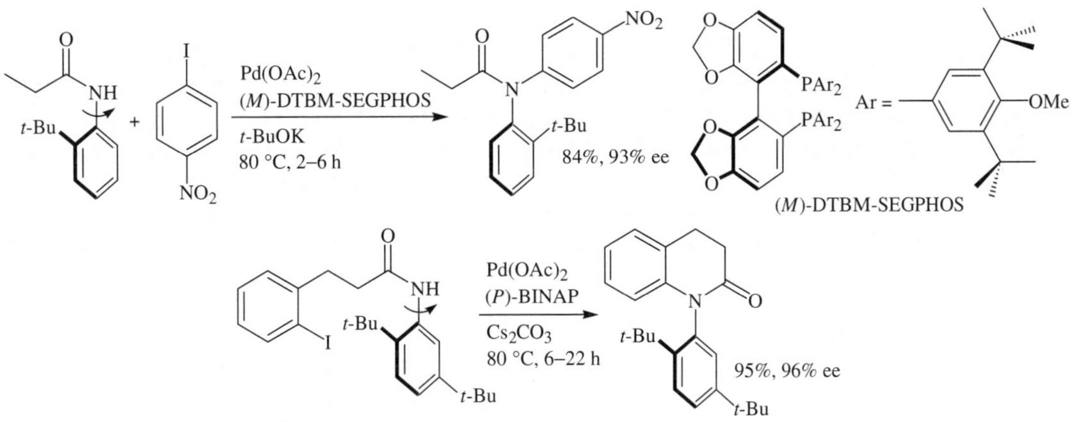

Scheme 6.38 Catalytic asymmetric *N*-arylation of anilides.

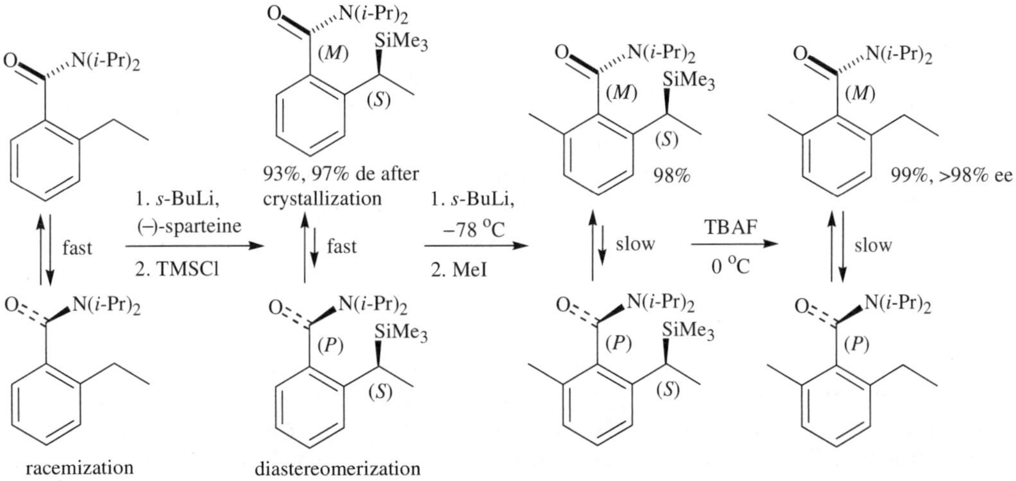

Scheme 6.39 Atropisomeric synthesis based on DTR of a stereolabile benzamide.

Clayden's group applied the principles of dynamic thermodynamic resolution (DTR) to atropisomeric synthesis of aryl amides, Chapter 7.6.[370] Asymmetric lithiation of rapidly racemizing *N,N*-diisopropyl-2-ethylbenzamide in the presence of (–)-sparteine followed by silylation produces the corresponding stereolabile (*S,M*)-trimethylsilyl atropisomer through asymmetric transformation of the second kind, Scheme 6.39. The presence of a chiral center in close proximity to the stereolabile chiral axis favors the *syn*-conformer which readily crystallizes from solution. This atropisomer is then trapped by *ortho*-lithiation and reaction with methyl iodide. If desirable, the TMS group can be removed with TBAF. Overall, resolution of *N,N*-diisopropyl-2-ethylbenzamide is achieved under thermodynamical control.[371]

6.4 CHIRALITY TRANSFER AND INTERCONVERSION OF CHIRAL ELEMENTS

The synthesis of chiral target molecules typically involves introduction of new stereogenic elements and manipulation of existing chirality, in particular when nonracemic starting materials from nature's chiral pool (carbohydrates, amino acids, α-hydroxy carboxylic acids, terpenes) are available. The displacement of a substituent attached to a chiral center of an enantiopure compound without concomitant racemization and diastereomerization is a key task in many asymmetric syntheses. This is routinely accomplished by bimolecular nucleophilic substitution when complete inversion of configuration is required. The stereochemical outcome of a stepwise replacement of a substituent attached to an asymmetric atom via an elimination-addition pathway, for instance an S_N1 reaction, is less predictable and partial racemization or diastereomerization is usually observed. Several strategies that allow manipulation of the absolute configuration at an asymmetric center including Seebach's concept of self-regeneration of stereocenters (SRS) are discussed in Chapter 6.5.

Nucleophilic substitutions may result in complete inversion or retention of the absolute configuration at an asymmetric carbon center but S_N1 and S_N2 reactions do not permit translocation of a stereocenter or conversion of central chirality to a chiral axis or another chiral element. Methods that enable chemists to relocate a chiral element or to transform a chiral element into another one have become powerful tools in asymmetric synthesis.[xiii] The term transfer of chirality has been coined for reactions in which at least one element of chirality is translocated from one site to another. Typically, this is an intramolecular process but the chiral element may also be exchanged between two molecules in an intermolecular reaction. In an intramolecular $[1,i]$-chirality transfer the stereogenic element is moved to a new site that is $(i-1)$ atoms away from its original position. By definition, transfer of chirality requires destruction of the original chiral element while it is formed somewhere else, preferably without compromising the stereochemical purity of the substrate.[xiv] This can often be achieved by stereospecific reactions such as S_N2' and S_E2' displacements or sigmatropic Cope, Wittig and Claisen rearrangements. Conceptually, interconversion of chirality elements is slightly different from chirality transfer, *i.e.*, the original chiral element is destroyed while *a different type* of chiral element is formed without loss of the stereochemical integrity of the molecule. Synthetic strategies based on chirality interconversion often provide access to non-racemic molecules that are difficult to obtain otherwise. Representative examples include conversion of central to axial chirality in asymmetric allene or atroposelective biaryl synthesis, *vide infra*. It should be noted that both concepts place emphasis on overall conservation of chirality, *i.e.*, the sterochemical purity of the product should not be compromised by simultaneous racemization or diastereomerization. The following chapters focus on reactions in which conservation of chirality is an inherent feature of chirality transfer or chiral element conversion rather than a result of stereocontrol due to the presence of other stereogenic elements within the substrate, or external control with chiral agents or catalysts. The term transfer of chirality is often inaccurately used to describe asymmetric induction during the course of a stereoselective reaction, for example enantioselective catalysis or diastereoselective synthesis. In these cases, the chiral element of a catalyst, reagent or substrate is not simultaneously destroyed and translocated, but remains intact while it is exploited for preferential formation of one stereoisomer over others.

6.4.1   Chirality Transfer in S_N2' and S_E2' Reactions

A bimolecular nucleophilic substitution at an allylic substrate involves either an S_N2 or an S_N2' mechanism.[372] An S_N2 reaction entails a nucleophilic displacement at the tetrahedral carbon without participation of the adjacent double bond. Alternatively, the nucleophile can approach the more distant sp^2-hybridized carbon atom in the alkene moiety and replace the leaving group through a simultaneous allylic rearrangement, Scheme 6.40. This S_N2' pathway prevails when a

---

[xiii] Since the original element of chirality is destroyed in these reactions, they are sometimes referred to as self-immolative processes.
[xiv] It is important to emphasize that only an *element of chirality* is destroyed at a specific site and simultaneously regenerated at another location, but the chirality of the whole molecule is maintained at all times.

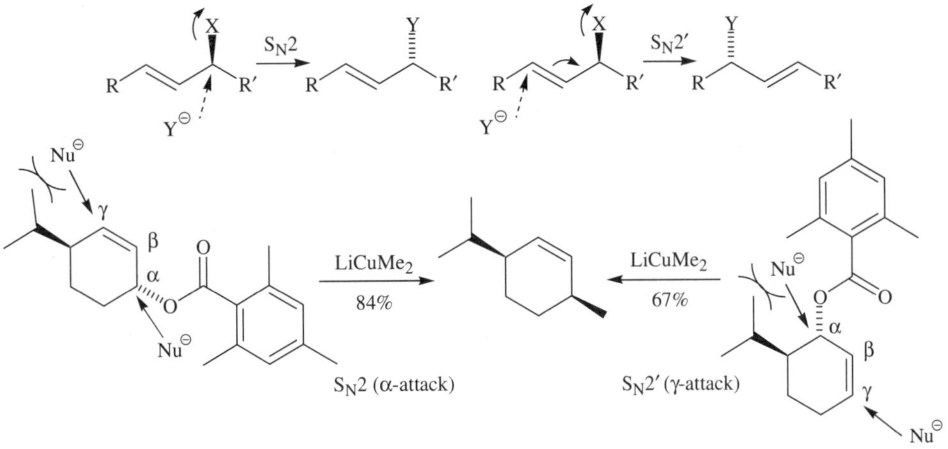

Scheme 6.40 Nucleophilic displacements at allylic substrates via S_N2 and S_N2' pathways.

direct attack at the tetrahedral center is sterically hindered and in transition metal-mediated allylic displacements, particularly when cyanocuprates are used. Asymmetric allylic substitutions commonly include the use of chiral catalysts,[373–383] chiral auxiliaries,[384–388] or nonracemic allylic electrophiles that undergo S_N2' reaction and simultaneous [1,3]-chirality transfer. Since the latter approach offers excellent regio- and stereocontrol, S_N2' reactions proceeding with high or even complete conservation of chirality have found widespread use in natural product synthesis.[389,390]

The ratio of the competing S_N2 and S_N2' pathways is generally determined by the structures of the substrate, the leaving group and the nucleophile. The presence of bulky groups at the γ-terminus impedes S_N2' attack and therefore favors S_N2 reaction whereas steric hindrance at the α-carbon favors the S_N2' pathway. Organocuprates, which are commonly used as nucleophiles in both S_N2 and S_N2' reactions, are known to coordinate strongly to the nitrogen in carbamoyl groups. This can give rise to leaving group-directed S_N2' displacement even in the presence of steric hindrance at the γ-position. In addition to regioselective assistance of a γ-attack, carbamate leaving groups affect the stereochemical outcome of the S_N2' reaction. Complexation of the cuprate generally directs the nucleophile to the same side of the leaving group, thus promoting *syn*-selectivity. Other leaving groups such as carboxylates, halides and sulfonates are usually replaced by an *anti*-S_N2' mechanism in which the nucleophile attacks the double bond at the side opposite to the leaving group.

As mentioned above, *anti*-S_N2' reactions between cuprates and readily available nonracemic allylic alcohol derivatives provide a powerful means for asymmetric carbon–carbon bond formation.[391–401] Major advantages of this transformation are excellent regio- and stereoselectivity and predictability of the stereochemical outcome of the [1,3]-chirality transfer. The cuprates can be generated *in situ* from Grignard, organolithium and organozinc reagents by transmetalation with copper salts, Scheme 6.41.[402,403] Yamamoto *et al.* were among the first to realize the potential of allylic substitutions for efficient generation of chiral quaternary carbon centers. Employing organocuprates in the reaction with enantiopure (*R*)-(*E*)-α-alkyl-γ-mesyloxy-α,β-enoates under cryogenic conditions they obtained (*R*)-(*E*)-α,α-dialkyl-β,γ-enoates in excellent yield and ee. Comparison of the most likely conformations suggests that allylic $A^{1,3}$ strain disfavors an arrangement that would lead to (*S*)-(*Z*)-isomers. It is assumed that the nucleophilic attack preferentially occurs at the major conformational isomer, thus producing the (*R*)-(*E*)-isomer by *anti*-S_N2' reaction with almost quantitative [1,3]-chirality transfer.

The *syn/anti*-selectivity of S_N2' reactions that fulfill the criteria of true chirality transfer, *i.e.*, the stereochemical outcome is not controlled by a chiral catalyst or auxiliary, is mostly determined by

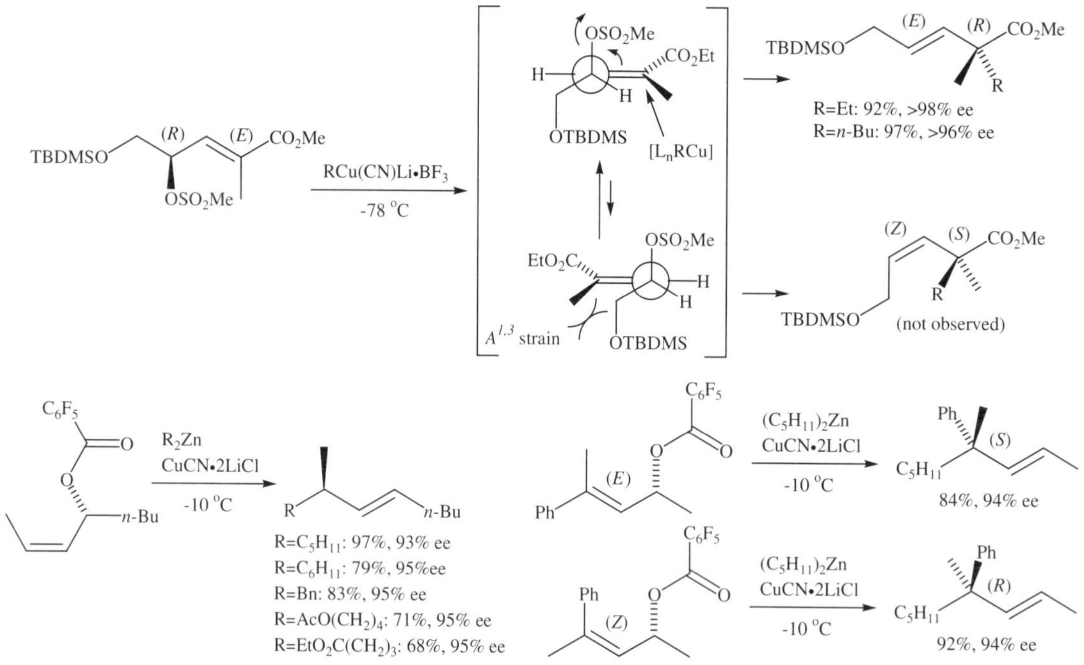

Scheme 6.41 Examples of *anti*-S$_N$2′ reactions.

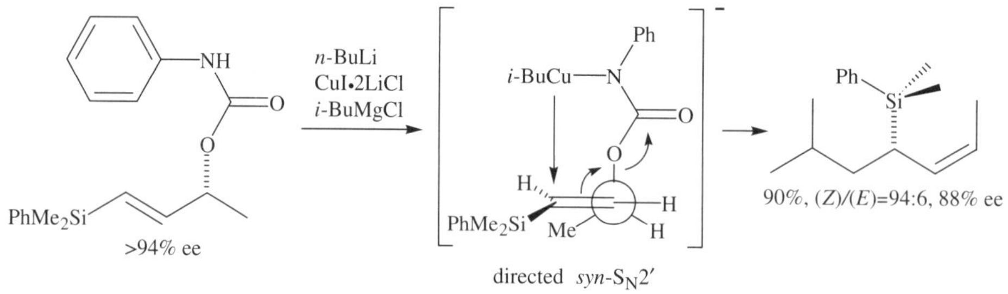

Scheme 6.42 Leaving group-directed alkylation of an allylic carbamate.

the substrate structure. Although *anti*-attack is quite common, the presence of bulky groups at the γ-terminus can significantly reduce the stereoselectivity of nucleophilic displacements.[404] To overcome this limitation, and when *syn*-substitution is desirable, nucleophile-directing leaving groups such as carbamates and benzothiazoles are used. The directing effect of these leaving groups is based on the affinity of organocopper species to the carbamoyl nitrogen or other heteroatoms. Formation of an intermediate complex between the cuprate and the allylic carbamate prior to the S$_N$2′ displacement governs the stereofacial attack of the nucleophile which approaches the γ-carbon atom from the side of the leaving group. Coordination of the cuprate to the carbamoyl nitrogen therefore leads to highly stereo- and regioselective *syn*-S$_N$2′ displacements, Scheme 6.42.[405–409]

Introduction of directing *ortho*-diarylphosphanylbenzoate leaving groups to copper-mediated S$_N$2′ reactions offers excellent stereocontrol, Scheme 6.43.[410] Coordination of the phosphine ligand to a cuprate prepared from a Grignard reagent and copper(I) bromide directs the nucleophlic attack to the *syn*-position with respect to the leaving group. The allylic substitution then proceeds with high stereoselectivity and regioselectivity (S$_N$2′/S$_N$2 > 99:1). The stereochemical outcome of the

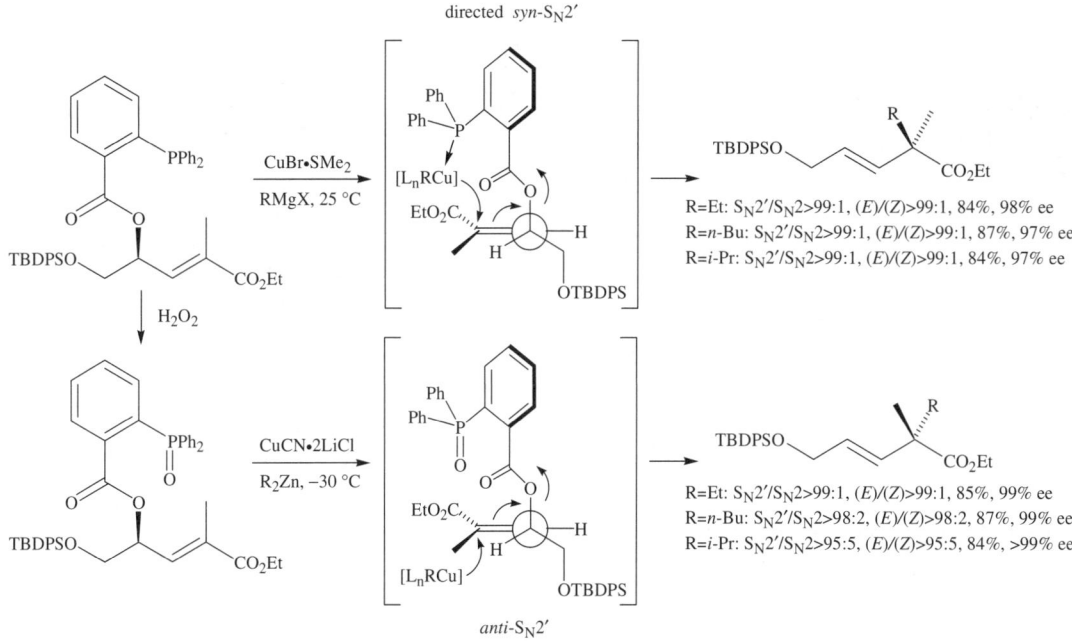

Scheme 6.43 Stereodivergent *syn-* and *anti-*allylic substitution using either a directing or a nondirecting leaving group.

[1,3]-chirality transfer can be reversed through oxidation of the directing group to a noncoordinating phosphine oxide. In the absence of a coordinating phosphine ligand, cuprates generated from organozinc reagents preferentially approach the alkene from the side opposite to the leaving benzoyl unit to minimize steric repulsion. This *anti-*S_N2' pathway therefore produces the opposite enantiomer. The results demonstrate how incorporation of a switchable directing/nondirecting leaving group affords stereodivergent access to multi-functionalized compounds with a quaternary chiral center.

Palladium-catalyzed allylic substitution is an important alternative to leaving group-directed *syn-*S_N2' reactions if racemization can be ruled out.[411] Initial formation of a cationic π-allyl palladium complex from a nonracemic allyl acetate and subsequent nucleophilic attack can proceed via a double-inversion mechanism with overall retention of configuration.[412,413] For example, allylic disilanyl ethers undergo palladium-catalyzed intramolecular disilylation followed by Peterson-type *syn-*elimination in the presence of a strong base.[xv] This reaction sequence involves formation of an intermediate four-membered *trans-*siloxane ring and yields chiral allylsilanes with almost perfect [1,3]-chirality transfer, Scheme 6.44.[414,415] Asymmetric alkylation of stereolabile allyl palladium complexes showing facile η^3-η^1-η^3-migration from one heterotopic face of the coordinating π-system to the other can be accomplished through dynamic kinetic asymmetric transformation, see Chapter 7.5.[416]

Relatively few examples of efficient asymmetric S_N2' reactions with heteroatom nucleophiles are known.[417,418] A noteworthy exception is the regioselective and stereospecific formation of Cbz-protected allylic amines by zirconium-mediated *syn-*S_N2' displacement of electron-rich allylic (trimethylsilyl)methyl ethers, Scheme 6.45.[419] The stereochemical outcome of this transformation has been explained by a chairlike transition state that accounts for the observed *syn-*selectivity, kinetic measurements and the well-known oxophilicity of zirconium imido complexes.

[xv] The *syn-*elimination occurs spontaneously but full conversion is observed upon addition of a base.

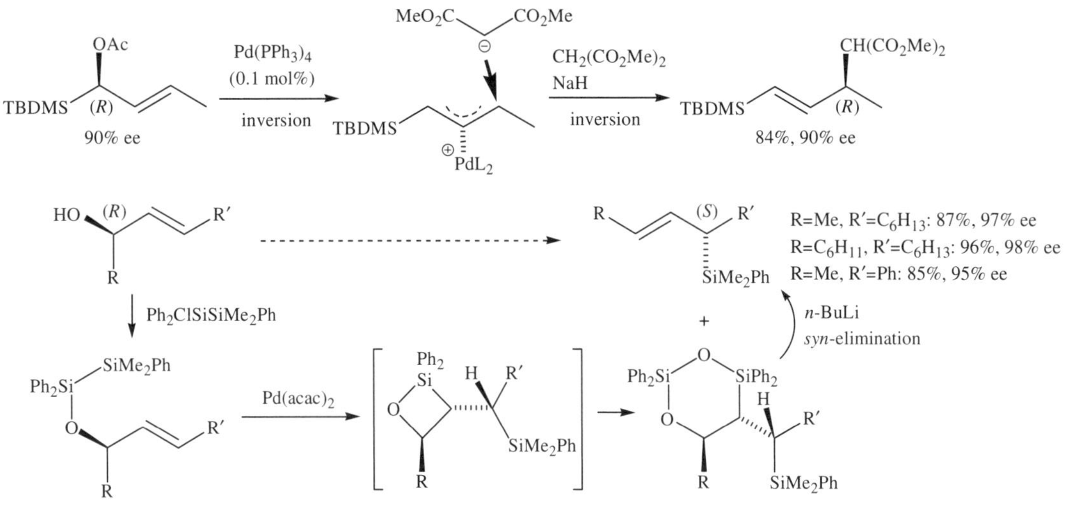

Scheme 6.44 Palladium-catalyzed alkylation of an allyl acetate with complete [1,3]-chirality transfer (top) and formation of allylsilanes through intramolecular disilylation of allyl alcohol derivatives (bottom).

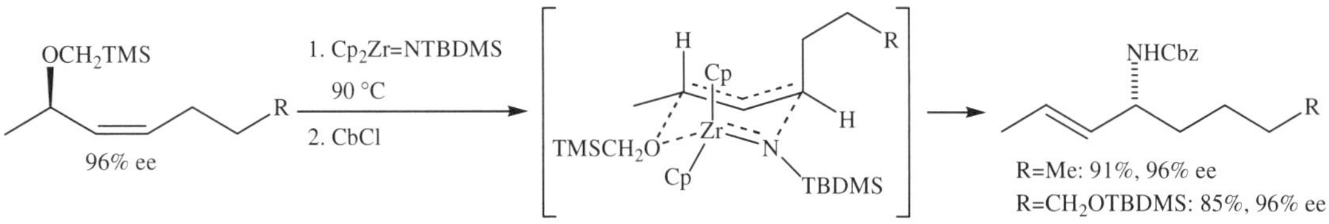

Scheme 6.45 Zirconium-mediated S_N2' displacement of allylic ethers.

Coordination of the (trimethylsilyl)methoxy functionality to the Lewis acidic zirconium complex generates a more effective leaving group and guides the nucleophilic imido moiety to the γ-position of the allylic system. Formation and cleavage of the two carbon–heteroatom bonds occur simultaneously at the same side of the allylic system.

Chirality transfer is a common feature of electrophilic substitutions such as S_E2' displacement of allylsilanes.[420] An interesting case is the Lewis acid-promoted electrophilic attack of an adamantyl cation at the *cis*- and *trans*-isomers of (R)-2-trimethylsilyl-3-pentene.[421] Electrophilic displacement at the (Z)-allylsilane gives (S)-(E)-2-adamantyl-3-pentene as the major product, but reaction of the (E)-isomer affords a mixture of three stereoisomers, Scheme 6.46. Stereochemical analysis of the products obtained from (R)-(E)-2-trimethylsilyl-3-pentene showed that (S)-(Z)-2-adamantyl-3-pentene is formed in 60% yield while the remaining 40% consist of a nonracemic mixture of the (E)-alkene. Since electrophilic substitutions often involve cationic intermediates, this has been attributed to *anti*-attack of the adamantyl cation at two preferentially populated conformations having either the allylic methyl or the hydrogen in the plane of the double bond. Addition to the former conformation gives a cationic intermediate that undergoes elimination to the major (S)-(Z)-diastereomer. Rotation about the single bond connecting the cationic carbon and the trimethylsilyl-derived group apparently occurs on the same time scale and produces the minor (S)-enantiomer of the *trans*-product after elimination. Electrophilic *anti*-addition to the other major conformer of the starting material displaying the allylic hydrogen in the plane of the double bond leads to the (R)-enantiomer of (E)-2-adamantyl-3-pentene. The corresponding cationic

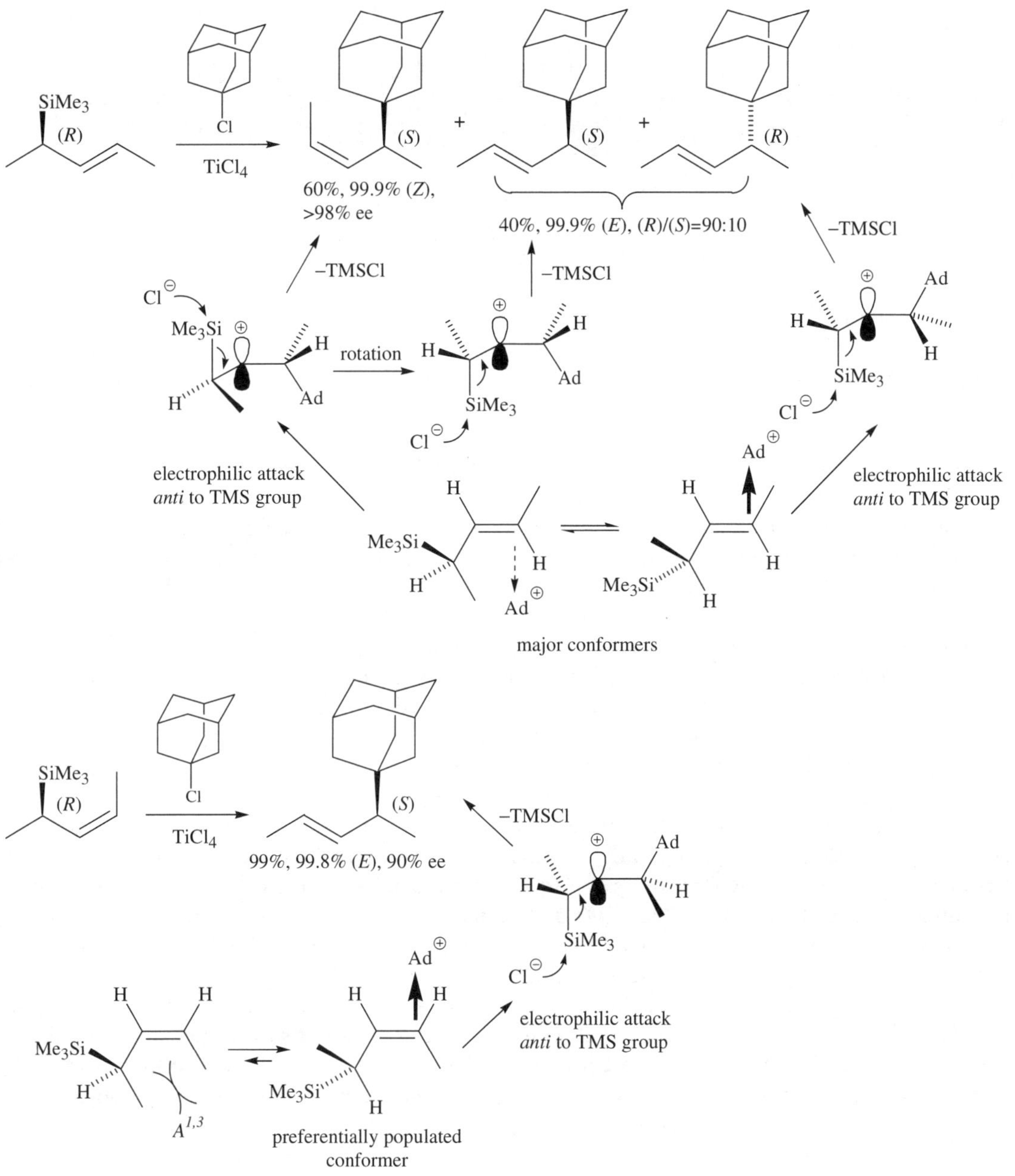

Scheme 6.46 Electrophilic *anti*-substitution of allylsilanes using adamantyl chloride and titanium tetrachloride.

intermediate occupies a relatively stable conformation with the bulky groups on opposite sides and is therefore less prone to rotation prior to elimination. In contrast, *anti*-S_E2' displacement of (R)-(Z)-2-trimethylsilyl-3-pentene furnishes only one major product. In this case, $A^{1,3}$ strain between the two methyl groups disfavors one conformer and electrophilic attack occurs preferentially at the major species bearing the hydrogen in the allylic plane. The cationic intermediate resides in the favored conformation during elimination and (S)-(E)-2-adamantyl-3-pentene is obtained almost exclusively.

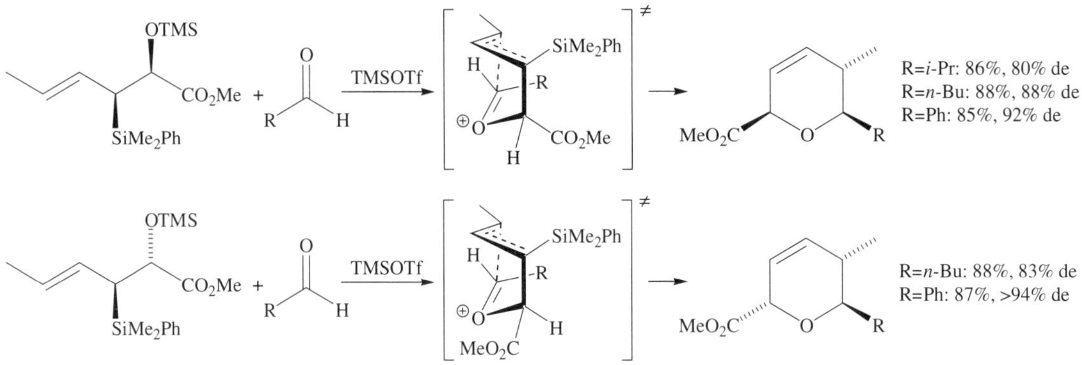

Scheme 6.47 Asymmetric synthesis of *cis*-2,6- and *trans*-2,6-dihydropyrans via intramolecular *anti*-S_E' reaction.

Lewis acid-catalyzed electrophilic substitutions at allylsilanes or allylstannanes with aldehydes are more complicated but usually follow an *anti*-S_E' mechanism.[422–428] The asymmetric synthesis of 2,6-disubstituted dihydropyrans and dihydropyridines from chiral (*E*)-crotylsilanes, such as methyl (2*S*,3*S*)-(4*E*)-3-(dimethylphenylsilyl)-2-(trimethylsiloxy)-4-hexenoate or its 2-amino derivative, and aldehydes formally constitutes a [4 + 2]-annulation but in fact involves an *anti*-S_E' pathway, Scheme 6.47.[429–431] It is generally assumed that trimethylsilyl triflate, a strong *O*-silylating agent, promotes the formation of an intermediate oxocarbenium ion that undergoes cyclization through a boatlike transition state having the dimethylphenylsilyl group in pseudoaxial orientation. An intramolecular *anti*-S_E' process then yields the *cis*-2,6-dihydropyran as the major diastereoisomer. Diastereomeric *trans*-2,6-dihydropyrans are obtained when (2*R*,3*S*)-hexenoates are applied to the formal [4 + 2]-annulation. By analogy with reactions with oxonium ions, chiral crotylsilanes show *anti*-S_E' displacement and simultaneous [1,3]-chirality transfer with thionium ions, which can be generated from dithioacetals in the presence of a Lewis acid, thus forming homoallylic thioethers.[432] The formal [4 + 2]-annulation of chiral crotylsilanes and electrophilic oxonium derivatives has been exploited by Panek's group for the total synthesis of a range of natural products including (+)-discodermolide,[433] the *C19-C28* fragment of phorboxazoles,[434] the *C1-C22* fragment of leucascandrolide A,[435,436] the *C1-C13* fragment of bistramide A,[437] the *C1a-C10* fragment of kendomycin,[438] methyl L-callipeltose,[439] and (–)-apicularen A.[440]

In some cases, S_N2' and S_E2' reactions proceed with simultaneous interconversion of chirality elements. Conversion of central to axial chirality via S_N2' reaction of substituted allyl and alkynyl substrates provides a unique entry to chiral allenes. On the other hand, nonracemic propargylic compounds can be obtained by allylic displacements from axially chiral allenes. These transformations are discussed in Chapters 6.4.5 and 6.4.6.

6.4.2 Rearrangements

Rearrangement reactions resulting in transfer of chirality with predictable stereochemical outcome are extremely useful for the construction of complex chiral frameworks. In particular, [2,3]-Wittig, [3,3]-Cope, [3,3]-Claisen and other stereospecific sigmatropic migrations with a highly ordered transition state have found numerous applications. Similarly to cycloadditions and electrocyclic reactions, sigmatropic rearrangements are pericyclic reactions which have been defined as processes "in which all first-order changes in bonding relationships take place in concert on a closed curve."[441] A sigmatropic [*i,j*]-rearrangement involves a shift of a σ-bond (carrying a substituent) across a system of conjugated double bonds to another site whose termini are (*i*–1) and (*j*–1) atoms away from the original location, Scheme 6.48.[442,443]

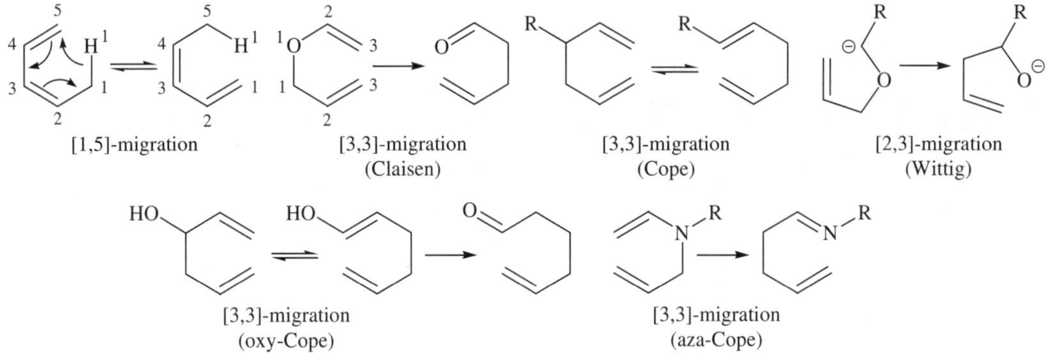

Scheme 6.48 Examples of sigmatropic [*i,j*]-rearrangements.

Frontier molecular orbitals and Woodward–Hoffmann rules have to be considered to determine whether a thermally or photochemically initiated rearrangement with $(4q+2)$ suprafacial and $4r$ antarafacial components is symmetry-allowed. With regard to the π-system, a rearrangement is considered suprafacial if the migrating group remains at the same side of the conjugated system. It is antarafacial if the bonds are made and cleaved on opposite sides. From the perspective of the migrating group, the process is suprafacial (antarafacial) if it moves with retention (inversion) of configuration. In general, thermal sigmatropic rearrangements involving $(4N+2)$ electrons are symmetry-allowed if both components react in a suprafacial fashion, and processes involving $4N$ electrons are less frequently observed because they require one antarafacial component. Both geometric and orbital symmetry constraints have stringent implications for reactivity and stereochemical outcome of sigmatropic rearrangements which often proceed with excellent conservation of chirality.

6.4.2.1 1,2-Chirality Transfer. Suprafacial [1,2]-shifts of an alkyl group to an adjacent carbocation are symmetry-allowed and show retention of configuration at the migrating atom, but suprafacial [1,2]-shifts to a carbanionic center such as the [1,2]-Wittig rearrangement are symmetry-forbidden and entail a stepwise mechanism.[444] The Wagner–Meerwein rearrangement has been identified as a synthetically useful stereospecific [1,2]-shift with predictable stereochemical outcome.[445–447] Various examples exhibiting [1,2]-chirality transfer play an important role in terpene chemistry.[448–450] Martinez described a highly enantiospecific electrophile-promoted Wagner–Meerwein rearrangement of a methylenenorbornanol derivative to enantiopure camphor

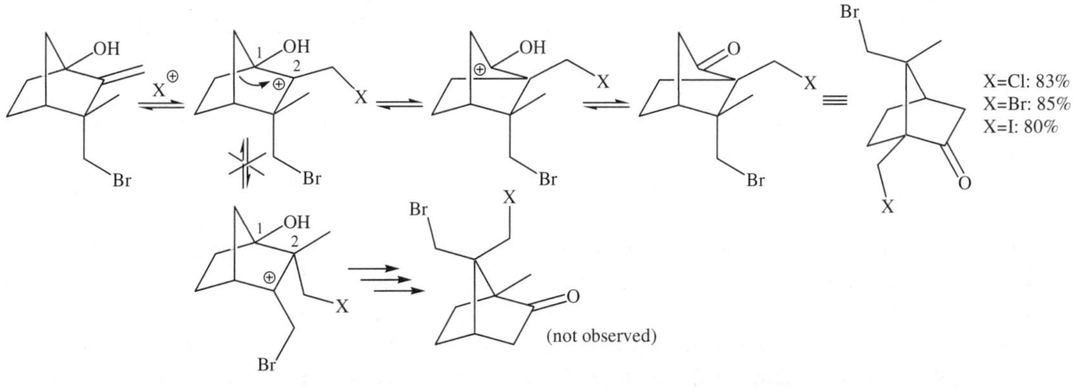

Scheme 6.49 Wagner–Meerwein rearrangement and [1,2]-chirality transfer.

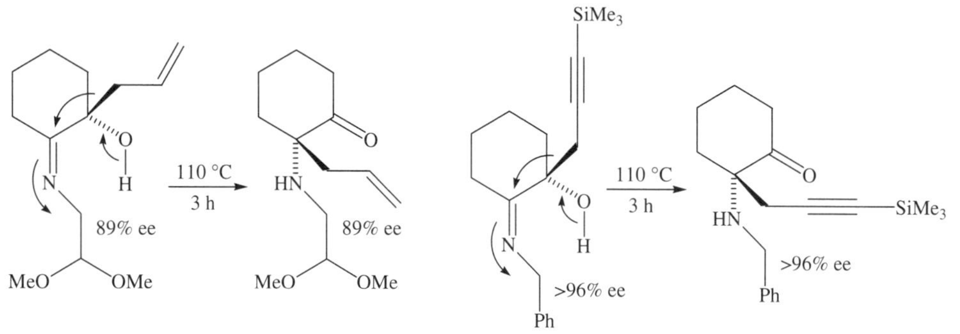

Scheme 6.50   [1,2]-Chirality transfer with α-hydroxy imines.

analogs, Scheme 6.49.[451] Interestingly, *exo*-migration of the neighbouring methyl group, the so-called Nametkin rearrangement, was not observed.

Thermal [1,2]-rearrangements of chiral α-hydroxy imines produce tertiary α-amino ketones.[452] This reaction includes migration of an allyl or propargyl group from an asymmetric α-carbon center to the *sp*²-hybridized carbon of an adjacent imine function. In many cases, the [1,2]-chirality transfer proceeds with complete conservation of chirality, Scheme 6.50. Palladium-catalyzed arylation of chiral allyl alcohols and subsequent [1,2]-chirality transfer has been reported to yield α-substituted ketones, albeit in low ee.[453] Nitrogen-to-carbon chirality transfer can be achieved by Stevens rearrangements and other [1,2]-shifts, see Chapter 6.4.4.

6.4.2.2   *1,3-Chirality Transfer.*

Electrophilic and nucleophilic allylic substitutions are powerful reactions because they combine synthetically useful carbon–carbon or carbon–heteroatom bond formation with [1,3]-chirality transfer, see Chapter 6.4.1. Sigmatropic rearrangements provide an equally viable alternative for stereospecific translocation of a chiral element across an allylic system. Among [3,3]-sigmatropic processes, the Claisen rearrangement is more synthetically useful than the Cope equivalent which is reversible and requires higher reaction temperatures.[454] Problems associated with the reversibility of Cope rearrangements can be overcome through introduction of a well-defined driving force that shifts the equilibrium to the desired product. This has been accomplished by incorporation of ring strain that is released upon rearrangement, ion stabilization in oxy- and aza-Cope reactions, and coupling of the Cope process with other reactions, for example in aza-Cope/Mannich tandem reactions. The high stereospecificity inherent to asymmetric Cope and Claisen rearrangements is generally explained by a chairlike transition state that is more stable than the corresponding boatlike conformation exhibiting destabilizing secondary orbital interactions, Figure 6.4. Because of the different geometry, antibonding orbital interactions are not possible in the energetically favored chairlike transition state. However, both reaction pathways are allowed by the Woodward–Hoffmann rules and can lead to the formation of diastereoisomeric mixtures.

The stereochemical outcome of sigmatropic rearrangements is usually highly stereospecific and predictable, and is often independent of remote chiral elements located outside the pericyclic system of the substrate. To predict the sense of chirality transfer one has to compare the relative stability of coexisting transition states. This requires consideration of the absolute configuration at chiral centers residing within the pericyclic system and the double bond geometry. For example, Claisen rearrangement of *cis*- and *trans*-isomers of (4*S*)-allyl vinyl ethers comprises destruction of the chiral center at *C*-4 and simultaneous formation of an asymmetric carbon in position 6. As a result of the strong preference for a chairlike transition state with equatorial substituents, the (*E*)-allyl vinyl

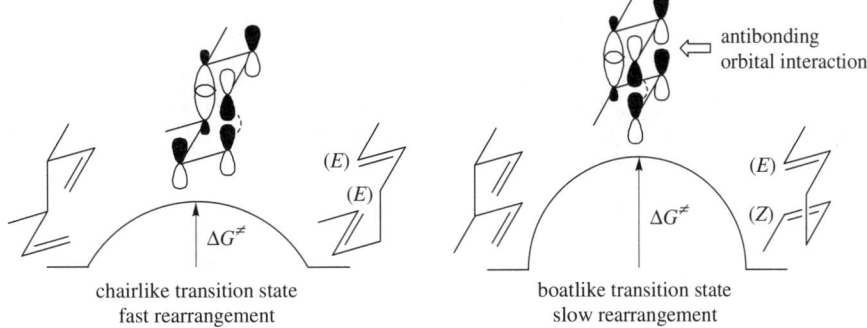

Figure 6.4   Chair- and boatlike transition states of [3,3]-sigmatropic rearrangements.

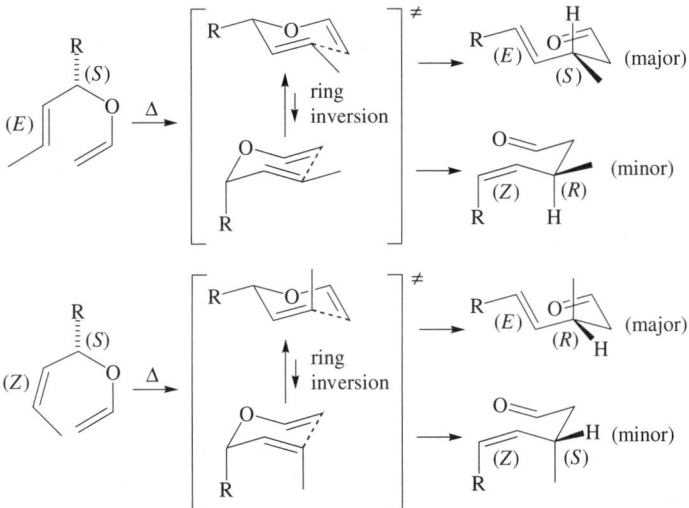

Scheme 6.51   Chairlike transition state geometries in Claisen rearrangements.

ether typically forms the (S)-(E)-product and the (Z)-allyl vinyl ether rearranges to the corresponding (R)-(E)-isomer, Scheme 6.51.

Asymmetric Claisen rearrangements with [1,3]-chirality transfer have been incorporated into total syntheses of challenging natural products including lentiginosine,[455] latifoline,[456] blasticidin S,[457] and others. The principles of the Claisen rearrangement have also been extended to aza- and thia-variations, and even [3,3]-sigmatropic allyl cyanate-to-isocyanate transformations have been reported, Scheme 6.52.[458–468]

The preparation of the allyl vinyl ether framework necessary for Claisen rearrangement and concurrent [1,3]-chirality transfer can be achieved with chiral nonracemic allylic alcohols and different enol trapping reagents. Orthoesters (Johnson variation),[469–474] N,N-dimethylacetamide dimethyl acetal (Eschenmoser variation),[475–477] and trimethylsilyl chloride (Ireland variation)[478–480] are frequently used for generation of reactive ketene acetal intermediates. These and other Claisen modifications have been incorporated into many natural product syntheses. For example, Heathcock *et al.* utilized both Eschenmoser's and Johnson's variation to convert (2S,3S)-(E)-2-methyl-1-(triphenylmethoxy)-4-hexen-3-ol to the corresponding (3S,6R)-(E)-2-heptenoate amide and ester via *in situ* formation of ketene acetal derivatives at 135–140 °C, Scheme 6.53.[481] The sigmatropic

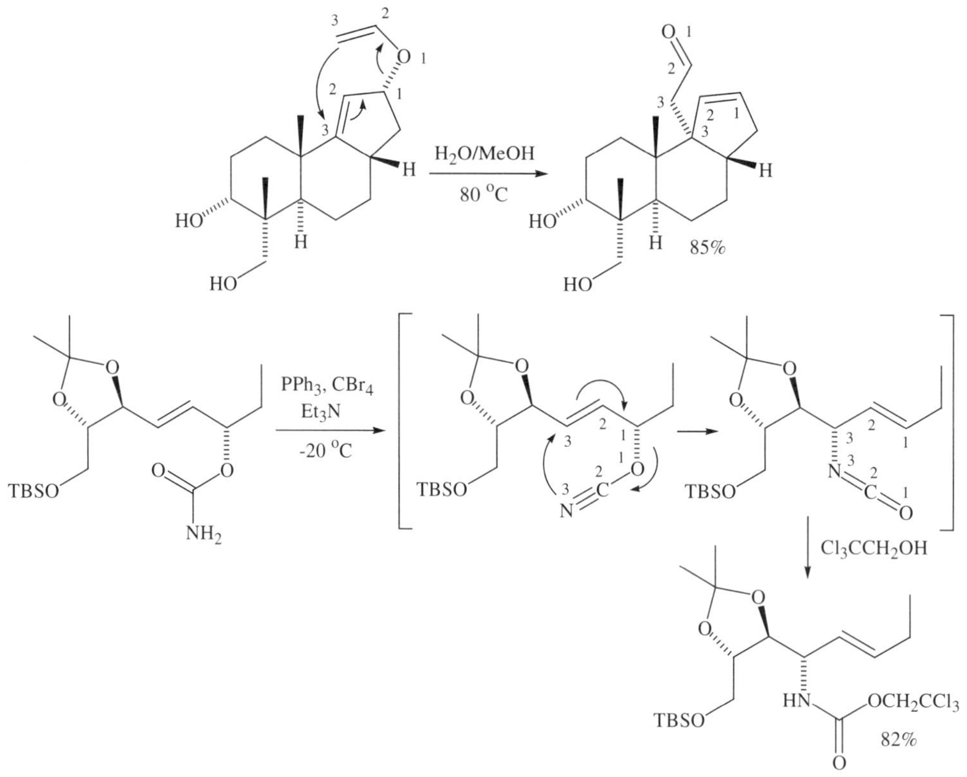

Scheme 6.52 [1,3]-Chirality transfer in Claisen rearrangements.

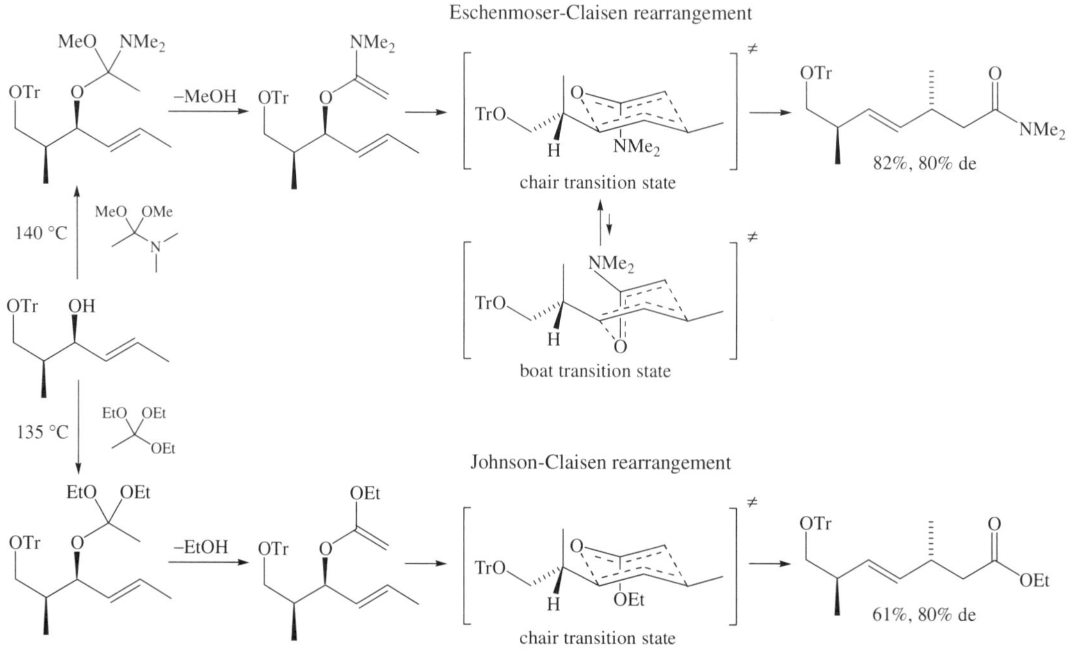

Scheme 6.53 [1,3]-Chirality transfer in Eschenmoser and Johnson Claisen rearrangements.

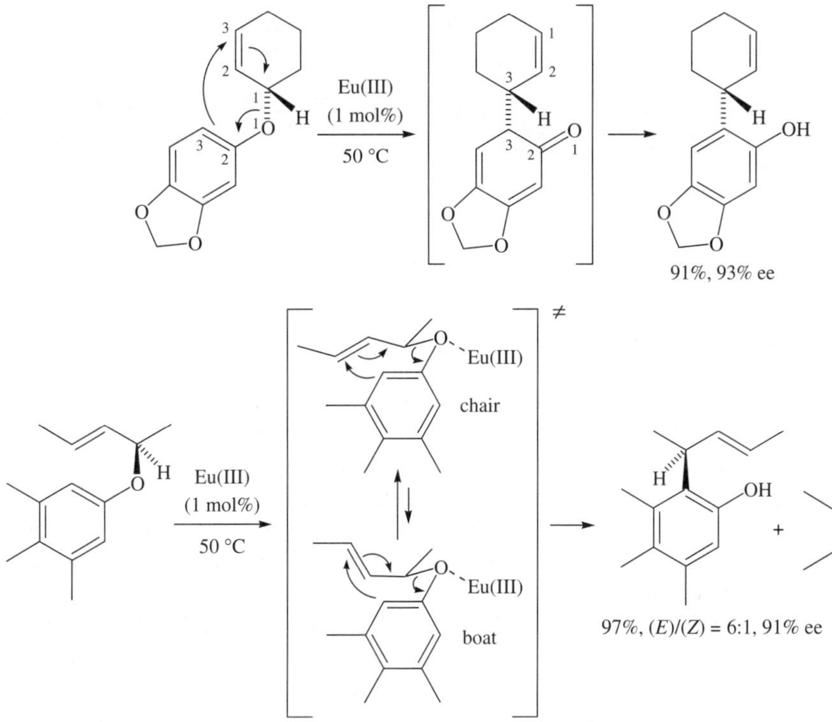

Scheme 6.54 Lewis acid-catalyzed aromatic Claisen rearrangement.

shifts proceed with good yield and stereospecificity. This has been explained by a favorable chair transition state having both alkyl substituents in equatorial positions, although a boatlike geometry that would produce the same stereoisomer is also feasible.

Claisen rearrangements are not restricted to aliphatic allyl ethers, but few asymmetric versions with allyl aryl ethers are known.[482] Aromatic Claisen rearrangements are often sluggish and of limited synthetic value due to substantial racemization, competing *ortho*- and *para*-migration, generation of intermediate allylic cations that migrate with low regio- and stereoselectivity, and other side reactions. The first aromatic Claisen rearrangement with effective [1,3]-chirality transfer was reported by Trost and Toste.[483] They found that europium-catalyzed migration in both cyclic and acyclic substrates proceeds with excellent enantioselectivity at ambient temperatures, although mixtures of (*E*)- and (*Z*)-alkenes are obtained with acyclic allyl aryl ethers, Scheme 6.54. The latter result has been attributed to two coexisting transition states possessing either a chairlike or a less favorable boatlike conformation.

A remarkable example of a retro Claisen rearrangement with [1,3]-chirality transfer is a key step in the total synthesis of the marine natural product (+)-laurenyne, Scheme 6.55.[484,485] Having developed a procedure for highly diastereoselective intramolecular *syn*-S$_N$2′ displacement of an allylic carbonate group by a diethyl malonate-derived enolate, Boeckman's group gained access to a stereochemically pure vinyl cyclobutane diester that was produced in 85% yield.[xvi] Reduction with lithium aluminum hydride followed by oxidation with Dess–Martin periodinane gave a vinyl cyclobutane dicarboxaldehyde that readily rearranged to the desired dihydrooxocene. The formation of the chiral eight-membered ring is thermodynamically favored due to release of ring strain in the cyclobutane precursor, which is the driving force for the retro Claisen rearrangement.

[xvi] The *syn*-S$_N$2′ cyclization of the diethyl malonate derivative constitutes another case of [1,3]-chirality transfer.

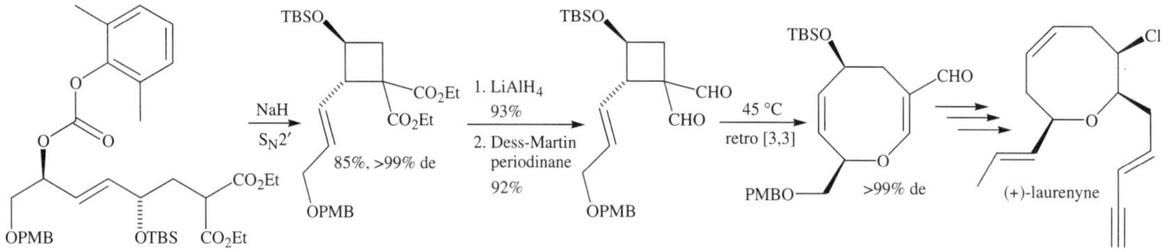

Scheme 6.55   Retro Claisen rearrangement of a strained vinyl cyclobutane dicarboxaldehyde.

Cope rearrangements are not as frequently employed in asymmetric synthesis as the Claisen variants, although highly stereospecific construction of complex carbon frameworks is feasible whenever a single transition state prevails. Impressive examples demonstrating the versatility and synthetic usefulness of Cope rerrangements include the synthesis of *cis*-decalins from unsaturated *endo*-bicyclo[2.2.2]octane derivatives,[486,487] and the diastereoselective synthesis of strychnan- and aspidospermatan-type alkaloids from a tryptophan derivative,[488–490] Scheme 6.56. The latter exploits a reaction cascade including a [3,3]-sigmatropic rearrangement and a Mannich cyclization. Aza- and oxy-Cope rearrangements afford synthetic entries to cyclooctanoids, terpenes and other natural products.[491–497] An intriguing case of an anionic oxy-Cope reaction with double [1,3]-chirality transfer was observed by Jung and Hudspeth.[498] The irreversible rearrangement of a benzofuran-derived norbornenyl system proceeds under relatively mild conditions and yields an important precursor of coronafacic acid. Tautomerization of the intermediate enol to the corresponding ketone shifts the equilibrium to the desired product. In this reaction both asymmetric centers at *C*-1 and *C*-1′ are destroyed while new chiral elements are generated at carbons 3 and 3′, Scheme 6.57. Knochel and coworkers developed a practical procedure for the synthesis of *trans*-1,2-diamino-3-cyclohexene based on two consecutive sigmatropic [3,3]-rearrangements of an allylic diimidate, Scheme 6.58.[499] Upon heating to 140 °C a double Overman rearrangement occurs with quantitative chirality transfer and the corresponding diamide is obtained in literally enantiopure form. Hydrolysis then generates the desired (*S,S*)-diaminocyclohexene.

According to the Woodward–Hoffmann rules, a thermally initiated sigmatropic rearrangement involving $4N$ electrons must include one antarafacial component. For example, a thermal [1,3]-sigmatropic shift of hydrogen across an allylic system is symmetry-allowed if the process is antarafacial, which means that the hydrogen must migrate from one side of the conjugated system to the other. This is not observed, since it is not feasible for geometric reasons. However, an atom of higher atomic number such as carbon can utilize both lobes of a *p*- or *sp³*-hybridized orbital and can therefore react in an antarafacial fashion. In this case a thermal [1,3]-shift involving $4N$ electrons is symmetry-allowed because it has one antarafacial component.[500] A [2πs + 2σa]-process with concomitant [1,3]-chirality transfer and inversion of configuration at the migrating carbon atom has been accomplished with deuterated bicyclo[3.2.0]-2-hepten-6-yl acetate at 300 °C, Scheme 6.59.[501]

The [2,3]-Wittig rearrangement of chiral allyl ethers is another powerful sigmatropic reaction exhibiting stereospecific transfer of chirality.[502,503] Similarly to Cope and Claisen reactions, the [2,3]-Wittig migration of deprotonated allylic ethers proceeds via a highly ordered cyclic transition state and the stereochemical outcome can often be predicted from the double bond geometry and absolute configuration of the substrate, in particular when allylic benzyl and propargylic ethers are present, Scheme 6.60.[xvii] The five-membered transition state is more flexible than the well-defined six-membered chairlike geometry of [3,3]-sigmatropic rearrangements, which somewhat reduces stereocontrol and the synthetic value of the [2,3]-migration.[504–507] The [2,3]-Wittig rearrangement is sometimes accompanied by the nonpericyclic [1,2]-Wittig shift. Although the [1,2]-rearrangement

---

[xvii] The Wittig rearrangement can be considered an orbital symmetry-allowed [2πs + 2σa + 2σa]-process.

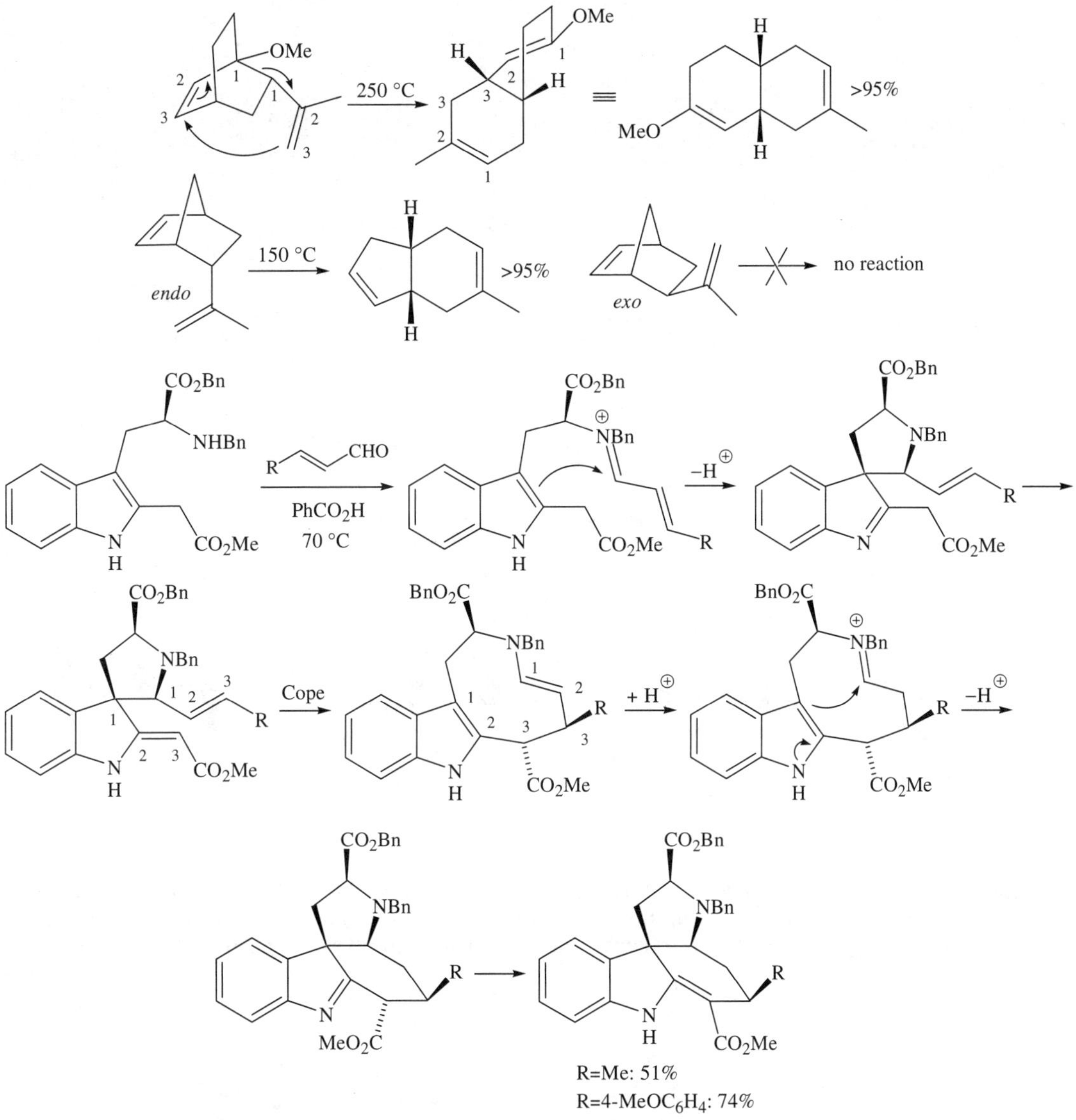

Scheme 6.56 Sigmatropic Cope rearrangements with [1,3]-chirality transfer.

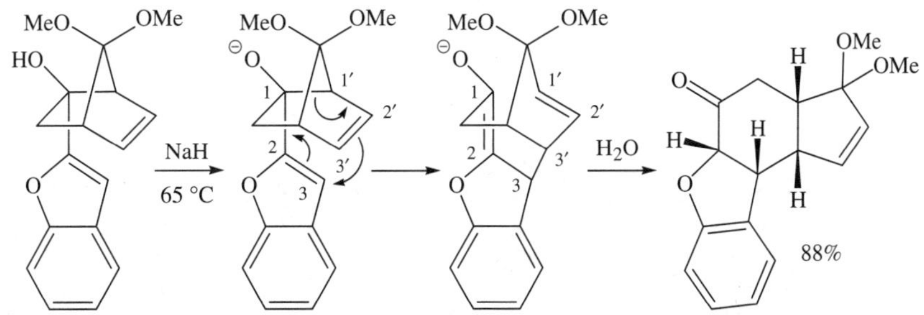

Scheme 6.57 Anionic oxy-Cope rearrangement with double [1,3]-chirality transfer.

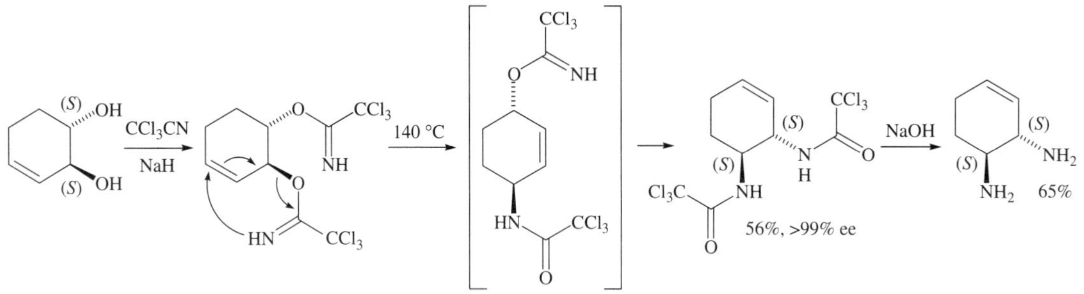

Scheme 6.58   Double Overman rearrangement.

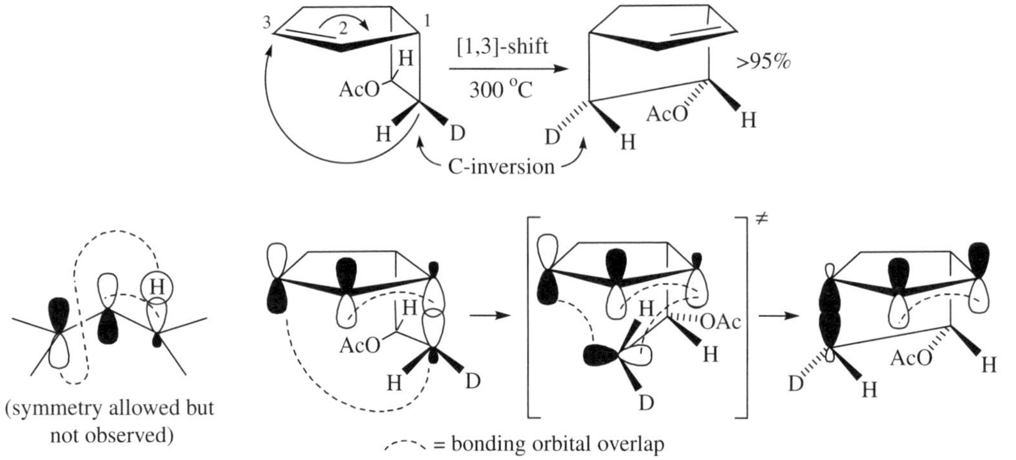

Scheme 6.59   Suprafacial sigmatropic [1,3]-rearrangement of deuterated bicyclo[3.2.0]-2-hepten-6-yl acetate with inversion of configuration at the migrating carbon atom (top). Orbital symmetry-allowed hydrogen and carbon [1,3]-shifts, and participating frontier molecular orbitals (bottom).

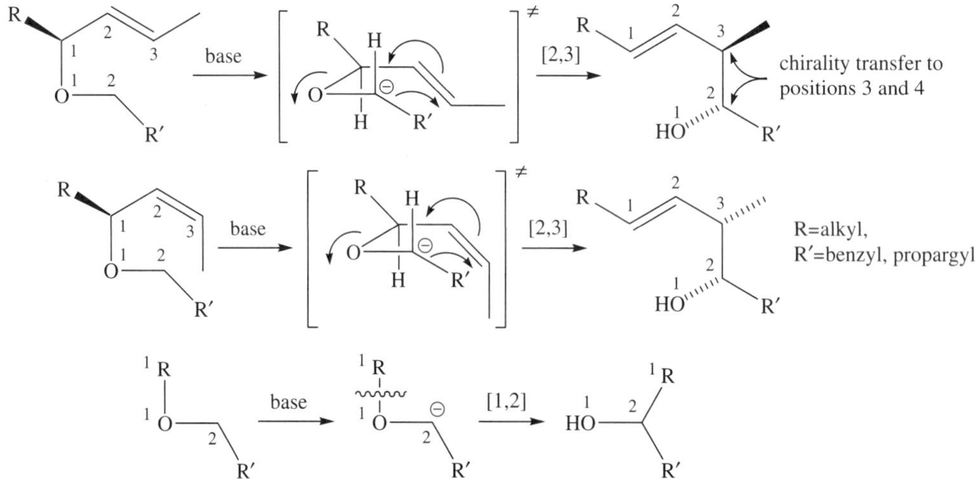

Scheme 6.60   Stereochemical outcome of the [2,3]-Wittig rearrangement (top) and the [1,2]-Wittig shift (bottom).

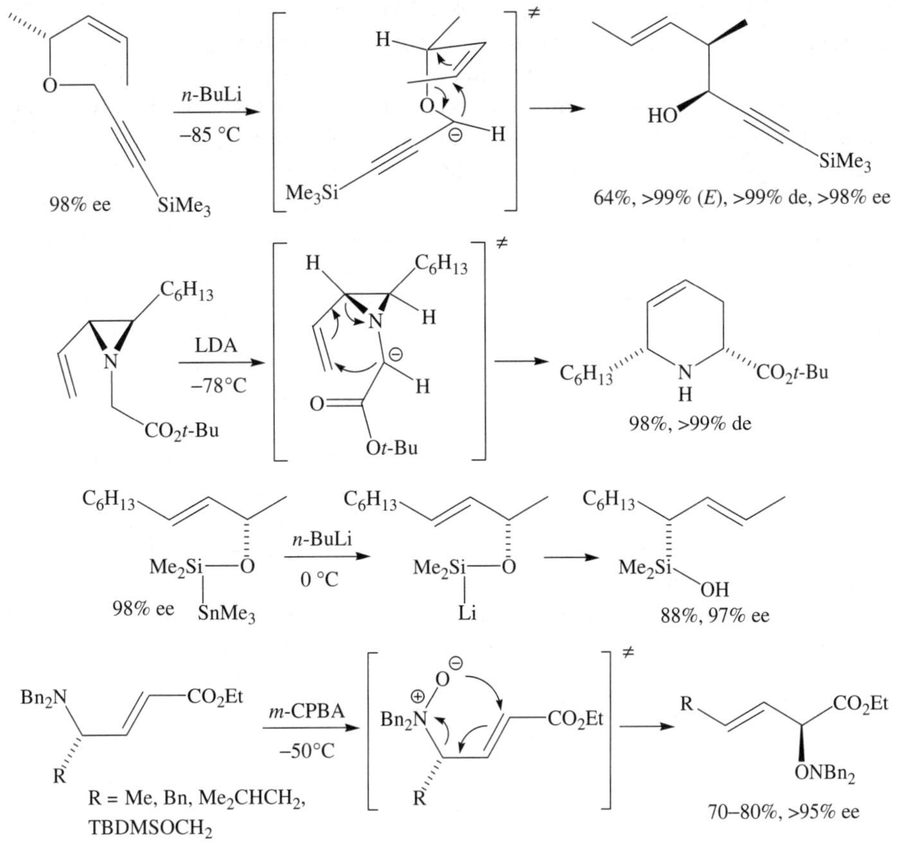

Scheme 6.61 Examples of [2,3]-Wittig and Meisenheimer rearrangements.

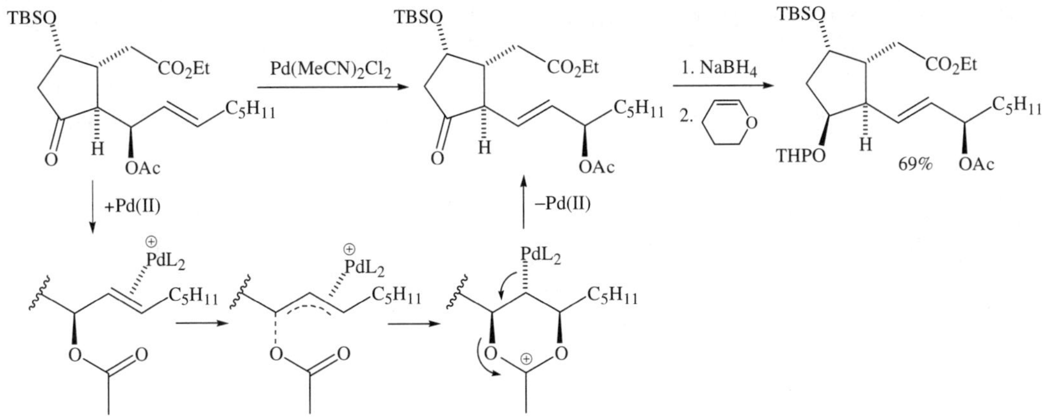

Scheme 6.62 Palladium(II)-catalyzed migration of an acetyl group across an allylic system.

involves a transient radical pair it can be highly stereoselective, but it is not as synthetically useful as the pericyclic [2,3]-variant.[508] The latter usually prevails over the [1,2]-shift when low temperatures are applied.[509,510] To date, an impressive range of [2,3]-Wittig rearrangements of allyl ethers has been reported, and sulfur-,[511–513] phosphorus-,[514] sila-,[515] and aza-variants[516–518] are also known,

Scheme 6.61.[519-532] The intrinsic stereospecificity of Wittig rearrangements of both allylic ethers and vinylaziridines has been exploited for the synthesis of *N*-heterocyclic alkaloids[533-535] and other natural products.[536-541] The Meisenheimer rearrangement of allylic tertiary amine oxides, which are readily prepared by oxidation of amines with *m*-chloroperbenzoic acid, involves a sigmatropic [2,3]-shift and offers stereospecific access to chiral hydroxylamines.[542-545]

Mercury(II)- and palladium(II)-mediated allylic rearrangements are somewhat reminiscent of sigmatropic shifts, and offer an additional entry to [1,3]-chirality transfer based on a nonpericyclic mechanism with an intermediate π-allyl transition metal complex.[546-549] The usefulness of transition metal-catalyzed allylic [1,3]-migrations for the total synthesis of prostaglandin congeners has been demonstrated by Danishefsky's group.[550] The palladium-catalyzed allylic transposition of the acetyl group across the allylic system shown in Scheme 6.62 is irreversible and strictly suprafacial. Diastereoselective reduction of the cyclopentanone moiety and protection of the corresponding alcohol furnishes the desired tetrahydropyranyl ether in 69% overall yield.

6.4.2.3   1,4-Chirality Transfer.   Sigmatropic Claisen rearrangements with simultaneous [1,4]-chirality transfer are quite common and have been implemented in the total synthesis of many natural products.[551-561] In an elegant approach to β-lactone enzyme inhibitors (−)-ebelactone A and B, Paterson's group utilized trimethylsilyl chloride to trap the kinetically favored ester enolate of (3*S*,4*S*,6*R*,7*R*)-3-[(*tert*-butyldimethylsilyl)oxy]-2-ethyl-4,6,8-trimethyl-7-(propanoyloxy)-1,8-nonadien-5-one.[562] The (*E*)-silyl ketene acetal undergoes stereospecific Ireland–Claisen rearrangement with [1,4]-chirality transfer and conversion of the original 1,2-*syn*-stereochemistry to a 1,5-*syn*-relationship, Scheme 6.63. As mentioned above, the stereochemical outcome of pericyclic rearrangements is generally controlled by the relative stability of competing chair- and boatlike transition states. By analogy with more flexible acyclic compounds, ring systems favor sigmatropic [3,3]-rearrangements with a chairlike transition state, unless this is not feasible due to steric or geometric constraints.[563,564] Dihydropyrans are known to rearrange through a boatlike transition state because a chairlike geometry would yield conformationally strained (*E*)-cyclohexenes.[565-567] Accordingly, the thermal rearrangement of lactonic silyl enolates derived from dihydropyrans and seven-membered analogs produces (*Z*)-cycloalkenes, Scheme 6.64.

Some examples of [1,4]-chirality transfer in Cope rearrangements of cyclic and acyclic systems are known.[568] Thermal rearrangement of (*R*)-(*E*)-3-methyl-3-phenylhepta-1,5-diene at 250 °C gives (*S*)-(*E*)-3-methyl-6-phenylhepta-1,5-diene in 87% yield and more than 94% ee. This reaction can occur via two chairlike transition states with the phenyl group either in equatorial or axial position. As expected, the latter is sterically less favored and the corresponding (*R*)-(*Z*)-isomer is formed in only

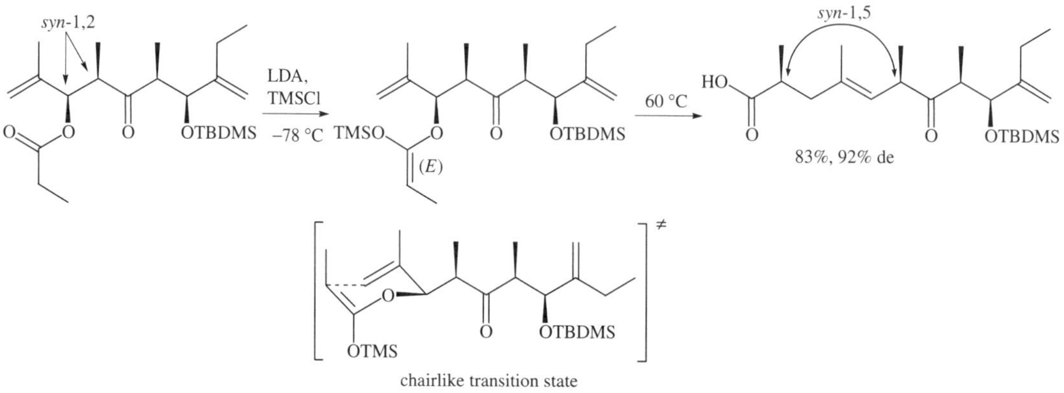

Scheme 6.63   Stereospecific Ireland–Claisen rearrangement and [1,4]-chirality transfer.

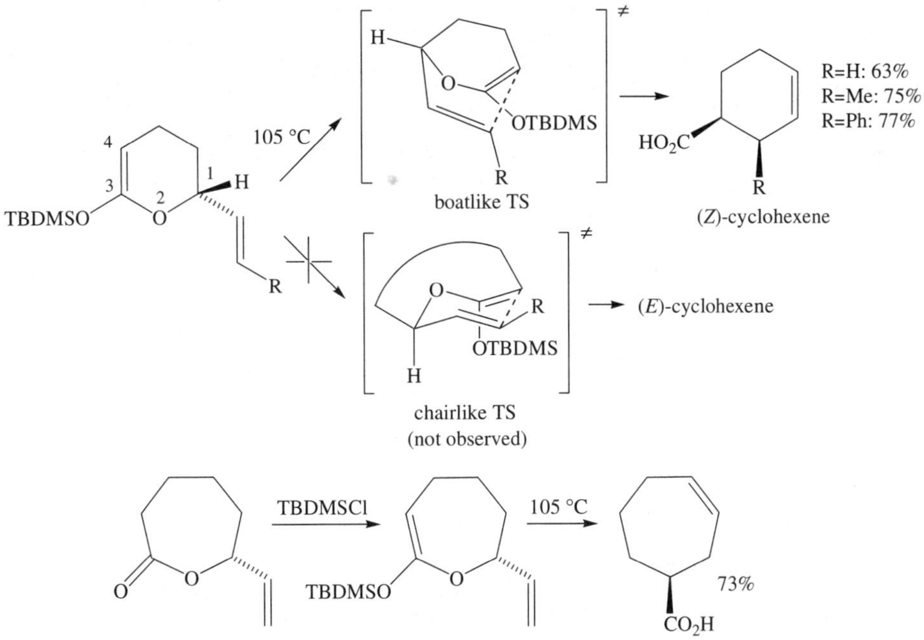

Scheme 6.64 Ireland–Claisen rearrangements with simultaneous [1,3]- and [1,4]-chirality transfer.

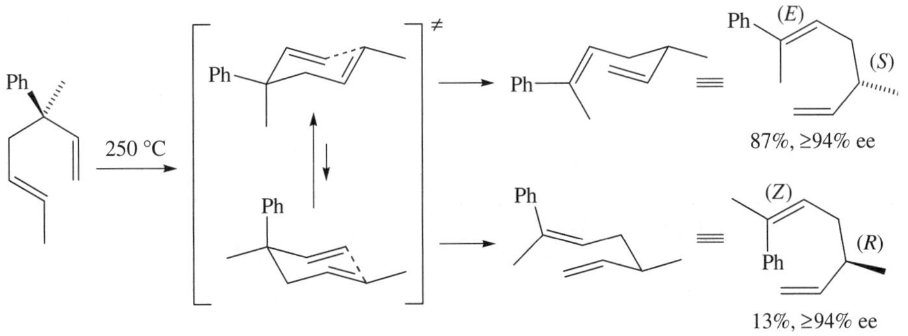

Scheme 6.65 Cope rearrangement of (*R*)-(*E*)-3-methyl-3-phenylhepta-1,5-diene.

13% yield, Scheme 6.65.[569] Paquette *et al.* studied the stereochemical course and [1,4]-chirality transfer in oxy-Cope shifts of cyclic molecules including norbornenyl derivatives, Scheme 6.66.[570,571] The oxy-variation occurs under considerably milder conditions than the Cope reaction and with excellent conservation of chirality. The rearrangement of (*R*)-4-(2-norbornenyl)-but-3-en-2-ol generates 59% of the *exo*-product in over 98% de. Apparently, the intermediate oxyanion group favors an equatorial position and *exo*-bond formation. The diastereoselectivity of the rearrangement decreases if an *exo*-attack involving a conformer with an equatorial oxyanion is not feasible. As a result, the oxy-Cope rearrangement of the (*S*)-diastereoisomer affords a similar yield but only 82% de.

As shown above, [2,3]-Wittig rearrangements often proceed with [1,3]-chirality transfer along the allylic system. Alternatively, the chiral information can be transferred one atom further across the newly formed σ-bond, Scheme 6.67.[572-575] Marshall *et al.* uncovered that sigmatropic [2,3]-rearrangements of propargylic ethers generate chiral allenes based on [1,4]-chirality transfer and conversion of central to axial chirality.[576-578]

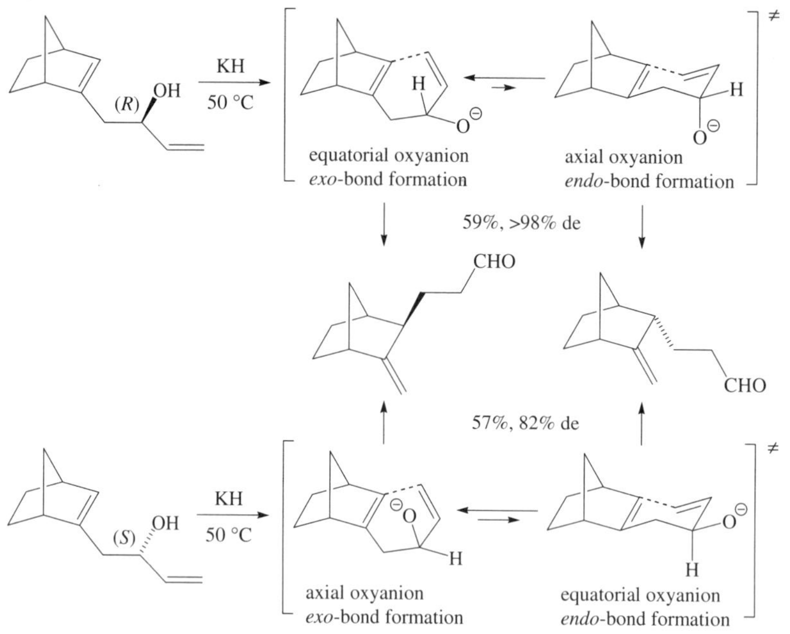

Scheme 6.66 Oxy-Cope rearrangement of diastereomeric norbornenyl derivatives.

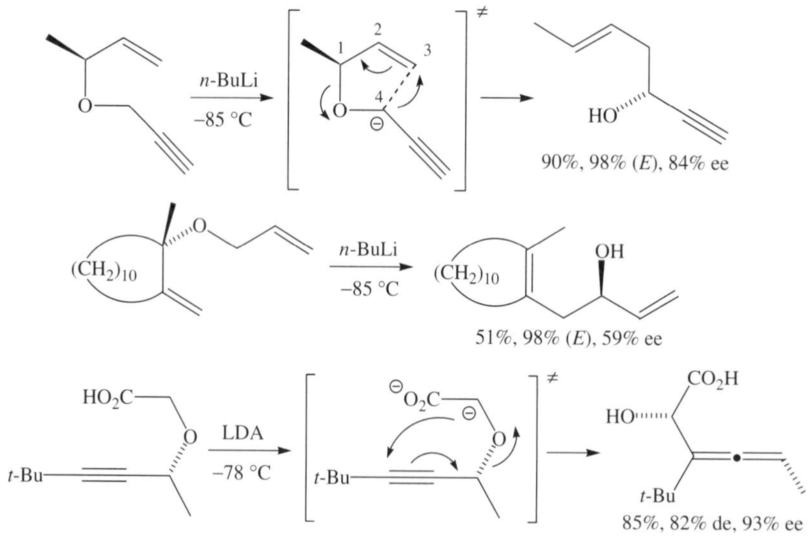

Scheme 6.67 [2,3]-Wittig rearrangements resulting in [1,4]-chirality transfer.

6.4.2.4 1,5-Chirality Transfer.   A symmetry-allowed thermal [1,5]-migration involving $(4N+2)$ electrons can have either two suprafacial or two antarafacial components.[xviii] Although [1,5]-hydrogen shifts are highly stereospecific they have found few synthetic applications, partly due to lack of reaction control.[579,580] For example, hydrogen migration from an asymmetric carbon center across two double bonds of the *s-cis*-diene framework of (*S*)-(2*E*,4*Z*)-2-deuterio-6-methyl-2,4-octadiene

---

[xviii] A thermal suprafacial [1,5]-shift of hydrogen and carbon constitutes a symmetry-allowed [4πs+2σs]-process. Accordingly, a carbon substituent migrates with retention of configuration in order to comply with Woodward–Hoffmann rules.

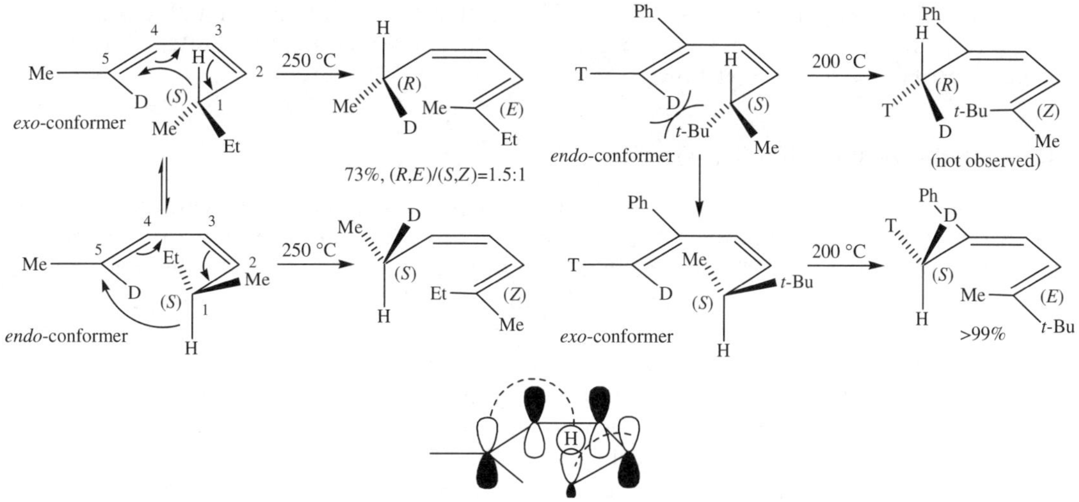

Scheme 6.68 Stereospecific suprafacial [1,5]-hydrogen shifts along *s-cis*-diene frameworks.

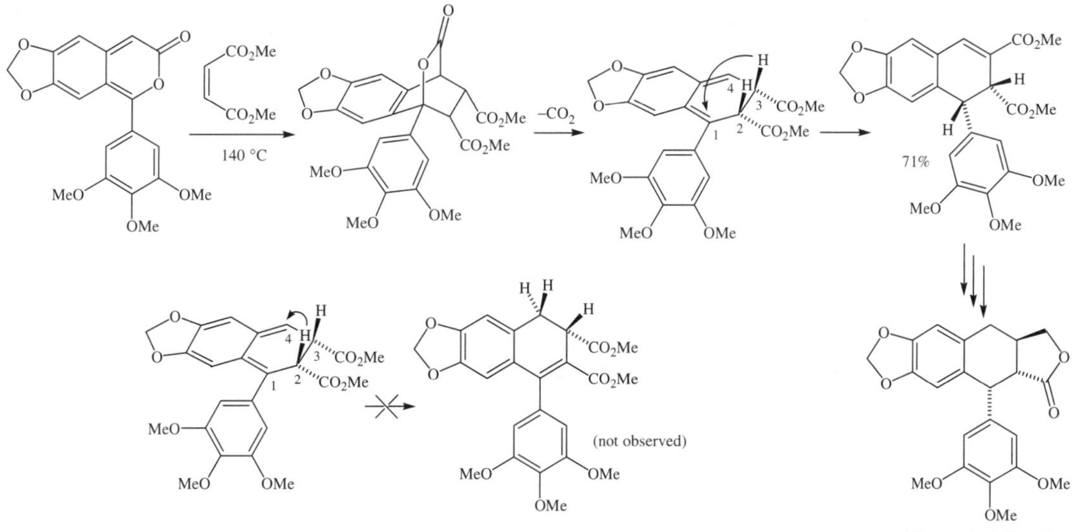

Scheme 6.69 Regioselective suprafacial hydrogen shift with [1,5]-chirality transfer.

engages two rapidly equilibrating conformers. The competing suprafacial hydrogen migrations form a mixture of diastereoisomers despite complete conservation of chirality during each [1,5]-sigmatropic shift, Scheme 6.68. Mulzer *et al.* showed that interconversion of *endo-* and *exo*-isomers can be avoided by incorporation of steric bulk into the asymmetric carbon center.[581] The presence of a sterically demanding *tert*-butyl group destabilizes the *endo*-conformation and exclusive sigmatropic rearrangement of the prevailing *exo*-isomer is observed.

A regioselective suprafacial [1,5]-hydrogen shift has been realized with an *o*-quinodimethane intermediate carrying *cis*-oriented methoxycarbonyl groups. Diels–Alder reaction of dimethyl maleate and a 1-(3,4,5-trimethoxyphenyl)-3*H*-2-benzopyran-3-one derivative is followed by carbon dioxide extrusion and subsequent migration of hydrogen from *C*-3 to the carbon bearing the phenyl ring.[582,583] This stereospecific hydrogen shift produces a *cis*-dihydronaphthalene, an important precursor of 4-deoxypodophyllotoxin and other chiral lignans, Scheme 6.69. Interestingly,

[1,5]-migration of the hydrogen at *C*-2 to *C*-4 is also symmetry-allowed but is not observed. Similar asymmetric [1,5]-hydride shifts have been reported for intramolecular Meerwein–Ponndorf–Verley/ Oppenauer reactions of hydroxyketones.[584,585]

6.4.3   Intermolecular Chirality Transfer

Most examples of chirality transfer are *intra*molecular processes, but chiral elements can also be relocated in *inter*molecular reactions. A quite intriguing case arises during asymmetric reduction of methyl benzoylformate with chiral 4-methyl-1,4-dihydropyridines.[586,587,xix] In this reaction, the chirality of (*S*)-*N*-benzyl-3-(hydroxymethyl)-4-methyl-1,4-dihydropyridine is transferred to the prochiral ester which is reduced to (*S*)-methyl mandelate, Scheme 6.70. As a result of the hydride transfer the central chirality of the reducing agent is destroyed as the achiral pyridinium salt is formed, but overall chirality is perfectly conserved in the newly formed α-hydroxy ester.

Intermolecular chirality transfer has been accomplished in asymmetric Meerwein–Ponndorf–Verley reactions[588–594] and reductions of ketones with chiral metal hydrides,[595,596] but moderate enantioselectivities are usually observed. In one of the more successful approaches, Maruoka's group utilized stoichiometric amounts of (*R*)-1-phenylethanol and other chiral secondary alcohols in the presence of an aluminum catalyst for reduction of 2-chloroacetophenone, Scheme 6.71.[597] The stereochemical outcome can be explained by the formation of a cyclic six-membered transition state in which the carbonyl substrate and the secondary alkoxide coordinate to the metal center.[598] The aluminate complex enhances the nucleophilicity of the chiral hydride donor while coordination

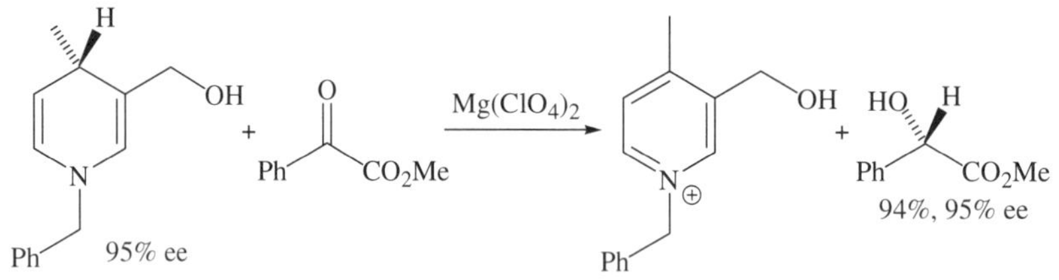

Scheme 6.70   Intermolecular chirality transfer with a chiral NADH mimic.

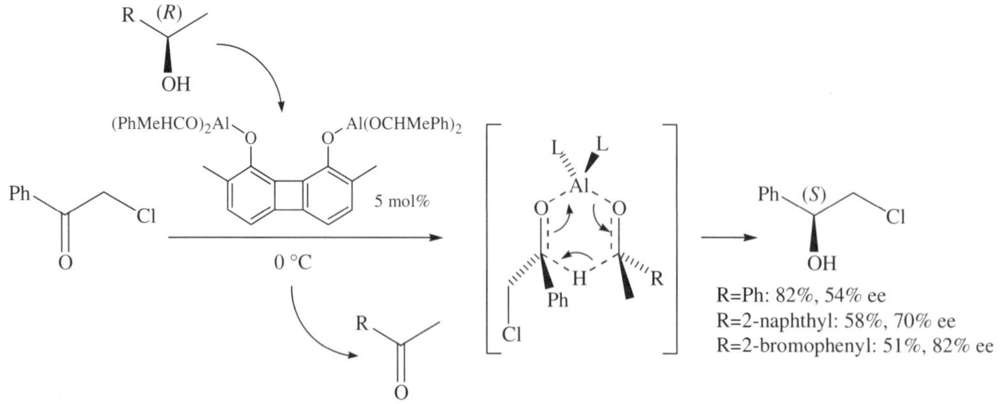

Scheme 6.71   Intermolecular chirality transfer in a Meerwein–Ponndorf–Verley reduction.

[xix] Dihydropyridines are frequently used as NADH analogs in biomimetic transformations.

of the ketone to the Lewis-acidic metal ion increases the electrophilicity of the hydride acceptor. To minimize steric repulsion, the largest groups of the two reagents preferentially reside on opposite sides of the cyclic transition state. This arrangement favors *Re*-facial attack at 2-chloroaceto-phenone and formation of (*S*)-2-chloro-1-phenylethanol.

6.4.4 Transfer of Stereogenicity Between Carbon and Heteroatoms

The transfer of chiral information between carbon and heteroatoms provides unique access to a range of chiral compounds. The translocation of stereogenicity is often accomplished with sigmatropic rearrangements exhibiting complete conservation of chirality. Nitrogen-to-carbon chirality transfer has been achieved by [1,2]- and [2,3]-rearrangements of chiral *N*-oxides and nitrogen ylides or via Stevens rearrangement of quaternary ammonium salts.[599–604] In accordance with the stereochemical course of sigmatropic [2,3]-migrations, allylic *N*-oxides undergo Meisenheimer rearrangements with concomitant transfer of stereogenicity from the configurationally stable nitrogen to a remote carbon atom, Scheme 6.72.[605,606] The thermally initiated concerted rearrangement of (*R*)-*trans*-crotyl-*N*-methyl-4-toluidine *N*-oxide gives (*S*)-*O*-methylvinylcarbinyl-*N*-methyl-4-tolylhydroxylamine in 87% yield. Comparison of two orbital symmetry-allowed doubly suprafacial five-membered transition states suggests that the absence of a nonbonding interaction between the 4-tolyl moiety and the allylic methyl group favors the pathway to the (*S*)-enantiomer.

Important alkaloids consisting of bicyclic pyrrolizidine, indolizidine and quinolizidine cores can be prepared from chiral ammonium ylides by stereospecific [1,2]- and [2,3]-migrations.[607–611] For example, copper carbenoids generated from vinylpyrrolidine-derived α-diazo ketones form spiro-fused ylides that generate bicyclic amines through [2,3]-migration with complete nitrogen-to-carbon chirality transfer, Scheme 6.73.[612] Since ylide formation preferentially occurs at the pyrrolidine side bearing the vinyl group, the two rapidly interconverting diastereomeric pyrrolidine-derived copper carbenoids afford an intermediate that further reacts to the corresponding (*R*)-(*Z*)-azabicyclo[6.3.0]undecene. This takes place through an *endo*-transition state having the vinyl group in equatorial position and pointing to the incipient eight-membered ring.

Chirality transfer based on destruction of stereogenicity at a sulfur atom and simultaneous generation of an asymmetric carbon center has been achieved with allylic and benzylic sulfonium ylides.[613–616] The formation of a reactive sulfur ylide from (*S*)-1-adamantylallylethylsulfonium tetrafluoroborate with potassium *tert*-butoxide entails sigmatropic [2,3]-rearrangement to a chiral

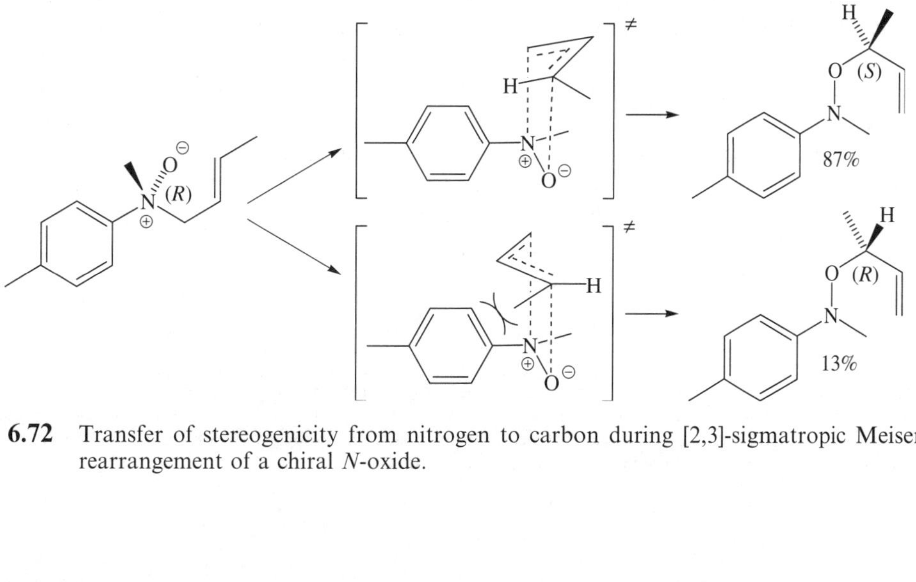

Scheme 6.72 Transfer of stereogenicity from nitrogen to carbon during [2,3]-sigmatropic Meisenheimer rearrangement of a chiral *N*-oxide.

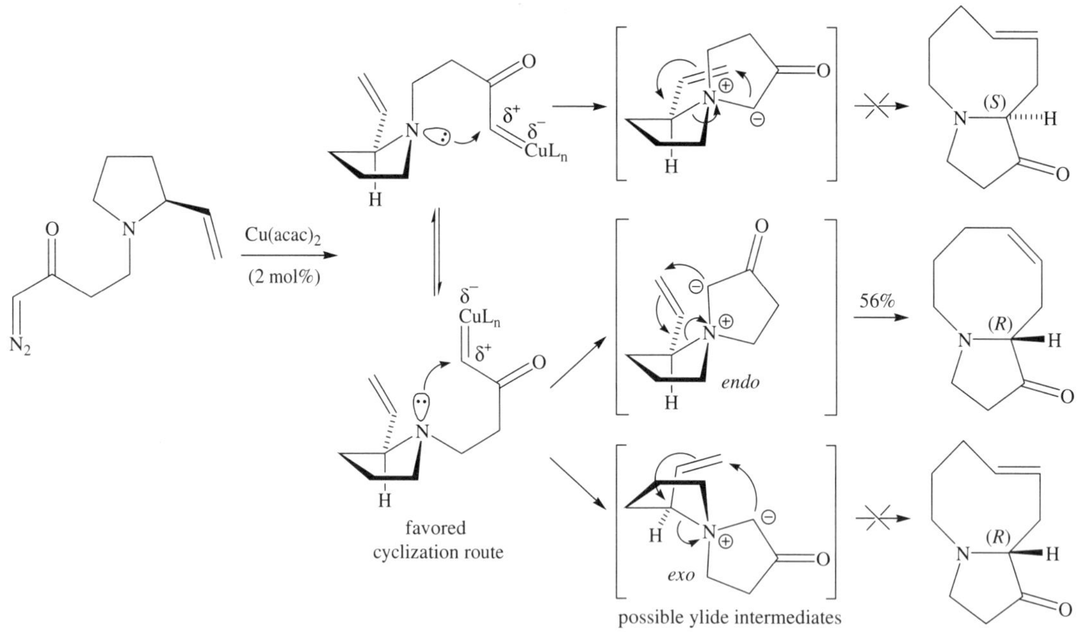

Scheme 6.73 Synthesis of an (R)-(Z)-azabicyclo[6.3.0]undecene via [2,3]-rearrangement of a spiro ammonium ylide.

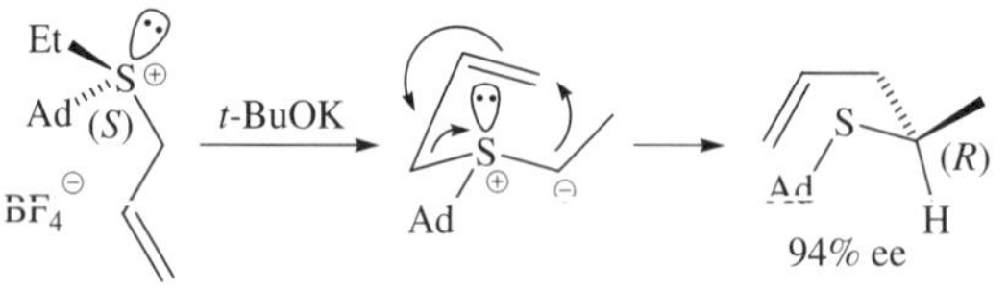

Scheme 6.74 Transfer of chirality from sulfur to carbon via a folded envelope conformation of an intermediate sulfonium ylide.

sulfide, Scheme 6.74. However, the scope of asymmetric synthesis with chiral sulfonium salts is limited because of the inherently low energy barrier to racemization.[xx]

Virtually complete chirality transfer from silicon to carbon is obtained with a strained allylic silacycle that can be prepared by rhodium(I)-catalyzed silylformylation of a propargylic (S,R_{Si})-silyl ether precursor, Scheme 6.75.[617,618] The five-membered silacycle is prone to spontaneous intramolecular allylation and forms a bicyclic siloxane that is readily cleaved to the (S,S)-1,5-*syn*-diol and *tert*-butyltrifluorosilane upon addition of TBAF. The reaction is solely controlled by the asymmetric silicon atom and proceeds without loss of stereochemical integrity. The presence of the chiral carbon center in the propargylic silyl ether has no influence on the stereochemical outcome since employment of the (S,S_{Si})-diastereomer in the same reaction sequence gives the (S,R)-1,5-*anti*-diol.[xxi]

[xx] Sulfonium salts undergo facile pyramidal S-inversion in solution at ambient temperatures.

[xxi] The introduction of the new chiral carbon center to the rearranged carbon skeleton can therefore be attributed to net silicon-to-carbon chirality transfer. However, the destruction of silicon stereogenicity and the generation of a new chiral carbon center do not occur simultaneously but in consecutive steps.

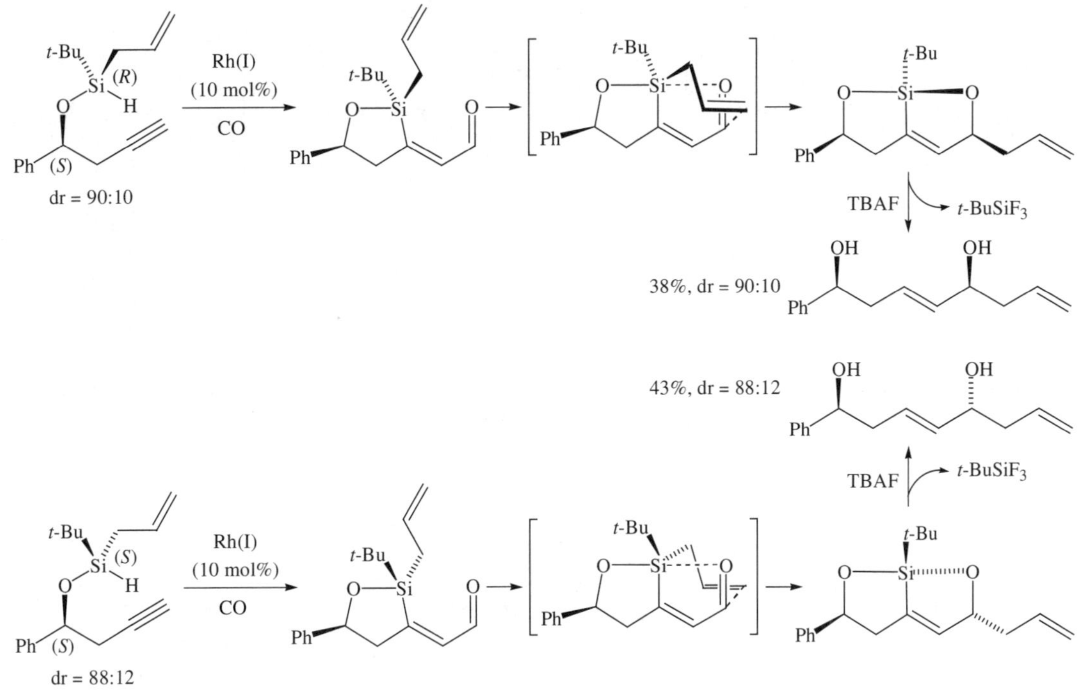

Scheme 6.75 Formation of 1,5-diols from strained allylic silacycles.

6.4.5 Conversion of Central Chirality to Other Chiral Elements

Atroposelective synthesis often relies on translation of central to axial chirality based on asymmetric induction with a temporary chiral template or asymmetric transformation of the first and second kind, see Chapters 6.2 and 6.3. A general drawback of these methodologies is the necessity of an auxiliary for introduction of axially chiral bias. In most cases this requires three steps: covalent attachment of the chiral auxiliary, atroposelective transformation, and removal of the auxiliary under carefully selected conditions to avoid concomitant racemization. A direct conversion of central chirality to an atropisomeric framework within a single reaction is a much more efficient synthetic alternative.[619] This has been accomplished through Lewis acid-promoted benzannulation of enantiopure (*S,S*)-diaryl-2,2-dichlorocyclopropylmethanols, Scheme 6.76.[620] Regioselective coordination of titanium tetrachloride to the (*S,S*)-dichlorocyclopropyl alcohol moiety initiates rotation of the adjacent phenyl ring to reduce steric interactions with the Lewis acid. As a result, the *ortho*-substituent *R'* preferentially occupies the opposite side of the chelation complex. Subsequent elimination of the hydroxyl group then generates a cationic intermediate that undergoes benzannulation to an axially chiral 1-phenyl-3,4-dihydronaphthalene. The two elements of central chirality in the (*S,S*)-diaryl-2,2-dichlorocyclopropylmethanol are destroyed as the chiral information is stored in the incipient biaryl axis. Finally, elimination of HCl produces the corresponding (*M*)-1-phenylnaphthalenes in good to high yields and without any sign of racemization. The reaction course is reminiscent of the concept of regeneration of stereogenicity via stereolabile chiral intermediates discussed in Chapter 6.5.3.

Interconversion of central to axial chirality in a single step without loss of stereochemical integrity has also been observed during mild oxidation of enantiomerically enriched (*S*)-4-(1-naphthyl)dihydroquinolines.[621] Addition of 1-naphthyllithium to an enantiomerically pure 3-quinolyl oxazoline and subsequent removal of the chiral auxiliary establishes a chiral center at *C*-4 in

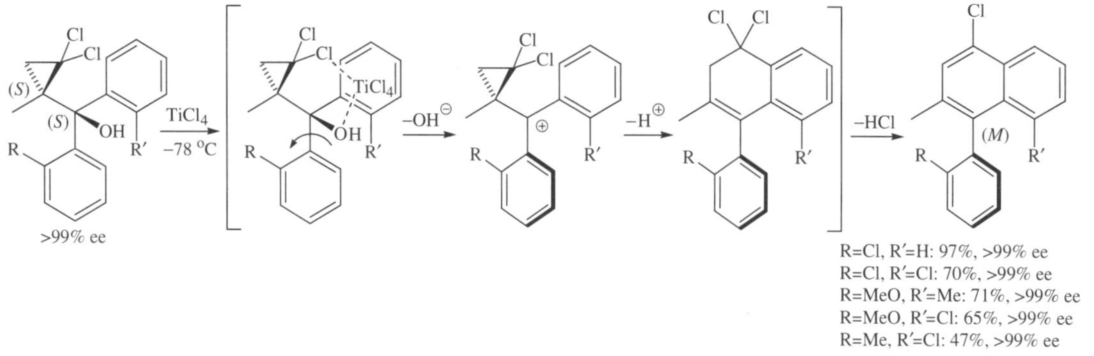

(S)

(S) OH

R R'

>99% ee

TiCl₄
−78 °C

−OH⁻

−H⁺

−HCl

(M)

R R'

R=Cl, R'=H: 97%, >99% ee
R=Cl, R'=Cl: 70%, >99% ee
R=MeO, R'=Me: 71%, >99% ee
R=MeO, R'=Cl: 65%, >99% ee
R=Me, R'=Cl: 47%, >99% ee

Scheme 6.76 Conversion of central to axial chirality via Lewis acid-promoted benzannulation of enantiopure (*S,S*)-diaryl-2,2-dichlorocyclopropylmethanols.

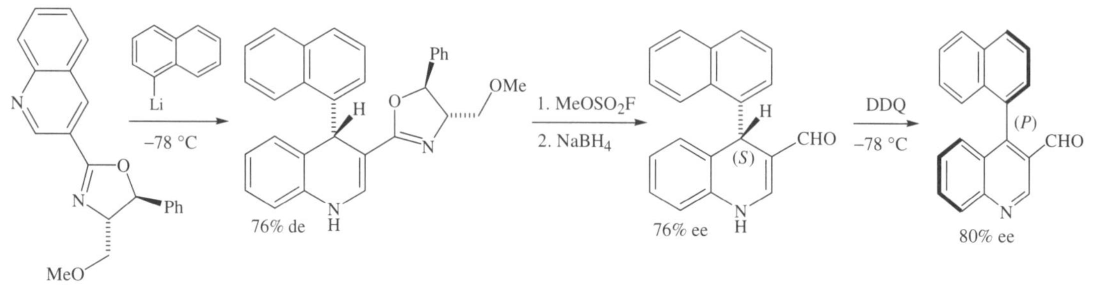

Li

−78 °C

76% de

1. MeOSO₂F
2. NaBH₄

(S) CHO

76% ee

DDQ
−78 °C

(P) CHO

80% ee

Scheme 6.77 Generation of axial chirality by oxidation of 4-(1-naphthyl)-1,4-dihydroquinoline-3-carboxaldehyde.

4-(1-naphthyl)-1,4-dihydroquinoline-3-carboxaldehyde. Oxidation of the dihydroquinoline with DDQ under cryogenic conditions then results in stereoselective transformation of the asymmetric center to a chiral axis. Because the naphthyl ring attached to the chiral center can freely rotate about the naphthyl–dihydroquinolyl bond the sense of axial chirality is not established prior to oxidation and this reaction constitutes a true example of chirality interconversion, Scheme 6.77.

Another viable approach to axially chiral biaryls entails dehydration of an enantiopure 1-naphthyl-1,2,3,4-tetrahydronaphthalene-1-ol that is readily available through ytterbium triflate-controlled diastereoselective addition of 2-methoxy-1-naphthylmagnesium bromide to 2-methyl-1-tetralone, Scheme 6.78.[622] The tertiary alcohol exists as a mixture of two interconverting conformers in a ratio of 1:6.3 in dichloromethane and 1:1.9 in benzene, corresponding to 86% and 34% de, respectively. Although axial chirality is already present prior to the dehydration step, trifluoroacetic anhydride-mediated elimination of water in these solvents gives (*M*)-3,4-dihydro-2'-methoxy-2-methyl-1,1'-binaphthalene in 95% or 82% ee, respectively, indicating that chirality interconversion coincides with dynamic kinetic asymmetric transformation. Dehydrogenation of the sample having 95% ee with DDQ at elevated temperature affords (*M*)-2'-methoxy-2-methyl-1,1'-binaphthalene in 83% ee due to partial racemization. Despite the attractive features of this strategy, few examples of natural product synthesis based on one-step interconversion of central to axial chirality have been reported to date.[623,624]

Since axially chiral allenes have been identified as versatile building blocks for asymmetric synthesis, and because of the abundance of allenic pharmacophores and subunits in natural products, the development of stereoselective methods providing access to this class of compounds

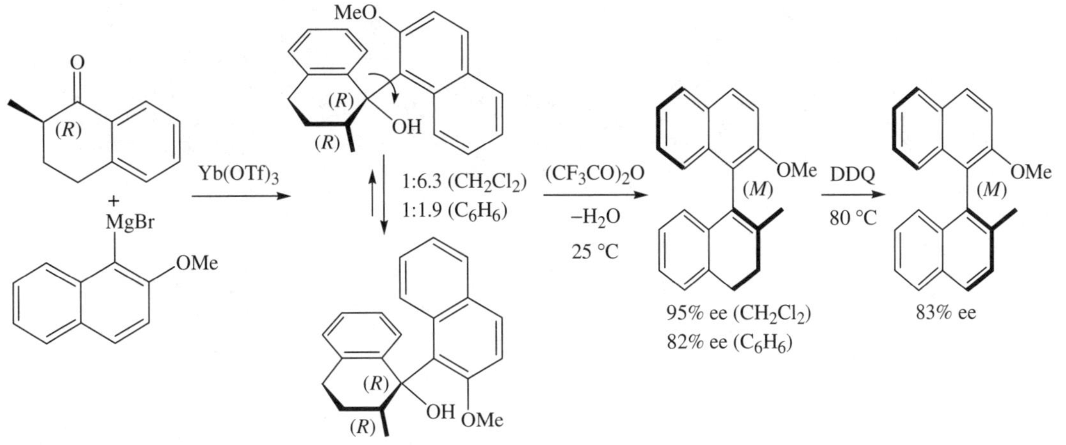

Scheme 6.78 Atroposelective formation of (*M*)-2′-methoxy-2-methyl-1,1′-binaphthalene.

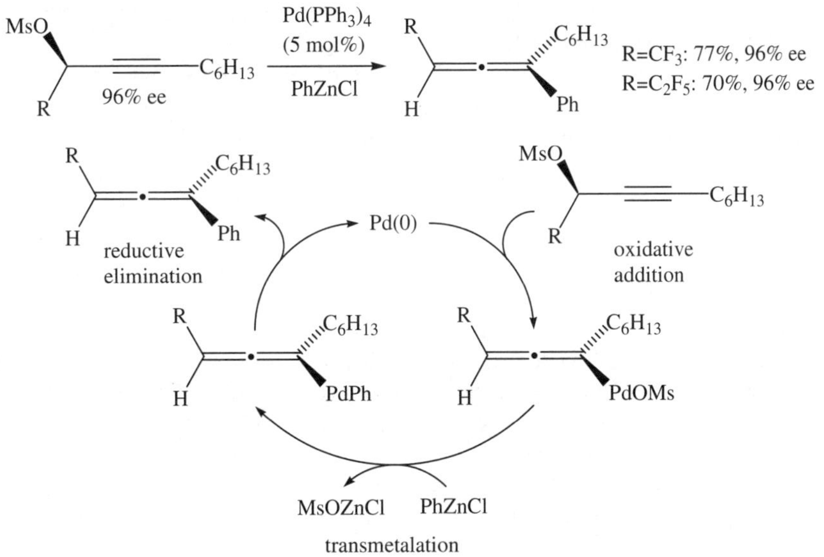

Scheme 6.79 Palladium-catalyzed synthesis of axially chiral allenes from propargylic mesylates.

has received considerable attention.[625–638] However, chirality conversion not only plays an important role in the synthesis of allenes, the chiral axis of allenic starting materials is also often utilized to gain access to compounds exhibiting central chirality, Chapter 6.4.6.[639] Copper- and palladium-mediated S_N2' attack of suitable nucleophiles at nonracemic chiral propargylic substrates provides axially chiral allenes.[640] An important example is the palladium-catalyzed reaction between propargylic mesylates bearing perfluorinated alkyl groups and phenylzinc chloride shown in Scheme 6.79.[641] The reaction course and interconversion of chirality elements can be explained by *anti*-S_N2'-type removal of the mesylate group.[642] Oxidative addition of the palladium catalyst initially forms an allenylpalladium species that undergoes transmetalation with the organozinc reagent and subsequent reductive elimination with retention of configuration. The final step produces an axially chiral allene while the catalytically active Pd(0) species is regenerated. It should be noted, however, that Pd-catalyzed S_N2' reactions of chiral propargylic or allylic alcohol

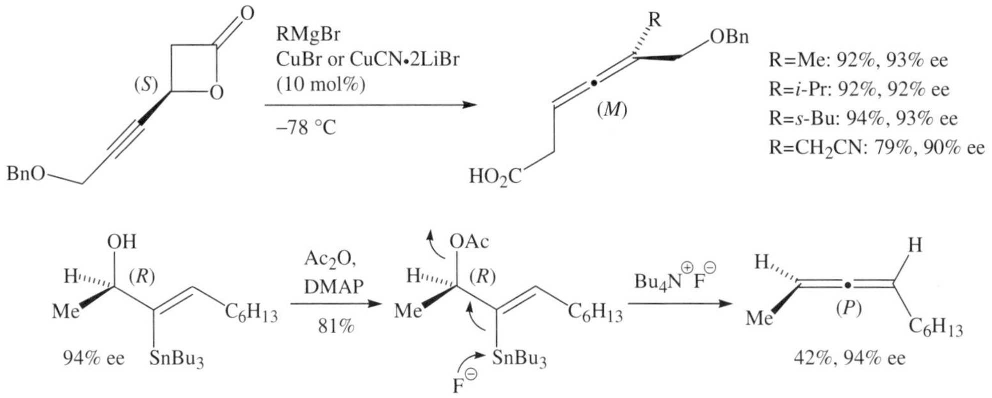

Scheme 6.80   Copper-catalyzed S_N2'-type ring opening of alkynyl-substituted β-lactones with Grignard reagents (top) and chirality conversion during stereoselective deoxystannylation of (*R*)-(*Z*)-3-tributylstannyl-3-decene-2-ol (bottom).

derivatives are frequently accompanied by partial or even complete racemization.[643,xxii] To date, S_N2' reactions of propargylic alcohol derivatives with organometallic reagents[644–647] or hydrides[648] and S_N''-type substitutions of enyne acetates[649,xxiii] are the most common methods for the synthesis of allenes based on central to axial chirality conversion. A conceptually different approach to axially chiral allenes involves tetrabutylammonium fluoride-promoted *anti*-elimination of tributyltin acetate from allylic esters, Scheme 6.80.[650]

Allenylcarbinols can be prepared via stereospecific [2,3]-Wittig rearrangement of propargylic ethers, see Chapter 6.4.2.3. Enantioselective lithiation of prochiral Cb-protected propargylic alcohols with butyllithium in the presence of (–)-sparteine followed by transmetalation and treatment with aldehydes or acetic acid offers another route to chiral allenes.[651] Deprotonation of alkynyl carbamates may occur with high enantiotopic selectivity, but trapping of the lithium salt with various electrophiles usually yields products in low enantiomeric purity. This is probably due to the low configurational stability of (–)-sparteine-derived propargylic complexes. In some cases, asymmetric transformation of the second kind furnishes a pure lithiated propargylic stereoisomer. Transmetalation of such a diastereomerically pure (–)-sparteine complex with titanium chlorotriisopropoxide has been reported to proceed with inversion of configuration, and treatment of the corresponding titanate with aldehydes affords a range of 4-hydroxyallenyl carbamates, Scheme 6.81.

Conversion of central to helical chirality has been realized through aromatic oxy-Cope rearrangement of (1*R*,2*S*,4*S*)-1-methoxybicyclo[2.2.2]oct-5-ene-2-(5,8-dimethyl-3-phenanthryl)-2-ol which can be prepared in 60% yield by addition of a phenanthryl Grignard complex to (1*R*,4*S*)-1-methoxybicyclo[2.2.2]oct-5-ene-2-one, Scheme 6.82.[652] The rearrangement is initiated by potassium bis(trimethylsilyl)amide in the presence of 18-crown-6 at 0 °C. The corresponding fused ring framework already displays a helical structure and after four further steps (*P*)-2-acetoxy-11,14-dimethyl[5]helicene is obtained in literally enantiopure form.

Cyclic *trans*-1,2-diols provide convenient access to nonracemic helical (*E*)-cycloalkenes. For example, preparative HPLC enantioseparation of 1,1,3,3,6,6-hexamethyl-1-sila-*trans*-4,5-cycloheptanediol, and conversion of the levorotatory (*S*,*S*)-enantiomer to the corresponding thionocarbonate, paved the way to (–)-(*R*)-(*E*)-1,1,3,3,6,6-hexamethyl-1-sila-4-cycloheptene through

---

xxii Racemization during Pd-catalyzed S_N2' reactions is almost negligible when perfluorinated alkyl groups are present in the propargylic substrate.

xxiii An S_N'' reaction is similar to an S_N' displacement but occurs with simultaneous rearrangement of two double bonds.

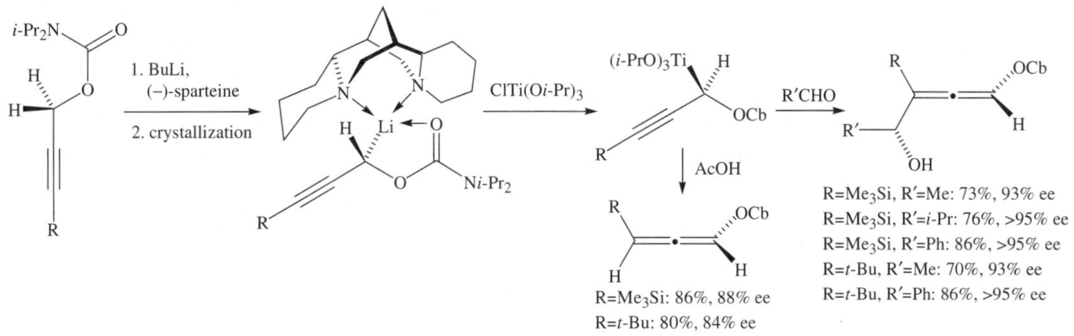

Scheme 6.81 Asymmetric synthesis of 4-hydroxyallenyl carbamates.

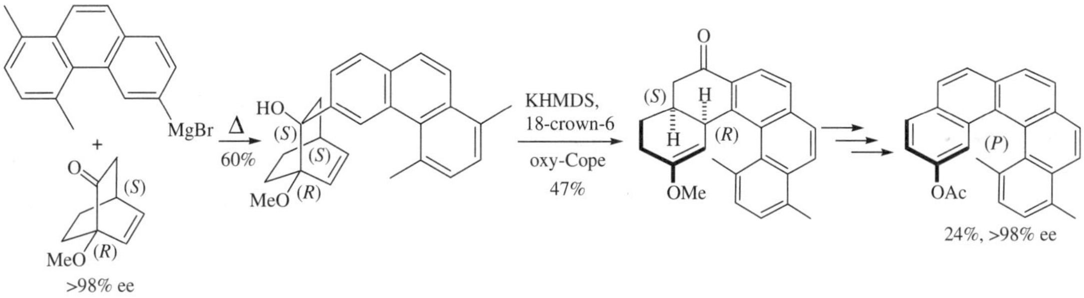

Scheme 6.82 Conversion of central to helical chirality via oxy-Cope rearrangement.

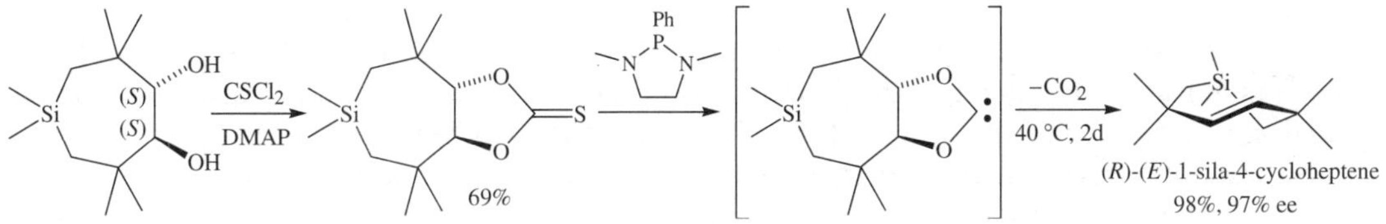

Scheme 6.83 Enantioselective synthesis of (*R*)-(*E*)-1,1,3,3,6,6-hexamethyl-1-sila-4-cycloheptene.

Corey–Winter elimination using 1,3-dimethyl-2-phenyl-1,3,2-diazaphospholidine for abstraction of sulfur.[653] Stereospecific *syn*-elimination of the intermediate carbene gave the helical (*R*)-cyclo-alkene in 98% yield and 97% ee, Scheme 6.83.[654] The opposite enantiomer was obtained from the dextrorotatory *trans*-1,2-diol by the same procedure.

6.4.6 Conversion of Axial Chirality to Other Chiral Elements

The advance of atroposelective methods described in Chapters 6.2 and 6.3 has led to a pool of readily available axially chiral compounds that serve as versatile synthetic building blocks. The development of reactions that allow one to transform axial chirality into other chiral elements has provided a powerful means for the total synthesis of natural products, *e.g.*, erythrinan alkaloids exhibiting a quaternary chiral center, Scheme 6.84.[655–661] Following a multiple step synthetic route based on Suzuki cross-coupling and HPLC enantioseparation, Matsumoto *et al.* prepared an enantiopure polysubstituted biphenyl bearing a *t*-Boc-protected amino function. Selective

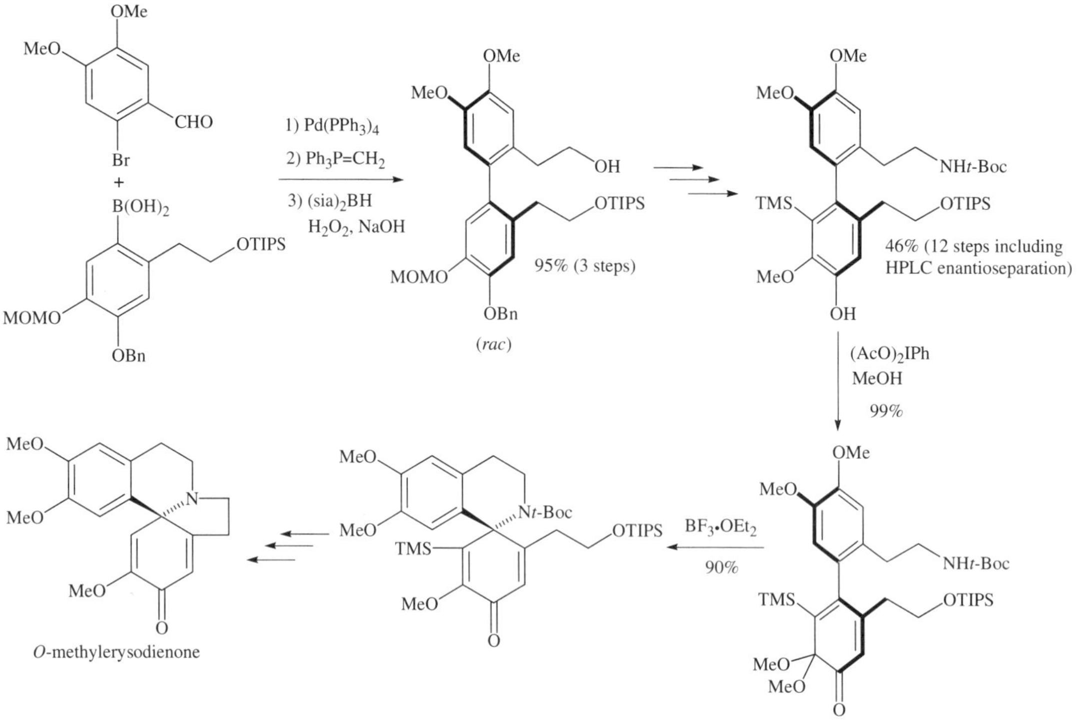

Scheme 6.84 Spirocyclization of an axially chiral quinone acetal.

oxidation in methanol gave an atropisomeric quinone acetal that was subjected to Lewis acid-catalyzed spirocyclization via intramolecular Michael-type addition. In this reaction sequence, the chiral axis of a tri-*ortho*-substituted biphenyl is destroyed during ring formation and a chiral spiro center is established without any sign of racemization. The desired spirotetrahydroisoquinoline was isolated in enantiopure form and transformed to *O*-methylerysodienone.

Curran and coworkers discovered that control of the stereodynamic properties of intermediate radicals prepared from axially chiral *ortho*-iodoacrylanilides is feasible and allows asymmetric cyclization coinciding with chirality conversion.[662–665] They determined the energy barrier to racemization of atropisomeric *N*-methyl-*N*-(2-iodo-4,6-dimethylphenyl)acrylamide as 128.9 kJ/mol, and employed enantiomerically enriched samples in Et$_3$B/O$_2$-initiated radical cyclizations using Bu$_3$SnH for chain propagation at room temperature, Scheme 6.85. Although the conformational stability of the intermediate radical is not known, comparison of the structure with *N*-alkyl-*N*-acyl-2-methyl-anilides, which possess rotational energy barriers of approximately 80 kJ/mol, suggests that racemization should occur rapidly at room temperature. Nevertheless, intramolecular trapping and lactam formation of the axially chiral *N*-aryl acrylamide radical appears to be considerably faster than rotation about the stereolabile chiral axis and radical cyclization of various *ortho*-iodoacryl-anilides yields chiral oxindoles in high enantiomeric excess.

The usefulness of allenes as synthetic building blocks stems from the exceptional combination of axial chirality and reactivity in regioselective and stereospecific cycloadditions, sigmatropic rearrangements and electrocyclic reactions.[666,667] A wide range of chiral carbocyclic and heterocyclic compounds has been prepared by [2 + 2]-, [4 + 2]-, [4 + 3]-, [2 + 2 + 1]-, and [2 + 2 + 2]cyclo-additions, and ene reactions.[668–684] The potential of this approach, the conversion of the axial chirality of allenes to other chiral elements, was recognized by Kanematsu and coworkers. This group found that the Lewis acid-promoted Diels–Alder reaction between a (−)-menthyl-derived

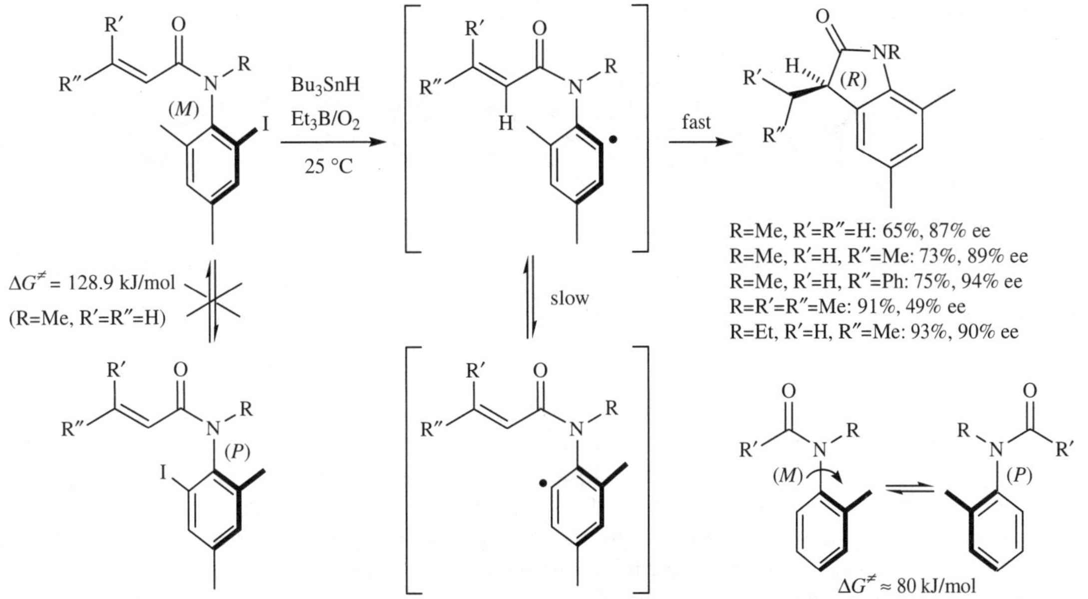

Scheme 6.85 Chirality conversion in radical cyclizations of *ortho*-iodoacrylanilides and racemization of *N*-alkyl-*N*-acyl-2-methylanilides.

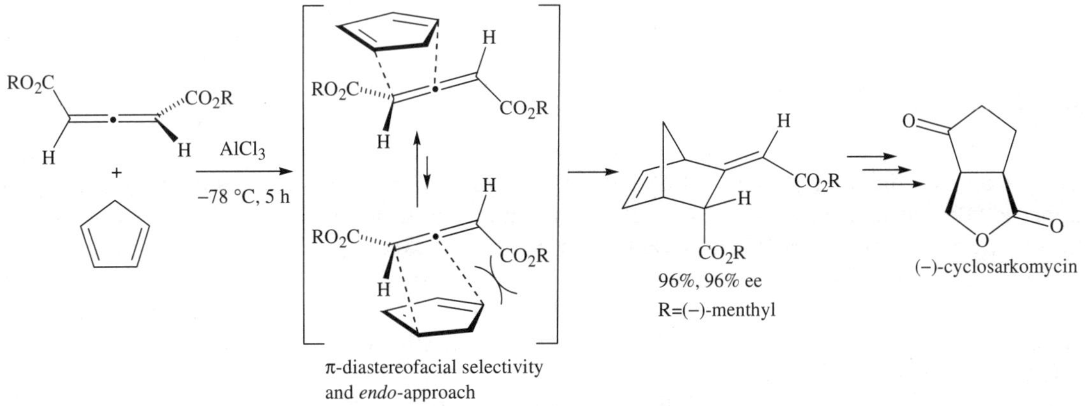

Scheme 6.86 Cycloaddition of allene-1,3-dicarboxylates and cyclopentadiene.

(*M*)-allene-1,3-dicarboxylate and cyclopentadiene produces a synthetically challenging bicyclo-[2.2.1]hept-2-ene derivative with an exocyclic (*Z*)-alkene moiety.[685–687,xxiv] The corresponding *endo*-adduct was obtained in almost enantiopure form and used for the synthesis of (−)-cyclosarkomycin, Scheme 6.86.

The availability of chiral allenes has generated increasing interest in the use of these versatile building blocks for the asymmetric synthesis of complex target compounds. For instance, introduction of a diene motif to an axially chiral allene sets the stage for fast Rh(I)-catalyzed intramolecular cycloaddition. This reaction furnishes diastereomerically pure bicyclic products

---

xxiv The high π-diastereofacial selectivity of this reaction is probably due to the axial chirality of the allene, although menthyl esters are known to be effective chiral auxiliaries in cycloadditions.

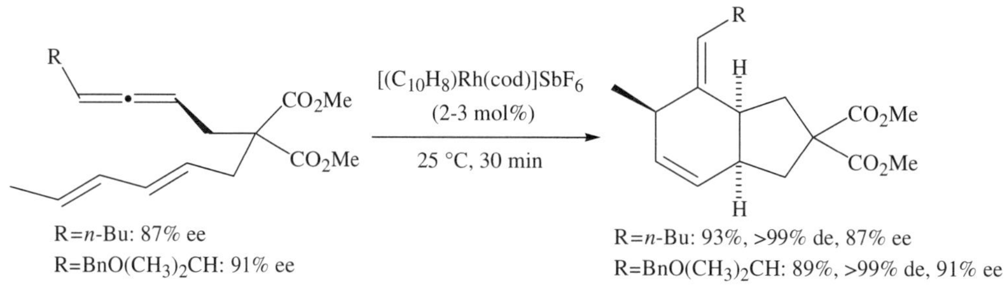

Scheme 6.87 Intramolecular Diels–Alder reaction of allenes.

exhibiting three stereocenters, Scheme 6.87.[688] Irradiation of chiral allenediynes in the presence of $CpCo(CO)_2$ produces tricyclic adducts through a sequence of intramolecular cycloadditions showing excellent regio-, chemo- and stereoselectivity.[689] It is assumed that initial $[2+2]$cyclo-addition generates a cobaltcyclopentadiene-derived intermediate that participates in a Diels–Alder reaction with the internal double bond of the allenic unit. Overall, this reaction sequence constitutes a $[2+2+2]$cycloaddition. The cyclization is controlled by steric interactions between the diene and the dienophilic allene which is preferentially approached at the readily accessible *Re*-face. The unstable cobalt-bridged cycloadduct then undergoes reductive elimination to a [6.6.5]tricyclic η^4-cobalt complex with an (*E*)-exocyclic double bond. Interestingly, formation of the corresponding (*Z*)-alkene via *Si*-face attack is not observed, Scheme 6.88.

Substituted allenes are excellent substrates for Claisen rearrangements resulting in [1,4]-chirality transfer and simultaneous translation of axial to central chirality.[690,691] Hoppe found that treatment of an enantiopure α-hydroxy allene with a ketene *N,O*-acetal gives a (*Z*)-vinyl ether that spontaneously forms an (*E,E*)-diene via [3,3]-sigmatropic rearrangement, Scheme 6.89. The stereochemical outcome of this reaction suggests that the favored transition state has a chairlike conformation with the benzyloxy-derived substituent in pseudoequatorial position. This arrangement avoids 1,3-diaxial repulsion with the dimethylamino group but it requires the vinyl moiety to approach the allene from the more sterically crowded face bearing the carbamate protecting group. Substrates with a propargylic chiral center are often employed in the synthesis of axially chiral allenes, see Chapter 6.4.5. On the other hand, the direction of chirality conversion can be reversed by stereoselective allylic substitutions at haloallenes with organometallic nucleophiles. Organocopper reagents are particularly useful for stereoselective *anti*-S_N2' displacements of 1,3-disubstituted 1-bromoallenes, Scheme 6.90.[692]

By analogy with S_N2' reactions of allylic, allenylic and propargylic structures that are susceptible to nucleophilic displacements, unsaturated silyl analogs react with electrophiles through an S_E' mechanism. These transformations proceed with efficient chirality conversion and afford synthetically useful chiral building blocks from allyl-, allenyl- or (allenylmethyl)silanes. Ito, Marshall and others demonstrated that electrophilic displacements of allenylsilanes with aldehydes yield chiral homopropargylic alcohols possessing two new chiral centers, Scheme 6.91.[693–696] Reaction of matched and mismatched pairs of chiral allenylsilanes and aldehydes in the presence of titanium tetrachloride produces diastereomeric stereotriads[xxv] with perfect substrate-based stereocontrol.[xxvi]

[xxv] Multifunctional stereotriads are invaluable precursors for the synthesis of polyketides which are important secondary metabolites of bacteria, fungi, plants, and animals.

[xxvi] In a reaction with substrate-based stereocontrol, the stereochemical outcome is completely controlled by the chirality of the substrate and is independent of the reagent structure. For comparison, see the concept of double stereodifferentiation described in Chapter 5.3.2.

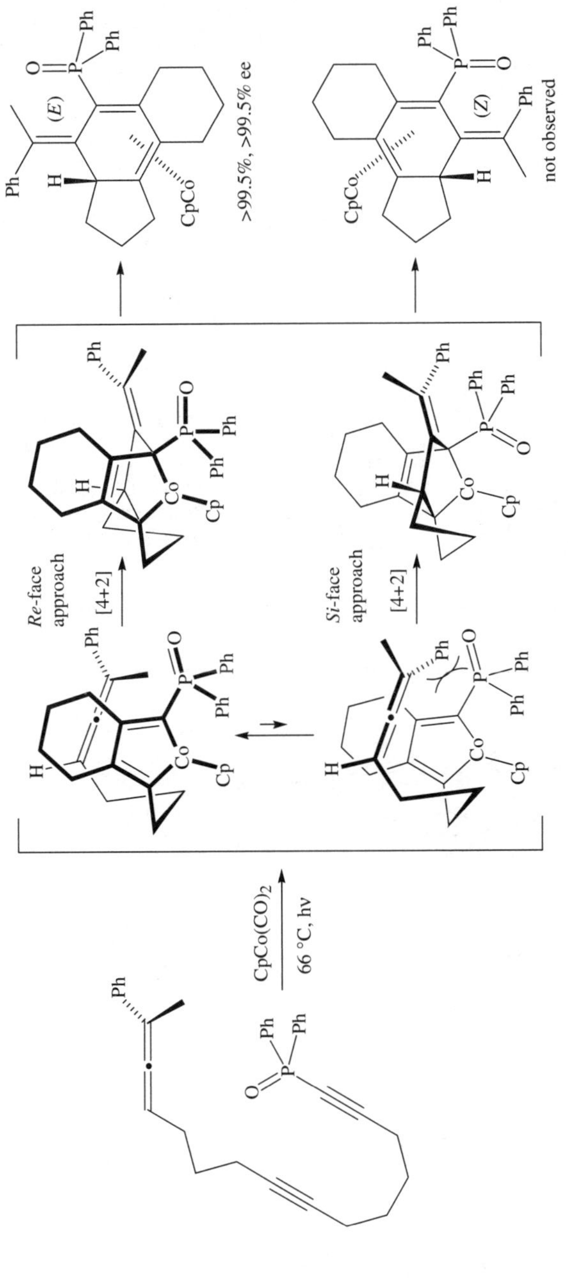

Scheme 6.88 Transformation of axial to central chirality in a cobalt-mediated [2 + 2 + 2]cycloaddition of allene diynes.

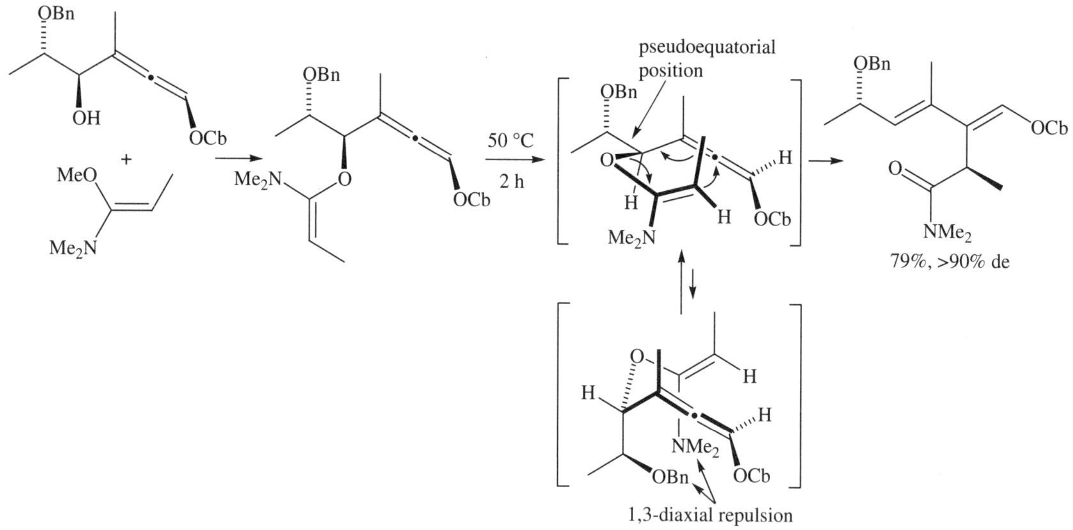

Scheme 6.89   Claisen rearrangement of a vinylic α-alkoxy allene.

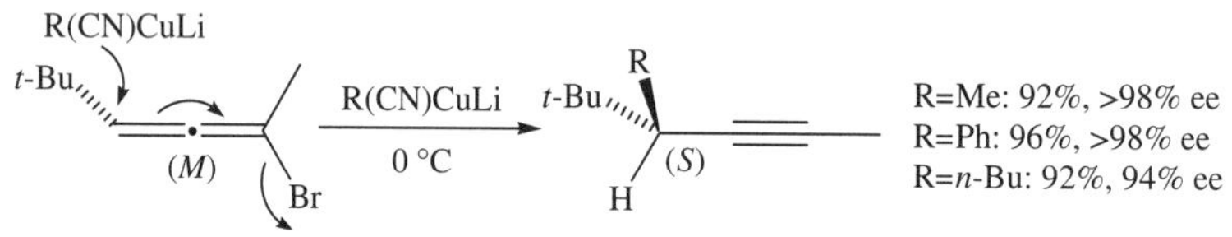

Scheme 6.90   Allylic *anti*-substitution of 1,3-disubstituted 1-bromoallenes with organocopper reagents.

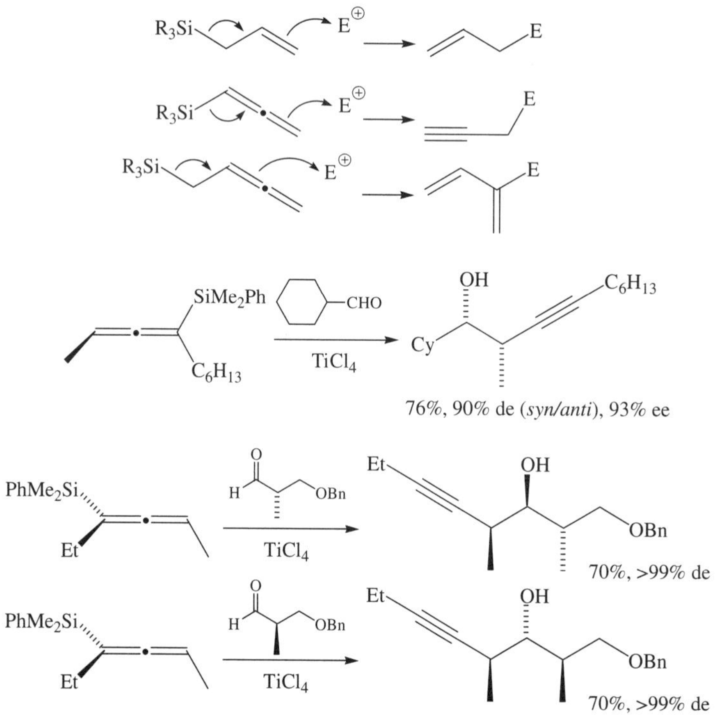

Scheme 6.91   Classification of S$_E'$ reactions with allyl-, allenyl- and (allenylmethyl)silanes (top), and asymmetric transformation of allenylsilanes to propargylic alcohols (bottom).

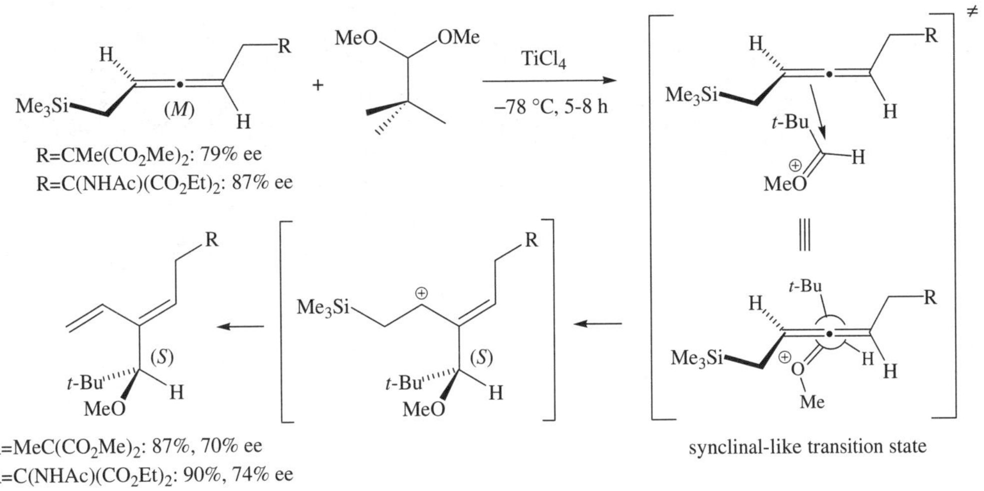

R=CMe(CO$_2$Me)$_2$: 79% ee
R=C(NHAc)(CO$_2$Et)$_2$: 87% ee

R=MeC(CO$_2$Me)$_2$: 87%, 70% ee
R=C(NHAc)(CO$_2$Et)$_2$: 90%, 74% ee

synclinal-like transition state

Scheme 6.92 Chirality interconversion in S$_E'$ reactions between (allenylmethyl)silanes and pivalaldehyde dimethyl acetal in the presence of a Lewis acid.

Ogasawara's and Hayashi's groups successfully employed axially chiral (allenylmethyl)silanes in asymmetric S$_E'$ reactions, Scheme 6.92.[697] Treatment of pivalaldehyde dimethyl acetal with stoichiometric amounts of titanium tetrachloride generates an electrophile that preferentially approaches the central carbon of (allenylmethyl)silanes from the less sterically hindered side. Interconversion of axial to central chirality occurs through an S$_E'$ pathway having a synclinal-like transition state in which the most sterically demanding *tert*-butyl group of the electrophile is placed in the least crowded location. The electrophile therefore approaches the allene with its *Re*-face to form an (*S*)-configured secondary ether, and spontaneous cleavage of the silyl group establishes a 1,3-diene unit.

Few tandem and multi-component reactions of axially chiral allenes are known.[698–703] Jamison discovered a highly regio- and enantioselective Ni-catalyzed three-component coupling procedure that utilizes nonracemic allenes, aromatic aldehydes and silanes. This reaction includes addition of an electrophilic aldehyde to the central *sp*-hybridized carbon and affords allylic silyl ethers in good yields and with high diastereo- and enantioselectivity, Scheme 6.93.[704] A remarkable feature is the unusual addition to the least nucleophilic *sp*-carbon in the allenic substrate and the surprising stereochemical outcome, which has prompted mechanistic studies including deuterium labeling with Et$_3$SiD.[xxvii] Results obtained with (*P*)-1-cyclohexyl-3-methylallene and benzaldehyde are in accordance with an initial approach of the nickel complex to the less substituted double bond from the more accessible face opposite to the cyclohexyl ring. Oxidative addition is followed by coordination of the prochiral aldehyde from the less sterically encumbered face opposite to the axial methyl group. The intermediate five-membered metallacycle reacts with the silane to give an η^3-allyl nickel complex that undergoes reductive elimination with retention of configuration at the isotope-labeled carbon.

Chiral pyrrolines, butenolides and dihydrofurans are abundant components of natural products and are important building blocks for asymmetric synthesis. These and other heterocycles are often prepared from chiral α-amino and α-hydroxy allenes or 2,3-allenoic acids via copper(II)-, silver(I)-, PhSeCl- or NBS-mediated cyclization with concomitant conversion of chirality.[705–709] In particular,

---

[xxvii] Multi-component coupling reactions of allenes usually occur at the more reactive *sp^2*-hybridized carbons.

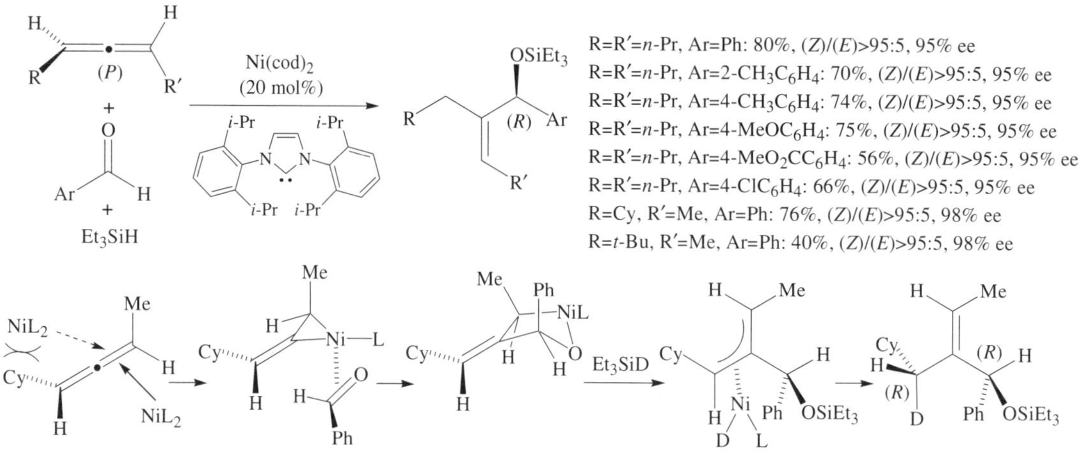

Scheme 6.93 Multi-component coupling of nonracemic allenes, aryl aldehydes and silanes using an *N*-heterocyclic carbene-derived nickel catalyst.

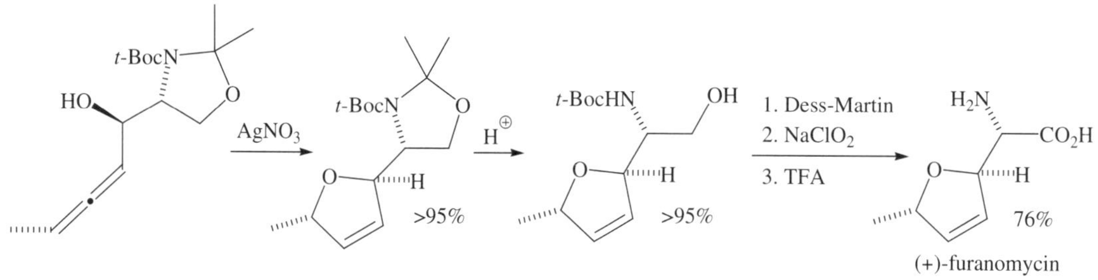

Scheme 6.94 Total synthesis of (+)-furanomycin.

the Ag(I)-catalyzed cyclization of α-allenic alcohols, which has been intensively investigated by Marshall and coworkers, has found many synthetic applications.[710–715] Standaert utilized this stereospecific reaction for construction of a *trans*-2,5-dihydrofuran ring that is easily derivatized to (+)-furanomycin, a natural amino acid found in *Streptomyces*, Scheme 6.94.[716] The introduction of carbophilic gold(III)-derived Lewis acids to catalytic cyclization of chiral allenes has led to additional synthetic entries towards dihydrofurans, pyrrolines and dihydrothiophenes, Scheme 6.95.[717,718] Cycloisomerization of α-hydroxy allenes in the presence of catalytic amounts of $AuCl_3$ at room temperature furnishes tri- and tetrasubstituted 2,5-dihydrofurans.[719] Mechanistic insights into this reaction, which can also be applied to α-amino allenes and α-thio allenes, have been obtained by NMR analysis and *ab initio* calculations. The soft Lewis acid first enhances the electrophilicity of one of the two allenic double bonds upon coordination. Stereoselective cyclization and demetalation of the intermediate aurate complex then destroy the chiral axis and generate a new chiral center.

Regio- and stereoselective halohydroxylation of allenic sulfoxides transforms axial to central chirality.[720] The mild reaction produces (*E*)-2-halo-1-phenylsulfinyl-1-alkene-3-ols in excellent yield and enantiopurity. A mechanism that is in agreement with the stereochemical outcome of the reaction between (+)-camphor-derived (*M*,S_S)-allenyl sulfoxide and iodine in the presence of water is given in Scheme 6.96. Accordingly, electrophilic addition of iodine forms a three-membered iodonium ring that is opened by intramolecular attack of the sulfinyl oxygen. This

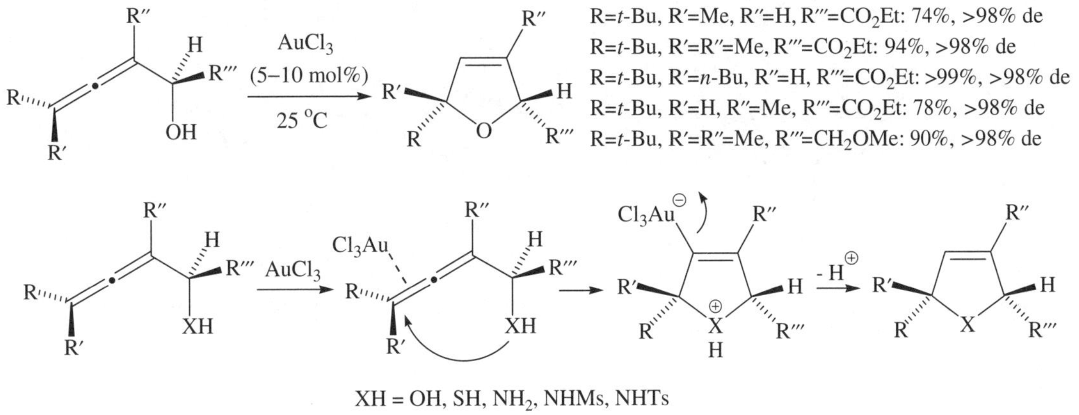

R=*t*-Bu, R′=Me, R″=H, R‴=CO$_2$Et: 74%, >98% de
R=*t*-Bu, R′=R″=Me, R‴=CO$_2$Et: 94%, >98% de
R=*t*-Bu, R′=*n*-Bu, R″=H, R‴=CO$_2$Et: >99%, >98% de
R=*t*-Bu, R′=H, R″=Me, R‴=CO$_2$Et: 78%, >98% de
R=*t*-Bu, R′=R″=Me, R‴=CH$_2$OMe: 90%, >98% de

XH = OH, SH, NH$_2$, NHMs, NHTs

Scheme 6.95 Gold(III)-catalyzed asymmetric cyclization of allenes.

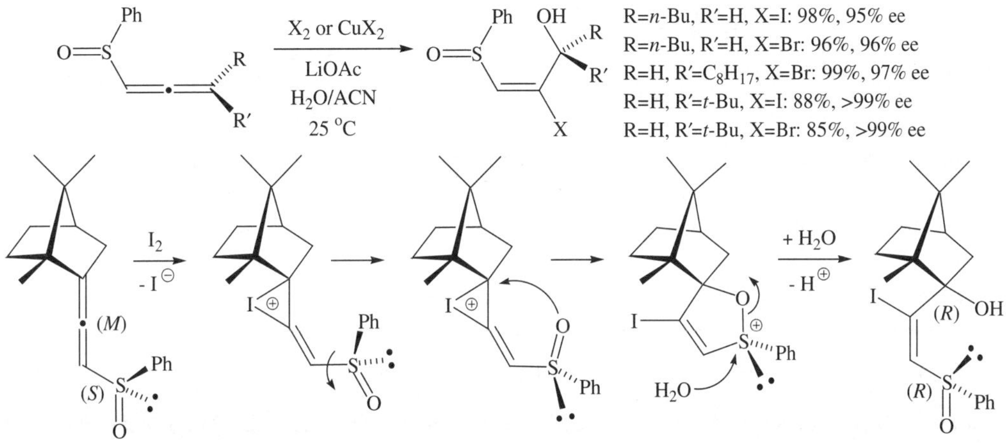

R=*n*-Bu, R′=H, X=I: 98%, 95% ee
R=*n*-Bu, R′=H, X=Br: 96%, 96% ee
R=H, R′=C$_8$H$_{17}$, X=Br: 99%, 97% ee
R=H, R′=*t*-Bu, X=I: 88%, >99% ee
R=H, R′=*t*-Bu, X=Br: 85%, >99% ee

Scheme 6.96 Asymmetric halohydroxylation of chiral allenyl phenyl sulfoxides.

S$_N$i reaction furnishes a heterocyclic five-membered ring exhibiting a vinyl iodide unit. Finally, ring opening through S$_N$2 attack of water at the positively charged chiral sulfur atom completes the halohydroxylation with concurrent inversion of configuration.

Helicenes have become popular synthetic targets due to the inherent chirality and intriguing optical and electronic properties of this class of compounds.[721–725,xxviii] Nozaki and coworkers developed a synthetic route towards hetero[7]helicenes that is based on conversion of axial to helical chirality during palladium-catalyzed *N*- or *O*-arylation and concomitant ring fusion of 4,4′-biphenanthryl-3,3′-diol derivatives, Scheme 6.97.[726] For example, cyclization via double *N*-arylation of enantiopure (*P*)-4,4′-biphenanthryl-3,3′-ylene dinonaflate and aniline in the presence of catalytic amounts of Pd$_2$(dba)$_3$ and Xantphos gives 94% of the corresponding (*P*)-aza[7]helicene in more than 99% ee. The stereospecific helicene formation requires long reaction times but proceeds with excellent yield and conservation of chirality. Several readily available axially chiral compounds have been employed in the synthesis of nonracemic helical compounds.[727–730] A synthetic route to thiahelicenes based on transition metal-mediated reductive cyclization of sterically crowded biaryl dialdehyde precursors has been described by Tanaka's group.[731] The key step of this approach is an

---

xxviii Helicenes are nonplanar *ortho*-annulated aromatic frameworks with a helical structure.

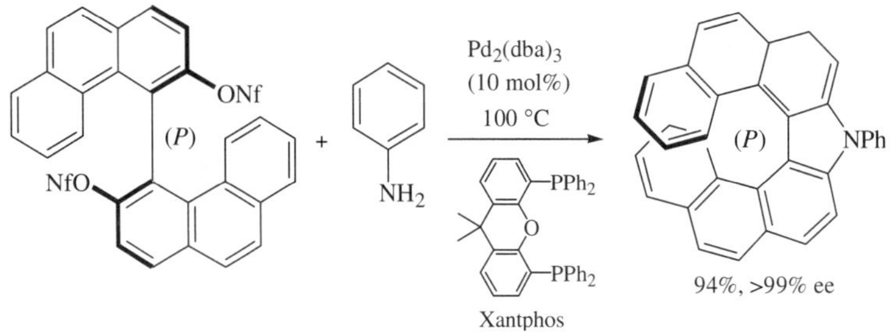

Scheme 6.97   Asymmetric synthesis of an aza[7]helicene via double *N*-arylation of enantiopure (*P*)-4,4′-biphenanthryl-3,3′-ylene dinonaflate.

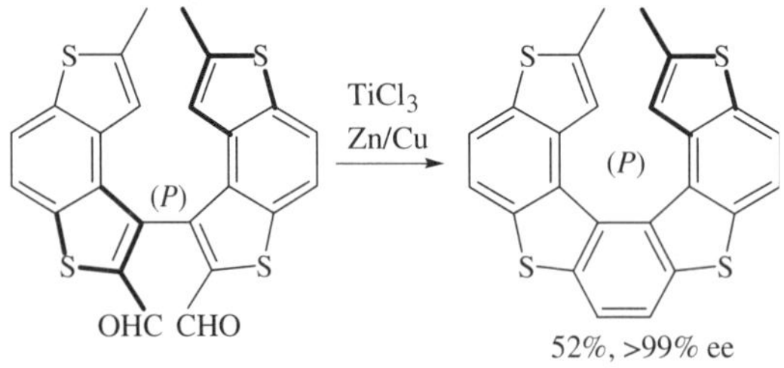

Scheme 6.98   Thia[7]helicene formation based on intramolecular McMurry coupling.

intramolecular McMurry coupling step that accomplishes ring closure and aromatization. Cyclization of enantiopure (*P*)-2,2′-diformyl-7,7′-dimethyl-1,1′-bi[benzo[1,2-*b*:4,3-*b*′]dithiophene gives a C$_2$-symmetric (*P*)-tetrathia[7]helicene in literally enantiopure form, Scheme 6.98.

6.4.7   Conversion of Planar Chirality to Other Chiral Elements

Few examples of conversion of a chiral plane to other elements of chirality have been reported to date.[732] Regio- and stereoselective ring opening of isopropyl (*R*)-1,12-dioxa[12](1,14) naphthalenophane-14-carboxylate with 2-methoxy-1-naphthylmagnesium bromide provides an atropisomeric (*P*)-1,1′-binaphthyl derivative.[733] The stereochemical outcome can be explained by preferential Grignard attack from the less sterically hindered side opposite to the naphthalenophane bridge. Chelation of the aryl Grignard by both the ester and *ortho*-methoxy group of the ansa compound initiates the regioselective nucleophilic displacement of the tether. Additional coordination of the 2-methoxy group of the introduced naphthyl moiety to magnesium then governs the generation of axial chirality during opening of the naphthalenophane bridge, Scheme 6.99.

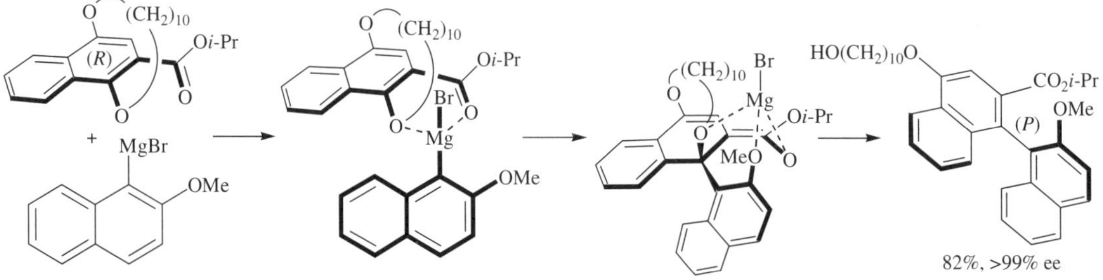

Scheme 6.99   Regio- and stereoselective opening of an ansa compound.

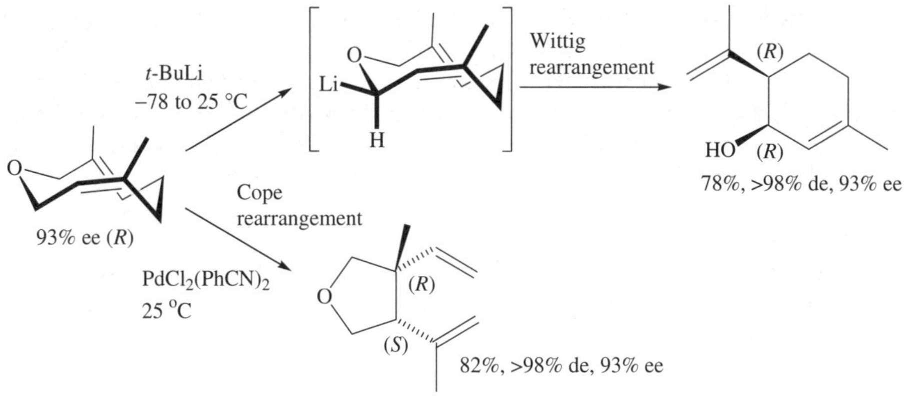

Scheme 6.100 Conversion of planar chirality of a cyclic ether to central chirality by [2,3]-Wittig and palladium-catalyzed Cope-type rearrangements.

Tomooka's group realized that macrocyclic unsaturated ethers are useful synthetic precursors of complex tetrahydrofurans and cyclohexenols.[734] Treatment of a chiral diallylic cyclic ether with *tert*-butyllithium under cryogenic conditions generates a reactive organolithium species that undergoes transannular [2,3]-Wittig rearrangement, Scheme 6.100. As a result, the planar chirality of the nine-membered ether ring is translated to two asymmetric carbon atoms. Alternatively, a trisubstituted tetrahydrofuran can be obtained from the same ether by Pd(II)-catalyzed Cope-type rearrangement. In both cases, chirality interconversion proceeds without measurable racemization.

6.5 SELF-REGENERATION OF STEREOGENICITY AND CHIRAL RELAYS

The advance of stereoselective synthesis has been fueled by the discovery of powerful asymmetric reactions and imaginative strategies. In particular, Seebach's concept of self-regeneration of stereocenters (SRS), and other methods that exploit temporary or transient chiral intermediates, provide an elegant entry to asymmetric synthesis of complex target compounds. The SRS approach allows manipulation of a chiral center without concomitant racemization by a four-step reaction sequence that involves formation of isolable chiral intermediates. The term "memory of chirality" has been coined for transformations of enantiopure compounds via transient and stereolabile chiral intermediates that can not be isolated. As Siegel and Cozzi have pointed out, the use of this stereochemical metaphor should be discouraged because it implies that the chirality of the molecule is temporarily lost and then somehow "remembered" later in the reaction.[735] In such a case, one would obtain either achiral products or a racemate. A closer look at reactions that supposedly show a "chiral memory" reveals that chirality is in fact conserved throughout the whole reaction, see Chapter 6.5.3. Diastereoselective synthesis with so-called chiral relays relies on fluxional auxiliaries that enhance the stereochemical information of an existing chiral element and allow effective transmission of chirality to a remote reaction site. Asymmetric synthesis with relays may be considered the diastereoselective variant of enantioselective catalysis with conformationally flexible achiral ligands, which is discussed in detail in Chapter 6.6.

6.5.1 Stereocontrolled Substitution at a Chiral Center

The synthesis of chiral compounds often requires modifications at an asymmetric carbon center. This can often be accomplished by a concerted reaction with predictable stereochemical outcome and without racemization or diastereomerization of precious starting materials. Bimolecular nucleophilic substitution allows stereocontrolled replacement of a substituent attached to an asymmetric carbon atom. For example, nucleophilic attack of an azide at a chiral secondary tosylate follows an S_N2 mechanism with inversion of configuration. Alternatively, the Mitsunobu reaction provides a venue for stereospecific S_N2 reactions of compounds that do not bear a good leaving group or when weak nucleophiles are used, Scheme 6.101.[736,737] Treatment of a carboxylic acid and an alcohol with diethyl azodicarboxylate and triphenylphosphine generates a nucleophilic carboxylate and an intermediate alkoxyphosphonium salt susceptible to S_N2 attack. The Mitsunobu reaction conveniently transforms a chiral secondary alcohol to the corresponding acylate with inversion of configuration.

Of course, interconversion of enantiomers can also be accomplished by other S_N2 displacements. A classic example is the transformation of a chiral alcohol to the tosylate and subsequent nucleophilic replacement by a hydroxide ion, Scheme 6.102. Alternatively, a reversal of the

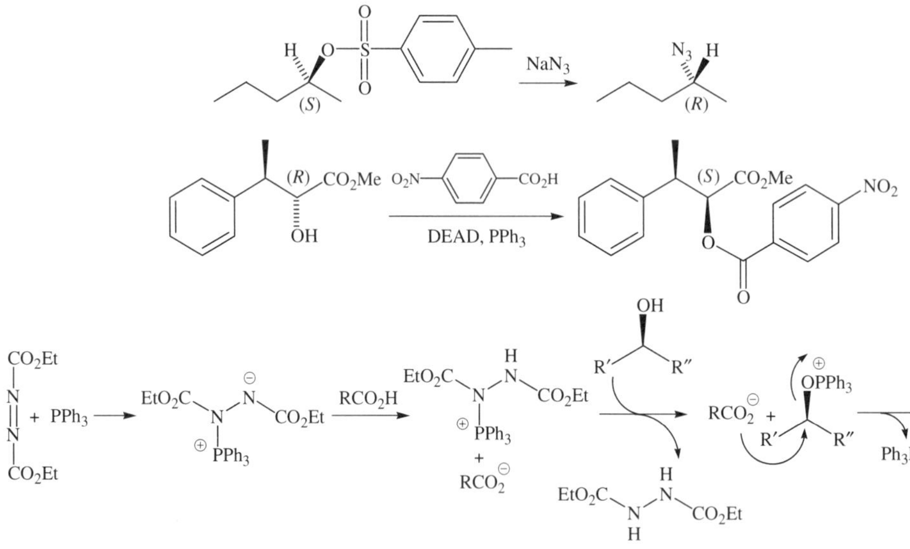

Scheme 6.101 Stereocontrolled transformations at a chiral center: nucleophilic substitution of a tosylate (top) and Mitsunobu reaction (bottom).

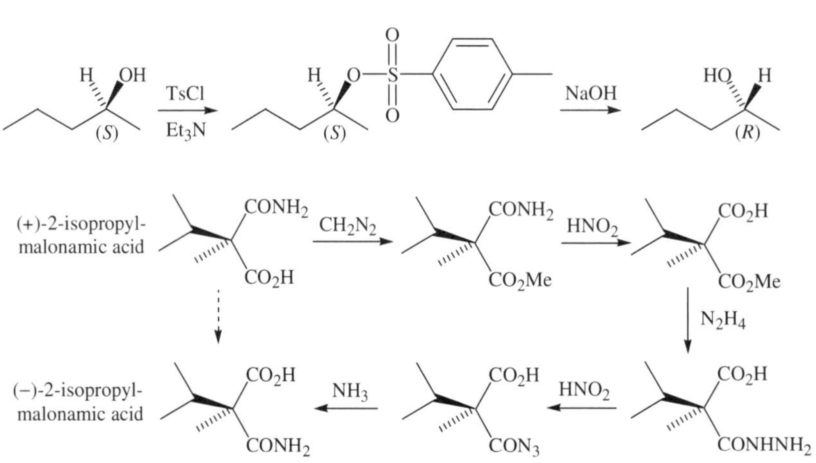

Scheme 6.102 Strategies for controlled interconversion of enantiomers.

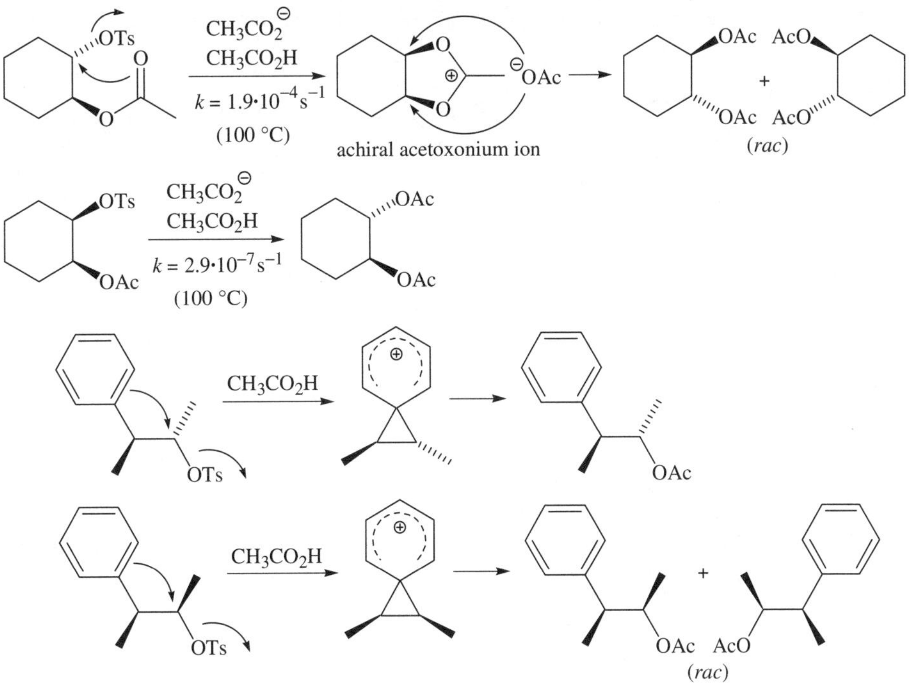

Scheme 6.103 Neighboring effects in nucleophilic displacements.

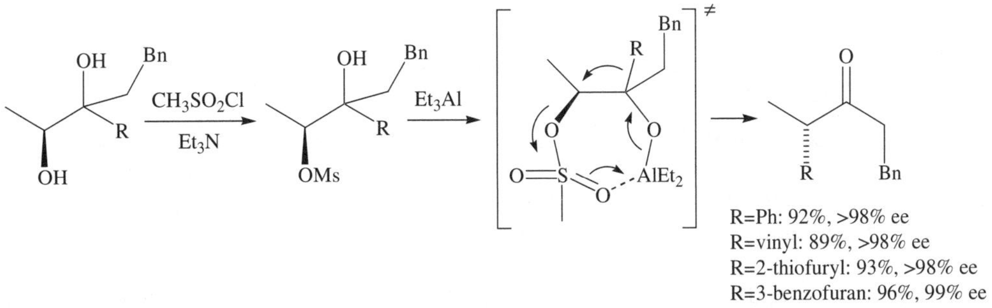

R=Ph: 92%, >98% ee
R=vinyl: 89%, >98% ee
R=2-thiofuryl: 93%, >98% ee
R=3-benzofuran: 96%, 99% ee

Scheme 6.104 Asymmetric Pinacol-type rearrangement.

absolute configuration at an asymmetric carbon atom with complete conservation of chirality is feasible through stepwise exchange of peripheral functionalities, *i.e.*, without direct involvement of any bonds at the asymmetric carbon atom. This has been demonstrated in a historic experiment conducted by Fischer and Brauns.[738] They were able to transform dextrorotatory 2-iso-propylmalonamic acid to the levorotatory enantiomer by stepwise interconversion of the primary amide and carboxylic acid functions, Scheme 6.102.[xxix] In the first step, the free carboxylic acid group was protected as the methyl ester. After hydrolysis of the primary amide function, the ester group was converted to an acyl azide using hydrazine and nitrous acid. Finally, treatment with ammonia completed the conversion of the originally free acid group to a primary amide. Systematic

---

[xxix] Fischer and Brauns carefully chose a route that (a) avoids achiral intermediates such as malonic acid and (b) allows chemical differentiation of the two peripheral carbon atoms of the chiral substrate during each step. All reactions were performed under mild conditions to exclude enolization and racemization.

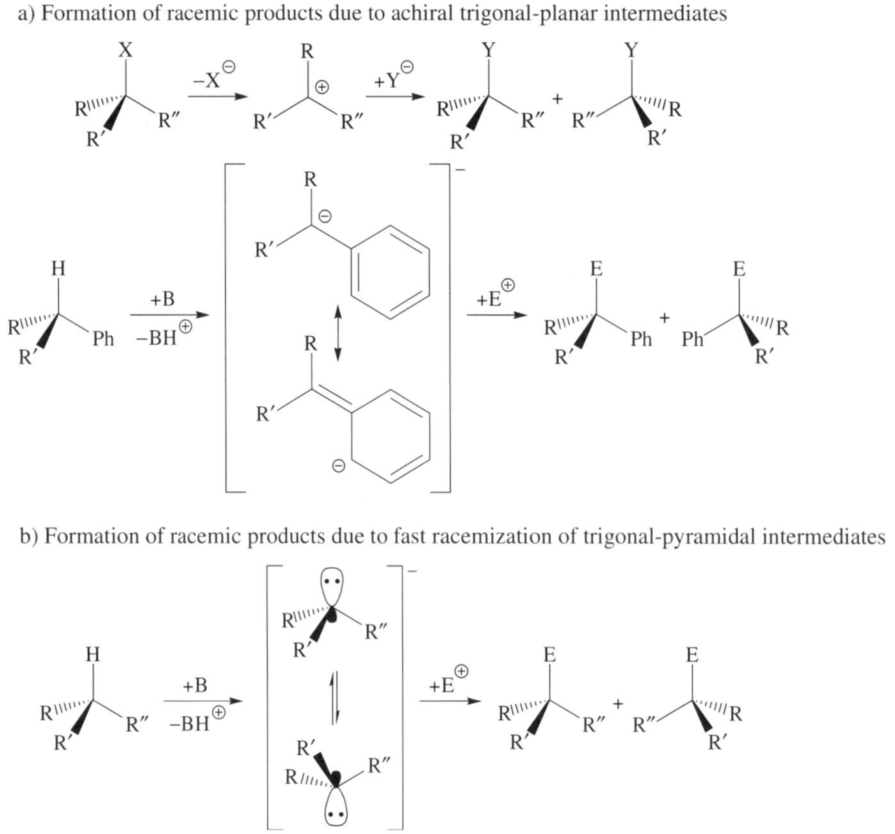

a) Formation of racemic products due to achiral trigonal-planar intermediates

b) Formation of racemic products due to fast racemization of trigonal-pyramidal intermediates

Scheme 6.105   Possible racemization courses of substitution reactions at asymmetric carbons involving either achiral or stereolabile chiral intermediates.

functional group transformations thus allowed conversion of (+)-2-isopropylmalonamic acid to the levorotatory enantiomer without manipulation of the bonds attached to the asymmetric carbon center.

The stereochemistry and reaction rate of nucleophilic substitutions are susceptible to anchimeric effects originating from participation of neighboring groups.[739–741] For example, nucleophilic displacement of the tosylate group of *cis*- and *trans*-2-acetoxycyclohexyl 4-toluenesulfonate by acetate requires considerably different reaction times. Solvolysis of the *trans*-isomer is approximately 700 times faster than the reaction of the *cis*-isomer, Scheme 6.103. This is generally attributed to participation of the neighboring acetoxy group in the reaction of the *trans*-isomer. The nucleophilic displacement in this isomer includes formation of an intermediate acetoxonium ion by intramolecular S_N2 attack. Such intramolecular assistance is geometrically impossible in the *cis*-isomer which explains its different reactivity in nucleophilic substitutions. The adjacent acetoxy group not only facilitates replacement of the tosylate group but also affects the stereochemical outcome of the reaction. Conversion of enantiopure *trans*-2-acetoxycyclohexyl 4-toluenesulfonate to the corresponding acetoxonium ion generates a symmetry plane and therefore results in the loss of chirality. Nucleophilic ring opening by an acetate anion occurs at either carbon atom with inversion of configuration and gives the racemic *trans*-diacetate. By contrast, the enantiopure *cis*-isomer undergoes S_N2 reaction with inversion of configuration to the *trans*-isomer without any loss of stereochemical integrity. Similar effects have been observed with 3-phenyl-2-butyl tosylates that

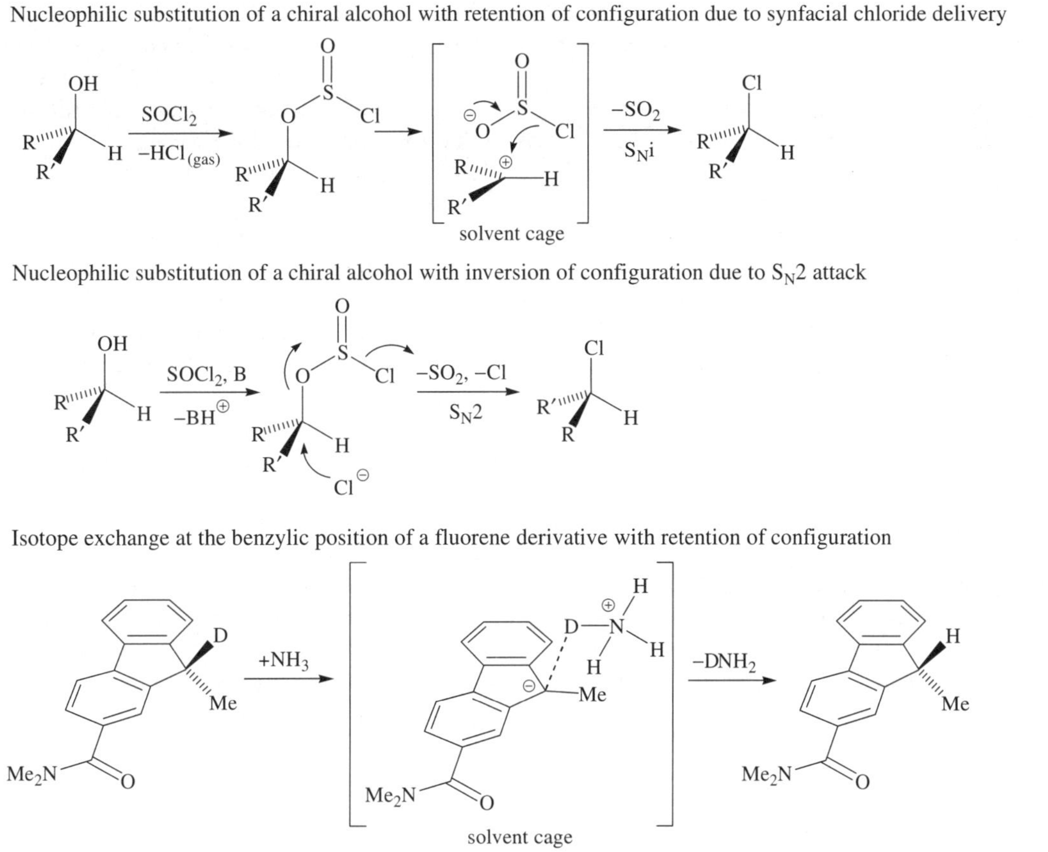

Nucleophilic substitution of a chiral alcohol with retention of configuration due to synfacial chloride delivery

Nucleophilic substitution of a chiral alcohol with inversion of configuration due to S_N2 attack

Isotope exchange at the benzylic position of a fluorene derivative with retention of configuration

Scheme 6.106 Concerted and dissociative displacements with trigonal-planar cationic and anionic intermediates.

show participation of the adjacent phenyl ring and S_Ni replacement of the tosylate prior to solvolysis.[742,743]

Nucleophilic displacements that occur during concerted rearrangements generally provide excellent stereocontrol, see also Chapter 6.4. For instance, Pinacol-type [1,2]-migrations allow stereospecific conversion of chiral diols to α-substituted ketones with complete conservation of chirality.[744–747] This has been demonstrated with regioselectively mesylated diols, Scheme 6.104.[748] Reaction of methansulfonyl chloride with chiral diols bearing adjacent tertiary and secondary alcohol groups proceeds faster at the latter position, thus setting the stage for an aluminum-catalyzed rearrangement. A variety of substituents with inherently high migratory aptitude such as vinyl, aryl and heteroaryl groups undergo enantiospecific [1,2]-migration from the carbon carrying the tertiary alcohol to the secondary sulfonate moiety. Importantly, the [1,2]-migration is not affected by the configuration at the tertiary alcohol group, probably due to the flexibility of the seven-membered transition state. The rearrangement proceeds with inversion of configuration at the mesylated carbon center and yields enantiopure α-substituted ketones. In fact, methane-sulfonylation of chiral diols and Lewis acid-catalyzed Pinacol rearrangement have been combined in a convenient one-pot procedure.

In contrast to stereospecific bimolecular nucleophilic displacements and rearrangements, two-step substitution pathways based on heterolytic or homolytic elimination of one substituent prior to incorporation of a new one often proceed via trigonalization of the reaction center, resulting in loss of stereochemical purity. This is generally attributed to formation of (a) achiral trigonal-planar cationic or π-conjugated radical and carbanionic intermediates, or (b) chiral unconjugated radical and anionic trigonal-pyramidal species that are prone to rapid racemization. For example, unimolecular nucleophilic substitutions usually involve an achiral carbocation while deprotonation at a chiral center generates a stereolabile carbanionic intermediate. Racemic products are ultimately obtained in both cases, Scheme 6.105. Some noteworthy exceptions to the scenarios outlined above have been developed into practical synthetic strategies. In particular, chiral organolithium intermediates have been utilized for asymmetric synthesis by Hoppe, Gawley, Still, Beak, and others, see Chapter 6.1. The interconversion of chiral alcohols to alkyl chlorides with thionyl chloride is known to follow an S_Ni-type mechanism with retention of configuration.[749] Treatment of an alcohol with thionyl chloride in the absence of a base generates a reactive chlorosulfinate ester while hydrogen chloride is released. Dissociation of this ester generates a short-lived ion pair consisting of a trigonal-planar cation and a chlorosulfite anion. The two reactive species are trapped within a solvent cage that effectively limits rotational and translational movements. Subsequent chlorination of the intermediate carbocation occurs with retention of configuration. The same reaction proceeds with inversion of configuration in the presence of pyridine or another base that forms a salt and thus retains nucleophilic chloride in solution. The change in the stereochemical outcome upon base addition is a consequence of an S_N2 attack by the free chloride anion on the chlorosulfinate ester, Scheme 6.106. Similarly, isotope exchange of a deuterated chiral fluorene derivative in the presence of ammonia proceeds with considerable retention of configuration.[750] Although one would expect the intermediate carbanion to display an achiral trigonal-planar geometry to maximize π-conjugation, the protonated product is obtained with high enantioselectivity. Apparently, the rate of D/H exchange is much faster than competing tumbling movements that would lead to the formation of racemic products.

6.5.2 Self-regeneration of Stereocenters

The stereoselective replacement of a substituent attached to an asymmetric atom via an elimination/addition pathway represents a major synthetic challenge. Seebach *et al.* developed a general strategy that utilizes a four-step sequence for stereoselective manipulation of a chiral center without concomitant racemization.[751] The first step involves diastereoselective introduction of a temporary chiral center to the enantiopure starting material. Then, the original chiral center of the substrate is trigonalized upon removal of one substituent. Importantly, the chiral information is not lost and remains stored in the temporarily incorporated asymmetric center that controls diastereoselective introduction of a new substituent to regenerate the original chiral center in the third step. The temporary chiral element is finally removed in the fourth step. Seebach coined the term self-regeneration of stereocenters (SRS) to describe the stepwise manipulation of a chiral center via intermediate trigonalization and subsequent diastereoselective regeneration in the presence of a second chiral element. The principles and usefulness of Seebach's SRS method are evident in the synthesis of the cell adhesion inhibitor BIRT-377 exhibiting an *N*-aryl substituted hydantoin structure and a quaternary chiral center, Scheme 6.107.[752–754] In this example, enantiopure (*R*)-*N*-*t*-Boc-alanine is converted to a diastereomerically pure *trans*-imidazolidinone with a pivalaldehyde-derived *N,N*-acetal unit. Diastereoselective formation of a rigid heterocyclic structure with a second chiral center thus accomplishes the first of the four steps required for SRS. Protection of the imidazolidinone using trifluoroacetic anhydride is followed by trigonalization of the asymmetric carbon in the amino acid unit in the presence of a strong base and stereoselective alkylation of the intermediate enolate with 4-bromobenzyl bromide, to realize the second and third steps of the SRS sequence. The alkylation occurs predominantly at the face of the enolate opposite to the shielding *tert*-butyl group of the chiral *N,N*-acetal unit. The synthesis of BIRT-377 is finally completed by removal of the temporary chiral center and conversion of the imidazolidinone to the desired

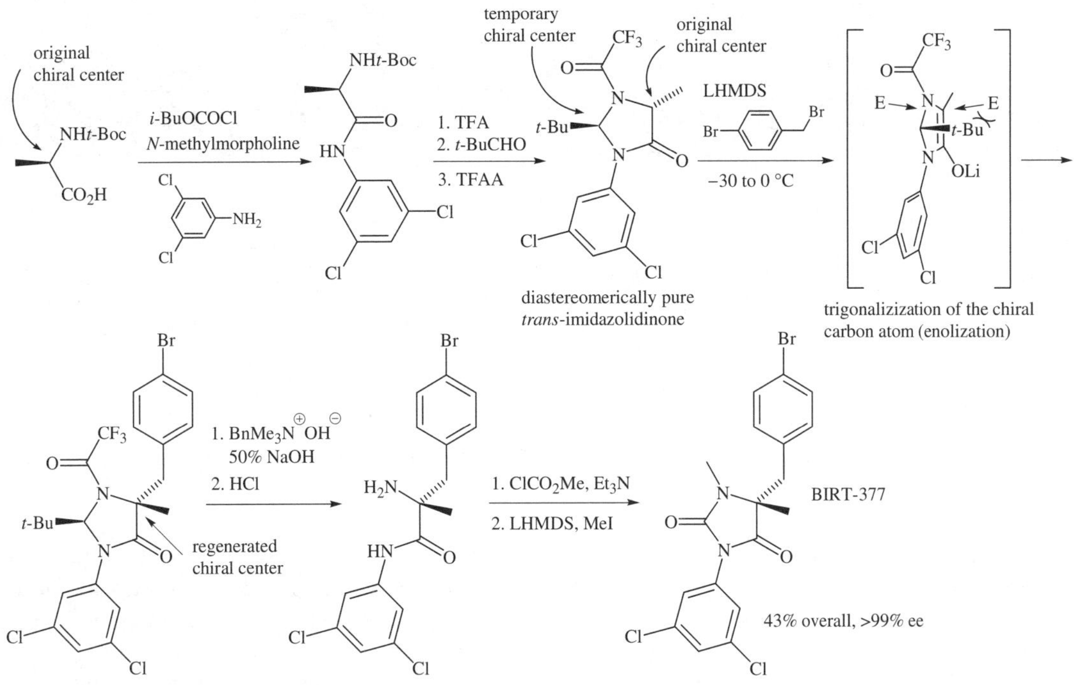

Scheme 6.107 Synthesis of (*R*)-*N*-arylsubstituted hydantoin BIRT-377 from (*R*)-*N*-*t*-Boc-alanine.

hydantoin bearing an α,α-disubstituted amino acid moiety. This route was found to be robust and cost-effective, and was successfully employed in the production of BIRT-377 on a multi-kilogram scale.

The overall success of SRS depends on the selectivity of two asymmetric reactions: introduction of the temporary chiral element and regeneration of the original chiral center. The enantiopurity of the product is determined by the individual diastereoselectivity of these steps. The preparation of diastereomerically pure imidazolidinones, oxazolidinones and dioxolanones from enantiopure α-amino and α-hydroxy acids, respectively, is a common feature of many SRS routes because enolization of these species generates rigid intermediates that undergo highly stereoselective alkylations with predictable asymmetric induction. Incorporation of the second chiral center into these heterocycles is usually accomplished with pivalaldehyde rather than benzaldehyde because the bulky *tert*-butyl group affords superior diastereoselectivity during both acetalization and regeneration of the original chiral center. Acetalizations with pivalaldehyde are often quantitative and unreacted excess of the volatile aldehyde can easily be removed from the product. Alternatively, prochiral ketones such as pinacolone and acetophenone can be used to generate a temporary chiral ketal center.[755,756] Selective formation of *cis*- and *trans*-isomers of imidazolidinones, oxazolidinones and dioxolanones is feasible by careful selection of reaction conditions. In many cases, but not exclusively, *cis*-substituted heterocycles are obtained by thermodynamically controlled acetalization of α-heterosubstituted carboxylic acid derivatives and *trans*-isomers are favored under kinetic control. The higher thermodynamic stability of *cis*-isomers compared to *trans*-isomers of oxazolidinones and imidazolidinones has been attributed to a combination of ring strain and $A^{1,3}$ strain involving the exocyclic amide or carbamate group.[757-759] Allylic strain can also have a significant impact on the stereochemical outcome of alkylation reactions of heterocyclic enolates. In particular, protecting *N*-acyl groups are frequently not just passive spectators but serve as chiral relays, see Chapter 6.5.4.[760,761]

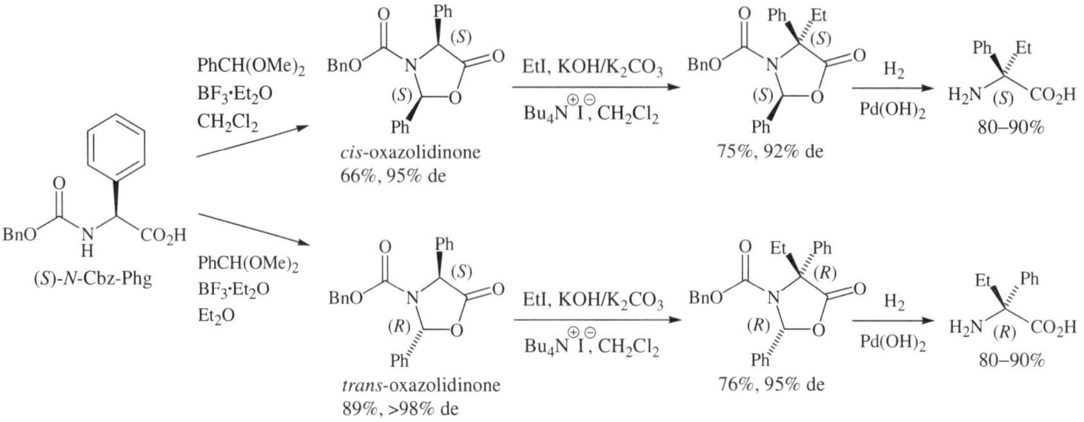

Scheme 6.108 Stereodivergent conversion of *N*-Cbz-protected (*S*)-phenylglycine to (*S*)- and (*R*)-α-ethyl-α-phenylglycine.

O'Donnell *et al.* found that treatment of *N*-Cbz-protected (*S*)-phenylglycine with benzaldehyde dimethyl acetal in the presence of a Lewis acid in dichloromethane yields a *cis*-oxazolidinone, whereas the corresponding *trans*-isomer is obtained when the reaction is carried out in diethyl ether, Scheme 6.108. Alkylation of the diastereomeric *cis*- and *trans*-oxazolidinones with ethyl iodide under catalytic phase-transfer conditions followed by hydrogenolysis furnishes (*S*)- and (*R*)-α-ethyl-α-phenylglycine, respectively.[762] Apparently the absolute configuration of the temporary chiral *N,O*-acetal center is controlled during ring closure simply by the choice of solvent, which gives rise to a convenient switch for selection of the stereochemical course of the SRS sequence.

By analogy with stereoselective transformation of readily available amino acids using imidazolidinones or oxazolidinones as key intermediates, formation of *cis*- or *trans*-dioxolanones from enantiopure α-hydroxy acids, for example tartaric, isovaleric, mandelic, citramalic, lactic, and malic acids, provides another entry to SRS. Chiral α-alkyl malates are important building blocks in biologically active natural compounds such as *Orchidaceae* alkaloids and *Orchidaceae* glycosides. Tietze's group employed (*R*)-malic acid in the synthesis of enantiopure α-alkyl malates, Scheme 6.109.[763] Acid-catalyzed acetalization of (*R*)-malic acid with pivalaldehyde furnished the corresponding *cis*-dioxolanone, which is thermodynamically favored over the *trans*-isomer, in high yield. Enolization with lithium bis(trimethylsilyl)amide and subsequent alkylation with dimethylallyl bromide was found to proceed with excellent stereocontrol. Because alkylation occurred predominantly at the face of the enolate opposite to the shielding *tert*-butyl group of the chiral *O,O*-acetal unit, the chiral center of the malic acid moiety was re-established with high diastereoselectivity. Subsequent hydrolysis and hydrogenation gave the desired α,α-disubstituted malic acid in enantiopure form. The principles of SRS have also been applied to the synthesis of 3,4-disubstituted β-lactams which are important subunits in many antiobiotics. Battaglia and coworkers discovered that (*S*)-lactic acid-derived dioxolanones are invaluable precursors of chiral β-lactams.[764–766] Treatment of (*S*)-lactic acid with pivalaldehyde gives the thermodynamically favored (2*S*,5*S*)-*cis*-dioxolanone in 94% de, Scheme 6.110. Addition of this dioxolanone-derived enolate to a solution of *N*-trimethylsilylphenyl aldimine generates the corresponding (1′*S*,2*S*,5*R*)-heterocycle in moderate yield but with remarkable diastereofacial control. Finally, MeMgBr-promoted cyclization of the (1′*S*,2*S*,5*R*)-amino lactone produces a (3*R*,4*S*)-lactam in 68% yield and 86% ee.

The general availability of inexpensive naturally occurring chiral building blocks, in particular enantiopure α-amino acids and α-hydroxy acids, has certainly contributed to the popularity of the SRS method for the synthesis of α,α-disubstituted α-heterocarboxylic acids, Figure 6.5.[767–781] Of

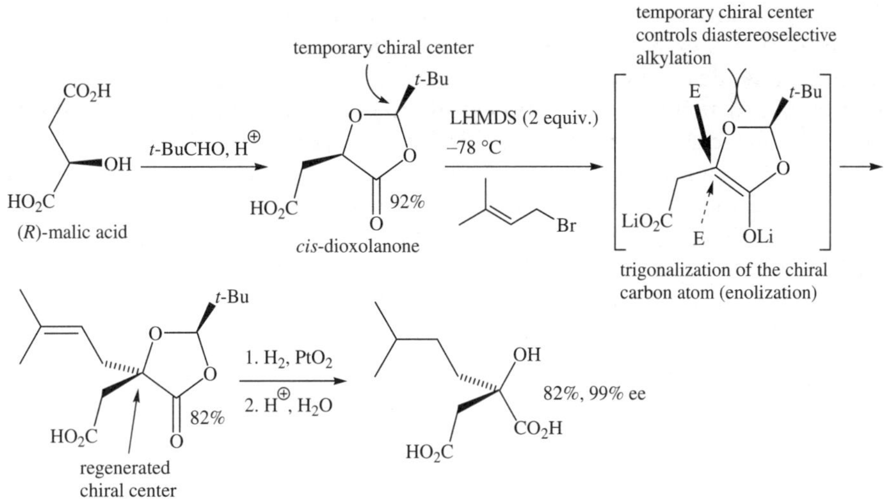

Scheme 6.109 Formation of an enantiopure α-alkyl malate.

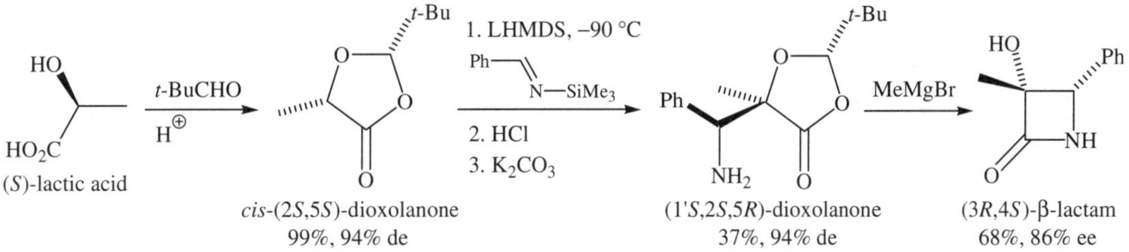

Scheme 6.110 Synthesis of a 3,4-disubstituted β-lactam from (*S*)-lactic acid.

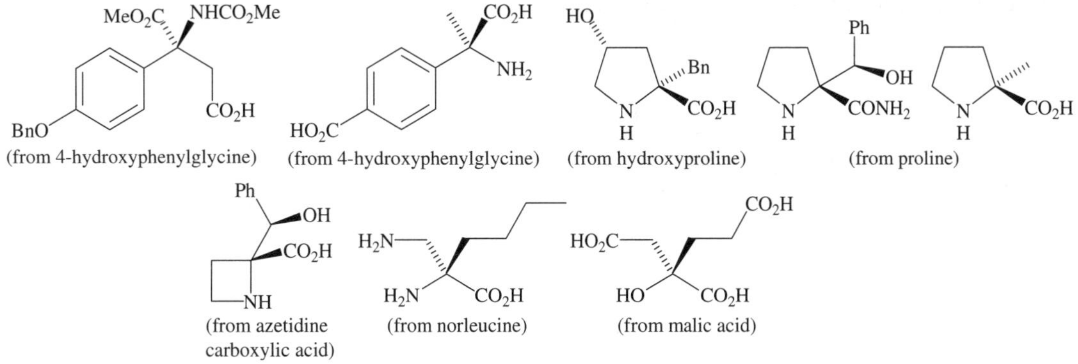

Figure 6.5 Selected examples of α,α-disubstituted α-heterocarboxylic acids synthesized from enantiopure α-amino acids and α-hydroxy acids.

course, this methodology is not limited to intermediate formation of imidazolidinones, oxazolidinones and dioxolanones. Heterobicyclic analogs derived from cyclic amino acids such as proline,[782,783] hydroxyproline,[784] azetidine carboxylic acid,[785] and oxazolopyrrolidine phosphonates,[786] C$_2$-symmetric pyrrolidinyl nitrones,[787] configurationally stable aziridines,[788] boron-amine adducts,[789–791] and oxazaborolidinones,[792–794] have also been explored.

Dispiroketals derived from α-hydroxy acids provide another interesting opportunity for diastereo-selective α-substitution. Ley *et al.* found that (*S*)-lactic acid and 1,1'-bis(dihydropyran) form a dispiroketal in high de due to anomeric stabilization.[795] Deprotonation of the diastereomerically pure dispirane obtained after recrystallization results in trigonalization of the originally asymmetric carbon atom in the lactic acid unit while the chiral information is stored in the two ketal moieties. Subsequent regeneration of the chiral center by treatment of the intermediate dispiro enolate with alkyl halides or aldehydes is highly diastereoselective and the α,α-disubstituted α-hydroxy carb-oxylic acids can be isolated by transketalization with an excess of ethylene glycol, Scheme 6.111.

The use of chiral heterocycles with endo- or exocyclic double bonds extends the SRS concept, described above for α-substitution of enolates, to stereoselective incorporation of substituents into the β- or γ-position of enoates via Michael additions, cycloadditions or radical reactions.[796–800] Seebach's group converted (*S*)-trifluorohydroxybutanoic acid into enantiopure (*R*)-2-*tert*-butyl-6-trifluoromethyl-1,3-dioxin-4-one exhibiting an endocyclic double bond, by acetalization, bromi-nation and amidine-promoted elimination, Scheme 6.112.[801] The dioxin-4-one was subjected to

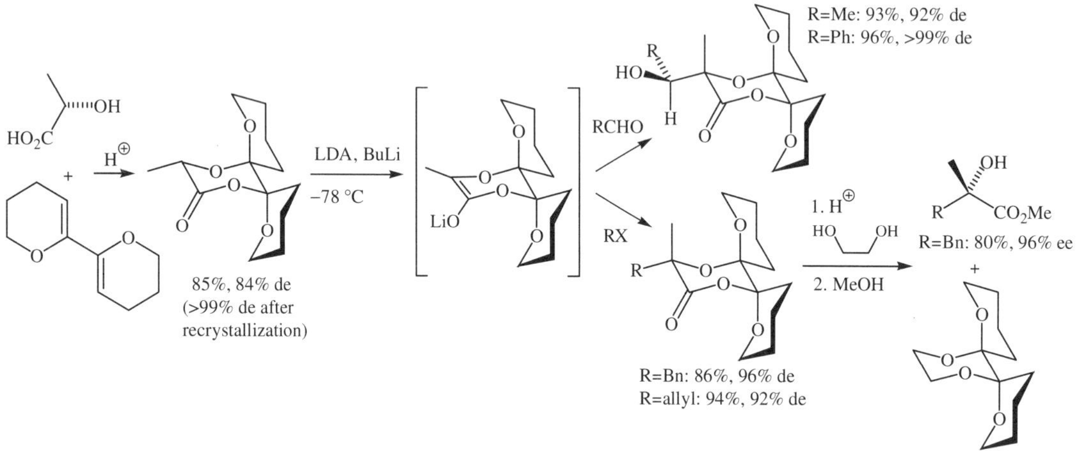

Scheme 6.111   Self-regeneration of stereocenters using dispiroketals.

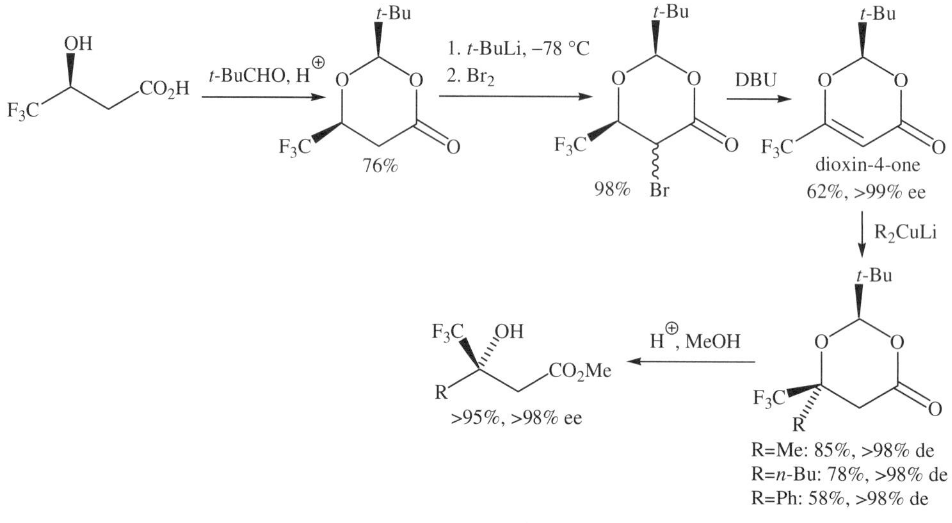

Scheme 6.112   Synthesis of β-substituted carboxylic acids using a 1,3-dioxin-4-one intermediate.

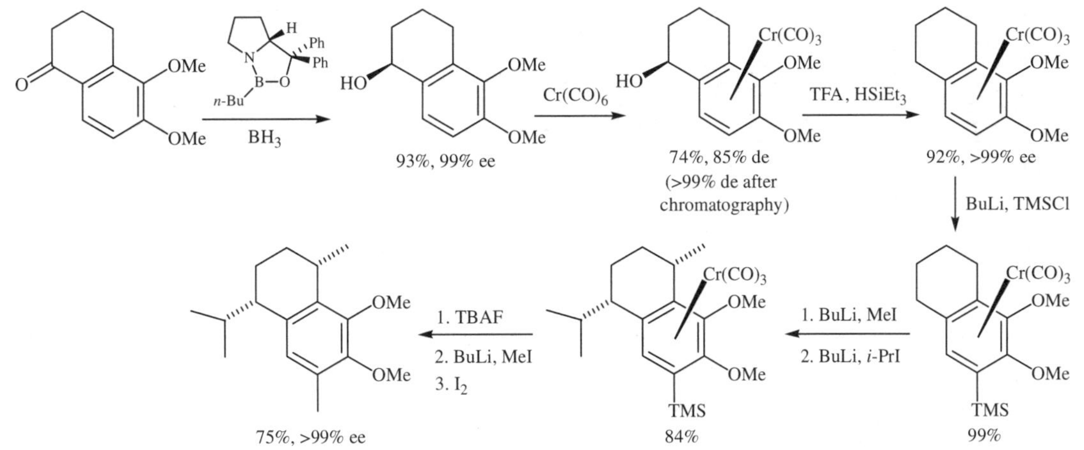

Scheme 6.113 Synthesis of (1*S*,4*S*)-7,8-dihydroxyclamenene via an η^6-arene Cr(CO)$_3$ complex.

Michael addition with organocuprates and then hydrolyzed to (*S*)-3-hydroxy-3-trifluoromethyl alkanoates.

The diastereoselective synthesis of the anti-infective sesquiterpene (1*S*,4*S*)-7,8-di-hydroxyclamenene shown in Scheme 6.113 involves formation of a temporary chiral plane.[802–804] In this reaction sequence, a benzylic (*S*)-alcohol is first prepared by enantioselective oxazaborolidine-catalyzed borane reduction of 1,2,3,4-tetrahydro-5,6-dimethoxy-1-naphthone. Complexation with chromium hexacarbonyl then gives the *syn*-adduct and the chiral alcohol group is reductively removed to generate an enantiopure η^6-aryl Cr(CO)$_3$ complex. Protection of the aryl position adjacent to the methoxy groups is necessary to ensure selective alkylation in both benzylic positions. The intermediate chiral η^6-arene tricarbonylchromium complex fulfills two functions. Temporary metal complexation enhances the acidity at the benzylic positions for mild deprotonation/alkylation and implements a bulky group shielding one π-face of the tetrahydronaphthalene ring for diastereoselective regeneration of the original central chirality. Benzylic alkylations occur at the face opposite to the tricarbonyl chromium unit and (1*S*,4*S*)-7,8-dihydroxyclamenene is obtained in 75% yield and without loss of enantiopurity after replacement of the trimethylsilyl group by a methyl substituent and oxidative decomplexation with iodine.

Enders and others used chiral η^3-allyl-derived transition metal complexes prepared from unsaturated substrates such as (*E*)-(*S*)-4-benzyloxypent-2-enoic methyl ester for asymmetric allylic substitution, Scheme 6.114.[805–810] Reaction of the chiral alkene with stoichiometric amounts of Fe$_2$(CO)$_9$ gives a neutral η^2-olefin tetracarbonyliron complex that is converted to an isolable η^3-allyl tetracarbonyliron tetrafluoroborate salt via acid-promoted removal of the benzyloxy group. As a result, the chirality at the asymmetric allylic carbon atom of the ligand is transferred to the chiral plane of the cationic metal complex. Nucleophilic addition of amines and subsequent oxidative cleavage of the iron complex with CAN regenerates the chiral center, providing (*E*)-(*S*)-4-amino-2-enoates with high regio- and stereoselectivity.[xxx]

The principle of self-regeneration of stereocenters has found widespread use in the synthesis of an impressive range of natural products including 10-methyltridecan-2-one,[811,812] amathaspiramides,[813] rhodomycinones,[814] frontalin,[815] myoporone,[816] eremantholide A,[817] indicine *N*-oxide,[818] mevalo-lactone,[819] brevianamide B,[820] desferrithiocin,[821] didehydromirabazole A,[822] and lactacystin,[823,824] Figure 6.6. Nevertheless, the scope and efficiency of the SRS concept is hampered by the inevitability

---

[xxx] The concept of transition metal-mediated allylic substitution has been further developed to powerful catalytic processes based on dynamic kinetic asymmetric transformation, see Chapter 7.5.

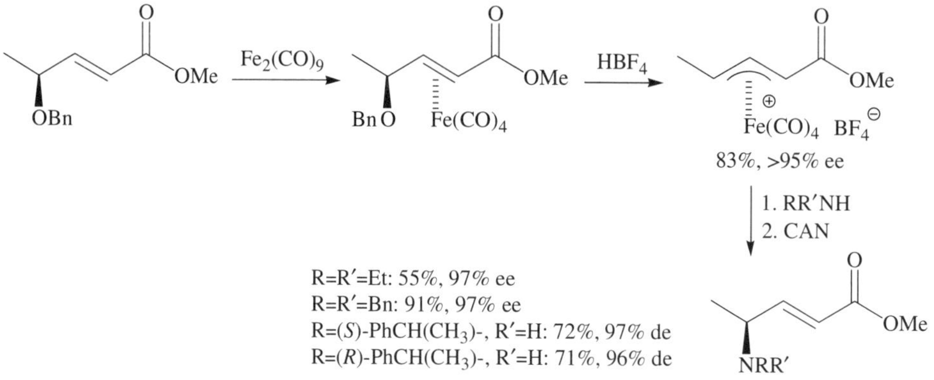

Scheme 6.114   Iron-mediated allylic substitution via a cationic η^3-allyl tetracarbonyliron complex.

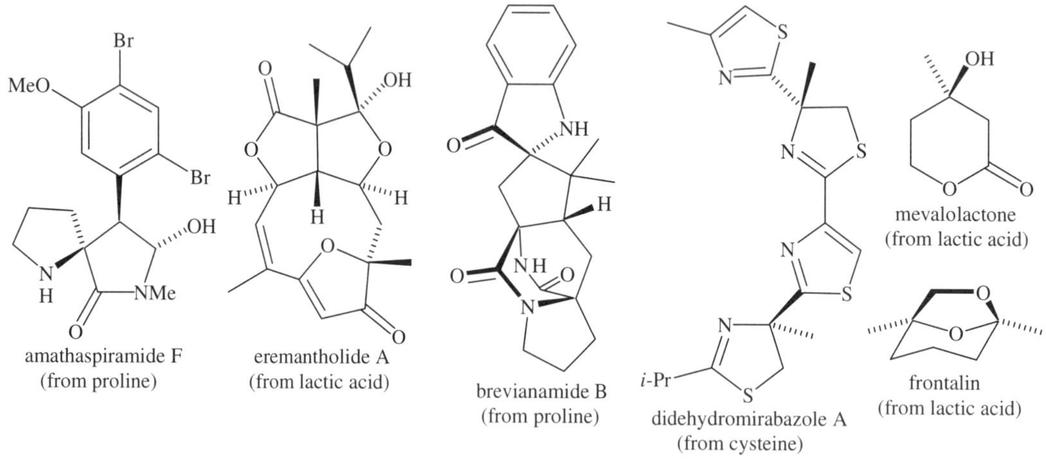

Figure 6.6   Structures of natural products prepared via self-regeneration of stereocenters.

of multiple laborious reaction and purification steps. Although formation of imidazolidinones, oxazolidinones and dioxolanones in more than 90% de is common, time-consuming purification by crystallization or chromatography is often necessary to obtain SRS products in high enantiopurity. Consequently, faster and more convenient alternatives based on one-step regeneration of stereogenicity with nonisolable chiral intermediates and chiral relays have been developed, *vide infra*.

6.5.3   Self-regeneration of Chiral Elements with Stereolabile Intermediates

In 1991, Fuji and coworkers explored the feasibility of asymmetric α-alkylation of chiral alkyl aryl ketones possessing an enolizable chiral center adjacent to the carbonyl function. An enantiomerically enriched 1-naphthyl ketone was prepared from the Weinreb amide of (*S*)-mandelic acid and 2,3-diethoxy-1-naphthyllithium and then subjected to deprotonation with potassium hydride in the presence of 18-crown-6. The enolate was trapped with various electrophiles at −78 °C in the absence of any additional chiral source, Scheme 6.115.[825] Although enolization generates a trigonalized carbon atom, alkylation did not afford racemic products. These findings were explained by an intermediate enolate that stores the central chirality of the starting material in the form of transient axial chirality. Fuji coined the term "memory of chirality" to emphasize that

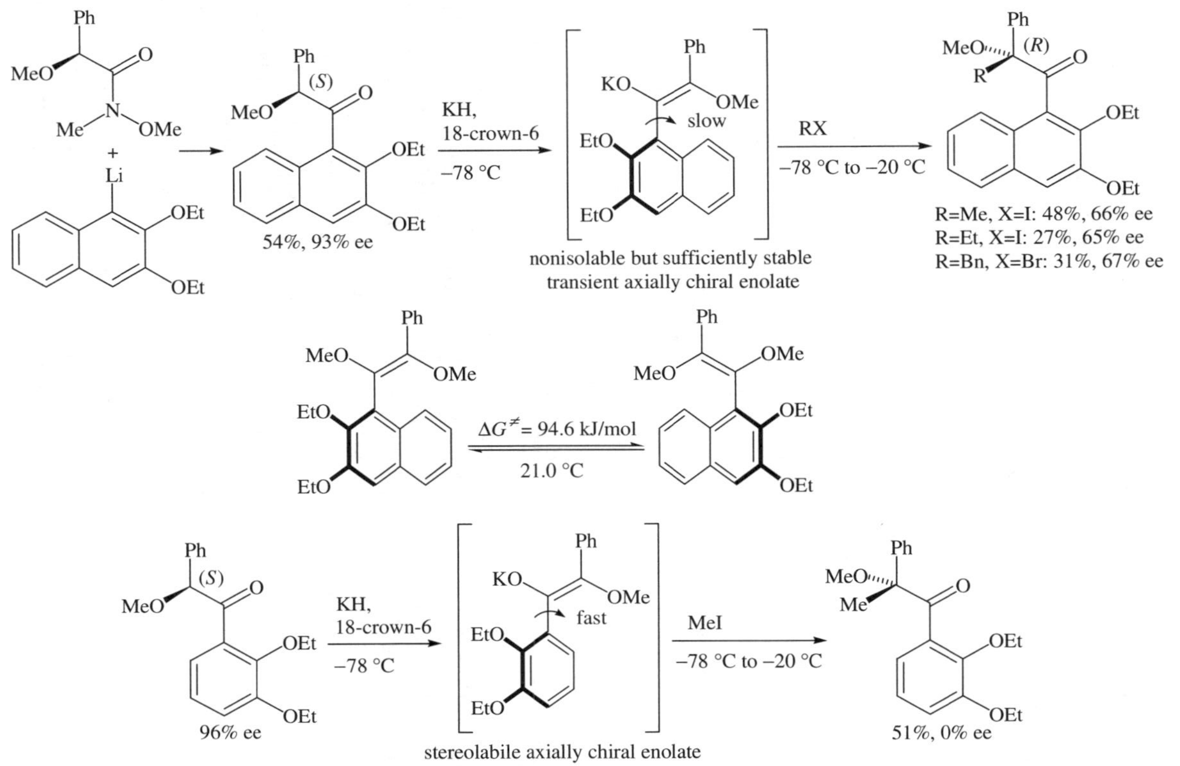

Scheme 6.115 Deprotonation of chiral aryl ketones and alkylation of a transient axially chiral enolate. Asymmetric regeneration of the stereocenter due to conformational stability of the intermediate naphthyl enolate (top) and racemization of the corresponding phenyl enolate (bottom).

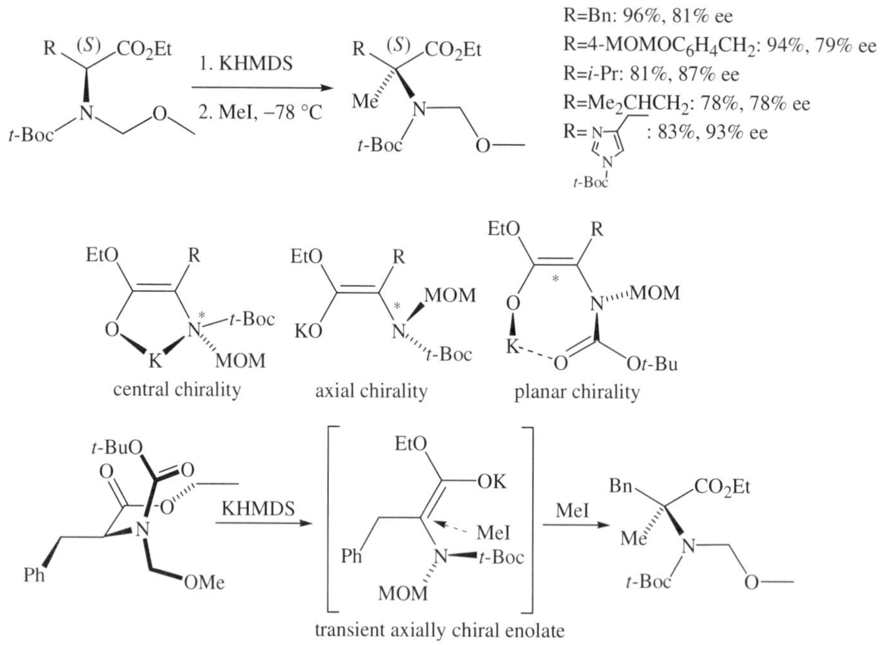

Figure 6.7 Methylation of *N*-MOM-*N*-*t*-Boc-protected amino esters (top), possible transient enolate structures (middle), and postulated reaction course (bottom).

the stereochemical information is temporarily stored at a different site in the intermediate. Since the molecule remains chiral at all times and does not undergo substantial racemization, the original chiral element can be regenerated with considerable selectivity upon reaction with an alkyl halide. For example, deprotonation of the (*S*)-mandelic acid-derived naphthyl ketone under cryogenic conditions and immediate addition of methyl iodide gives the (*R*)-enantiomer in 48% yield and 66% ee.[xxxi] When the corresponding phenyl ketone is employed in the same reaction, a racemic product is obtained because enantioconversion of the stereolabile axially chiral enolate intermediate is faster than the trapping reaction. Apparently, self-regeneration of a stereocenter is feasible within a single reaction if the transient chiral species does not undergo spontaneous racemization. Unfortunately, the term "memory of chirality" has led to the false impression that chiral information is lost and somehow regenerated later in the reaction. In fact, molecular chirality is conserved throughout the reaction while only a stereogenic element is temporarily lost.[xxxii] It has been pointed out by Carlier that this scenario involves *transient* storage of the chiral information at another site of a nonisolable intermediate. This is conceptually different from Seebach's SRS strategy which requires introduction of a *temporary* chiral element for control of a separate diastereoselective reaction.[826]

Fuji's group then applied the concept of self-regeneration of stereocenters with nonisolable chiral intermediates to the asymmetric synthesis of α-alkylated amino acids.[827–830] They found that *N*-MOM-*N*-*t*-Boc-protected amino esters undergo enolization and subsequent methylation under cryogenic conditions with excellent selectivity. The results clearly demonstrate the usefulness of this methodology, Figure 6.7. One can assume formation of three different transient enolates exhibiting

---

[xxxi] The formation of an intermediate atropisomeric enolate was confirmed by isolation of small amounts of the *O*-methylated side product. Kinetic studies revealed that this *O*-methyl enolate has a rotational energy barrier of 94.6 kJ/mol at 21.0 °C. It is therefore reasonable to assume that the intermediate axially chiral enolate shown in Scheme 6.115 is sufficiently stable to racemization under cryogenic conditions and therefore affords *C*-alkylation products in 65–67% ee.

[xxxii] Chirality is a molecular property whereas stereogenicity refers to an atom or group within a molecule.

either central, axial or planar chirality. Computational analysis favors a transient axially chiral enolate that is formed by deprotonation of the most populated amino ester conformation. Alkylation then occurs from the less sterically hindered side and affords the quaternary carbon center with retention of configuration. Kawabata *et al.* employed β-branched α-amino acids in similar enolization/alkylation sequences and observed that the stereoselectivity of the transformation is decisively controlled by the transient chiral *C–N* axis of the intermediate enolate, while the central chirality at the β-carbon has only a minor effect on the stereochemical outcome of the reaction.[831] Kawabata, Fuji and others have successfully utilized this concept for asymmetric alkylation of chiral amides,[832] enantioselective formation of azacyclic amino acids[833] and for intramolecular conjugate addition of enolates, producing pyrrolidine-, piperidine- and tetrahydroisoquinoline-derived chiral heterocycles.[834,835] In closely related studies, Stoodley's and González-Muñiz' groups discovered that complex thiazolidines and 4-alkyl-4-carboxy-2-azetidinones can be prepared by ring closure of transient axially chiral enolates generated from amino esters under basic conditions, Scheme 6.116.[836–843]

Clayden's group developed a multi–step route for stereoselective functionalization of aryl amides combining features of Fuji's and Seebach's SRS methodologies, Scheme 6.117.[844] In the first two of

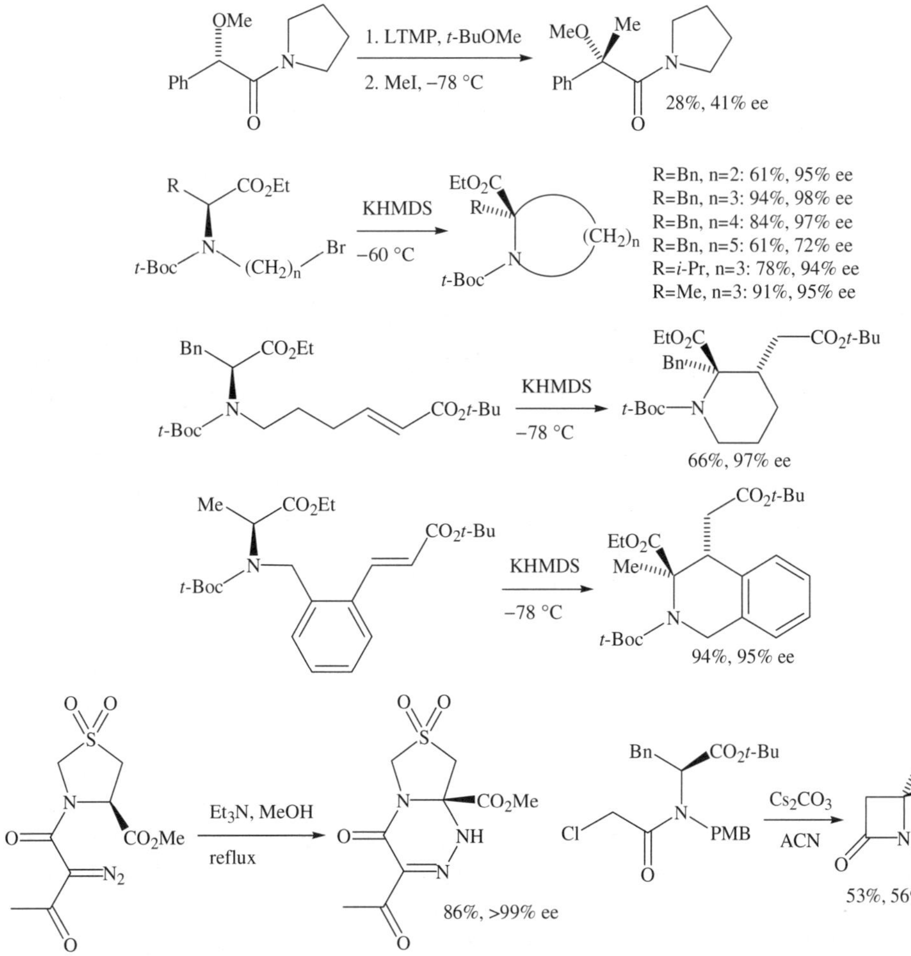

Scheme 6.116 Synthesis of amino esters and amides via transient axially chiral enolates.

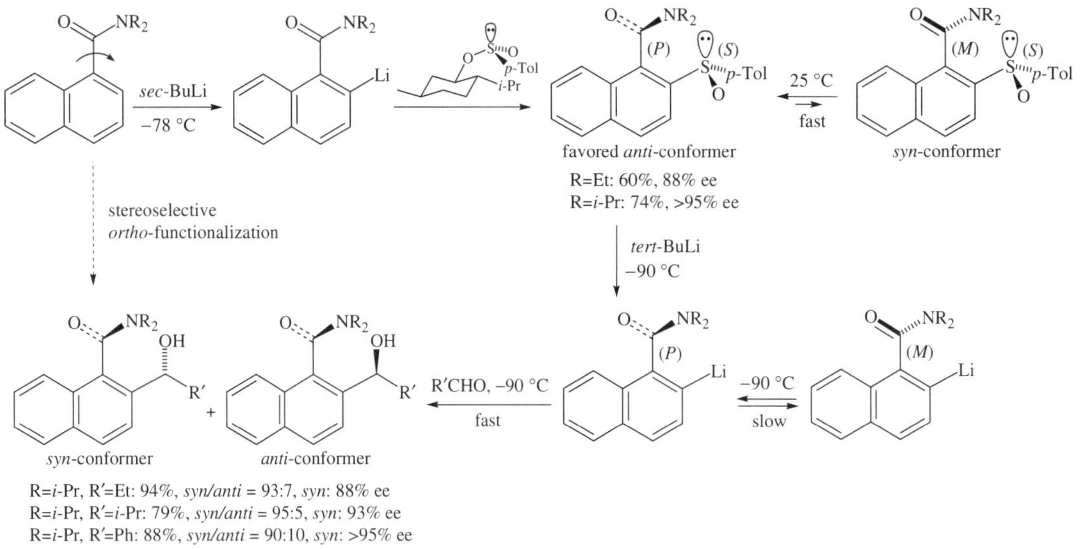

Scheme 6.117 Stereoselective *ortho*-functionalization of 1-naphthamides.

four steps, regioselective lithiation of tertiary aryl amides and treatment with (1*R*,2*S*,5*R*,*S*$_S$)-(−)-menthyl toluenesulfinate gives rise to amido sulfoxides. According to NMR analysis, incorporation of the *p*-tolyl sulfoxide group into the *ortho*-position of aromatic amides favors formation of a single diastereomeric conformer. Due to facile rotation about the axially chiral aryl amide bond at room temperature, the amide group readily adopts an *anti*-orientation relative to the sulfoxide unit in order to minimize dipole–dipole interactions. The (*S*)-sulfoxide group thus induces formation of the (*P*)-atropisomer. The (*S*,*P*)-isomer is then subjected to sulfoxide/lithium exchange at −90 °C and stereoselective reaction with an aldehyde. Although the chiral sulfoxide group is removed upon lithiation, the previously induced axial chirality is maintained because rotation about the aryl amide axis is very slow under cryogenic conditions. The stereochemical information of the replaced chiral sulfoxide group is thus temporarily stored in the axially chiral aryl amide moiety which governs the diastereoselective addition to aldehydes.[xxxiii]

The unique dynamic stereochemistry of chiral benzodiazepines offers another opportunity for self-regeneration of stereogenicity. Diazepam and other glycine-derived 1,4-benzodiazepin-2-ones lacking a chiral center exist in the form of axially chiral enantiomers that usually show rapid interconversion due to facile ring flipping at room temperature.[845–847] The stability to racemization of these seven-membered ring systems depends on the bulkiness of the substituent attached to the amide nitrogen. The presence of a *tert*-butyl group gives rise to isolable enantiomers whereas *N*-methyl and *N*-isopropyl analogs afford energy barriers to racemization of only 75.3 and 88.3 kJ/mol, respectively, Scheme 6.118. The synthesis of 1,4-benzodiazepin-2-ones from chiral amino acids introduces a chiral center which renders the diazepine conformations diastereomeric. The diastereomeric ratio is controlled by the absolute configuration at the asymmetric carbon atom: a substituent at *C*-3 preferentially occupies the pseudoequatorial position and (3*S*)-benzodiazepines adopt (*M*)-helicity.[848,849,xxxiv] Carlier *et al.* exploited Fuji's concept for the synthesis of 1,4-benzodiazepin-2-ones with a quaternary chiral center.[850,851] Deprotonation of an (*S*)-alanine-derived 1,4-benzodiazepin-2-one and trapping of the enolate intermediate with an electrophile

---

[xxxiii] In contrast to Seebach's SRS, only one intermediate (the naphthalic amido sulfoxide) of this four-step reaction sequence has to be isolated.

[xxxiv] The sense of helicity can be described based on the *R*(amide)-*N*(1)-*C*(7)-*C*(8) dihedral angle.

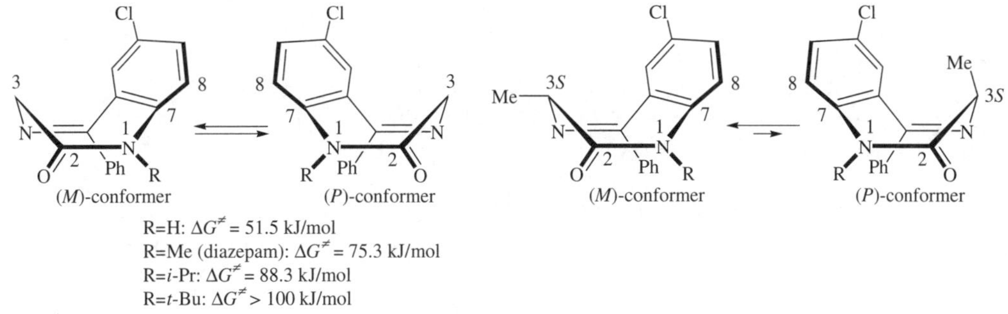
(M)-conformer
(P)-conformer
(M)-conformer
(P)-conformer
R=H: ΔG≠ = 51.5 kJ/mol
R=Me (diazepam): ΔG≠ = 75.3 kJ/mol
R=i-Pr: ΔG≠ = 88.3 kJ/mol
R=t-Bu: ΔG≠ > 100 kJ/mol

Scheme 6.118 Interconversion barriers of enantiomeric conformations of glycine-derived 1,4-benzodiazepin-2-ones (left) and diastereomerization of (*S*)-alanine derivatives (right).

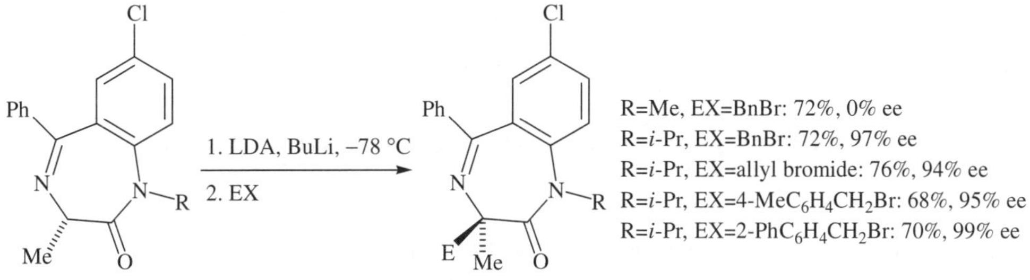
1. LDA, BuLi, −78 °C
2. EX
R=Me, EX=BnBr: 72%, 0% ee
R=i-Pr, EX=BnBr: 72%, 97% ee
R=i-Pr, EX=allyl bromide: 76%, 94% ee
R=i-Pr, EX=4-MeC6H4CH2Br: 68%, 95% ee
R=i-Pr, EX=2-PhC6H4CH2Br: 70%, 99% ee

Scheme 6.119 Enantioselective alkylation of 1,4-benzodiazepin-2-ones.

proceeds with retention of the original stereochemistry and gives 3,3-dialkylated products in up to 99% ee, Scheme 6.119. Although the stereogenic center is essentially lost due to enolization,[xxxv] chirality is conserved in the diazepine ring. The diastereofacial alkylation of the enolate is therefore controlled by the transient axial chirality and the asymmetric center is regenerated with high selectivity. It is crucial that the axially chiral benzodiazepine enolate does not racemize via ring flipping during the reaction. This can be avoided when enolization of 1,4-benzodiazepin-2-ones carrying an isopropyl group at the amide nitrogen is carried out under cryogenic conditions. Since the corresponding methyl derivative is prone to fast equilibration at −78 °C, one observes formation of racemic products. Nevertheless, stereogenicity can be selectively regenerated even with conformationally labile *N*-methyl 1,4-benzodiazepin-2-ones at −109 °C if the enolate intermediate is trapped within a few seconds in the presence of benzyl bromide, or by instantaneous deuteration using deuterated methanol as solvent.[852]

Few examples of self-regeneration of a stereocenter via chiral radical intermediates are known. Schmalz and coworkers found that benzylic radicals derived from arene tricarbonylchromium complexes are best described as a 17-valence electron species bearing an exocyclic double bond.[853] They rationalized that enantioselective reduction of readily available 1-arylalkanol-derived tricarbonylchromium complexes would generate an intermediate radical exhibiting a chiral plane. Central chirality could thus be temporarily stored in the chiral plane and regenerated through rapid formation of the corresponding anion and treatment with an electrophile. Based on computational calculations, the energy barrier to racemization of such a radical chromium complex was estimated as 55.3 kJ/mol which corresponds to a half-life of approximately one minute at −78 °C. Indeed, careful reduction of an (*R*)-1-phenylethanol-derived chromium complex with two

[xxxv] Enolization of 1,4-benzodiazepin-2-ones affords a trigonal-planar carbon atom which has been confirmed by DFT calculations.

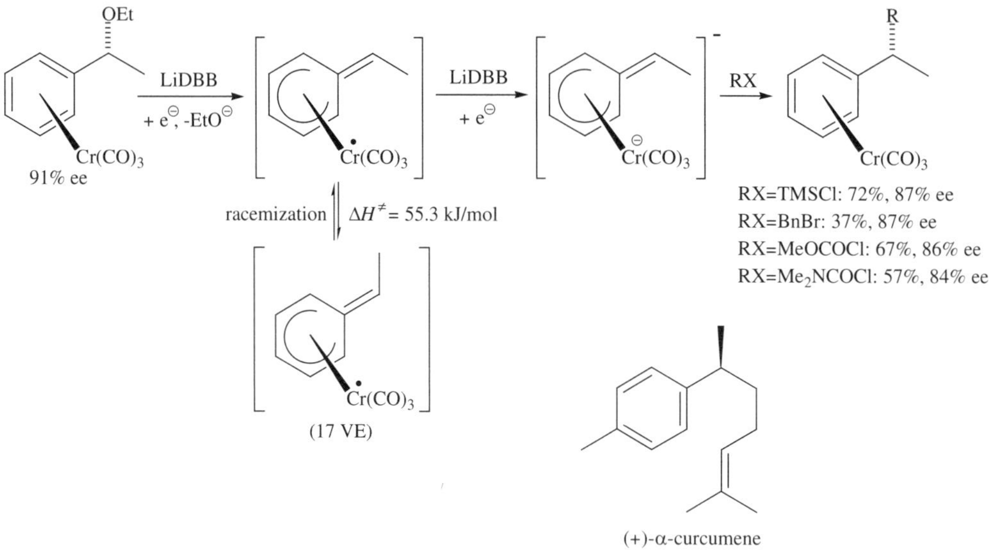

Scheme 6.120 Stereoselective radical reaction of a tricarbonylchromium complex derived from (*R*)-1-phenylethanol.

equivalents of lithium 4,4'-di-*tert*-butylbiphenyl (LiDBB) at $-78\,^{\circ}$C and subsequent treatment with various electrophiles gave the desired products with high stereoselectivity, Scheme 6.120.[854] Apparently, generation of the radical intermediate, reduction to the benzylic anion and alkylation or acylation are considerably faster than competing racemization pathways. This method has been used for the synthesis of the sesquiterpene (+)-α-curcumene. Regeneration of a chiral center through formation of a planar chiral intermediate in a stereospecific tandem oxy-Cope/ene reaction has also been reported.[855]

The stereoselective displacement of a substituent attached to a chiral center via an elimination/ addition sequence does not necessarily require transient or temporary storage of stereochemical information in the form of another chiral element. In many cases racemization can be avoided at low temperatures.[xxxvi] For example, stabilization of the absolute configuration of chiral organo-lithium compounds by a proximate carbamoyl group has been used extensively for racemization-free synthesis under cryogenic conditions, see Chapter 6.1. Rychnovsky's group recognized the remarkable stereochemical stability of tetrahydropyranyl radicals. While acyclic chiral radicals undergo rapid racemization due to inversion barriers below $4\,$kJ/mol,[856] cyclic analogs show significantly enhanced half-lives.[857,858] Racemization of tetrahydropyranyl radicals requires ring inversion which is likely to have an energy barrier between 20 and $40\,$kJ/mol. Decarboxylation or reductive decyanation of enantiopure tetrahyropyranyl-derived carboxylic acids or nitriles produce an anomeric-stabilized chiral radical that can be trapped at $-78\,^{\circ}$C, Scheme 6.121.[859] Similar concepts have been employed in stereoselective transannular cyclizations of intermediate cyclo-decenyl radicals to bicyclo[5.3.0]decanes,[860] cyclizations of photochemically generated diradi-cals,[861,862] and in-cage recombinations of radical pairs obtained by photolytic cleavage of aryl ethers and esters in *n*-alkanes and polyethylene films.[863,864] The stereoselectivity observed in the latter case is a consequence of templating effects, *i.e.*, reaction cavities control tumbling and translational movements of short-lived radical intermediates.[865–867]

[xxxvi] In the context of asymmetric synthesis, the evaluation of the configurational and conformational stability of a chiral intermediate requires consideration of reaction conditions and time scale, as the knowledge of the *relative* rate of isomerization and trapping reactions is sufficient.

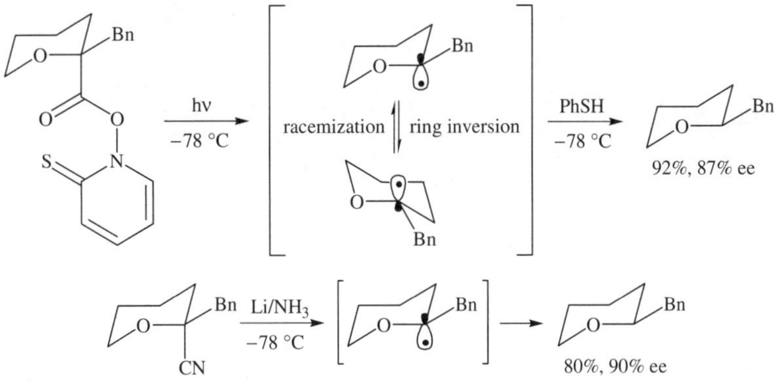

Scheme 6.121 Photolytic and reductive generation of a chiral tetrahydropyranyl radical and enantioselective trapping under cryogenic conditions.

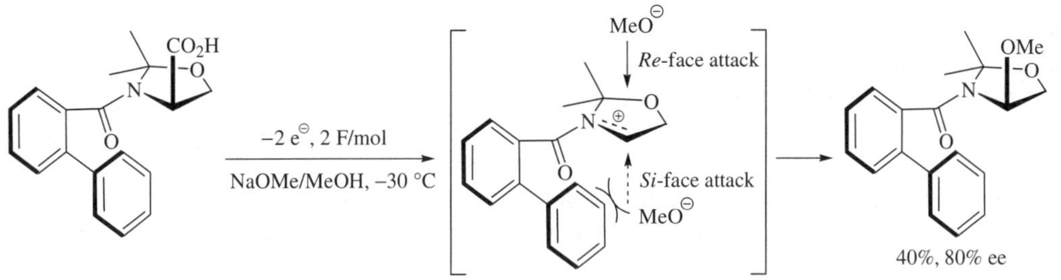

Scheme 6.122 Regeneration of a stereogenic center involving an electrochemically generated acyliminium ion.

Few examples of regeneration of chiral centers with carbocation intermediates have been reported.[868,869] Oxidative decarboxylation of *N*-2-phenylbenzoyl serine, threonine or *allo*-threonine *N,O*-ketals, and subsequent addition of methoxide to the intermediate iminium ion, proceeds with retention of configuration.[870] Electrochemical oxidation of the *N*-2-phenylbenzoyl serine *N,O*-ketal in the presence of sodium methoxide gives the corresponding α-methoxylated product in 40% yield and 80% ee, Scheme 6.122. To account for the stereochemical outcome of this reaction, enantiofacial shielding of the *Si*-face of the iminium ion by the 2-phenylbenzoyl group has been proposed. Computational studies of the serine-derived *N,O*-ketal suggest that the *N*-2-phenylbenzoyl unit resides preferentially beneath the carboxylic acid group. Oxidative decarboxylation of the most stable conformer forms a chiral iminium ion species that slowly racemizes due to hindered rotation about the amide bond. Nucleophilic methoxide addition therefore occurs at the freely accessible *Re*-face resulting in the observed retention of configuration.

6.5.4 Chiral Relays

The success of asymmetric synthesis is determined by the difference in the free energy of the reaction products (thermodynamic control) or by the difference in the free activation energy of competing reaction pathways (kinetic control). The extent of asymmetric induction depends on how effectively chirality can be expressed in close proximity to the reaction center. Several groups have recognized the usefulness of fluxional auxiliaries that convey stereochemical information of an existing chiral element to a remote reaction site. The underlying principle of this strategy is to utilize intramolecular or interligand-mediated relays for chiral amplification and enhanced stereochemical

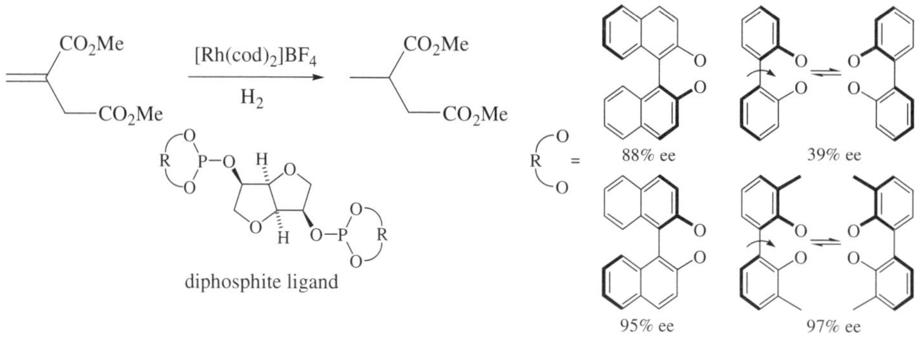

Scheme 6.123 Asymmetric hydrogenation of dimethyl itaconate using conformationally stable and stereo-labile chiral diphosphite ligands.

control of an asymmetric reaction. The advantage of stereodynamic auxiliaries has been demonstrated by Reetz and coworkers. They attached diphosphite ligands carrying either two conformationally stable BINOL units or two stereolabile biphenyl moieties to a rigid sugar-derived template, Scheme 6.123.[871] The chiral ligands were then used for rhodium-catalyzed asymmetric hydrogenation of dimethyl itaconate. Using the diastereomeric (*P*)- and (*M*)-BINOL-derived diphosphites they obtained the chiral ester in 88% and 95% ee, respectively, and with opposite absolute configuration. These findings imply that the stereochemical outcome of this reaction is overwhelmingly controlled by the axially chiral components of the catalyst and not by the chiral sugar backbone. Replacement of the BINOL portions by stereolabile 2,2′-biphenol groups decreased the ee to 39% but incorporation of 3,3′-dimethyl-2,2′-biphenol into the diphosphite ligand gave a rhodium catalyst that produced the diester in 97% ee. The excellent selectivity is probably due to stabilization of one distinct chiral conformation of the biphenyl moiety. Accordingly, intraligand chiral amplification conveys asymmetric induction from the bicyclic sugar backbone via the biphenyl phosphite motif to the reaction site and thus forms a powerful asymmetric catalyst.[xxxvii]

Exploiting Schöllkopf's bislactimether method for the synthesis of enantiopure amino acids,[872] Davies *et al.* were among the first to systematically design chiral relays for diastereofacial reaction control, Scheme 6.124.[873–876] They concluded from molecular modeling studies that deprotonation of (3*S*)-*N*,*N*′-bis(4′-methoxybenzyl)-3-isopropylpiperazine-2,5-dione would produce a planar enolate structure that preferentially populates a conformation with the proximal benzyl moiety in *anti*-position relative to the isopropyl group at the chiral center. To reduce steric interactions, the proximal and distal benzyl groups reside on opposite sides of the ring plane. The principal idea is that the chiral information of the (*S*)-valine-derived auxiliary is amplified by both benzyl groups and effectively relayed to the reaction center to impede alkylation at the *Si*-face of the enolate. Indeed, reaction of the enolate with various electrophiles gives *trans*-alkylation products with excellent diastereoselectivity. Comparison of the results obtained with Schöllkopf's (*S*)-valine-derived diketopiperazine auxiliary uncovers the significant enhancement in diastereoselectivity which is transmitted by the communicating benzyl groups. Davies' chiral relay approach provides a useful tool for the synthesis of (*R*)-α-amino acids (or (*S*)-α-amino acids if the dibenzylated diketopiperazine is derived from (*R*)-valine) and the auxiliary can be recovered after completion of the reaction by hydrolysis of the diketopiperazone.

[xxxvii] The assumption of a single prevalent catalytic species seems reasonable but this is not necessarily the case. The high enantioselectivity can also be a consequence of coexisting (and rapidly interconverting) diastereomeric rhodium complexes, one of which affords excellent stereoselectivity and considerably higher catalytic activity.

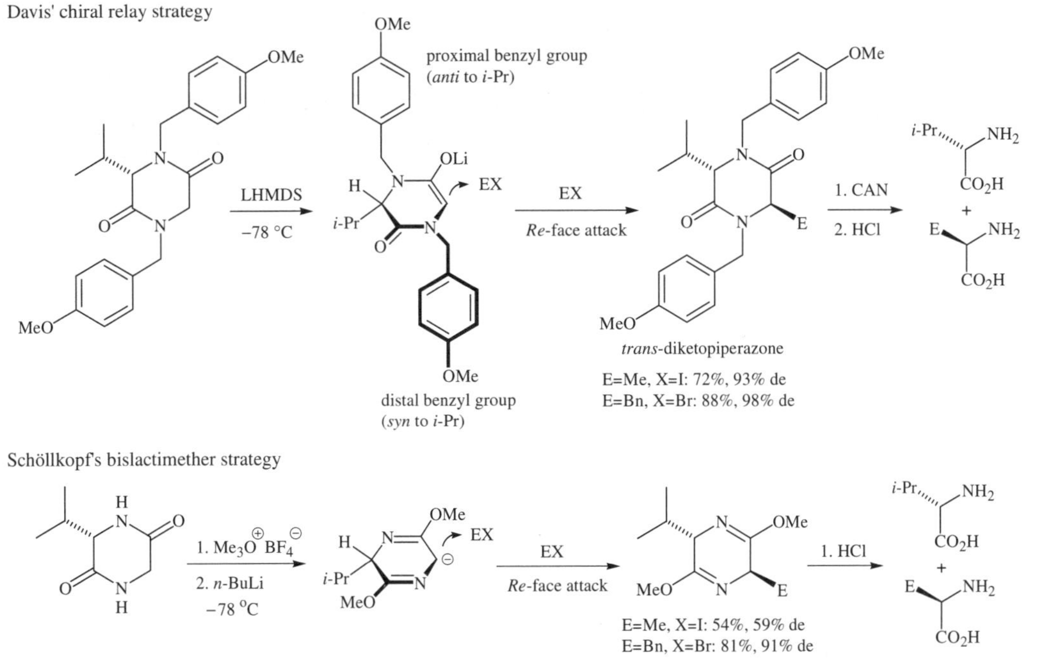

Davis' chiral relay strategy

trans-diketopiperazone

E=Me, X=I: 72%, 93% de
E=Bn, X=Br: 88%, 98% de

Schöllkopf's bislactimether strategy

E=Me, X=I: 54%, 59% de
E=Bn, X=Br: 81%, 91% de

Scheme 6.124 Synthesis of α-amino acids using Davies' chiral relay. Schöllkopf's bislactimether strategy is shown for comparison.

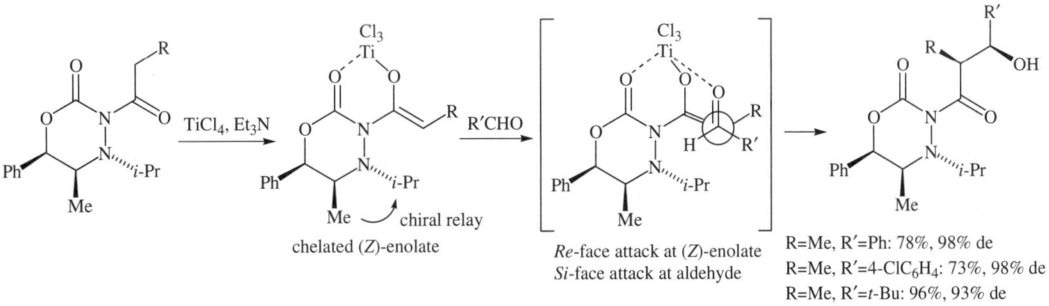

R=Me, R′=Ph: 78%, 98% de
R=Me, R′=4-ClC$_6$H$_4$: 73%, 98% de
R=Me, R′=t-Bu: 96%, 93% de

Scheme 6.125 Diastereoselective aldol reaction with an intramolecular chiral relay.

Hitchcock applied a conceptually similar approach in diastereoselective titanium(IV)-mediated aldol reactions of a (1R,2S)-norephedrine-derived N^3-acyl-N^4-isopropyloxadiazinone with aldehydes, Scheme 6.125.[877–879] The chirality of the (1R,2S)-norephedrine portion is relayed to the N^3-enolate via the stereogenic N^4-nitrogen atom carrying an isopropyl group that selectively shields the *Si*-face of the titanium (*Z*)-enolate. The Lewis acidic titanium(IV) chloride provides chelation control and enhances the electrophilicity of the coordinating aldehyde. The stereoselective outcome of the aldol reaction is in agreement with a chairlike Zimmermann–Traxler transition state in which the *Re*-face of the enolate undergoes carbon–carbon bond formation with the *Si*-face of the prochiral aldehyde.

Sibi's group used a configurationally labile auxiliary to amplify the chiral induction of a C$_2$-symmetric bisoxazoline-copper(II) complex in the Diels–Alder reaction of butenoyl-pyrazolidinones with cyclopentadiene, Scheme 6.126.[880,881] The configuration at the rapidly interconverting tertiary amino group in the pyrazolidinone ring is effectively controlled by the chiral bisoxazoline

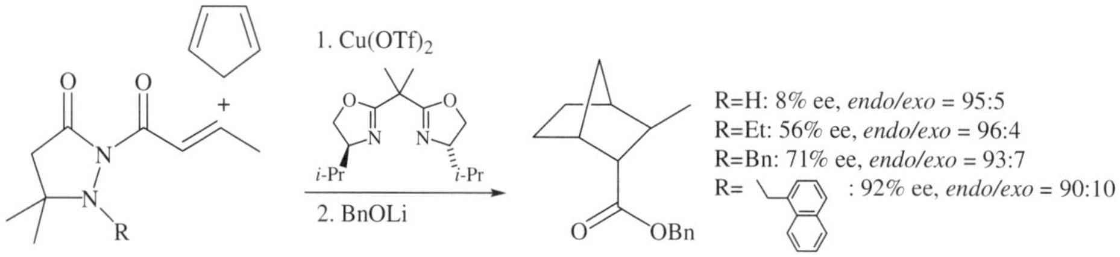

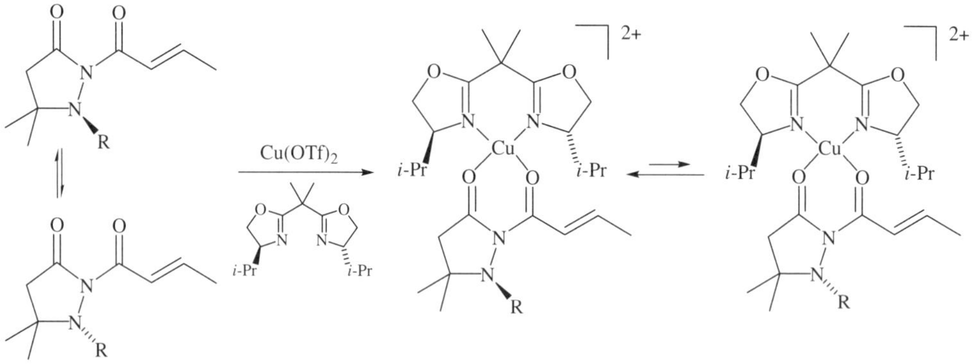

Scheme 6.126 Template-mediated chiral amplification using a configurationally labile pyrazolidinone auxiliary.

ligand of the chiral Lewis acid. Replacement of the nitrogen substituent with hydrogen results in almost complete loss of stereoselectivity, indicating that the π-facial selectivity is not directly induced by the chiral bisoxazoline ligand. It is therefore assumed that the amino function in the pyrazolidinone adopts a single chiral configuration in the presence of the chiral Lewis acid and relays the chiral information to the adjacent dienophile. The selectivity of the reaction is independent of the size of the residues attached to the bisoxazoline ligand and the ee of the cycloadduct improves to 92% with increasing bulkiness of the amino substituent. The direct impact of the steric demand of the amino group on the stereoselectivity is in agreement with the proposed template-mediated chiral amplification. Pyrazolidinone templates have also been used as effective chirality relays in conjugate additions of hydroxylamines and tributyltin-propagated isopropyl radical reactions.[882,883]

Other auxiliaries that possess stereolabile units suited to chiral amplification and relay of asymmetric induction have been employed in enantioselective alkylations of aldehydes with diethylzinc,[884] Diels–Alder reactions[885,886] and Michael additions.[887] The principles of configurational dynamics of nitrogen-based relays outlined above have been exploited for azasugar synthesis and the first preparation of adenophorine, an α-glycosidase-specific inhibitor.[888] Clayden's group used the relay concept for transmission of chiral information across an array of conformationally interlocked axially chiral aromatic amide moieties to afford impressive asymmetric induction at a remote reaction center, see Chapter 8.1.[889–894]

6.6 ASYMMETRIC CATALYSIS WITH STEREOLABILE LIGANDS

Combinatorial synthesis and high-throughput screening assays, as well as computational and iterative optimization methods, have produced a plethora of chiral catalysts that can be applied in numerous asymmetric reactions. In many cases, however, catalytic activity and asymmetric induction of a chiral catalyst need to be fine-tuned to achieve high turnover numbers in addition to satisfactory enantioselectivity and product yield. This is routinely accomplished by optimization of temperature and solvent effects, and extensive screening of other reaction parameters. The striking sensitivity of many catalytic processes to small amounts of additives has culminated in the development of asymmetric activation and chiral poisoning strategies.[895–900] In particular, conformationally labile axially chiral ligands and fluxional achiral additives that can populate chiral conformations have been reported to increase both the enantioselectivity and efficiency of chiral catalysts.[901]

As has been discussed in Chapter 2.2, achiral ligands can exist as a fluxional mixture of achiral and enantiomeric conformers. Interactions with another chiral compound can render the equienergetic and equally populated chiral conformations diastereomeric. Since diastereomers differ in energy, the ligand is likely to preferentially occupy one chiral conformation. The presence of an enantiopure compound can therefore induce a conformational bias in an achiral ligand resulting in amplification of chirality. This concept has been exploited for asymmetric catalysis. One can imagine that a stereodynamic achiral ligand populates a distinctive chiral conformation upon coordination to a chiral catalyst consisting of a metal ion and an enantiopure ligand. The chirality of the metal complex is then amplified and this may increase asymmetric induction during catalysis. At the same time, accommodation of the stereodynamic ligand by the chiral metal complex changes the coordination sphere of the metal which can affect the rate of individual catalytic steps including substrate coordination, product formation and dissociation of the product-catalyst complex. This can further improve catalytic properties. For example, a fast release of the product facilitates regeneration of the catalyst and ultimately increases turnover frequency. The prospect of increasing both enantioselectivity and efficiency of a catalyst with inexpensive and readily available additives is attractive, compared with laborious fine-tuning of the chiral ligand structure.[902,903]

Alternatively, the structure of a fluxional metal complex comprising rapidly interconverting enantiomeric species can be controlled by coordination of a nonracemizing enantiopure ligand. This concept of asymmetric activation was introduced by Mikami, and has been successfully applied to catalysts derived from conformationally unstable chiral ligands such as BIPHEP, Figure 6.8.[904]

(a) Asymmetric activation via nonselective formation of equilibrating diastereomers

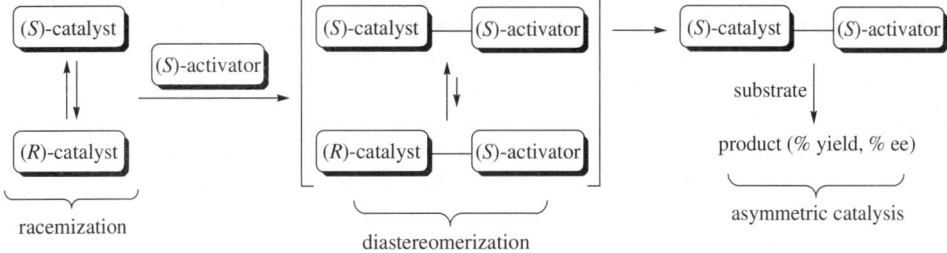

(b) Asymmetric activation via selective formation of a single diastereomer and concomitant interconversion of the remaining enantiomer

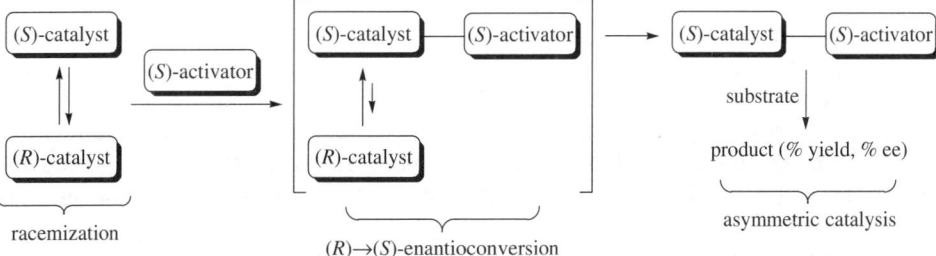

Figure 6.8 Asymmetric activation of a stereolabile racemic catalyst with a nonracemic activator.

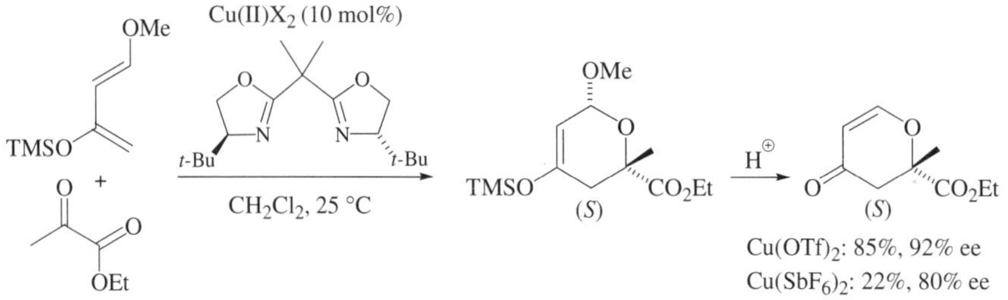

Scheme 6.127 Counteranion effects on the enantioselectivity and yield of the chiral Lewis acid-catalyzed hetero-Diels–Alder reaction of Danishefsky's diene and ethyl pyruvate.

The interactions between a stereolabile racemic catalyst and a nonracemic chiral activator can be classified into two ultimate scenarios: (a) nonselective formation of an equimolar mixture of diastereomeric species that equilibrate towards a single catalytically active diastereomer, and (b) selective coordination of the chiral activator to one enantiomer of the metal complex and concomitant conversion of the remaining enantiomer to the same diastereomerically pure catalyst. Although formation of a single diastereomer through asymmetric activation of a racemic stereolabile catalyst is desirable, a mixture of diastereomeric catalytic species with individual enantioselectivity and turnover frequency is obtained in many cases.

The introduction of achiral additives to asymmetric catalysis can have multi-faceted mechanistic implications leading to increased stereoselectivity and faster turnover. Increased selectivity is usually attributed to amplification of the chiral environment of the catalyst, preferential formation of one well-defined nonfluxional catalyst structure, or selective poisoning of an undesired catalytic species that favors an alternative reaction pathway with lower stereoselectivity. Enhanced turnover frequency is often explained by facilitated dissociation of inactive or less reactive oligomeric complexes to a catalytically active monomeric species, faster release of the product from the catalyst in the presence of an achiral additive, and conformational changes in the catalyst structure and geometry resulting in increased activity. The presence of additives can completely alter the mechanistic pathway and outcome of an asymmetric reaction. An impressive example is the poorly understood counterion effect of triflate and hexafluoroantimonate anions on the hetero-Diels–Alder reaction of α-keto esters and activated dienes, Scheme 6.127.[905–908,xxxviii] Jørgensen and coworkers found that bisoxazoline-derived copper(II) triflates provide dihydropyranones in significantly higher yield and ee than their hexafluoroantimonate analogs. Both anions are generally considered noncoordinating counterions, but these findings confirm that the impact of triflate and hexafluoroantimonate ions on chiral Lewis acid catalysis is by no means negligible.[909,910]

6.6.1 Stereodynamic Achiral Ligands

Several stereodynamic bidentate ligands that effectively amplify chirality at a metal center have been reported to improve both selectivity and efficiency of asymmetric catalysis. One of the first successful examples was reported by Katsuki and coworkers who employed chiral *N*-oxides and manganese complexes bearing flexible achiral salen ligands in the asymmetric epoxidation of olefins. This important reaction was discovered by Jacobsen and Katsuki who independently developed chiral (salen)manganese catalysts for asymmetric epoxidation of nonfunctionalized alkenes.[911,912] Katsuki recognized that pyridine *N*-oxide increases the enantioselectivity and life time of chiral (salen)Mn catalysts which are believed to form reactive intermediate

---

[xxxviii] Dihydropyranones are important building blocks for natural product synthesis of carbohydrates, pheromones, insect toxins, antitumor agents, antibiotics, and anti-inflammatory sesterterpenoids.

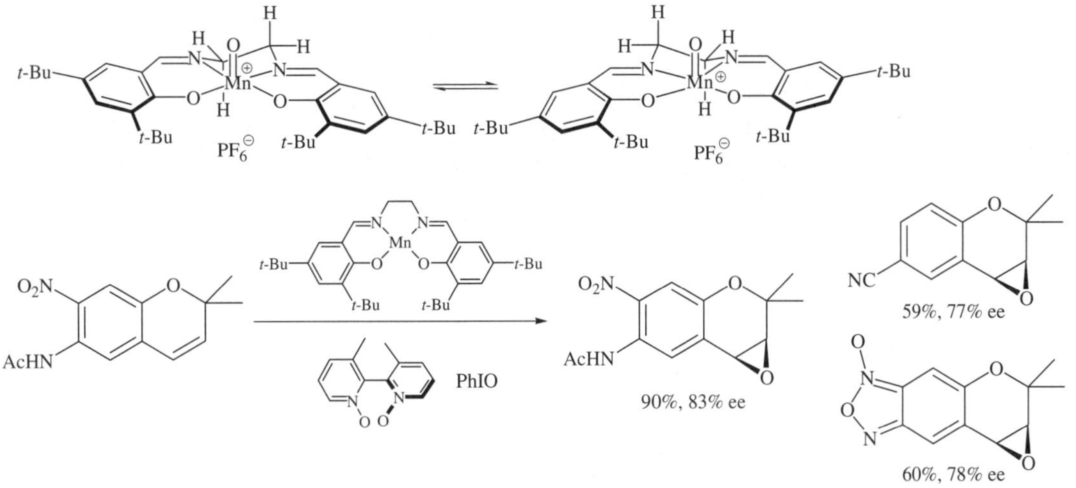

Scheme 6.128 Enantiomeric stepped conformations of interconverting (salen)oxomanganese(v) complexes (top) and asymmetric epoxidation of a 2,2-dimethylchromene using an achiral (salen)Mn catalyst in the presence of axially chiral 3,3′-dimethyl-2,2′-bipyridine *N,N*′-dioxide (bottom).

oxomanganese(v) complexes possessing a nonplanar stepped conformation.[913–917] It is generally assumed that the stepped conformation of the salen ligand plays a crucial role in the stereochemical course of this reaction. Katsuki hypothesized that achiral (salen)Mn catalysts may exist in the form of rapidly interconverting enantiomeric stepped conformations that can be rendered diastereomeric in the presence of a chiral Lewis base. Screening of a range of achiral (salen)Mn(III) complexes and chiral Lewis bases gave only moderate results when amino alcohols, alcohols or amines such as (−)-sparteine were used.[918,919] In contrast, the asymmetric epoxidation of 2,2-dimethylchromenes proceeds with remarkable enantioselectivity when catalytic amounts of enantiopure 3,3′-dimethyl-2,2′-bipyridine *N,N*′-dioxide are added.[920] The (salen)manganese-catalyzed reaction of 6-acetamido-2,2-dimethyl-7-nitrochromene gives the corresponding epoxide in 83% ee in the presence of the *N,N*′-dioxide, Scheme 6.128. This has been explained by coordination of the atropisomeric *N,N*′-dioxide to the metal center and selective stabilization of one chiral salen conformation. This interligand chiral amplification ultimately leads to high asymmetric induction during olefin epoxidation. Nguyen and coworkers applied the same concept to asymmetric olefin cyclopropanation using an achiral (salen)ruthenium(II) complex in the presence of catalytic amounts of chiral sulfoxides.[921]

Walsh and coworkers realized the advantage of screening inexpensive and readily available stereodynamic achiral ligands for optimization of asymmetric catalysis. They systematically examined the influence of achiral additives on the asymmetric induction in titanium- and zinc-mediated alkylations of aromatic aldehydes.[922–924] The employment of achiral diamine or diimine activators significantly altered the stereochemical outcome of the (*P*)-3,3′-diphenyl BINOL-catalyzed alkylation of benzaldehyde with diethylzinc. In the absence of a stereodynamic additive, (*S*)-1-phenylpropanol was obtained in only 44% ee. However, the alcohol was produced in ee's ranging from 96% (*R*) to 75% (*S*) when an activator was added. The dramatic effect on enantioselectivity can be attributed to amplification of the chiral environment of the (*P*)-BINOL-derived zinc catalyst upon coordination of the diamine or diimine. Achiral ligands such as *cis*-diaminocyclohexane derivatives can access equienergetic enantiomeric chair conformations that easily interconvert via ring inversion. Coordination of such a ligand to a (*P*)-3,3′-diphenyl BINOL-zinc complex gives rise to diastereomeric conformations that exist in nonequimolar amounts. The preferential population of one chiral conformation of the *cis*-diaminocyclohexane ligand therefore

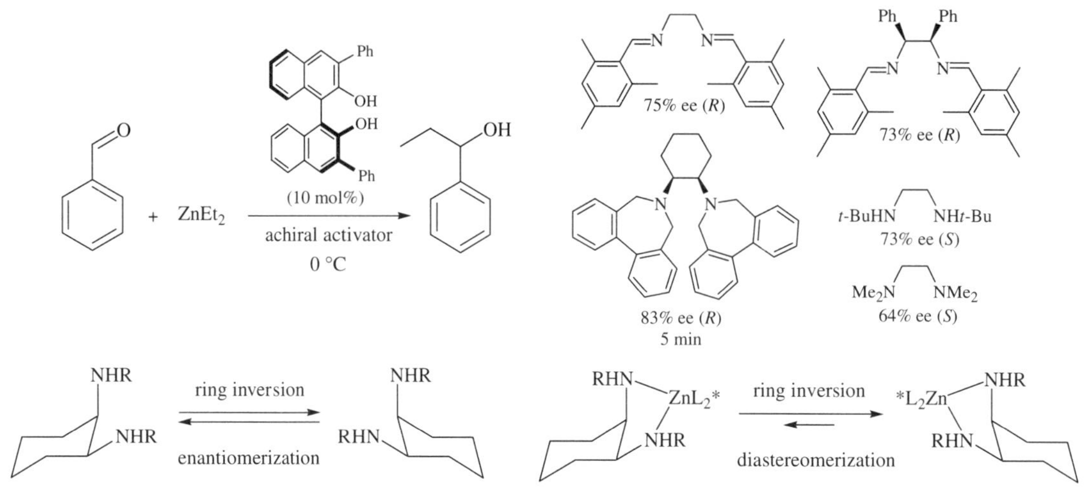

Scheme 6.129   (*P*)-3,3′-Diphenyl BINOL-catalyzed alkylation of benzaldehyde and structures of achiral activators (top). The absolute configuration and enantiomeric excess of 1-phenylpropanol obtained in the presence of a diamine or diimine additive is given under the structure of the activator. Enantiomeric and diastereomeric conformations of a diaminocyclohexane-derived ligand and chiral amplification of a BINOL-derived zinc catalyst, ZnL$_2$*, upon coordination (bottom).

amplifies the asymmetric environment of the catalyst and enhances asymmetric induction. In addition, the presence of some diamines or diimines gives rise to ligand-accelerated catalysis. For example, ethylation of benzaldehyde with diethylzinc using (*P*)-3,3′-diphenyl BINOL combined with a meso diamine carrying flexible pendant biphenyl groups is completed within 5 minutes at 0 °C. This reaction proceeds much more slowly when it is conducted without the achiral ligand, Scheme 6.129. This approach is conceptually different from Mikami's asymmetric activation methodology depicted in Figure 6.8. Mikami's concept is based on rapidly interconverting chiral catalysts, for example racemic BIPHEP-derived transition metal complexes, that are converted to a diastereomerically enriched or pure catalytic species through addition of an enantiopure activator. In contrast, Walsh's approach utilizes stereolabile achiral activators to optimize the performance of a chiral enantiopure catalyst that is inherently stable to racemization.

6.6.2   Stereolabile Axially Chiral Ligands

Axially chiral biaryls exist in the form of enantiomeric ground state conformations with dihedral angles typically above 30°.[xxxix] The energy barrier to rotation about the chiral axis of biaryls mainly depends on the number and size of *ortho*-substituents, see Chapter 3.3.1. Because of severe steric hindrance to rotation, 2,2′-di-*tert*-butylbiphenyl is stable to racemization and the enantiomers can be isolated.[925,926] Incorporation of smaller groups gives rise to 2,2′-disubstituted biaryls that undergo facile rotation about the chiral axis and rapid enantioconversion at room temperature.[927] Stereolabile 2,2′-disubstituted biphenyls are inherently chiral but the enantiomers can not be isolated at 25 °C. Mikami's group recognized the usefulness of conformationally unstable biphenol-derived ligands in asymmetric catalysis. The selectivity of the (biphenoxide)Ti(TADDOLate)-catalyzed alkylation of 3,5-bis(trifluoromethyl)benzaldehyde with methyltitanium triisopropoxide depends strongly on the

---

[xxxix] Biphenyl has a dihedral angle of 42° in the gas phase and 20–30° in solution. The angle between the aryl rings generally increases with the size of the *ortho*-substituents when intramolecular interactions such as hydrogen bonding are negligible.

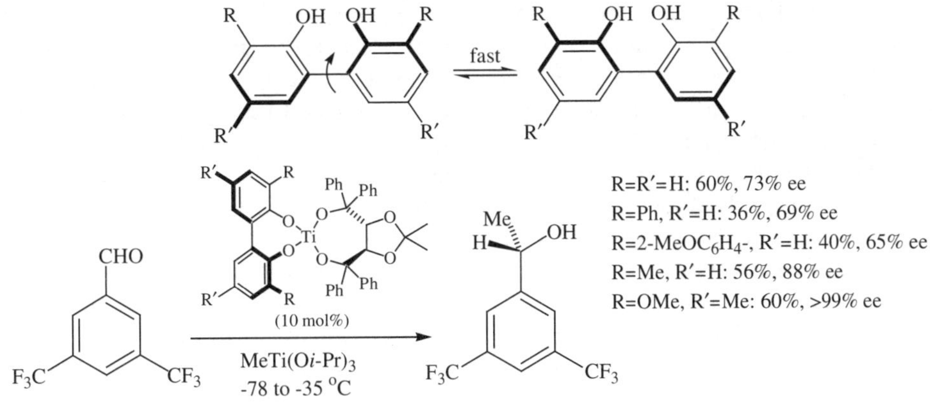

Scheme 6.130 Enantiomerization of biphenols and (biphenoxide)Ti(TADDOLate)-catalyzed methylation of 3,5-bis(trifluoromethyl)benzaldehyde with MeTi(Oi-Pr)$_3$.

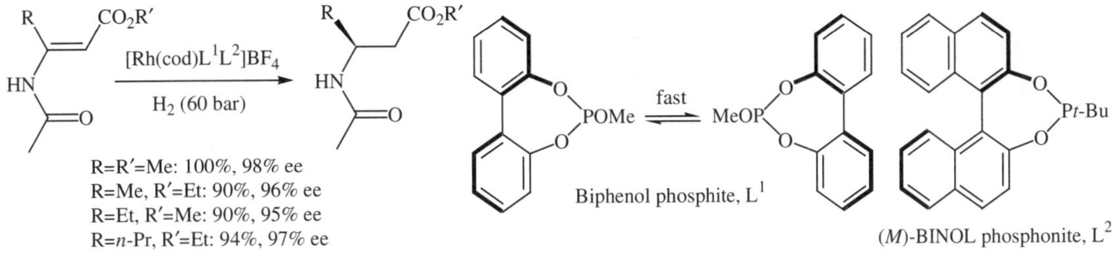

Scheme 6.131 Hydrogenation of β-acylamino acrylates with an (*M*)-BINOL-derived Rh catalyst bearing a conformationally flexible biphenol phosphite ligand.

substitution pattern of the biphenol additive, Scheme 6.130.[928] While 1-[3′,5′-bis(trifluoro-methyl)phenyl]ethanol is produced in 60% yield and 73% ee in the presence of 2,2′-biphenol, enantioselectivity increases dramatically to more than 99% when 3,3′-dimethoxy-5,5′-dimethyl-2,2′-biphenol is used. It has been hypothesized that this impressive increase in enantioselectivity is due to formation of a highly effective Ti(TADDOLate)-derived catalytic species exhibiting the coordinating biphenol ligand in one exclusively populated chiral conformation. This assumption is supported by molecular mechanics suggesting that the (*M*)-3,3′-dimethoxy-5,5′-dimethyl-2,2′-biphenoxide-Ti[(*R*,*R*)-TADDOLate] complex is 15 kJ/mol more stable than the (*P*)-biphenoxide-Ti[(*R*,*R*)-TADDOLate] diastereomer.[xl] Other groups have utilized stereolabile axially chiral biphenols to improve the efficacy of asymmetric ene reactions[929] and Baeyer–Villiger oxidations.[930]

Reetz and Li examined a series of rhodium complexes consisting of monodentate BINOL- and fluxional biphenol-derived ligands to uncover an effective catalyst for hydrogenation of β-acylamino acrylates.[931] Extensive screening revealed that combination of an (*M*)-BINOL phosphonite and a conformationally labile biphenol phosphite provides a rhodium species that allows reduction of β-acylamino acrylates to the corresponding β-amino acids with excellent enantioselectivity, Scheme 6.131. Mechanistic studies proved the coexistence of two rapidly interconverting diastereomers, of which the homochiral (*M*,*M*)-form is assumed to afford higher enantioselectivity and catalytic activity.[932,933]

[xl] Alternatively, the high enantioselectivity of the methylation can be attributed to two coexisting diastereomeric (biphen-oxide)Ti(TADDOLate) complexes that have considerably different catalytic activity.

Another class of flexible ligands that is known to improve the performance of chiral catalysts comprises 2,2′-bis(diarylphosphino)biphenyls such as BIPHEP and DM-BIPHEP. Noyori, Mikami and others introduced these chelating diphosphines as conformationally flexible analogs of BINAP.[934–937] NMR spectroscopic studies revealed that addition of (*S,S*)-diphenylethylene-diamine (DPEN) to racemic (DM-BIPHEP)RuCl$_2$(DMF)$_n$ initially produces a 1:1 mixture of diastereomeric ruthenium(II) precatalysts, which slowly changes within three hours to a 3:1 ratio in favor of the (*P*)-DM-BIPHEP-derived complex. Treatment of these species with base and hydrogen generates catalytically active ruthenium(0) complexes that were applied in the reduction of ketones. Because of the slow equilibration process, Mikami *et al.* were able to employ different dia-stereomeric catalyst mixtures in the hydrogenation of 1-acetylnaphthalene. They found that the enantiomeric excess of the (*R*)-alcohol product steadily improves with increasing diastereomeric purity of the catalyst. Hydrogenation of 1-acetylnaphthalene using 0.4 mol% of a 1:1 dia-stereomeric ruthenium catalyst mixture gave (*R*)-1-(1′-naphthyl)ethanol in more than 99% yield and 63% ee. The enantiomeric excess increased to 84% when a 3:1 catalyst mixture was used under the same conditions, Scheme 6.132. Apparently, coordination of (*S,S*)-DPEN to (DM-BIPHEP)RuCl$_2$ results in deracemization of the diphosphine ligand and effectively amplifies the chiral environment of the ruthenium complex. It is important to remember that enantioselective synthesis requires kinetic reaction control and that the stereochemical outcome is determined by the individual stereoselectivity and turnover frequency of each catalytic species present. Mikami's studies show that the thermodynamically favored ruthenium complex bearing (*S,S*)-DPEN and

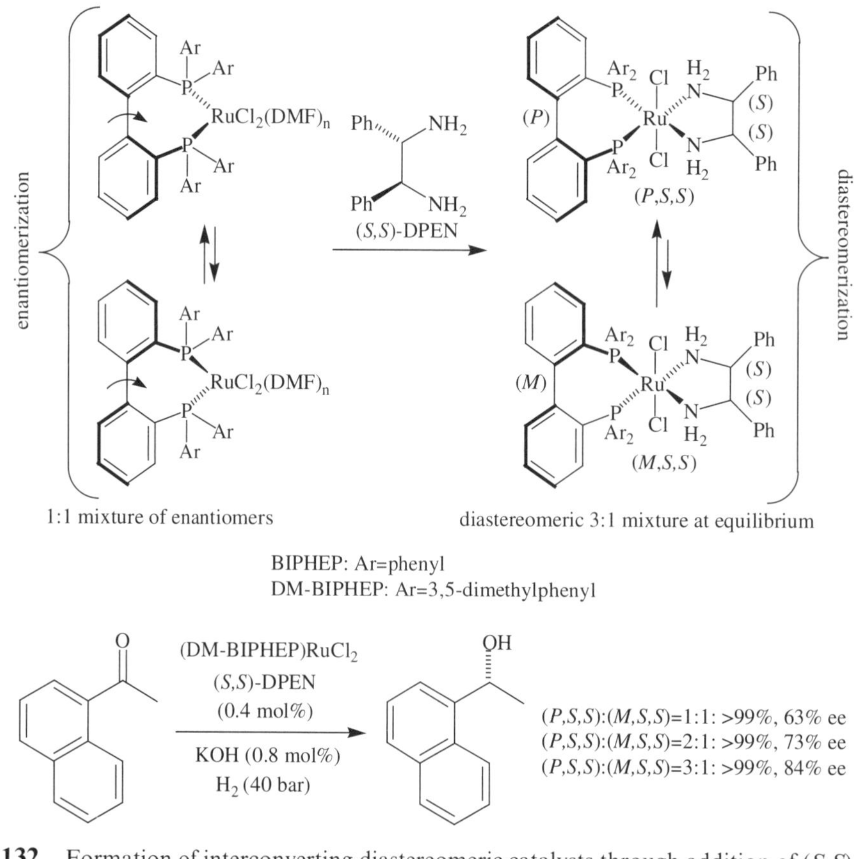

BIPHEP: Ar=phenyl
DM-BIPHEP: Ar=3,5-dimethylphenyl

(*P,S,S*):(*M,S,S*)=1:1: >99%, 63% ee
(*P,S,S*):(*M,S,S*)=2:1: >99%, 73% ee
(*P,S,S*):(*M,S,S*)=3:1: >99%, 84% ee

Scheme 6.132 Formation of interconverting diastereomeric catalysts through addition of (*S,S*)-DPEN to a racemic mixture of (DM-BIPHEP)RuCl$_2$(DMF)$_n$ (top) and enantioselective hydrogenation of 1-acetylnaphthalene using different catalyst ratios (bottom).

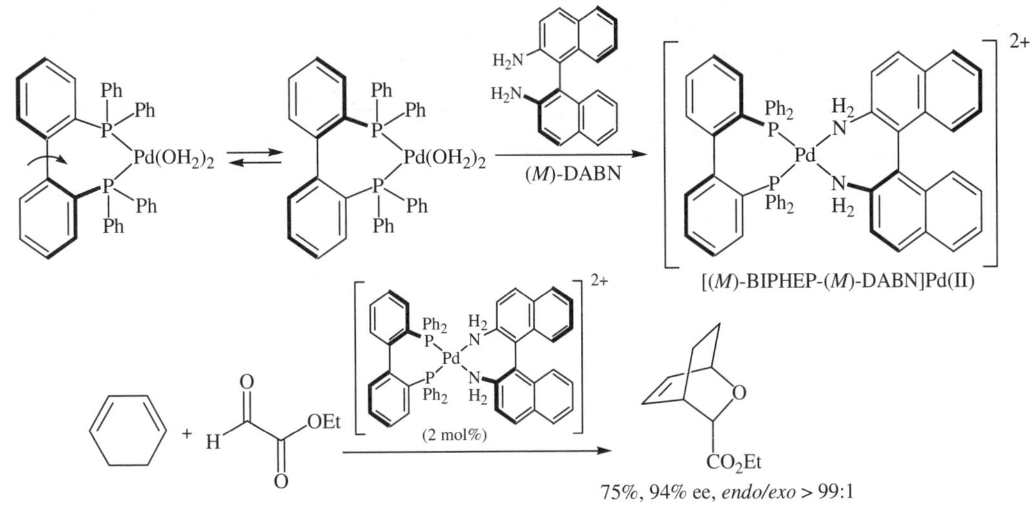

Scheme 6.133   [(*M*)-BIPHEP-(*M*)-DABN]Pd(II)-catalyzed hetero-Diels–Alder reaction of ethyl glyoxylate and cyclohexadiene.

(*P*)-DM-BIPHEP affords higher enantioselectivity and reaction rates than the diastereomeric (*M*)-DM-BIPHEP-derived catalyst.

The concept of asymmetric activation of racemic BIPHEP-derived metal complexes with chiral amines has been extended to other reactions.[938,939] For example, (*M*)-2,2'-diamino-1,1'-binaphthyl (DABN) forms a Lewis-acidic [(*M*)-BIPHEP-(*M*)-DABN]Pd(II) catalyst exhibiting both high enantioselectivity and activity in the hetero-Diels–Alder reaction of ethyl glyoxylate and cyclohexadiene, Scheme 6.133. The energy barrier to rotation of BIPHEP is 92 kJ/mol at 125 °C, and one can expect that this di-*ortho*-substituted biphenyl undergoes rapid racemization at room temperature.[940] It is important to realize, however, that coordination of BIPHEP to a metal center can significantly enhance the stability to rotation about the chiral axis. This would explain the slow equilibration towards a thermodynamically favored 3:1 ratio of diastereomeric (*S*,*S*)-DPEN-derived ruthenium complexes, *vide supra*.[941] The mechanism and extent of metal-mediated conformational stabilization of coordinating 2,2'-bis(diarylphosphino)biphenyls is difficult to predict, and it depends on a variety of factors including electron configuration, coordination number and geometry of the metal complex, as well as electronic and steric properties of ancillary ligands. It is assumed that interconversion of metal-coordinated 2,2'-bis(diarylphosphino)biphenyl conformers occurs via intermediate dissociation of one phosphine group, subsequent rotation and recoordination to the metal center.[942] Mikami's group discovered that the conformational stability of (BIPHEP)rhodium(I) complexes depends critically on the presence and nature of ancillary diamino ligands.[943] They found that BIPHEP coordinating to (DPEN)rhodium(I) does not rotate about the chiral axis, while the corresponding (DABN)rhodium complex and amine-free (BIPHEP)Rh(I) show atropisomerization at room temperature. Asymmetric activation of racemic (DM-BIPHEP)Rh(I) with (*M*)-DABN generates diastereomerically pure [(*M*)-DM-BIPHEP-(*M*)-DABN]Rh(I) which has been successfully employed in enantioselective cyclization of 1,6-enynes, Scheme 6.134. Gagné exploited the conformational stability of platinum-coordinated BIPHEP to prepare [(*M*)-BIPHEP]Pt(OTf)$_2$ in 98% enantiomeric excess from racemic [(*M*)-BIPHEP]PtCl$_2$ via isolation of [(*M*)-BIPHEP]Pt[(*P*)-Binolate] and ligand exchange with triflic acid.[944] The high enantiopurity and considerable stability to racemization of [(*M*)-BIPHEP]platinum(II) complexes at room temperature for at least eight hours was confirmed by ^{31}P NMR measurements using (*S*,*S*)-DPEN as chiral shift reagent. The chiral catalyst shows high enantio- and *endo*-selectivity in the Diels–Alder

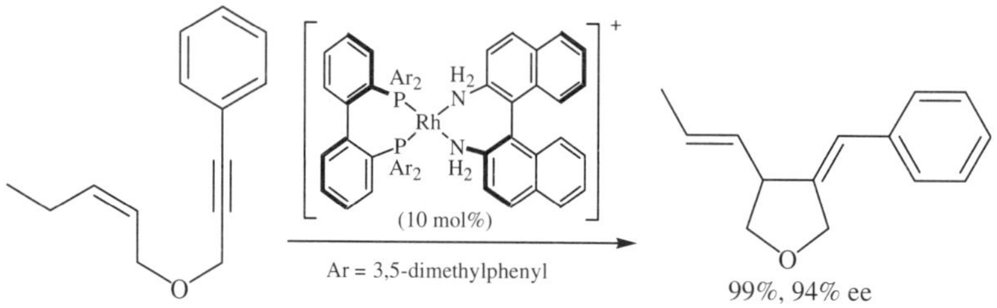

Scheme 6.134   [(*M*)-DM-BIPHEP-(*M*)-DABN]Rh(I)-catalyzed cyclization of 1,6-enynes.

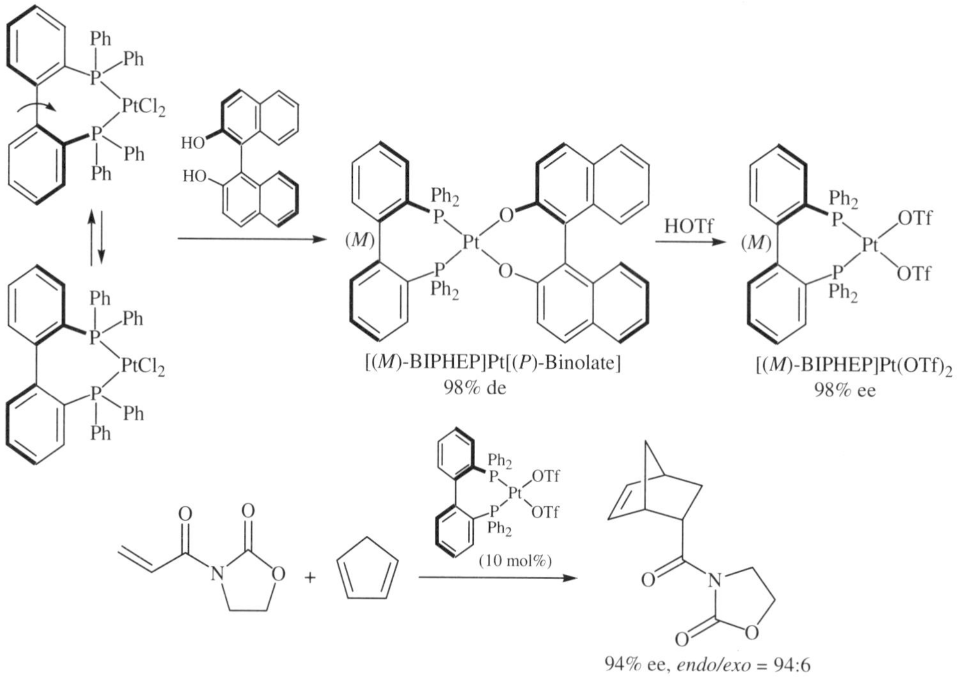

Scheme 6.135   Preparation of [(*M*)-BIPHEP]PtCl₂, and enantioselective Diels–Alder reaction of cyclo-pentadiene and 3-acryloyl-oxazolidin-2-one.

reaction of cyclopentadiene and 3-acryloyl-oxazolidin-2-one, Scheme 6.135. Both Mikami's and Gagné's groups have successfully employed isolated [(*M*)-BIPHEP]palladium and platinum complexes in hetero-Diels–Alder and carbonyl ene reactions.[945–948]

Axially chiral 1,1'-diphenyl-3,3',4,4'-tetramethyl-2,2'-biphosphole (BIPHOS) and other biphos-pholes exist as a mixture of three pairs of rapidly interconverting enantiomers in solution due to the presence of a stereolabile chiral axis and two chiral phosphorus atoms that undergo facile pyramidal inversion.[949,xli] Fortunately, BIPHOS crystallizes as a conglomerate consisting of (*M*,*S*ₚ,*S*ₚ)- and (*P*,*R*ₚ,*R*ₚ)-enantiomers that can be isolated under cryogenic conditions, and the ligands are stable to isomerization upon coordination to palladium. The enantiopure (BIPHOS)Pd complex catalyzes the asymmetric allylic substitution of 1,3-diphenyl-prop-2-enyl acetate with sodium dimethyl malonate and the corresponding homoallylic diester is obtained in 93% yield and 80% ee, Scheme 6.136.

Although preparation of enantiopure transition metal complexes derived from BIPHEP or other stereolabile ligands is feasible, formation of asymmetric catalysts from inherently stable atrop-isomeric diphosphines remains a much more attractive strategy. The Diels–Alder reaction depicted

---

[xli] The facile *P*-pyramidal inversion in 1-phenylphospholes is a consequence of a resonance-stabilized planar transition state.

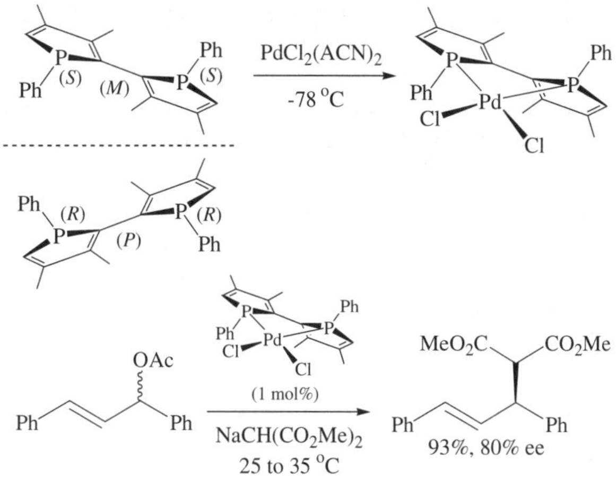

Scheme 6.136 Structures of enantiomeric (M,S_P,S_P)- and (P,R_P,R_P)-BIPHOS, preparation of the $[(M,S_P,S_P)$-BIPHOS]Pd catalyst, and asymmetric allylic substitution of 1,3-diphenyl-prop-2-enyl acetate with sodium dimethyl malonate.

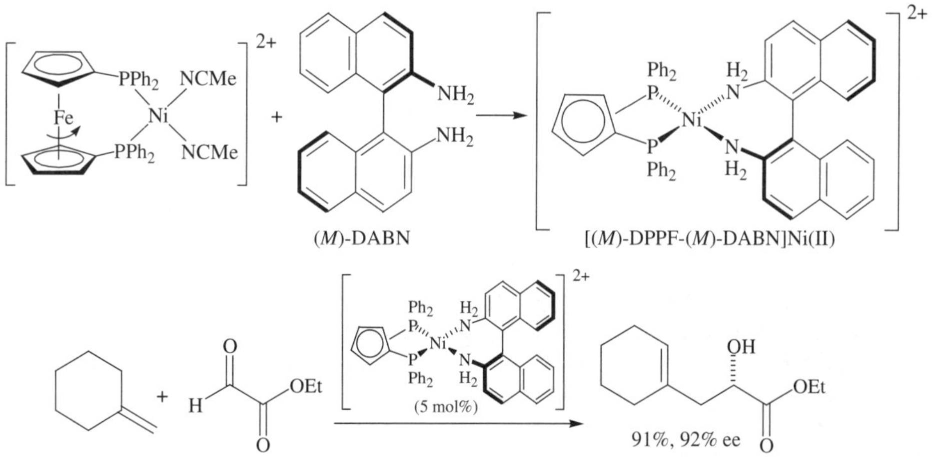

Scheme 6.137 Formation of a diastereomerically pure $[(M)$-DPPF-(M)-DABN]Ni complex using (M)-DABN as chirality controller (top) and glyoxylate-ene reaction (bottom).

in Scheme 6.135 has also been accomplished with readily available $[(M)$-BINAP]platinum and palladium catalysts, providing the cycloadduct in up to 98% ee.[950] The studies conducted by Mikami's and Gagné's groups discussed above demonstrate that the stereodynamic properties of BIPHEP limit its use as a conformationally stable diphosphine ligand that has the stereochemical integrity of BINAP, while the slow interconversion rate restricts its use as a perfectly flexible activator for amplification of asymmetric induction of a chiral catalyst. Mikami therefore introduced the conformationally less stable bis(diphenylphosphino)ferrocene (DPPF), a highly flexible diphosphine ligand forming axially chiral staggered conformations upon coordination to a transition metal. In contrast to (BIPHEP)platinum and palladium complexes, the chiral conformations of metal-coordinated DPPF can be effectively controlled with enantiopure diamines, and equilibration is completed within seconds.[951] Deracemization of a (DPPF)Ni complex with (M)-DABN produces the homochiral diastereomer which catalyzes the glyoxylate-ene reaction of methylene-cyclohexane with high enantioselectivity, Scheme 6.137.

6.7 STEREOSELECTIVE SYNTHESIS IN THE SOLID STATE

The success of asymmetric synthesis is directly tied to the control of both the stereodynamics and the relative orientation of reactants. In solution, molecules exist as a mixture of rapidly interconverting conformational isomers that constantly undergo tumbling motions. A fundamental strategy of asymmetric synthesis is to limit the number of coexisting conformers and competing transition states that ultimately lead to the formation of different stereoisomers and thus compromise selectivity. As discussed in Chapter 5, this is often achieved by chelation control and use of C_2-symmetric catalysts. For the same reason, superior enantio- and diastereoselectivities are generally obtained with rigid and cyclic compounds that possess less rotational freedom than acyclic analogs.

A well-defined fixation of the three-dimensional structure and relative orientation of organic compounds is often elusive in solution, but this can be achieved in the solid state.[952–954] Solid-state synthesis is attractive for several reasons: it is interesting from environmental and sustainability standpoints because it eliminates the use of solvents, and it often affords stereochemically pure products in literally quantitative yield. Weak interactions including hydrogen bonding and π-stacking have been exploited to build supramolecular assemblies of unsaturated substrates that participate in stereo- and regioselective photodimerizations.[955] Topochemical polymerization of unsaturated carboxylic acids has been accomplished through photochemical and γ-radiation-induced reactions using ammonium carboxylates as solid-state templates for control of polymer tacticity.[956–960] Similarly, stereocontrolled solid-state photodimerization of ammonium salts of *trans*-cinnamic acid or *trans*-cinnamamide carboxylates allows stereoselective formation of truxinic acids and truxinamides.[961–963] The combination of the principles of molecular recognition and supramolecular self-assembly of bifunctional building blocks has paved the way for powerful template-assisted solid-state reactions.[964–970] The well-defined regularity, order and orientation of organic molecules in crystalline materials give rise to distinguished stereo- and regiochemical selectivity and excellent yields.[971–976] Nevertheless, few enantioselective solid-state reactions have been reported to date.[977–985] Cocrystallization of coumarin or thiocoumarin and C_2-symmetric (*S,S*)-bis(hydroxydiphenylmethyl)-2,2-dimethyl-1,3-dioxacyclopentane, which can be prepared from tartaric acid, pre-organizes two coumarin molecules in the chiral environment of the diol template. Irradiation of the coumarin-derived cocrystal with a 400 W high-pressure mercury lamp results in asymmetric single-crystal-to-single-crystal [2 + 2]cycloaddition providing the enantiopure *anti*-head-to-head dimer in 89% yield, Scheme 6.138.[986,987]

Enantiopure ammonium salts have been used to arrange achiral carboxylic acids in an asymmetric solid-state environment for enantioselective photoreactions and topotactic single-crystal-to-single-crystal transformation.[988–990] Similarly, inclusion complexes of prochiral guest molecules and hosts such as cyclodextrins or tartaric acid derivatives have been used for stereoselective

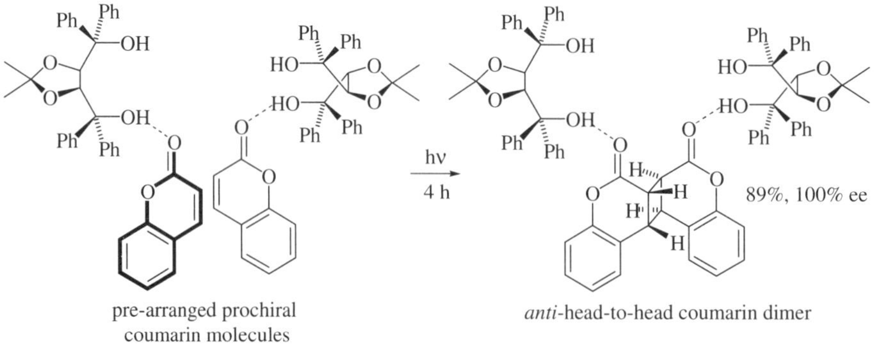

pre-arranged prochiral
coumarin molecules

anti-head-to-head coumarin dimer

89%, 100% ee

Scheme 6.138   Stereoselective solid-state synthesis of an *anti*-head-to-head coumarin dimer using a tartaric acid-derived template.

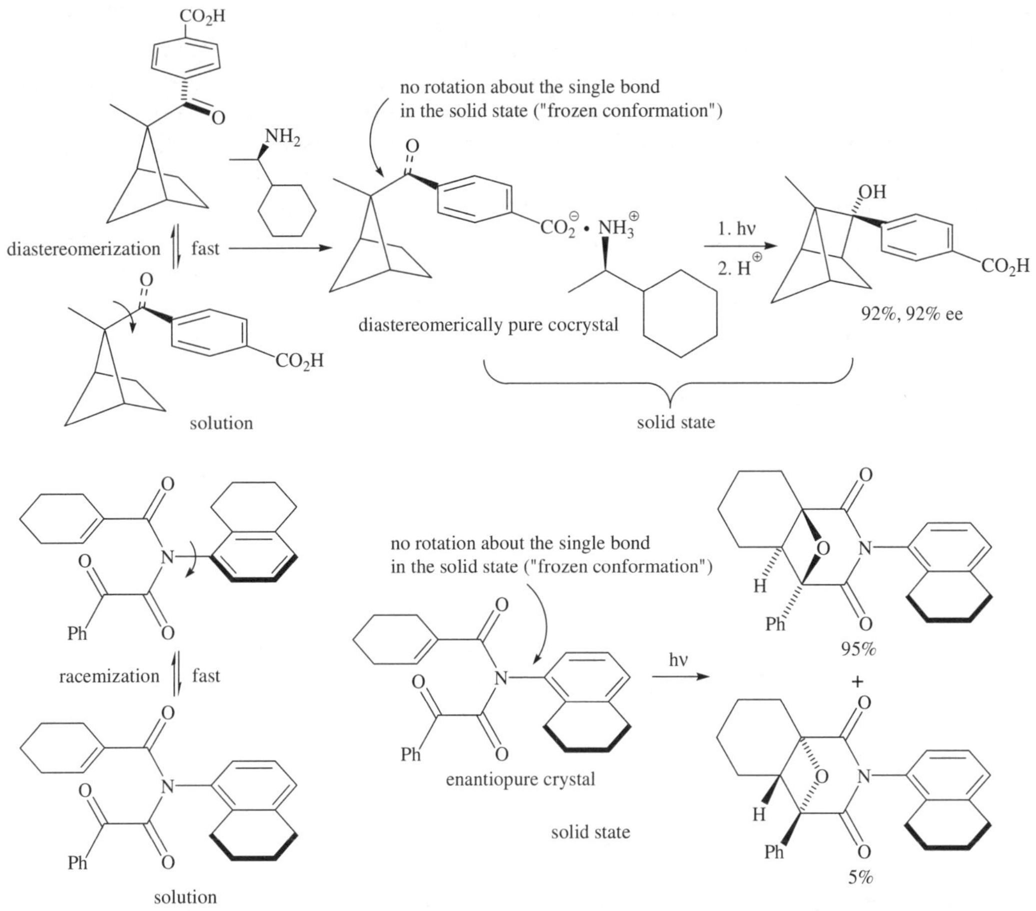

Scheme 6.139 Asymmetric solid-state reactions of "frozen" chiral conformations.

synthesis within the chiral cavity.[991–994] The scope of this approach is limited because the chiral recognition mechanism of cyclodextrins and the reactivity and stereoselectivity of organic reactions proceeding within the hydrophobic interior are very difficult to control, and the sense of chiral induction can not be predicted.

Stereoselective intramolecular reactions of compounds that exist as a mixture of rapidly inter-converting chiral conformations in solution have been realized in the solid state. The conversion of equilibrating stereoisomers to a single chiral conformer is usually achieved by cocrystallization with a chiral template or spontaneous resolution based on crystallization-induced asymmetric transformation, see Chapter 7.3. In these cases, the stereodynamic freedom of a fluxional compound is reduced to a single "frozen" chiral conformation which can be exploited for asymmetric synthesis.[995–999] Scheffer observed that rapidly interconverting mixtures of enantiomeric keto acids form diastereomerically pure ammonium salts with chiral amines via asymmetric transformation of the second kind. Irradiation of the crystal initiates an asymmetric photoreaction that transforms the "frozen" chiral conformer to a single configurationally stable isomer. For example, crystallization of an *endo*-bicyclo[2.1.1]hexane-derived keto acid with (*R*)-1-cyclohexylethylamine affords one conformational diastereomer that undergoes asymmetric Yang photocyclization, thus generating the corresponding chiral alcohol in 90% ee at 92% conversion, Scheme 6.139.[1000–1003,xlii] Another example of an asymmetric intramolecular reaction using a "frozen" chiral conformation

---

xlii The same ammonium salt forms a racemate when the photocyclization is carried out in solution.

of an otherwise stereolabile compound was reported by Sakamoto and coworkers.[1004] They observed that spontaneous resolution of the interconverting enantiomers of *N*-cyclohexenecarbonyl-*N*-(5,6,7,8-tetrahydronaphthalen-1-yl)benzoyl formamide locks the molecule into a single geometry that yields a bicyclic chiral oxetane by asymmetric [2 + 2]photocycloaddition in the solid state.

Despite the attractive features of these examples, planning of asymmetric solid-state synthesis remains a difficult task. Once the frozen conformation of a molecule or the arrangement of two reactants in a crystal has been elucidated by X-ray analysis, the stereochemical outcome of an intra- or intermolecular reaction is quite predictable. However, a major problem is that the structure and properties of supramolecular assemblies of organic compounds relies mainly on weak interactions including hydrogen bonding, dipole–dipole interactions, and *CH*/π- and π-stacking forces.[1005–1010] It is therefore difficult to systematically embed prochiral substrates with a specific relative orientation into a chiral solid-state environment. The design and growth of cocrystals consisting of different compounds is particularly challenging, since the supramolecular architecture and molecular arrangement is determined by a complex synergism of noncovalent forces, molecular shape, conformational flexibility, and stoichiometry of the individual building blocks.[1011,1012]

REFERENCES

1. Letsinger, R. L. *J. Am. Chem. Soc.* **1950**, *72*, 4842.
2. Curtin, D. Y.; Koehl Jr., W. J. *J. Am. Chem. Soc.* **1962**, *84*, 1967-1973.
3. Nozaki, H.; Aratani, T.; Toraya, T.; Noyori, R. *Tetrahedron* **1971**, *27*, 905-913.
4. Basu, A.; Beak, P. *J. Am. Chem. Soc.* **1996**, *118*, 1575-1576.
5. Hoppe, D.; Carstens, A. *Tetrahedron* **1994**, *50*, 6097-6108.
6. Pearson, W. H.; Lindbeck, A. C.; Kampf, J. W. *J. Am. Chem. Soc.* **1993**, *115*, 2622-2636.
7. Chan, P. C.-M.; Chong, J. M. *J. Org. Chem.* **1988**, *53*, 5584-5586.
8. Basu, A.; Thayumanavan, S. *Angew. Chem., Int. Ed.* **2002**, *41*, 716-738.
9. Li, X.; Schenkel, L. B.; Kozlowski, M. C. *Org. Lett.* **2000**, *2*, 875-878.
10. Harrison, J. R.; O'Brien, P.; Porter, D. W.; Smith, N. M. *Chem. Commun.* **2001**, 1202-1203.
11. Hermet, J.-P. R.; Porter, D. W.; Dearden, M. J.; Harrison, J. R.; Koplin, T.; O'Brien, P.; Parmene, J.; Tyurin, V.; Whitwood, A. C.; Gilday, J.; Smith, N. M. *Org. Biomol. Chem.* **2003**, *1*, 3977-3988.
12. Wilkinson, J. A.; Rossington, S. B.; Ducki, S.; Leonard, J.; Hussain, N. *Tetrahedron: Asymm.* **2004**, *15*, 3011-3013.
13. Dearden, M. J.; McGrath, M. J.; O'Brien, P. *J. Org. Chem.* **2004**, *69*, 5789-5792.
14. Phuan, P.-W.; Ianni, J. C.; Kozlowski, M. C. *J. Am. Chem. Soc.* **2004**, *126*, 15473-15479.
15. O'Brien, P.; Wiberg, K. B.; Bailey, W. F.; Hermet, J.-P. R.; McGrath, M. J. *J. Am. Chem. Soc.* **2004**, *126*, 15480-15489.
16. Morita, Y.; Tokuyama, H.; Fukuyama, T. *Org. Lett.* **2005**, *7*, 4337-4340.
17. Hoppe, D. *Angew. Chem., Int. Ed. Engl.* **1984**, *23*, 932-948.
18. Özlügedik, M.; Kristensen, J.; Wibbeling, B.; Fröhlich, R.; Hoppe, D. *Eur. J. Org. Chem.* **2002**, 414-427.
19. Özlügedik, M.; Kristensen, J.; Reuber, J.; Fröhlich, R.; Hoppe, D. *Synthesis* **2004**, 2303-2316.
20. Özlügedik, M.; Ünaldi, S.; Wibbeling, B.; Hoppe, D. *Adv. Synth. Catal.* **2005**, *347*, 1627-1631.
21. Dieter, R. K.; Chen, N.; Watson, R. T. *Tetrahedron* **2005**, *61*, 3221-3230.
22. Gawley, R. E.; Zhang, Q. *Tetrahedron* **1994**, *50*, 6077-6088.
23. Meyers, A. I.; Guiles, J.; Warmus, J. S.; Gonzales, M. A. *Tetrahedron Lett.* **1991**, *32*, 5505-5508.
24. Hoffmann, R. W.; Rühl, T.; Chemla, F.; Zahneisen, T. *Liebigs Ann. Chem.* **1992**, 719-724.
25. Beak, P.; Du, H. *J. Am. Chem. Soc.* **1993**, *115*, 2516-2518.

26. Derwing, C.; Hoppe, D. *Synthesis* **1996**, 149-154.
27. Park, Y. S.; Boys, M. L.; Beak, P. *J. Am. Chem. Soc.* **1996**, *118*, 3757-3758.
28. Basu, A.; Beak, P. *J. Am. Chem. Soc.* **1996**, *118*, 1575-1576.
29. Basu, A.; Gallagher, D. J.; Beak, P. *J. Org. Chem.* **1996**, *61*, 5718-5719.
30. Clayden, J.; Pink, J. H. *Tetrahedron Lett.* **1997**, *38*, 2561-2564.
31. Clayden, J.; Pink, J. H. *Tetrahedron Lett.* **1997**, *38*, 2565-2568.
32. Thayumanavan, S.; Basu, A.; Beak, P. *J. Am. Chem. Soc.* **1997**, *119*, 8209-8216.
33. Thayumanavan, S.; Lee, S.; Liu, C.; Beak, P. *J. Am. Chem. Soc.* **1994**, *116*, 9755-9756.
34. Elworthy, T. R.; Meyers, A. I. *Tetrahedron* **1994**, *50*, 6089-6096.
35. Still, W. C.; Sreekumar, C. *J. Am. Chem. Soc.* **1980**, *102*, 1201-1202.
36. McGarvey, G. J.; Kimura, M. *J. Org. Chem.* **1982**, *47*, 5422-5424.
37. Tomooka, K.; Igarashi, T.; Watanabe, M.; Nakai, T. *Tetrahedron Lett.* **1992**, *33*, 5795-5798.
38. Manabe, S. *Chem. Commun.* **1997**, 737-738.
39. Kawasaki, T.; Kimachi, T. *Tetrahedron* **1999**, *55*, 6847-6862.
40. Gibson, S. E.; Ham, P.; Jefferson, G. R. *Chem. Commun.* **1998**, 123-124.
41. Tomooka, K.; Komine, N.; Nakai, T. *Chirality* **2000**, *12*, 505-509.
42. Barrett, I. M.; Breeden, S. W. *Tetrahedron: Asymm.* **2004**, *15*, 3015-3017.
43. Sasaki, M.; Higashi, M.; Masu, H.; Yamaguchi, K.; Takeda, K. *Org. Lett.* **2005**, *7*, 5913-5915.
44. Marshall, J. A.; Lebreton, J. *J. Am. Chem. Soc.* **1988**, *110*, 2925-2931.
45. Marshall, J. A.; Lebreton, J. *J. Org. Chem.* **1988**, *53*, 4108-4112.
46. Hodgson, D. M.; Gras, E. *Angew. Chem., Int. Ed.* **2002**, *41*, 2376-2378.
47. Hodgson, D. M.; Maxwell, C. R.; Miles, T. J.; Paruch, E.; Stent, M. A. H.; Matthews, I. R.; Wilson, F. X.; Witherington, J. *Angew. Chem., Int. Ed.* **2002**, *41*, 4313-4316.
48. O'Brien, P.; Rosser, C. M.; Caine, D. *Tetrahedron Lett.* **2003**, *44*, 6613-6615.
49. Hodgson, D. M.; Stent, M. A. H.; Stefane, B.; Wilson, F. X. *Org. Biomol. Chem.* **2003**, *1*, 1139-1150.
50. Hoppe, D.; Zschage, O. *Angew. Chem., Int. Ed. Engl.* **1989**, *28*, 69-71.
51. Hoppe, D.; Hintze, F.; Tebben, P. *Angew. Chem., Int. Ed. Engl.* **1990**, *29*, 1422-1424.
52. Marsch, M.; Harms, K.; Zschage, O.; Hoppe, D.; Boche, G. *Angew. Chem., Int. Ed. Engl.* **1991**, *30*, 321-323.
53. Zschage, O.; Hoppe, D. *Tetrahedron* **1992**, *48*, 5657-5666.
54. Zschage, O.; Hoppe, D. *Tetrahedron* **1992**, *48*, 8389-8392.
55. Hoppe, D.; Paetow, M.; Hintze, F. *Angew. Chem., Int. Ed. Engl.* **1993**, *32*, 394-396.
56. Hoppe, D.; Hense, T. *Angew. Chem., Int. Ed. Engl.* **1997**, *36*, 2282-2316.
57. Deiters, A.; Mück-Lichtenfeld, C.; Fröhlich, R.; Hoppe, D. *Org. Lett.* **2000**, *2*, 2415-2418.
58. Schultz-Fademrecht, C.; Wibbeling, B.; Fröhlich, R.; Hoppe, D. *Org. Lett.* **2001**, *3*, 1221-1224.
59. Deiters, A.; Hoppe, D. *J. Org. Chem.* **2001**, *66*, 2842-2849.
60. Christoph, G.; Hoppe, D. *Org. Lett.* **2002**, *4*, 2189-2192.
61. Gralla, G.; Wibbeling, B.; Hoppe, D. *Org. Lett.* **2002**, *3*, 2193-2195.
62. Zeng, W.; Fröhlich, R.; Hoppe, D. *Tetrahedron* **2005**, *61*, 3281-3287.
63. Brandau, S.; Hoppe, D. *Tetrahedron* **2005**, *61*, 12244-12255.
64. Reuber, J.; Fröhlich, R.; Hoppe, D. *Org. Lett.* **2004**, *6*, 783-786.
65. Seppi, M.; Kalkofen, R.; Reupohl, J.; Fröhlich, R.; Hoppe, D. *Angew. Chem., Int. Ed.* **2004**, *43*, 1423-1427.
66. Derwing, C.; Frank, H.; Hoppe, D. *Eur. J. Org. Chem.* **1999**, 3519-3524.
67. Hoppe, D.; Carstens, A.; Krämer, T. *Angew. Chem., Int. Ed. Engl.* **1990**, *29*, 1424-1425.
68. Carstens, A.; Hoppe, D. *Tetrahedron* **1994**, *50*, 6097-6108.
69. Derwing, C.; Hoppe, D. *Synthesis* **1996**, 149-154.
70. Hammerschmidt, F.; Hanninger, A. *Chem. Ber.* **1995**, *128*, 1069-1077.

71. Hammerschmidt, F.; Hanninger, A.; Vollenkle, H. *Chem. Eur. J.* **1997**, *3*, 1728-1732.

72. Hammerschmidt, F.; Hanninger, A.; Simov, B.; Vollenkle, H.; Werner, A. *Eur. J. Org. Chem.* **1999**, 3511-3518.

73. Beak, P.; Zajdel, W. J.; Reitz, D. B. *Chem. Rev.* **1984**, *84*, 471-523.

74. Pearson, W. H.; Lindbeck, A. C. *J. Am. Chem. Soc.* **1991**, *113*, 8546-8548.

75. Tsunoda, T.; Fujiwara, K.; Yamamoto, Y.; Ito, S. *Tetrahedron Lett.* **1991**, *32*, 1975-1978.

76. Burchat, A. F.; Chong, J. M.; Park, S. B. *Tetrahedron Lett.* **1993**, *34*, 51-54.

77. Coldham, I.; Hufton, R.; Snowden, D. J. *J. Am. Chem. Soc.* **1996**, *118*, 5322-5323.

78. Katritzky, A. R.; Qi, M. *Tetrahedron* **1998**, *54*, 2647-2668.

79. Gawley, R. E.; Zhang, Q. *J. Am. Chem. Soc.* **1993**, *115*, 7515-7516.

80. Coldham, I.; Hufton, R.; Snowden, D. J. *J. Am. Chem. Soc.* **1996**, *118*, 5322-5323.

81. Coldham, I.; Vennall, G. P. *Chem. Commun.* **2000**, 1569-1570.

82. Gawley, R. E.; Zhang, Q. *J. Org. Chem.* **1995**, *60*, 5763-5769.

83. Kerrick, S. T.; Beak, P. *J. Am. Chem. Soc.* **1991**, *113*, 9708-9710.

84. Chong, J. M.; Park, S. B. *J. Org. Chem.* **1992**, *57*, 2220-2222.

85. Beak, P.; Kerrick, S. T.; Wu, S.; Chu, J. *J. Am. Chem. Soc.* **1994**, *116*, 3231-3239.

86. Park, Y. S.; Boys, M. L.; Beak, P. *J. Am. Chem. Soc.* **1996**, *118*, 3757-3758.

87. Weisenburger, G. A.; Beak, P. *J. Am. Chem. Soc.* **1996**, *118*, 12218-12219.

88. Park, Y. S.; Weisenburger, G. A.; Beak, P. *J. Am. Chem. Soc.* **1997**, *119*, 10537-10538.

89. Faibish, N. C.; Park, Y. S.; Lee, S.; Beak, P. *J. Am. Chem. Soc.* **1997**, *119*, 11561-11570.

90. Pippel, D. J.; Weisenburger, G. A.; Wilson, S. R.; Beak, P. *Angew. Chem., Int. Ed.* **1998**, *37*, 2522-2524.

91. Whisler, M. C.; Vaillancourt, L.; Beak, P. *Org. Lett.* **2000**, *2*, 2655-2658.

92. Gross, K. M. B.; Beak, P. *J. Am. Chem. Soc.* **2001**, *123*, 315-321.

93. Coldham, I.; Copley, R. C. B.; Haxell, T. F. N.; Howard, S. *Org. Lett.* **2001**, *3*, 3799-3801.

94. Pippel, D. J.; Weisenburger, G. A.; Faibish, N. C.; Beak, P. *J. Am. Chem. Soc.* **2001**, *123*, 4919-4927.

95. Whisler, M. C.; Beak, P. *J. Org. Chem.* **2003**, *68*, 1207-1215.

96. Ashweek, N. J.; Coldham, I.; Haxell, T. F. N.; Howard, S. *Org. Biomol. Chem.* **2003**, *1*, 1532-1544.

97. Kim, D. D.; Lee, S. J.; Beak, P. *J. Org. Chem.* **2005**, *70*, 5376-5386.

98. Schlosser, M.; Limat, D. *J. Am. Chem. Soc.* **1995**, *117*, 12342-12343.

99. Weisenburger, G. A.; Faibish, N. C.; Pippel, D. J.; Beak, P. *J. Am. Chem. Soc.* **1999**, *121*, 9522-9530.

100. Johnson, T. A.; Curtis, M. D.; Beak, P. *J. Am. Chem. Soc.* **2001**, *123*, 1004-1005.

101. Johnson, T. A.; Curtis, M. D.; Beak, P. *Org. Lett.* **2002**, *4*, 2747-2749.

102. Johnson, T. A.; Jang, D. O.; Slafer, B. W.; Curtis, M. D.; Beak, P. *J. Am. Chem. Soc.* **2002**, *124*, 11689-11698.

103. Jang, D. O.; Kim, D. D.; Pyun, D. K.; Beak, P. *Org. Lett.* **2003**, *5*, 4155-4157.

104. Wu, S.; Lee, S.; Beak, P. *J. Am. Chem. Soc.* **1996**, *118*, 715-721.

105. Aggarwal, V. K. *Angew. Chem., Int. Ed. Engl.* **1994**, *33*, 175-177.

106. Nakamura, S.; Nakagawa, R.; Watanabe, Y.; Toru, T. *Angew. Chem., Int. Ed.* **2000**, *39*, 353-355.

107. Nakamura, S.; Nakagawa, R.; Watanabe, Y.; Toru, T. *J. Am. Chem. Soc.* **2000**, *122*, 11340-11347.

108. Nakamura, S.; Furutani, A.; Toru, T. *Eur. J. Org. Chem.* **2002**, 1690-1695.

109. Nakamura, S.; Ito, Y.; Wang, L.; Toru, T. *J. Org. Chem.* **2004**, *69*, 1581-1589.

110. Gais, H.-J.; Hellmann, G.; Günther, H.; Lopez, F.; Lindner, H. J.; Braun, S. *Angew. Chem., Int. Ed. Engl.* **1989**, *28*, 1025-1027.

111. Raabe, G.; Gais, H.-J.; Fleischhauer, J. *J. Am. Chem. Soc.* **1996**, *118*, 4622-4630.

112. Gais, H.-J.; van Gumpel, M.; Schleusner, M.; Raabe, G.; Runsink, J.; Vermeeren, C. *Eur. J. Org. Chem.* **2001**, 4275-4303.
113. Hoffmann, R. W.; Bewersdorf, M. *Liebigs Ann. Chem.* **1992**, 643-653.
114. Hoffmann, R. W.; Klute, W. *Chem. Eur. J.* **1996**, *2*, 694-700.
115. Biellmann, J. F.; Vicens, J. J. *Tetrahedron Lett.* **1978**, *19*, 467-470.
116. Yamamoto, Y.; Maruyama, K. *J. Chem. Soc., Chem. Commun.* **1980**, 239-240.
117. Boche, G. *Angew. Chem., Int. Ed. Engl.* **1989**, *28*, 277-297.
118. Kaiser, B.; Hoppe, D. *Angew. Chem., Int. Ed. Engl.* **1995**, *34*, 323-325.
119. Hoffmann, R.; Koberstein, R. *J. Chem. Soc., Perkin Trans. 2* **2000**, 595-602.
120. Hoppe, D.; Kaiser, B.; Stratmann, O.; Fröhlich, R. *Angew. Chem., Int. Ed. Engl.* **1997**, *36*, 2784-2786.
121. Marr, F.; Fröhlich, R.; Hoppe, D. *Org. Lett.* **1999**, *1*, 2081-2083.
122. Stratmann, O.; Kaiser, B.; Fröhlich, R.; Meyer, O.; Hoppe, D. *Chem. Eur. J.* **2001**, *7*, 423-435.
123. Marr, F.; Fröhlich, R.; Wibbeling, B.; Diedrich, C.; Hoppe, D. *Eur. J. Org. Chem.* **2002**, 2970-2988.
124. Otte, R.; Fröhlich, R.; Wibbeling, B.; Hoppe, D. *Angew. Chem., Int. Ed.* **2005**, *44*, 5492-5496.
125. Hammerschmidt, F.; Hanninger, A. *Chem. Ber.* **1995**, *128*, 823-830.
126. Hammerschmidt, F.; Schmidt, S. *Chem. Ber.* **1996**, *129*, 1503-1508.
127. Hoffmann, R. W.; Bewersdorf, M. *Chem. Ber.* **1991**, *124*, 1259-1264.
128. Rao, A. V. R.; Gurjar, M. K.; Reddy, K. L.; Rao, A. S. *Chem. Rev.* **1995**, *95*, 2135-2167.
129. Nicolaou, K. C.; Boddy, C. N. C.; Bräse, S.; Winssinger, N. *Angew. Chem., Int. Ed.* **1999**, *38*, 2096-2152.
130. Baudoin, O.; Guéritte, F. *Stud. Nat. Prod. Chem.* **2003**, *29*, 355-418.
131. Rizzacasa, M. A. *Stud. Nat. Prod. Chem.* **1998**, *20*, 407-455.
132. Dagne, E.; Steglich, W. *Phytochemistry* **1984**, *23*, 1729-1731.
133. Fukuyama, Y.; Asakawa, J. *J. Chem. Soc., Perkin Trans. 1* **1991**, 2737-2741.
134. Ito, C.; Thoyama, Y.; Omura, M.; Kajiura, I.; Furukawa, H. *Chem. Pharm. Bull.* **1993**, *41*, 2096-2100.
135. Bringmann, G.; Tasler, S.; Endress, H.; Kraus, J.; Messer, K.; Wohlfarth, M.; Lobin, W. *J. Am. Chem. Soc.* **2001**, *123*, 2703-2711.
136. Rosini, C.; Franzini, L.; Raffaelli, A.; Salvadori, P. *Synthesis* **1992**, 503-517.
137. McCarthy, M.; Guiry, P. J. *Tetrahedron* **2001**, *57*, 3809-3844.
138. Takita, R.; Yakura, K.; Ohshima, T.; Shibasaki, M. *J. Am. Chem. Soc.* **2005**, *127*, 13760-13761.
139. Yamagiwa, N.; Qin, H.; Matsunaga, S.; Shibasaki, M. *J. Am. Chem. Soc.* **2005**, *127*, 13419-13427.
140. Hatano, M.; Ikeno, T.; Miyamoto, T.; Ishihara, K. *J. Am. Chem. Soc.* **2005**, *127*, 10776-10777.
141. Brunel, J. M. *Chem. Rev.* **2005**, *105*, 857-897.
142. Miyashita, A.; Yasuda, A.; Takaya, H.; Toriumi, K.; Ito, T.; Souchi, T.; Noyori, R. *J. Am. Chem. Soc.* **1980**, *102*, 7932-7934.
143. Noyori, R.; Takaya, H. *Acc. Chem. Res.* **1990**, *23*, 345-350.
144. Noyori, R. *Adv. Synth. Catal.* **2003**, *345*, 15-32.
145. Yanagisawa, A.; Touge, T.; Arai, T. *Angew. Chem., Int. Ed.* **2005**, *44*, 1546-1548.
146. Berthod, M.; Mignani, G.; Woodward, G.; Lemaire, M. *Chem. Rev.* **2005**, *105*, 1801-1836.
147. Schmid, R.; Cereghetti, M.; Heiser, B.; Schönholzer, P.; Hansen, H.-J. *Helv. Chim. Acta* **1988**, *71*, 897-929.
148. Ogawa, C.; Sugiura, M.; Kobayashi, S. *Angew. Chem., Int. Ed.* **2004**, *43*, 6419-6493.
149. Kocovsky, P.; Vyskocil, S.; Smrcina, M. *Chem. Rev.* **2003**, *103*, 3213-3245.
150. Aikawa, K.; Mikami, K. *Angew. Chem., Int. Ed.* **2003**, *42*, 5458-5461.

151. Ogawa, C.; Sugiura, M.; Kobayashi, S. *Angew. Chem., Int. Ed.* **2004**, *43*, 6491-6493.

152. Ihori, Y.; Yamashita, Y.; Ishitani, H.; Kobayashi, S. *J. Am. Chem. Soc.* **2005**, *127*, 15528-15535.

153. Ding, K.; Li, X.; Ji, B.; Guo, H.; Kitamura, M. *Curr. Org. Synth.* **2005**, *2*, 499-545.

154. Evans, D. A.; Thomson, R. J.; Franco, F. *J. Am. Chem. Soc.* **2005**, *127*, 10816-10817.

155. Maruoka, K.; Ooi, T. *Chem. Rev.* **2003**, *103*, 3013-3028.

156. Ooi, T.; Maruoka, K. *Acc. Chem. Res.* **2004**, *37*, 526-533.

157. Shirakawa, S.; Yamamoto, K.; Kitamura, M.; Ooi, T.; Maruoka, K. *Angew. Chem., Int. Ed.* **2005**, *44*, 625-628.

158. Kitamura, M.; Shirakawa, S.; Maruoka, K. *Angew. Chem., Int. Ed.* **2005**, *44*, 1549-1551.

159. Ooi, T.; Takeuchi, M.; Kato, D.; Uematsu, Y.; Tayama, E.; Sakai, D.; Maruoka, K. *J. Am. Chem. Soc.* **2005**, *127*, 5073-5083.

160. Malkov, A. V.; Orsini, M.; Pernazza, D.; Muir, K. W.; Langer, V.; Meghani, P.; Kocovsky, P. *Org. Lett.* **2002**, *4*, 1047-1049.

161. Malkov, A. V.; Dufkova, L.; Farrugia, L.; Kocovsky, P. *Angew. Chem., Int. Ed.* **2003**, *42*, 3674-3677.

162. Malkov, A. V.; Bell, M.; Orsini, M.; Pernazza, D.; Massa, A.; Herrmann, P.; Meghani, P.; Kocovsky, P. *J. Org. Chem.* **2003**, *68*, 9659-9668.

163. Malkov, A. V.; Bell, M.; Fabiomassimo, C.; Kocovsky, P. *Org. Lett.* **2005**, *7*, 3219-3222.

164. Nakajima, M.; Saito, M.; Shiro, M.; Hashimoto, S. *J. Am. Chem. Soc.* **1998**, *120*, 6419-6420.

165. Nakajima, M.; Saito, M.; Uemura, M.; Hashimoto, S. *Tetrahedron Lett.* **2002**, *43*, 8827-8829.

166. Shimada, T.; Kina, A.; Ikeda, S.; Hayashi, T. *Org. Lett.* **2002**, *4*, 2799-2801.

167. Denmark, S. E.; Fan, Y. *J. Am. Chem. Soc.* **2002**, *124*, 4233-4235.

168. Denmark, S. E.; Fu, J. *Chem. Commun.* **2003**, 167-170.

169. Jiao, Z.; Feng, X.; Liu, B.; Chen, F.; Zhang, G.; Jiang, J. *Eur. J. Org. Chem.* **2003**, 3818-3826.

170. Shimada, T.; Kina, A.; Hayashi, T. *J. Org. Chem.* **2003**, *68*, 6329-6337.

171. Chen, F.-X.; Qin, B.; Feng, X.; Zhang, G.; Jiang, Y. *Tetrahedron* **2004**, *60*, 10449-10460.

172. Denmark, S. E.; Fan, Y.; Eastgate, M. D. *J. Org. Chem.* **2005**, *70*, 5235-5248.

173. Mei, X.; Wolf, C. *Chem. Commun.* **2004**, 2078-2079.

174. Mei, X.; Wolf, C. *J. Am. Chem. Soc.* **2004**, *126*, 14736-14737.

175. Tumambac, G. E.; Wolf, C. *Org. Lett.* **2005**, *7*, 4045-4048.

176. Bringmann, G.; Mortimer, A. J. P.; Keller, P. A.; Gresser, M. J.; Garner, J.; Breuning, M. *Angew. Chem., Int. Ed.* **2005**, *44*, 5384-5427.

177. Bringmann, G.; Reichert, Y.; Kane, V. V. *Tetrahedron* **2004**, *60*, 3539-3574.

178. Nelson, T. D.; Meyers, A. I. *Tetrahedron Lett.* **1994**, *35*, 3259-3262.

179. Nelson, T. D.; Meyers, A. I. *J. Org. Chem.* **1994**, *59*, 2577-2580.

180. Wolf, C.; König, W. A.; Roussel, C. *Liebigs Ann.* **1995**, 781-786.

181. Wolf, C.; Hochmuth, D. H.; König, W. A.; Roussel, C. *Liebigs Ann.* **1996**, 357-363.

182. Miyano, S.; Tobita, M.; Hashimoto, H. *Bull. Chem. Soc. Jpn.* **1981**, *54*, 3522-3526.

183. Miyano, S.; Handa, S.; Shimizu, K.; Tagami, K.; Hashimoto, H. *Bull. Chem. Soc. Jpn.* **1984**, *57*, 1943-1947.

184. Miyano, S.; Fukushima, H.; Handa, S.; Ito, H.; Hashimoto, H. *Bull. Chem. Soc. Jpn.* **1988**, *61*, 3249-3254.

185. Lipshutz, B. H.; Siegmann, K.; Garcia, E.; Kayser, F. *J. Am. Chem. Soc.* **1993**, *115*, 9276-9282.

186. Lipshutz, B. H.; Kayser, F.; Liu, Z.-P. *Angew. Chem., Int. Ed. Engl.* **1994**, *33*, 1842-1844.

187. Kyasnoor, R. V.; Sargent, M. V. *Chem. Commun.* **1998**, 2713-2714.

188. Lipshutz, B. H.; James, B.; Vance, S.; Carrico, I. *Tetrahedron Lett.* **1997**, *38*, 753-756.

189. Lipshutz, B. H.; Shin, Y.-J. *Tetrahedron Lett.* **1998**, *39*, 7017-7020.

190. Lipshutz, B. H.; Buzard, D. J.; Olsson, C.; Noson, K. *Tetrahedron* **2004**, *60*, 4443-4449.

191. Noyori, R.; Tomino, I.; Tanimoto, Y. *J. Am. Chem. Soc.* **1979**, *101*, 3129-3131.

192. Smrcina, M.; Lorenc, M.; Hanus, V.; Kocovsky, P. *Synlett* **1991**, 231-232.

193. Lipshutz, B. H.; Shin, Y.-J. *Tetrahedron Lett.* **2000**, *41*, 9515-9521.

194. Bringmann, G.; Keller, P. A.; Rölfing, K. *Synlett* **1994**, 423-424.

195. Rawal, V. H.; Florjancic, A. S.; Singh, S. P. *Tetrahedron Lett.* **1994**, *35*, 8985-8988.

196. Sugimura, T.; Yamada, H.; Inoue, S.; Tai, A. *Tetrahedron: Asymm.* **1997**, *8*, 649-655.

197. Spring, D. R.; Krishnan, S.; Schreiber, S. L. *J. Am. Chem. Soc.* **2000**, *122*, 5656-5657.

198. Spring, D. R.; Krishnan, S.; Blackwell, H. E.; Schreiber, S. L. *J. Am. Chem. Soc.* **2002**, *124*, 1355-1363.

199. Baker, R. W.; Kyasnoor, R. V.; Sargent, M. V.; Skelton, B. W.; White, A. H. *Aust. J. Chem.* **2000**, *53*, 487-506.

200. Lipshutz, B. H.; Shin, Y.-J.; Kayser, F. *Tetrahedron Lett.* **1994**, *35*, 5567-5570.

201. Lin, G.-Q.; Zhong, M. *Tetrahedron: Asymm.* **1997**, *8*, 1369-1372.

202. Bowie Jr., A. L.; Hughes, C. C.; Trauner, D. *Org. Lett.* **2005**, *7*, 5207-5209.

203. Michaud, G.; Billiard, M.; Ricard, L.; Genêt, J.-P.; Marinetti, A. *Chem. Eur. J.* **2002**, *8*, 3327-3330.

204. Madec, J.; Michaud, G.; Genêt, J.-P.; Marinetti, A. *Tetrahedron: Asymm.* **2004**, *15*, 2253-2261.

205. Qiu, L.; Wu, J.; Chan, S.; Au-Yeung, T. T.-L.; Ji, J.-X.; Guo, R.; Pai, C.-C.; Zhou, Z.; Li, X.; Fan, Q.-H. Chan, A. S. C. *Proc. Natl. Acad. Sci.* **2004**, *101*, 5815-5820.

206. Feldman, K. S.; Ensel, S. M. *J. Am. Chem. Soc.* **1993**, *115*, 1162-1163.

207. Feldman, K. S.; Ensel, S. M. *J. Am. Chem. Soc.* **1994**, *116*, 3357-3366.

208. Feldman, K. S.; Ensel, S. M.; Minard, R. D. *J. Am. Chem. Soc.* **1994**, *116*, 1742-1745.

209. Feldman, K. S.; Lawlor, M. D. *J. Am. Chem. Soc.* **2000**, *122*, 7396-7397.

210. Evans, D. A.; Dinsmore, C. J. *Tetrahedron Lett.* **1993**, *34*, 6029-6032.

211. Evans, D. A.; Wood, M. R.; Trotter, B. W.; Richardson, T. I.; Barrow, J. C.; Katz, J. L. *Angew. Chem., Int. Ed.* **1998**, *37*, 2700-2704.

212. Nicolaou, K. C.; Chen, D. Y.-K.; Huang, X.; Ling, T.; Bella, M.; Snyder, S. A. *J. Am. Chem. Soc.* **2004**, *126*, 12888-12896.

213. Robin, J.-P.; Landais, Y. *J. Org. Chem.* **1988**, *53*, 224-226.

214. Ward, R. S.; Hughes, D. D. *Tetrahedron* **2001**, *57*, 4015-4022.

215. Kramer, B.; Averhoff, A.; Waldvogel, S. R. *Angew. Chem., Int. Ed.* **2002**, *41*, 2981-2982.

216. Meyers, A. I. *J. Org. Chem.* **2005**, *70*, 6137-6151.

217. Meyers, A. I.; Meier, A.; Rawson, D. J. *Tetrahedron Lett.* **1992**, *33*, 853-856.

218. Moorlag, H.; Meyers, A. I. *Tetrahedron Lett.* **1993**, *34*, 6989-6992.

219. Moorlag, H.; Meyers, A. I. *Tetrahedron Lett.* **1993**, *34*, 6993-6996.

220. Meyers, A. I.; Nelson, T. D.; Moorlag, H.; Rawson, D. J.; Meier, A. *Tetrahedron* **2004**, *60*, 4459-4473.

221. Meyers, A. I.; Flisak, J. R.; Aitken, R. A. *J. Am. Chem. Soc.* **1987**, *109*, 5446-5452.

222. Leighton, B. N.; Rizzacasa, M. A. *J. Org. Chem.* **1995**, *60*, 5702-5705.

223. Warshawsky, A. M.; Meyers, A. I. *J. Am. Chem. Soc.* **1990**, *112*, 8090-8099.

224. Nelson, T. D.; Meyers, A. I. *J. Org. Chem.* **1994**, *59*, 2577-2580.

225. Nelson, T. D.; Meyers, A. I. *J. Org. Chem.* **1994**, *59*, 2655-2658.

226. Andrus, M. B.; Asgari, D.; Sclafani, J. A. *J. Org. Chem.* **1997**, *62*, 9365-9368.

227. Meyers, A. I. *J. Heterocycl. Chem.* **1998**, *35*, 991-1002.

228. Meyers, A. I.; Willemsen, J. J. *Chem. Commun.* **1997**, 1573-1574.

229. Meyers, A. I.; Willemsen, J. J. *Tetrahedron* **1998**, *54*, 10493-10511.

230. Coleman, R. S.; Grant, E. B. *J. Am. Chem. Soc.* **1995**, *117*, 10889-10904.

231. Baudoin, O.; Décor, A.; Cesario, M.; Guéritte, F. *Synlett* **2003**, 2009-2012.

232. Broutin, P.-E.; Colobert, F. *Org. Lett.* **2003**, *5*, 3281-3284.

233. Lipshutz, B. H.; Keith, J. M. *Angew. Chem., Int. Ed.* **1999**, *38*, 3530-3533.

234. Suzuki, T.; Hotta, H.; Hattori, T.; Miyano, S. *Chem. Lett.* **1990**, 807-810.

235. Baker, R. W.; Pocock, G. R.; Sargent, M. V.; Twiss, E. *Tetrahedron: Asymm.* **1993**, *4*, 2423-2426.

236. Hattori, T.; Koike, N.; Miyano, S. *J. Chem. Soc., Perkin Trans. I* **1994**, 2273-2282.

237. Baker, R. W.; Hockless, D. C. R.; Pocock, G. R.; Sargent, M. V.; Skelton, B. W.; Sobolev, A. N.; Twiss, E. *J. Chem. Soc., Perkin Trans. I* **1995**, 2615-2629.

238. Kamikawa, K.; Uemura, M. *Synlett* **2000**, 938-949.

239. Kamikawa, K.; Tachibana, A.; Sugimoto, S.; Uemura, M. *Org. Lett.* **2001**, *3*, 2033-2036.

240. Kamikawa, K.; Watanabe, A.; Daimon, M.; Uemura, M. *Tetrahedron* **2000**, *56*, 2325-2337.

241. Watanabe, T.; Tanaka, Y.; Shoda, R.; Sakamoto, R.; Kamikawa, K.; Uemura, M. *J. Org. Chem.* **2004**, *69*, 4152-4158.

242. Uemura, M.; Nishimura, H.; Kamikawa, K.; Nakayama, K.; Hayashi, Y. *Tetrahedron Lett.* **1994**, *35*, 1909-1912.

243. Kamikawa, K.; Uemura, M. *J. Chem. Soc., Chem. Commun.* **1994**, 2697-2698.

244. Kamikawa, K.; Watanabe, T.; Uemura, M. *J. Org. Chem.* **1996**, *61*, 1375-1384.

245. Nelson, S. G.; Hilfiker, M. A. *Org. Lett.* **1999**, *1*, 1379-1382.

246. Tanaka, Y.; Sakamoto, T.; Kamikawa, K.; Uemura, M. *Synlett* **2003**, 519-521.

247. Kamikawa, K.; Uemura, M. *Tetrahedron Lett.* **1996**, *37*, 6359-6362.

248. Li, X.; Yang, J.; Kozlowski, M. C. *Org. Lett.* **2001**, *3*, 1137-1140.

249. Kozlowski, M. C.; Li, X.; Carroll, P. J.; Xu, Z. *Organometallics* **2002**, *21*, 4513-4522.

250. Mulrooney, C. A.; Li, X.; DiVirgilio, E. S.; Kozlowski, M. C. *J. Am. Chem. Soc.* **2003**, *125*, 6856-6857.

251. Li, X.; Hewgley, J. B.; Mulrooney, C. A.; Yang, J.; Kozlowski, M. C. *J. Org. Chem.* **2003**, *68*, 5500-5511.

252. Smrcina, M.; Polakova, J.; Vyskocil, S.; Kocovsky, P. *J. Org. Chem.* **1993**, *58*, 4534-4538.

253. Osa, T.; Kashiwaga, Y.; Yanagisawa, Y.; Bobbit, J. M. *J. Chem. Soc., Chem. Commun.* **1994**, 2535-2537.

254. Irie, R.; Masutani, K.; Katsuki, T. *Synlett* **2000**, 1433-1436.

255. Nakajima, M.; Kanayama, K.; Miyoshi, I.; Hashimoto, S.-I. *Tetrahedron Lett.* **1995**, *36*, 9519-9520.

256. Nakajima, M.; Miyoshi, I.; Kanayama, K.; Hashimoto, S.-I.; Noji, M.; Koga, K. *J. Org. Chem.* **1999**, *64*, 2264-2271.

257. Kim, K. H.; Lee, D.-W.; Lee, Y.-S.; Ko, D.-H.; Ha, D.-C. *Tetrahedron* **2004**, *60*, 9037-9042.

258. Colletti, S. L.; Halterman, R. L. *Tetrahedron Lett.* **1989**, *30*, 3513-3516.

259. Harris, J. M.; McDonald, R.; Vederas, J. C. *J. Chem. Soc., Perkin Trans. I.* **1996**, 2669-2674.

260. Baudoin, O. *Eur. J. Org. Chem.* **2005**, 4223-4229.

261. Hayashi, T.; Hayashizaki, K.; Kiyoi, T.; Ito, Y. *J. Am. Chem. Soc.* **1988**, *110*, 8153-8156.

262. Hayashi, T.; Hayashizaki, K.; Ito, Y. *Tetrahedron Lett.* **1989**, *30*, 215-218.

263. Cammidge, A. N.; Crépy, K. V. L. *Chem. Commun.* **2000**, 1723-1724.

264. Jensen, J. F.; Johannsen, M. *Org. Lett.* **2003**, *5*, 3025-3028.

265. Cammidge, A. N.; Crépy, K. V. L. *Tetrahedron* **2004**, *60*, 4377-4386.

266. Nicolaou, K. C.; Li, H.; Boddy, C. N. C.; Ramanjulu, J. M.; Yue, T.-Y.; Natarajan, S.; Chu, X.-J.; Bräse, S.; Rübsam, F. *Chem. Eur. J.* **1999**, *5*, 2584-2601.

267. Castanet, A.-S.; Colobert, F.; Broutin, P.-E.; Obringer, M. *Tetrahedron: Asymm.* **2002**, *13*, 659-665.

268. Mikami, K.; Miyamoto, T.; Hatano, M. *Chem. Commun.* **2004**, 2082-2083.

269. Yin, J.; Buchwald, S. L. *J. Am. Chem. Soc.* **2000**, *122*, 12051-12052.

270. Saito, S.; Kano, T.; Muto, H.; Nakadai, M.; Yamamoto, H. *J. Am. Chem. Soc.* **1999**, *121*, 8943-8944.

271. Kano, T.; Ohyabu, Y.; Saito, S.; Yamamoto, H. *J. Am. Chem. Soc.* **2002**, *124*, 5365-5373.

272. Gutnov, A.; Heller, B.; Fischer, C.; Drexler, H.-J.; Spannenberg, A.; Sundermann, B.; Sundermann, C. *Angew. Chem., Int. Ed.* **2004**, *43*, 3795-3797.

273. Bradley, A.; Motherwell, W. B.; Ujjainwalla, F. *Chem. Commun.* **1999**, 917-918.

274. Tanaka, K.; Nishida, G.; Ogino, M.; Hirano, M.; Noguchi, K. *Org. Lett.* **2005**, *7*, 3119-3121.

275. Shibata, T.; Fujimoto, T.; Yokota, K.; Takagi, K. *J. Am. Chem. Soc.* **2004**, *126*, 8382-8383.

276. Bao, J.; Wulff, W. D.; Fumo, M. J.; Grant, E. B.; Heller, D. P.; Whitcomb, M. C. Yeung, S.-M. *J. Am. Chem. Soc.* **1996**, *118*, 2166-2181.

277. Vorogushin, A. V.; Wulff, W. D.; Hansen, H.-J. *J. Am. Chem. Soc.* **2002**, *124*, 6512-6513.

278. Anderson, J. C.; Cran, J. W.; King, N. P. *Tetrahedron Lett.* **2003**, *44*, 7771-7774.

279. Harada, T.; Ueda, S.; Yoshida, T.; Inoue, A.; Takeuchi, M.; Ogawa, N.; Oku, A. *J. Org. Chem.* **1994**, *59*, 7575-7576.

280. Harada, T.; Ueda, S.; Tuyet, T. M. T.; Inoue, A.; Fujita, K.; Takeuchi, M.; Ogawa, N.; Oku, A.; Shiro, M. *Tetrahedron* **1997**, *53*, 16663-16678.

281. Tuyet, T. M. T.; Harada, T.; Hashimoto, K.; Hatsuda, M.; Oku, A. *J. Org. Chem.* **2000**, *65*, 1335-1343.

282. Harada, T.; Tuyet, T. M. T.; Oku, A. *Org. Lett.* **2000**, *2*, 1319-1322.

283. Hayashi, T.; Niizuma, S.; Kamikawa, T.; Suzuki, N.; Uozumi, Y. *J. Am. Chem. Soc.* **1995**, *117*, 9101-9102.

284. Kamikawa, T.; Uozumi, Y.; Hayashi, T. *Tetrahedron Lett.* **1996**, *37*, 3161-3164.

285. Kamikawa, T.; Hayashi, T. *Tetrahedron* **1999**, *55*, 3455-3466.

286. Engelhardt, L. M.; Leung, W.-P.; Raston, C.-L.; Salem, G.; Twiss, P.; White, A. H. *J. Chem. Soc., Dalton Trans.* **1988**, 2403-2409.

287. Matsumoto, T.; Konegawa, T.; Nakamura, T.; Suzuki, K. *Synlett* **2002**, 122-124.

288. Capozzi, G.; Ciampi, C.; Delogu, G.; Menichetti, S.; Nativi, C. *J. Org. Chem.* **2001**, *66*, 8787-8792.

289. Feldman, K. S.; Eastman, K. J.; Lessene, G. *Org. Lett.* **2002**, *4*, 3525-3528.

290. Penhoat, M.; Levacher, V.; Dupas, G. *J. Org. Chem.* **2003**, *68*, 9517-9520.

291. Edwards, D. J.; Pritchard, R. G.; Wallace, T. W. *Tetrahedron Lett.* **2003**, *44*, 4665-4668.

292. Shimada, T.; Kina, A.; Hayashi, T. *J. Org. Chem.* **2003**, *68*, 6329-6337.

293. Korenaga, T.; Aikawa, K.; Terada, M.; Kawauchi, S.; Mikami, K. *Adv. Synth. Catal.* **2001**, *343*, 284-288.

294. Bringmann, G.; Menche, D. *Acc. Chem. Res.* **2001**, *34*, 615-624.

295. Bringmann, G.; Hinrichs, J.; Pabst, T.; Henschel, P.; Peters, K.; Peters, E.-M. *Synlett* **2001**, 155-167.

296. Bringmann, G.; Menche, D.; Mühlbacher, J.; Reichert, M.; Saito, N.; Pfeiffer, S. S.; Lipshutz, B. H. *Org. Lett.* **2002**, *4*, 2833-2836.

297. Bringmann, G.; Vitt, D. *J. Org. Chem.* **1995**, *60*, 7674-7681.

298. Bringmann, G.; Güssregen, S.; Vitt, D.; Stowasser, R. *J. Mol. Model.* **1998**, *4*, 165-175.

299. Bringmann, G.; Hartung, T. *Angew. Chem., Int. Ed. Engl.* **1992**, *31*, 761-762.

300. Bringmann, G.; Hartung, T. *Tetrahedron* **1993**, *49*, 7891-7902.

301. Seebach, D.; Jaeschke, G.; Gottwald, K.; Matsuda, K.; Formisano, R.; Chaplin, D. A.; Breuning, M.; Bringmann, G. *Tetrahedron* **1997**, *53*, 7539-7556.

302. Bringmann, G.; Breuning, M. *Tetrahedron: Asymm.* **1999**, *10*, 385-390.

303. Bringmann, G.; Breuning, M.; Walter, R.; Wuzik, A.; Peters, K.; Peters, E.-M. *Eur. J. Org. Chem.* **1999**, 3047-3055.

304. Bringmann, G.; Breuning, M.; Tasler, S.; Endress, H.; Ewers, C. L. J.; Göbel, L.; Peters, K.; Peters, E.-M.; von Schnering, H. G.; Burschka, C. *Chem. Eur. J.* **1999**, *5*, 3029-3038.

305. Bringmann, G.; Hartung, T.; Göbel, L.; Schupp, O.; Ewers, C. L. J.; Schöner, Zagst, R.; Peters, K.; von Schnering, H. G.; Burschka, C. *Liebigs Ann.* **1992**, 225-232.

306. Bringmann, G.; Breuning, M.; Henschel, P.; Hinrichs, J. *Org. Synth.* **2002**, *79*, 72-83.

307. Abe, H.; Takeda, S.; Fujita, T.; Nishioka, K.; Takeuchi, Y.; Harayama, T. *Tetrahedron Lett.* **2004**, *45*, 2327-2329.

308. Molander, G. A.; George, K. M.; Monovich, L. G. *J. Org. Chem.* **2003**, *68*, 9533-9540.

309. Bringmann, G.; Pabst, T.; Henschel, P.; Kraus, J.; Peters, K.; Peters, E.-M.; Rycroft, D. S.; Connolly, J. *J. Am. Chem. Soc.* **2000**, *122*, 9127-9133.

310. Bringmann, G.; Menche, D. *Angew. Chem., Int. Ed.* **2001**, *40*, 1687-1690.

311. Bringmann, G.; Ochse, M. *Synlett* **1998**, 1294-1296.

312. Bringmann, G.; Holenz, J.; Weirich, R.; Rübenacker, M.; Funke, C.; Boyd, M. R.; Gulakowski, R. J.; François, G. *Tetrahedron* **1998**, *54*, 497-512.

313. Bringmann, G.; Wuzik, A.; Breuning, P.; Henschel, K.; Peters, K.; Peters, E.-M. *Tetrahedron: Asymm.* **1999**, *10*, 3025-3031.

314. Bringmann, G.; Pfeifer, R.-M.; Rummey, C.; Hartner, K.; Breuning, P. *J. Org. Chem.* **2003**, *68*, 6859-6863.

315. Bringmann, G.; Pfeifer, R.-M.; Schreiber, P.; Hartner, K.; Kocher, N.; Brun, R.; Peters, K.; Peters, E.-M.; Breuning, M. *Tetrahedron* **2004**, *60*, 6335-6344.

316. Kamikawa, K.; Furusyo, M.; Uno, T.; Sato, Y.; Konoo, A.; Bringmann, G.; Uemura, M. *Org. Lett.* **2001**, *3*, 3667-3670.

317. Kamikawa, K.; Norimura, K.; Furusyo, M.; Uno, T.; Sato, Y.; Konoo, A.; Bringmann, G.; Uemura, M. *Organometallics* **2003**, *22*, 1038-1046.

318. Cho, Y.-H.; Kina, A.; Shimada, T.; Hayashi, T. *J. Org. Chem.* **2004**, *69*, 3811-3823.

319. Shimada, T.; Cho, Y.-H.; Hayashi, T. *J. Am. Chem. Soc.* **2002**, *124*, 13396-13397.

320. Clayden, J. *Angew. Chem., Int. Ed. Engl.* **1997**, *36*, 949-951.

321. Curran, D. P.; Qi, H.; Geib, S. J.; DeMello, N. C. *J. Am. Chem. Soc.* **1994**, *116*, 3131-3132.

322. Clayden, J.; Westlund, N.; Wilson, F. X. *Tetrahedron Lett.* **1996**, *37*, 5577-5580.

323. Hughes, A. D.; Price, D. A.; Shishkin, O.; Simpkins, N. S. *Tetrahedron Lett.* **1996**, *37*, 7607-7610.

324. Clayden, J.; Pink, J. *Tetrahedron Lett.* **1997**, *38*, 2561-2564.

325. Kitagawa, O.; Izawa, H.; Taguchi, T. *Tetrahedron Lett.* **1997**, *38*, 4447-4450.

326. Clayden, J.; Darbyshire, M.; Pink, J.; Westlund, N.; Wilson, F. X. *Tetrahedron Lett.* **1997**, *38*, 8587-8590.

327. Kitagawa, O.; Izawa, H.; Sato, K.; Dobashi, A.; Taguchi, T. *J. Org. Chem.* **1998**, *63*, 2634-2640.

328. Fujita, M.; Kitagawa, O.; Izawa, H.; Dobashi, A.; Fukuya, H.; Taguchi, T. *Tetrahedron Lett.* **1999**, *40*, 1949-1952.

329. Clayden, J.; Westlund, N.; Wilson, F. X. *Tetrahedron Lett.* **1999**, *40*, 3329-3330.

330. Curran, D. P.; Geib, S.; DeMello, N. *Tetrahedron* **1999**, *55*, 5681-5704.

331. Clayden, J.; Westlund, N.; Wilson, F. X. *Tetrahedron Lett.* **1999**, *40*, 7883-7887.

332. Curran, D. P.; Liu, W.; Chen, C. H.-T. *J. Am. Chem. Soc.* **1999**, *121*, 11012-11013.

333. Kitagawa, O.; Momose, S.-I.; Fushimi, Y.; Taguchi, T. *Tetrahedron Lett.* **1999**, *40*, 8827-8831.

334. Clayden, J.; Westlund, N.; Beddoes, R. L.; Helliwell, M. *J. Chem. Soc., Perkin Trans 1.* **2000**, 1351-1361.

335. Clayden, J.; McCarthy, C.; Westlund, N.; Frampton, C. S. *J. Chem. Soc., Perkin Trans 1.* **2000**, 1363-1378.

336. Clayden, J.; Westlund, N.; Frampton, C. S. *J. Chem. Soc., Perkin Trans 1.* **2000**, 1379-1385.

337. Clayden, J.; Helliwell, M.; McCarthy, C.; Westlund, N. *J. Chem. Soc., Perkin Trans 1.* **2000**, 3232-3249.

338. Clayden, J.; McCarthy, C.; Cumming, J. C. *Tetrahedron Lett.* **2000**, *41*, 3279-3283.

339. Clayden, J.; Kenworthy, M. N.; Youssef, L. H.; Helliwell, M. *Tetrahedron Lett.* **2000**, *41*, 5171-5175.

340. Clayden, J.; Helliwell, M.; Pink, J. H.; Westlund, N. *J. Am. Chem. Soc.* **2001**, *123*, 12449-12457.

341. Ach, D.; Reboul, V.; Metzner, P. *Eur. J. Org. Chem.* **2003**, 3398-3406.

342. Curran, D. P.; Chen, C. H.-T.; Geib, S. J.; Lapierre, A. J. B. *Tetrahedron* **2004**, *60*, 4413-4424.

343. Bennett, D. J.; Blake, A. J.; Cooke, P. A.; Godfrey, C. R. A.; Pickering, P. L.; Simpkins, N. S.; Walker, M. D.; Wilson, C. *Tetrahedron* **2004**, *60*, 4491-4511.

344. Chen, Y.; Smith, M. D.; Shimizu, K. D. *Tetrahedron Lett.* **2001**, *42*, 7185-7187.

345. Dai, W.-M.; Yeung, K. K. Y.; Liu, J.-T.; Zhang, Y.; Williams, I. D. *Org. Lett.* **2002**, *4*, 1615-1618.

346. Mino, T.; Tanaka, Y.; Yabusaki, T.; Okumura, D.; Sakamato, M.; Fujita, T. *Tetrahedron: Asymm.* **2003**, *14*, 2503-2506.

347. Dai, W.-M.; Yeung, K. K. Y.; Wang, Y. *Tetrahedron* **2004**, *60*, 4425-4430.

348. Clayden, J.; Pink, J. H. *Angew. Chem., Int. Ed.* **1998**, *37*, 1937-1939.

349. Bragg, R. A. Clayden, J. *Org. Lett.* **2000**, *2*, 3351-3354.

350. Bragg, R. A.; Clayden, J.; Morris, G. A.; Pink, J. L. *Chem. Eur. J.* **2002**, *8*, 1279-1289.

351. Clayden, J.; Pink, J. H.; Yasin, S. A. *Tetrahedron Lett.* **1998**, *39*, 105-108.

352. Clayden, J. *Synlett* **1998**, 810-816.

353. Clayden, J.; Westlund, N.; Wilson, F. X. *Tetrahedron Lett.* **1999**, *40*, 3331-3334.

354. Clayden, J.; Lai, L. W.; Helliwell, M. *Tetrahedron: Asymm.* **2001**, *12*, 695-698.

355. Clayden, J.; Lund, A.; Youssef, L. H. *Org. Lett.* **2001**, *3*, 4133-4136.

356. Clayden, J.; Lund, A.; Vallverdu, L.; Helliwell, M. *Nature* **2004**, *431*, 966-971.

357. Betson, M. S.; Clayden, J.; Lam, H. K.; Helliwell, M. *Angew. Chem., Int. Ed.* **2005**, *44*, 1241-1244.

358. Thayumanavan, S.; Beak, P.; Curran, D. P. *Tetrahedron Lett.* **1996**, *37*, 2899-2902.

359. Bowles, P.; Clayden, J.; Helliwell, M.; McCarthy, C.; Tomkinson, M.; Westlund, N. *J. Chem. Soc., Perkin Trans 1.* **1997**, 2607-2616.

360. Kitagawa, O.; Kohriyama, M.; Taguchi, T. *J. Org. Chem.* **2002**, *67*, 8682-8684.

361. Terauchi, J.; Curran, D. P. *Tetrahedron: Asymm.* **2003**, *14*, 587-592.

362. Clayden, J. *Chem. Commun.* **2004**, 127-135.

363. Bennett, D. J.; Pickering, P. L.; Simpkins, N. S. *Chem. Commun.* **2004**, 1392-1393.

364. Koide, H.; Uemura, M. *Chem. Commun.* **1998**, 2483-2484.

365. Koide, H.; Hata, T.; Yoshihara, K.; Kamikawa, K.; Uemura, M. *Tetrahedron* **2004**, *60*, 4527-4541.

366. Hata, T.; Koide, H.; Uemura, M. *Synlett* **2000**, 1145-1147.

367. Hata, T.; Koide, H.; Taniguchi, N.; Uemura, M. *Org. Lett.* **2000**, *2*, 1907-1910.

368. Kitagawa, O.; Takahashi, M.; Yoshikawa, M.; Taguchi, T. *J. Am. Chem. Soc.* **2005**, *127*, 3676-3677.

369. Chan, V.; Kim, J. G.; Jimeno, C.; Carroll, P. J.; Walsh, P. J. *Org. Lett.* **2004**, *6*, 2051-2053.

370. Clayden, J.; Stimson, C. C.; Keenan, M. *Synlett* **2005**, 1716-1720.

371. Clayden, J.; Johnson, P.; Pink, J. H.; Helliwell, M. *J. Org. Chem.* **2000**, *65*, 7033-7040.

372. Yamamoto, Y. *Formation of C-C Bonds by Reactions Involving Olefinic Double Bonds*, In Helmchen, G.; Hoffmann, R. W.; Mulzer, J.; Schaumann, E. (Eds.) *Methods of Organic Chemistry (Houben-Weyl)*, Vol. E21b, Thieme, Stuttgart, 1995, pp. 2011-2040.

373. Karlström, A. S. E.; Huerta, F. F.; Meuzelaar, G. J.; Bäckvall, J.-E. *Synlett* **2001**, 923-926.

374. Alexakis, A.; Malan, C.; Lea, L.; Benhaim, C.; Fournioux, X. *Synlett* **2001**, 927-930.

375. Dübner, F.; Knochel, P. *Angew. Chem., Int. Ed.* **1999**, *38*, 379-381.

376. Malda, H.; van Zijl, A. W.; Arnold, L. A.; Feringa, B. L. *Org. Lett.* **2001**, *3*, 1169-1171.

377. Luchaco-Cullis, C. A.; Mizutani, H.; Murphy, K. E.; Hoveyda, A. H. *Angew. Chem., Int. Ed.* **2001**, *40*, 1456-1460.

378. Alexakis, A.; Croset, K. *Org. Lett.* **2002**, *4*, 4147-4149.

379. Tissot-Croset, K.; Polet, D.; Alexakis, A. *Angew. Chem., Int. Ed.* **2004**, *43*, 2426-2428.

380. Diéguez, M.; Pàmies, O.; Claver, C. *J. Org. Chem.* **2005**, *70*, 3363-3368.

381. Hu, X.-P.; Chen, H. L.; Zheng, Z. *Adv. Synth. Catal.* **2005**, *347*, 541-548.

382. Leitner, A.; Shekhar, S.; Pouy, M. J.; Hartwig, J. F. *J. Am. Chem. Soc.* **2005**, *127*, 15506-15514.

383. Ito, H.; Kawakami, C.; Sawamura, M. *J. Am. Chem. Soc.* **2005**, *127*, 16034-16035.

384. Yamamoto, Y.; Yamamoto, S.; Yatagai, H.; Maruyama, K. *J. Am. Chem. Soc.* **1980**, *102*, 2318-2325.

385. Nakamura, E.; Sekiya, K.; Arai, M.; Aoki, S. *J. Am. Chem. Soc.* **1989**, *111*, 3091-3093.

386. Denmark, S. E.; Marble, L. K. *J. Org. Chem.* **1990**, *55*, 1984-1986.

387. Marino, J. P.; Viso, A.; Lee, J.-D. *J. Org. Chem.* **1997**, *62*, 645-653.

388. Spino, C.; Beaulieu, C.; Lafreniere, J. *J. Org. Chem.* **2000**, *65*, 7091-7097.

389. Helmchen, G.; Ernst, M; Paradies, G. *Pure Appl. Chem.* **2004**, *76*, 495-506.

390. Uenishi, J.; Ohmi, M. *Angew. Chem., Int. Ed.* **2005**, *44*, 2756-2760.

391. Goering, H. L.; Singleton Jr., V. D. *J. Org. Chem.* **1983**, *48*, 1531-1533.

392. Ibuka, T.; Nakao, T.; Nishii, S.; Yamamoto, Y. *J. Am. Chem. Soc.* **1986**, *108*, 7420-7422.

393. Ibuka, T.; Tanaka, M.; Nishii, S.; Yamamoto, Y. *J. Chem. Soc., Chem. Commun.* **1987**, 1596-1598.

394. Marshall, J. A.; Trometer, J. D.; Blough, B. E.; Crute, T. D. *J. Org. Chem.* **1988**, *53*, 4274-4282.

395. Ibuka, T.; Akimoto, N.; Tanaka, M.; Nishii, S.; Yamamoto, Y. *J. Org. Chem.* **1989**, *54*, 4055-4061.

396. Ibuka, T.; Tanaka, M.; Nishii, S.; Yamamoto, Y. *J. Am. Chem. Soc.* **1989**, *111*, 4864-4872.

397. Yamamoto, Y.; Chounan, Y.; Tanaka, M.; Ibuka, T. *J. Org. Chem.* **1992**, *57*, 1024-1026.

398. Wipf, P.; Henniger, T. C.; Geib, S. J. *J. Org. Chem.* **1998**, *63*, 6088-6089.

399. Spino, C.; Beaulieu, C. *J. Am. Chem. Soc.* **1998**, *120*, 11832-11833.

400. Agami, C.; Couty, F.; Mathieu, H.; Pilot, C. *Tetrahedron Lett.* **1999**, *40*, 4539-4542.

401. Belelie, J. L.; Chong, J. M. *J. Org. Chem.* **2001**, *66*, 5552-5555.

402. Calaza, M. I.; Hupe, E.; Knochel, P. *Org. Lett.* **2003**, *5*, 1059-1061.

403. Harrington-Frost, N.; Leuser, H.; Calaza, M. I.; Kneisel, F. F.; Knochel, P. *Org. Lett.* **2003**, *5*, 2111-2114.

404. Corey, E. J.; Boaz, N. W. *Tetrahedron Lett.* **1984**, *25*, 3063-3066.

405. Goering, H. L.; Kantner, S. S.; Tseng, C. C. *J. Org. Chem.* **1983**, *48*, 715-721.

406. Calo, V.; Lopez, L.; Carlucci, W. F. *J. Chem. Soc., Perkin Trans. 1* **1983**, 2953-2956.

407. Valverde, S.; Bernabe, M.; Garcia-Ochoa, S.; Gomez, A. M. *J. Org. Chem.* **1990**, *55*, 2294-2298.

408. Smitrovich, J. H.; Woerpel, K. A. *J. Am. Chem. Soc.* **1998**, *120*, 12998-12999.

409. Smitrovich, J. H.; Woerpel, K. A. *J. Org. Chem.* **2000**, *65*, 1601-1614.

410. Breit, B.; Demel, P.; Studte, C. *Angew. Chem., Int. Ed.* **2004**, *43*, 3786-3789.

411. Trost, B. M.; Ceschi, M. A.; König, B. *Angew. Chem., Int. Ed. Engl.* **1997**, *36*, 1486-1489.

412. Uenishi, J.; Ohmi, M.; Ueda, A. *Tetrahedron: Asymm.* **2005**, *16*, 1299-1303.

413. Sakaguchi, K.; Yamada, T.; Ohfune, Y. *Tetrahedron Lett.* **2005**, *46*, 5009-5012.

414. Suginome, M.; Matsumoto, A.; Ito, Y. *J. Am. Chem. Soc.* **1996**, *118*, 3061-3062.

415. Suginome, M.; Iwanami, T.; Matsumoto, A.; Ito, Y. *Tetrahedron: Asymm.* **1997**, *8*, 859-862.

416. Trost, B. M.; Krische, M. J.; Radinov, R.; Zanoni, G. *J. Am. Chem. Soc.* **1996**, *118*, 6297-6298.

417. Magid, R. M.; Fruchey, O. S. *J. Am. Chem. Soc.* **1977**, *99*, 8368-8370.

418. Magid, R. M. *Tetrahedron* **1980**, *36*, 1901-1930.

419. Lalic, G.; Blum, S. A.; Bergmann, R. G. *J. Am. Chem. Soc.* **2005**, *127*, 16790-16791.

420. Fleming, I.; Barbero, A.; Walter, D. *Chem. Rev.* **1997**, *97*, 2063-2192.

421. Buckle, M. J. C.; Fleming, I.; Gil, S. *Tetrahedron Lett.* **1992**, *33*, 4479-4482.

422. Hayashi, T.; Konishi, M.; Ito, H.; Kumada, M. *J. Am. Chem. Soc.* **1982**, *104*, 4962-4963.

423. Hayashi, T.; Konishi, M.; Kumada, M. *J. Am. Chem. Soc.* **1982**, *104*, 4963-4965.

424. Mikami, K.; Kimura, Y.; Kishi, N.; Nakai, T. *J. Org. Chem.* **1983**, *48*, 281-282.

425. Denmark, S. E.; Almstead, N. G. *J. Org. Chem.* **1994**, *59*, 5130-5132.

426. Denmark, S. E.; Hosoi, S. *J. Org. Chem.* **1994**, *59*, 5133-5135.

427. Suginome, M.; Iwanami, T.; Ito, Y. *J. Org. Chem.* **1998**, *63*, 6096-6097.

428. Suginome, M.; Iwanami, T.; Yamamoto, A.; Ito, Y. *Synlett* **2001**, 1042-1045.

429. Masse, C. E.; Dakin, L. A.; Knight, B. S.; Panek, J. S. *J. Org. Chem.* **1997**, *62*, 9335-9338.

430. Huang, H.; Panek, J. S. *J. Am. Chem. Soc.* **2000**, *122*, 9836-9837.

431. Huang, H.; Spande, T. F.; Panek, J. S. *J. Am. Chem. Soc.* **2003**, *125*, 626-627.

432. Liu, P. L.; Binnun, E. D.; Schaus, J. V.; Valentino, N. M.; Panek, J. S. *J. Org. Chem.* **2002**, *67*, 1705-1707.

433. Arefolov, A.; Panek, J. S. *J. Am. Chem. Soc.* **2005**, *127*, 5596-5603.

434. Huang, H.; Panek, J. S. *Org. Lett.* **2001**, *3*, 1693-1696.

435. Dakin, L. A.; Panek, J. S. *Org. Lett.* **2003**, *5*, 3995-3998.

436. Su, Q.; Panek, J. S. *Angew. Chem., Int. Ed.* **2005**, *44*, 1223-1225.

437. Lowe, J. T.; Panek, J. S. *Org. Lett.* **2005**, *7*, 3231-3234.

438. Lowe, J. T.; Panek, J. S. *Org. Lett.* **2005**, *7*, 1529-1532.

439. Huang, H.; Panek, J. S. *Org. Lett.* **2003**, *5*, 1991-1993.

440. Su, Q.; Panek, J. S. *J. Am. Chem. Soc.* **2004**, *126*, 2425-2430.

441. Woodward, R. B.; Hoffmann, R. *The Conservation of Orbital Symmetry*, Verlag Chemie, Weinheim, 1970.

442. Woodward, R. B.; Hoffmann, R. *J. Am. Chem. Soc.* **1965**, *87*, 2511-2513.

443. Fleming, I. *Frontier Molecular Orbitals and Organic Chemical Reactions*, Wiley, London, 1976, p. 98.

444. Shono, T.; Fujita, K.; Kumai, S. *Tetrahedron Lett.* **1973**, *14*, 3123-3126.

445. Paquette, L. A.; Elmore, S. W.; Combrink, K. D.; Hickey, E. R.; Rogers, R. D. *Helv. Chim. Acta* **1992**, *75*, 1755-1771.

446. Trost, B. M.; Yasukata, T. *J. Am. Chem. Soc.* **2001**, *123*, 7162-7163.

447. Xin, G.; Paquette, L. A. *J. Org. Chem.* **2005**, *70*, 315-320.

448. Money, T.; Palme, M. H. *Tetrahedron: Asymm.* **1993**, *4*, 2363-2370.

449. Martinez, A. G.; Vilar, E. T.; Fraile, A. G.; de la Moya Cerero, S.; Maroto, B. L. *Tetrahedron: Asymm.* **2002**, *12*, 3325-3327.

450. De la Cerero, S.; Martinez, A. G.; Vilar, E. T.; Fraile, A. G.; Maroto, B. L. *J. Org. Chem.* **2003**, *68*, 1451-1458.

451. Martinez, A. G.; Vilar, E. T.; Fraile, A. G.; de la Moya Cerero, S.; Morillo, C. D.; Morillo, R. P. *J. Org. Chem.* **2004**, *69*, 7348-7351.

452. Compain, P.; Gore, J.; Vatele, J.-M. *Tetrahedron* **1996**, *52*, 6647-6664.

453. Smadja, W.; Czernecki, S.; Ville, G.; Georgoulis, C. *Organometallics* **1987**, *6*, 166-169.

454. Nubbemeyer, U. *Synthesis* **2003**, 961-1008.

455. Ichikawa, Y.; Ito, T.; Isobe, M. *Chem. Eur. J.* **2005**, *11*, 1949-1957.

456. Drutu, I.; Krygowski, E. S.; Wood, J. L. *J. Org. Chem.* **2001**, *66*, 7025-7029.

457. Ichikawa, Y.; Hirata, K.; Ohbayashi, M.; Isobe, M. *Chem. Eur. J.* **2004**, *10*, 3241-3251.

458. Boeckman Jr., R. K.; Blum, D. M.; Arthur, S. D. *J. Am. Chem. Soc.* **1979**, *101*, 5060-5062.
459. Takahashi, T.; Yamada, H.; Tsuji, J. *J. Am. Chem. Soc.* **1981**, *103*, 5259-5261.
460. Zielinski, J.; Li, H. T.; Djerassi, C. *J. Org. Chem.* **1982**, *47*, 620-625.
461. Nonoshita, K.; Banno, H.; Maruoka, K.; Yamamoto, H. *J. Am. Chem. Soc.* **1990**, *112*, 316-322.
462. Grieco, P. A.; Brandes, E. B.; McCann, S.; Clark, J. D. *J. Org. Chem.* **1999**, *54*, 5849-5851.
463. Diederich, M.; Nubbemeyer, U. *Chem. Eur. J.* **1996**, *2*, 894-900.
464. Wood, J. L.; Moniz, G. A. *Org. Lett.* **1999**, *1*, 371-374.
465. Trost, B. M.; Schroeder, G. M. *J. Am. Chem. Soc.* **2000**, *122*, 3785-3786.
466. Sakaguchi, K.; Suzuki, H.; Ohfune, Y. *Chirality* **2001**, *13*, 357-365.
467. Sauer, E. L. O.; Barriault, L. *J. Am. Chem. Soc.* **2004**, *126*, 8569-8575.
468. Xu, Q.; Rozners, E. *Org. Lett.* **2005**, *7*, 2821-2824.
469. Chan, K.-K.; Specian Jr., A. C.; Saucy, G. *J. Org. Chem.* **1978**, *43*, 3435-3440.
470. Stork, G.; Takahashi, T.; Kawamoto, I. Suzuki, T. *J. Am. Chem. Soc.* **1978**, *100*, 8272-8273.
471. Hiyama, T.; Kobayashi, K.; Fujita, M. *Tetrahedron Lett.* **1984**, *25*, 4959-4962.
472. Takahashi, T.; Jinbo, Y.; Titamura, K.; Tsuji, J. *Tetrahedron Lett.* **1984**, *25*, 5921-5924.
473. Heneghan, M.; Procter, G. *Synlett* **1992**, 489-490.
474. McKinney, J. A.; Eppley, D. F.; Keenan, R. M. *Tetrahedron Lett.* **1994**, *35*, 5985-5988.
475. Hill, R. K.; Soman, R.; Swada, S. *J. Org. Chem.* **1972**, *37*, 3737-3740.
476. Heathcock, C. H.; Finkelstein, B. L. *J. Chem. Soc., Chem. Commun.* **1983**, 919-920.
477. Russell, A. T.; Procter, G. *Tetrahedron Lett.* **1987**, *28*, 2041-2044.
478. Still, W. C.; Schneider, M. J. *J. Am. Chem. Soc.* **1977**, *99*, 948-950.
479. Curran, D. P.; Suh, Y.-G. *Tetrahedron Lett.* **1984**, *25*, 4179-4182.
480. Fehr, C.; Galindo, J. *Angew.Chem., Int. Ed.* **2000**, *39*, 569-573.
481. Heathcock, C. H.; Finkelstein, B. L.; Jarvi, E. T.; Radel, P. A.; Hadley, C. R. *J. Org. Chem.* **1988**, *53*, 1922-1942.
482. Ito, H.; Taguchi, T. *Chem. Soc. Rev.* **1999**, *28*, 43-50.
483. Trost, B. M.; Toste, F. D. *J. Am. Chem. Soc.* **1998**, *120*, 815-816.
484. Boeckman Jr., R. K.; Reeder, M. R. *J. Org. Chem.* **1997**, *62*, 6456-6457.
485. Boeckman Jr., R. K.; Zhang, J.; Reeder, M. R. *Org. Lett.* **2002**, *4*, 3891-3894.
486. Johnson, A. P.; Rahman, M. *Tetrahedron Lett.* **1974**, *15*, 359-362.
487. Adames, G.; Grigg, R.; Grover, J. N. *Tetrahedron Lett.* **1974**, *15*, 363-366.
488. Kuehne, M. E.; Xu, F. *J. Org. Chem.* **1997**, *62*, 7950-7960.
489. Kuehne, M. E.; Xu, F. *J. Org. Chem.* **1998**, *63*, 9427-9433.
490. Kuehne, M. E.; Xu, F. *J. Org. Chem.* **1998**, *63*, 9434-9439.
491. Paquette, L. A.; Ladouceur, G. *J. Org. Chem.* **1989**, *54*, 4278-4279.
492. Boeckman Jr., R. K.; Springer, D. M.; Alessi, T. R. *J. Am. Chem. Soc.* **1989**, *111*, 8284-8286.
493. Wei, S.-Y.; Tomooka, K.; Nakai, T. *J. Org. Chem.* **1991**, *56*, 5973-5974.
494. Paquette, L. A.; Maynard, G. D. *J. Am. Chem. Soc.* **1992**, *114*, 5018-5027.
495. Wei, S.-Y.; Tomooka, K.; Nakai, T. *Tetrahedron* **1993**, *49*, 1025-1042.
496. Deng, W.; Overman, L. E. *J. Am. Chem. Soc.* **1994**, *116*, 11241-11250.
497. Gauvreau, D.; Barriault, L. *J. Org. Chem.* **2005**, *70*, 1382-1388.
498. Jung, M. E.; Hudspeth, J. P. *J. Am. Chem. Soc.* **1980**, *102*, 2463-2464.
499. Demay, S.; Kotschy, A.; Knochel, P. *Synthesis* **2001**, 863-866.
500. Baldwin, J. E.; Fleming, R. H. *J. Am. Chem. Soc.* **1972**, *94*, 2140-2142.
501. Berson, J. A.; Nelson, G. L. *J. Am. Chem. Soc.* **1967**, *89*, 5503-5504.
502. Nakai, T.; Mikami, K. *Chem. Rev.* **1986**, *86*, 885-902.
503. Mikami, K.; Nakai, T. *Synthesis* **1991**, 594-604.
504. Wu, Y.-D.; Houk, K. N.; Marshall, J. A. *J. Org. Chem.* **1990**, *55*, 1421-1423.
505. Verner, E. J.; Cohen, T. *J. Am. Chem. Soc.* **1992**, *114*, 375-377.

506. Hoffmann, R.; Brückner, R. *Angew.Chem., Int. Ed. Engl.* **1992**, *31*, 647-649.

507. Tomooka, K.; Igarashi, T.; Watanabe, M.; Nakai, T. *Tetrahedron Lett.* **1992**, *33*, 5795-5798.

508. Kitagawa, O.; Momose, S.; Yamada, Y.; Shiro, M.; Taguchi, T. *Tetrahedron Lett.* **2001**, *42*, 4865-4868.

509. Wipf, P. In: Trost, B. M.; Fleming I. (Eds.) *Comprehensive Organic Synthesis. Selectivity, Strategy, and Efficiency in Modern Organic Chemistry*, Pergamon, Oxford, 1991, Vol. 5, pp. 827-873.

510. Capriati, V.; Florio, S.; Ingrosso, G.; Granito, C.; Troisi, L. *Eur. J. Org. Chem.* **2002**, 478-484.

511. Andrews, G.; Evans, D. A. *Tetrahedron Lett.* **1972**, *13*, 5121-5124.

512. Hoffmann, R. W. *Angew. Chem., Int. Ed. Engl.* **1979**, *18*, 563-572.

513. Miller, J. G.; Kurz, W.; Untch, K. G.; Stork, G. *J. Am. Chem. Soc.* **1974**, *96*, 6774-6775.

514. Liron, F.; Knochel, P. *Chem. Commun.* **2004**, 304-305.

515. Kawachi, A.; Maeda, H.; Nakamura, H.; Doi, N.; Tamao, K. *J. Am. Chem. Soc.* **2001**, *123*, 3143-3144.

516. Ahman, J.; Somfai, P. *Tetrahedron Lett.* **1995**, *36*, 303-306.

517. Aggarwal, V. K.; Fang, G.; Charmant, J. P. H.; Meek, G. *Org. Lett.* **2003**, *5*, 1757-1760.

518. Anderson, J. C.; Ford, J. G.; Whiting, M. *Org. Biomol. Chem.* **2005**, *3*, 3734-3748.

519. Tsai, D. J. S.; Midland, M. M. *J. Org. Chem.* **1984**, *49*, 1842-1843.

520. Sayo, N.; Azuma, K.; Mikami, K.; Nakai, T. *Tetrahedron Lett.* **1984**, *25*, 565-568.

521. Keegan, D. S.; Midland, M. M.; Werley, R. T.; Mcloughlin, J. I. *J. Org. Chem.* **1991**, *56*, 1185-1191.

522. Midland, M. M.; Kwon, Y. C. *Tetrahedron Lett.* **1985**, *26*, 5017-5020.

523. Midland, M. M.; Kwon, Y. C. *Tetrahedron Lett.* **1985**, *26*, 5021-5024.

524. Tsai, D. J.-S.; Midland, M. M. *J. Am. Chem. Soc.* **1985**, *107*, 3915-3918.

525. Uchikawa, M.; Katsuki, T.; Yamaguchi, M. *Tetrahedron Lett.* **1986**, *27*, 4581-4582.

526. Doyle, M. P.; Bagheri, V.; Harn, N. K. *Tetrahedron Lett.* **1988**, *29*, 5119-5122.

527. Tong, X.; Kallmerten, J. *Synlett* **1992**, 845-846.

528. Mulzer, J.; List, B. *Tetrahedron Lett.* **1994**, *35*, 9021-9024.

529. Konno, T.; Umetani, H.; Kitazume, T. *J. Org. Chem.* **1997**, *62*, 137-150.

530. Li, A.-H.; Dai, L.-X.; Aggarwal, V. K. *Chem. Rev.* **1997**, *97*, 2341-2372.

531. Tsubuki, M.; Kamata, T.; Nakatani, M.; Yamazaki, K.; Matsui, T.; Honda, T. *Tetrahedron: Asymm.* **2000**, *11*, 4725-4736.

532. Itoh, T.; Kudo, K. *Tetrahedron Lett.* **2001**, *42*, 1317-1320.

533. Ahman, J.; Somfai, P. *J. Am. Chem. Soc.* **1994**, *116*, 9781-9782.

534. Coldham, I.; Collis, A. J.; Mould, R. J.; Rathmell, R. E. *Tetrahedron Lett.* **1995**, *36*, 3557-3560.

535. Ahman, J.; Jarevang, T.; Somfai, P. *J. Org. Chem.* **1996**, *61*, 8148-8159.

536. Castedo, L.; Granja, J. R.; Mourino, A. *Tetrahedron Lett.* **1985**, *26*, 4959-4960.

537. Koreeda, M.; Ricca, D. J. *J. Org. Chem.* **1986**, *51*, 4090-4092.

538. Balestra, M.; Wittman, M. D.; Kallmerten, J. *Tetrahedron Lett.* **1988**, *29*, 6905-6908.

539. Eshelman, J. E.; Epps, J. L.; Kallmerten, J. *Tetrahedron Lett.* **1993**, *34*, 749-752.

540. Cywin, C. L.; Kallmerten, J. *Tetrahedron Lett.* **1993**, *34*, 1103-1106.

541. Bohnstedt, A. C.; Prasad, J. V. N. V.; Rich, D. H. *Tetrahedron Lett.* **1993**, *34*, 5217-5220.

542. Yamamoto, Y.; Oda, J.; Inouye, Y. *J. Org. Chem.* **1976**, *41*, 303-306.

543. Reetz, M. T.; Lauterbach, E. H. *Tetrahedron Lett.* **1991**, *32*, 4481-4482.

544. Davies, S. G.; Smyth, G. D. *Tetrahedron: Asymm.* **1996**, *7*, 1001-1004.

545. Blanchet, J.; Bonin, M.; Micouin, L.; Husson, H.-P. *Tetrahedron Lett.* **2000**, *41*, 8279-8283.

546. Grieco, P. A.; Takigawa, T.; Bongers, S. L.; Tanaka, H. *J. Am. Chem. Soc.* **1980**, *102*, 7588-7590.

547. Overman, L. E. *Angew. Chem., Int. Ed. Engl.* **1984**, *23*, 579-586.

548. Mehmandoust, M.; Petit, Y.; Larcheveque, M. *Tetrahedron Lett.* **1992**, *33*, 4313-4316.

549. Sugiura, M.; Yanagisawa, M.; Nakai, T. *Synlett* **1995**, 447-448.

550. Danishefsky, S. J.; Cabal, M. P.; Chow, K. *J. Am. Chem. Soc.* **1989**, *111*, 3456-3457.

551. Heathcock, C. H.; Jarvi, E. T.; Rosen, T. *Tetrahedron Lett.* **1984**, *25*, 243-246.

552. Schreiber, S. L.; Smith, D. B. *J. Org. Chem.* **1989**, *54*, 9-10.

553. Herold, P.; Duthaler, R.; Ribs, G.; Angst, C. *J. Org. Chem.* **1989**, *54*, 1178-1185.

554. Paquette, L. A.; Sweeney, T. J. *J. Org. Chem.* **1990**, *55*, 1703-1704.

555. Paquette, L. A.; Sweeney, T. J. *Tetrahedron* **1990**, *46*, 4487-4502.

556. Burke, S. D.; Porter, W. J.; Rancourt, J.; Kaltenbach, R. F. *Tetrahedron Lett.* **1990**, *31*, 5285-5288.

557. Paterson, I.; Hulme, A. N.; Wallace, D. J. *Tetrahedron Lett.* **1991**, *32*, 7601-7604.

558. Andersen, M. W.; Hildebrandt, B.; Dahmann, G.; Hoffmann, R. W. *Chem. Ber.* **1991**, *124*, 2127-2139.

559. Morimoto, Y.; Mikami, S.; Kuwabe, H.; Shirahama, H. *Tetrahedron Lett.* **1991**, *32*, 2909-2912.

560. Kim, D.; Kim, J. I. *Tetrahedron Lett.* **1995**, *36*, 5035-5036.

561. Sudau, A.; Münch, W.; Nubbemeyer, U. *J. Org. Chem.* **2000**, *65*, 1710-1720.

562. Paterson, I.; Hulme, A. N. *J. Org. Chem.* **1995**, *60*, 3288-3300.

563. Nagatsuma, M.; Shirai, N.; Sayo, N.; Nakai, T. *Chem. Lett.* **1984**, 1393-1396.

564. Grattan, T. J.; Whitehurst, J. S. *J. Chem. Soc., Perkin Trans. 1* **1990**, 11-18.

565. Büchi, G.; Powell Jr., J. E. *J. Am. Chem. Soc.* **1970**, *92*, 3126-3133.

566. Childers Jr., W. E.; Pinnick, H. W. *J. Org. Chem.* **1984**, *49*, 5276-5277.

567. Danishefsky, S.; Funk, R. L.; Kerwin Jr., J. K. *J. Am. Chem. Soc.* **1980**, *102*, 6889-6891.

568. Lee, E.; Shin, I.-J.; Kim, T.-S. *J. Am. Chem. Soc.* **1990**, *112*, 260-264.

569. Overman, L. E.; Jacobsen, E. N. *J. Am. Chem. Soc.* **1982**, *104*, 7225-7231.

570. Paquette, L. A.; Maynard, G. D. *Angew. Chem., Int. Ed. Engl.* **1991**, *30*, 1368-1370.

571. Paquette, L. A.; Maynard, G. D. *J. Am. Chem. Soc.* **1992**, *114*, 5018-5027.

572. Sayo, N.; Shirai, F.; Nakai, T. *Chem. Lett.* **1984**, 255-258.

573. Marshall, J. A.; Jenson, T. M. *J. Org. Chem.* **1984**, *49*, 1707-1712.

574. Midland, M. M.; Gabriel, J. *J. Org. Chem.* **1985**, *50*, 1143-1144.

575. Mikami, K.; Kawamoto, K.; Nakai, T. *Tetrahedron Lett.* **1987**, *26*, 5799-5802.

576. Marshall, J. A.; Robinson, E. D.; Zapata, A. *J. Org. Chem.* **1989**, *54*, 5854-5855.

577. Marshall, J. A.; Wang, X.-J. *J. Org. Chem.* **1990**, *55*, 2995-2996.

578. Marshall, J. A.; Wang, X.-J. *J. Org. Chem.* **1991**, *56*, 4913-4918.

579. Roth, W. R.; König, J.; Stein, K. *Chem. Ber.* **1970**, *103*, 426-439.

580. Hastings, C. A.; Ringgenberg, J. D.; Carreira, E. M. *Tetrahedron Lett.* **1997**, *38*, 8789-8792.

581. Dehnhardt, C.; McDonald, M.; Lee, S.; Floss, H. G.; Mulzer, J. *J. Am. Chem. Soc.* **1999**, *121*, 10848-10849.

582. Jones, D. W.; Thompson, A. M. *J. Chem. Soc., Chem. Commun.* **1987**, 1797-1798.

583. Jones, D. W.; Thompson, A. M. *J. Chem. Soc., Chem. Commun.* **1988**, 1095-1096.

584. Fujita, M.; Takarada, Y.; Sugimura, T.; Tai, A. *Chem. Commun.* **1997**, 1631-1632.

585. Nishide, K.; Node, M. *Chirality* **2002**, *14*, 759-767.

586. Ohno, A.; Ikeguchi, M.; Kimura, T.; Oka, S. *J. Am. Chem. Soc.* **1979**, *101*, 7036-7040.

587. Meyers, A. I.; Oppenlaender, T. *J. Am. Chem. Soc.* **1986**, *108*, 1989-1996.

588. Doering, W. E.; Young, R. W. *J. Am. Chem. Soc.* **1950**, *72*, 631.

589. Newman, P.; Rutkin, P.; Mislow, K. *J. Am. Chem. Soc.* **1958**, *80*, 465-473.

590. Eliel, E. L.; Nasipuri, D. *J. Org. Chem.* **1965**, *30*, 3809-3814.

591. Morrison, J. D.; Ridgway, R. W. *J. Org. Chem.* **1974**, *39*, 3107-3110.

592. Warnhoff, E. W.; Reynolds-Warnhoff, P.; Wong, M. Y. H. *J. Am. Chem. Soc.* **1980**, *102*, 5956-5957.

593. Seebach, D.; Plattner, D. A.; Beck, A. K.; Wang, Y. M.; Hunziker, D. *Helv. Chim. Acta* **1992**, *75*, 2171-2209.

594. Campbell, E. J.; Zhou, H.; Nguyen, S. T. *Org. Lett.* **2001**, *3*, 2391-2393.

595. Giacomelli, G.; Menicagli, R.; Lardicci, L. *J. Org. Chem.* **1974**, 39, 1757-1758.

596. Morrison, J. D.; Tomaszewski, J. E.; Mosher, H. S.; Dale, J.; Miller, D.; Elsenbaumer, R. L. *J. Am. Chem. Soc.* **1977**, *99*, 3167-3168.

597. Ooi, T.; Miura, T.; Maruoka, K. *Angew. Chem., Int. Ed.* **1998**, *37*, 2347-2349.

598. De Graauw, C. F.; Peters, J. A.; van Bekkum, H.; Huskens, J. *Synthesis* **1994**, 1007-1017.

599. Hill, R. K.; Chan, T.-H. *J. Am. Chem. Soc.* **1966**, *88*, 866-867.

600. Brewster, J. H.; Jones Jr., R. S. *J. Org. Chem.* **1969**, *34*, 354-358.

601. Gawley, R. E.; Zhang, Q.; Campagna, S. *J. Am. Chem. Soc.* **1995**, *117*, 11817-11818.

602. Glaeske, K. W.; West, F. G. *Org. Lett.* **1999**, *1*, 31-33.

603. Saba, A. *Tetrahedron Lett.* **2003**, *44*, 2895-2898.

604. Muroni, D.; Saba, A.; Culeddu, N. *Tetrahedron: Asymm.* **2004**, *15*, 2609-2614.

605. Moriwaki, M.; Yamamoto, Y.; Oda, J.; Inouye, Y. *J. Org. Chem.* **1976**, *41*, 300-303.

606. Buston, J. E. H.; Coldham, I.; Mulholland, K. R. *Synlett* **1997**, 322-324.

607. Vedejs, E.; Hagen, J. P.; Roach, B. L.; Spear, K. L. *J. Org. Chem.* **1978**, *43*, 1185-1190.

608. West, F. G.; Naidu, B. N. *J. Am. Chem. Soc.* **1994**, *116*, 8420-8421.

609. Wright, D. L.; Weekly, R. M.; Groff, R.; McMills, M. C. *Tetrahedron Lett.* **1996**, *37*, 2165-2168.

610. Naidu, B. N.; West, F. G. *Tetrahedron* **1997**, *53*, 16565-16574.

611. Vanecko, J. A.; West, F. G. *Org. Lett.* **2002**, *4*, 2813-2816.

612. Clark, J. S.; Hodgson, P. B.; Goldsmith, M. D.; Blake, A. J.; Cooke, P. A.; Street, L. J. *J. Chem. Soc., Perkin Trans. 1* **2001**, 3325-3337.

613. Trost, B. M.; Hammen, R. F. *J. Am. Chem. Soc.* **1973**, *95*, 962-964.

614. Campbell, S. J.; Darwish, D. *Can. J. Chem.* **1976**, *54*, 193-201.

615. Kurth, M. J.; Hasan, T. S.; Olmstead, M. M. *J. Org. Chem.* **1990**, 55, 2286-2288.

616. Cagle, P. C.; Arif, A. M.; Gladysz, J. A. *J. Am. Chem. Soc.* **1994**, *116*, 3655-3656.

617. Schmidt, D. R.; O'Malley, S. J.; Leighton, J. L. *J. Am. Chem. Soc.* **2003**, *125*, 1190-1191.

618. Zhang, X.; Houk, K. N.; Leighton, J. L. *Angew. Chem., Int. Ed.* **2005**, *44*, 938-941.

619. Vorogushin, A. V.; Wulff, W. D.; Hansen, H.-J. *J. Am. Chem. Soc.* **2002**, *124*, 6512-6513.

620. Nishii, Y.; Wakasugi, K.; Koga, K.; Tanabe, Y. *J. Am. Chem. Soc.* **2004**, *126*, 5358-5359.

621. Meyers, A. I.; Wettlaufer, D. G. *J. Am. Chem. Soc.* **1984**, *106*, 1135-1136.

622. Hattori, T.; Date, M.; Sakurai, K.; Morohashi, N.; Kosugi, H.; Miyano, S. *Tetrahedron Lett.* **2001**, *42*, 8035-8038.

623. Warnhoff, E. W.; Valverde Lopez, S. *Tetrahedron Lett.* **1967**, *28*, 2723-2727.

624. Wanjohi, J. M.; Yenesew, A.; Midiwo, J. O.; Heydenreich, M.; Peter, M. G.; Dreyer, M.; Reichert, M.; Bringmann, G. *Tetrahedron* **2005**, *61*, 2667-2674.

625. Zimmer, R.; Dinesh, C. U.; Nandanan, E.; Khan, F. A. *Chem. Rev.* **2000**, *100*, 3067-3125.

626. Marshall, J. A. *Chem. Rev.* **2000**, *100*, 3163-3186.

627. Sweeney, Z. K.; Salsman, J. L.; Andersen, R. A.; Bergman, R. G. *Angew. Chem., Int. Ed.* **2000**, *39*, 2339-2343.

628. Hashmi, A. S. K. *Angew. Chem., Int. Ed.* **2000**, *39*, 3590-3593.

629. Yamazaki, J.; Watanabe, T.; Tanaka, K. *Tetrahedron: Asymm.* **2001**, *12*, 669-675.

630. Hiroi, K.; Kato, F. *Tetrahedron* **2001**, *57*, 1543-1550.

631. Marshall, J. A.; Chobanian, H. R.; Yanik, M. M. *Org. Lett.* **2001**, *3*, 3369-3372.

632. Marshall, J. A.; Adams, N. D. *J. Org. Chem.* **2002**, *67*, 733-740.

633. Marshall, J. A.; Ellis, K. C. *Org. Lett.* **2003**, *5*, 1729-1732.

634. Marshall, J. A.; Piettre, A.; Paige, M. A.; Valeriote, F. *J. Org. Chem.* **2003**, *68*, 1771-1779.

635. Marshall, J. A.; Schaaf, G. M. *J. Org. Chem.* **2003**, *68*, 7428-7432.

636. Marshall, J. A.; Ellis, K. *Tetrahedron Lett.* **2004**, *45*, 1351-1353.

637. Sherry, B. D.; Toste, F. D. *J. Am. Chem. Soc.* **2004**, *126*, 15978-15979.

638. Marshall, J. A.; Mulhearn, J. *J. Org. Lett.* **2005**, *7*, 1593-1596.

639. Hoffmann-Röder, A.; Krause, N. *Angew. Chem., Int. Ed.* **2002**, *41*, 2933-2935.

640. Eliel, E. L.; Lynch, J. E.; Kenan Jr., W. R. *Tetrahedron Lett.* **1987**, *28*, 4813-4816.

641. Konno, T.; Tanikawa, M.; Ishihara, T.; Yamanaka, H. *Chem. Lett.* **2000**, 1360-1361.

642. Elsevier, C. J.; Stehouwer, P. M.; Westmijze, H.; Vermeer, P. *J. Org. Chem.* **1983**, *48*, 1103-1105.

643. Ohno, H.; Miyamura, K.; Tanaka, T.; Oishi, S.; Toda, A.; Takemoto, Y.; Fujii, N.; Ibuka, T. *J. Org. Chem.* **2002**, *67*, 1359-1367.

644. Marek, I.; Mangeney, P.; Alexakis, A.; Normant, J. F. *Tetrahedron Lett.* **1986**, *27*, 5499-5502.

645. Alexakis, A.; Marek, I.; Mangeney, P.; Normant, J. F. *Tetrahedron Lett.* **1989**, *30*, 2387-2390.

646. Wan, Z.; Nelson, S. G. *J. Am. Chem. Soc.* **2000**, *122*, 10470-10471.

647. Spino, C.; Fréchette, S. *Tetrahedron Lett.* **2000**, *41*, 8033-8036.

648. Claesson, A.; Olsson, L.-I. *J. Am. Chem. Soc.* **1979**, *101*, 7302-7311.

649. Krause, N.; Purpura, M. *Angew. Chem., Int. Ed.* **2000**, *39*, 4355-4356.

650. Konoike, T.; Araki, Y. *Tetrahedron Lett.* **1992**, *33*, 5093-5096.

651. Schultz-Fademrecht, C.; Wibbeling, B.; Fröhlich, R.; Hoppe, D. *Org. Lett.* **2001**, *3*, 1221-1224.

652. Ogawa, Y.; Toyama, M.; Karikomi, M.; Seki, K.; Haga, K.; Uyehara, T. *Tetrahedron Lett.* **2003**, *44*, 2167-2170.

653. Krebs, A. W.; Thölke, B.; Pforr, K.-I.; König, W. A.; Scharwächter, K.; Grimme, S.; Vögtle, F.; Sobanski, A.; Schramm, J.; Hormes, J. *Tetrahedron: Asymm.* **1999**, *10*, 3483-3492.

654. Corey, E. J.; Hopkins, P. B. *Tetrahedron* **1982**, *23*, 1979-1982.

655. Amadji, M.; Cahard, D.; Moriggi, J.-D.; Toupet, L.; Plaquevent, J.-C. *Tetrahedron: Asymm.* **1998**, *9*, 1657-1660.

656. Ohmori, K.; Kitamura, M.; Suzuki, K. *Angew. Chem., Int. Ed.* **1999**, *38*, 1226-1229.

657. Kitamura, M.; Ohmori, K.; Kawase, T.; Suzuki, K. *Angew. Chem., Int. Ed.* **1999**, *38*, 1229-1232.

658. Taniguchi, N.; Hata, T.; Uemura, M. *Angew. Chem., Int. Ed.* **1999**, *38*, 1232-1235.

659. Ohmori, K.; Tamiya, M.; Kitamura, M.; Kato, H.; Oorui, M.; Suzuki, K. *Angew. Chem., Int. Ed.* **2005**, *44*, 3871-3874.

660. Baker, R. W.; Kyasnoor, R. V.; Sargent, M. V. *Tetrahedron Lett.* **1999**, *40*, 3475-3478.

661. Yasui, Y.; Suzuki, K.; Matsumoto, T. *Synlett* **2004**, 619-622.

662. Curran, D. P.; Liu, W.; Chen, C. H.-T. *J. Am. Chem. Soc.* **1999**, *121*, 11012-11013.

663. Curran, D. P.; Chen, C. H.-T.; Geib, S. J.; Lapierre, A. J. B. *Tetrahedron* **2004**, *60*, 4413-4424.

664. Petit, M.; Geib, S. J.; Curran, D. P. *Tetrahedron* **2004**, *60*, 7543-7552.

665. Petit, M.; Lapierre, A. J. B.; Curran, D. P. *J. Am. Chem. Soc.* **2005**, *127*, 14994-14995.

666. Gibbs, R. A.; Okamura, W. H. *J. Am. Chem. Soc.* **1988**, *110*, 4062-4063.

667. Hu, H.; Smith, D.; Cramer, R. E.; Tius, M. A. *J. Am. Chem. Soc.* **1999**, *121*, 9895-9896.

668. Bertrand, M.; Roumestant, M. L.; Sylvestre-Panthet, P. *Tetrahedron Lett.* **1981**, *22*, 3589-3590.

669. Becker, D.; Nagler, M.; Harel, Z.; Gillon, A. *J. Org. Chem.* **1983**, *48*, 2584-2590.

670. Trost, B. M.; Tour, J. M. *J. Am. Chem. Soc.* **1988**, *110*, 5231-5233.

671. Gibbs, R.; Bartels, K.; Lee, R. W. K.; Ojamura, W. H. *J. Am. Chem. Soc.* **1989**, *111*, 3717-3725.

672. Pasto, D. J.; Sugi, K. D. *J. Org. Chem.* **1991**, *56*, 3795-3801.

673. Carreira, E. M.; Hastings, C. A.; Shepard, M. S.; Yerkey, L. A.; Millward, D. B. *J. Am. Chem. Soc.* **1994**, *116*, 6622-6630.

674. Yeo, S.-K.; Hatae, N.; Seki, M.; Kanematsu, K. *Tetrahedron* **1995**, *51*, 3499-3506.

675. Wender, P. A.; Jenkins, T. E.; Suzuki, S. *J. Am. Chem. Soc.* **1995**, *117*, 1843-1844.

676. Llerena, D.; Aubert, C.; Malacria, M. *Tetrahedron Lett.* **1996**, *37*, 7027-7030.

677. Shepard, M. S.; Carreira, E. M. *J. Am. Chem. Soc.* **1997**, *119*, 2597-2605.

678. Urabe, H.; Takeda, T.; Hideura, D.; Sato, F. *J. Am. Chem. Soc.* **1997**, *119*, 11295-11305.

679. Wender, P. A.; Glorius, F.; Husfeld, C. O.; Langkopf, E.; Love, J. A. *J. Am. Chem. Soc.* **1999**, *121*, 5348-5349.

680. Brummond, K. M.; Kerekes, A. D.; Wan, H. *J. Org. Chem.* **2002**, *67*, 5156-5163.

681. Rameshkumar, C.; Hsung, R. P. *Angew. Chem., Int. Ed.* **2004**, *43*, 615-618.

682. Danh, T. T.; Bocian, W.; Kozerski, L.; Szczukiewicz, P.; Frelek, J.; Chmielewski, M. *Eur. J. Org. Chem.* **2005**, 429-440.

683. Daidouji, K.; Fuchibe, K.; Akiyama, T. *Org. Lett.* **2005**, *7*, 1051-1053.

684. Jung, M. E.; Min, S.-J. *J. Am. Chem. Soc.* **2005**, *127*, 10834-10835.

685. Aso, M.; Ikeda, I.; Kawabe, T.; Shiro, M.; Kanematsu, K. *Tetrahedron Lett.* **1992**, *33*, 5787-5790.

686. Ikeda, I.; Kanematsu, K. *J. Chem. Soc., Chem. Commun.* **1995**, 453-454.

687. Ikeda, I.; Honda, K.; Osawa, E.; Shiro, M.; Aso, M.; Kanematsu, K. *J. Org. Chem.* **1996**, *61*, 2031-2037.

688. Trost, B. M.; Fandrick, D. R.; Dinh, D. C. *J. Am. Chem. Soc.* **2005**, *127*, 14186-14187.

689. Buisine, O.; Aubert, C.; Malacria, M. *Synthesis* **2000**, 985-989.

690. Hoppe, D.; Gonschorrek, C.; Egert, E.; Schmidt, D. *Angew. Chem., Int. Ed. Engl.* **1985**, *24*, 700-701.

691. Behrens, U.; Wolff, C.; Hoppe, D. *Synthesis* **1991**, 644-646.

692. Corey, E. J.; Boaz, N. W. *Tetrahedron Lett.* **1984**, *25*, 3059-3062.

693. Suginome, M.; Matsumoto, A.; Ito, Y. *J. Org. Chem.* **1996**, *61*, 4884-4885.

694. Marshall, J. A.; Adams, N. D. *J. Org. Chem.* **1997**, *62*, 8976-8977.

695. Marshall, J. A.; Maxson, K. *J. Org. Chem.* **2000**, *65*, 630-633.

696. Han, J. W.; Tokunaga, N.; Hayashi, T. *J. Am. Chem. Soc.* **2001**, *123*, 12915-12916.

697. Ogasawara, M.; Ueyama, K.; Nagano, T.; Mizuhata, Y.; Hayashi, T. *Org. Lett.* **2003**, *5*, 217-219.

698. Anwar, U.; Grigg, R.; Rasparini, M.; Sridharan, V. *Chem. Commun.* **2000**, 645-646.

699. Ha, Y.-H.; Kang, S.-K. *Org. Lett.* **2002**, *4*, 1143-1146.

700. Kang, S.-K.; Yoon, S.-K. *Chem. Commun.* **2002**, 2634-2635.

701. Montgomery, J.; Song, M. *Org. Lett.* **2002**, *4*, 4009-4011.

702. Hopkins, C. D.; Malinakova, H. C. *Org. Lett.* **2004**, *6*, 2221-2224.

703. Wu, M.-S.; Rayabarapu, D. K.; Cheng, C.-H. *J. Am. Chem. Soc.* **2003**, *125*, 12426-12427.

704. Ng, S.-S.; Jamison, T. F. *J. Am. Chem. Soc.* **2005**, *127*, 7320-7321.

705. Ma, S.; Wu, S. *J. Org. Chem.* **1999**, *64*, 9314-9317.

706. Dieter, R. K.; Yu, H. *Org. Lett.* **2001**, *3*, 3855-3858.

707. Ma, S.; Wu, S. *Chem. Commun.* **2001**, 441-442.

708. Yoshida, M.; Fujita, M.; Ihara, M. *Org. Lett.* **2003**, *5*, 3325-3427.

709. Horvath, A.; Benner, J.; Bäckvall, J.-E. *Eur. J. Org. Chem.* **2004**, 3240-3243.

710. Marshall, J. A.; Wang, X. *J. Org. Chem.* **1990**, *55*, 2995-2996.

711. Marshall, J. A.; Wang, X. *J. Org. Chem.* **1991**, *56*, 4913-4918.

712. Marshall, J. A.; Pinney, K. G. *J. Org. Chem.* **1993**, *58*, 7180-7184.

713. Marshall, J. A.; Bartley, G. S. *J. Org. Chem.* **1994**, *59*, 7169-7171.

714. Marshall, J. A.; Yu, R. H.; Perkins, J. F. *J. Org. Chem.* **1995**, *60*, 5550-5555.

715. Marshall, J. A.; Wallace, E. M. *J. Org. Chem.* **1997**, *62*, 367-371.

716. VanBrunt, M. P.; Standaert, R. F. *Org. Lett.* **2000**, *2*, 705-708.

717. Morita, N.; Krause, N. *Org. Lett.* **2004**, *6*, 4121-4123.

718. Hoffmann-Röder, A.; Krause, N. *Org. Biomol. Chem.* **2005**, *3*, 387-391.

719. Hoffmann-Röder, A.; Krause, N. *Org. Lett.* **2001**, *3*, 2537-2538.

720. Ma, S.; Ren, H.; Wei, Q. *J. Am. Chem. Soc.* **2003**, *125*, 4817-4830.

721. Sudhakar, A.; Katz, T. J. *J. Am. Chem. Soc.* **1986**, *108*, 179-181.

722. Carreno, M. C.; Garcia-Cerrada, S.; Urbano, A. *J. Am. Chem. Soc.* **2001**, *123*, 7929-7930.

723. Urbano. A. *Angew. Chem., Int. Ed.* **2003**, *42*, 3986-3989.

724. Carreno, M. C.; Gonzalez-Lopez, M.; Urbano, A. *Chem. Commun.* **2005**, 611-613.

725. Stara, I. G.; Alexandrova, Z.; Teply, F.; Sehnal, P.; Stary, I.; Saman, D.; Budesinsky, M.; Cvacka, J. *Org. Lett.* **2005**, *7*, 2547-2550.

726. Nakano, K.; Hidehira, Y.; Takahashi, K.; Hiyama, T.; Nozaki, K. *Angew. Chem., Int. Ed.* **2005**, *44*, 7136-7138.

727. Bestman, H. J.; Both, W. *Angew. Chem., Int. Ed. Engl.* **1972**, *11*, 296.

728. Stara, I. G.; Stary, I.; Tichy, M.; Zavada, J.; Hanus, V. *J. Am. Chem. Soc.* **1994**, *116*, 5084-5088.

729. Rajca, A.; Miyasaka, M.; Pink, M.; Wang, H.; Rajca, S. *J. Am. Chem. Soc.* **2004**, *126*, 15211-15222.

730. Wigglesworth, T. J.; Sud, D.; Norsten, T. B.; Lekhi, V. S.; Branda, N. R. *J. Am. Chem. Soc.* **2005**, *127*, 7272-7273.

731. Tanaka, K.; Suzuki, H.; Osuga, H. *J. Org. Chem.* **1997**, *62*, 4465-4470.

732. Cope, A. C.; Moore, W. R.; Bach, R. D.; Winkler, H. J. S. *J. Am. Chem. Soc.* **1970**, *92*, 1243-1247.

733. Hattori, T.; Koike, N.; Okaishi, Y.; Miyano, S. *Tetrahedron Lett.* **1996**, *37*, 2057-2060.

734. Tomooka, K.; Komine, N.; Fujiki, D.; Nakai, T.; Yanagitsuru, S. *J. Am. Chem. Soc.* **2005**, *127*, 12182-12183.

735. Cozzi, F.; Siegel, J. S. *Org. Biomol. Chem.* **2005**, *3*, 4296-4298.

736. Hughes, D. L. *Org. Reac.* **1992**, *42*, 335-656.

737. Nicolaou, K. C.; Boddy, C. N. C.; Natarajan, S.; Yue, T.-Y.; Li, H.; Bräse, S.; Ramanjulu, J. M. *J. Am. Chem. Soc.* **1997**, *119*, 3421-3422.

738. Fischer, E.; Brauns, F. *Ber. Dtsch. Chem. Ges.* **1914**, *47*, 3181-3193.

739. Winstein, S.; Hess, H. V.; Buckles, R. E. *J. Am. Chem. Soc.* **1942**, *64*, 2796- 2801.

740. Winstein, S.; Hanson, C.; Grunwald, E. *J. Am. Chem. Soc.* **1948**, *70*, 812-816.

741. Winstein, S.; Grunwald, E.; Buckles, R. E.; Hanson, C. *J. Am. Chem. Soc.* **1948**, *70*, 816-821.

742. Cram, D. J. *J. Am. Chem. Soc.* **1949**, *71*, 3863-3870.

743. Cram, D. J. *J. Am. Chem. Soc.* **1952**, *74*, 2129-2137.

744. Suzuki, K.; Tomooka, K.; Katayama, E.; Matsumoto, T.; Tsuchihashi, G. *J. Am. Chem. Soc.* **1986**, *108*, 5221-5229.

745. Nemoto, H.; Miyata, J.; Hakamata, H.; Fukumoto, K. *Tetrahedron Lett.* **1995**, *36*, 1055-1058.

746. Nemoto, H.; Miyata, J.; Hakamata, H.; Nagamochi, M.; Fukumoto, K. *Tetrahedron Lett.* **1995**, *36*, 5511-5522.

747. Shinohara, T.; Suzuki, K. *Tetrahedron Lett.* **2002**, *43*, 6937-6940.

748. Shionhara, T.; Suzuki, K. *Synthesis* **2003**, 141-146.

749. Schreiner, P. R.; von Schleyer, R.; Hill, R. K. *J. Org. Chem.* **1994**, *59*, 1849-1854.

750. Cram, D. J. *Fundamentals of Carbanion Chemistry*, Academic New Press, New York, 1965, pp. 85-135.

751. Seebach, D.; Sting, A. R.; Hoffmann, M. *Angew. Chem., Int. Ed. Engl.* **1996**, *35*, 2708-2748.

752. Yee, N. K. *Org. Lett.* **2000**, *2*, 2781-2783.

753. Napolitano, E.; Farina, V. *Tetrahedron Lett.* **2001**, *42*, 3231-3234.

754. Yee, N. K.; Nummy, L. J.; Frutos, R. P.; Song, J. J.; Napolitano, E.; Byrne, D. P.; Jones P.-J.; Farina, V. *Tetrahedron: Asymm.* **2003**, *14*, 3495-3501.

755. Neveux, M.; Seiller, B.; Hagedorn, F.; Bruneau, C.; Dixneuf, P. H. *J. Organomet. Chem.* **1993**, *451*, 133-138.

756. Ortholand, J.-Y.; Vicart, N.; Greiner, A. *J. Org. Chem.* **1995**, *60*, 1880-1884.

757. Naef, R.; Seebach, D. *Helv. Chim. Acta* **1985**, *68*, 135-143.

758. Seebach, D.; Aebi, J. D.; Naef, R.; Weber, T. *Helv. Chim. Acta* **1985**, *68*, 144-154.

759. Seebach, D.; Amatsch, B.; Amstutz, R.; Beck, A. K.; Doler, M.; Egli, M.; Fitzi, R.; Gautschi, M.; Herradön, B.; Hidber, P. C.; Irwin, J. J.; Locher, R.; Maestro, M.; Maetzke, T.; Mouriño, A.; Pfammatter, E.; Plattner, D. A.; Schickli, C.; Schweizer, W. B.; Seiler, P.; Stucky, G.; Petter, W.; Escalante, J.; Juaristi, E.; Quintana, D.; Miravitlles, C.; Molins, E. *Helv. Chim. Acta* **1992**, *75*, 913-934.

760. Lamatsch, B.; Seebach, D. *Helv. Chim. Acta* **1992**, *75*, 1095-1110.

761. Suzuki, K.; Seebach, D. *Liebigs Ann. Chem.* **1992**, 51-61.

762. O'Donnell, M. J.; Fang, Z.; Ma, X.; Huffman, J. C. *Heterocycles* **1997**, *46*, 617-630.

763. El Bialy, S. A. A.; Braun, H.; Tietze, L. F. *Eur. J. Org. Chem.* **2005**, 2965-2972.

764. Barbaro, G.; Battaglia, A.; Guerrini, A.; Bertucci, C. *Tetrahedron: Asymm.* **1997**, *8*, 2527-2531.

765. Barbaro, G.; Battaglia, A.; Di Giuseppe, F.; Giorgianni, P.; Guerrini, A.; Bertucci, C.; Geremia, S. *Tetrahedron: Asymm.* **1999**, *10*, 2765-2773.

766. Barbaro, G.; Battaglia, A.; Guerrini, A. *J. Org. Chem.* **1999**, *64*, 4643-4651.

767. Seebach, D.; Weber, T. *Tetrahedron Lett.* **1983**, *24*, 3315-3318.

768. Seebach, D.; Weber, T. *Helv. Chim. Acta* **1984**, *67*, 1650-1661.

769. Aebi, J. D.; Seebach, D. *Helv. Chim. Acta* **1985**, *68*, 1507-1518.

770. Seebach, D.; Fadel, A. *Helv. Chim. Acta* **1985**, *68*, 1243-1250.

771. Calderari, G. Seebach, D. *Helv. Chim. Acta* **1985**, *68*, 1592-1604.

772. Fadel, A.; Salaün, J. *Tetrahedron Lett.* **1987**, *28*, 2243-2246.

773. Strijtveen, B.; Kellogg, R. M. *Tetrahedron* **1987**, *43*, 5039-5054.

774. Seebach, D.; Aebi, J. D.; Gander-Coquoz, M.; Naef, R. *Helv. Chim. Acta* **1987**, *70*, 1194-1216.

775. Gander-Coquoz, M.; Seebach, D. *Helv. Chim. Acta* **1988**, *71*, 224-236.

776. Beck, A. K.; Seebach, D. *Chimia* **1988**, *42*, 142-144.

777. Pfammatter, E.; Seebach, D. *Liebigs Ann. Chem.* **1991**, 1323-1336.

778. Jones, R. C. F.; Crockett, A. K.; Rees, D. C.; Gilbert, I. H. *Tetrahedron: Asymm.* **1994**, *5*, 1661-1664.

779. Ma, D.; Tian, H. *J. Chem. Soc., Perkin Trans. 1* **1997**, 3493-3496.

780. Ma, G.; Palmer, D. R. J. *Tetrahedron Lett.* **2000**, *41*, 9209-9212.

781. Ma, D.; Ding, K.; Tian, H.; Wang, B.; Cheng, D. *Tetrahedron: Asymm.* **2002**, *13*, 961-969.

782. Seebach, D.; Naef, R. *Helv. Chim. Acta* **1981**, *64*, 2704-2708.

783. Seebach, D.; Boes, M.; Naef, R.; Schweitzer, W. B. *J. Am. Chem. Soc.* **1983**, *105*, 5390-5398.

784. Weber, T.; Seebach, D. *Helv. Chim. Acta* **1985**, *68*, 1655-1661.

785. Seebach, D.; Vettiger, T.; Müller, H.-D.; Plattner, D. A.; Petter, W. *Liebigs Ann. Chem.* **1990**, 687-695.

786. Amedjikouh, M.; Westerlund, K. *Tetrahedron Lett.* **2004**, *45*, 5175-5177.

787. Einhorn, J.; Einhorn, C.; Ratajczak, F.; Gautier-Luneau, I.; Pierre, J.-L. *J. Org. Chem.* **1997**, *62*, 9385-9388.

788. Häner, R.; Olano, B.; Seebach, D. *Helv. Chim. Acta* **1987**, *70*, 1676-1693.

789. Ferey, V.; Le Gall, T.; Mioskowski, C. *J. Chem. Soc., Chem. Commun.* **1995**, 487-489.

790. Ferey, V.; Toupet, L.; Le Gall, T.; Mioskowski, C. *Angew. Chem., Int. Ed. Engl.* **1996**, *35*, 430-432.

791. Ferey, V.; Toupet, L.; Le Gall, T.; Mioskowski, C. *Angew. Chem., Int. Ed.* **2000**, *39*, 430-432.

792. Vedejs, E.; Fields, S. C.; Schrimpf, M. R. *J. Am. Chem. Soc.* **1993**, *115*, 11612-11613.

793. Vedejs, E.; Fields, S. C.; Lin, S.; Schrimpf, M. R. *J. Org. Chem.* **1995**, *60*, 3028-3034.

794. Vedejs, E.; Fields, S. C.; Hayashi, R.; Hitchcock, S. R.; Powell, D. R.; Schrimpf, M. R. *J. Am. Chem. Soc.* **1999**, *121*, 2460-2470.

795. Boons, G.-J.; Downham, R.; Kim, K. S.; Ley, S. V.; Woods, M. *Tetrahedron* **1994**, *50*, 7157-7176.

796. Seebach, D.; Zimmermann, J.; Gysel, U.; Ziegler, R.; Ha, T.-K. *J. Am. Chem. Soc.* **1988**, *110*, 4763-4772.

797. Stucky, G.; Seebach, D. *Chem. Ber.* **1989**, *122*, 2365-2375.

798. Konopelski, J. P.; Chu, K. S.; Negrete, G. R. *J. Org. Chem.* **1991**, *56*, 1355-1357.

799. Roush, W. R.; Brown, B. B. *J. Org. Chem.* **1992**, *57*, 3380-3387.

800. Organ, M. G.; Froese, R. D. J.; Goddard, J. D.; Taylor, N. J.; Lange, G. L. *J. Am. Chem. Soc.* **1994**, *116*, 3312-3323.

801. Gautschi, M.; Seebach, D. *Angew. Chem., Int. Ed. Engl.* **1992**, *31*, 1083-1085.

802. Schmalz, H.-G.; Millies, B.; Bats, J. W.; Dürner, G. *Angew. Chem., Int. Ed. Engl.* **1992**, *31*, 631-633.

803. Schmalz, H.-G.; Hollander, J.; Arnold, M.; Dürner, G. *Tetrahedron Lett.* **1993**, *34*, 6259-6262.

804. Schmalz, H.-G.; Arnold, M.; Hollander, J.; Dürner, G. *Angew. Chem., Int. Ed. Engl.* **1994**, *33*, 109-111.

805. Hayashi, T.; Konishi, M.; Kumada, M. *J. Chem. Soc., Chem. Commun.* **1984**, 107-108.

806. Koot, W.-J.; Hiemstra, H.; Speckamp, W. N. *J. Chem. Soc., Chem. Commun.* **1993**, 156-158.

807. Enders, D.; Finkam, M. *Synlett* **1993**, 401-403.

808. Enders, D.; Jandeleit, B.; Raabe, G. *Angew. Chem., Int. Ed. Engl.* **1994**, *33*, 1949-1951.

809. Hopman, J. C. P.; Hiemstra, H.; Speckkamp, W. N. *J. Chem. Soc., Chem. Commun.* **1995**, 617-618.

810. Enders, D.; von Berg, S.; Jandeleit, B. *Synlett* **1996**, 18-20.

811. Enders, D.; Jandeleit, B. *Tetrahedron* **1995**, *51*, 6273-6284.

812. Enders, D.; Jandeleit, B. *Liebigs Ann.* **1995**, 1173-1176.

813. Hughes, C. C.; Trauner, D. *Angew. Chem., Int. Ed.* **2002**, *41*, 4556-4559.

814. Krohn, K.; Hamann, I. *Liebigs Ann. Chem.* **1988**, 949-953.

815. Naef, R.; Seebach, D. *Liebigs Ann. Chem.* **1983**, 1930-1936.

816. Enders, D.; Jandeleit, B. *Synthesis* **1994**, 1327-1330.

817. Boeckman Jr., R. K.; Yoon, S. K.; Heckendorn, D. K. *J. Am. Chem. Soc.* **1991**, *113*, 9682-9684.

818. Ogawa, T.; Niwa, H.; Yamada, K. *Tetrahedron* **1993**, *49*, 1571-1578.

819. Krohn, K.; Meyer, A. *Liebigs Ann. Chem.* **1994**, 167-174.

820. Williams, R. M.; Glinka, T.; Kwast, E. *J. Am. Chem. Soc.* **1988**, *110*, 5927-5929.

821. Mulqueen, G. C.; Pattenden, G.; Whiting, D. A. *Tetrahedron* **1993**, *49*, 5359-5364.

822. Boyce, R. J.; Mulqueen, G. C.; Pattenden, G. *Tetrahedron Lett.* **1994**, *35*, 5705-5708.

823. Corey, E. J.; Reichard, G. A. *J. Am. Chem. Soc.* **1992**, *114*, 10677-10678.

824. Uno, H.; Baldwin, J. E.; Russell, A .T. *J. Am. Chem. Soc.* **1994**, *116*, 2139-2140.

825. Kawabata, T.; Yahiro, K.; Fuji, K. *J. Am. Chem. Soc.* **1991**, *113*, 9694-9696.

826. Zhao, H.; Hsu, D. C.; Carlier, P. R. *Synthesis* **2005**, 1-16.

827. Kawabata, T.; Wirth, T.; Yahiro, K.; Suzuki, H.; Fuji, K. *J. Am. Chem. Soc.* **1994**, *116*, 10809-10810.

828. Kawabata, T.; Suzuki, H.; Nagae, Y.; Fuji, K. *Angew. Chem., Int. Ed.* **2000**, *39*, 2155-2157.

829. Kawabata, T.; Kawakami, S.-P.; Fuji, K. *Tetrahedron Lett.* **2002**, *43*, 1465-1467.

830. Kawabata, T.; Kawakami, S.-P.; Shimada, S.; Fuji, K. *Tetrahedron* **2003**, *59*, 965-974.

831. Kawabata, T.; Chen, J.; Suzuki, H.; Fuji, K. *Synthesis* **2005**, 1368-1377.

832. Kawabata, T.; Öztürk, O.; Chen, J.; Fuji, K. *Chem. Commun.* **2003**, 162-163.

833. Kawabata, T.; Kawakami, S.; Majumdar, S. *J. Am. Chem. Soc.* **2003**, *125*, 13012-13013.

834. Kawabata, T.; Fuji, K. *Top. Stereochem.* **2003**, *23*, 175-205.

835. Kawabata, T.; Majumdar, S.; Tsubaki, K.; Monguchi, D. *Org. Biomol. Chem.* **2005**, *3*, 1609-1611.

836. Beagley, B.; Betts, M. J.; Pritchard, R. G.; Schofield, A.; Stoodley, R. J.; Vohra, S. *Chem. Commun.* **1991**, 924-925.

837. Beagley, B.; Betts, M. J.; Pritchard, R. G.; Schofield, A.; Stoodley, R. J.; Vohra, S. *J. Chem. Soc., Perkin Trans. 1* **1993**, 1761-1770.

838. Betts, M. J.; Pritchard, R. G.; Schofield, A.; Stoodley, R. J.; Vohra, S. *J. Chem. Soc., Perkin Trans. I.* **1999**, 1067-1072.

839. Brewster, A. G.; Frampton, C. S.; Jayatissa, J.; Mitchell, M. B.; Stoodley, R. J.; Vohra, S. *Chem. Commun.* **1998**, 299-300.

840. Brewster, A. G.; Jayatissa, J.; Mitchell, M. B.; Schofield, A.; Stoodley, R. J. *Tetrahedron Lett.* **2002**, *43*, 3919-3922.

841. Gerona-Navarro, G.; Bonache, M. A.; Herranz, R.; García-López, M. T.; González-Muñiz, R. *J. Org. Chem.* **2001**, *66*, 3538-3547.

842. Bonache, M. A.; Gerona-Navarro, G.; Martin-Martinez, M.; García-López, M. T.; López, P.; Cativiela, C.; González-Muñiz, R. *Synlett* **2003**, 1007-1011.

843. Bonache, M. A.; Gerona-Navarro, G.; Garcia-Aparicio, C.; Alias, M.; Martin-Martinez, M.; García-López, M. T.; Lopez, P.; Cativiela, C.; González-Muñiz, R. *Tetrahedron: Asymm.* **2003**, *14*, 2161-2169.

844. Clayden, J.; Stimson, C. C.; Keenan, M. *Synlett* **2005**, 1716-1720.

845. Linscheid, P.; Lehn, J.-M. *Bull. Chim. Soc. Fr.* **1967**, 992-997.

846. Konwal, A.; Snatzke, G.; Alebic-Kolbah, T.; Kajfez, F.; Rendic, S.; Sunjic, V. *Biochem. Pharm.* **1979**, *28*, 3109-3113.

847. Gilman, N. W.; Rosen, P.; Earley, J. V.; Cook. C.; Todaro, L. J. *J. Am. Chem. Soc.* **1990**, *112*, 3969-3978.

848. Sunjic, V.; Lisini, A.; Sega, A.; Kovac, T.; Kajfez, F.; Ruscic, B. *J. Heterocycl. Chem.* **1979**, *16*, 757-761.

849. Paizs, B.; Simonyi, M. *Chirality* **1999**, *11*, 651-658.

850. Carlier, P. R.; Zhao, H.; DeGuzman, J.; Lam, P. C.-H. *J. Am. Chem. Soc.* **2003**, *125*, 11482-11483.

851. Lam, P. C.-H.; Carlier, P. R. *J. Org. Chem.* **2005**, *70*, 1530-1538.

852. Carlier, P. R.; Lam, P. C.-H.; DeGuzman, J.; Zhao, H. *Tetrahedron: Asymm.* **2005**, *16*, 2998-3002.

853. Pfletschinger, A.; Dargel, T. K.; Schmalz, H.-G.; Koch, W. *Chem. Eur. J.* **1999**, *5*, 537-545.

854. Schmalz, H.-G.; de Koning, C. B.; Bernicke, D.; Siegel, S.; Pfletschinger, A. *Angew. Chem., Int. Ed.* **1999**, *38*, 1620-1623.

855. Gauvreau, D.; Barriault, L. *J. Org. Chem.* **2005**, *70*, 1382-1388.

856. Griller, D.; Ingold, K. U.; Krusic, P. J.; Fischer, H. *J. Am. Chem. Soc.* **1978**, *100*, 6750-6752.

857. Simamura, O. *Top. Stereochem.* **1969**, *4*, 1-37.

858. Roberts, B. P.; Steel, A. J. *J. Chem. Soc., Perkin Trans. 2* **1992**, 2025-2029.

859. Buckmelter, A. J.; Kim, A. I.; Rychnovsky, S. D. *J. Am. Chem. Soc.* **2000**, *122*, 9386-9390.

860. Dalgard, J. E.; Rychnovsky, S. D. *Org. Lett.* **2004**, *6*, 2713-2716.

861. Giese, B.; Wettstein, P.; Stähelin, C.; Barbosa, F.; Neuberger, M.; Zehnder, M.; Wessig, P. *Angew. Chem., Int. Ed.* **1999**, *38*, 2586-2587.

862. Griesbeck, A. G.; Kramer, W.; Lex, J. *Angew. Chem., Int. Ed.* **2001**, *40*, 577-579.
863. Xu, J.; Weiss, R. G. *Org. Lett.* **2003**, *5*, 3077-3080.
864. Xu, J.; Weiss, R. G. *J. Org. Chem.* **2005**, *70*, 1243-1252.
865. Turro, N. J. *Chem. Commun.* **2002**, 2279-2293.
866. Bhanthumnavin, W.; Bentrude, W. G. *J. Org. Chem.* **2001**, *66*, 980-990.
867. Ellison, M. E.; Ng, D.; Dang, H.; Garcia-Garibay, M. A. *Org. Lett.* **2003**, *5*, 2531-2534.
868. Starling, S. M.; Vonwiller, S. C. *Tetrahedron Lett.* **1997**, *38*, 2159-2162.
869. Matsumara, Y.; Shirakawa, Y.; Satoh, Y.; Umino, M.; Tanaka, T.; Maki, T.; Onomura, O. *Org. Lett.* **2000**, *2*, 1689-1691.
870. Wanyoike, G. N.; Onomura, O.; Maki, T.; Matsumara, Y. *Org. Lett.* **2002**, *4*, 1875-1877.
871. Reetz, M. T.; Neugebauer, T. *Angew. Chem., Int. Ed.* **1999**, *38*, 179-181.
872. Deng, C.; Groth, U.; Schöllkopf, U. *Angew. Chem., Int. Ed. Engl.* **1981**, *20*, 798-799.
873. Bull, S. D.; Davies, S. G.; Fox, D. J.; Sellers, T. G. R. *Tetrahedron: Asymm.* **1998**, *9*, 1483-1487.
874. Bull, S. D.; Davies, S. G.; Fox, D. J.; Garner, A. C.; Sellers, T. G. R. *Pure Appl. Chem.* **1998**, *70*, 1501-1508.
875. Bull, S. D.; Davies, S. G.; Epstein, S. W.; Leech, M. A.; Ouzman, J. V. A. *J. Chem. Soc., Perkin Trans. 1* **1998**, 2321-2330.
876. Bull, S. D.; Davies, S. G.; Epstein, S. W.; Ouzman, J. V. A. *Chem. Commun.* **1998**, 659-660.
877. Casper, D. M.; Burgeson, J. R.; Esken, J. M.; Ferrence, G. M.; Hitchcock, S. R. *Org. Lett.* **2002**, *4*, 3739-3742.
878. Casper, D. M.; Hitchcock, S. R. *Tetrahedron: Asymm.* **2003**, *14*, 517-521.
879. Hitchcock, S. R.; Casper, D. M.; Vaughn, J. F.; Finefield, J. M.; Ferrence, G. M.; Esken, J, M, *J. Org. Chem.* **2004**, *69*, 714-718.
880. Sibi, M. P.; Venkatram, L. *Angew. Chem., Int. Ed.* **1999**, *38*, 8444-8445.
881. Sibi, M. P.; Venkatram, L.; Liu, M.; Jasperse, C. P. *J. Am. Chem. Soc.* **2001**, *123*, 8444-8445.
882. Sibi, M. P.; Liu, M. *Org. Lett.* **2001**, *3*, 4181-4184.
883. Sibi, M. P.; Prabagaran, N. *Synlett* **2004**, 2421-2424.
884. Sibi, M. P.; Stanley, L. M. *Tetrahedron: Asymm.* **2004**, *15*, 3353-3356.
885. Quaranta, L.; Corminboeuf, O.; Renaud, P. *Org. Lett.* **2002**, *4*, 39-42.
886. Corminboeuf, O.; Quaranta, L.; Renaud, P.; Liu, M.; Jasperse, C. R.; Sibi, M. P. *Chem. Eur. J.* **2003**, *9*, 29-35.
887. Malkov, A. V.; Hand, J. B.; Koèovsky, P. *Chem. Commun.* **2003**, 1948-1949.
888. Maughan, M. A. T.; Davies, I. G.; Claridge, T. D. W.; Courtney, S.; Hay, P.; Davies, B. G. *Angew. Chem., Int. Ed.* **2003**, *42*, 3788-3792.
889. Clayden, J.; Pink, J. H.; Yasin, S. A. *Tetrahedron Lett.* **1998**, *39*, 105-108.
890. Clayden, J. *Synlett* **1998**, 810-816.
891. Clayden, J.; Westlund, N.; Wilson, F. X. *Tetrahedron Lett.* **1999**, *40*, 3331-3334.
892. Clayden, J.; Lai, L. W.; Helliwell, M. *Tetrahedron: Asymm.* **2001**, *12*, 695-698.
893. Clayden, J. *Chem. Commun.* **2004**, 127-135.
894. Clayden, J.; Lund, A.; Vallverdu, L.; Helliwell, M. *Nature* **2004**, *431*, 966-971.
895. Vogl, E. M.; Groeger, H.; Shibasaki, M. *Angew. Chem., Int. Ed.* **1999**, *38*, 1570-1577.
896. Mikami, K.; Matsukawa, S. *Nature* **1997**, *385*, 613-615.
897. Mikami, K.; Terada, M.; Korenaga, T.; Matsumoto, Y.; Ueki, M.; Angelaud, R. *Angew. Chem., Int. Ed.* **2000**, *39*, 3532-3556.
898. Mikami, K.; Korenaga, T.; Yusa, Y.; Yamanaka, M. *Adv. Synth. Catal.* **2003**, *345*, 246-254.
899. Faller, J. W.; Lavoie, A.; Parr, J. *Chem. Rev.* **2003**, *103*, 3345-3367.

900. Mikami, K.; Yamanaka, M. *Chem. Rev.* **2003**, *103*, 3369-3340.

901. Vogl, E. M.; Gröger, H.; Shibasaki, M. *Angew. Chem., Int. Ed.* **1999**, *38*, 1571-1577.

902. Walsh, P. J. *Acc. Chem. Res.* **2003**, *36*, 739-749.

903. Walsh, P. J.; Lurain, A. E.; Balsells, J. *Chem. Rev.* **2003**, *103*, 3297-3344.

904. Mikami, K.; Aikawa, K.; Yusa, Y.; Jodry, J. J.; Yamanaka, M. *Synlett* **2002**, 1561-1578.

905. Johannsen, M.; Yao, S.; Jørgensen, K. A. *Chem. Commun.* **1997**, 2169-2170.

906. Yao, S.; Johannsen, M.; Audrain, H.; Hazell, R. G.; Jørgensen, K. A. *J. Am. Chem. Soc.* **1998**, *120*, 8599-8605.

907. Jørgensen, K. A.; Johannsen, M.; Yao, S.; Audrian, H.; Thorhauge, J. *Acc. Chem. Res.* **1999**, *32*, 605-613.

908. Wolf, C.; Fadul, Z.; Hawes, P. A.; Volpe, E. C. *Tetrahedron: Asymm.* **2004**, *15*, 1987-1993.

909. Evans, D. A.; Kozlowski, M. C.; Burgey, C. S.; MacMillan, D. W. C. *J. Am. Chem. Soc.* **1997**, *119*, 7893-7894.

910. Johnson, J. S.; Evans, D. A. *Acc. Chem. Res.* **2000**, *33*, 325-335.

911. Zhang, W.; Loebach, J. L.; Wilson, S. R.; Jacobsen, E. N. *J. Am. Chem. Soc.* **1990**, *112*, 2801-2803.

912. Irie, R.; Noda, K.; Ito, Y.; Matsumoto, N.; Katsuki, T. *Tetrahedron Lett.* **1990**, *31*, 7345-7348.

913. Irie, R.; Ito, Y.; Katsuki, T. *Synlett* **1991**, 265-266.

914. Norrby, P.-O. *J. Am. Chem. Soc.* **1995**, *117*, 11035-11036.

915. Hamada, T.; Fukuda, T.; Imanishi, H.; Katsuki, T. *Tetrahedron* **1996**, *52*, 515-530.

916. Finney, N. S.; Popisil, P. J.; Chang, S.; Palucki, M. Konsler, R. G.; Hansen, K. B.; Jacobsen, E. N. *Angew. Chem., Int. Ed. Engl.* **1997**, *36*, 1720-1723.

917. Feichtinger, D.; Plattner, D. A. *Angew. Chem., Int. Ed. Engl.* **1997**, *36*, 1718-1719.

918. Hashihayata, T.; Ito, Y.; Katsuki, T. *Synlett* **1996**, 1079-1081.

919. Hashihayata, T.; Ito, Y.; Katsuki, T. *Tetrahedron* **1997**, *53*, 9541-9552.

920. Miura, K.; Katsuki, T. *Synlett* **1999**, 783-785.

921. Miller, J. A.; Gross, B. A.; Zhuravel, M. A.; Jin, W.; Nguyen, S. T. *Angew. Chem., Int. Ed.* **2005**, *44*, 3885-3889.

922. Balsells, J.; Walsh, P. J. *J. Am. Chem. Soc.* **2000**, *122*, 1802-1803.

923. Costa, A. M.; Jimeno, C.; Gavenonis, J.; Carroll, P. J.; Walsh, P. J. *J. Am. Chem. Soc.* **2002**, *124*, 6929-6941.

924. Davis, T. J.; Balsells, J.; Carroll, P. J.; Walsh, P. J. *Org. Lett.* **2001**, *3*, 2161-2164.

925. Wolf, C.; König, W. A.; Roussel, C. *Liebigs Ann.* **1995**, 781-786.

926. Wolf, C.; Hochmuth, D. H.; König, W. A.; Roussel, C. *Liebigs Ann.* **1996**, 357-363.

927. Wolf, C. *Chem. Soc. Rev.* **2005**, *34*, 595-608.

928. Ueki, M.; Matsumoto, Y.; Jodry, J. J.; Mikami, K. *Synlett* **2001**, 1889-1892.

929. Chavarot, M.; Bryne, J. J.; Chavant, P. Y.; Guindet, P.; Vallée, Y. *Tetrahedron: Asymm.* **1998**, *9*, 3889-3894.

930. Bolm, C.; Beckmann, O. *Chirality*, **2000**, *12*, 523-525.

931. Reetz, M. T.; Li, X. *Angew. Chem., Int. Ed.* **2005**, *44*, 2959-2962.

932. Reetz, M. T.; Neugebauer, T. *Angew. Chem., Int. Ed.* **1999**, *38*, 179-181.

933. Blackmond, D. G.; Rosner, T.; Neugebauer, T.; Reetz, M. T. *Angew. Chem., Int. Ed.* **1999**, *38*, 2196-2199.

934. Ohkuma, T.; Doucet, H.; Pham, T.; Mikami, K.; Korenaga, T.; Terada, M.; Noyori, R. *J. Am. Chem. Soc.* **1998**, *120*, 1086-1087.

935. Mikami, K.; Korenaga, T.; Terada, M.; Ohkuma, T.; Pham, T.; Noyori, R. *Angew. Chem., Int. Ed.* **1999**, *38*, 495-497.

936. Aikawa, K.; Mikami, K. *Angew. Chem., Int. Ed.* **2003**, *42*, 5455-5458.

937. Aikawa, K.; Mikami, K. *Angew. Chem., Int. Ed.* **2003**, *42*, 5458-5461.

938. Mikami, K.; Aikawa, K.; Yamanaka, M. *Pure Appl. Chem.* **2004**, *76*, 537-540.

939. Mikami, K.; Kakuno, H.; Aikawa, K. *Angew. Chem., Int. Ed.* **2005**, *44*, 7257-7260.

940. Desponds, O.; Schlosser, M. *Tetrahedron Lett.* **1996**, *37*, 47-48.

941. Tudor, M. D.; Becker, J. J.; White, P. S.; Gagné, M. R. *Organometallics* **2000**, *19*, 4376-4384.

942. Yamanaka, M.; Mikami, K. *Organometallics* **2002**, *21*, 5847-5851.

943. Mikami, K.; Kataoka, S.; Yusa, Y. Aikawa, K. *Org. Lett.* **2004**, *6*, 3699-3701.

944. Becker, J. J.; White, P. S.; Gagné, M. R. *J. Am. Chem. Soc.* **2001**, *123*, 9478-9479.

945. Koh, J. H.; Larsen, A. L.; Gagné, M. R. *Org. Lett.* **2001**, *3*, 1233-1236.

946. Mikami, K.; Aikawa, K.; Yusa, Y. *Org. Lett.* **2002**, *4*, 91-94.

947. Mikami, K.; Aikawa, K.; Yusa, Y. *Org. Lett.* **2002**, *4*, 95-97.

948. Mikami, K.; Yusa, Y.; Aikawa, K.; Hatano, M. *Chirality* **2003**, *15*, 105-107.

949. Tissot, O.; Gouygou, M.; Dallemer, F.; Daran, J.-C.; Balavoine, G. G. A. *Angew. Chem., Int. Ed.* **2001**, *40*, 1076-1078.

950. Ghosh, A. K.; Matsuda, H. *Org. Lett.* **1999**, *1*, 2157-2159.

951. Mikami, K.; Aikawa, K. *Org. Lett.* **2002**, *4*, 99-101.

952. Tanaka, K.; Toda, F. *Chem. Rev.* **2000**, *100*, 1025-1074.

953. Braga, D.; Grepioni, F. *Angew. Chem., Int. Ed.* **2004**, *43*, 4002-4011.

954. Gao, X.; Friscic, T.; MacGillivray, L. R. *Angew. Chem., Int. Ed.* **2004**, *43*, 232-236.

955. Feldman, K. S.; Campbell, R. F. *J. Org. Chem.* **1995**, *60*, 1924-1925.

956. Matsumoto, A.; Odani, T.; Chikada, M.; Sada, K.; Miyata, M. *J. Am. Chem. Soc.* **1999**, *121*, 11122-11129.

957. Matsumoto, A.; Katayama, K.; Odani, T.; Oka, K.; Tashiro, K.; Saragai, S.; Nakamoto, S. *Macromolecules* **2000**, *33*, 7786-7792.

958. Matsumoto, A.; Nagahama, S.; Odani, T. *J. Am. Chem. Soc.* **2000**, *122*, 9109-9119.

959. Matsumoto, A.; Sada, K.; Tashiro, K.; Miyata, M.; Tsubouchi, T.; Tanaka, T.; Odani, T.; Nagahama, S.; Tanaka, T.; Inoue, K.; Saragai, S.; Nakamoto, S. *Angew. Chem., Int. Ed.* **2002**, *41*, 2502-2505.

960. Nagahama, S.; Inoue, K.; Sada, K.; Miyata, M.; Matsumoto, A. *Cryst. Growth Des.* **2003**, *3*, 247-256.

961. Ito, Y.; Borecka, B.; Trotter, J.; Scheffer, J. R. *Tetrahedron Lett.* **1995**, *36*, 6083-6086.

962. Ito, Y.; Hosomi, H.; Ohba, S. *Tetrahedron* **2000**, *56*, 6833-6844.

963. Ohba, S.; Hosomi, H.; Ito, Y. *J. Am. Chem. Soc.* **2001**, *123*, 6349-6352.

964. Koshima, H.; Ding, K.; Chisaka, Y.; Matsuura, T. *J. Am. Chem. Soc.* **1996**, *118*, 12059-12065.

965. Koshima, H.; Nakagawa, T.; Matsuura, T. *Tetrahedron Lett.* **1997**, *38*, 6063-6066.

966. Koshima, H.; Ding, K.; Chisaka, Y.; Matsuura, T.; Miyahara, I.; Hirotsu, K. *J. Am. Chem. Soc.* **1997**, *119*, 10317-10324.

967. MacGillivray, L. R.; Reid, J. L.; Ripmeester, J. A. *J. Am. Chem. Soc.* **2000**, *122*, 7817-7818.

968. Amirsakis, D. G.; Garcia-Garibay, M. A.; Rowan, S. J.; Stoddart, J. F.; White, A. J. P.; Williams, D. J. *Angew. Chem., Int. Ed.* **2001**, *40*, 4256-4261.

969. Friscic, T.; MacGillivray, L. R. *Chem. Commun.* **2003**, 1306-1307.

970. Ouyang, X.; Fowler, F. W.; Lauher, J. W. *J. Am. Chem. Soc.* **2003**, *125*, 12400-12401.

971. Tashiro, K.; Kamae, T.; Kobayashi, M.; Matsumoto, A.; Yokoi, K.; Aoki, S. *Macromolecules* **1999**, *32*, 2449-2454.

972. Tanaka, T.; Matsumoto, A. *J. Am. Chem. Soc.* **2002**, *124*, 9676-9677.

973. Matsumoto, A.; Tanaka, T.; Tsubouchi, T.; Tashiro, K.; Saragai, S.; Nakamato, S. *J. Am. Chem. Soc.* **2002**, *124*, 8891-8902.

974. Matsumoto, A.; Oshita, S.; Fujioka, D. *J. Am. Chem. Soc.* **2002**, *124*, 13749-13756.

975. Wu, D.-Y.; Chen, B.; Fu, X.-G.; Wu, L.-Z.; Zhang, L.-P.; Tung, C.-H. *Org. Lett.* **2003**, *5*, 1075-1077.

976. Caronna, T.; Liantonio, R.; Logothetis, T. A.; Metrangolo, P.; Pilati, T.; Resnati, G. *J. Am. Chem. Soc.* **2004**, *126*, 4500-4501.

977. Kaftory, M.; Yagi, M.; Tanaka, K.; Toda, F. *J. Org. Chem.* **1988**, *53*, 4391-4393.

978. Toda, F.; Akai, H. *J. Org. Chem.* **1990**, *55*, 3446-3447.

979. Toda, F.; Tanaka, K.; Sato, J. *Tetrahedron: Asymm.* **1993**, *4*, 1771-1774.

980. Toda, F.; Miyamoto, H.; Takeda, K.; Matsugawa, R.; Maruyama, N. *J. Org. Chem.* **1993**, *58*, 6208-6211.

981. Koshima, H.; Nakagawa, T.; Matsuura, T. *Tetrahedron Lett.* **1997**, *38*, 6063-6066.

982. Ohba, S.; Hosomi, H.; Tanaka, K.; Miyamoto, H.; Toda, F. *Bull. Chem. Soc. Jpn* **2000**, *73*, 2075-2085.

983. Scheffer, J. R.; Wang, L. *J. Phys. Org. Chem.* **2000**, *13*, 531-538.

984. Kohmoto, S.; Ono, Y.; Masu, H.; Yamaguchi, K.; Kishikawa, K. I.; Yamamoto, M. *Org. Lett.* **2001**, *3*, 4153-4155.

985. Chen, S.; Patrick, B. O.; Scheffer, J. R. *J. Org. Chem.* **2004**, *69*, 2711-2718.

986. Tanaka, K.; Toda, F.; Mochizuki, E.; Yasui, N; Kai, Y.; Miyahara, I.; Hirotsu, K. *Angew. Chem., Int. Ed.* **1999**, *38*, 3523-3525.

987. Ito, Y.; Hosomi, H.; Ohba, S. *Tetrahedron* **2000**, *56*, 6833-6844.

988. Leibovitch, M.; Olovsson, G.; Scheffer, J. R.; Trotter, J. *J. Am. Chem. Soc.* **1997**, *119*, 1462-1463.

989. Cheung, E.; Kang, T.; Raymond, J. R.; Scheffer, J. R.; Trotter, J. *Tetrahedron Lett.* **1999**, *40*, 8729-8732.

990. Scheffer, J. R. *Can. J. Chem.* **2001**, *79*, 349-357.

991. Toda, F.; Miyamoto, H.; Kikuchi, S. *J. Chem. Soc., Chem. Commun.* **1995**, 621-622.

992. Toda, F.; Miyamoto, H.; Kanemoto, K.; Tanaka, K.; Takahashi, Y.; Takenaka, Yi. *J. Org. Chem.* **1999**, *64*, 2096-2102.

993. Toda, F. *Pure Appl. Chem.* **2001**, *73*, 1137-1145.

994. Shailaja, J.; Karthikeyan, S.; Ramamurthy, V. *Tetrahedron Lett.* **2002**, *43*, 9335-9339.

995. Vestergren, M.; Gustafsson, B.; Davidsson, O.; Hakansson, M. *Angew. Chem., Int. Ed.* **2000**, *39*, 3435-3437.

996. Sakamoto, M.; Sekine, N.; Miyoshi, H.; Mino, T.; Fujita, T. *J. Am. Chem. Soc.* **2000**, *122*, 10210-10211.

997. Vestergren, M.; Eriksson, J.; Hakansson, M. *Chem. Eur. J.* **2003**, *9*, 4678-4686.

998. Johansson, A.; Hakansson, M. *Chem. Eur. J.* **2005**, *11*, 5238-5248.

999. Sakamoto, M. Unosawa, A.; Kobaru, S.; Saito, A.; Mino, T.; Fujita, T. *Angew. Chem., Int. Ed.* **2005**, *44*, 5523-5526.

1000. Leibovitch, M.; Olovvson, G.; Scheffer, J. R.; Trotter, J. *J. Am. Chem. Soc.* **1998**, *120*, 12755-12769.

1001. Patrick, B. O.; Scheffer, J. R.; Scott, C. *Angew. Chem., Int. Ed.* **2003**, *42*, 3775-3777.

1002. Chong, K. C. W.; Scheffer, J. R. *J. Am. Chem. Soc.* **2003**, *125*, 4040-4041.

1003. Xia, W.; Scheffer, J. R.; Patrick, B. O. *CrystEngComm.* **2005**, *7*, 728-730.

1004. Sakamoto, M.; Iwamoto, T.; Nono, N.; Ando, M.; Arai, W.; Mino, T.; Fujita, T. *J. Org. Chem.* **2003**, *68*, 942-946.

1005. Kane, J. J.; Liao, R.-F.; Lauher, J. W.; Fowler, F. W. *J. Am. Chem. Soc.* **1995**, *117*, 12003-12004.

1006. Koshima, H.; Nakagawa, T.; Matsuura, T.; Miyamoto, H.; Toda, F. *J. Org. Chem.* **1997**, *62*, 6322-6325.

1007. Gdaniec, M.; Jankowski, W.; Milewska, M. J.; Polonski, T. *Angew. Chem., Int. Ed.* **2003**, *42*, 3903-3906.

1008. Crihfield, A.; Hartwell, J.; Phelps, D.; Walsh, R. B.; Harris, J. L.; Payne, J. F.; Pennington, W. T.; Hanks, T. W. *Cryst. Growth Des.* **2003**, *3*, 313-320.
1009. Chu, Q.; Swenson, D. C.; MacGillivray, L. R. *Angew. Chem., Int. Ed.* **2005**, *44*, 3569-3572.
1010. Nishio, M. *CrystEngComm.* **2004**, *6*, 130-158.
1011. Aakeroy, C. B.; Beatty, A. M.; Helfrich, B. A. *J. Am. Chem. Soc.* **2002**, *124*, 14425-14432.
1012. Mei, X.; Wolf, C. *Eur. J. Org. Chem.* **2004**, 4340-4347.

Asymmetric Resolution and Transformation of Chiral Compounds under Thermodynamic and Kinetic Control

Many pharmaceuticals, nutrients, flavors, and fragrances such as naproxen, esomeprazole, vancomycin, vitamin C, aspartame, menthol, carvone, and limonene are chiral. Biomedical and nutritional applications generally require the use of chiral compounds in enantiopure form, which has fueled the development of ingenious synthetic strategies and numerous asymmetric reactions. Despite the advance of asymmetric synthesis, chiral compounds are often obtained as racemates or nonracemic mixtures that need additional purification. Preparative separation of enantiomers can be accomplished by chromatography on a chiral stationary phase, preferential crystallization with seed crystals and chiral additives, selective crystallization of diastereomeric salts, and kinetic resolution. A major drawback of these methods is the inherently poor atom economy because the maximum yield is limited to 50%.[1] This dilemma may be alleviated by enantioconvergent synthesis, deracemization or through repetition of a laborious sequence of preparative enantioseparation and external racemization of the undesired enantiomer.[2]

The term deracemization has been coined to describe conversion of a racemate to either enantiomer; this is often confused with desymmetrization, enantioconvergent synthesis or kinetic resolution, Figure 7.1. A desymmetrization reduces the number of symmetry elements in a molecule. In many cases, stereoselective destruction of a symmetry plane or inversion center of a meso compound provides a useful entry to asymmetric synthesis of chiral compounds. In a kinetic resolution one enantiomer of a racemate is selectively derivatized, while the other one is less reactive and is therefore not, or only slightly, consumed at 50% conversion. In the case of enantioconvergent synthesis, both enantiomers of a racemic starting material are used to prepare a single chiral product which may require several synthetic steps.

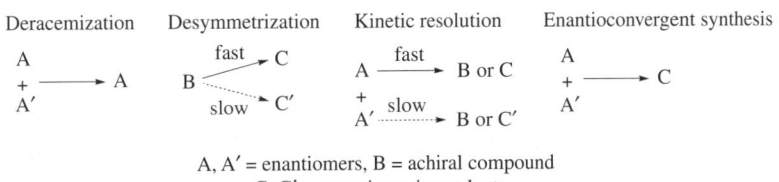

Figure 7.1   Deracemization, desymmetrization, kinetic resolution, and enantioconvergent synthesis.

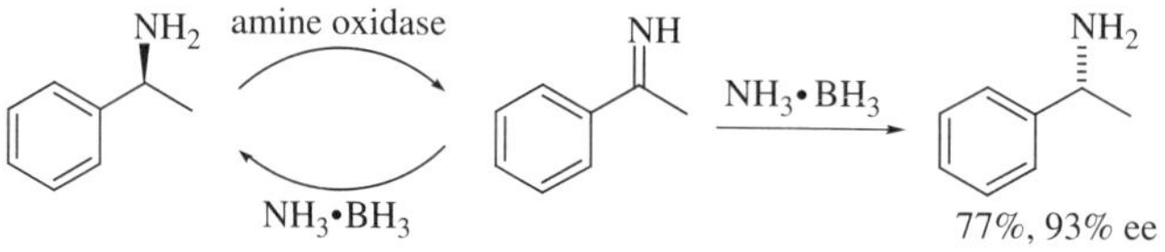

Scheme 7.1 Deracemization of α-methylbenzylamine.

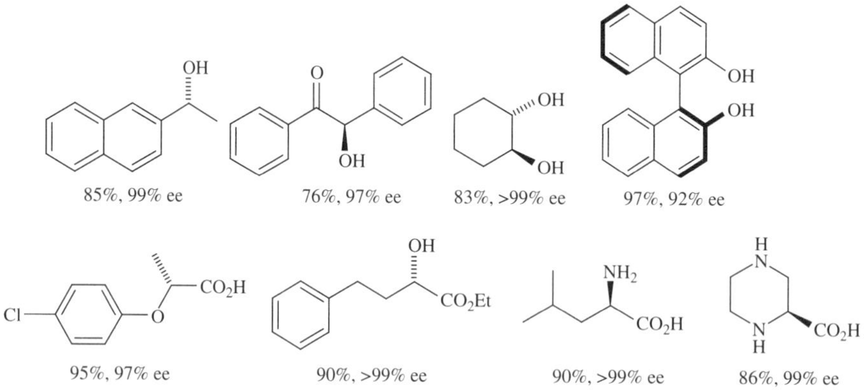

Figure 7.2 Structures of compounds obtained by deracemization.

An elegant approach to deracemization of chiral amines has been reported by Turner.[3] Treatment of racemic α-methylbenzylamine with an enantioselective amine oxidase results in selective conversion of the (*S*)-enantiomer to the corresponding achiral imine which is reduced *in situ* with ammonia borane.[i] The nonselective chemical reduction produces equal amounts of the enantiomers of α-methylbenzylamine. However, incorporation of both reactions into a one-pot procedure allows continuous recycling of the (*S*)-enantiomer until the entire mixture is deracemized to (*R*)-α-methyl-benzylamine, Scheme 7.1. Deracemization is frequently achieved by formation of achiral intermediates in combination with enantioselective biocatalytic processes that exploit isolated enzymes or even intact cells.[4] Since quantitative deracemization eliminates cumbersome preparative enantio-separation and presents an atom-economical solution to asymmetric synthesis with racemic starting materials, it has been applied to many classes of chiral compounds including amino acids,[5–7] amines,[8] alcohols,[9,10] diols,[11–14] carboxylic acids,[15,16] esters,[17] aryl alkanes,[18] biaryls,[19] and transition metal complexes,[20] Figure 7.2.

Desymmetrization of achiral molecules provides another useful entry to asymmetric synthesis. In most cases, meso compounds possessing a mirror plane are desymmetrized with a chiral reagent or in the presence of a chiral catalyst or enzyme.[21,22] However, an increasing number of desymmetrization procedures applicable to centrosymmetric compounds has emerged.[23] A renowned example is the asymmetric ring opening of achiral epoxides with trimethylsilyl azide. In particular, Jacobsen's (salen)chromium(III) complex catalyzes ring opening of a series of readily available epoxides with high enantioselectivity and affords azido silyl ethers that can easily be hydrolyzed to azido alcohols, Scheme 7.2.[24]

A representative case of enantioconvergent synthesis involving an enzyme-catalyzed kinetic resolution step of a racemic tricyclic diol has been reported by Yoshida and Ogasawara, Scheme 7.3.[25] Lipase-mediated transesterification of the racemic starting material using vinyl acetate as acyl

---

[i]The enzyme used does not recognize (*R*)-α-methylbenzylamine and selectively oxidizes the (*S*)-enantiomer.

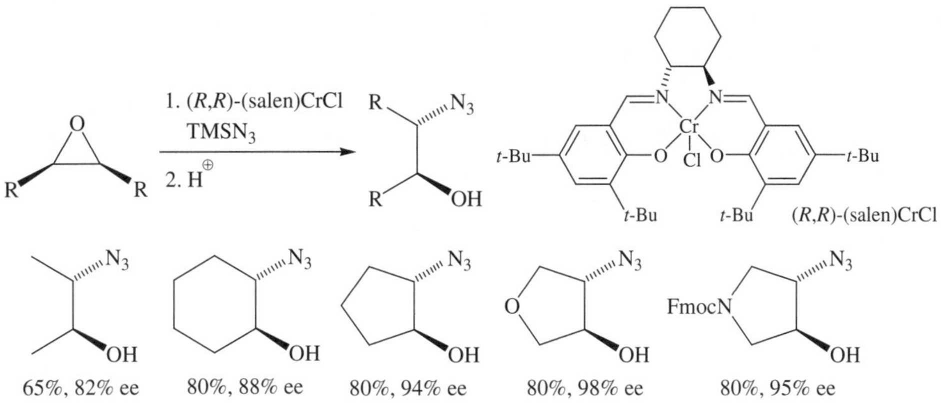

Scheme 7.2 Desymmetrization of meso epoxides with (salen)Cr(III)Cl.

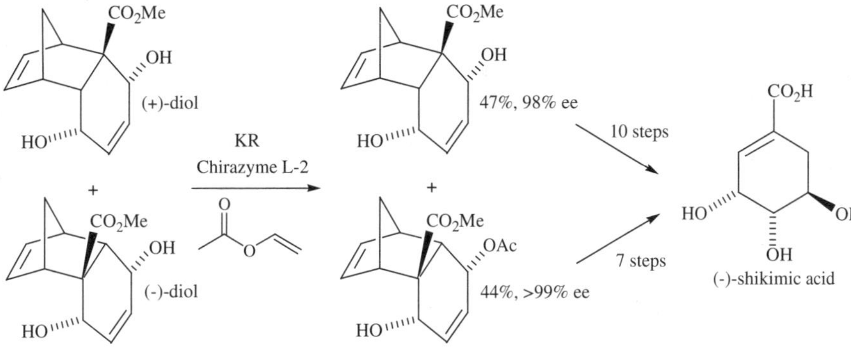

Scheme 7.3 Enantioconvergent synthesis of (−)-shikimic acid based on kinetic resolution.

donor produced a mixture of the remaining dextrorotatory diol and the acetate of the (−)-enantiomer. This reaction constitutes a kinetic resolution (KR) which is discussed in more detail in Chapter 7.4.1. The two compounds were then separated and employed in individual synthetic routes to prepare (−)-shikimic acid. Although (−)-shikimic acid can be prepared from both enantiomers of the racemic starting material, the enantioconvergent approach shown above requires many synthetic steps. To limit the total number of reactions, it is desirable to establish enantioconvergence as early as possible.[ii] In many cases, coupling of racemization of a stereolabile chiral compound with an enantiomer-differentiating reaction in the presence of a chiral catalyst or reagent allows quantitative transformation of a racemic mixture to an enantiopure compound within a single process. Important atom-economical synthetic strategies that incorporate interconverting chiral starting materials or intermediates into a one-pot procedure are dynamic kinetic resolution (DKR), dynamic kinetic asymmetric transformation (DYKAT) and dynamic thermodynamic resolution (DTR). The success and popularity of these broadly applicable methodologies clearly emphasize the usefulness of stereolabile chiral compounds in asymmetric synthesis.

[ii] The strategy towards (−)-shikimic acid shown in Scheme 7.3 requires a total of 18 steps including kinetic resolution because enantioconvergence is established at the end of each synthetic branch. If enantioconvergence were achieved in the first step, for example by deracemization to the (−)-diol, one could synthesize the final product from the racemate in just nine steps.

7.1 SCOPE AND PRINCIPLES OF ASYMMETRIC RESOLUTION
AND TRANSFORMATION

The equilibration of a stereoisomeric mixture to a single stereoisomer has traditionally been referred to as asymmetric transformation of the first or second kind. If such a process is not coupled with a separation step, *e.g.*, crystallization and extraction, it is called asymmetric transformation of the first kind, whereas the term asymmetric transformation of the second kind is reserved for processes that combine isomerization and concomitant separation. Both strategies involve interconversion of stereolabile chiral compounds under thermodynamic control. Asymmetric transformation of the first kind is based on the relative thermodynamic stability of interconverting diastereomers or diastereomeric complexes. Asymmetric transformation of the second kind entails physical separation and depends on conglomerate formation of interconverting enantiomers, and different solubility or partitioning between two immiscible phases in the case of stereolabile diastereomers.[iii] Asymmetric transformations of the first and second kind are generally aimed at the recovery of one stereoisomer of the starting materials and are conceptually related to separation techniques such as crystallization and chromatography. Asymmetric transformation is often the ultimate method of choice, and many examples have been incorporated into asymmetric syntheses of natural products, *vide infra*.

As has been mentioned above, the scope of kinetic resolution (KR) is limited due to the inherently low yield of 50% but it can be the most effective method when enantioseparation by chromatography or crystallization is not successful. According to the 1996 IUPAC recommendation,[26] KR is defined as "the achievement of partial or complete resolution by virtue of unequal rates of reaction of the enantiomers in a racemate with a chiral agent (reagent, catalyst or solvent)." Kinetic resolutions are usually stopped at 50% conversion, and are used to recover the less reactive enantiomer. Alternatively, the objective of KR can be to obtain the enantiopure or enantioenriched reaction product. In other words, kinetic resolutions can be aimed either at purification of enantiomers or at asymmetric synthesis.

The coupling of kinetically controlled enantiomerization or diastereomerization with a stereoisomer-differentiating reaction in a one-pot process constitutes a dynamic kinetic resolution or a dynamic kinetic asymmetric transformation, and allows quantitative conversion of a stereoisomeric mixture to a single chiral product, see Chapters 7.4 and 7.5. The term dynamic thermodynamic resolution (DTR) has been coined for the combination of asymmetric transformation of the first or second kind, *i.e.*, thermal equilibration of interconverting stereoisomers, with a subsequent synthetic step under conditions that exclude isomerization, Chapter 7.6. In contrast to KR, which is often used for preparative *separation* of a racemate by consumption of one enantiomer in a stereoselective reaction, DKR, DYKAT and DTR are always aimed at asymmetric *synthesis*.

7.2 ASYMMETRIC TRANSFORMATION OF THE FIRST KIND

Asymmetric transformation of the first kind refers to an equilibration process resulting in the enrichment of one stereoisomer from a racemic or diastereomeric mixture of stereolabile chiral compounds without concomitant separation. This concept comprises deracemization of enantiomers in the presence of a chiral solvating agent and interconversion of diastereomers under thermodynamical control. Pirkle and Reno discovered base-promoted deracemization of stereolabile enantiomers of *N*-(3,5-dinitrobenzoyl)leucine butyl thioester in the presence of substoichiometric amounts of (*R*)-11-undecenyl *N*-(2-naphthyl)alaninate.[27] The transformation of the racemic starting material to a mixture exhibiting the (*R*)-leucine thioester in 78% ee is a consequence of

---

[iii] Isomerization due to diastereoselective crystallization is based on the different stability of the crystal lattices formed by the interconverting isomers. This method provides the less soluble diastereoisomer which is not necessarily the more thermodynamically stable one.

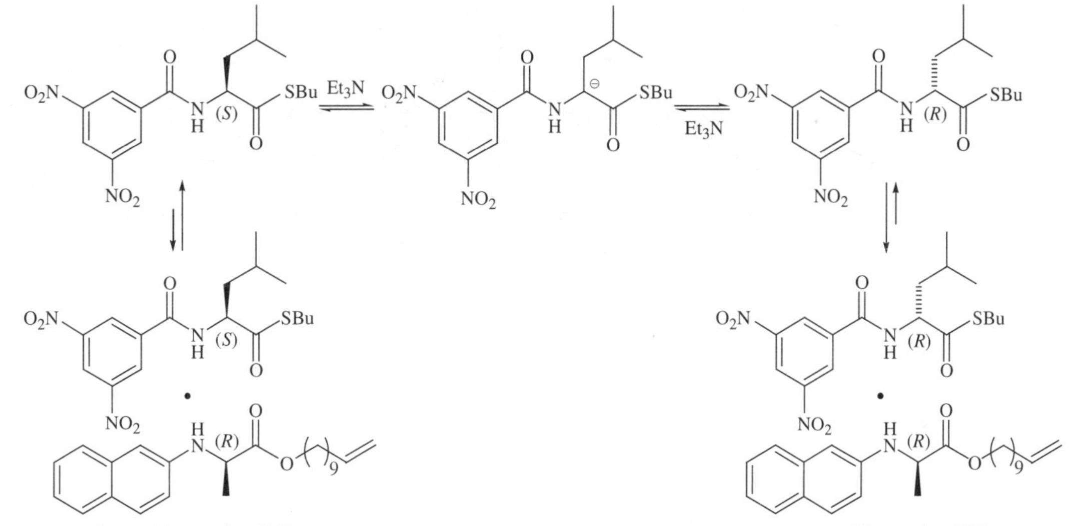

Scheme 7.4 Deracemization of *N*-(3,5-dinitrobenzoyl)leucine butyl thioester through asymmetric transformation of the first kind.

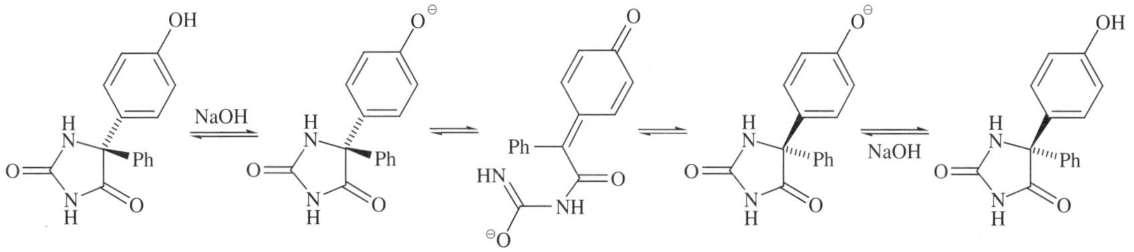

Scheme 7.5 Base-promoted interconversion of hydantoin enantiomers.

differential complexation and formation of diastereomeric complexes with different thermodynamic stability, Scheme 7.4. It has been proposed that three simultaneous interactions, *i.e.*, two hydrogen bonds in addition to π-π-stacking, occur during complexation of the (*R*)-leucine thioester with the (*R*)-alaninate, while the heterochiral complex is less stable because it can only accommodate two of these interactions at the same time. Riedner and Vogel applied the same concept to a stereolabile 5-(4-hydroxyphenyl)-5-phenylhydantoin bearing an asymmetric quaternary carbon atom.[28] Equilibration of the racemic hydantoin mixture in the presence of sodium hydroxide and brucine as the chiral complexing agent is highly stereoselective and permits isolation of the (*S*)-enantiomer in excellent enantiopurity. It is assumed that the enantioconversion involves ring opening of the nonenolizable hydantoin and intramolecular Michael addition of an achiral intermediate, Scheme 7.5.

The deracemization processes described above are based on equilibration of noncovalent diastereomeric adducts derived from a stereolabile racemate and an enantiopure resolving agent. Interconversion of covalent diastereomers is a practical alternative and has been used to enrich stereoisomers of a variety of chiral compounds including amino acids,[29] oxazinones[30] and transition metal complexes,[31] Figure 7.3. Diastereomerization processes including mutarotation and epimerization are prime examples of asymmetric transformation of the first kind, see Chapters 3.1.3 and 3.1.4.

Asymmetric transformation of the first kind provides a convenient means of enhancing the diastereomeric ratio of axially chiral compounds. Meyers' group described a remarkable example

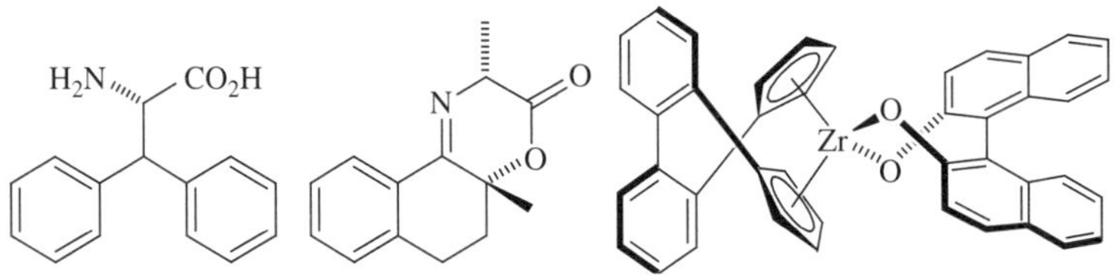

Figure 7.3   Structures of chiral compounds prepared by asymmetric transformation of the first kind.

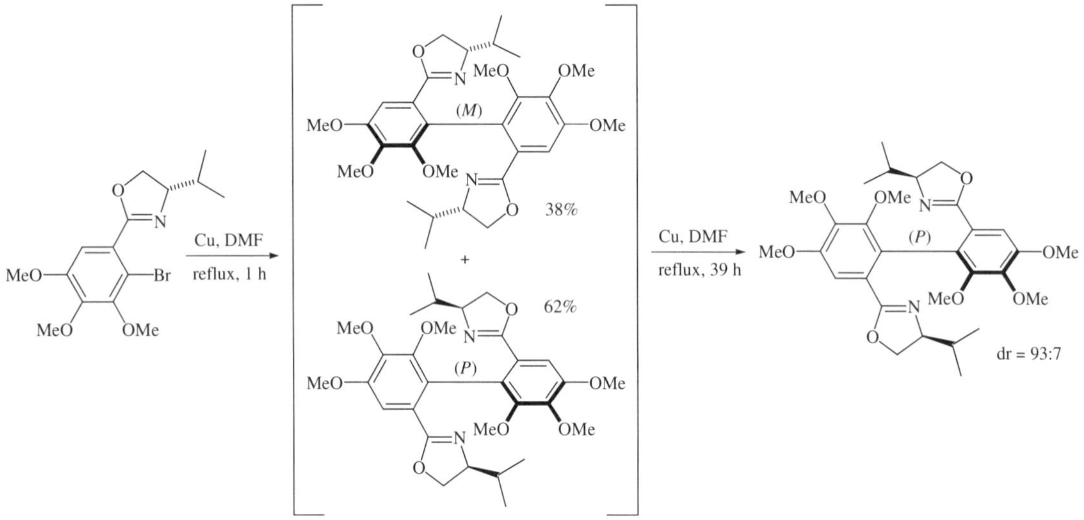

Scheme 7.6   Atroposelective biaryl synthesis based on asymmetric transformation of the first kind.

that is based on oxazoline-mediated asymmetric Ullmann coupling of aryl bromides and subsequent thermodynamically controlled transformation to a diastereomerically enriched biaryl, Scheme 7.6.[32–34] The coupling reaction occurs within one hour generating both biaryl atropisomers with low diastereoselectivity. Continuous refluxing of the interconverting isomers in DMF for a further 39 hours increases the initial diastereomeric ratio from 62:38 to 93:7, favoring the desired (*P*)-atropisomer which is a key intermediate for the total synthesis of permethylated tellimagrandin, an important ellagitannin.[35] Asymmetric transformation of the first kind has also been exploited for deracemization of *ortho*-dihydroxylated biaryl ligands VANOL and VAPOL in the presence of (−)-sparteine-derived copper(II) complexes,[36,37] and for the synthesis of the aglycon of the antibiotic glycopeptide vancomycin.[38–41] Evans and coworkers prepared a bicyclic tetrapeptide vancomycin precursor exhibiting the unnatural (*M*)-actinoidinic acid moiety. They recognized that the conformation of the actinoidinic acid unit is controlled by the global aglycon structure rather than by proximate chiral elements, and that rotation about the central bond between aryl rings *A* and *B* proceeds at ambient temperature favoring formation of the (*P*)-axis. Indeed, thermal equilibration at 55 °C resulted in atropisomerselective interconversion of the two diastereomeric tetrapeptides providing the desired axial stereochemistry in 90% de, Scheme 7.7. Boger *et al.* developed a different approach to vancomycin, in which the tricyclic aglycon is obtained by thermally controlled stereomutation of the axially chiral biaryl moiety and the two chiral aryl ether planes. Asymmetric transformation of the first kind has thus become a key feature in the synthesis of vancomycin, teicoplanin and other antibiotics. Atroposelective synthesis with stereodynamic compounds is discussed in more detail in Chapters 6.2 and 6.3.

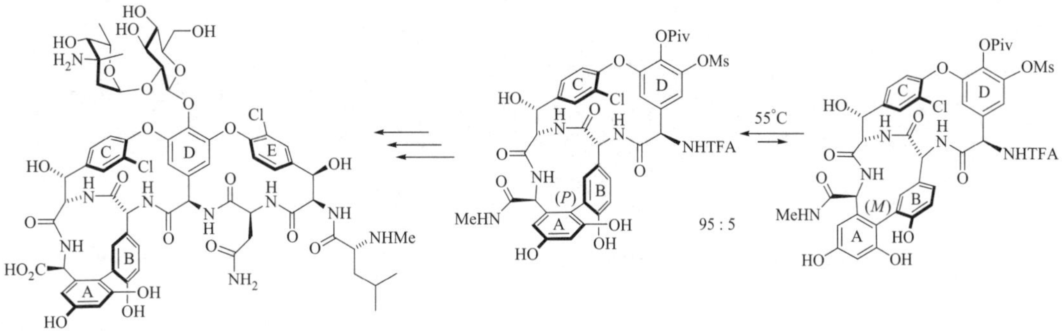

Scheme 7.7 Structure of vancomycin (left) and asymmetric transformation favoring the formation of the natural (*P*)-diastereoisomer of the tetrapeptide precursor (right).

7.3 ASYMMETRIC TRANSFORMATION OF THE SECOND KIND

Deracemization of stereolabile chiral compounds that undergo enantioconversion at elevated temperature can be achieved through repetitive enantioseparation based on chromatography or crystallization and racemization of the undesired enantiomer upon heating. For example, isolation of each enantiomer of a chiral compound by HPLC allows selective off-column racemization of the undesired fraction which is then employed in a second resolution process to yield another 25% of the preferred enantiomer. The remaining 25% of the undesired enantiomer can then be converted to the racemate and subjected to a third separation process, and so on. Theoretically, deracemization based on five separation and four racemization steps affords 96.9% of the desired enantiomer. Although either enantiomer of a stereolabile compound can be enriched through this procedure, a series of separation and off-column racemization steps is required. Enantioenrichment of a racemate is more conveniently accomplished through asymmetric transformation of the second kind, which is defined as equilibration of a mixture of stereoisomers towards a stereochemically enriched or pure sample with concomitant separation. The stereochemical outcome is controlled by the stereoselective crystallization, extraction or chromatographic separation step. Asymmetric transformation of the second kind eliminates the need for cumbersome separation and recycling of the undesired enantiomer by external racemization, and several economically viable one-pot procedures are known.

7.3.1 Crystallization-induced Asymmetric Transformation

Crystallization of a mixture of stereolabile enantiomers in solution produces either a racemic precipitate or a conglomerate. In the latter case, and when enantioconversion occurs on the crystallization time scale, spontaneous or crystal seed-induced crystallization from solution or from the melt may lead to quantitative deracemization and formation of enantiopure solids. Crystallization-induced asymmetric transformation is sometimes referred to as crystallization-induced dynamic resolution (CIDR) or total spontaneous resolution, and can also be observed with interconverting diastereoisomers.[42,43] While deracemization of enantiomers requires conglomerate formation, the success of crystallization-induced asymmetric transformation of interconverting diastereomers depends solely on the relative thermodynamic stability of the corresponding crystal lattices. Effective asymmetric transformation of racemic narwedine to either enantiomer based on total spontaneous resolution has been reported by Shieh and Carlson, Scheme 7.8.[44] Addition of 2.5% of enantiopure seed crystals to a supersaturated ethanolic solution of racemic narwedine, an important tertiary alkaloid, initiates crystallization of a single enantiomer in 84% yield. The total

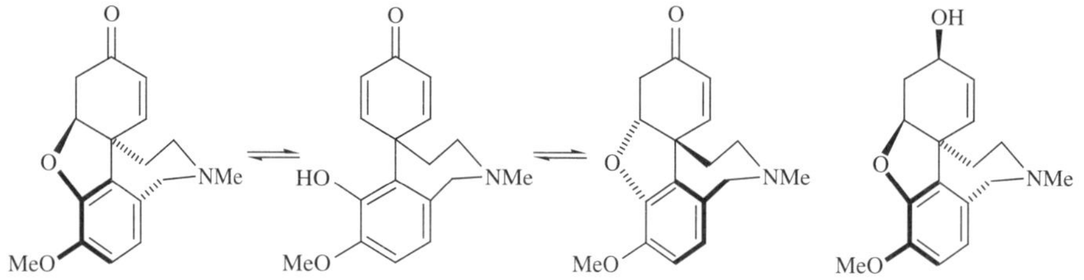

Scheme 7.8 Enantiomerization of narwedine via an achiral intermediate (left) and structure of galanthamine (right).

spontaneous resolution of narwedine can also be induced by addition of enantiopure seed crystals of a structural analog such as galanthamine. Hydrogen/deuterium exchange experiments suggest that the enantiomers of narwedine racemize via reversible retro Michael addition and subsequent ring closure of the achiral intermediate.

Asymmetric transformation of the second kind has been incorporated into several synthetic routes to α-amino acids.[45–49] Broxterman and coworkers coupled a diastereoselective Strecker reaction using pivalaldehyde, sodium cyanide and (R)-phenylglycine amide with crystallization-induced asymmetric transformation to produce the corresponding α-amino nitrile in 93% yield and 98% de, Scheme 7.9.[50] It has been suggested that the interconversion proceeds through reversible diastereofacial addition of HCN to an intermediate imine. Subsequent hydolysis and hydrogenolysis of the amino nitrile finally gave (S)-*tert*-leucine, and the same concept has been used to prepare α-methyl-DOPA. Jochims *et al.* accomplished deracemization of α-amino nitriles via asymmetric transformation of the second kind based on Dimroth's principle,[51] *i.e.*, interconversion of solid chiral stereoisomers through an equilibrium process that occurs in solution, Scheme 7.9. They prepared an ethanolic suspension consisting of equimolar amounts of diastereomeric amygdalates derived from an aliphatic or aromatic racemic α-amino nitrile and (R)-mandelic acid. The mixture was then converted to a stereochemically pure diastereomer by stirring at room temperature and the chiral auxiliary was recovered by extraction.

Asymmetric transformation of amino acids based on *in situ* racemization of N-acylamino derivatives or Schiff base intermediates has become a popular method.[52–62] Virtually every amino acid has been employed in deracemication studies by using aldehydes or ketones to generate a stereolabile Schiff base that undergoes crystallization in the presence of a chiral additive, such as camphersulfonic and tartaric acid, or enantiopure crystal seeds of the desired enantiomer, Figure 7.4.[63] Villani and coworkers observed racemization of the hydrochloride salt of 4-chlorophenylalanine methyl ester in the presence of catalytic amounts of salicylaldehyde in refluxing methanol. Addition of one equivalent of (S,S)-tartaric acid generates an insoluble diastereomeric salt, thus driving the equilibrium of the interconverting enantiomers towards the (R)-amino ester which can be isolated in 68% yield and 96% ee, Scheme 7.10.[64]

Vedejs explored crystallization-induced asymmetric transformation of diastereomeric oxazaborolidinone complexes derived from amino acids or their amidino-protected derivatives, Scheme 7.11.[65–67] Appropriate substitution at boron affords a stereolabile chiral center showing epimerization due to reversible cleavage of the boron–nitrogen or boron–oxygen bond. Slow evaporation of the solvent from a mixture of stereolabile oxazaborolidinone diastereoisomers derived from (S)-phenylalanine results in crystallization of the *anti*-diastereoisomer, which increases the diastereomeric ratio from 3.1:1 in solution to 39:1 in the solid state. In this case, the more stable diastereoisomer in solution undergoes preferential crystallization. However, the predominant species in solution must not necessarily be the one that precipitates because asymmetric transformation of the second kind is solely driven by crystallization of the stereoisomer that forms the more stable crystal lattice. Oxazaborolidinone complexes derived from boranes with

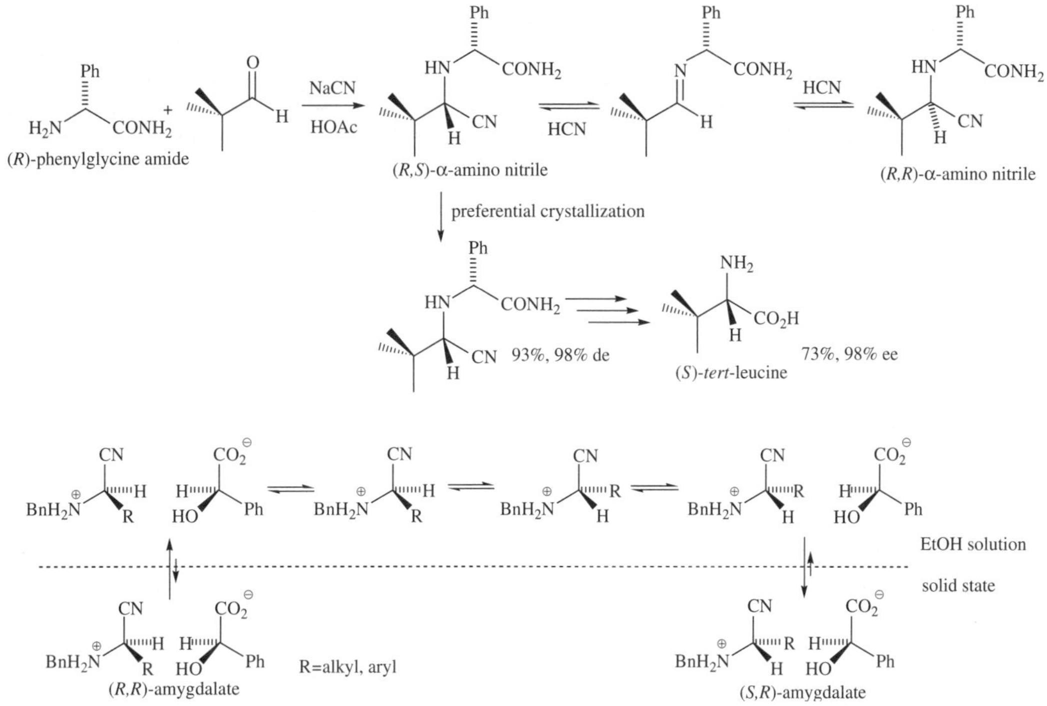

Scheme 7.9 Synthesis of (*S*)-*tert*-leucine via crystallization-induced asymmetric transformation of an intermediate α-amino nitrile (top) and deracemization of α-amino nitriles using Dimroth's principle (bottom).

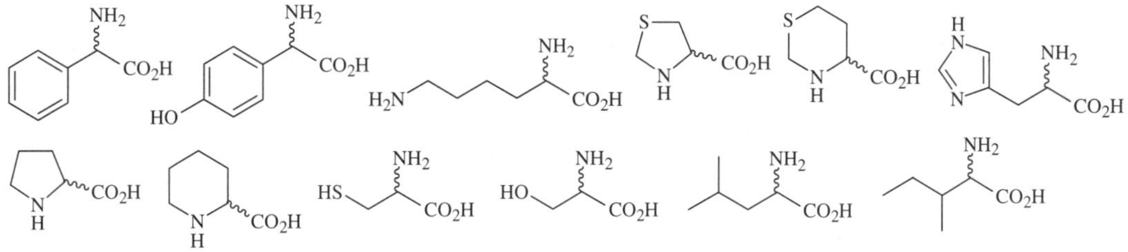

Figure 7.4 Structures of amino acids that have been purified by crystallization-induced asymmetric transformation.

electron-withdrawing substituents can be exploited for asymmetric synthesis based on the principles of self-regeneration of stereocenters (SRS), see Chapter 6.5.2.[68] Vedejs' group uncovered that oxazaborolidinones are configurationally stable under cryogenic conditions and allow base-promoted enolization of the amino acid moiety and subsequent asymmetric alkylation, Scheme 7.11.[69] In this reaction sequence, the sense of chirality at the boron atom is first determined by crystallization-induced asymmetric transformation of diastereomeric borate complexes bearing an α-amidino derivative of an amino acid such as (*S*)-phenylalanine. The chiral center of the amino acid moiety is then destroyed by deprotonation and the chiral borate unit is used to control the stereoselective alkylation of the enolate with propyl bromide to regenerate the asymmetric carbon atom. In the final step, hydrolysis of the borate complex releases (*R*)-2-propylphenylalanine in 81% yield and 99.5% ee. This approach circumvents racemization of (*S*)-phenylalanine and provides asymmetric access to quaternary amino acids. A similar strategy involving crystallization-induced

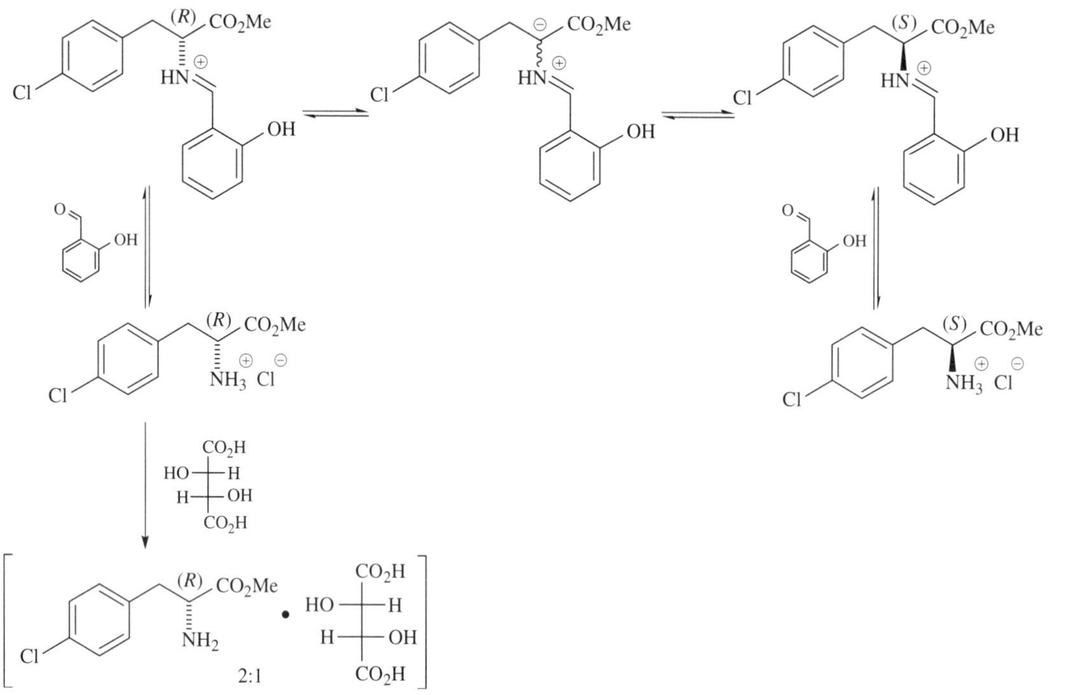

Scheme 7.10   Asymmetric transformation of 4-chlorophenylalanine methyl ester.

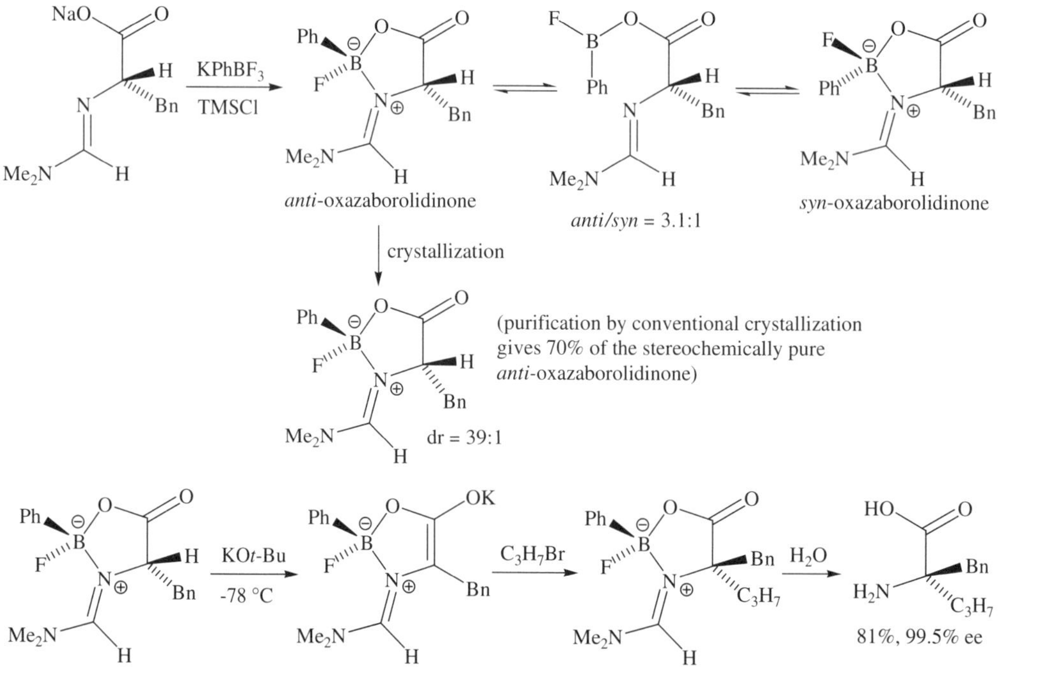

Scheme 7.11   Asymmetric transformation of borate complexes bearing an amino acid ligand (top) and self-regeneration of the original stereocenter during alkylation of an oxazaborolidinone-derived enolate (bottom).

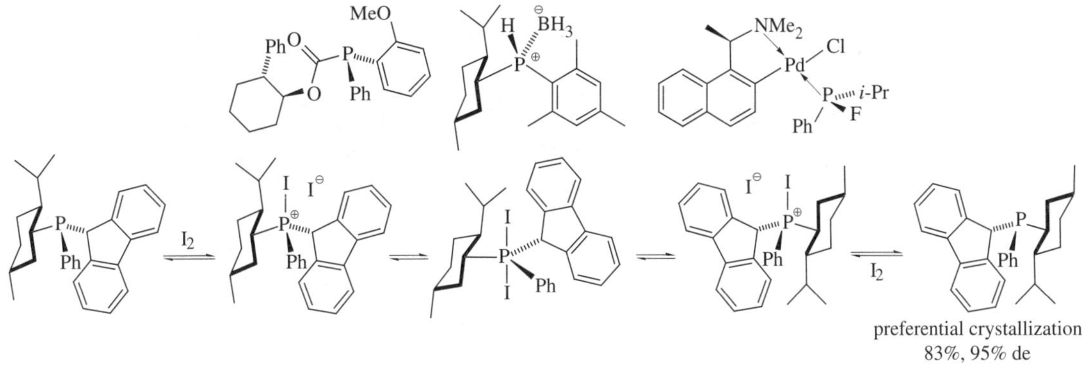

Scheme 7.12 Chiral phosphines employed in asymmetric transformation of the second kind (top) and iodine-catalyzed interconversion of diastereomeric tertiary phosphines (bottom).

asymmetric transformation of diastereomeric oxazolidinones derived from (*S*)-alanine followed by enolization and asymmetric alkylation based on SRS has been reported.[70]

Asymmetric transformations with tertiary acylphosphines, secondary phosphines and fluorophosphines exhibiting an energy barrier to pyramidal inversion between 80 and 100 kJ/mol have been described.[71–74] Since tertiary phosphines containing three alkyl or aryl groups have pyramidal inversion barriers exceeding 120 kJ/mol and are therefore configurationally stable at ambient temperature, Vedejs and Donde developed a crystallization-induced asymmetric transformation procedure based on iodine-catalyzed equilibration of diastereomeric tertiary phosphines. The interconversion process presumably involves a pentavalent trigonal bipyramidal adduct, Scheme 7.12.[75] Conversion of a 3:1 mixture of a menthyl-derived diastereomeric phosphine to a highly enriched isomer was achieved through slow solvent evaporation over 30 hours. The diastereomer that forms the more stable crystal lattice, *i.e.*, the less soluble isomer, was isolated in 83% yield and 95% de.

Asymmetric transformation of several chiral carboxylic acids including mandelic acid and naproxen has been reported.[76] Zwanenburg *et al.* discovered that 2-arylpropionic acids show amidine-promoted racemization and stereoselective crystallization from the melt in the presence of enantiopure α-methylbenzylamine.[77] Kiau and coworkers developed a procedure for deracemization of α-substituted carboxylic acids that interconvert under mild conditions and form diastereomeric ammonium salts, Scheme 7.13.[78] Crystallization-induced asymmetric transformation of racemic 2-bromo-4-(3,4,4-trimethyl-2,5-dioxo-1-imidazolidine)butanoic acid occurs upon addition of (1*R*,2*S*)-2-amino-1,2-diphenylethanol and catalytic amounts of tetrabutylammonium bromide. The configurational instability of other α-halogenated carboxylic acids in the presence of a quaternary ammonium salt has been utilized for crystallization-induced epimerization of diastereomeric imidazolidinone analogs.[79]

Komatsu and Awano developed an attractive synthetic route to 2-deoxy-α-D-ribosyl-1-phosphate, a key intermediate for the synthesis of 2-deoxynucleosides, Scheme 7.14.[80] Treatment of readily available 3,5-di-*O*-(4-chlorobenzoyl)-2-deoxy-α-D-ribosyl-1-chloride with orthophosphoric acid produces an almost equimolar mixture of anomeric sugar 1-phosphates that rapidly interconvert in acetonitrile. Selective crystallization shifts the epimeric equilibrium to the desired α-anomer. The crystallization-controlled epimerization increases the anomeric ratio to 89:11 and subsequent removal of the 4-chlorobenzoyl residues and crystallization yields 2-deoxy-α-D-ribosyl-1-phosphate in 98.8% anomeric purity. Silverberg *et al.* incorporated anomerization of a 2,3-di-*O*-benzyl-4,6-*O*-ethylidene-α,β-D-glucopyranose unit and concomitant crystallization of the desired β-anomer into a multi-kilogram scale synthesis of the anticancer drug etoposide.[81] Crystallographic

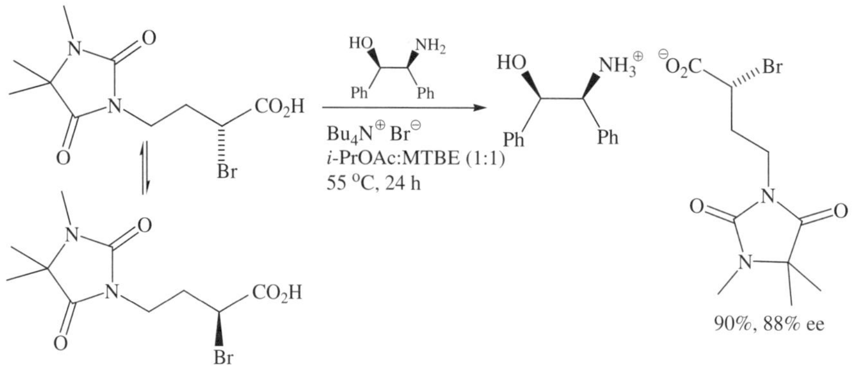

Scheme 7.13 Crystallization-induced deracemization of an α-substituted carboxylic acid.

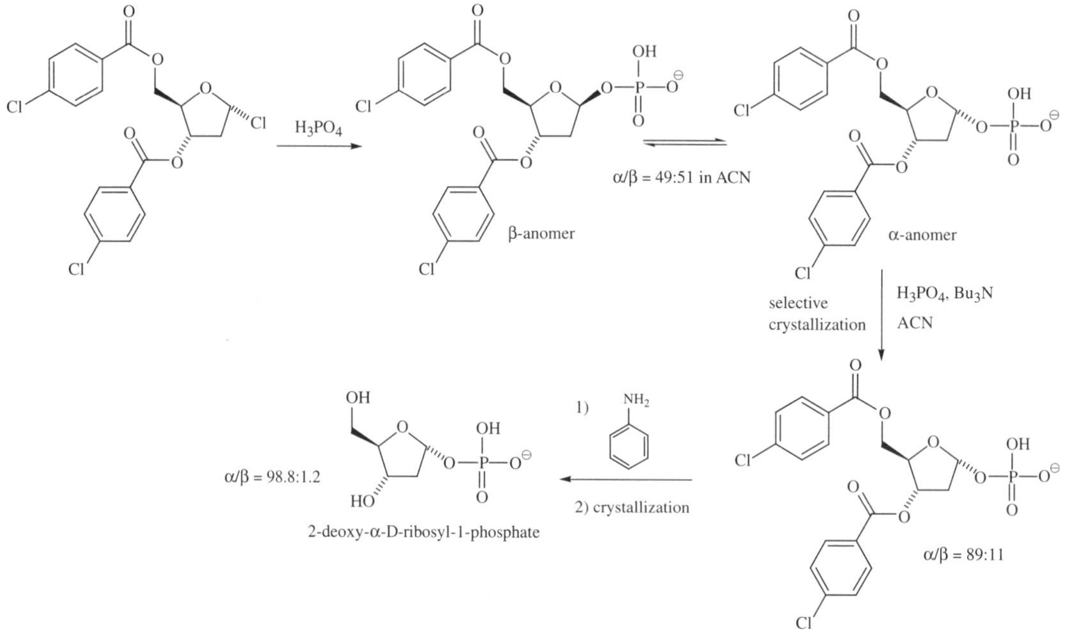

Scheme 7.14 Synthesis of 2-deoxy-α-D-ribosyl-1-phosphate by crystallization-induced epimerization.

analysis revealed that the β-isomer of the carbohydrate derivative forms a more stable crystal lattice than the α-anomer, and thus overcomes the anomeric effect which favors formation of the latter.

Zwanenburg's group described a process that affords the cephalosporin antibiotic cefadroxil through clathration-induced asymmetric transformation, Scheme 7.15.[82,iv] They combined pyridoxal-mediated epimerization of a mixture of cefadroxil and one of its diastereoisomers, *epi*-cefadroxil, with selective formation of an inclusion complex containing the desired stereoisomer in a cocrystal with 2,7-dihydroxynaphthalene. Epimerization in the presence of pyridoxal and concomitant clathrate formation increased the diastereomeric purity of cefadroxil from 63 to 94%. Another clathrate-induced asymmetric transformation of stereolabile *N*-nitrosamines that undergo racemization in solution at room temperature has been accomplished with

---

[iv] A clathrate is an assembly consisting of a molecule which is trapped in a cage formed by a different molecule.

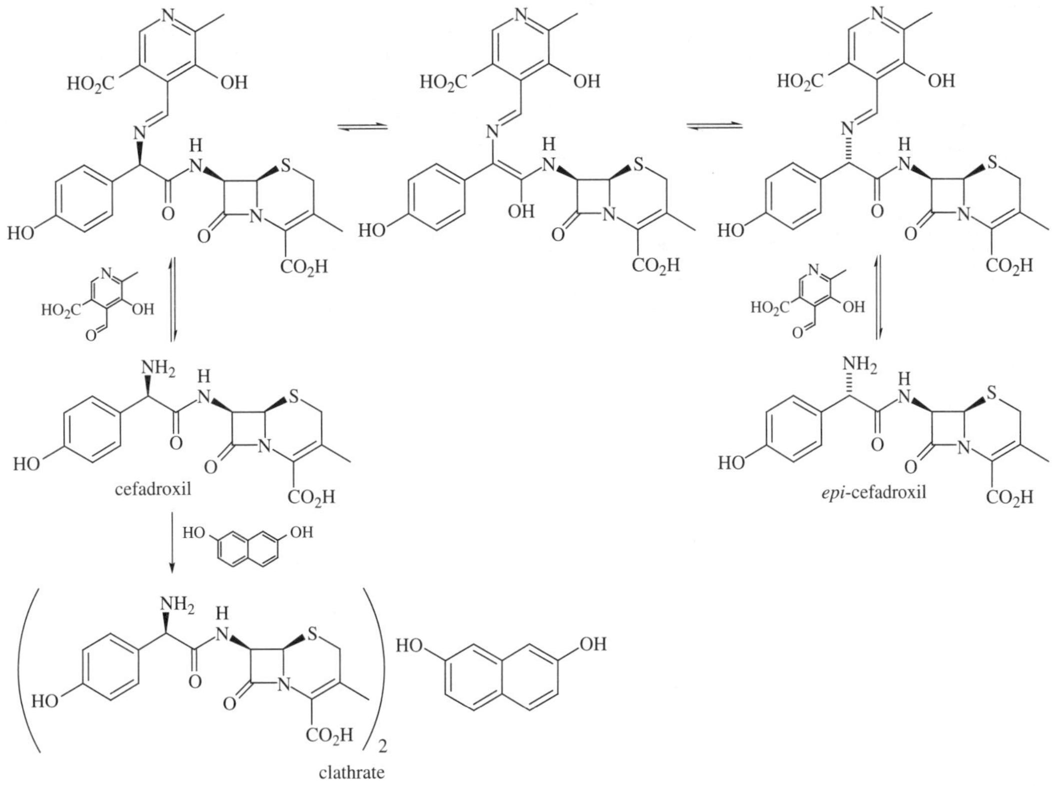

Scheme 7.15 Clathration-induced asymmetric transformation of cefadroxil.

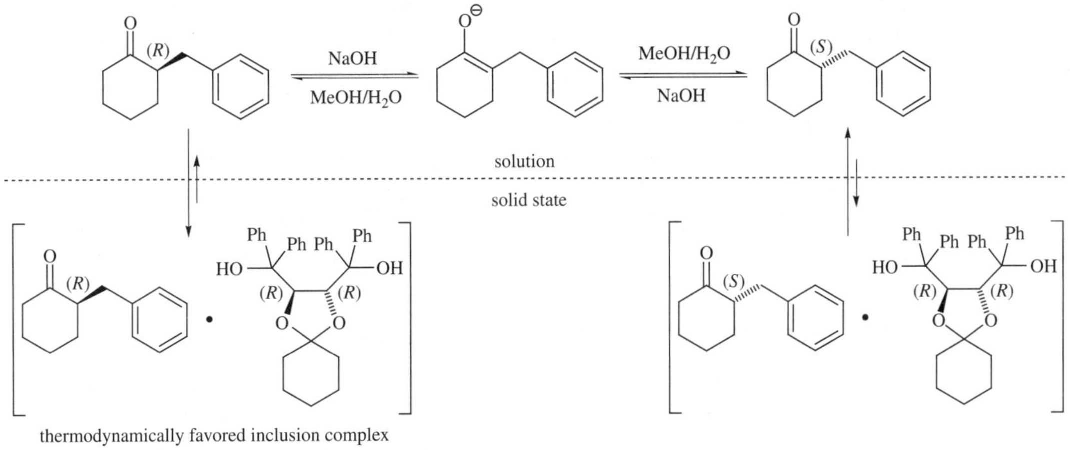

Scheme 7.16 Thermodynamically controlled deracemization of 2-benzylcyclohexanone based on enantio-conversion and formation of insoluble diastereomeric TADDOL complexes.

(*R,R*)-TADDOLs as crystal host.[83] In a conceptually similar approach that exploits stereoselective partitioning, deracemization of a solution of 2-substituted cyclohexanones is achieved by complexation with insoluble chiral host molecules.[84,85] For example, sodium hydroxide-catalyzed enantioconversion of 2-benzylcyclohexanone in aqueous methanolic solution can be coupled with

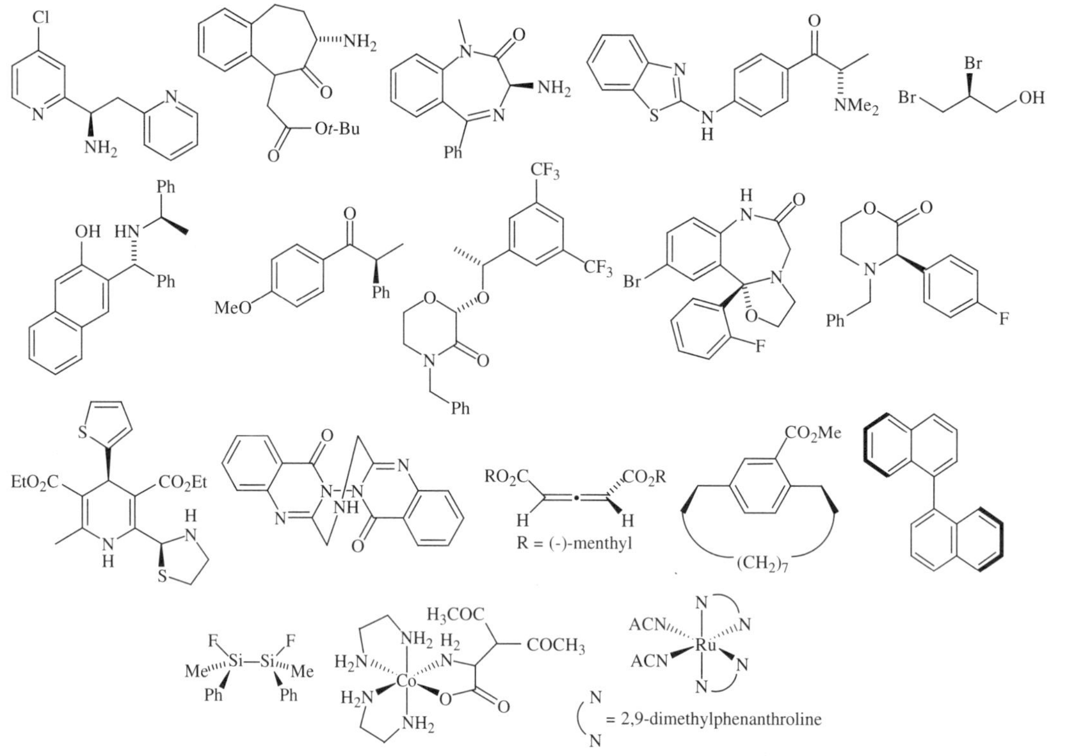

Figure 7.5   Compounds obtained by asymmetric transformation of the second kind.

enantioselective extraction of the (*R*)-enantiomer into a solid phase consisting of an (*R*,*R*)-TADDOL complexing agent, Scheme 7.16.[v] Formation of the thermodynamically favored homochiral inclusion complex of the chiral host and 2-benzylcyclohexanone causes deracemization, and the (*R*)-enantiomer is obtained in 74% ee after two days of equilibration.

In addition to the examples given above, asymmetric transformation of the second kind has been applied to a wide range of chiral compounds. This resolution technique provides invaluable access to chiral amines and amino azepinones,[86–94] alcohols,[95] amino alcohols,[96] aldehydes and ketones,[97,98] acetals,[99,100] *N,O*-ketals,[101] oxazinones,[102] thiazoles,[103] atropisomeric biaryl lactones and diquinazolinones,[104,105] allenes,[106,107] paracyclophanes,[108,109] binaphthyls and oligonaphthalenes,[110–112] helicenes,[113] meso silanes,[114,115] and chiral transition metal complexes,[116–119] Figure 7.5.

7.3.2   Asymmetric Transformation based on Chromatographic Separation

Few chromatographic processes that involve asymmetric transformation of the first or second kind to produce enantioenriched or enantiopure compounds have been reported. The possibility of deracemization based on chromatographic enantioseparation coupled to on-column enantioconversion was first mentioned by Pirkle.[120] Wolf and coworkers described a chromatographic procedure that integrates several racemization and enantioseparation steps into one HPLC process.[121] They determined the rotational energy barrier of 2,2′-diiodobiphenyl as 96.6 kJ/mol and observed that the enantiomers can be separated by HPLC on microcrystalline triacetyl cellulose with an enantioselectivity factor, α, of 5.7. In this case, high chromatographic enantioselectivity is

---

[v]TADDOLs generally consist of two adjacent *trans*-diarylhydroxymethyl groups attached to a 1,3-dioxolane ring.

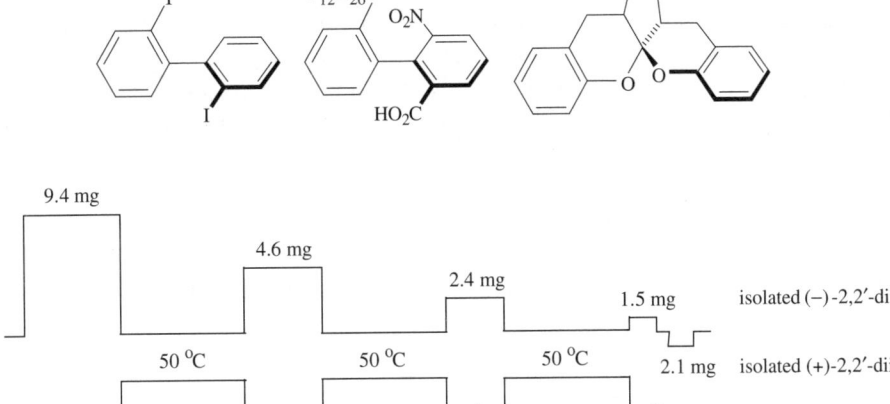

Figure 7.6 Structures of stereolabile compounds that have been employed in preparative chromatographic on-column racemization and separation processes (top). Chromatographic conditions for the deracemization of 20 mg of 2,2′-diiodobiphenyl using microcrystalline triacetyl cellulose as chiral stationary phase (bottom). A total of 17.9 mg of the (−)-enantiomer was collected in four fractions and 2.1 mg of the (+)-enantiomer was recovered.

combined with a large capacity factor for the more strongly retained (+)-enantiomer. The remarkable chromatographic selectivity and suitable capacity for the dextrorotatory enantiomer, combined with the propensity of this atropisomer to rapid rotation about the chiral axis at ambient temperatures, allowed enantioseparation and enantioenrichment by a series of on-column racemization/elution cycles, Figure 7.6. Once the less retained (−)-enantiomer was eluted from the column, the chromatographic separation process was stopped and the column temperature was increased from 8 to 50 °C for on-column racemization of the remaining (+)-2,2′-diiodobiphenyl. After cooling to 8 °C, the flow was re-adjusted to 2 mL/min to continue the enantioseparation process until elution of a further 25% of the (−)-enantiomer was completed. On-column racemization and chromatographic isolation of the less strongly retained enantiomer were repeated two more times to afford a total of 90% of the (−)-enantiomer and 10% of the finally eluted (+)-enantiomer. This procedure allows almost quantitative conversion of a racemate to a single enantiomer and eliminates the need for repetitive injection and off-column racemization. Okamoto's group later described a deracemization process that exploits on-column enantioconversion of a stereolabile chiral spiro compound.[122] Reversible [1,6]-ring opening on a Chiralcel OD column at stopped flow occurred at 40 °C and produced an enantiomeric excess of 31% favoring the more retained enantiomer at thermal equilibrium.[vi] The use of two Chiralcel OD columns coupled in series allowed separation of the enantiomers and selective on-column racemization of the less retained enantiomer at 40 °C in the second column at stopped flow, while the other enantiomer was still on the first column which was cooled to 0 °C. After continuation of the chromatographic

---

[vi] This is an example of asymmetric transformation of the first kind. Formation of diastereomeric complexes between the interconverting enantiomers and the chiral stationary phase apparently results in significant enantioenrichment.

process through both columns at low temperature, the separated enantiomers were collected in a ratio of 4:1. These examples demonstrate that coupling of enantioselective chromatography and thermal on-column deracemization of one enantiomer allows incorporation of enantiomeric enrichment and separation into a single operation. Following Wolf's and Okamoto's work, Lindner *et al.* conducted stopped-flow HPLC and CEC experiments to evaluate the stereodynamics of axially chiral 2′-dodecyl-6-nitrobiphenyl-2-carboxylic acid. They observed on-column deracemization of the racemate due to asymmetric transformation of the first kind on a quinine-derived chiral stationary phase, providing the atropisomers in 14% ee after chromatographic elution.[123]

7.4 KINETIC RESOLUTION AND DYNAMIC KINETIC RESOLUTION

Enantiomers are usually separated by enantioselective chromatography and selective crystallization of covalent, or preferably noncovalent, diastereomeric adducts formed with enantiopure resolving agents. When conventional separation attempts are unsuccessful, kinetic resolution (KR) can be a useful alternative. However, a common shortcoming of these resolution techniques is that only 50% of one enantiomer can be obtained from a racemate. Integration of racemization and KR into one process results in synthetically more attractive dynamic kinetic resolution (DKR), which in principle allows quantitative conversion of a racemic mixture to a single chiral product. By definition, both KR and DKR are confined to kinetically controlled reactions and are therefore conceptually different from asymmetric transformation of the first and second kind.

7.4.1 Kinetic Resolution

Kinetic resolution is based on the different reactivity of enantiomers during derivatization with a chiral reagent, or with an achiral reagent in the presence of an asymmetric catalyst.[vii] Because one enantiomer of a racemate reacts more rapidly than the other due to formation of diastereomeric transition states, it can be selectively transformed to a new compound if the reaction is conducted under kinetic control. For example, a racemate consisting of enantiomers A and A' can be resolved to a mixture of the remaining enantiomer A and product B' in the presence of a suitable chiral catalyst (or reagent) if $k_A \ll k_{A'}$. In this scenario, the catalyst and A' constitute a matched pair whereas A gives rise to a mismatched pair, see Chapter 5.3.2. Since A' reacts much more rapidly than A, one can recover the latter if the reaction is not allowed to proceed to completion. The two possible products B and B' can be enantiomers, as indicated in Figure 7.7, but this is not always the case. The effectiveness of KR depends on the selectivity factor, s, which is defined as the ratio of the rate constants $k_{A'}$ and k_A. The enantioselectivity of a KR is a function of the conversion c.

$$s = \frac{k_{A'}}{k_A} = e^{\Delta\Delta G^{\neq}/RT} = \frac{\ln\left[(1-c)(1-\text{ee})\right]}{\ln\left[(1-c)(1+\text{ee})\right]} \tag{7.1}$$

where ee is the enantiomeric excess of the remaining enantiomer A.

As a rule of thumb, a selectivity factor of 100 or above is required for effective resolution of both A and B'.[viii] Usually, either A or B' (but not both) can be obtained in sufficient enantiomeric excess if s is less than 100. The enantiomeric purities of the product and the remaining starting material are necessarily interdependent and it is impossible to optimize the results for both A and B'. The ideal

---

[vii] The term kinetic resolution was originally restricted to separation of enantiomeric substrates aimed at the recovery of one enantiomer while the other is consumed in an enantiomer-differentiating reaction. However, the principles of KR are often exploited for asymmetric synthesis (and not for resolution), which is most obvious in the case of parallel kinetic resolution, Chapter 7.4.1.3. In accordance with the classifications made in the original literature, processes that have been introduced as KR are presented in this chapter.

[viii] A nonselective reaction has an *s*-value of 1.

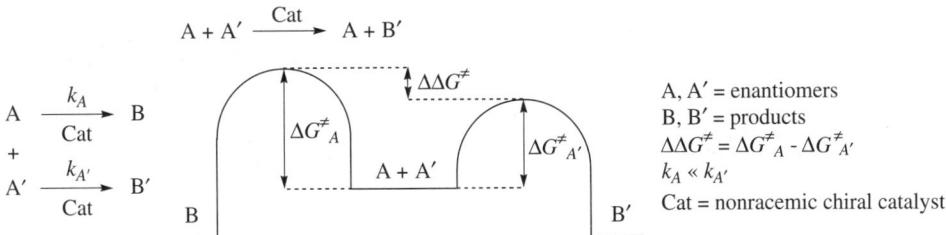

Figure 7.7 Kinetic resolution of *A* and *A'* using a chiral catalyst.

conversion, *c*, of a given KR process is governed by the selectivity, *s*, and by the choice between optimization of the enantiomeric purity of either *A* or *B'*. For example, optimization of the ee of the remaining enantiomer *A* often requires one to exceed 50% conversion, which compromises the ee of product *B'* and also the yield of *A*. To maximize the enantiopurity of the reaction product *B'*, kinetic resolutions are usually stopped before 50% conversion is reached. In accordance with the selectivity, *s*, an enantioselectivity factor, *E*, has been defined for enzymatic kinetic resolutions:

$$E = \left(\frac{k_{cat}}{K_M}\right)_R \bigg/ \left(\frac{k_{cat}}{K_M}\right)_S = \frac{\ln\left[(1-c)(1-ee)\right]}{\ln\left[(1-c)(1+ee)\right]} \tag{7.2}$$

where K_M is the Michaelis–Menten constant.

7.4.1.1 Enzyme-catalyzed Kinetic Resolution. Since Pasteur discovered the first kinetic resolution in 1858 by treating tartaric acid with fermenting yeast, a plethora of enzyme-catalyzed processes with transferases, oxidoreductases and hydrolases has been reported.[124–126] Kinetic resolution of amino acids has received considerable attention due to potential biomedical applications and their usefulness as chiral synthons or auxiliaries in asymmetric synthesis. Many examples, including industrial applications using acylases, amidases, hydantoinases, and esterases, can be found in the literature.[127–131]

In particular, lipases have been exploited for kinetic resolution of alcohols, carboxylic acids and other classes of compounds because this family of enzymes tolerates a wide range of unnatural substrates, maintains catalytic activity and selectivity in organic solvents, does not require a cofactor, and is readily available.[132] Racemic alcohols are frequently resolved by lipase-catalyzed transesterification using vinyl acetate or isopropenyl acetate as acyl donor. The advantage of these reagents is that thermodynamically stable and easily removable side products (acetaldehyde and acetone) are obtained. For example, *Candida antarctica* lipase-catalyzed KR of propargylic alcohols in the presence of vinyl acetate yields a mixture of approximately 50% of (*S*)-propargylic acetate, similar amounts of the remaining (*R*)-propargylic alcohol, and acetaldehyde, Scheme 7.17.[133] The transfer of the acetyl group is followed by spontaneous tautomerization of vinyl alcohol to acetaldehyde, which provides the driving force for the esterification process. The enantiomers of numerous chiral primary,[134–137] secondary[138–146] and tertiary alcohols[147,148] including hydroxy butenolides,[149] spiropentane and dispiroheptane methanols,[150] allenes,[151] and fluorohydrins[152] have successfully been resolved by lipase-catalyzed transesterification, Figure 7.8. A continuous-flow KR of 1-phenylethanol using immobilized lipase and vinyl acetate in super-critical carbon dioxide has been reported.[153] Enzymatic KR of alcohols with lipases is quite practical but other methods such as sulfatase-catalyzed hydrolysis of *sec*-alkyl sulfates are viable alternatives.[154]

Several lipases, including *Candida antarctica* lipase (CAL), have been employed in kinetic resolutions of carboxylic acids,[155] esters,[156,157] diols,[158,159] lactams,[160] amino alcohols,[161,162] amino acids and esters,[163–167] oxaziridines,[168] and amines, Figure 7.9.[169] Gotor and others reported CAL-catalyzed amidation of racemic 1,2-diaminocyclohexane to the corresponding (*R*,*R*)-bisamidoester

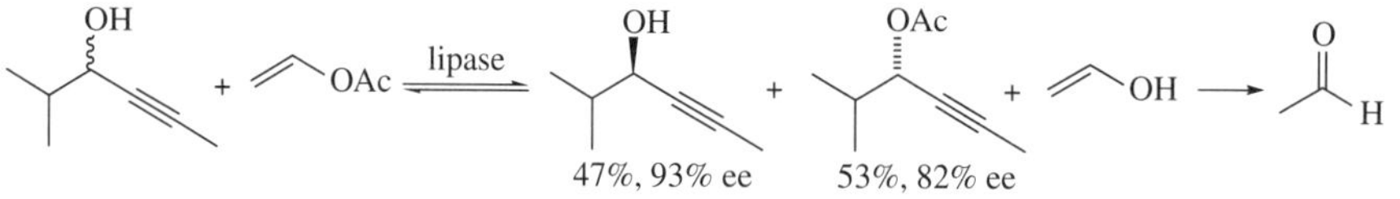

Scheme 7.17 Lipase-catalyzed KR of propargylic alcohols with vinyl acetate.

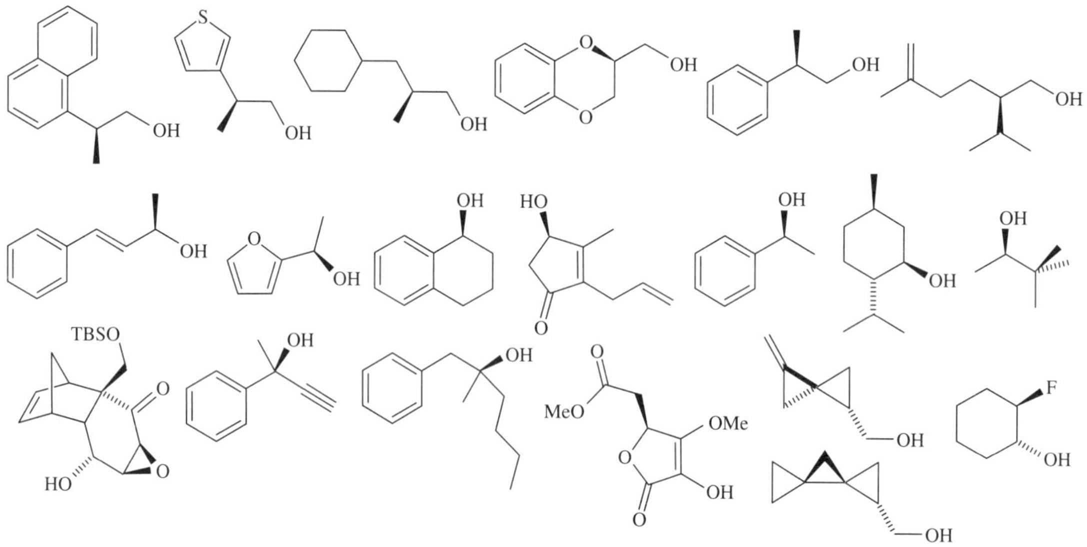

Figure 7.8 Structures of alcohols resolved by lipase-catalyzed transesterification.

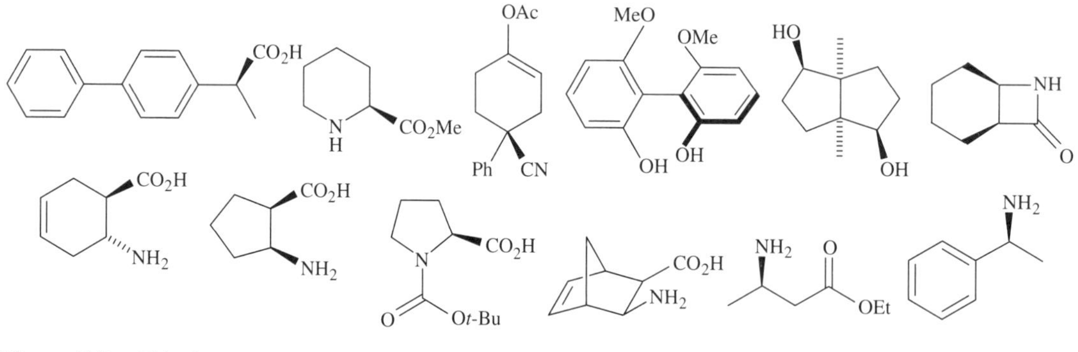

Figure 7.9 Chiral compounds resolved by lipase-catalyzed KR.

with dimethyl malonate as acyl donor, Scheme 7.18.[170–172] This KR involves two consecutive amidations with *E*-values of 45 and 68, respectively. Through careful optimization of the amount of acyl donor and substrate conversion, a process was developed that affords 43% of essentially enantiopure (*S,S*)-1,2-diaminocyclohexane and 57% of the (*R,R*)-bisamidoester in 97% ee, while only minute amounts of the intermediate monoamidoester are formed. Enzymatic and microbial KR has been applied to 2-hydroxy carboxylic acids,[173] amino acids,[174] and even unnatural substrates including metallocenes exhibiting a chiral plane.[175] A general drawback of enzymatic resolution is the inherently high substrate specificity. Reetz and others have shown that this

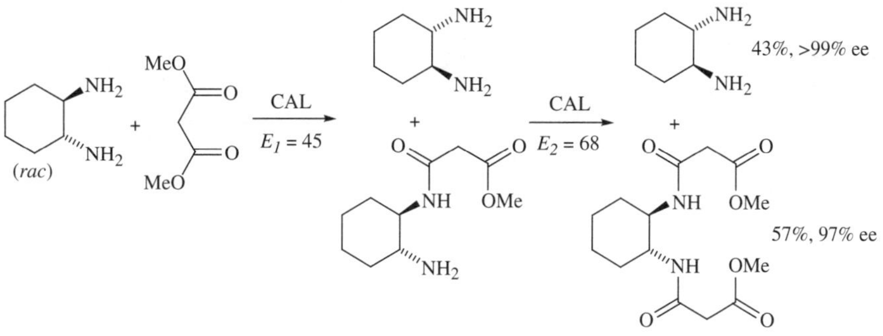

Scheme 7.18 Enzymatic transesterification of 1,2-diaminocyclohexane with CAL.

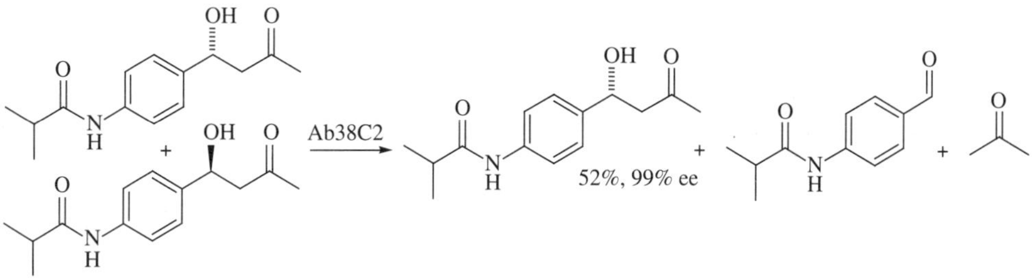

Scheme 7.19 Kinetic resolution of β-hydroxy ketones by Ab38C2-catalyzed retro aldol reaction.

limitation can be overcome by bioengineering of enzymatic properties via site-specific mutagenesis or directed evolution based on random mutagenesis and protein expression in conjunction with high-throughput screening.[176–178] Alternatively, KR can be achieved with antibody-catalyzed aldol reactions, transesterifications and ester hydrolyses.[179–186] The antibody Ab38C2 has been known to convert prochiral aldehydes in the presence of acetone to β-hydroxy ketones with excellent enantioselectivity. According to the principle of microscopic reversibility, Ab38C2 should not only be a useful catalyst for asymmetric aldol reactions but also applicable to KR of aldols. Indeed, treatment of racemic β-hydroxy ketones with Ab38C2 furnishes literally enantiopure starting material due to enantioselective retro aldol reaction, Scheme 7.19.[187]

Imaginative enantioconvergent processes that overcome the impractical limitation of 50% theoretical yield inherent to traditional KR have been described. Faber and others coupled biocatalytic KR of racemic epoxides with nonenzymatic hydrolysis exhibiting complementary stereoselectivity to achieve quantitative formation of highly enantioenriched vicinal diols.[188–191] Resolution of racemic *cis*-2,3-epoxyheptane by epoxide hydrolase from bacterial *Nocardia* EH1 results in selective conversion of the (2*S*,3*R*)-enantiomer to (2*R*,3*R*)-heptane-2,3-diol. Kinetic studies and labeling experiments with H$_2$^{18}O revealed that under basic conditions the remaining (2*R*,3*S*)-epoxide undergoes regioselective hydrolytic ring opening at *C*-3 to the same (2*R*,3*R*)-diol, Scheme 7.20. The combination of KR and enantioconvergent hydrolysis of the remaining enantiomer produces (2*R*,3*R*)-heptane-2,3-diol in virtually quantitative amounts and 97% ee. This concept has been utilized for the synthesis of a variety of chiral vicinal diols from racemic epoxides.[192–195] Chadha and Baskar developed a biocatalytic deracemization procedure for α-hydroxy esters such as ethyl 2-hydroxy-4-phenylbutanoate and ethyl mandelate using intact cells of *Candida parapsilosis*.[196] It is assumed that the enantioconvergent stereoinversion involves oxidation of the (*R*)-α-hydroxy ester to the corresponding 2-keto ester, which is probably catalyzed

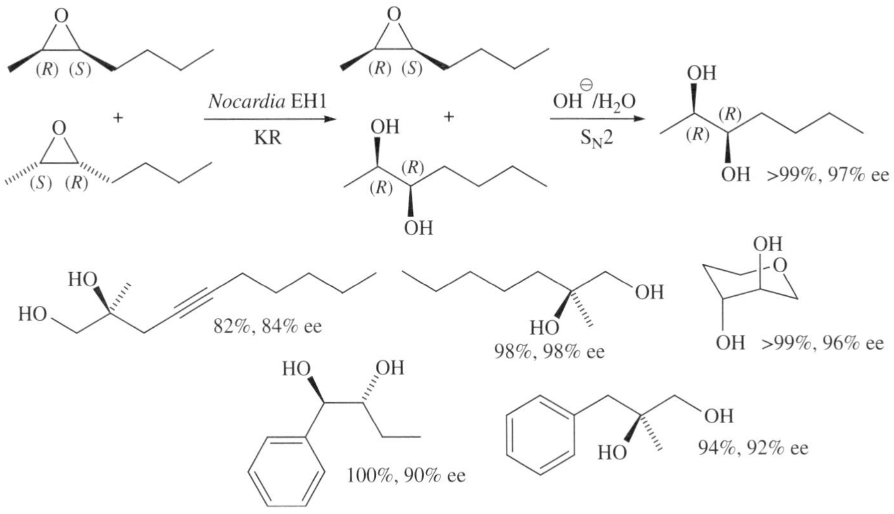

Scheme 7.20 Kinetic resolution and enantioconvergent *in situ* hydrolysis of racemic *cis*-2,3-epoxyheptane (top) and structures of vicinal diols obtained from epoxide precursors based on KR and subsequent hydrolysis (bottom).

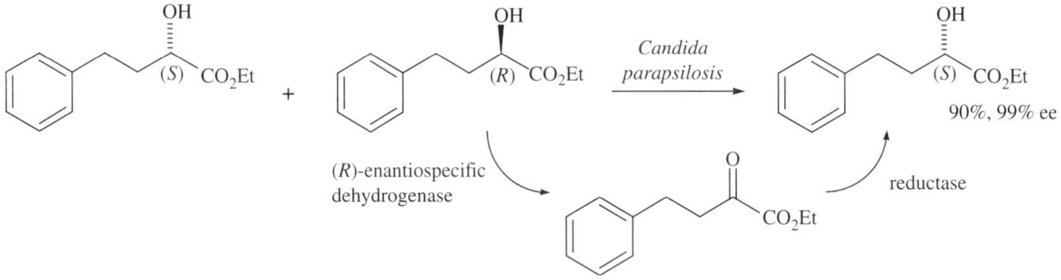

Scheme 7.21 Biocatalytic deracemization of ethyl 2-hydroxy-4-phenylbutanoate with *Candida parapsilosis*.

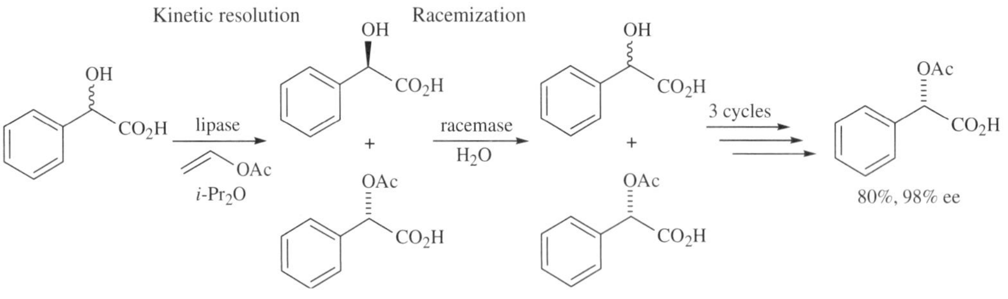

Scheme 7.22 Two-step enzymatic KR and racemization of mandelic acid.

by an NAD$^+$-dependent dehydrogenase. The achiral intermediate is then converted to the (*S*)-enantiomer in the presence of an NADPH-dependent reductase, Scheme 7.21.

Deracemization can also be achieved by combination of KR and external racemization of the isolated undesired enantiomer. The racemate is then recycled into another KR process, and so on.[197] Strauss and Faber showed that such a laborious approach can be incorporated into a one-pot procedure.[198] For example, deracemization of mandelic acid is feasible by lipase-catalyzed

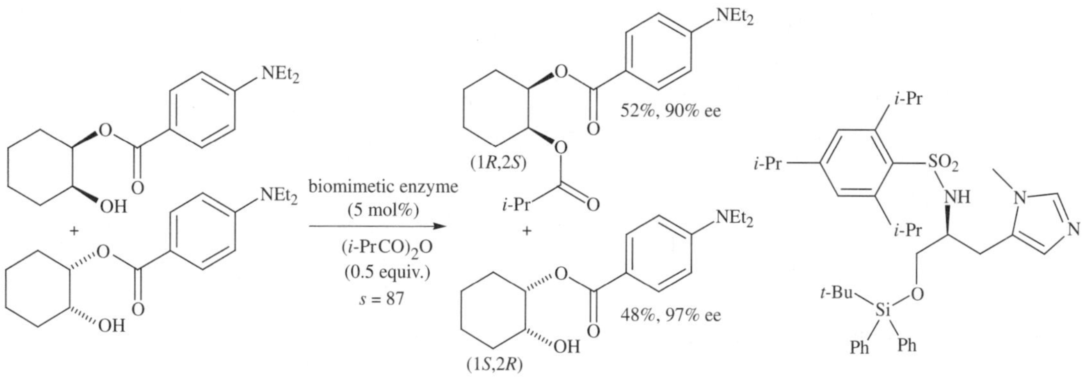

Scheme 7.23 Biomimetic KR of *cis*-1-(4-diethylaminobenzoyl)-2-hydroxyhexane (left) and structure of the artificial enzyme used (right).

acetylation in diisopropyl ether followed by enzymatic racemization of (*R*)-mandelic acid in aqueous solution. Although the two enzymes require different reaction conditions and have to be removed after each step, the desired transesterification product, (*S*)-*O*-acetyl mandelic acid, and the remaining substrate, (*R*)-mandelic acid, do not have to be separated. Employment of racemic mandelic acid in four KR/racemization sequences affords (*S*)-*O*-acetyl mandelic acid in 98% ee and 80% yield, Scheme 7.22.[ix]

7.4.1.2 Nonenzymatic Kinetic Resolution. The attractive features of lipase-catalyzed biotransformations have intensified the search for biomimetic catalysts applicable to KR. In particular, acylase[199–204] and hydrolase[205] mimics have been employed in the resolution of chiral alcohols and esters. Ishihara *et al.* developed a simple (*S*)-histidine-derived artificial acylase for KR of secondary alcohols, Scheme 7.23.[206] Treatment of *cis*-1-(4-diethylaminobenzoyl)-2-hydroxyhexane with 0.5 equivalents of isobutyric anhydride in the presence of 5 mol% of the artificial enzyme allowed preferential acylation of the (1*R*,2*S*)-alcohol. The (1*S*,2*R*)-enantiomer was recovered in 97% ee at 52% conversion which corresponds to a selectivity factor, *s*, of 87.

Many enantioselective catalysts and chiral auxiliaries originally developed for asymmetric synthesis are suitable to kinetic resolution.[207,208] One of the most impressive examples, that certainly rivals enzymatic methods, is the application of Sharpless' asymmetric epoxidation (AE) of allylic alcohols with *tert*-butyl hydroperoxide (TBHP) in the presence of catalytic amounts of Ti(O*i*-Pr)$_4$ and diisopropyl tartrate (DIPT)[209] to KR of chiral substrates, Scheme 7.24.[210–216] The matched pair consisting of the (*S*)-allylic alcohol and the L-(+)-tartrate-derived catalyst generates an epoxy alcohol, whereas the mismatched pair, *i.e.*, L-(+)-DIPT/Ti(O*i*-Pr)$_4$ and the (*R*)-alcohol, is substantially less reactive. The (*R*)-alcohol can therefore be recovered in excellent enantiomeric excess at approximately 50% conversion. In general, (*E*)-1,2- and 2,2-disubstituted allylic alcohols provide better results than (*Z*)-1,2-disubstituted substrates and selectivities, *s*, up to 700 have been observed. The Sharpless KR protocol has been applied to a variety of other alcohols including furyl alcohols,[217–219] cyanohydrins[220] and amino alcohols[221,222] which are converted to *N*-oxides. It is noteworthy that Sharpless' KR has been employed in the total synthesis of (+)-grandisol,[223] koromicin,[224] (+)-methynolide,[225] and laulimalide,[226] a component of the aggregation pheromone of the male cotton boll weevil.

Since chiral cyclic allylic alcohols are not good candidates for Sharpless' AE, Noyori and coworkers developed a complementary method based on (BINAP)ruthenium-catalyzed

---

[ix] The development of a more practical dynamic kinetic resolution of mandelic acid, based on simultaneous use of lipase and mandelate racemase in one reaction mixture, is not possible because the enzymes require incompatible reaction conditions.

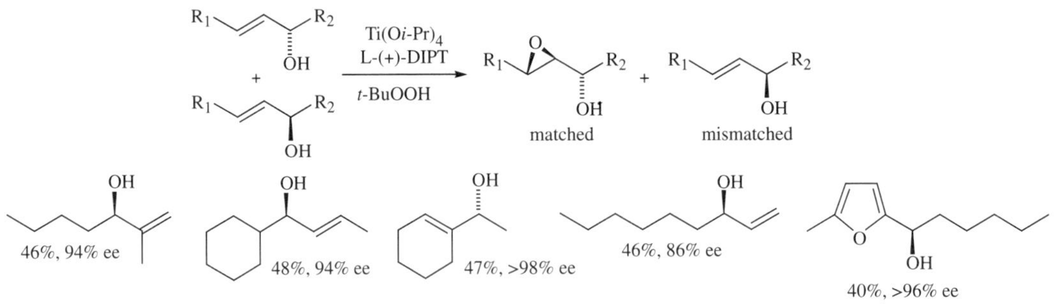

Scheme 7.24 Sharpless' AE of secondary allylic alcohols.

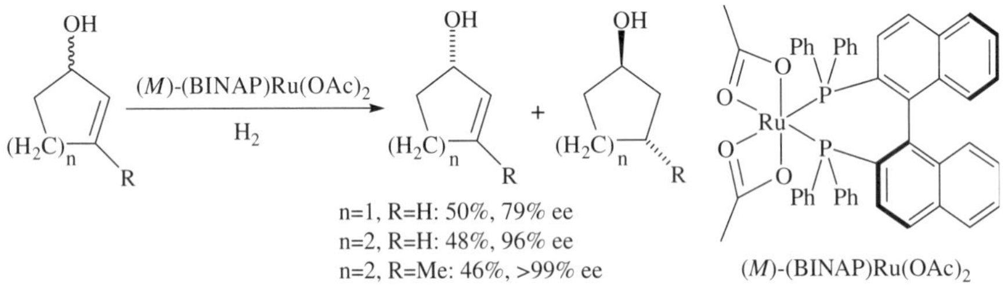

Scheme 7.25 Noyori's reductive KR of cyclic allylic alcohols.

enantioselective hydrogenation.[227] Using ruthenium diacetate in the presence of (*M*)-BINAP, they recovered unreacted (*S*)-allylic alcohols in remarkable ee at approximately 50% conversion, Scheme 7.25. A drawback of this procedure is that it generates a mixture of saturated and unsaturated alcohols, which complicates isolation of the unreacted enantiomer. This problem can be avoided with an oxidative Ru-catalyzed KR protocol generating an achiral ketone that is more easily separated from the remaining alcohol.[228,229] Following the pioneering work of Evans and Vedejs, who utilized stoichiometric amounts of chiral acylating reagents for KR of alcohols,[230,231] Fu *et al.* introduced a broadly applicable planar chiral ferrocene-derived DMAP acylation catalyst that affords excellent results with benzylic, propargylic and allylic secondary alcohols, Scheme 7.26.[232,233] Since then, several catalytic processes that allow kinetic resolution of chiral alcohols have been developed.[234–245,x]

Nonfunctionalized terminal epoxides can be conveniently prepared in racemic form from readily available alkene precursors, but separation of enantiomers by classical crystallization methods or enantioselective chromatography remains a difficult task. Jacobsen and coworkers therefore employed a (salen)chromium(III) complex, which can also be used for desymmetrization of meso epoxides, in the kinetic resolution of chiral epoxides.[246] A range of terminal and 2,2′-disubstituted epoxides undergo [(*R*,*R*)-(salen)]CrN_3-catalyzed ring opening in the presence of $TMSN_3$ with selectivity factors as high as 280, Scheme 7.27.[247,248] This reaction produces chiral 1-azido-2-siloxy derivatives that can be transformed to synthetically versatile amino alcohols. Other nucleophiles such as indoles are also suitable to (salen)Cr-catalyzed KR of epoxides.[249] Since the corresponding (salen)cobalt(III) complex catalyzes epoxide ring opening with water, Jacobsen refined the method described above to a hydrolytic kinetic resolution (HKR) process exhibiting selectivities, *s*, up to 630, Scheme 7.28.[250–257] A broad variety of functionalized terminal epoxides has been resolved to

[x] In many cases, however, chiral secondary alcohols can be prepared with superior atom economy by enantioselective reduction of ketones or alkylation of aldehydes.

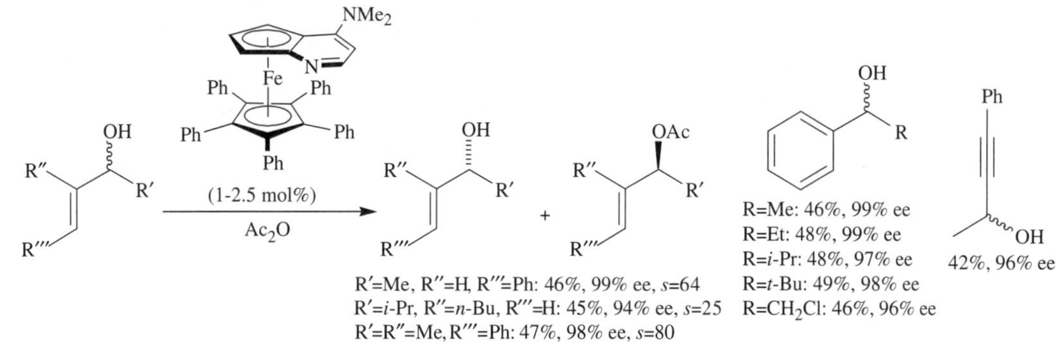

Scheme 7.26 Enantioselective acylation of allylic and propargylic alcohols.

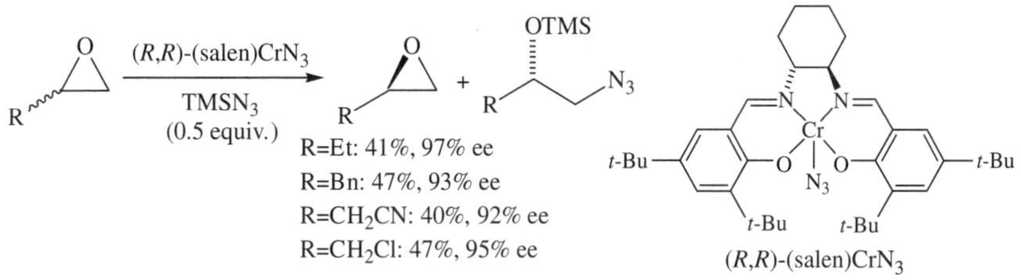

Scheme 7.27 Jacobsen's KR of terminal epoxides with [(*R*,*R*)-(salen)]Cr(III).

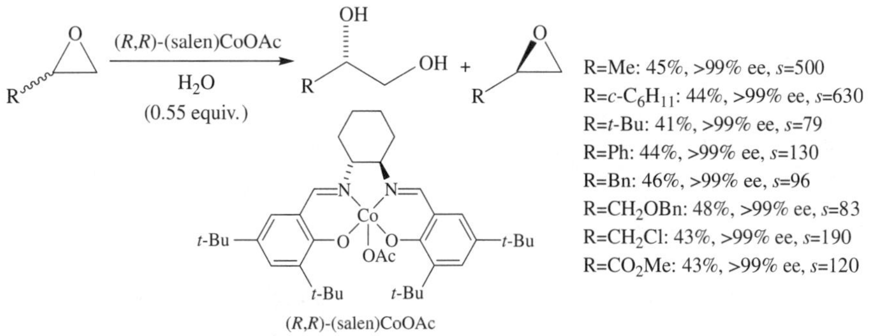

Scheme 7.28 HKR of terminal epoxides with (*R*,*R*)-(salen)cobalt(III).

literally enantiopure materials by this method.[258,259] The usefulness of KR is most apparent when enantiopure compounds can not be obtained by other means. Chiral epoxides can be prepared through powerful asymmetric reactions such as Sharpless epoxidation of allylic alcohols, Jacobsen epoxidation of conjugated olefins with a (salen)Mn(II) catalyst, and Shi epoxidation of nonfunctionalized alkenes with a readily available fructose-derived catalyst.[260–262] Despite the broad success of these methods, enantiopure epoxides are usually not obtained without further purification. Since simple epoxides tend to be liquid and do not undergo strong interactions with commonly used chiral resolving agents or chiral stationary phases, separation by enantioselective chromatography, crystallization of diastereomeric adducts or preferential crystallization is often not feasible. Accordingly, HKR of epoxides, which are very useful intermediates in natural product synthesis, has become a viable alternative to traditional resolution methods. Hydrolytic KR has

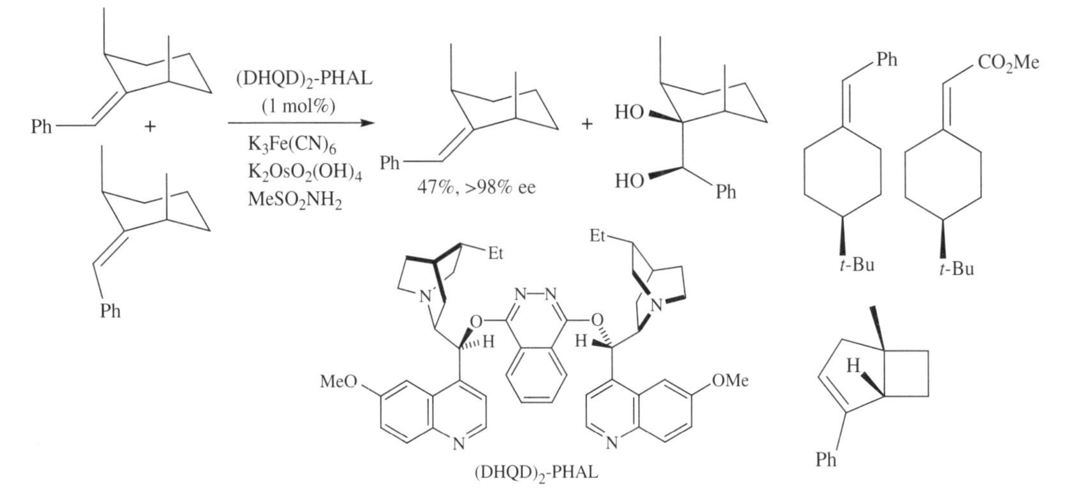

Scheme 7.29 KR of chiral trisubstituted olefins based on Sharpless AD and structures of recovered enantiomers.

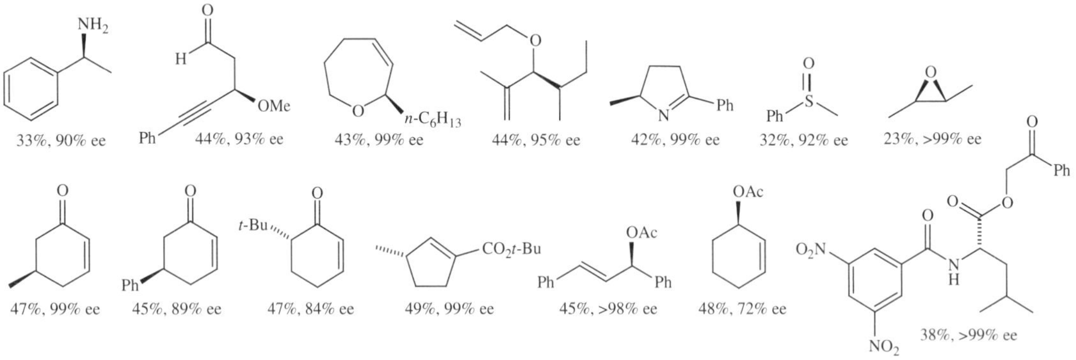

Figure 7.10 Structures of enantiomers obtained by kinetic resolution.

been incorporated into the total synthesis of epothilone,[263] laulimalide,[264] fostriecin,[265] (–)-pyreno-phorin,[266] carquinostatin A,[267] bryostatins,[268] ulapalide,[269] and (–)-mycalolide.[270]

Sharpless' renowned cinchona alkaloid-catalyzed asymmetric dihydroxylation (AD) of olefins has been extended to kinetic resolution.[271] Although cinchona alkaloids are usually sensitive to existing chirality in alkenes,[xi] AD has successfully been applied to a few substrates, including chiral fullerenes and axially chiral alkenes, which are very difficult to resolve by other means, Scheme 7.29.[272–277] Oxidative KR of alkenes via selective epoxidation with either Shi's fructose-derived catalyst[278] or (salen)manganese complexes[279,280] has also been reported. Some alkenes, *e.g.*, 1-substituted 1,2-dihydronaphthalenes, have been resolved by reductive KR based on hydrobora-tion.[281,282] Other practical protocols involving either a chiral catalyst or a chiral auxiliary have been developed for alkanes,[283] alkyl halides,[284] amines,[285–288] alcohols,[289–292] aldehydes,[293] ethers,[294–298] aziridines,[299] imines and pyrrolines,[300–302] sulfoxides,[303,304] epoxides,[305–307] allenes,[308,309] ke-tones,[310–312] esters,[313–319] carbamates,[320–322] and lactones,[323,324] Figure 7.10.

[xi] This is an example of a reaction course with significant reagent-based stereocontrol, see Chapter 5.3.2.

The principles of KR can also be exploited for analytical purposes. Horeau employed an excess of racemic 2-phenylbutyric anhydride or the corresponding acyl chloride to convert substoichiometric amounts of a chiral alcohol of unknown configuration to a mixture of diastereomeric esters. Kinetic resolution gives rise to preferential esterification of one anhydride enantiomer while the other one is enantiomerically enriched. Horeau found that hydrolysis of the remaining nonracemic reagent and subsequent polarimetric analysis of the carboxylic acid mixture can be used to deduce the absolute configuration of the resolving alcohol. This method has been applied to a variety of chiral secondary alcohols and primary alcohols that are chiral only by virtue of isotopic substitution (RCHDOH).[325,326] Siuzdak, Finn and coworkers developed a high-throughput screening method for determination of the enantiomeric composition of amines and alcohols based on kinetic resolution with pseudoenantiomeric mass-tagged chiral acylating agents in conjunction with mass spectrometric analysis.[327]

7.4.1.3 Parallel Kinetic Resolution. The outcome of KR is determined by the rate of the two competing reactions and conversion of the starting material. A major shortcoming of KR is that the enantiomeric purity of the product decreases at conversion close to 50%, when the concentration and consequently the rate of consumption of the more reactive enantiomer is reduced while the *relative* reaction rate of the less reactive enantiomer increases significantly. Because of the interdependence of the ee of the product and the remaining starting material, it is impossible to maximize the enantiomeric purity of both. For example, to optimize the purity of the less reactive enantiomer one must exceed 50% conversion and thus compromise the ee of the product. Since the ee of the remaining enantiomer steadily increases with conversion, one can not maximize both enantiomeric purity and recovery of the substrate.

To overcome these limitations, Vedejs and Chen introduced a strategy called parallel kinetic resolution (PKR). In PKR both enantiomers are consumed simultaneously by two independent reactions that ideally proceed with similar rates, Figure 7.11.[328] In contrast to KR, both enantiomers of the starting material are converted to nonenantiomeric products by different reactions with complementary enantiodifferentiation and one can obtain products with substantially improved enantiomeric purity at 100% conversion, even in the case of relatively low selectivity factors. Vedejs and Chen utilized pseudoenantiomeric DMAP-derived resolving agents for dual kinetic resolution of secondary arylalkyl alcohols. Although the individual stereoselectivity factors of the two chiral *N*-alkoxycarbonylpyridinium salts shown in Scheme 7.30 for the enantiomers of 1-(1-naphthyl)ethanol are only 41 and 42, respectively, the corresponding carbonates are produced in 88% and 95% ee at 98% conversion. For comparison, the recovery of 49% of one enantiomer in 95% ee by traditional KR requires a stereoselectivity factor of at least 125.

In theory, PKR of enantiomers *A* and *A'* based on two independent reactions with a complementary selectivity, *s*, of 49 gives products *B* and *C* in 96% ee at quantitative conversion. Formation of a mixture containing product *B* and the remaining starting material *A'* ($k_A \gg k_{A'}$) in the same enantiomeric excess by traditional KR would require a selectivity factor of 200 at 50% conversion. As a rule of thumb, PKR is advantageous when the selectivity factor is below 150 and when dynamic kinetic resolution or KR coupled with enantioconvergent *in situ* transformation of the less reactive enantiomer is not feasible.[329]

Based on the structural relationship of the products, PKR has been classified into chemo-, regio- and stereodivergent processes.[330] The PKR of a secondary alcohol with stoichiometric amounts of

$$A \xrightarrow{k_A} B$$
$$+$$
$$A' \dashrightarrow{k_{A'}} B'$$

A, A' = enantiomers
B, B' = (enantiomeric) products
$k_A \gg k_{A'}$

$$A \xrightarrow{k_A} B$$
$$+$$
$$A' \xrightarrow{k_{A'}} C$$

A, A' = enantiomers
B, C = nonenantiomeric products
$k_A \approx k_{A'}$

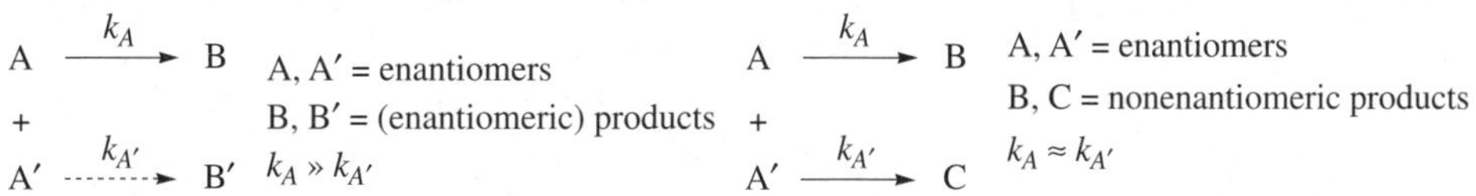

Figure 7.11 Principles of KR (left) and PKR (right).

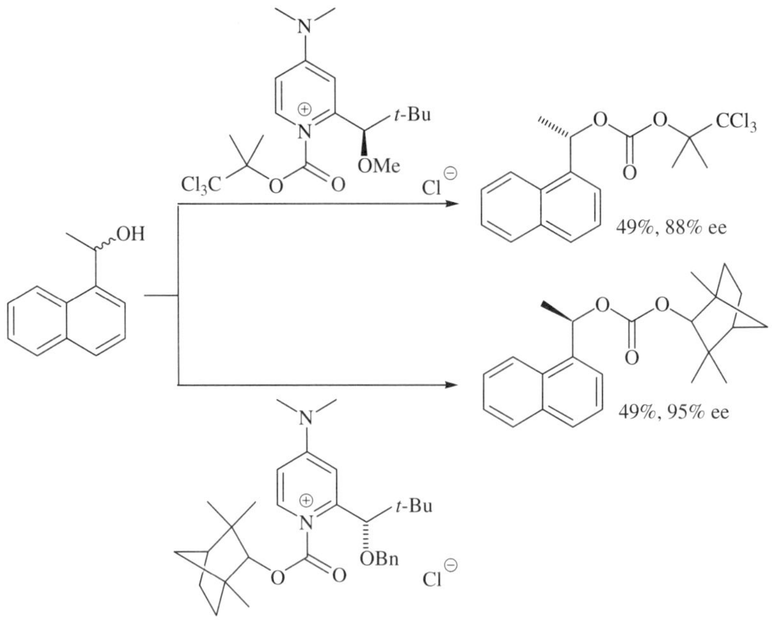

Scheme 7.30   Parallel kinetic resolution of a secondary arylalkyl alcohol using stoichiometric amounts of quasienantiomeric resolving agents.

two chiral *N*-alkoxycarbonylpyridinium salts discussed above affords nonisomeric products and thus constitutes a chemodivergent process. Several other catalytic chemodivergent resolutions are known.[331–335] Davies and coworkers developed a PKR protocol for methyl 5-alkyl-cyclopentene-1-carboxylates that is based on conjugate addition of pseudoenantiomeric lithium *N*-α-methyl-benzylamide-derived salts. They also demonstrated the feasibility of dual KR of structurally related 3-alkyl-cyclopentene-1-carboxylates and 2-amino-5-*tert*-butyl-cyclopentene-1-carboxylates.[336–338] Parallel kinetic resolution of an activated ester of 2-phenylpropionate has been achieved with stoichiometric amounts of pseudoenantiomeric oxazolidinones.[339] It must be emphasized that PKR gives rise to a mixture of nonenantiomeric products that still need to be separated, which is not necessarily a trivial task. The practicality of PKR can be improved when stereoselective chemodivergent resolution is combined with derivatizing agents exhibiting quite different chromatographic properties.[340] Tanaka and Fu observed that racemic 4-alkynals can be kinetically resolved through enantioselective cyclization with a DUPHOS-derived rhodium complex. This procedure yields a chiral cyclopentenone in 93% ee and recovered starting material in 72% ee.[341] Further investigation and catalyst screening revealed that chemodivergent PKR of the same starting material gives rise to a chiral cyclobutanone if a BINAP-derived rhodium complex is used, Scheme 7.31.[342] This finding is remarkable because of the surprising effect of the chiral catalyst structure on the reactivity and fate of the enantiomeric alkynals, and the simultaneous formation of two synthetically challenging cycloalkanones. Chemodivergent PKR can also produce a mixture of a chiral and an achiral product. For example, treatment of a 3-oxo-7,7-(ethylenedioxy)bicyclo[3.3.0]octane-2-carboxylate with baker's yeast leads to reduction of one enantiomer to the corresponding *syn*-hydroxy ester, while the other enantiomer is transformed to an achiral 7,7-(ethylenedioxy)bicyclo[3.3.0]octane-3-one by enzymatic hydrolysis and subsequent decarboxylation.[343] Apparently, each enantiomer of this β-keto ester is recognized by a different enzyme, *i.e.*, an oxidoreductase and an esterase, Scheme 7.32.

Regiodivergent resolutions either involve competing reaction pathways that give rise to regioisomers or examples of positional selectivity, in which one molecule displays similar functional

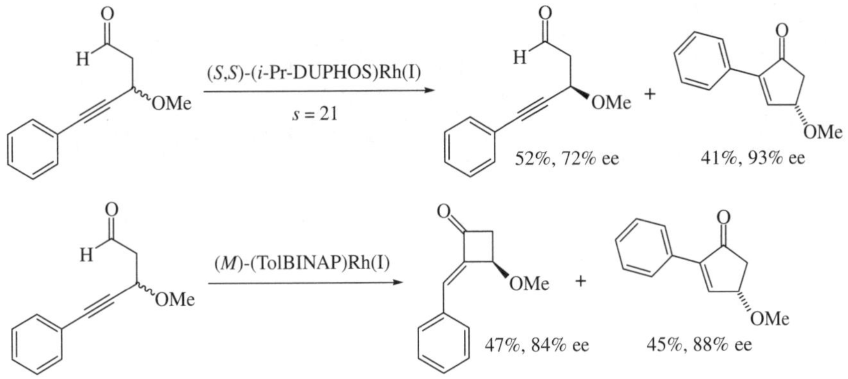

Scheme 7.31   KR (top) and PKR (bottom) of racemic 4-alkynals with chiral rhodium catalysts.

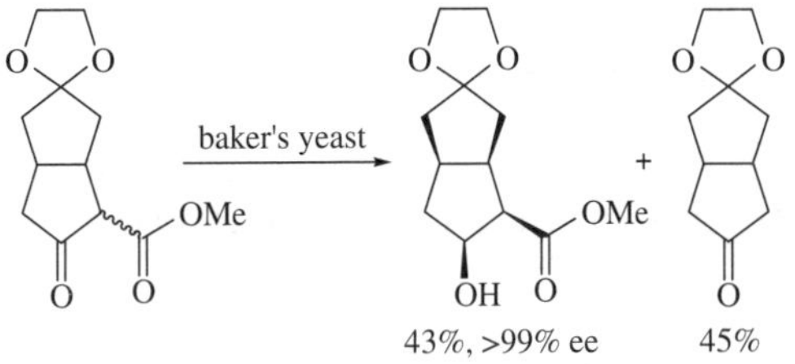

Scheme 7.32   Biocatalytic chemodivergent PKR of 3-oxo-7,7-(ethylenedioxy)bicyclo[3.3.0]octane-2-carboxylate.

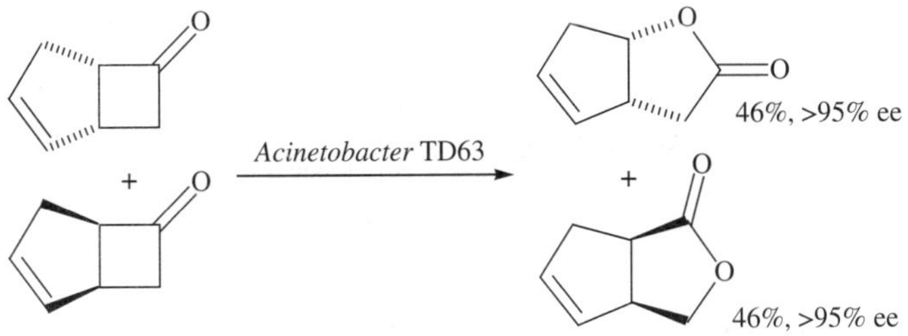

Scheme 7.33   Regiodivergent PKR of bicyclo[3.2.0]hept-2-en-6-one with *Acinetobacter* TD63.

groups that can participate in the same reaction. This resolution strategy has been applied in a variety of reactions including catalytic aerobic Baeyer–Villiger oxidation of cyclobutanones,[344] Sharpless epoxidation,[345,346] cinchona alkaloid-catalyzed alcoholysis of chiral succinic anhydrides,[347] and zirconocene-catalyzed ring opening of dihydrofurans.[348] Furstoss observed regiodivergent biocatalytic Baeyer–Villiger oxidation of bicyclic ketones with whole cells of *Acinetobacter* cultures.[349] For instance, racemic bicyclo[3.2.0]hept-2-en-6-one is almost quantitatively converted to two regioisomeric lactones by incubation with *Acinetobacter* TD63, Scheme 7.33. It has been hypothesized that both oxidations are catalyzed by the same monooxygenase. The enzyme differentiates between the enantiomeric ketones and favors migration of different carbon–carbon bonds during Baeyer–Villiger oxidation due to a change in the orientation of peroxidic intermediates in the active site of the monooxygenase.

Martin *et al.* employed Rh$_2$[(5S)-MEPY]$_4$, a useful catalyst originally designed by Doyle for asymmetric cyclopropanation of prochiral alkenes,[350] in the resolution of a racemic diallyl

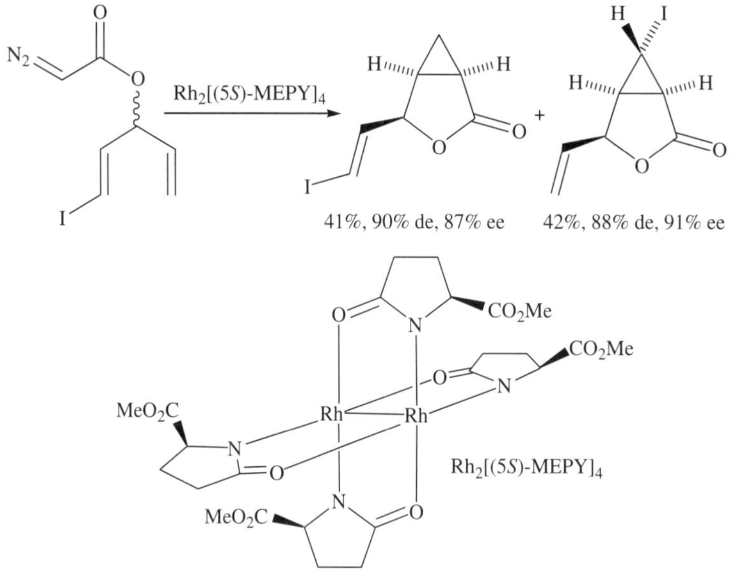

Scheme 7.34   Rh$_2$[(5S)-MEPY]$_4$-catalyzed intramolecular cyclopropanation of a racemic diallyl diazo-acetate.

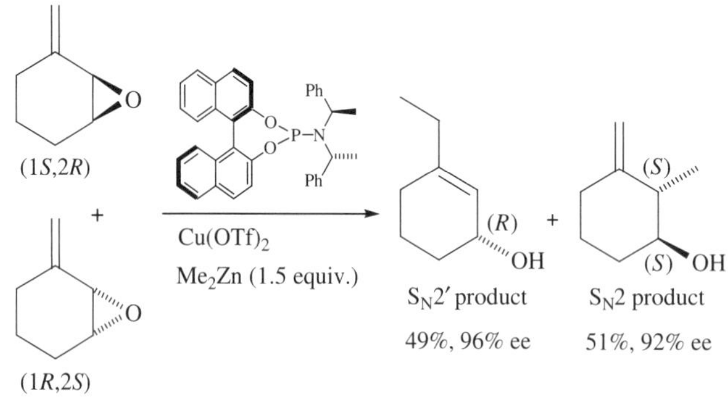

Scheme 7.35   Regiodivergent PKR of 1,2-epoxy-3-methylenecyclohexane with dimethylzinc catalyzed by an (M,R,R)-phosphoramidite-derived copper complex.

diazoacetate possessing two distinct double bonds.[351] The chiral rhodium complex catalyzes cyclization of each enantiomer and activates a different double bond with excellent *endo*-, enantio- and diastereoselectivity. This procedure gives access to multifunctionalized bicyclic compounds, Scheme 7.34. Pineschi and coworkers utilized Feringa's (M,R,R)-phosphoramidite-derived copper complex for regiodivergent addition of dialkylzinc reagents to vinyl oxiranes, Scheme 7.35.[352,353] The complex effectively differentiates between the enantiomers of 1,2-epoxy-3-methylene-cyclohexane and catalyzes S$_N$2 and S$_N$2′ reaction with organozinc compounds. While treatment of 1,2-epoxy-3-methylenecyclohexane with 0.5 equivalents of dimethyl- or diethylzinc establishes a KR, the use of excess amounts of dialkylzinc results in PKR and provides two constitutional isomers. The (1R,2S)-epoxide forms an (R)-allylic alcohol in 49% and 96% ee based on (M,R,R)-phosphoramidite copper-catalyzed S$_N$2′ attack and the (1S,2R)-enantiomer undergoes S$_N$2 reaction to an (S,S)-alcohol having an exocyclic double bond.

The complete conversion of a racemate to a mixture of diastereoisomeric products by complementary enantiomer-differentiating reactions constitutes a stereodivergent PKR. This has been

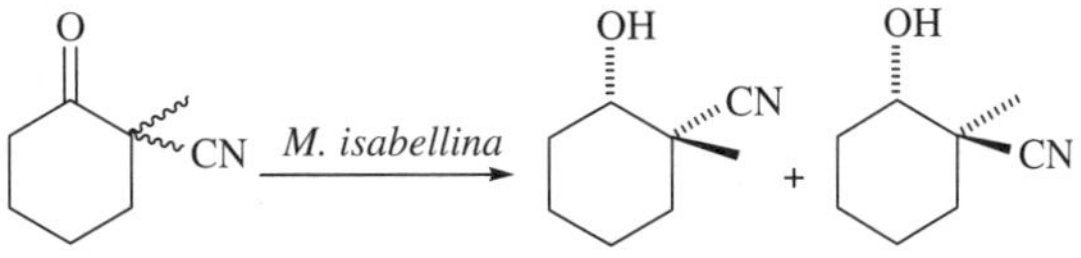

58%, 73% ee   42%, >99% ee

Scheme 7.36   Diastereodivergent PKR of 1-methyl-2-oxocyclohexane carbonitrile with fungus *Mortierella isabellina* NRRL1757.

realized in Horner–Wadsworth–Emmons reactions[354,355] and in a oxazaborolidine-catalyzed reduction of a steroidal ketone.[356] Alternatively, one can employ isolated enzymes or whole cells that promote selective biotransformation of enantiomers to diastereomeric products. This has been demonstrated with enzymatic reductions of racemic ketones[357] and cyanations of chiral aldehydes.[358] Dehli and Gotor examined the use of microorganisms for diastereodivergent bioreduction of racemic 1-methyl-2-oxocycloalkane carbonitrile, Scheme 7.36.[359] Incubation of a culture of fungus *Mortierella isabellina* NRRL1757 gave a mixture of the diastereomeric β-hydroxy nitriles with 99% and 73% ee, respectively.

7.4.2 Dynamic Kinetic Resolution

Dynamic kinetic resolution (DKR) is based on rapid racemization of a chiral compound in conjunction with irreversible conversion of one enantiomer to a chiral product. Since dynamic kinetic resolutions obey the Curtin–Hammett principle, the ratio of the products formed by the enantiomer-differentiating step is determined by the difference in the free activation energies, $\Delta\Delta G^{\neq}$. In contrast to kinetic resolution, DKR can provide quantitative amounts of enantiopure materials and the ee of the product does not change with conversion. A basic requirement for efficient DKR is a selectivity factor, E or s, of at least 30. In addition, the rate constant for racemization, k_{rac}, should be higher than the rate constant for the faster of the two conversions, $k_{A'}$, or at least 10 times higher than k_A in cases of very high selectivity. The energy profile of a typical DKR obeying Curtin–Hammett kinetics is illustrated in Figure 7.12. In this example, the catalyst and A' constitute a matched pair, whereas A gives rise to a mismatched pair, see Chapter 5.3.2. As has been pointed out for kinetic resolutions, products B and B' can be enantiomeric but this is not a prerequisite for DKR, *vide infra*.

Asymmetric transformation of the first or second kind and DKR are conceptually different methods. While the former is based on equilibration and thermodynamic reaction control, the latter takes advantage of different reactivities of stereolabile enantiomers and therefore requires kinetic control. Dynamic KR is aimed at complete conversion of a racemate to an enantiopure product and is closely related to asymmetric *synthesis*.[xii] In contrast, asymmetric transformations of the first and second kind are used for *resolution* of stereoisomers. Because of the inherently high atom economy and enantioconvergence, numerous DKR procedures have been developed to date and applied in the synthesis of complex natural products. Only racemizations proceeding under mild reaction conditions compatible with the consecutive resolution step are suitable for DKR. Most common DKR processes are based on thermal racemization, base- or acid-catalyzed racemization, racemization via redox reactions, enzymatic racemization, and formation of stereolabile intermediates such as organolithium or transition metal complexes. By analogy with KR, the

---

[xii] The term DKR was originally reserved for processes involving interconversion of enantiomers. Since the same principles can be exploited for stereoselective transformation of interconverting diastereomers, the original definition is often not strictly followed in the literature. However, the latter is better described as dynamic kinetic asymmetric transformation, see Chapter 7.5.

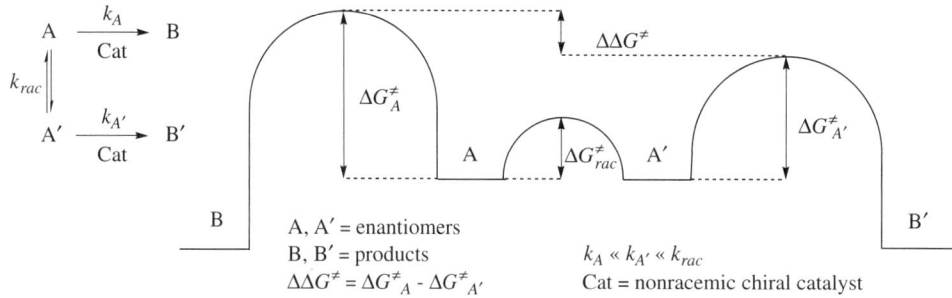

Figure 7.12 Illustration of dynamic kinetic resolution using a chiral catalyst.

resolution step can involve bioorganic methods or the use of nonenzymatic chiral catalysts or reagents.

7.4.2.1 Enzyme-catalyzed Dynamic Kinetic Resolution. Many dynamic kinetic resolutions involve transition metal-catalyzed racemization coupled to an enantiomer-differentiating biotransformation.[360–365] Because lipases accept a broad variety of substrates, operate in organic solvents and do not require cofactors, the use of these readily available enzymes greatly facilitates screening of KR conditions compatible with fast substrate racemization. Transition metal-catalyzed racemization and lipase-catalyzed resolution have been combined for DKR of secondary alcohols,[366–370] monoprotected 1,2-diols, hydroxy acetals and hydroxy esters,[371] amines,[372] and allylic alcohols or acetates.[373–375] Pàmies and Bäckvall incorporated lipase-catalyzed KR and ruthenium-catalyzed racemization of δ-hydroxy esters into a one-pot process.[376] Using lipase PS-C, *i.e.*, immobilized lipase isolated from *Pseudomonas cepacia*, exhibiting E values greater than 360 and 4-chlorophenyl acetate as acyl donor in the presence of a ruthenium complex, they obtained (*R*)-δ-acetoxy esters in up to 92% and 98% ee, Scheme 7.37.[xiii] Similar procedures have been developed for a range of secondary alcohols,[377–379] β-halo alcohols,[380] β-azido alcohols,[381] α-hydroxy esters,[382] β-hydroxy esters,[383] β-hydroxy nitriles,[384] β-hydroxy alkanephosphonates,[385] and γ-hydroxy esters and amides.[386,387]

In the case of allylic esters, one can use tetrakis(triphenylphosphine)palladium(0) and lipase from *Candida antarctica* (CAL) and *Pseudomonas cepacia* (PCL) to achieve racemization and concomitant resolution based on selective ester cleavage, Scheme 7.38.[388] The lipase differentiates between the enantiomers and transfers the acetyl unit from the (*R*)-substrate to isopropyl alcohol. The racemization probably proceeds via nucleophilic acetyl displacement and formation of an intermediate π-allylpalladium species.[389] Addition of 1,1'-bis(diphenylphosphino)ferrocene to the palladium catalyst effectively reduces side reactions such as acetate elimination, and high yields and ee's are obtained, albeit at the expense of long reaction times. Alternatively, reduction of prochiral ketones to rapidly racemizing secondary alcohol intermediates has been combined with enantioselective transesterification to afford chiral acetates.[390,391] Enzymatic acylation is not restricted to alcohols. Palladium-catalyzed hydrogenation of ketoximes generates amines that undergo transition metal-catalyzed racemization and enantioselective acylation in the presence of *Candida antarctica* lipase B (CALB).[392] Kita's group introduced a lipase-catalyzed domino process that includes DKR of chiral alcohols exhibiting a diene moiety and an acyl donor providing an electron-deficient dienophile for subsequent intramolecular Diels–Alder reaction.[393] For example, 3-vinyl-cyclohex-2-en-1-ol undergoes rapid ruthenium-catalyzed racemization and resolution with a functionalized ethoxyvinyl ester in the presence of CALB, Scheme 7.39. The combination of racemization, resolution and cyclization provides a powerful one-pot procedure that furnishes

---

[xiii] See Chapter 3.2.8 for the mechanism of ruthenium-catalyzed racemization of secondary alcohols.

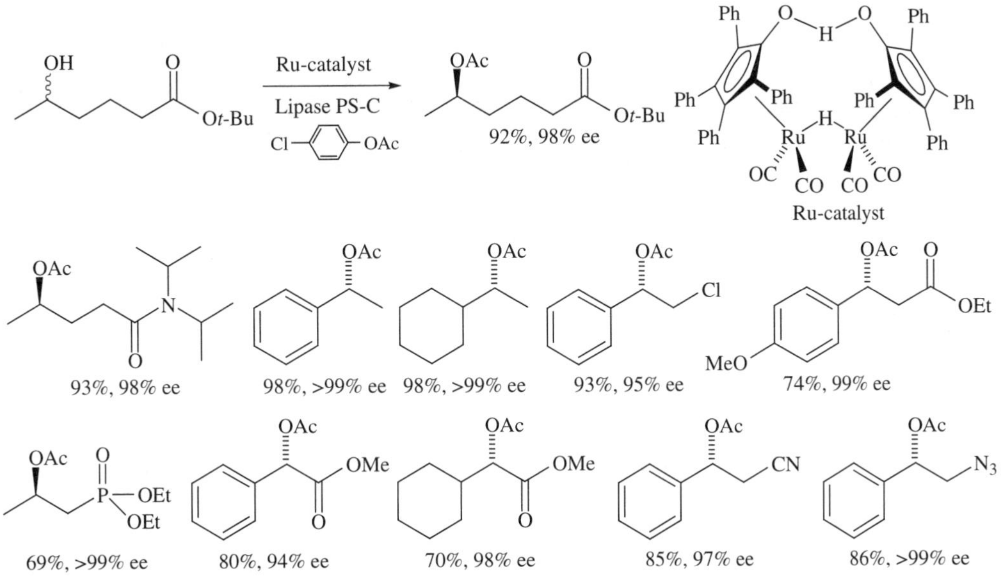

Scheme 7.37 Bäckvall's DKR of δ-hydroxy esters based on Ru-catalyzed alcohol racemization and lipase-controlled resolution.

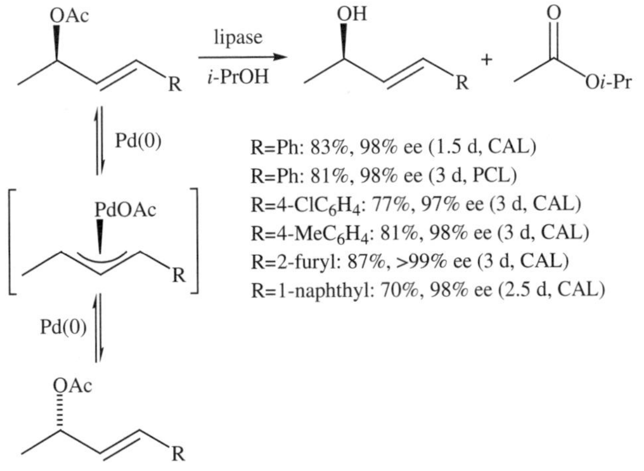

Scheme 7.38 Pd-catalyzed racemization and enzymatic resolution of allylic acetates.

complex tricyclic lactones with four chiral centers. Incorporation of the same lipase and acyl donor into DKR of cyclic α-hydroxy nitrones followed by intramolecular 1,3-dipole cycloaddition gives chiral tetrahydrofuro[3,4-c]isoxazoles.[394]

Lipase-catalyzed KR has been coupled to other types of racemization processes including ring opening and subsequent cyclication of N-acyl hemiaminals,[395] dissociation and recombination of cyanohydrins in the presence of amberlite,[396,397] zeolite-mediated alcohol racemization,[398] and base-promoted interconversion of chiral oxazolones and oxazolines.[399–401] Chiral thioesters possess fairly acidic α-protons and racemize under mild conditions.[402] A large-scale DKR protocol that takes advantage of base-catalyzed racemization of an isoxazoline-derived thioester was reported by Pesti.[403] Deprotonation of the adjacent thioester moiety facilitates racemization via a retro

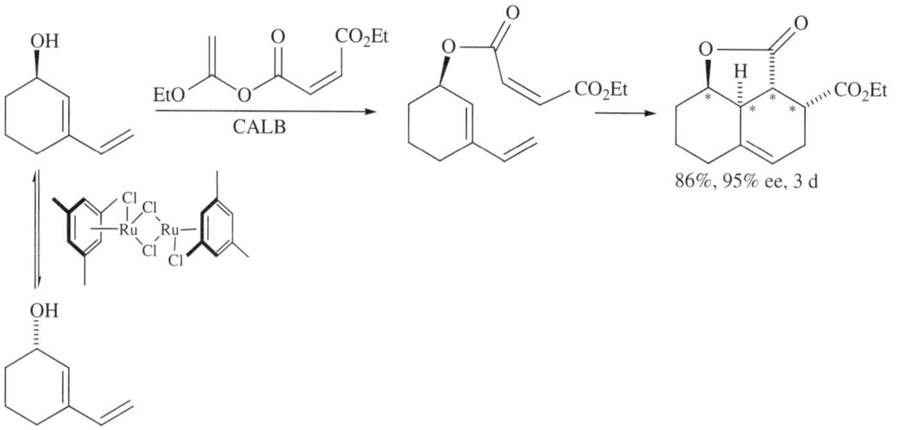

Scheme 7.39   Domino DKR of 3-vinylcyclohex-2-en-1-ol involving racemization, resolution and intramolecular cycloaddition.

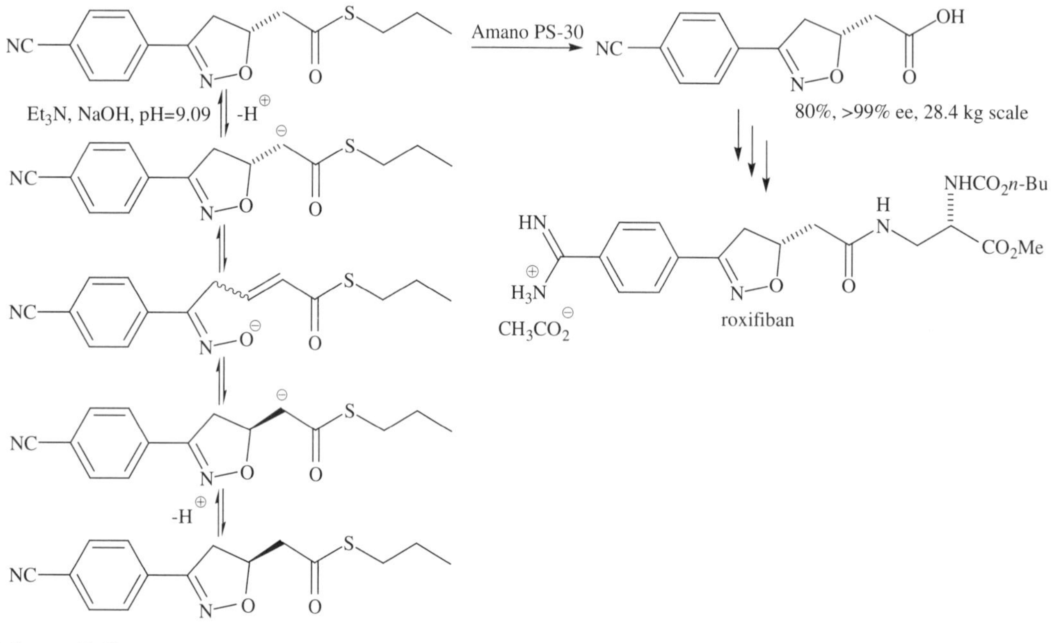

Scheme 7.40   DKR of an isoxazoline thioester.

Michael/Michael addition sequence followed by Amano PS 30 lipase-catalyzed resolution, Scheme 7.40. The isolated isoxazoline is a key intermediate in the synthesis of roxifiban, a promising candidate for the treatment of thrombotic diseases. This industial application highlights the efficiency and robustness of DKR.

Acetylation of interconverting hemithioacetals yields α-acetoxy sulfides from aldehydes and thiols.[404] Racemic hemithioacetals are readily formed *in situ* and can be resolved with *Pseudomonas fluorescens* lipase. Dynamic KR is accomplished by selective formation of the (*S*)-α-acetoxy sulfide because the (*R*)-hemithioacetal is not recognized by the enzyme and racemizes via silicon dioxide-promoted dissociation and subsequent recombination, Scheme 7.41. Racemization of α-bromo and α-chloro esters via reversible halide displacement has been coupled to enantioselective enzymatic

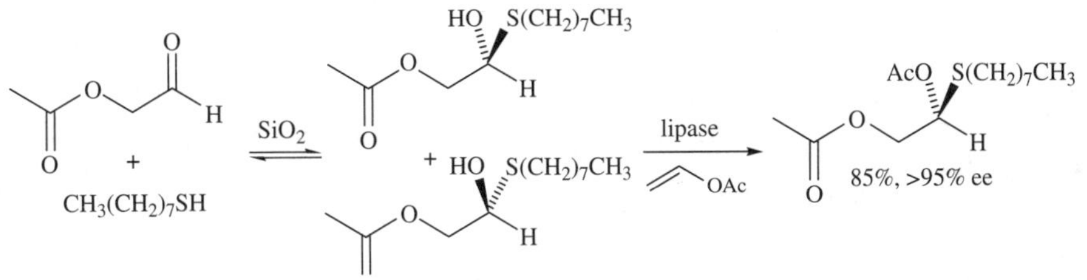

Scheme 7.41 Synthesis of α-acetoxy sulfides through lipase-catalyzed acetylation of interconverting hemithioacetals.

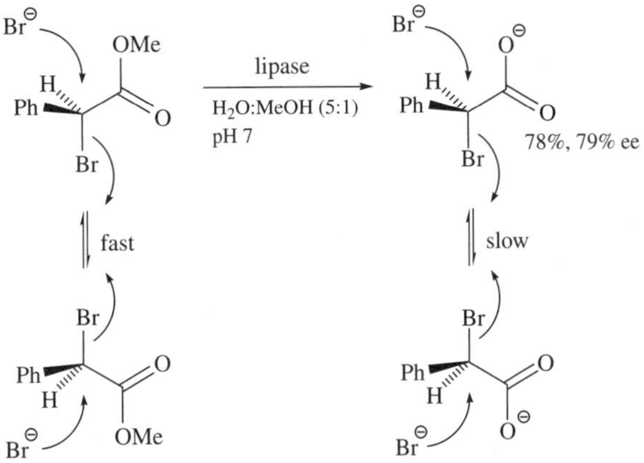

Scheme 7.42 DKR based on bromide displacement and enzymatic hydrolysis.

hydrolysis.[405,406] At neutral pH, a similar S_N2 attack at the corresponding α-halo carboxylate is relatively slow due to electrostatic repulsion between the negatively charged substrate and the halide. Dynamic kinetic resolution of methyl 2-bromo-2-phenylacetate using *Candida rugosa* lipase and polymer-linked phosphonium bromide as the halide source generates (*S*)-2-bromo-2-phenyl-acetic acid in 79% ee, Scheme 7.42. Although the product is considerably more stable to racemization than the racemic starting materials, the enantiopurity decreases over time and the reaction has to be stopped after 4.5 hours at only 78% conversion to maintain an enantiomeric excess of at least 79%.

Pyridoxal-catalyzed racemization of phenylglycine methyl ester via Schiff base formation and enantioselective ammonolysis with *Candida antarctica* lipase B provides (*R*)-phenylglycine amide in 88% ee at 85% conversion, Scheme 7.43.[407] The phenylglycine ester reacts with pyridoxal to an intermediate Schiff base that racemizes in the presence of ammonia, while the (*R*)-amide generated by lipase-catalyzed resolution precipitates and is no longer available for imine formation and racemization. Dynamic KR of amino acids based on pyridoxal 5-phosphate- and 4-chlorobenzal-dehyde-catalyzed racemization has also been reported.[408,409]

The applicability of intact microorganisms and enzymes other than lipases to DKR further expands the application spectrum of this technique to a wide range of chiral building blocks. Important examples are alcohols,[410] hydantoins,[411,412] 2-oxo-cycloalkanecarbonitriles,[413] β-keto esters, amides and lactones,[414–417] 2-benzenesulfonyl cycloalkanones,[418] and β-hydroxy esters.[419] Furstoss' group utilized base-catalyzed racemization of 2-benzyloxymethylcyclopentanone and resolution due to microbiological Baeyer–Villiger oxidation with recombinant *E. coli* for the synthesis of (*R*)-6-benzyloxymethyltetrahydropyran-2-one, Scheme 7.44.[420,421] Since racemic ox-azol-5(4*H*)-ones, thiazolin-5-ones and hydantoins are readily prepared and susceptible to enantio-conversion under basic conditions or in the presence of a racemase, they have become useful

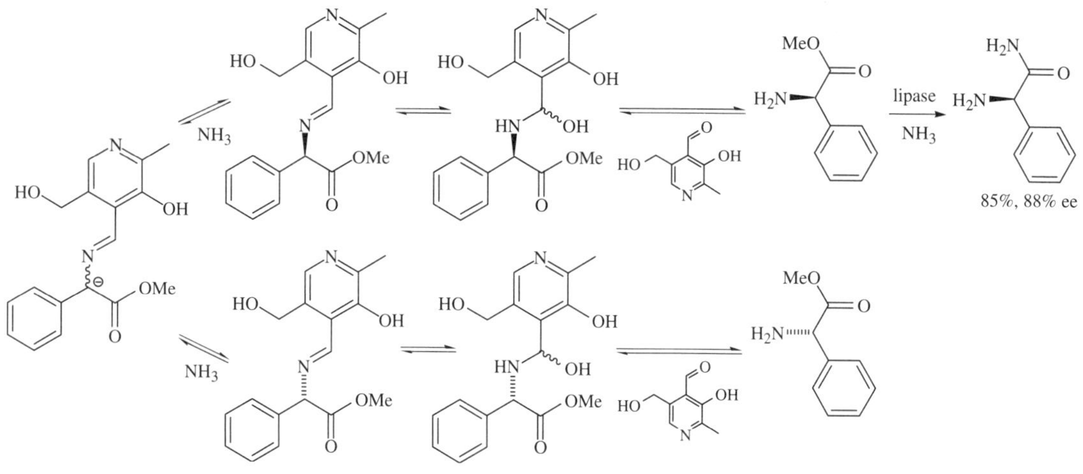

Scheme 7.43 DKR of phenylglycine methyl ester via Schiff base-catalyzed racemization.

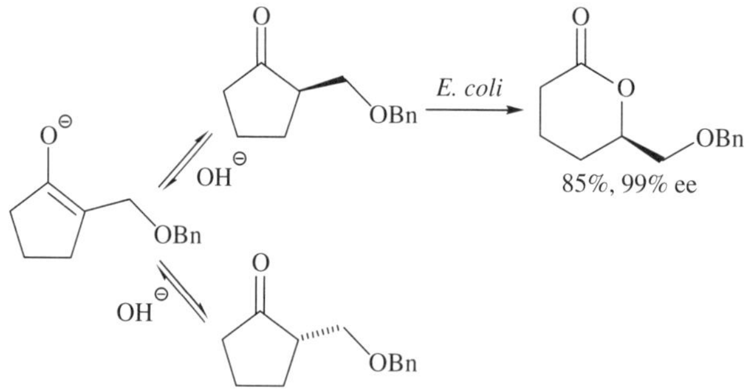

Scheme 7.44 Microbiological Baeyer–Villiger oxidation and DKR of 2-benzyloxymethylcyclopentanone.

substrates for asymmetric synthesis of α-amino acids.[422] The availability of complementary enzymes such as (R)- and (S)-hydantoinase significantly enhances the practicality of DKR. Combination of racemization of a 5-substituted hydantoin with hydantoinase-catalyzed hydrolytic ring opening can afford either enantiomer of the corresponding N-carbamoyl amino acid. Treatment with nitrous acid or addition of a carbamoylase then generates the free amino acid, Scheme 7.45. A similar approach allows conversion of racemic oxazol-5(4H)-ones and thiazolin-5-ones to N-protected amino acids.

Several examples of DKR involving enzyme catalysis in both the racemization and the resolution step are known. In most cases, a combination of a hydrolase and a racemase is used to produce (R)-amino acid derivatives.[423–425] Epibromohydrin racemizes via formation of 1,3-dibromo-2-propanol and participates in enzymatic resolution in the presence of sodium bromide, sodium azide and haloalcohol dehalogenase HheC isolated from *Agrobacterium radiobacter* AD1. The enzyme catalyzes both reactions, interconversion of the enantiomers of epibromohydrin via a meso haloalcohol and enantioselective nucleophilic epoxide ring opening with azide.[426] Through careful optimization of pH and concentration of the azide and bromide nucleophiles, a procedure providing (S)-1-azido-3-bromo-2-propanol in 94% ee was developed, Scheme 7.46. The suitability of dual enzyme DKR to industrial production has been demonstrated by incorporation of a

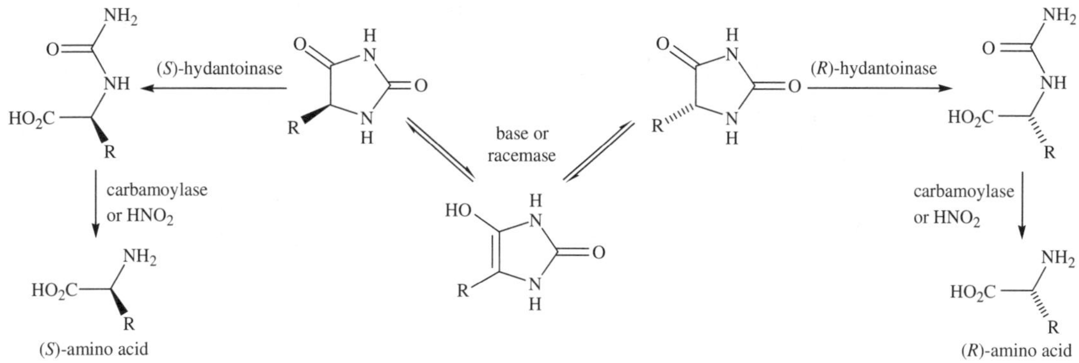

Scheme 7.45 Asymmetric synthesis of α-amino acids.

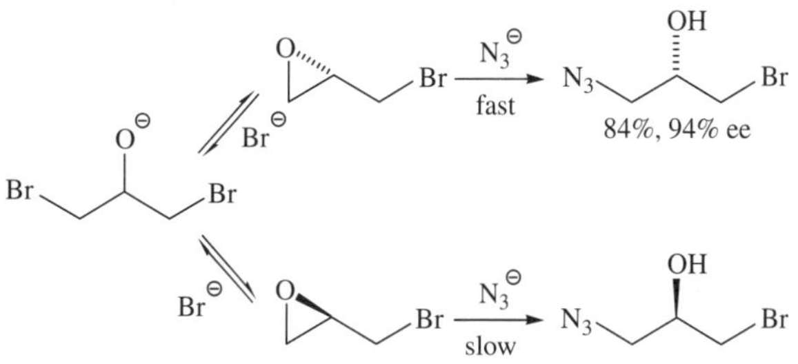

Scheme 7.46 Haloalcohol dehalogenase-catalyzed DKR of epibromohydrin.

racemase and a hydantoinase or acylase into the large scale synthesis of several natural and unnatural (*S*)- and (*R*)-amino acids.[427,428]

7.4.2.2 Nonenzymatic Dynamic Kinetic Resolution.

Weygand *et al.* were probably first to recognize the potential of DKR as early as 1966, but it took another 30 years before the concept and scope of this methodology received significant attention.[429] Since then, the development of many dynamic kinetic resolutions has been inspired by some of the KR processes discussed in Chapter 7.4.1, and by the well-known propensity of azlactones and other important chiral building blocks to racemization.[430–433]

In 1989, Noyori and coworkers reported that chiral β-keto esters carrying either α-alkyl or α-amido substituents readily racemize under conditions compatible with (BINAP)RuBr$_2$-catalyzed asymmetric hydrogenation.[434,435] The combination of enantioconversion and stereoselective reduction in a single operation has proved invaluable for asymmetric synthesis of β-hydroxy esters. The (*R*)-enantiomer of methyl cyclopentanone-2-carboxylate is selectively reduced by the (*M*)-BINAP-derived ruthenium catalyst and is constantly regenerated from the less reactive (*S*)-enantiomer via enolization.[xiv] As a result, racemic methyl cyclopentanone-2-carboxylate is converted to the corresponding (1*R*,2*R*)-*trans*-β-hydroxy ester which is obtained in 98% de and 92% ee, Scheme 7.47. Dynamic KR of α-substituted β-keto esters has been incorporated as a key step into the total synthesis of several natural products including biphenomycin A[436] and roxaticin.[437] This procedure has been further refined and successfully extended to β-keto phosphonates, α-substituted ketones and piperidinones.[438–442] A remarkable example is the quantitative reduction of 2-phenylcyclohexanone to (1*S*,2*S*)-*cis*-2-phenylcyclohexanol. This reaction proceeds at room temperature and requires only 1 mmol% of *trans*-RuCl$_2$[(*M*)-TolBINAP][(*R*,*R*)-DPEN],

---

[xiv] The (*R*)-β-keto ester and (*M*)-BINAP)RuBr$_2$ constitute a matched pair, see Chapter 5.3.2.

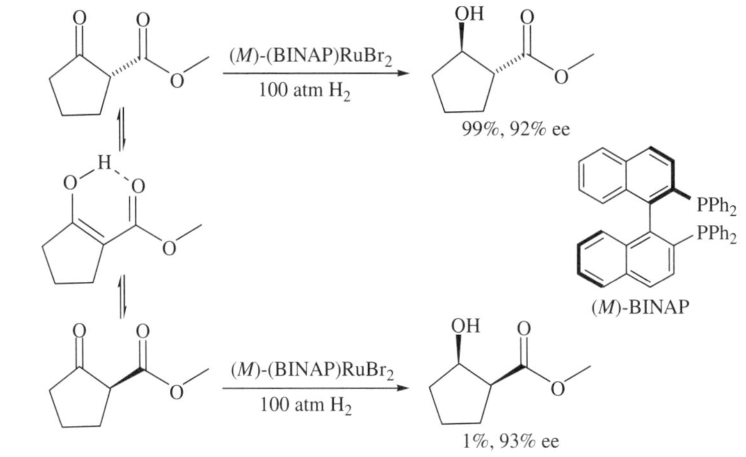

Scheme 7.47   Noyori's resolution of methyl cyclopentanone-2-carboxylate.

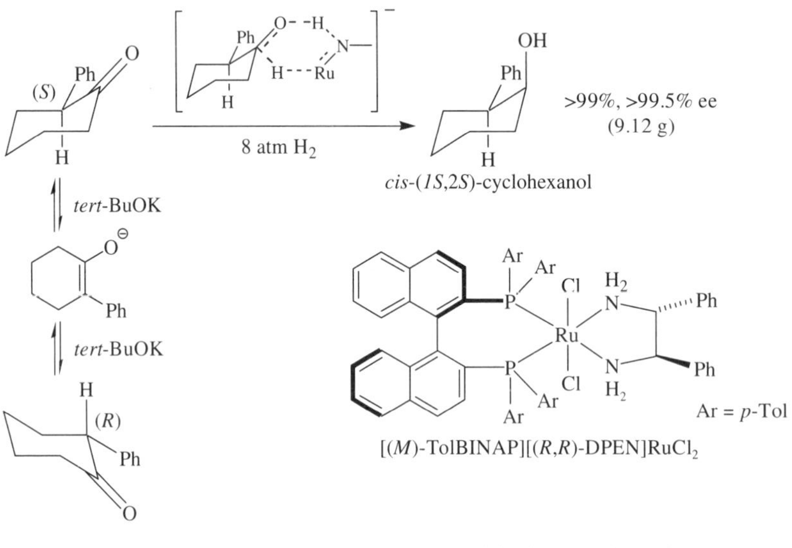

Scheme 7.48   [(*M*)-TolBINAP][(*R*,*R*)-DPEN]RuCl$_2$-catalyzed hydrogenation of 2-phenylcyclohexanone.

relatively low hydrogen pressure and potassium *tert*-butoxide to promote enolization. It is assumed that the ruthenium complex selectively recognizes (*S*)-2-phenylcyclohexanone in its most heavily populated chair conformation. Diastereofacial reduction then proceeds via equatorial attack and generates the *cis*-alcohol, Scheme 7.48. Both Genêt's[443–450] and Hamada's groups[451–453] applied a series of chiral phosphine-derived ruthenium and iridium catalysts to stereoselective hydrogenation of configurationally unstable α-amino-β-keto esters and α-chloro-β-keto esters, Figure 7.13. Alternatively, transition metal-catalyzed DKR of stereolabile β-keto esters,[454–456] ketones[457,458] and 3,5-dialkylcyclopentenones[459] can be achieved with formic acid, borohydrides or poly(methylhydrosiloxane) as the hydrogen source.

Noyori introduced a two-step reduction sequence that allows conversion of achiral diaryl ketones such as benzil and its *para*-substituted derivatives to chiral 1,2-diols based on DKR of a stereolabile α-hydroxy ketone intermediate, Scheme 7.49.[460] The [(*S*,*S*)-TSDPEN](η^6-*p*-cymene)RuCl-catalyzed reduction of benzil with formic acid as hydride source affords

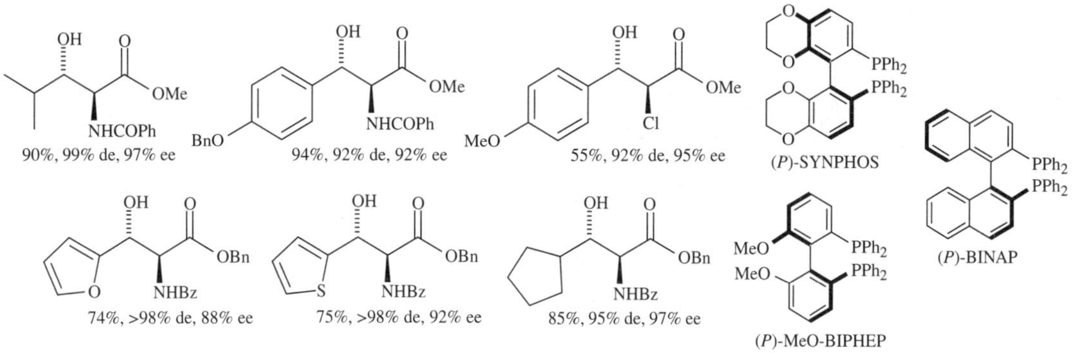

Structures with values below:

90%, 99% de, 97% ee

94%, 92% de, 92% ee

55%, 92% de, 95% ee

(*P*)-SYNPHOS

(*P*)-BINAP

74%, >98% de, 88% ee

75%, >98% de, 92% ee

85%, 95% de, 97% ee

(*P*)-MeO-BIPHEP

Figure 7.13 Compounds prepared by asymmetric hydrogenation of α-amino-β-keto esters and α-chloro-β-keto esters (left) and structures of chiral ligands used (right).

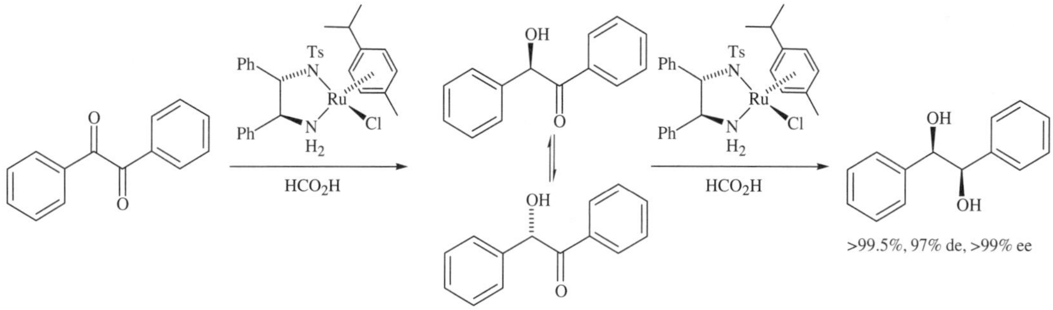

>99.5%, 97% de, >99% ee

Scheme 7.49 Asymmetric reduction of benzil to (*R*,*R*)-hydrobenzoin.

hydrobenzoin with excellent enantio- and diastereoselectivity. The reaction involves intermediate formation of rapidly racemizing benzoin. The (*S*,*S*)-TSDPEN-derived ruthenium complex selectively recognizes the (*R*)-enantiomer of benzoin and furnishes quantitative amounts of (*R*,*R*)-hydrobenzoin.

Few DKR protocols with stereolabile Grignard reagents have been reported.[461] Kumada and coworkers described an asymmetric cross-coupling procedure applicable to aryl and alkenyl bromides, Scheme 7.50.[462] They found that vinyl bromide and 1-phenylethylmagnesium chloride give (*R*)-3-phenyl-1-butene in 95% yield and 66% ee when a nickel catalyst derived from (*S*)-*N*,*N*-dimethyl-1-[(*R*)-2-(diphenylphosphino)ferrocenyl]ethylamine (PPFA) is used. Apparently, 1-phenylethylmagnesium chloride undergoes spontaneous racemization and the chiral (ferrocenylphosphine)nickel complex selectively incorporates the (*S*)-enantiomer into the cross-coupling cycle during transmetalation.[xv] Other important nonreductive transition metal-catalyzed dynamic kinetic resolutions include Jacobsen's ring opening of epichlorohydrin with TMSN₃ in the presence of (salen)Cr(III)N₃[463] and Gais' palladium-catalyzed asymmetric synthesis of thioesters from racemic allylic esters with potassium thioacetate.[464]

With the advance of asymmetric organocatalysis, several transition metal-free DKR procedures have become available. The usefulness of this approach is demonstrated by the synthesis of a wide range of multifunctionalized compounds. Noteworthy examples include (*S*)-proline-catalyzed intermolecular aldol reaction of enolizable and thus stereolabile aldehydes,[465] and regiospecific

---

[xv] The enantioselectivity observed can also be a consequence of nonstereoselective transmetalation followed by conversion of diastereomeric nickel species to the (*S*)-1-phenylethylnickel complex due to asymmetric transformation of the first kind.

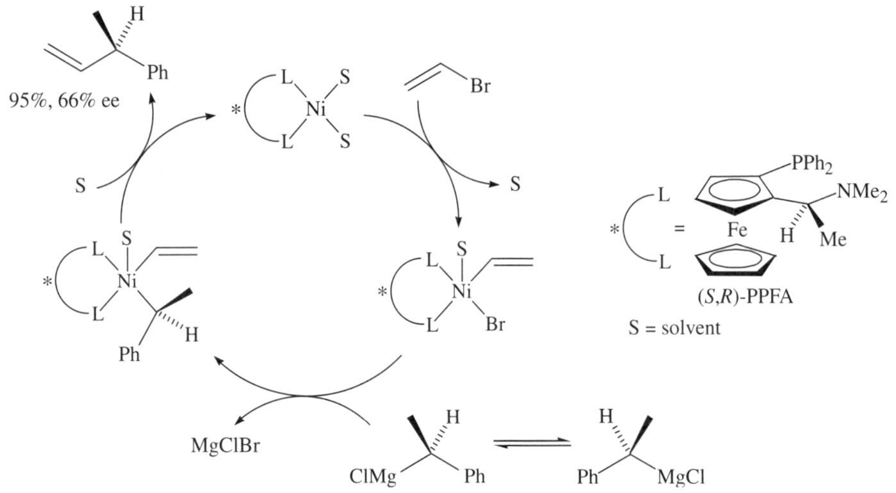

Scheme 7.50   Racemization of 1-phenylethylmagnesium chloride and resolution by [(*S*,*R*)-PPFA]NiCl$_2$-catalyzed cross-coupling with vinyl bromide.

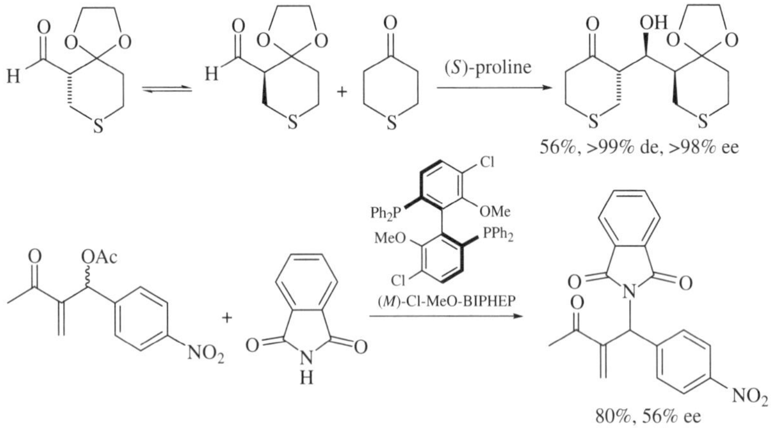

Scheme 7.51   Organocatalytic intermolecular aldol reaction (top) and allylic amination of Morita–Baylis–Hilman acetates (bottom).

allylic amination of Morita–Baylis–Hilman acetates in the presence of catalytic amounts of (*M*)-Cl-MeO-BIPHEP,[466] Scheme 7.51.

It has been known for a long time that azlactones undergo facile enolization and concomitant racemization. The low configurational stability has hampered a more extensive use of this readily available chiral building block for asymmetric synthesis. Nevertheless, azlactones are useful precursors of proteinogenic and unnatural α-amino acids and have been employed in many DKR efforts. Various enzymatic[467–469] and nonenzymatic methods[470–472] that are somewhat limited by narrow substrate scope or long reaction times are known. Berkessel introduced a thiourea-derived bifunctional organocatalyst that allows DKR of azlactones in the presence of allyl alcohol within 24 hours, Scheme 7.52.[473,474] This procedure has proved superior to other non-enzymatic methods and provides *N*-acylated amino esters with good to high enantioselectivities and yields. Based on NMR spectroscopic studies, it is assumed that the thiourea unit of the catalyst activates the azlactone for enolization and subsequent racemization through hydrogen bonding

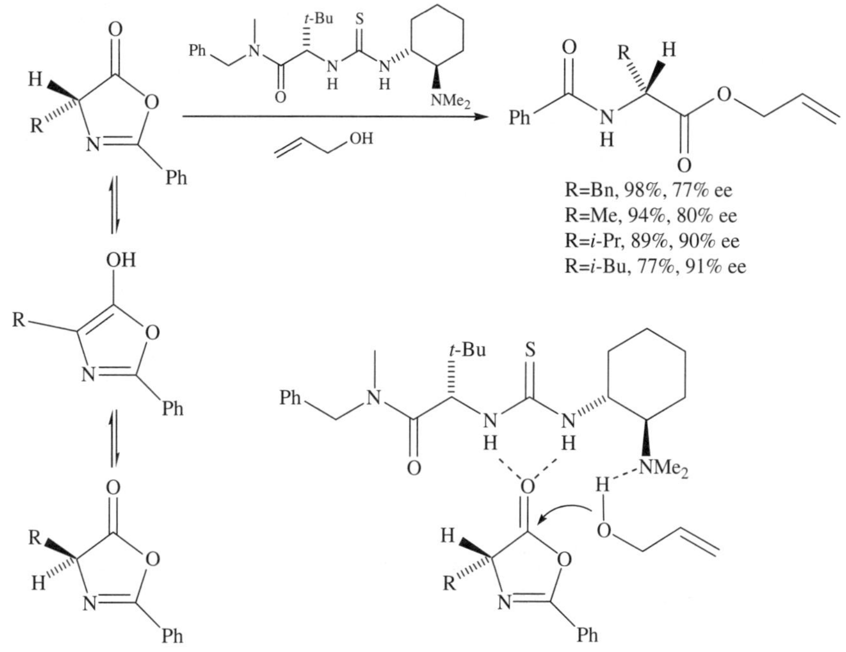

Scheme 7.52 DKR of azlactones with a thiourea catalyst.

while the terminal amino group serves as a Brønsted base and promotes alcoholytic ring opening. In a related approach, Deng *et al.* utilized the cinchona alkaloid (DHQD)$_2$AQN as a dual catalyst, accelerating both racemization and enantioselective alcoholytic ring opening of α-amino acid *N*-carboxyanhydrides and 1,3-dioxolane-2,4-diones, respectively.[475,476] The selectivity and time efficiency of this organocatalytic DKR is quite remarkable. In some cases, α-aryl amino acids and α-hydroxy carboxylic acids are obtained in excellent yield and high enantiomeric purity within a few hours, Scheme 7.53.

The vast majority of dynamic kinetic resolutions involves racemization of compounds possessing a configurationally labile chiral center. However, DKR is also applicable to stereolabile compounds that display other elements of chirality. Walsh demonstrated that interconverting atropisomeric 1-(2-formyl)naphthamides and benzamide analogs can be resolved by (*S*)-proline-catalyzed asymmetric aldol reaction when acetone is used as both reagent and solvent, Scheme 7.54.[477] Due to the low steric hindrance to rotation about the chiral aryl–amide bond, racemization of 1-formyl naphthamides is fast at room temperature. The rotational energy barrier is significantly enhanced upon formation of the aldol product and conformationally stable atropisomers are obtained. The (*S*)-proline-catalyzed aldol addition preferentially occurs at the *Si*-face of the aldehyde when the bulky dialkylamino moiety of the amide group resides on the opposite side of the arene plane. Accordingly, the reaction favors formation of the (*M*,*S*)-diastereomer.

Dynamic KR with enantiomer-differentiating reagents is also feasible, although catalytic processes are generally preferred.[xvi] Norton *et al.* discovered that stereolabile zircona aziridines, which can be prepared from *N*-substituted aryl amines and Cp$_2$ZrCl$_2$, undergo kinetic resolution in the presence of stoichiometric amounts of a C$_2$-symmetric carbonate, Scheme 7.55.[478,479] Interconversion of the aziridine enantiomers has been attributed to formation of an intermediate η^3-1-aza-allylzirconocene hydride complex. Apparently, the (*S*)-zircona aziridine reacts more rapidly

---

[xvi] Dynamic KR with a chiral reagent usually yields diastereomers. However, the use of a chiral reagent that differentiates between interconverting enantiomers does not always introduce a new chiral element to the substrate, see Scheme 7.57.

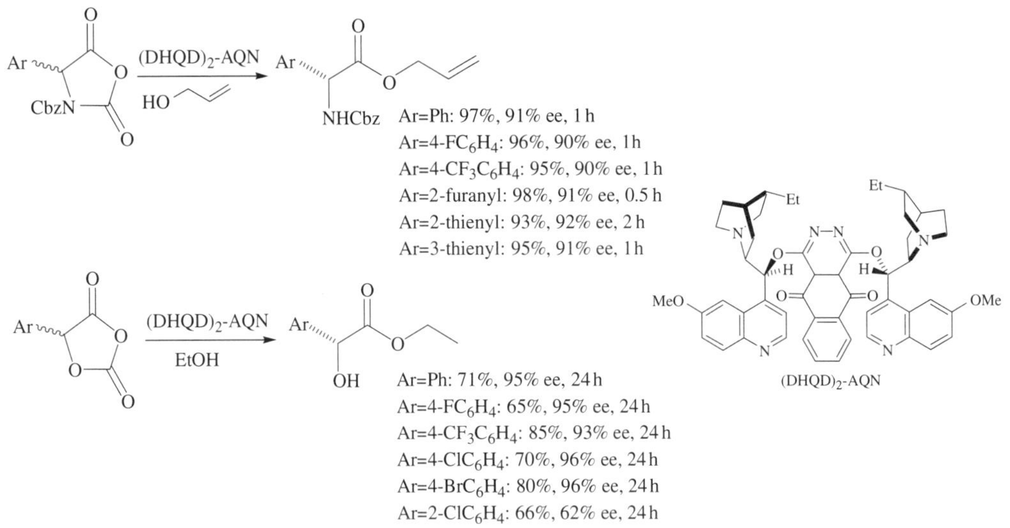

Scheme 7.53 Asymmetric synthesis of α-aryl amino acids and α-hydroxy carboxylic acids via cinchona alkaloid-catalyzed racemization and alcoholysis.

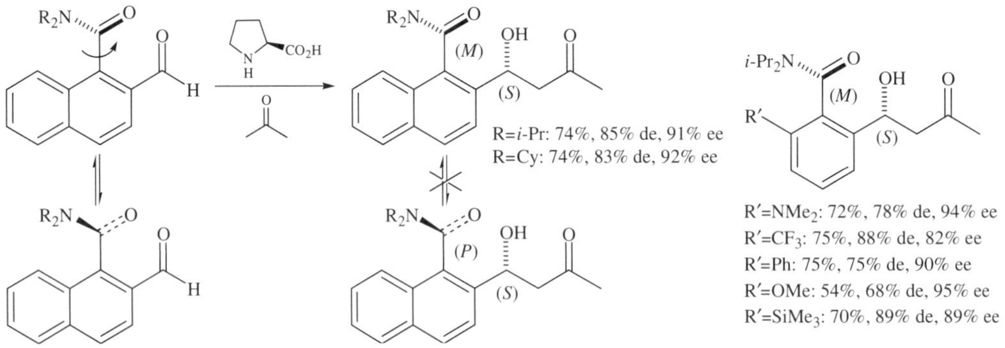

Scheme 7.54 DKR of atropisomeric amides.

with the resolving (*R,R*)-diphenylethylene carbonate to an (*R,R,R*)-metallacycle. Mild cleavage of the homochiral metallacycle releases the corresponding α-amino ester without concomitant racemization and the chiral auxiliary can be recovered. Activation of *N*-phthalylamino acids with dicyclohexylcarbodiimide and 4-dimethylaminopyridine generates stereolabile acyl 4-dimethyl-aminopyridinium enantiomers that can be kinetically resolved by stereoselective esterification with stoichiometric amounts of a chiral alcohol. When (*S*)-α-methylpantolactone is used as chiral resolving reagent one obtains the corresponding (*S,S*)-amino esters in up to 80% de within 15 hours, Scheme 7.56.[480] Similarly, DKR of interconverting chiral *N*-acyl hemiaminals has been achieved by transesterification with an axially chiral amide as enantiomer-differentiating acylating reagent.[481]

Mikami and Yoshida described a regio- and enantioselective samarium(II)-mediated reduction/protonation sequence of a propargylic phosphate ester in the presence of catalytic amounts of tetrakis(triphenylphosphine)palladium and one equivalent of a chiral alcohol, Scheme 7.57.[482] This cascade process converts the propargylic substrate to an axially chiral allenic ester in only 10 minutes. It is assumed that oxidative addition of the racemic phosphate to palladium(0) probably

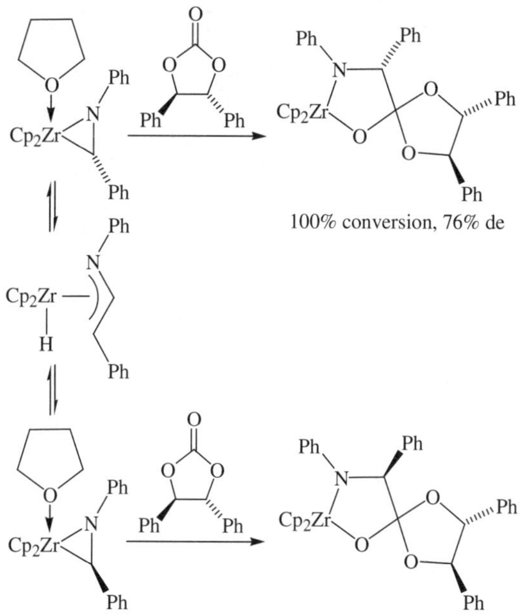

Scheme 7.55 DKR of zircona aziridines using (*R*,*R*)-diphenylethylene carbonate as chiral resolving reagent.

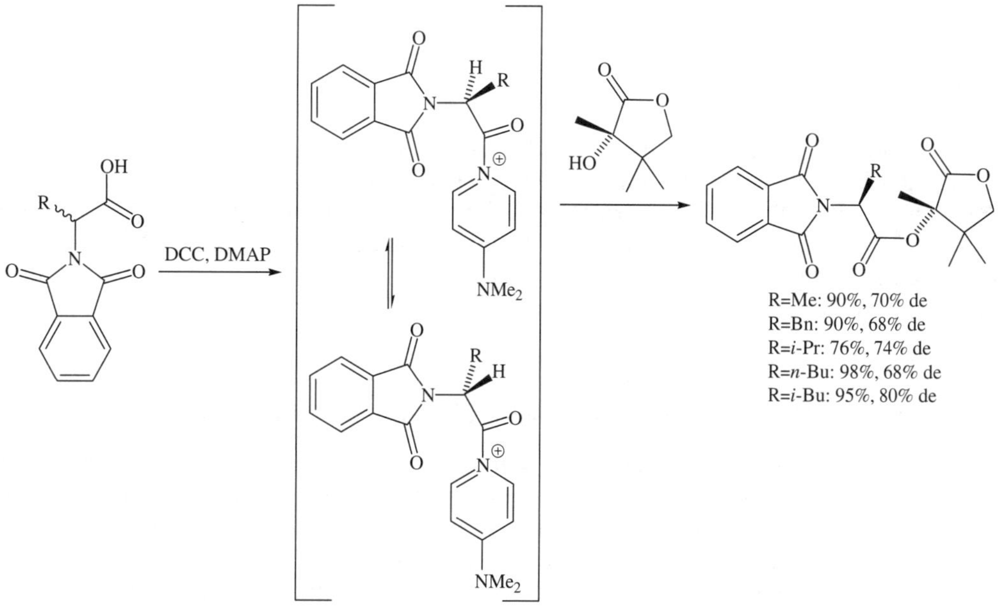

R=Me: 90%, 70% de
R=Bn: 90%, 68% de
R=*i*-Pr: 76%, 74% de
R=*n*-Bu: 98%, 68% de
R=*i*-Bu: 95%, 80% de

Scheme 7.56 DKR of activated *N*-phthalylamino acids with (*S*)-α-methylpantolactone.

affords an allenylpalladium(II) species that undergoes transmetalation to rapidly interconverting allenylsamarium(III) isomers. Because of the substantial carbanionic character of this stereolabile allenyl intermediate, dynamic kinetic resolution can be achieved through enantioselective protonation with pantolactone or other chiral alcohols.

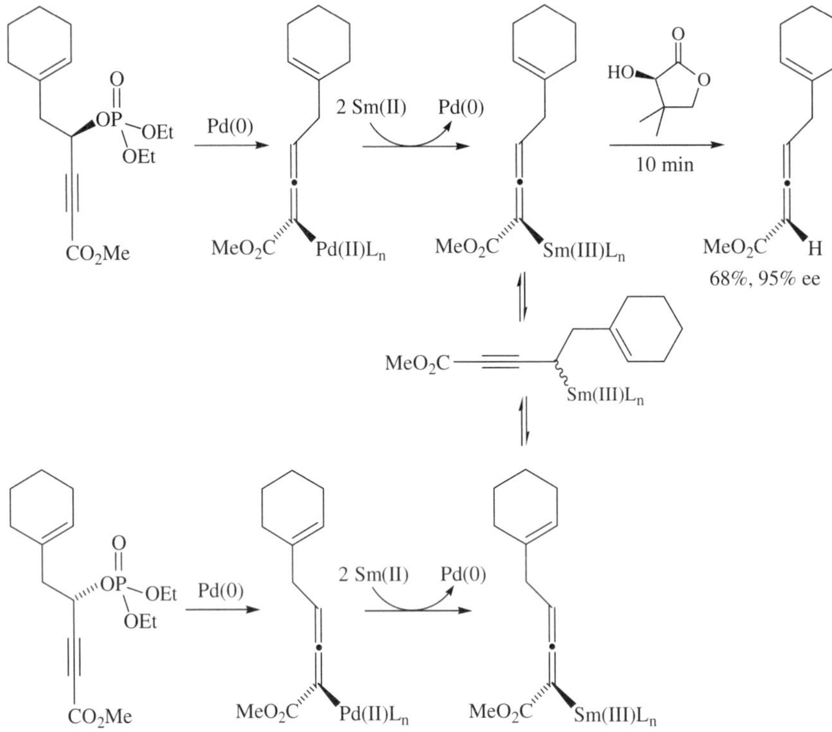

Scheme 7.57   Regio- and enantioselective SmI_2-mediated reduction/protonation sequence of a propargylic phosphate ester.

Other important DKR processes include the asymmetric Horner–Wadsworth–Emmons reaction of interconverting chiral α-amino aldehydes with chiral phosphonates,[483] enantiodivergent synthesis of sulfinate esters in the presence of either peptides,[484] cinchona alkaloids[485] or diacetone-D-glucose,[486,487] reduction of an enolizable diltiazem precursor with sodium borohydride and (S)-*tert*-leucine,[488] and oxidation of stereolabile chiral chlorophosphines and phospholes by (R,R)-bis(*tert*-butylphenylphosphinoyl)disulfide.[489]

7.5   DYNAMIC KINETIC ASYMMETRIC TRANSFORMATION

The concepts of dynamic kinetic resolution (DKR) and dynamic kinetic asymmetric transformation (DYKAT) are closely related and these terms are often confused in the literature. Both methodologies involve kinetic reaction control – in contrast to asymmetric transformation of the first and second kind and dynamic thermodynamic resolution (DTR). According to the original definition, the term DKR is strictly confined to one-pot procedures in which deracemization is coupled to an irreversible enantiomer-differentiating reaction. The combination of diastereomerization and stereoselective transformation is excluded by this definition and is better referred to as dynamic kinetic asymmetric transformation.[xvii] The purpose of resolution of a

---

[xvii] Some reactions that involve interconversion of diastereomers have been introduced as DKR in the literature. Although it has become common practice to extend the original definition of DKR to diastereomerization, these processes are more appropriately described as dynamic kinetic asymmetric transformation. This term has been consistently applied to asymmetric reactions based on sequential nucleophilic displacements at stereolabile π-allyl metal complexes. Following the classifications made in the original literature, these reactions are discussed in this chapter.

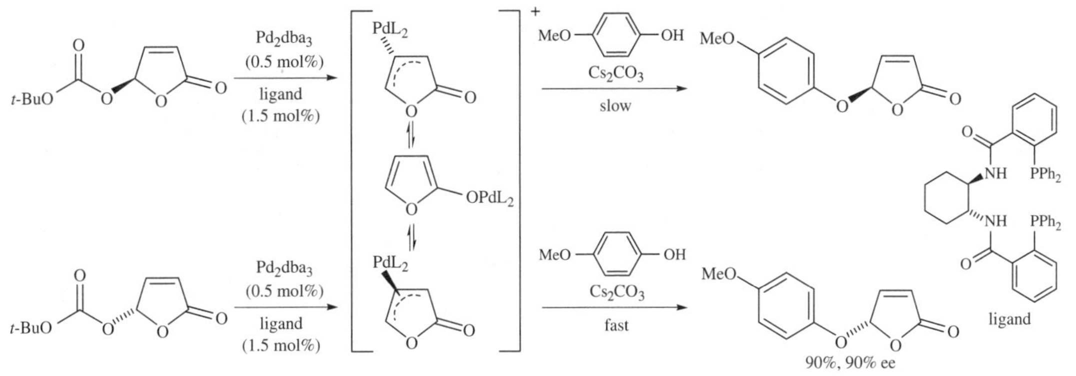

Scheme 7.58 Trost's palladium-catalyzed DYKAT of acyloxybutenolides.

racemic mixture is to recover at least one enantiomer. This is obvious in the case of preparative enantioseparation based on chromatography or crystallization, and with kinetic resolutions that are used to facilitate isolation of the less reactive enantiomer after approximately 50% conversion. In contrast, DKR is always aimed at quantitative conversion of a racemate to a single stereoisomer of a new compound and neither of the enantiomers is recovered because enantioconversion is combined with an asymmetric reaction that consumes the whole starting material. One can therefore argue that DYKAT is in principle a more appropriate term than DKR because it points more accurately to asymmetric synthesis rather than to resolution.

Trost and coworkers recognized that palladium undergoes facile η^3-η^1-η^3-migration from one enantiotopic face of a π-system to the other, and introduced efficient chiral palladium catalysts for asymmetric allylic alkylation (AAA) of a wide range of substrates.[490,491] This reaction has been successfully applied to racemic acyloxybutenolides, Scheme 7.58.[492] Ionization of racemic γ-*tert*-butoxycarbonyloxy-2-butenolide in the presence of a chiral palladium complex generates diastereomeric π-allyl metal complexes that rapidly interconvert via an intermediate aromatic η^1-complex. The stereochemical outcome of the subsequent nucleophilic attack of 4-methoxyphenoxide is governed by the chiral Pd complex which favors formation of an (S)-phenoxybutenolide in 90% ee. In accordance with DKR, a fundamental prerequisite for stereoselective DYKAT is that the diastereomerization of the intermediate π-allylpalladium complexes is faster than the stereoselective trapping with phenoxide. Palladium-catalyzed dynamic kinetic asymmetric transformation of acyloxybutenolides and 3,4,5,6-tetrahydroxylated cyclohexenes, so-called conduritols, affords invaluable synthetic access to important natural products such as aflatoxins, cyclophellitol and brefeldin A.[493–496]

Trost's group also developed palladium-catalyzed DYKAT of vinyl epoxides using alcohols as nucleophile and triethylborane as cocatalyst to generate vinyl glycidols which are versatile chiral building blocks.[497] Formation of a borate complex enhances the reactivity of the alcohol and controls the subsequent transfer of the nucleophile to the more highly substituted allylic carbon. The borate thus determines the regioselectivity of the reaction while enantioselectivity is induced by the chiral palladium complex, Scheme 7.59. Importantly, incorporation of terminal double bonds into the glycidol backbone provides an entry to enantiopure oxygen heterocycles and unnatural nucleosides via ruthenium-catalyzed ring closing metathesis.[498] Asymmetric allylic alkylation of racemic epoxides can be accomplished with other nucleophiles such as enolates, imides or carbonates[499] and has been employed in the total synthesis of malyngolide,[500] vigabatrin and ethambutol,[501] macrocyclic bisindolyl maleimides,[502] and the cyclopentyl core of viridinomycin.[503] Cheeseman and coworkers combined Pd-catalyzed DYKAT of 3,4-epoxy-1-butene using sodium bicarbonate as nucleophile with subsequent hydrolysis of the intermediate cyclic carbonate in the

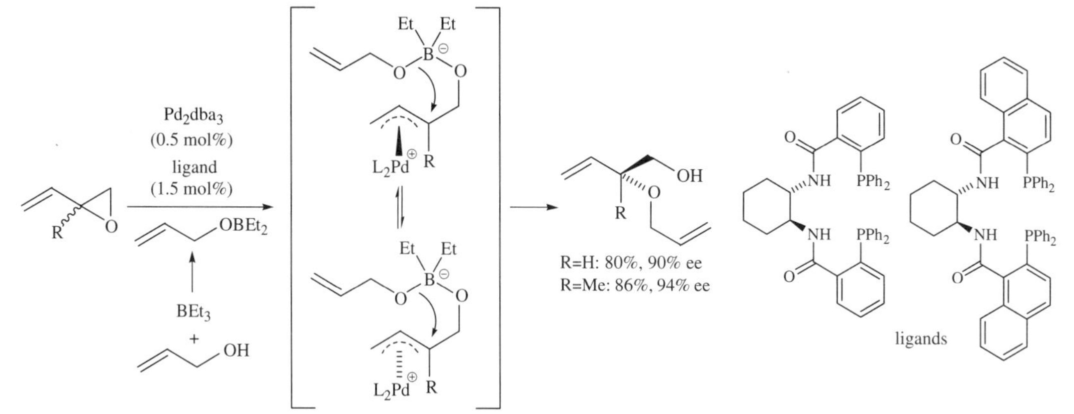

Scheme 7.59   Asymmetric synthesis of vinyl glycidols.

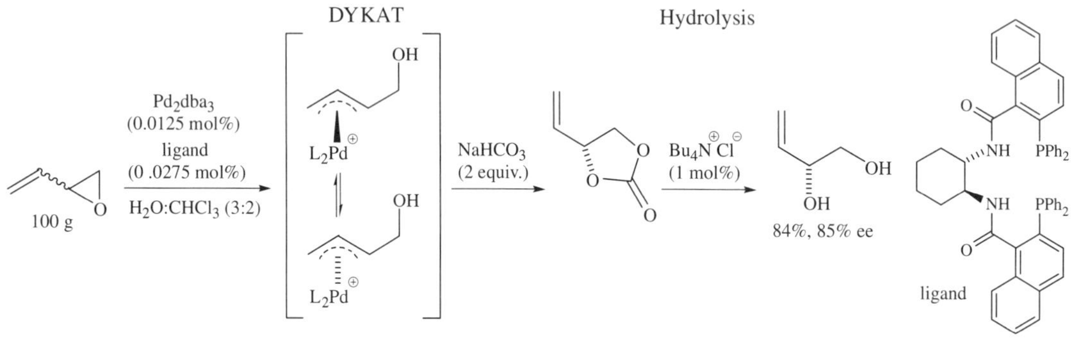

Scheme 7.60   DYKAT of 3,4-epoxy-1-butene and phase-transfer-catalyzed hydrolysis of the intermediate cyclic carbonate.

presence of a phase-transfer catalyst. This one-pot reaction sequence yields (2R)-3-butene-1,2-diol in 85% ee on a 100 g scale, Scheme 7.60.[504] Palladium-, molybdenum- and titanium-catalyzed AAA of vinylaziridines[505,506] and allylic alcohol derivatives[507–511] has been employed in the synthesis of a variety of chiral compounds including imidazolidinones, indenyl pivalate, 3-phthalimido-4,5-*O*-isopropylidenedioxycyclopentene, and indanes exhibiting a quaternary chiral carbon atom, Figure 7.14. The general usefulness of this approach has been demonstrated by incorporation of some of these chiral building blocks into the total synthesis of tipranavir and pseudodistomin D.

Dynamic kinetic resolution and dynamic kinetic asymmetric transformation are particularly attractive when starting materials are only available in racemic form or in unsatisfactory stereo-isomeric purity. The incorporation of *in situ* isomerization and stereoisomer-differentiating transformation into a single operation allows stereoconvergent, atom-economical asymmetric synthesis of many complex target compounds. For example, the Baylis–Hillman reaction gives convenient access to racemic β-hydroxy-α-methylene esters, ketones, nitriles, and other synthetically versatile building blocks. The preparation of enantiopure Baylis–Hillman products is quite challenging and despite remarkable progress one often has to resort to a synthetic strategy that utilizes racemic materials. Trost *et al.* therefore introduced palladium-catalyzed DYKAT of racemic Baylis–Hillman carbonates as an elegant alternative, Scheme 7.61.[512] The carbonates are readily prepared by treatment of β-hydroxy-α-methylene esters and nitriles with methyl chloroformate, and nucleophilic phenoxides or naphthoxides can be used for highly enantioselective trapping of the interconverting allylpalladium η³-complexes. This is important because the 4-methoxyphenyl ether

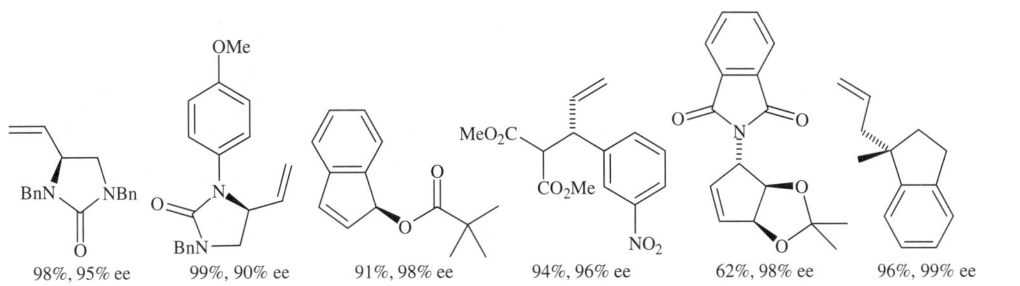

98%, 95% ee   99%, 90% ee   91%, 98% ee   94%, 96% ee   62%, 98% ee   96%, 99% ee

Figure 7.14 Compounds prepared by transition metal-catalyzed DYKAT of vinyl aziridines and allylic alcohol derivatives.

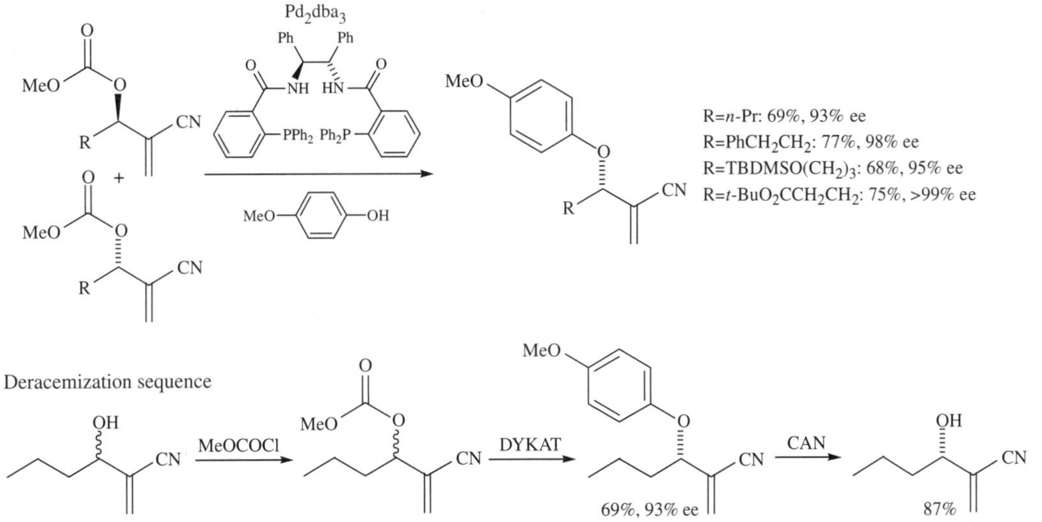

R=*n*-Pr: 69%, 93% ee
R=PhCH$_2$CH$_2$: 77%, 98% ee
R=TBDMSO(CH$_2$)$_3$: 68%, 95% ee
R=*t*-BuO$_2$CCH$_2$CH$_2$: 75%, >99% ee

Deracemization sequence

69%, 93% ee   87%

Scheme 7.61 DYKAT of Baylis–Hillman adducts (top) and deracemization (bottom).

group can easily be removed by mild oxidation with CAN to release the Baylis–Hillman adduct. Formation of the carbonate followed by DYKAT with 4-methoxyphenol and oxidative cleavage thus completes a deracemization cycle. Palladium-catalyzed AAA of Baylis–Hillman products with nucleophilic aliphatic and aromatic alcohols has been implemented into the total synthesis of furaquinocin A, B and E,[513,514] and hippospongic acid A.[515]

Many asymmetric transformations involving diastereomerization rather than racemization are conceptually very close to the processes discussed in Chapter 7.4.2. For instance, Bäckvall's one-pot transformation of *syn/anti*-mixtures of secondary diols to enantiopure diacetates requires the same ruthenium complex for interconversion of stereoisomers and a similar lipase as the parental DKR, see Schemes 7.37 and 7.62 for comparison. Because a mixture of diastereomers (versus a racemic or nonracemic mixture of enantiomers) is employed, this reaction is better described as a dynamic kinetic asymmetric transformation.[516,517] Ruthenium- and lipase-catalyzed DYKAT of 1,4-diols to enantiomerically enriched γ-acetoxy ketones has also been described.[518]

Amat and coworkers prepared bicyclic δ-lactams from racemic γ-aryl δ-oxoesters utilizing phenylglycinol as chiral auxiliary, Scheme 7.63.[519] According to NMR spectroscopic analysis, condensation of methyl 5-oxo-4-phenylpentanoate and the (*R*)-phenylglycine-derived amino

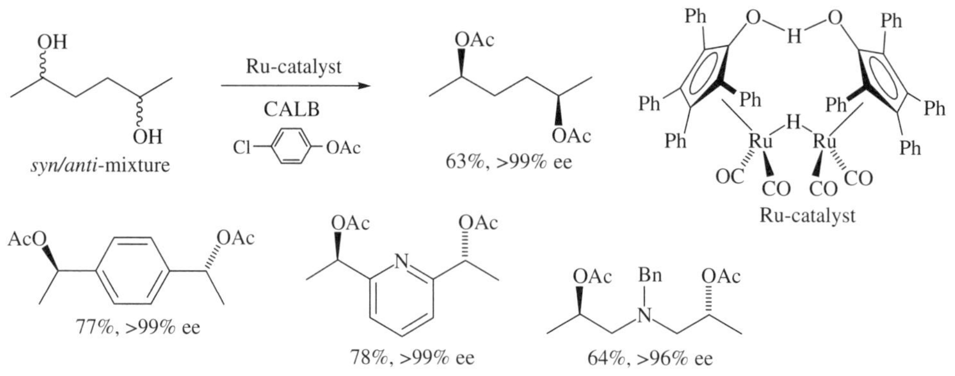

Scheme 7.62   DYKAT of secondary diols based on Ru-catalyzed diastereomerization and *Candida antarctica* lipase B-catalyzed resolution.

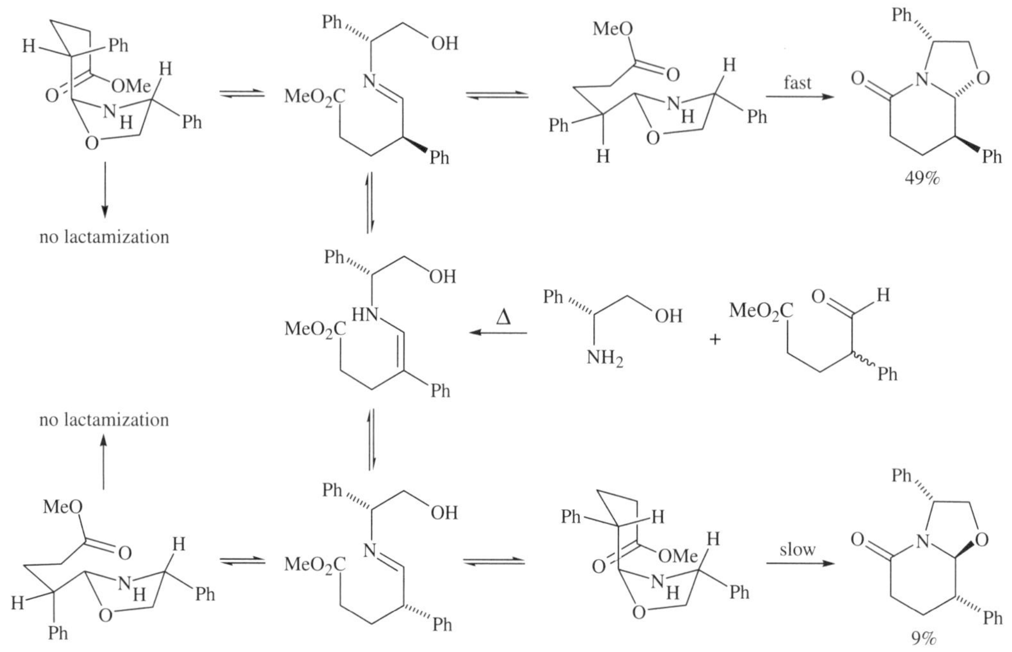

Scheme 7.63   Diastereoslective lactamization of methyl 5-oxo-4-phenylpentanoate with (*R*)-phenylglycinol as chiral auxiliary.

alcohol generates diastereomeric imines that interconvert via a common enamine intermediate. In addition to diastereomerization, the imines can produce four equilibrating oxazolidines through reversible ring closure. Subsequent lactamization is possible only for the two oxazolidines that can access a transition state bearing the aryl substituent of the incipient six-membered lactam ring in equatorial position. Less steric hindrance to nucleophilic attack of the nitrogen at the ester group in one of the diastereomeric oxazolidines results in stereoselective lactamization. Similar diastereoselective one-pot syntheses of other polysubstituted lactams involving phenylglycinol-promoted DYKAT of δ-keto esters and δ-oxo diesters have been reported.[520,521] Nucleophilic substitution of configurationally labile α-bromo and α-iodo carboxylic acid derivatives exhibiting a chiral auxiliary

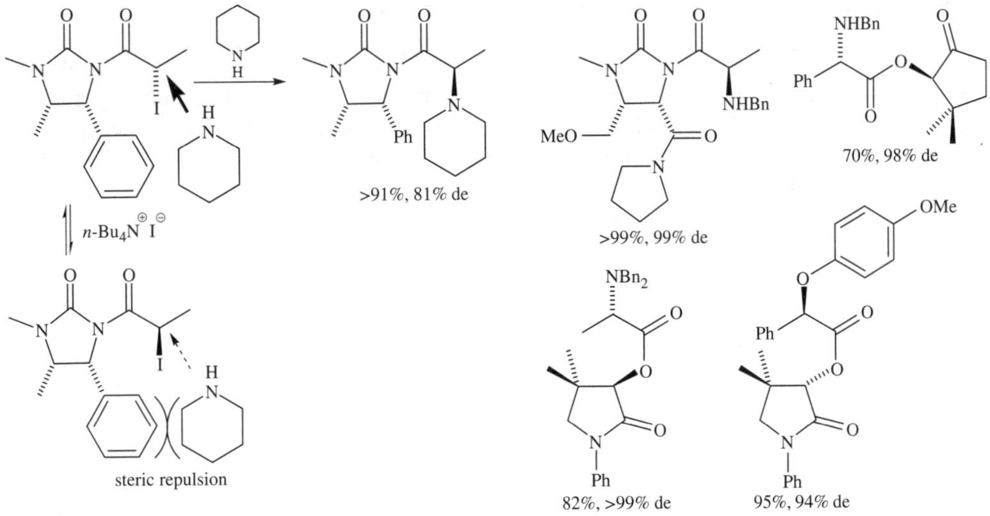

Scheme 7.64 Diastereoselective transformation of a stereolabile α-iodoacylimidazolidinone and structures of α-substituted carboxylic acid derivatives obtained by DYKAT.

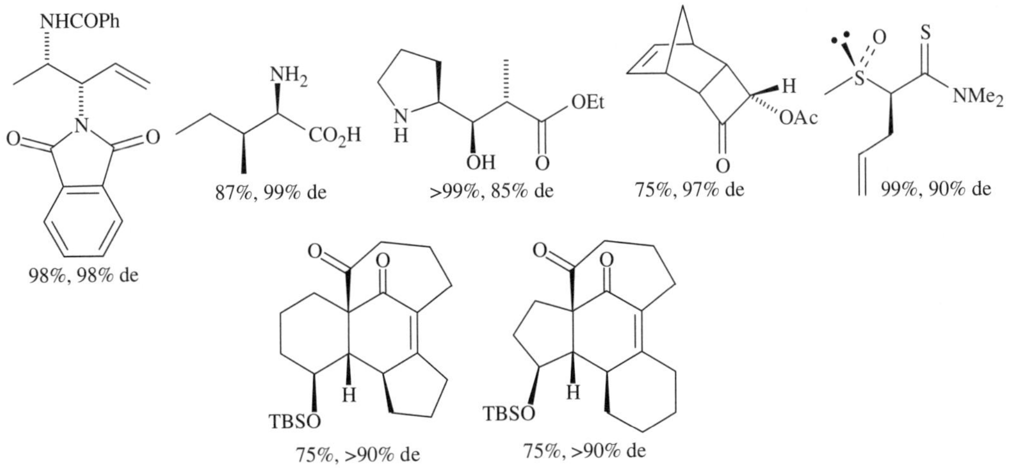

Figure 7.15 Chiral compounds synthesized by DYKAT.

that controls the stereochemical outcome of the S_N2 displacement has been studied to some extent. Diastereomerization due to consecutive inversions is usually induced by addition of a halide salt or base in polar solvents, Scheme 7.64. Typical chiral auxiliaries used for DYKAT of α-halo carboxylic acid derivatives are imidazolidinones,[522–525] hydroxy lactones,[526] hydroxy pyrrolidinones,[527–529] and oxazolidinones.[530]

To date, a variety of multifunctional chiral compounds including vicinal diamines,[531] amino acids such as D-*allo*-isoleucine,[532] α-substituted hydrazones,[533] pyrrolidine-derived β-hydroxy-α-methyl esters,[534] α-acetoxy ketones,[535] α-sulfinyl thioamides,[536] and polycyclic bridgehead enones[537] has been prepared by dynamic kinetic asymmetric transformation, Figure 7.15.

7.6   DYNAMIC THERMODYNAMIC RESOLUTION

Asymmetric synthesis based on DKR and DYKAT utilizes rapidly interconverting enantiomers or diastereomers that are consumed in a stereoisomer-differentiating reaction in the presence of a chiral catalyst or reagent. Both DKR and DYKAT obey the Curtin–Hammett principle, and the product ratio is solely determined by the difference in the free energy of the diastereomeric transition states of the irreversible derivatization step. Beak and others have developed an alternative strategy that is based on a one-pot procedure with two independently controlled steps: equilibration of diastereomeric adducts derived from stereolabile enantiomers and a chiral complexing agent, and subsequent transformation under conditions that exclude enantioconversion, Figure 7.16.[538] They coined the term dynamic thermodynamic resolution (DTR) to emphasize the difference between this approach and kinetically controlled DKR and DYKAT. A dynamic thermodynamic resolution is a two-stage process, in which the equilibrium between two interconverting diastereomeric adducts, $A \cdot B$ and $A' \cdot B$, is first established and then locked at lower temperature before the nonequilibrating species react to the enantiomeric products D and D'. The first stage of DTR determines the relative population of the diastereomeric adducts and resembles an asymmetric transformation of the first or second kind. In contrast to DKR and DYKAT, the rates of stereoisomer interconversion, k_{diast} and k'_{diast}, are smaller than the rates of the subsequent reaction step, k_c and k'_c. Furthermore, the final product ratio is determined by the difference in the Gibbs free energy, ΔG, of the intermediate diastereomeric adducts according to the Boltzmann equation. The individual rates of the reactions between the diastereomeric adducts $A \cdot B$ and $A' \cdot B$ and reagent C may affect the stereochemical outcome at low conversion, but they are irrelevant at quantitative conversion.[xviii] According to Beak's original definition, only processes providing enantioenriched or enantiopure products via thermal equilibration of intermediate diastereomeric noncovalent complexes constitute a DTR. Asymmetric reactions that involve equilibration of interconverting stereoisomers and *in situ* formation of diastereomeric products are conceptually similar but are formally excluded by this definition.[539–542] In other words, DTR has originally been restricted to enantioselective synthesis.

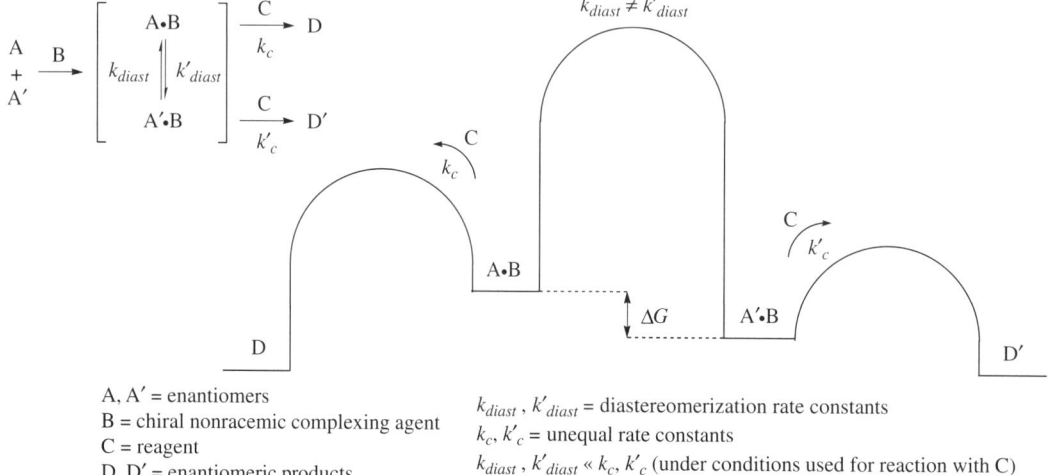

A, A' = enantiomers
B = chiral nonracemic complexing agent
C = reagent
D, D' = enantiomeric products

k_{diast}, k'_{diast} = diastereomerization rate constants
k_c, k'_c = unequal rate constants
k_{diast}, k'_{diast} ≪ k_c, k'_c (under conditions used for reaction with C)

Figure 7.16   Illustration of dynamic thermodynamic resolution.

[xviii] By contrast, the product ratio in DKR and DYKAT processes is independent of the conversion. Kinetic analysis can therefore be used to distinguish between DKR, DYKAT and DTR.

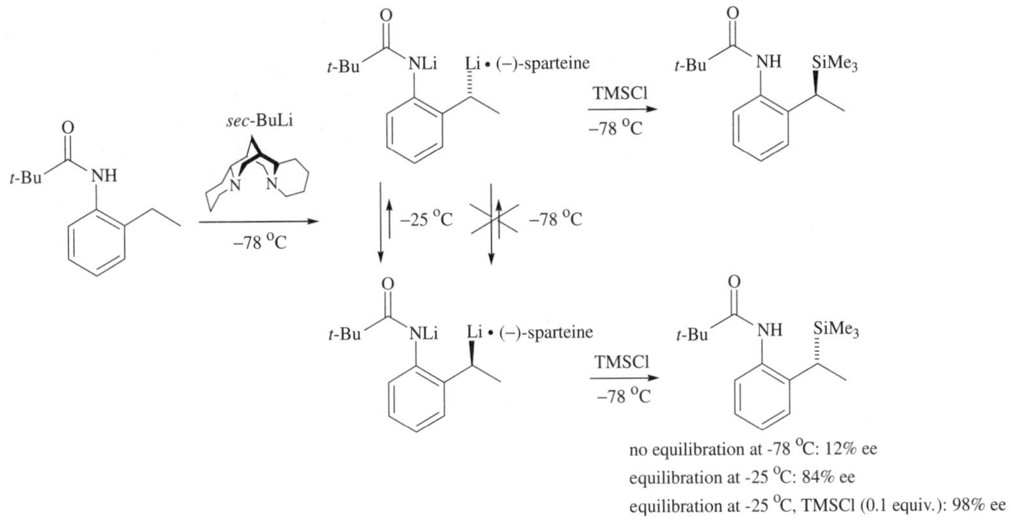

no equilibration at -78 °C: 12% ee

equilibration at -25 °C: 84% ee

equilibration at -25 °C, TMSCl (0.1 equiv.): 98% ee

Scheme 7.65 DTR of *N*-pivaloyl 2-ethylaniline.

Many dynamic thermodynamic resolutions involve formation of organolithium species in the presence of (−)-sparteine or another chiral complexing agent and concomitant trapping of the thermodynamically favored diastereomeric complex with a suitable electrophile.[543] A representative case that clearly illustrates the concept of DTR is shown in Scheme 7.65.[544–546] The treatment of *N*-pivaloyl 2-ethylaniline with two equivalents of *sec*-butyllithium generates a dilithium species that forms diastereomeric adducts with (−)-sparteine at −78 °C. An increase in the temperature to −25 °C allows rapid interconversion and equilibration of the diastereomeric complexes. Finally, the reaction mixture is cooled back to −78 °C and the chiral carbanions that are configurationally stable under these conditions are trapped with trimethylsilyl chloride. The silylation proceeds with inversion of configuration and the corresponding (*R*)-silane is obtained in 84% ee. In contrast, the chiral silane is produced in only 12% ee when the reaction is carried out entirely at −78 °C. Without a temporary increase in temperature, the diastereomeric sparteine complexes can not establish a thermodynamic equilibrium and the low enantiomeric excess obtained is probably a consequence of enantiotopic deprotonation of the prochiral substrate by (−)-sparteine. Addition of only 10% of trimethylsilyl chloride at −78 °C (after completion of the equilibration process at −25 °C) allows trapping of the chiral organolithium species at low conversion. In this case, the enantiomeric excess of the product improves dramatically to 98% ee, indicating that the more populated complex reacts more rapidly to the final product than the thermodynamically disfavored one. These findings prove that the stereochemical outcome is due to DTR, and that it is possible to combine DTR and kinetic resolution.

Beak's group employed other carboxamides and carbamates in similar DTR sequences.[547,548] Benzylic lithiation in β-position to an activating amide group affords a dipole-stabilized carbanion that in principle can enter three different reaction routes. Formation of a configurationally stable lithium salt and subsequent reaction with an electrophile would establish an enantiotopos-differentiating reaction in which asymmetric induction occurs during the deprotonation step. Alternatively, the deprotonation step can result – under appropriate reaction conditions – in the formation of a stereolabile lithium salt which may establish either a DKR or a DTR in the presence of (−)-sparteine or another complexing alkaloid. The latter was found to be the predominant pathway during (−)-sparteine-controlled lithiation and subsequent substitution of *N*-(2-phenylethyl)isobutyramide, Scheme 7.66.[549] Treatment of this carboxamide with

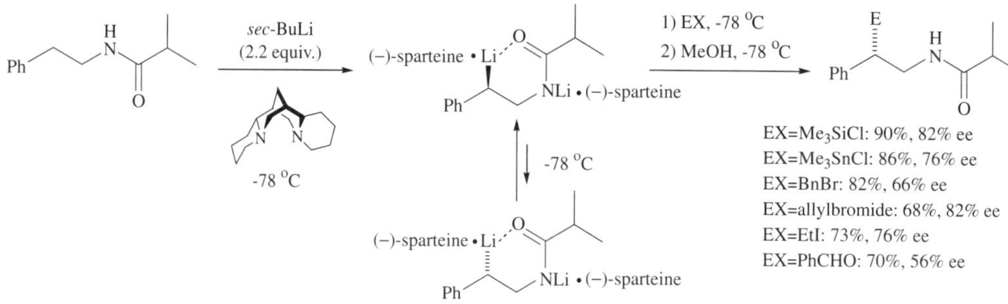

Scheme 7.66 Asymmetric synthesis of 2-substituted 2-phenylethylamines by DTR.

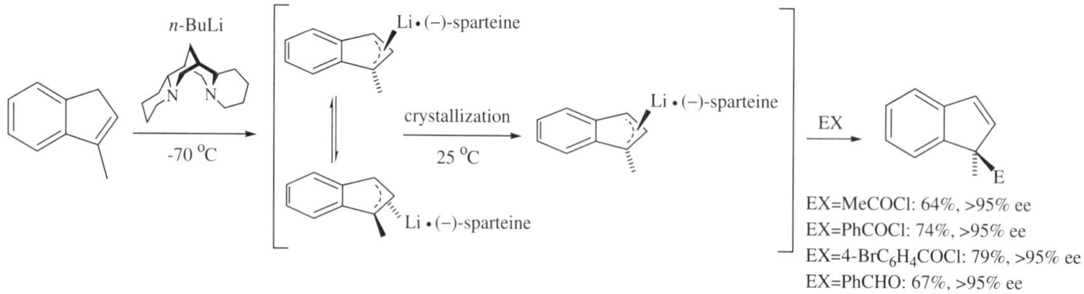

Scheme 7.67 DTR of lithium indenides based on asymmetric transformation of the second kind.

sec-butyllithium, sparteine and various electrophiles gives β-substituted amides in 68 to 90% yield and up to 82% ee.

Hoppe *et al.* reported that lithiation and successive (−)-sparteine-controlled DTR of carbamates and thiocarbamates proceeds with retention of configuration.[550–553] The same concept has been applied to lithium indenides. Deprotonation of 1-methylindene with *n*-butyllithium/(−)-sparteine at −70 °C was followed by an increase in the reaction temperature until crystallization occured. Reaction between the lithium indenide/sparteine precipitate with acyl chlorides or aldehydes gave (*R*)-1-substituted indenenes in good yields and high enantiomeric excess, Scheme 7.67.[554] The high ee can be explained by DTR based on thermodynamically controlled asymmetric transformation of the second kind. In other words, the stereochemical outcome is not determined by the relative stability of the equilibrating diastereomeric lithium salts in solution but is solely a consequence of crystallization-induced dynamic resolution.[xix]

Toru and coworkers utilized a variety of chiral chelating agents, including (−)-sparteine, C_2-symmetric bisoxazolines and 1,2-diaminocyclohexane derivatives, for DTR of configurationally labile α-lithiated thiazolidines,[555,556] dithioacetals,[557] α-sulfenyl carbanions,[558] and α-selenoorgano-lithium compounds.[559] Interestingly, DTR based on lithiation of benzyl 2-pyridyl sulfide and subsequent reaction with benzophenone or other electrophiles in the presence of a bisoxazoline ligand proceeds with inversion of configuration at the carbanion center, Scheme 7.68.[560] In contrast, diastereomeric bisoxazoline-derived adducts of α-lithiated benzyl phenyl sulfide prepared under similar conditions undergo significantly faster interconversion, thus establishing DYKAT with retention of configuration. The enantioconvergent synthesis of chiral phosphine–borane complexes via interconverting (−)-sparteine-derived diastereomers reported by Livinghouse *et al.*

[xix] In CIDR, the prevalent diastereomer is the one that forms the more stable crystal lattice but it is not necessarily thermodynamically favored in solution.

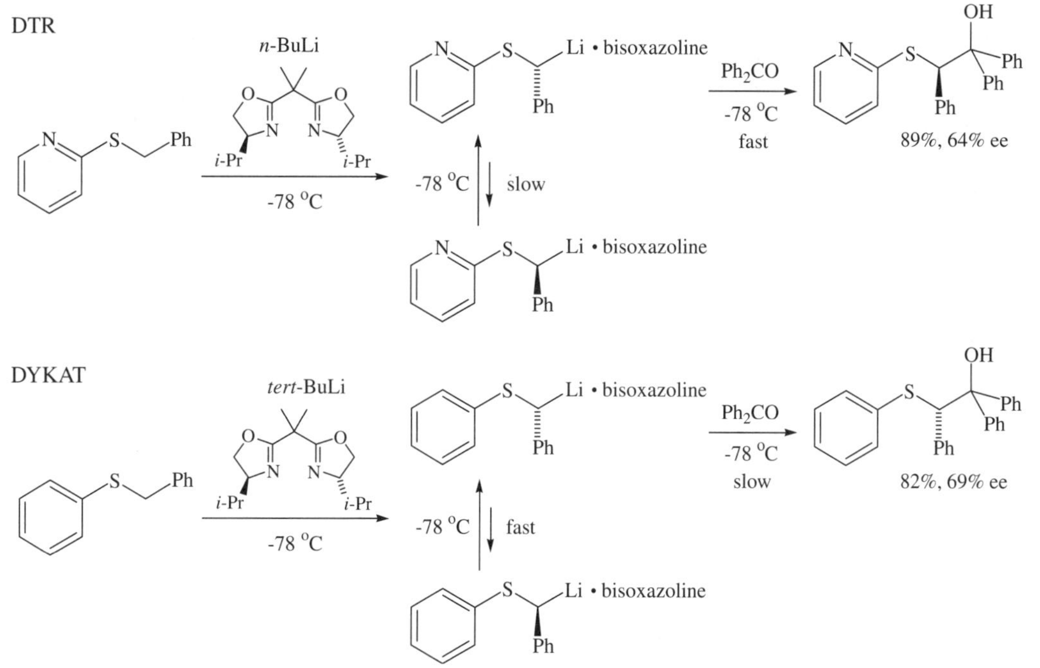

Scheme 7.68 DTR of benzyl 2-pyridyl sulfide (top) and DYKAT of benzyl phenyl sulfide (bottom).

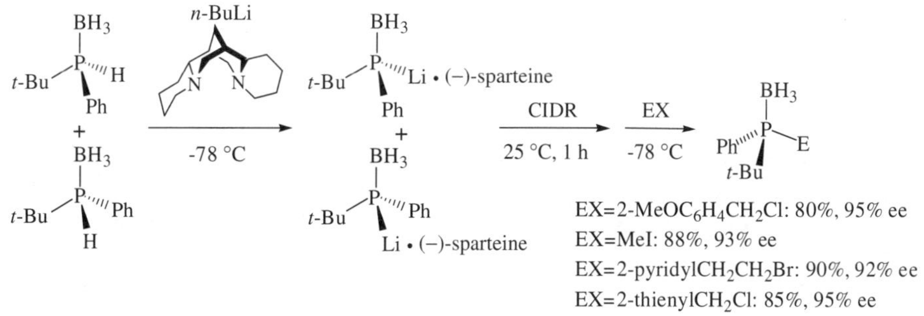

Scheme 7.69 Sparteine-mediated DTR of racemic dialkyl phosphine-borane complexes.

is another case that can be attributed to either DTR or DYKAT.[561,562] Lithiation of a racemic dialkyl phosphine–borane adduct in the presence of (–)-sparteine at −78 °C and equilibration at room temperature was accompanied by the formation of a suspension. Cooling of the reaction mixture and addition of an electrophile at −78 °C furnished chiral phosphines in excellent yield and ee, Scheme 7.69. Equilibration at 25 °C proved essential to the enantioselectivity of the process and only moderate ee's were observed when the temperature of the reaction mixture was not temporarily raised above 0 °C. The latter observation and the formation of a suspension during the equilibration step clearly show that DTR is operative, probably involving crystallization-induced dynamic resolution.

Following Beak's and Hoppe's pioneering work, several groups have utilized DTR to convert achiral or racemic chiral starting materials to enantiomerically enriched carbamates,[563] pyrrolidines,[564,565] β-hydroxy nitriles,[566] amino acids,[567] α-hydroxy carboxylic acid derivatives,[568,569] and 1,2-diols,[570] Figure 7.17. Clayden *et al.* applied the principles of DTR to the synthesis of axially

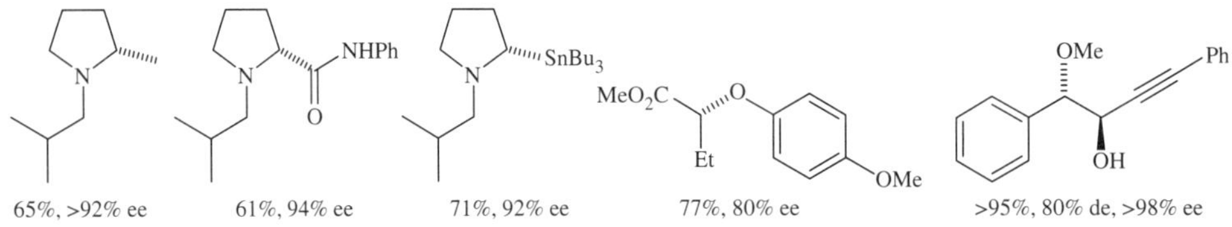

Figure 7.17 Structures of compounds obtained by DTR.

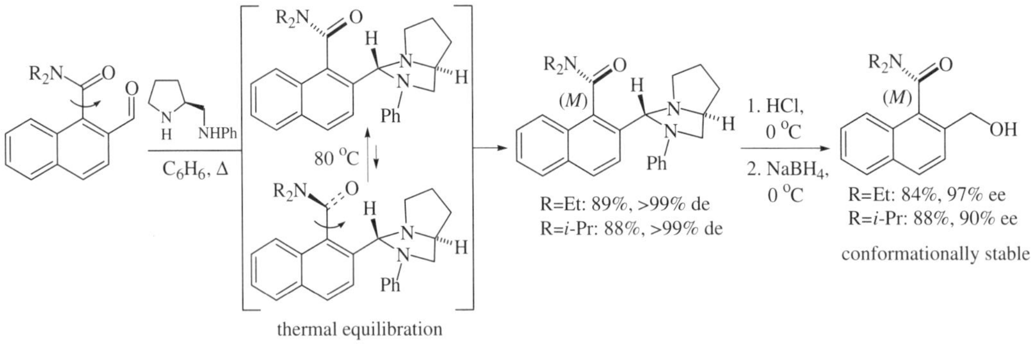

Scheme 7.70 Atroposelective synthesis of axially chiral naphthamides based on thermal equilibration of
interconverting diastereomeric aminals.

chiral aryl amides.[571–576] For example, refluxing of *N,N*-dialkyl-2-formyl-1-naphthamides with
stoichiometric amounts of an (*S*)-proline-derived diamine in benzene gave diastereomerically pure
aminals, Scheme 7.70. The atroposelectivity of this reaction sequence is established during thermal
equilibration of diastereomeric aminals at 80 °C (asymmetric transformation of the first kind) and
does not originate from DKR of rapidly interconverting 2-substituted 1-naphthamide enantiomers.
The chiral auxiliary was removed by acidic hydrolysis, and subsequent reduction with sodium
borohydride at low temperatures gave conformationally stable (*M*)-2-hydroxymethyl naphthamides in up to 97% ee. This method provides atropisomeric naphthamides with excellent
stereocontrol but requires at least three steps to complete the resolution sequence: introduction
of a resolving reagent, thermal equilibration of diastereomeric atropisomers and final removal of
the chiral auxiliary. The first two steps of the transformation shown in Scheme 7.70 can be
accomplished within a single operation, but more convenient one-pot DTR procedures that
eliminate isolation and purification of intermediates are often feasible. It is noteworthy that
Vedejs' crystallization-induced asymmetric transformation of diastereomeric oxazaborolidinone
complexes and subsequent conversion to chiral amino acids via asymmetric alkylation, discussed in
Chapter 7.3.1, is closely related to the DTR processes discussed above.

REFERENCES

1. Trost, B. M. *Angew. Chem., Int. Ed. Engl.* **1995**, *34*, 259-281.
2. Strauss, U. T.; Felfer, U.; Faber, K. *Tetrahedron: Asymm.* **1999**, *10*, 107-117.
3. Alexeeva, M.; Enright, A.; Dawson, M. J.; Mahmoudian, N. J.; Turner, N. J. *Angew. Chem., Int. Ed.* **2002**, *41*, 3177-3180.
4. Duhamel, L.; Duhamel, P.; Plaquevent, J.-C. *Tetrahedron: Asymm.* **2004**, *15*, 3653-3691.
5. Duhamel, L.; Plaquevent, J.-C. *J. Am. Chem. Soc.* **1978**, *100*, 7415-7416.
6. Alexandre, F.-R.; Pantaleone, D. P.; Taylor, P. P.; Fotheringham, I. G.; Ager, D. J.; Turner, N. J. *Tetrahedron Lett.* **2002**, *43*, 707-710.

7. Beard, T. M.; Turner, N. J. *Chem. Commun.* **2002**, 246-247.

8. Carr, R.; Alexeeva, M.; Enright, A.; Eve, T. S. C.; Dawson, M. J.; Turner, N. J. *Angew. Chem., Int. Ed.* **2003**, *42*, 4807-4810.

9. Allan, G. R.; Carnell, A. J. *J. Org. Chem.* **2001**, *66*, 6495-6497.

10. Demir, A. S.; Hamamci, H.; Sesenoglu, O.; Neslihanoglu, R.; Asikoglu, B.; Capanoglu, D. *Tetrahedron Lett.* **2002**, *43*, 6447-6449.

11. Harada, T.; Shintani, T.; Oku, A. *J. Am. Chem. Soc.* **1995**, *117*, 12346-12347.

12. Davis, A. P. *Angew. Chem., Int. Ed. Engl.* **1997**, *36*, 591-594.

13. Page, P. C. B.; Carnell, A. J.; McKenzie, M. J. *Synlett* **1998**, 774-776.

14. Goswami, A.; Mirfakhrae, K. D.; Patel, R. N. *Tetrahedron: Asymm.* **1999**, *10*, 4239-4244.

15. Kato, D.-I.; Mitsuda, S.; Ohta, H. *Org. Lett.* **2002**, *4*, 371-373.

16. Kato, D.-I.; Mitsuda, S.; Ohta, H. *J. Org. Chem.* **2003**, *68*, 7234-7242.

17. Chadha, A.; Baskar, B. *Tetrahedron: Asymm.* **2002**, *13*, 1461-1464.

18. Prat, L.; Dupas, G.; Duflos, J.; Queguiner, G.; Bourguignon, J.; Levacher, V. *Tetrahedron Lett.* **2001**, *42*, 4515-4518.

19. Zhang, Y.; Yeung, S.-M.; Wu, H.; Heller, D. P.; Wu, C.; Wulff, W. D. *Org. Lett.* **2003**, *5*, 1813-1816.

20. Hamada, T.; Ohtsuka, H.; Sakaki, S. *Chem. Lett.* **2000**, 364-365.

21. Willis, M. C. *J. Chem. Soc., Perkin Trans. 1* **1999**, 1765-1784.

22. Garcia-Urdiales, E.; Alfonso, I.; Gotor, V. *Chem. Rev.* **2005**, *105*, 313-354.

23. Anstiss, M.; Holland, J. M.; Nelson, A.; Titchmarsh, J. R. *Synlett* **2003**, 1213-1220.

24. Martinez, L. E.; Leighton, J. L.; Carsten, D. H.; Jacobsen, E. N. *J. Am. Chem. Soc.* **1995**, *117*, 5897-5898.

25. Yoshida, N.; Ogasawara, K. *Org. Lett.* **2000**, *2*, 1461-1463.

26. Moss, G. P. *Pure Appl. Chem.* **1996**, *68*, 2193-2122.

27. Pirkle, W. H.; Reno, D. S. *J. Am. Chem. Soc.* **1987**, *109*, 7189-7190.

28. Riedner, J.; Vogel, P. *Tetrahedron: Asymm.* **2004**, *15*, 2657-2660.

29. Tararov, V. I.; Savel'eva, T. F.; Kutznetsov, N. Y.; Ikonnikov, N. S.; Orlova, S. A.; Belekon, Y. N.; North, M. *Tetrahedron: Asymm.* **1997**, *8*, 79-83.

30. Solladie-Cavallo, A.; Sedy, O.; Salisova, M.; Schmitt, M. *Eur. J. Org. Chem.* **2002**, 3042-3049.

31. Ringwald, M.; Stürmer, R.; Brintzinger, H. H. *J. Am. Chem. Soc.* **1999**, *121*, 1524-1527.

32. Nelson, T. D.; Meyers, A. I. *Tetrahedron Lett.* **1993**, *34*, 3061-3062.

33. Nelson, T. D.; Meyers, A. I. *Tetrahedron Lett.* **1994**, *35*, 3259-3262.

34. Degnan, A. P.; Meyers, A. I. *J. Am. Chem. Soc.* **1999**, *121*, 2762-2769.

35. Nelson, T. D.; Meyers, A. I. *J. Org. Chem.* **1994**, *59*, 2577-2580.

36. Zhang, Y.; Yeung, S.-M.; Wu, H.; Heller, D. P.; Wu, C.; Wulff, W. D. *Org. Lett.* **2003**, *5*, 1813-1816.

37. Yu, S.; Rabalakos, C.; Mitchell, W. D.; Wulff, W. D. *Org. Lett.* **2005**, *7*, 367-369.

38. Evans, D. A.; Wood, M. R.; Trotter, B. W.; Richardson, T. I.; Barrow, J. C.; Katz, J. L. *Angew. Chem., Int. Ed.* **1998**, *37*, 2700-2704.

39. Evans, D. A.; Dinsmore, C. J.; Watson, P. S.; Wood, M. R.; Richardson, T. I.; Trotter, B. W.; Katz, J. L. *Angew. Chem., Int. Ed.* **1998**, *37*, 2704-3708.

40. Boger, D. L.; Miyazaki, S.; Kim, S. H.; Wu, J. H.; Loiseleur, O.; Castle, S. L. *J. Am. Chem. Soc.* **1999**, *121*, 3226-3227.

41. Boger, D. L.; Miyazaki, S.; Kim, S. H.; Wu, J. H.; Castle, S. L.; Loiseleur, O.; Jin, Q. *J. Am. Chem. Soc.* **1999**, *121*, 10004-10011.

42. Eliel, E. L.; Wilen, S. H. *Stereochemistry of Organic Compounds*, Wiley, New York, 1994, pp. 297-464.

43. Caddick, S.; Jenkins, K. *Chem. Soc. Rev.* **1996**, 447-456.

44. Shieh, W.-C.; Carlson, J. A. *J. Org. Chem.* **1994**, *59*, 5463-5465.

45. Shiraiwa, T.; Furukawa, T.; Tsuchida, T.; Sakata, S.; Sunami, M.; Kurokawa, H. *Bull. Chem. Soc. Jpn.* **1991**, *64*, 3729-3731.

46. Shiraiwa, T.; Baba, Y.; Miyazaki, H.; Sakata, S.; Kawamura, S. Uehara, M.; Kurokawa, H. *Bull. Chem. Soc. Jpn.* **1993**, *66*, 1430-1437.

47. Kolarovic, A.; Berkes, D.; Baran, P.; Povazanec, F. *Tetrahedron Lett.* **2001**, *42*, 2579-2582.

48. Kolarovic, A.; Berkes, D.; Baran, P.; Povazanec, F. *Tetrahedron Lett.* **2005**, *46*, 975-978.

49. Moseley, J. D.; Williams, B. J.; Owen, S. N.; Verrier, H. M. *Tetrahedron: Asymm.* **1996**, *7*, 3351-3352.

50. Boesten, W. H. J.; Seerden, J.-P. G.; de Lange, B.; Dielemans, H. J. A.; Elsenberg, H. L. M.; Kaptein, B.; Moody, H. M.; Kellogg, R. M.; Broxterman, Q. B. *Org. Lett.* **2001**, *3*, 1121-1124.

51. Hassan, N.; Bayer, E.; Jochims, J. C. *J. Chem. Soc., Perkin Trans. 1* **1998**, 3747-3757.

52. Hongo, C.; Yamada, S.; Chibata, I. *Bull. Chem. Soc. Jpn.* **1981**, *54*, 3286-3290.

53. Hongo, C.; Yamada, S.; Chibata, I. *Bull. Chem. Soc. Jpn.* **1981**, *54*, 3291-3295.

54. Yamada, S.; Hongo, C.; Yoshioka, R.; Chibata, I. *J. Org. Chem.* **1983**, *48*, 843-846.

55. Hongo, C.; Yoshioka, R.; Tohyama, M.; Yamada, S.; Chibata, I. *Bull. Chem. Soc. Jpn.* **1983**, *56*, 3744-3747.

56. Hongo, C.; Yoshioka, R.; Tohyama, M.; Yamada, S.; Chibata, I. *Bull. Chem. Soc. Jpn.* **1984**, *57*, 1328-1330.

57. Hongo, C.; Tohyama, M.; Yoshioka, R.; Yamada, S.; Chibata, I. *Bull. Chem. Soc. Jpn.* **1985**, *58*, 433-436.

58. Tabushi, I.; Kuroda, Y.; Yamada, M. *Tetrahedron Lett.* **1987**, *28*, 5695-5698.

59. Shiraiwa, T.; Shinjo, K.; Kurokawa, H. *Chem. Lett.* **1989**, 1413-1414.

60. Shiraiwa, T.; Kataoka, K.; Sakata, S.; Kurokawa, H. *Bull. Chem. Soc. Jpn.* **1989**, *62*, 109-113.

61. Shiraiwa, T.; Shinjo, K.; Kurokawa, H. *Bull. Chem. Soc. Jpn.* **1991**, *64*, 3251-3255.

62. Miyzaki, H.; Ohta, A.; Kawakatsu, N.; Waki, Y.; Gogun, Y.; Shiraiwa, T.; Kurokawa, H. *Bull. Chem. Soc. Jpn.* **1993**, *66*, 536-540.

63. Ebbers, E. J.; Ariaans, G. J. A.; Houbiers, J. P. M.; Bruggink, A.; Zwanenburg, B. *Tetrahedron* **1997**, *53*, 9417-9476.

64. Marynoff, C. A.; Scott, L.; Shah, R. D.; Villani Jr., F. J. *Tetrahedron: Asymm.* **1998**, *9*, 3247-3250.

65. Vedejs, E.; Chapman, R. W.; Fields, S. C.; Lin, S.; Schrimpf, M. R. *J. Org. Chem.* **1995**, *60*, 3020-3027.

66. Vedejs, E.; Fields, S. C.; Lin, S.; Schrimpf, M. R. *J. Org. Chem.* **1995**, *60*, 3028-3034.

67. Vedejs, E.; Chapman, R. W.; Lin, S.; Müller, M.; Powell, D. R. *J. Am. Chem. Soc.* **2000**, *122*, 3047-3052.

68. Seebach, D.; Sting, A. R.; Hoffmann, M. *Angew. Chem., Int. Ed. Engl.* **1996**, *35*, 2708-2748.

69. Vedejs, E.; Fields, S. C.; Hayashi, R.; Hitchcock, S. R.; Powell, D. R.; Schrimpf, M. R. *J. Am. Chem. Soc.* **1999**, *121*, 2460-2470.

70. Napolitano, E.; Farina, V. *Tetrahedron Lett.* **2001**, *42*, 3231-3234.

71. Bader, A.; Pabel, M.; Wild, S. B. *J. Chem. Soc., Chem. Commun.* **1994**, 1405-1406.

72. Pabel, M.; Willis, A. C.; Wild, S. B. *Inorg. Chem.* **1996**, *35*, 1244-1249.

73. Bader, A.; Pabel, M.; Willis, A. C.; Wild, S. B. *Inorg. Chem.* **1996**, *35*, 3874-3877.

74. Vedejs, E.; Donde, Y. *J. Am. Chem. Soc.* **1997**, *119*, 9293-9294.

75. Vedejs, E.; Donde, Y. *J. Org. Chem.* **2000**, *65*, 2337-2343.

76. Lopez, F. J.; Ferrino, S. A.; Reyes, M. S.; Roman, R. *Tetrahedron: Asymm.* **1997**, *8*, 2497-2500.

77. Ebbers, E. J.; Ariaans, G. J. A.; Bruggink, A.; Zwanenburg, B. *Tetrahedron: Asymm.* **1999**, *10*, 3701-3718.

78. Kiau, S.; Discordia, R. P.; Madding, G.; Okuniewicz, F. J.; Rosso, V.; Venit, J. J. *J. Org. Chem.* **2004**, *69*, 4256-4261.

79. Caddick, S.; Jenkins, K. *Tetrahedron Lett.* **1996**, *37*, 1301-1304.

80. Komatsu, H.; Awano, H. *J. Org. Chem.* **2002**, *67*, 5419-5421.

81. Silverberg, L. J.; Dillon, J. L.; Vemishetti, P.; Sleezer, P. D.; Discordia, R. P.; Hartung, K. B. *Org. Proc. Res. Dev.* **2000**, *4*, 34-42.

82. Kemperman, G. J.; Zhu, J.; Klunder, A. J. H.; Zwanenburg, B. *Org. Lett.* **2000**, *2*, 2829-2831.

83. Olszewska, T.; Milewska, M. J.; Gdaniec, M.; Maluszynska, H.; Polonski, T. *J. Org. Chem.* **2001**, *66*, 501-506.

84. Tsunoda, T.; Kaku, H.; Nagaku, M.; Okuyama, E. *Tetrahedron Lett.* **1997**, *38*, 7759-7760.

85. Kaku, H.; Takaoka, S.; Tsunoda, T. *Tetrahedron* **2002**, *58*, 3401-3407.

86. Reider, P. J.; Davis, P.; Hughes, D. L.; Grabowski, E. J. J. *J. Org. Chem.* **1987**, *52*, 955-957.

87. Wilen, S. H.; Qi, J. Z. *J. Org. Chem.* **1991**, *56*, 485-487.

88. Negi, S.; Matsukura, M.; Mizuno, M.; Miyake, K.; Minami, N. *Synthesis* **1996**, 991-996.

89. Shieh, W.-C.; Carlson, J. A.; Zaunius, G. M. *J. Org. Chem.* **1997**, *62*, 8271-8272.

90. Shi, Y.-J.; Wells, K. M.; Pye, P. J.; Choi, W.-B.; Churchill, H. R. O.; Lynch, J. E.; Maliakal, A.; Sager, J. W.; Rossen, K.; Volante, R. P.; Reider, P. J. *Tetrahedron* **1999**, *55*, 909-918.

91. Aelterman, W.; Lang, Y.; Willemsens, B.; Vervest, I.; Leurs, S.; de Knaep, F. *Org. Proc. Res. Dev.* **2001**, *5*, 467-471.

92. Singh, J.; Kronenthal, D. R.; Schwinden, M.; Godfrey, J. D.; Fox, R.; Vawter, E. J.; Zhang, B.; Kissick, T. P.; Patel, B.; Mneimne, O.; Humora, M.; Papaioannou, C. G.; Szymanski, W.; Wong, M. K. Y.; Chen, C. K.; Heikes, J. E.; DiMarco, J. D.; Qiu, J.; Desphande, R. P.; Gougoutas, J. Z.; Mueller, R. H. *Org. Lett.* **2003**, *5*, 3155-3158.

93. Barrett, R.; Caine, D. M.; Cardwell, K. S.; Cooke, J. W. B.; Lawrence, R. M.; Scott, P.; Sjolin, A. *Tetrahedron Asym.* **2003**, *14*, 3627-3631.

94. Jakubec, P.; Berkes, D.; Povazanec, F. *Tetrahedron Lett.* **2004**, *45*, 4755-4758.

95. Brunetto, G.; Gori, S.; Fiaschi, R.; Napolitano, E. *Helv. Chim. Acta* **2002**, *85*, 3785-3791.

96. Cimarelli, C.; Mazzanti, A.; Palmieri, G.; Volpini, E. *J. Org. Chem.* **2001**, *66*, 4759-4765.

97. Chandrasekhar, S.; Ravindranath, M. *Tetrahedron Lett.* **1989**, *30*, 6207-6208.

98. Kosmrlj, J.; Weigel, L. O.; Evans, D. A.; Downey, C. W.; Wu, J. *J. Am. Chem. Soc.* **2003**, *125*, 3208-3209.

99. Pye, P. J.; Rossen, K.; Weissman, S. A.; Maliakal, A.; Reamer, R. A.; Ball, R.; Tsou, N. T.; Volante, R. P.; Reider, P. J. *Chem. Eur. J.* **2002**, *8*, 1372-1376.

100. Brands, K. M. J.; Payack, J. F.; Rosen, J. D.; Nelson, T. D.; Candelario, A.; Huffman, M. A.; Zhao, M. M.; Li, J.; Craig, B.; Song, Z. J.; Tschaen, D. M.; Hansen, K.; Devine, P. N.; Pye, P. J.; Rossen, K.; Dormer, P. G.; Reamer, R. A.; Welch, C. J.; Mathre, D. J.; Tsou, N. N.; McNamara, J. M.; Reider, P. J. *J. Am. Chem. Soc.* **2003**, *125*, 2129-2135.

101. Okada, Y.; Takebayashi, T.; Hashimoto, M.; Kasuga, S.; Sato, S.; Tamura, C. *J. Chem. Soc., Chem. Commun.* **1983**, 784-785.

102. Alabaster, R. J.; Gibson, A. W.; Johnson, S. A.; Edwards, J. S.; Cottrell, I. F. *Tetrahedron: Asymm.* **1997**, *8*, 447-450.

103. Marchalin, S.; Cvopova, K.; Kriz, M.; Baran, P.; Oulyadi, H.; Daich, A. *J. Org. Chem.* **2004**, *69*, 4227-4237.

104. Lange, J.; Burzlaff, H.; Bringmann, G.; Schupp, O. *Tetrahedron* **1995**, *51*, 9361-9366.

105. Coogan, M. P.; Hibbs, D. E.; Smart, E. *Chem. Commun.* **1999**, 1991-1992.

106. Naruse, Y.; Watanabe, H.; Ishiyama, Y.; Yoshida, T. *J. Org. Chem.* **1997**, *62*, 3862-3866.

107. Node, M.; Nishide, K.; Fujiwara, T.; Ichihashi, S. *Chem. Commun.* **1998**, 2363-2364.

108. Kanomata, N.; Ochiai, Y. *Tetrahedron Lett.* **2001**, *42*, 1045-1048.

109. Ueda, T.; Kanomata, N.; Machida, H. *Org. Lett.* **2005**, *7*, 2365-2368.

110. Wilson, K. R.; Pincock, R. E. *J. Am. Chem. Soc.* **1975**, *97*, 1474-1478.

111. Smrcina, M.; Lorenc, M.; Hanus, V.; Sedmera, P.; Kocovsky, P. *J. Org. Chem.* **1992**, *57*, 1917-1920.

112. Tsubaki, K.; Miura, M.; Morikawa, H.; Tanaka, H.; Kawabata, T.; Furuta, T.; Tanaka, K.; Fuji, K. *J. Am. Chem. Soc.* **2003**, *125*, 16200-16201.

113. Wynberg, H.; Groen, M. B. *J. Am. Chem. Soc.* **1968**, *90*, 5339-5341.

114. Trankler, K. A.; Wyman, D. S.; Corey, J. Y.; Katz, E. E.; Rath, N. P. *Organometallics* **2001**, *20*, 5139-5148.

115. Trankler, K. A.; Corey, J. Y.; Rath, N. P. *J. Organomet. Chem.* **2003**, *686*, 66-74.

116. Boyle Jr., W. J.; Sifniades, S.; van Peppen, J. F. *J. Org. Chem.* **1979**, 44, 4841-4847.

117. Bendahl, L.; Hammershøi, A.; Jensen, D. K.; Larsen, S.; Riisager, A.; Sargeson, A.; Sørenson, H. O. *J. Chem. Soc., Dalton Trans.* **2002**, 3054-3064.

118. Yamanari, K.; Ito, R.; Yamamoto, S.; Konno, T.; Fuyuhiro, A.; Kobayashi, M.; Arakawa, R. *Dalton Trans.* **2003**, 380-386.

119. Hamelin, O.; Pecaut, J.; Fontecave, M. *Chem. Eur. J.* **2004**, *10*, 2548-2554.

120. Pirkle, W. H.; Welch, C. J.; Zych, A. J. *J. Chromatogr.* **1993**, *648*, 101-109.

121. Wolf, C.; König, W. A.; Roussel, C. *Chirality* **1995**, *7*, 610-611.

122. Lorenz, K.; Yashima, E.; Okamoto, Y. *Angew. Chem., Int. Ed.* **1998**, *37*, 1922-1925.

123. Tobler, E.; Lämmerhofer, M.; Mancicni, G.; Lindner, W. *Chirality* **2001**, *13*, 641-647.

124. Kagan, H. B.; Fiaud, J. C. *Top. Stereochem.* **1988**, *18*, 249-330.

125. Santaniello, E.; Ferraboschi, P.; Grisenti, P.; Manzocchi, A. *Chem. Rev.* **1992**, *92*, 1071-1140.

126. Hu, S.; Tat, D.; Martinez, C. A.; Yazbeck, D. R.; Tao, J. *Org. Lett.* **2005**, *7*, 4329-4331.

127. Roper, J. M.; Bauer, D. P. *Synthesis* **1983**, 1041-1043.

128. Chenault, H. K.; Dahmer, J.; Whitesides, G. M. *J. Am. Chem. Soc.* **1989**, *111*, 6354-6364.

129. Williams, R. M.; Hendrix, J. A. *Chem. Rev.* **1992**, *92*, 889-917.

130. Tomisaka, K.; Ishida, Y.; Konishi, K.; Aida, T. *Chem. Commun.* **2001**, 133-134.

131. Landis, B. H.; Mullins, P. B.; Karen, E.; Wang, P. T. *Org. Proc. Res. Dev.* **2002**, *6*, 539-546.

132. Ghanem, A.; Aboul-Enein, H. Y. *Chirality* **2005**, *17*, 1-15.

133. Raminelli, C.; Comasseto, J. V.; Andrade, L. H.; Porto, A. L. M. *Tetrahedron: Asymm.* **2004**, *15*, 3117-3122.

134. Nordin, O.; Nguyen, B.; Vorde, C.; Hedenstrom, E.; Hogberg, H.-E. *J. Chem. Soc., Perkin Trans. 1* **2000**, 367-376.

135. Hirose, K.; Naka, H.; Ohashi, S.; Naemura, K.; Tobe, Y. *Tetrahedron: Asymm.* **2000**, *11*, 1199-1210.

136. Kawasaki, M.; Goto, M.; Kawabata, S.; Kometani, T. *Tetrahedron: Asymm.* **2001**, *12*, 585-596.

137. Sakai, T.; Matsuda, A.; Korenaga, T.; Ema, T. *Bull. Chem. Soc. Jpn.* **2003**, *76*, 1819-1821.

138. Bierstedt, A.; Stoelting, J.; Froehlich, R.; Metz, P. *Tetrahedron: Asymm.* **2001**, *12*, 3399-3407.

139. Ghanem, A.; Schurig, V. *Chirality* **2001**, *13*, 118-123.

140. Ghanem, A.; Schurig, V. *Tetrahedron: Asymm.* **2001**, *12*, 2761-2766.

141. Vorlova, S.; Bornscheuer, U. T.; Gatfield, I.; Hilmer, J.-M.; Bertram, H.-J.; Schmid, R. D. *Adv. Synth. Catal.* **2002**, *344*, 1152-1155.

142. Lindner, E.; Ghanem, A.; Warad, I.; Eichle, K.; Mayer, E.; Schurig, V. *Tetrahedron: Asymm.* **2003**, *14*, 1045-1053.

143. Ghanem, A.; Schurig, V. *Tetrahedron: Asymm.* **2003**, *14*, 57-62.

144. Ghanem, A.; Schurig, V. *Tetrahedron: Asymm.* **2003**, *14*, 2547-2555.

145. Ghanem, A.; Schurig, V. *Org. Biomol. Chem.* **2003**, *1*, 1282-1291.

146. Mehta, G.; Islam, K. *Tetrahedron. Lett.* **2004**, *45*, 7683-7687.

147. Chen, S. T.; Fang, J. M. *J. Org. Chem.* **1997**, *62*, 4349-4357.

148. Krishna, S. H.; Persson, M.; Bornscheuer, U. T. *Tetrahedron: Asymm.* **2002**, *13*, 2693-2696.

149. Kirschner, A.; Langer, P.; Bornscheuer, U. T. *Tetrahedron: Asymm.* **2004**, *15*, 2871-2874.

150. De Meijere, A.; Khlebnikov, A. F.; Kozhushkov, S. I.; Kostikov, R. R.; Schreiner, P. R.; Wittkopp, A.; Rinderspacher, C.; Menzel, H.; Yufit, D. S.; Howard, J. A. K. *Chem. Eur. J.* **2002**, *8*, 828-842.
151. Xu, D.; Li, Z.; Ma, S. *Tetrahedron: Asymm.* **2003**, *14*, 4657-3666.
152. Wölker, D.; Haufe, G. *J. Org. Chem.* **2002**, *67*, 3015-3021.
153. Matsuda, T.; Watanabe, K.; Harada, T.; Nakamura, K.; Arita, Y.; Misumi, Y.; Ichikawa, S.; Ikariya, T. *Chem. Commun.* **2004**, 2286-2287.
154. Pogorevc, M.; Kroutil, W.; Wallner, S. R.; Faber, K. *Angew. Chem., Int. Ed.* **2002**, *41*, 4052-4054.
155. Ceynowa, J.; Rauchfkeisz, M. *J. Mol. Catal. B: Enzymatic* **2002**, *23*, 43-51.
156. Allan, G. R.; Carnell, A. J.; Kroutil, W. *Tetrahedron Lett.* **2001**, *42*, 5959-5962.
157. Liljeblad, A.; Lindborg, J.; Kanerva, A.; Katajisto, J.; Kanerva, L. T. *Tetrahedron Lett.* **2002**, *43*, 2471-2474.
158. Theil, F. *Methods Biotechnol.* **2001**, *15*, 277-289.
159. Sanfilippo, C.; Nicolosi, G.; Delogu, G.; Fabbri, D.; Dettori, M. A. *Tetrahedron: Asymm.* **2003**, *14*, 3267-3270.
160. Forro, E.; Fülöp, F. *Org. Lett.* **2003**, *5*, 1209-1212.
161. Gotor, V.; Brieva, R.; Rebolledo, F. *J. Chem. Soc., Chem. Commun.* **1988**, 757-958.
162. Bevinakatti, H. S.; Newadkar, R. V. *Tetrahedron: Asymm.* **1990**, *1*, 579-582.
163. Kanerva, L. T.; Csomos, P.; Sundholm, O.; Bernath, G.; Fueloep, F. *Tetrahedron: Asymm.* **1996**, *7*, 1705-1716.
164. Gedey, S.; Liljeblad, A.; Lazar, L.; Fulop, F.; Kanerva, L. T. *Tetrahedron: Asymm.* **2001**, *12*, 105-110.
165. Kurokawa, M.; Shindo, T.; Suzuki, M.; Nakajima, N.; Ishihara, K.; Sugai, T. *Tetrahedron: Asymm.* **2003**, *14*, 1323-1333.
166. Gyarmati, Z. C.; Liljeblad, A.; Rintola, M.; Bernath, G.; Kanerva, L. T. *Tetrahedron: Asymm.* **2003**, *14*, 3805-3814.
167. Pousset, C.; Callens, R.; Haddad, M. Larcheveque, M. *Tetrahedron: Asymm.* **2004**, *15*, 3407-3412.
168. Bucciarelli, M.; Forni, A.; Moretti, I.; Prati, F. *J. Chem. Soc., Chem. Commun.* **1988**, 1614-1615.
169. Schmid, A.; Dordick, S. J.; Hauer, B.; Kiener, A.; Wubbolts, M.; Witholt, B. *Nature* **2001**, *409*, 258-268.
170. Alfonso, I.; Astorga, C.; Rebolledo, F.; Gotor, V. *Chem. Commun.* **1996**, 2471-2472.
171. Alfonso, I.; Rebolledo, F.; Gotor, V. *Chem. Eur. J.* **2000**, *6*, 3331-3338.
172. Tumambac, G. E.; Wolf, C. *Org. Lett.* **2005**, *7*, 4045-4048.
173. Skopan, H.; Günther, H.; Simon, H. *Angew. Chem., Int. Ed. Engl.* **1987**, *26*, 128-130.
174. Izumi, Y.; Chibata, I.; Itoh, T. *Angew. Chem., Int. Ed. Engl.* **1978**, *17*, 176-183.
175. Top, S.; Jaouen, G.; Gillois, J.; Baldoli, C.; Maiorana, S. *J. Chem. Soc., Chem. Commun.* **1988**, 1284-1285.
176. Reetz, M. T.; Zonta, A.; Schimossek, K.; Liebeton, K.; Jaeger, K.-E. *Angew. Chem., Int. Ed. Engl.* **1997**, *36*, 2830-2832.
177. Reetz, M. T.; Wilensek, S.; Zha, D.; Jaeger, K.-E. *Angew. Chem., Int. Ed.* **2001**, *40*, 3589-3591.
178. Reetz, M. T.; Torre, C.; Eipper, A.; Lohmer, R.; Hermes, M.; Brunner, B.; Maichele, A.; Bocola, M.; Arand, M.; Cronin, A.; Genzel, Y.; Archelas, A.; Furstoss, R. *Org. Lett.* **2004**, *6*, 177-180.
179. Napper, A. D.; Benkovic, S. J.; Tramontano, A.; Lerner, R. A. *Science* **1987**, *237*, 1041-1043.
180. Janda, K. D.; Benkovic, S. J.; Lerner, R. A. *Science* **1989**, *244*, 437-440.
181. Pollack, S. J.; Hsiun, P.; Schultz, P. G. *J. Am. Chem. Soc.* **1989**, *111*, 5961-5962.
182. Tanaka, F.; Kinoshita, K.; Tanimura, R.; Fujii, I. *J. Am. Chem. Soc.* **1996**, *118*, 2332-2339.

183. Wade, H.; Scanlan, T. S. *J. Am. Chem. Soc.* **1996**, *118*, 6510-6511.

184. Gijsen, H. J. M.; Qiao, L.; Fitz, W.; Wong, C. H. *Chem. Rev.* **1996**, *96*, 443-473.

185. Lo, C.-H. L.; Wentworth Jr., P.; Jung, K. W.; Yoon, J.; Ashley, J. A.; Janda, K. D. *J. Am. Chem. Soc.* **1997**, *119*, 10251-10252.

186. Flanagan, M. E.; Jacobsen, J. R.; Sweet, E.; Schultz, P. G. *J. Am. Chem. Soc.* **1996**, *118*, 6078-6079.

187. Zhong, G.; Shabat, D.; List, B.; Anderson, J.; Sinha, S. C.; Lerner, R. A.; Barbas III, C. F. *Angew. Chem., Int. Ed.* **1998**, *37*, 2481-2484.

188. Mitsuda, S.; Umemura, T.; Hirohara, H. *Appl. Microbiol. Biotechnol.* **1988**, *29*, 310-315.

189. Danda, H.; Nagatomi, T.; Maehara, A.; Umemura, T. *Tetrahedron* **1991**, *47*, 8701-8716.

190. Wallner, A.; Mang, H.; Glueck, S. M.; Steinreiber, A.; Mayer, S. F.; Faber, K. *Tetrahedron: Asymm.* **2003**, *14*, 2427-2432.

191. Kroutil, W.; Mischitz, M.; Faber, K. *J. Chem. Soc., Perkin Trans. 1* **1997**, 3629-3636.

192. Belucci, G.; Berti, G.; Catelani, G.; Mastrorilli, E. *J. Org. Chem.* **1981**, *46*, 5148-5150.

193. Belucci, G.; Chiappe, C.; Cordoni, A. *Tetrahedron: Asymm.* **1996**, *7*, 197-202.

194. Orru, R. V. A.; Mayer, S. F.; Kroutil, W.; Faber, K. *Tetrahedron* **1998**, *54*, 859-874.

195. Ueberbacher, B. J.; Osprian, I.; Mayer, S. F.; Faber, K. *Eur. J. Org. Chem.* **2005**, 1266-1270.

196. Chadha, A.; Baskar, B. *Tetrahedron: Asymm.* **2002**, *13*, 1461-1464.

197. Kamphuis, J.; Boesten, W. H. J.; Kaptein, B.; Hermes, H. F. M.; Sonke, T.; Broxterman, Q. B.; van den Tweel, W. J. J.; Schoemaker, H. E. In: Collins, A. N.; Sheldrake, G. N.; Crosby, J. (Eds.) *Chirality in Industry*, Wiley, New York, 1992, pp. 187-208.

198. Strauss, U. T.; Faber, K. *Tetrahedron: Asymm.* **1999**, *10*, 4079-4081.

199. Kawabata, T.; Nagato, M.; Takasu, K.; Fuji, K. *J. Am. Chem. Soc.* **1997**, *119*, 3169-3170.

200. Copeland, G. T.; Jarvo, E. R.; Miller, S. J. *J. Org. Chem.* **1998**, *63*, 6784-6785.

201. Sano, T.; Imai, K.; Ohashi, K.; Oriyama, T. *Chem. Lett.* **1999**, 265-266.

202. Jarvo, E. R.; Copeland, G. T.; Papaioannou, N.; Bonitatebus Jr., P. J.; Miller, S. J. *J. Am. Chem. Soc.* **1999**, *121*, 11638-11643.

203. Miller, S. J.; Copeland, G. T.; Papaioannou, N.; Horstmann, T. E.; Ruel, E. M. *J. Am. Chem. Soc.* **1998**, *120*, 1629-1630.

204. Copeland, G. T.; Miller, S. J. *J. Am. Chem. Soc.* **2001**, *123*, 6496-6502.

205. Dro, C.; Bellemin-Laponnaz, S.; Welter, R.; Gade, L. H. *Angew. Chem., Int. Ed.* **2004**, *43*, 4479-4482.

206. Ishihara, K.; Kosugi, Y.; Akakura, M. *J. Am. Chem. Soc.* **2004**, *126*, 12212-12213.

207. Keith, J. M.; Larrow, J. F.; Jacobsen, E. N. *Adv. Synth. Catal.* **2001**, *343*, 5-26.

208. Robinson, D. E. J. E.; Bull, S. D. *Tetrahedron: Asymm.* **2003**, *14*, 1407-1446.

209. Katsuki, T.; Sharpless, K. B. *J. Am. Chem. Soc.* **1980**, *102*, 5974-5976.

210. Martin, V. S.; Woodard, S. S.; Katsuki, T.; Yamada, Y.; Ikeda, M.; Sharpless, K. B. *J. Am. Chem. Soc.* **1981**, *103*, 6237-6240.

211. Hanson, R. M.; Sharpless, K. B. *J. Org. Chem.* **1986**, *51*, 1922-1925.

212. Kitano, Y.; Matsumoto, T.; Sato, F. *J. Chem. Soc., Chem. Commun.* **1986**, 1323-1325.

213. Kitano, Y.; Matsumoto, T.; Takeda, Y.; Sato, F. *J. Chem. Soc., Chem. Commun.* **1986**, 1732-1733.

214. Kitano, Y.; Matsumoto, T.; Sato, F. *Tetrahedron* **1988**, *44*, 4073-4086.

215. Gao, Y.; Hanson, R. M.; Klunder, J. M.; Ko, S. Y.; Masamune, H. Sharpless, K. B. *J. Am. Chem. Soc.* **1987**, *109*, 5765-5780.

216. Carlier, P. R.; Mungati, W. S.; Schröder, G.; Sharpless, K. B. *J. Am. Chem. Soc.* **1988**, *110*, 2978-2979.

217. Kobayashi, Y.; Kusakabe, M.; Kitano, Y.; Sato, F. *J. Org. Chem.* **1988**, *53*, 1586-1587.

218. Honda, T.; Sano, N.; Kanai, K. *Heterocycles* **1995**, *41*, 425-429.

219. Yang, Z.-C.; Jiang, X.-B.; Wang, Z.-M.; Zhou, W.-S. *J. Chem. Soc., Perkin Trans. 1* **1997**, 317-321.

220. Black, P. J.; Jenkins, K.; Williams, J. M. J. *Tetrahedron: Asymm.* **2002**, *13*, 317-323.

221. Miyano, S.; Lu, L. D.-L.; Viti, S. M.; Sharpless, K. B. *J. Org. Chem.* **1983**, *48*, 3608-3611.

222. Miyano, S.; Lu, L. D.-L.; Viti, S. M.; Sharpless, K. B. *J. Org. Chem.* **1985**, *50*, 4350-4360.

223. Hamon, D. G.; Tuck, K. L. *J. Org. Chem.* **2000**, *65*, 7839-7846.

224. Kobayashi, Y.; Yoshida, S.; Nukayama, Y. *Eur. J. Org. Chem.* **2001**, 1873-1881.

225. Cossy, J.; Bauer, D.; Bellosta, V. *Tetrahedron* **2002**, *58*, 5909-5922.

226. Mulzer, J.; Öhler, E. *Angew. Chem., Int. Ed.* **2001**, *40*, 3842-3846.

227. Kitamura, M.; Kasahara, I.; Manabe, K.; Noyori, R.; Takaya, H. *J. Org. Chem.* **1988**, *53*, 708-710.

228. Hashiguchi, S.; Fujii, A.; Takehara, J.; Ikariya, T.; Noyori, R. *J. Am. Chem. Soc.* **1995**, *117*, 7562-7563.

229. Hashiguchi, S.; Fujii, A.; Haack, K.-J.; Matsumura, K.; Ikariya, T.; Noyori, R. *Angew. Chem., Int. Ed. Engl.* **1997**, *36*, 288-290.

230. Evans, D. A.; Anderson, J. C.; Taylor, M. K. *Tetrahedron Lett.* **1993**, *34*, 5563-5566.

231. Vedejs, E.; Chen, X. *J. Am. Chem. Soc.* **1996**, *118*, 1809-1810.

232. Ruble, J. C.; Latham, H. A.; Fu, G. C. *J. Am. Chem. Soc.* **1997**, *119*, 1492-1493.

233. Bellemin-Lapponaz, S.; Tweddell, J.; Ruble, J. C.; Breitling, F. M.; Fu, G. C. *Chem. Commun.* **2000**, 1009-1010.

234. Nishibayashi, Y.; Takei, I.; Uemura, S.; Hidai, M. *Organometallics* **1999**, *18*, 2291-2293.

235. Sekar, G.; Nishiyama, H. *J. Am. Chem. Soc.* **2001**, *123*, 3603-3604.

236. Jensen, D. R.; Pugsley, J. S.; Sigman, M. S. *J. Am. Chem. Soc.* **2001**, *123*, 7475-7476.

237. Ferreira, E. M.; Stoltz, B. M. *J. Am. Chem. Soc.* **2001**, *123*, 7725-7726.

238. Mueller, J. A.; Jensen, D. R.; Sigman, M. S. *J. Am. Chem. Soc.* **2002**, *124*, 8202-8203.

239. Mikami, K.; Yusa, Y.; Korenaga, T. *Org. Lett.* **2002**, *4*, 1643-1645.

240. Sun, W.; Wang, H.; Xia, C.; Li, J.; Zhao, P. *Angew. Chem., Int. Ed.* **2003**, *42*, 1042-1044.

241. Mandal, S. K.; Sigman, M. S. *J. Org. Chem.* **2003**, *68*, 7535-7537.

242. Birman, V. B.; Uffman, E. W.; Jiang, H.; Li, X.; Kilbane, C. J. *J. Am. Chem. Soc.* **2004**, *126*, 12226-12227.

243. Chen, S. L.; Hu, Q.-Y.; Loh, T.-P. *Org. Lett.* **2004**, *6*, 3365-3367.

244. Suzuki, Y.; Yamauchi, K.; Muramatsu, K.; Sato, M. *Chem. Commun.* **2004**, 2770-2771.

245. Matsumura, Y.; Maki, T.; Tsurumaki, K.; Onomura, O. *Tetrahedron Lett.* **2004**, *45*, 9131-9134.

246. Martinez, L. E.; Leighton, J. L.; Carsten, D. H.; Jacobsen, E. N. *J. Am. Chem. Soc.* **1995**, *117*, 5897-5898.

247. Larrow, J. F.; Schaus, S. E.; Jacobsen, E. N. *J. Am. Chem. Soc.* **1996**, *118*, 7420-7421.

248. Lebel, H.; Jacobsen, E. N. *Tetrahedron Lett.* **1999**, *40*, 7303-7306.

249. Bandini, M.; Cozzi, P. G.; Melchiorre, P.; Umani-Ronchi, A. *Angew. Chem., Int. Ed.* **2004**, *43*, 84-87.

250. Tokunaga, M.; Larrow, J. F.; Kakiuchi, F.; Jacobsen, E. N. *Science* **1997**, *277*, 936-938.

251. Furrow, M. E.; Schaus, S. E.; Jacobsen, E. N. *J. Org. Chem.* **1998**, *63*, 6776-6777.

252. Annis, D. A.; Jacobsen, E. N. *J. Am. Chem. Soc.* **1999**, *121*, 4147-4154.

253. Jacobsen, E. N. *Acc. Chem. Res.* **2000**, *33*, 421-431.

254. Schaus, S. E.; Brandes, B. D.; Larrow, J. F.; Tokunaga, M.; Hansen, K. B.; Gould, A. E.; Furrow, M. E.; Jacobsen, E. N. *J. Am. Chem. Soc.* **2002**, *124*, 1307-1315.

255. Ready, J. M.; Jacobsen, E. N. *J. Am. Chem. Soc.* **2001**, *123*, 2687-2688.

256. Ready, J. M.; Jacobsen, E. N. *Angew. Chem., Int. Ed.* **2002**, *41*, 1374-1377.

257. Larrow, J. F.; Hemberger, K. E.; Jasmin, S.; Kabir, H.; Morel, P. *Tetrahedron: Asymm.* **2003**, *14*, 3589-3592.

258. Wu, M. H.; Hansen, K. B.; Jacobsen, E. N. *Angew. Chem., Int. Ed.* **1999**, *38*, 2012-2014.

259. Ready, J. M.; Jacobsen, E. N. *J. Am. Chem. Soc.* **1999**, *121*, 6086-6087.

260. Jacobsen, E. N.; Zhang, W.; Muci, A. R.; Ecker, J. R.; Deng, L. *J. Am. Chem. Soc.* **1991**, *113*, 7063-7064.

261. Zhang, W.; Loebach, J. L.; Wilson, S. R.; Jacobsen, E. N. *J. Am. Chem. Soc.* **1990**, *112*, 2801-2803.

262. Wang, Z.-X.; Tu, W.; Zhang, J.-R.; Shi, Y. *J. Am. Chem. Soc.* **1997**, *119*, 11224-11235.

263. Liu, Z.-Y.; Chen, Z.-C.; Yu, C.-Z.; Wang, R.-F.; Zhang, R.-Z.; Huang, C.-S.; Yan, Z.; Cao, D.-R.; Sun, J.-B.; Li, G. *Chem. Eur. J.* **2002**, *8*, 3747-3756.

264. Dorling, E. K.; Öhler, E.; Mantoulidis, A.; Mulzer, J. *Synlett* **2001**, 1105-1108.

265. Chavez, D. E.; Jacobsen, E. N. *Angew. Chem., Int. Ed.* **2001**, *40*, 3667-3670.

266. Liu, P.; Panek, S. *J. Am. Chem. Soc.* **2000**, *122*, 1235-1236.

267. Knölker, H.-J.; Baum, E.; Reddy, K. R. *Tetrahedron Lett.* **2000**, *41*, 1171-1174.

268. Yadav, J. S.; Bandyopadhyay, A.; Kunwar, A. C. *Tetrahedron Lett.* **2001**, *42*, 4907-4911.

269. Celatka, C. A.; Panek, J. S. *Tetrahedron Lett.* **2002**, *43*, 7043-7046.

270. Fürstner, A.; Thiel, O. R.; Ackermann, L. *Org. Lett.* **2001**, *3*, 449-451.

271. Jacobsen, E. N.; Marko, I.; Mungall, W. M.; Schroder, G.; Sharpless, K. B. *J. Am. Chem. Soc.* **1988**, *110*, 1968-1970.

272. Van Nieuwenhze, M. S.; Sharpless, K. B. *J. Am. Chem. Soc.* **1993**, *115*, 7864-7865.

273. Hawkins, J. M.; Meyer, A. *Science* **1993**, *260*, 1918-1920.

274. Hawkins, J. M.; Nambu, M. *J. Am. Chem. Soc.* **1994**, *116*, 7642-7645.

275. Corey, E. J.; Noe, M. C.; Guzman-Perez, A. *J. Am. Chem. Soc.* **1995**, *117*, 10817-10824.

276. Christie, H. S.; Hamon, D. P. G.; Tuck, K. L. *Chem. Commun.* **1999**, 1989-1990.

277. Rios, R.; Jimeno, C.; Carroll, P. J.; Walsh, P. J. *J. Am. Chem. Soc.* **2002**, *124*, 10272-10273.

278. Frohn, M.; Zhou, X.; Zhang, J.-R.; Tang, Y.; Shi, Y. *J. Am. Chem. Soc.* **1999**, *121*, 7718-7719.

279. Van der Velde, S. L.; Jacobsen, E. N. *J. Org. Chem.* **1995**, *60*, 5380-5381.

280. Linker, T.; Rebien, F.; Toth, G.; Simon, A.; Kraus, J.; Bringmann, G. *Chem. Eur. J.* **1998**, *4*, 1944-1951.

281. Brown, H. C.; Ayyangar, N. R.; Zweifel, G. *J. Am. Chem. Soc.* **1964**, *86*, 397-403.

282. Maeda, K.; Brown, J. M. *Chem. Commun.* **2002**, 310-311.

283. Johnson, T. H.; Baldwin, T. F.; Klein, K. C. *Tetrahedron Lett.* **1979**, *20*, 1191-1192.

284. Meyers, A. I.; Kamata, K. *J. Am. Chem. Soc.* **1976**, *98*, 2290-2294.

285. Wiesner, K.; Jay, E. W. K.; Tsai, T. Y. R.; Demerson, C.; Jay, L.; Kanno, T.; Krepinsky, J.; Vilim, A.; Wu, C. S. *Can. J. Chem.* **1972**, *50*, 1925-1943.

286. Arai, S.; Bellemin-Lapponaz, S.; Fu, G. C. *Angew. Chem., Int. Ed.* **2001**, *40*, 234-236.

287. Arseniyadis, S.; Valleix, A.; Wagner, A.; Mioskowski, C. *Angew. Chem., Int. Ed.* **2004**, *43*, 3314-3317.

288. Arseniyadis, S.; Subhash, P. V.; Valleix, A.; Wagner, A.; Mioskowski, C. *Chem. Commun.* **2005**, 3310-3312.

289. Hashiguchi, S.; Fujii, A.; Haack, K.J.; Matsumura, K.; Ikariya, T.; Noyori, R. *Angew. Chem., Int. Ed. Engl.* **1997**, *36*, 288-290.

290. Noyori, R.; Yamakawa, M.; Hashiguchi, S. *J. Org. Chem.* **2001**, *66*, 7931-7944.

291. Birman, V. B.; Jiang, H. *Org. Lett.* **2005**, *7*, 3445-3447.

292. Rendler, S.; Auer, G.; Oestreich, M. *Angew. Chem., Int. Ed.* **2005**, *44*, 7620-7624.

293. Tanaka, K.; Fu, G. C. *J. Am. Chem. Soc.* **2002**, *124*, 10296-10297.

294. Morken, J. P.; Didiuk, M. T.; Visser, M. S.; Hoveyda, A. H. *J. Am. Chem. Soc.* **1994**, *116*, 3123-3124.

295. Visser, M. S.; Heron, N. M.; Didiuk, M. T.; Sagal, J. F.; Hoveyda, A. H. *J. Am. Chem. Soc.* **1996**, *118*, 4291-4298.

296. Alexander, J. B.; La, D. S.; Cefalo, D. R.; Hoveyda, A. H.; Schrock, R. R. *J. Am. Chem. Soc.* **1998**, *120*, 2499-2500.

297. La, D. S.; Alexander, J. B.; Cefalo, D. R.; Graf, D. D.; Hoveyda, A. H.; Schrock, R. R. *J. Am. Chem. Soc.* **1998**, *120*, 9720-9721.

298. Aoyama, H.; Tokunaga, M.; Kiyosu, J.; Iwasawa, T.; Obora, Y.; Tsuji, Y. *J. Am. Chem. Soc.* **2005**, *127*, 10474-10475.

299. Calet, S.; Urso, F.; Alper, H. *J. Am. Chem. Soc.* **1989**, *111*, 931-934.

300. Viso, A.; Lee, N. E.; Buchwald, S. L. *J. Am. Chem. Soc.* **1994**, *116*, 9373-9374.

301. Yun, J.; Buchwald, S. L. *J. Org. Chem.* **2000**, *65*, 767-774.

302. Yun, J.; Buchwald, S. L. *Chirality* **2000**, *12*, 476-478.

303. Sun, J.; Zhu, C.; Dai, Z.; Yang, M.; Pan, Y.; Hu, H. *J. Org. Chem.* **2004**, *69*, 8500-8503.

304. Jia, X.; Li, X.; Xu, L.; Li, Y.; Shi, Q.; Au-Yeung, T. T.-L.; Yip, C. W.; Yao, X.; Chan, A. S. C. *Adv. Synth. Catal.* **2004**, *346*, 723-726.

305. Schurig, V.; Betschinger, F. *Bull. Soc. Chi. Fr.* **1994**, *131*, 555-560.

306. Södergren, M. J.; Bertilsson, S. K.; Andersson, P. G. *J. Am. Chem. Soc.* **2000**, *122*, 6610-6618.

307. Gayet, A.; Bertilsson, S.; Andersson, P. G. *Org. Lett.* **2002**, *4*, 3777-3779.

308. Sweeney, Z. K.; Salsman, J. L.; Andersen, R. A.; Bergmann, R. G. *Angew. Chem., Int. Ed.* **2000**, *39*, 2339-2343.

309. Chemla, F.; Ferreira, F. *J. Org. Chem.* **2004**, *69*, 8244-8250.

310. Naasz, R.; Arnold, L. A.; Minnaard, A. J.; Feringa, B. L. *Angew. Chem., Int. Ed.* **2001**, *40*, 927-930.

311. Urbaneja, L. M.; Alexakis, A.; Krause, N. *Tetrahedron Lett.* **2002**, *43*, 7887-7890.

312. Fehr, C.; Galindo, J.; Etter, O. *Eur. J. Org. Chem.* **2004**, 1953-1957.

313. Ramdeehul, S.; Dierkes, P.; Aguado, R.; Kamer, P. C. J.; van Leeuwen, P. W. N. M.; Osborn, J. A. *Angew. Chem., Int. Ed.* **1998**, *37*, 3118-3121.

314. Trost, B. M.; Hembre, E. J. *Tetrahedron Lett.* **1999**, *40*, 219-222.

315. Gilbertson, S. R.; Lan, P. *Org. Lett.* **2001**, *3*, 2237-2240.

316. Bailey, S.; Davies, S. G.; Smith, A. D.; Withey, J. M. *Chem. Commun.* **2002**, 2910-2911.

317. Snyder, S. E.; Pirkle, W. H. *Org. Lett.* **2002**, *4*, 3283-3286.

318. Lüssem, B. J.; Gais, H.-J. *J. Am. Chem. Soc.* **2003**, *125*, 6066-6067.

319. Faller, J. W.; Wilt, J. C.; Parr, J. *Org. Lett.* **2004**, *6*, 1301-1304.

320. Woltering, M. J.; Fröhlich, R.; Wibbeling, B.; Hoppe, D. *Synlett* **1998**, 797-800.

321. Weber, B.; Schwerdtfeger, J.; Fröhlich, R.; Göhrt, A.; Hoppe, D. *Synthesis* **1999**, 1915-1924.

322. Laqua, H.; Fröhlich, R.; Wibbeling, B.; Hoppe, D. *J. Organomet. Chem.* **2001**, *624*, 96-104.

323. Wegener, B.; Hansen, M.; Winterfeldt, E. *Tetrahedron: Asymm.* **1993**, *4*, 345-350.

324. Lim, S. H.; Beak, P. *Org. Lett.* **2002**, *4*, 2657-2660.

325. Horeau, A.; Nouaille, A. *Tetrahedron Lett.* **1966**, *7*, 3953-3959.

326. Horeau, A. *Determination of the Configuration of Secondary Alcohols by Partial Resolution* In: Kagan, H. B. (Ed.) *Stereochemistry, Fundamentals and Methods*, Vol. 3, Georg Thieme Verlag, Stuttgart, 1977, chapter 3.

327. Guo, J.; Wu, J.; Siuzdak, G.; Finn, M. G. *Angew. Chem., Int. Ed.* **1999**, *38*, 1755-1758.

328. Vedejs, E.; Chen, X. *J. Am. Chem. Soc.* **1997**, *119*, 2584-2585.

329. Eames, J. *Angew. Chem., Int. Ed.* **2000**, *39*, 885-888.

330. Dehli, J. R.; Gotor, V. *Chem. Soc. Rev.* **2002**, *31*, 365-370.

331. Königsberger, K.; Alphand, V.; Furstoss, R.; Griengl, H. *Tetrahedron Lett.* **1991**, *32*, 499-500.

332. Mischitz, M.; Faber, K. *Tetrahedron Lett.* **1994**, *35*, 81-84.

333. Cardona, F.; Valenza, S.; Goti, A.; Brandi, A. *Eur. J. Org. Chem.* **1999**, 1319-1323.

334. Vedejs, E.; Rozners, E. *J. Am. Chem. Soc.* **2001**, *123*, 2428-2429.

335. Al-Sehemi, A. G.; Atkinson, R. S.; Meades, C. K. *Chem. Commun.* **2001**, 2684-2685.

336. Davies, S. G.; Diez, D.; El Hammouni, M. M.; Christopher G. A.; Garrido, N. M.; Long, M. J. C.; Morrison, R. M.; Smith, A. D.; Sweet, M. J.; Withey, J. M. *Chem. Commun.* **2003**, 2410-2411.

337. Davies, S. G.; Garner, A. C.; Long, M. J. C.; Smith, A. D.; Sweet, M. J.; Withey, J. M. *Org. Biomol. Chem.* **2004**, *2*, 3355-3362.

338. Davies, S. G.; Garner, A. C.; Long, M. J. C.; Morrison, R. M.; Roberts, P. M.; Savory, E. D.; Smith, A. D.; Sweet, M. J.; Withey, J. M. *Org. Biomol. Chem.* **2005**, *3*, 2762-2775.

339. Coumbarides, G. S.; Eames, J.; Flinn, A.; Northen, J.; Yohannes, Y. *Tetrahedron Lett.* **2005**, *46*, 2897-2902.

340. Liao, L.-A.; Zhang, F.; Dmintrenko, O.; Bach, R. D.; Fox, J. M. *J. Am. Chem. Soc.* **2004**, *126*, 4490-4491.

341. Tanaka, K.; Fu, G. C. *J. Am. Chem. Soc.* **2002**, *124*, 10296-10297.

342. Tanaka, K.; Fu, G. C. *J. Am. Chem. Soc.* **2003**, *125*, 8078-8079.

343. Brooks, D. W.; Wilson, M.; Webb, M. *J. Org. Chem.* **1987**, *52*, 2244-2248.

344. Bolm, C.; Schlingloff, G. *J. Chem. Soc., Chem. Commun.* **1995**, 1247-1248.

345. Yang, Z.-C.; Jiang, X.-B.; Wang, Z.-M.; Zhou, W-S. *J. Chem. Soc., Chem. Commun.* **1995**, 2389-2390.

346. Honda, T.; Sano, N.; Kanai, K. *Heterocycles* **1995**, *41*, 425-429.

347. Chen, Y.; Deng, L. *J. Am. Chem. Soc.* **2001**, *123*, 11302-11303.

348. Visser, M. S.; Hoveyda, A. H. *Tetrahedron* **1995**, *51*, 4383-4394.

349. Alphand, V.; Furstoss, R. *J. Org. Chem.* **1992**, *57*, 1306-1309.

350. Doyle, M. P.; Forbes, D. C. *Chem Rev.* **1998**, *98*, 911-936.

351. Martin, S. F.; Spaller, M. R.; Liras, S.; Hartmann, B. *J. Am. Chem. Soc.* **1994**, *116*, 4493-4494.

352. Bertozzi, F.; Crotti, P.; Macchia, F.; Pineschi, M.; Feringa, B. L. *Angew. Chem., Int. Ed.* **2001**, *40*, 930-932.

353. Pineschi, M.; Del Moro, F.; Crotti, P.; Di Bussolo, V.; Macchia, F. *J. Org. Chem.* **2004**, *69*, 2009-2105.

354. Rein, T.; Kann, N.; Kreuder, R.; Gangloff, B.; Reiser, O. *Angew. Chem., Int. Ed. Engl.* **1994**, *33*, 556-558.

355. Pedersen, T. M.; Hansen, E. L.; Kane, J.; Rein, T.; Helquist, P.; Norrby, P.-O.; Tanner, D. *J. Am. Chem. Soc.* **2001**, *123*, 9738-9742.

356. Kurosu, M.; Kishi, Y. *J. Org. Chem.* **1998**, *63*, 6100-6101.

357. Davies, J.; Jones, J. B. *J. Am. Chem. Soc.* **1979**, *101*, 5405-5410.

358. Bianchi, P.; Roda, G.; Riva, S.; Danieli, B.; Zabelinskaja-Mackova, A.; Griengl, H. *Tetrahedron* **2001**, *57*, 2213-2220.

359. Dehli, J. R.; Gotor, V. *J. Org. Chem.* **2002**, *67*, 1716-1718.

360. Stürmer, R. *Angew. Chem., Int. Ed. Engl.* **1997**, *36*, 1173-1174.

361. Huerta, F. F.; Minidis, A. B. E.; Bäckvall, J.-E. *Chem. Soc. Rev.* **2001**, *30*, 321-331.

362. Kim, M.-J.; Ahn, Y.; Park, J. *Curr. Opin. Biotechnol.* **2002**, *13*, 578-587.

363. Pàmies, O.; Bäckvall, J. E. *Chem. Rev.* **2003**, *103*, 3247-3261.

364. Pàmies, O.; Bäckvall, J. E. *Trends Biotechnol.* **2004**, *22*, 130-135.

365. Turner, N. J. *Curr. Opin. Chem. Biol.* **2004**, *8*, 114-119.

366. Dinh, P. M.; Howarth, J. A.; Hudnott, A. R.; Williams, J. M. J.; Harris, W. *Tetrahedron Lett.* **1996**, *37*, 7623-7626.

367. Koh, J. H.; Jung, M.; Kim, M.-J.; Park, J. *Tetrahedron Lett.* **1999**, *40*, 6281-6284.

368. Choi, J. H.; Kim, Y. H.; Nam, S. H.; Shin, S. T.; Kim, M.-J.; Park, J. *Angew. Chem., Int. Ed.* **2002**, *41*, 2373-2376.

369. Choi, J. H.; Choi, Y. K.; Kim, Y. H.; Park, E. S.; Kim, E. J.; Kim, M.-J.; Park, J. *J. Org. Chem.* **2004**, *69*, 1972-1977.

370. Riermeier, T. H.; Gross, P.; Monsees, A.; Hoff, M.; Trauthwein, H. *Tetrahedron Lett.* **2005**, *46*, 3403-3406.

371. Kim, M.-J.; Choi, Y. K.; Choi, M. Y.; Kim, M.; Park, J. *J. Org. Chem.* **2001**, *66*, 4736-4738.
372. Reetz, M.; Schimossek, K. *Chimia* **1996**, *50*, 668-669.
373. Allen, J. V.; Williams, J. M. J. *Tetrahedron Lett.* **1996**, *37*, 1859-1862.
374. Choi, Y. K.; Suh, J. H.; Lee, D.; Lim, I. T.; Jung, J. Y.; Kim, M. J. *J. Org. Chem.* **1999**, *64*, 8423-8424.
375. Lee, D.; Huh, E. A.; Kim, M.-J.; Jung, H. M.; Koh, J. H.; Park, J. *Org. Lett.* **2000**, *2*, 2377-2379.
376. Pàmies, O.; Bäckvall, J. E. *J. Org. Chem.* **2002**, *67*, 1261-1265.
377. Larsson, A. L. E.; Persson, B. A.; Bäckvall, J. E. *Angew. Chem., Int. Ed. Engl.* **1997**, *36*, 1211-1212.
378. Persson, B. A.; Larsson, A. L. E.; Le Ray, M.; Bäckvall, J. E. *J. Am. Chem. Soc.* **1999**, *121*, 1645-1650.
379. Martin-Matute, B.; Edin, M.; Bogar, K.; Bäckvall, J. E. *Angew. Chem., Int. Ed.* **2004**, *43*, 6535-6539.
380. Pàmies, O.; Bäckvall, J. E. *J. Org. Chem.* **2002**, *67*, 9006-9010.
381. Pàmies, O.; Bäckvall, J. E. *J. Org. Chem.* **2001**, *66*, 4022-4025. Correction: *J. Org. Chem.* **2002**, *67*, 1418.
382. Huerta, F. F.; Laxmi, Y. R. S.; Bäckvall, J. E. *Org. Lett.* **2000**, *2*, 1037-1040.
383. Huerta, F. F.; Bäckvall, J. E. *Org. Lett.* **2001**, *3*, 1209-1212.
384. Pàmies, O.; Bäckvall, J. E. *Adv. Synth. Catal.* **2002**, *344*, 947-952.
385. Pàmies, O.; Bäckvall, J. E. *J. Org. Chem.* **2003**, *68*, 4815-4818.
386. Runmo, A.-B. L.; Pàmies, O.; Faber, K.; Bäckvall, J. E. *Tetrahedron Lett.* **2002**, *43*, 2983-2986.
387. Fransson, A.-B. L.; Borén, L.; Pàmies, O.; Bäckvall, J. E. *J. Org. Chem.* **2005**, *70*, 2582-2587.
388. Choi, Y. K.; Suh, J. H.; Lee, D.; Lim, I. T.; Jung, J. Y.; Kim, M.-J. *J. Org. Chem.* **1999**, *64*, 8423-8424.
389. Granberg, K. L.; Bäckvall, J. E. *J. Am. Chem. Soc.* **1992**, *114*, 6858-6863.
390. Jung, H. M.; Koh, J. H.; Kim, M.-J.; Park, J. *Org. Lett.* **2000**, *2*, 409-411.
391. Jung, H. M.; Koh, J. H.; Kim, M.-J.; Park, J. *Org. Lett.* **2000**, *2*, 2487-2490.
392. Choi, Y.; Kim, M. J.; Ahn, Y.; Kim, M. J. *Org. Lett.* **2001**, *3*, 4099-4101.
393. Akai, S.; Tanimoto, K.; Kita, Y. *Angew. Chem., Int. Ed.* **2004**, *43*, 1407-1410.
394. Akai, S.; Tanimoto, K.; Kanao, Y.; Omura, S.; Kita, Y. *Chem. Commun.* **2005**, 2369-2371.
395. Paizs, C.; Tähtinen, P.; Lundell, K.; Poppe, L.; Irimie, F.-D.; Kanerva, L. T. *Tetrahedron: Asymm.* **2003**, *14*, 1895-1904.
396. Inagaki, M.; Hiratake, J.; Nishioka, T.; Oda, J. *J. Am. Chem. Soc.* **1991**, *113*, 9360-9361.
397. Paizs, C.; Tähtinen, P.; Lundell, K.; Poppe, L.; Irimie, F.-D.; Kanerva, L. T. *Tetrahedron: Asymm.* **2003**, *14*, 1895-1904.
398. Wuyts, S.; de Temmerman, K.; de Vos, D.; Jacobs, P. A. *Chem. Commun.* **2003**, 1928-1929.
399. Turner, N. J.; Winterman, J. R. *Tetrahedron Lett.* **1995**, *36*, 1113-1116.
400. Brown, S. A.; Parker, M.-C.; Turner, N. J. *Tetrahedron: Asymm.* **2000**, *11*, 1687-1690.
401. Lin, H.-Y.; Tsai, S.-W. *J. Mol. Catal. B* **2003**, *24-25*, 111-120.
402. Tan, D. S.; Gunter, M. M.; Drueckhammer, D. G. *J. Am. Chem. Soc.* **1995**, *117*, 9093-9094.
403. Pesti, J. A.; Yin, J.; Zhang, L. H.; Anzalone, L.; Waltermire, R. E.; Ma, P.; Gorko, E.; Confalone, P. N.; Fortunak, J.; Silverman, C.; Blackwell, J.; Chung, J. C.; Hrytsak, M. D.; Cooke, M.; Powell, L.; Ray, C. *Org. Proc. Res. Dev.* **2004**, *8*, 22-27.
404. Brand, S.; Jones, M. F.; Rayner, C. M. *Tetrahedron Lett.* **1995**, *36*, 8493-8496.
405. Jones, M. M.; Williams, J. M. J. *Chem. Commun.* **1998**, 2519-2520.
406. Laugthon, L.; Williams, J. M. J. *Synthesis* **2001**, 943-946.
407. Wegman, M. A.; Hacking, M. A. P. J.; Rops, J.; Pereira, P.; van Rantwijk, F.; Sheldon, R. A. *Tetrahedron: Asymm.* **1999**, *10*, 1739-1750.

408. Chen, S.-T.; Huang, W.-H.; Wang, K.-T. *J. Org. Chem.* **1994**, *59*, 7580-7581.

409. Parmar, V. S.; Singh, A.; Bisht, K. S.; Kumar, N.; Belekon, Y. N.; Kochetkov, K. A.; Ikonnikov, N. S.; Orlova, S. A.; Tararov, V. I.; Saveleva, T. F. *J. Org. Chem.* **1996**, *61*, 1223-1227.

410. Kim, M.-J.; Chung, Y. I.; Choi, Y. K.; Lee, H. K.; Kim, D.; Park, J. *J. Am. Chem. Soc.* **2003**, *125*, 11494-11495.

411. Garcia, M. J.; Azerad, R. *Tetrahedron: Asymm.* **1997**, *8*, 85-92.

412. Suzuki, M.; Yamazaki, T.; Ohta, H.; Shima, K.; Ohi, K.; Nishiyama, S.; Sugai, T. *Synlett* **2000**, 189-192.

413. Dehli, J. R.; Gotor, V. *J. Org. Chem.* **2002**, *67*, 6816-6819.

414. Fantin, G.; Fogagnolo, M.; Giovannini, P. P.; Medici, A.; Pedrini, P. *Tetrahedron* **1996**, *52*, 3547-3552.

415. Danchet, S.; Bigot, C.; Buisson, D.; Azerad, R. *Tetrahedron: Asymm.* **1997**, *8*, 1735-1739.

416. Ji, A.; Wolberg, M.; Hummel, W.; Wandrey, C.; Müller, M. *Chem. Commun.* **2001**, 57-58.

417. Quiros, M.; Rebolledo, F.; Gotor, V. *Tetrahedron: Asymm.* **1999**, *10*, 473-486.

418. Maguire, A. R.; O'Riordan, N. *Tetrahedron Lett.* **1999**, *40*, 9285-9288.

419. Rodriguez, S.; Schroeder, K. T.; Kayser, M. M.; Stewart, J. D. *J. Org. Chem.* **2000**, *65*, 2586-2587.

420. Berezina, N.; Alphand, V.; Furstoss, R. *Tetrahedron: Asymm.* **2002**, *13*, 1953-1955.

421. Gutierrez, M.-C.; Furstoss, R.; Alphand, V. *Adv. Synth. Catal.* **2005**, *347*, 1051-1059.

422. Stecher, H.; Faber, K. *Synthesis* **1997**, 1-16.

423. Tokuyama, S.; Hatano, K. *Appl. Microbiol. Biotechnol.* **1996**, *44*, 774-777.

424. Wakayama, M.; Yoshimune, K.; Hirose, Y. Moriguchi, M. *J. Mol. Catal. B* **2003**, *23*, 71-85.

425. Asano, Y.; Yamaguchi, S. *J. Am. Chem. Soc.* **2005**, *127*, 7696-7697.

426. Spelberg, J. H. L.; Tang, L.; Kellog, R. M.; Janssen, D. B. *Tetrahedron: Asymm.* **2004**, *15*, 1095-1102.

427. Schulze, B.; Wubbolts, M. G. *Curr. Opin. Biotechnol.* **1999**, *10*, 6609-6615.

428. May, O.; Verseck, S.; Bommarius, A.; Drauz, K. *Org. Proc. Res. Dev.* **2002**, *6*, 452-457.

429. Weygand, F.; Steglich, W.; de la Lama, X. B. *Tetrahedron* **1966**, *22* (*Suppl. 8*), 9-13.

430. Noyori, R.; Tokunaga, M.; Kitamura, M. *Bull. Chem. Soc. Jpn.* **1995**, *68*, 36-55.

431. Caddick, S.; Jenkins, K. *Chem. Soc. Rev.* **1996**, 447-456.

432. Pellissier, H. *Tetrahedron* **2003**, *59*, 8291-8327.

433. Vedejs, E.; Jure, M. *Angew. Chem., Int. Ed.* **2005**, *44*, 3974-4001.

434. Noyori, R.; Ikeda, T.; Ohkuma, T.; Widhalm, M.; Kitamura, M.; Takaya, H.; Akutagawa, S.; Sayo, N.; Saito, T.; Taketomi, T.; Kumobayashi, H. *J. Am. Chem. Soc.* **1989**, *111*, 9134-9135.

435. Kitamura, M.; Tokunaga, M.; Noyori, R. *J. Am. Chem. Soc.* **1993**, *115*, 144-152.

436. Schmidt, U.; Leitenberger, V.; Griesser, H.; Schmidt, J.; Meyer, R. *Synthesis* **1992**, 1248-1254.

437. Rychnovsky, S. D.; Hoye, R. C. *J. Am. Chem. Soc.* **1994**, *116*, 1753-1765.

438. Kitamura, M.; Tokunaga, M.; Noyori, R. *J. Am. Chem. Soc.* **1995**, *117*, 2931-2932.

439. Ohkuma, T.; Ooka, H.; Yamakawa, M.; Ikariya, T.; Noyori, R. *J. Org. Chem.* **1996**, *61*, 4872-4873.

440. Matsumoto, T.; Murayama, T.; Mitsuhashi, S.; Miura, T.; *Tetrahedron Lett.* **1999**, *40*, 5043-5046.

441. Ohkuma, T.; Ishii, D.; Takeno, H. Noyori, R. *J. Am. Chem. Soc.* **2000**, *122*, 6510-6511.

442. Ohkuma, T.; Li, J.; Noyori, R. *Synthesis* **2004**, 1383-1386.

443. Genêt, J.-P.; Pinel, C.; Mallart, S.; Juge, S.; Thorimbert, S.; Laffitte, J. A. *Tetrahedron: Asymm.* **1991**, *2*, 555-567.

444. Genêt, J.-P.; de Andrade, M. C. C.; Ratovelomanama-Vidal, V. *Tetrahedron Lett.* **1995**, *36*, 2063-2066.

445. Coulon, E.; de Andrade, C. C.; Ratovelomanama-Vidal, V.; Genêt, J.-P. *Tetrahedron Lett.* **1998**, *39*, 6467-6470.

446. Girard, A.; Greck, C.; Ferroud, D.; Genêt, J.-P. *Tetrahedron Lett.* **1996**, *37*, 7967-7970.

447. Mordant, C.; de Andrade, C. C.; Touati, R.; Ratovelomanama-Vidal, V.; Hassine, B. B.; Genêt, J.-P. *Synthesis* **2003**, 2405-2409.

448. Mordant, C.; Dünkelmann, P.; Ratovelomanama-Vidal, V.; Genêt, J.-P. *Chem. Commun.* **2004**, 1296-1297.

449. Labeeuw, O.; Phansavath, P.; Genêt, J.-P. *Tetrahedron: Asymm.* **2004**, *15*, 1899-1908.

450. Mordant, C.; Dünkelmann, P.; Ratovelomanama-Vidal, V.; Genêt, J.-P. *Eur. J. Org. Chem.* **2004**, 3017-3026.

451. Makino, K.; Okamoto, N.; Hara, O.; Hamada, Y. *Tetrahedron: Asymm.* **2001**, *12*, 1757-1762.

452. Makino, K.; Goto, T.; Hiroki, Y.; Hamada, Y. *Angew. Chem., Int. Ed.* **2004**, *43*, 882-884.

453. Makino, K.; Hiroki, Y.; Hamada, Y. *J. Am. Chem. Soc.* **2005**, *127*, 5784-5785.

454. Ohtsuka, Y.; Miyazaki, D.; Ikeno, T.; Yamada, T. *Org. Lett.* **2001**, *3*, 2543-2546.

455. Mohar, B.; Valleix, A.; Desmurs, J.-R.; Felemez, M.; Wagner, A.; Mioskowski, C. *Chem. Commun.* **2001**, 2572-2573.

456. Eustache, F.; Dalko, P. I.; Cossy, J. *Org. Lett.* **2002**, *4*, 1263-1265.

457. Sugi, K. D.; Nagata, T.; Yamada, T.; Mukaiyama, T. *Chem. Lett.* **1996**, 1081-1082.

458. Alcock, N. J.; Mann, I.; Peach, P.; Wills, M. *Tetrahedron: Asymm.* **2002**, *13*, 2485-2490.

459. Jurkauskas, V.; Buchwald, S. L. *J. Am. Chem. Soc.* **2002**, *124*, 2892-2893.

460. Murata, K.; Okano, K.; Miyagi, M.; Iwane, H.; Noyori, R.; Ikariya, T. *Org. Lett.* **1999**, *1*, 1119-1121.

461. Lloyd-Jones, G. C.; Butts, C. P. *Tetrahedron* **1998**, *54*, 901-914.

462. Hayashi, T.; Konishi, M.; Fukushima, M.; Mise, T.; Kagotani, M.; Tajika, M.; Kumada, M. *J. Am. Chem. Soc.* **1982**, *104*, 180-186.

463. Schaus, S. E.; Jacobsen, E. N. *Tetrahedron Lett.* **1996**, *37*, 7937-7940.

464. Lüssem, B. J.; Gais, H.-J. *J. Org. Chem.* **2004**, *69*, 4041-4052.

465. Ward, D. E.; Jheengut, V.; Akinnusi, O. T. *Org. Lett.* **2005**, *7*, 1181-1184.

466. Cho, C.-W.; Kong, J.-R.; Krische, M. J. *Org. Lett.* **2004**, *6*, 1337-1339.

467. Crich, J. Z.; Brieva, R.; Marquart, P.; Gu, R. L.; Flemming, S.; Sih, C. J. *J. Org. Chem.* **1993**, *58*, 3252-3258.

468. Turner, N. J.; Winterman, J. R.; McCague, R.; Parratt, J. S.; Taylor, S. J. C. *Tetrahedron Lett.* **1995**, *36*, 1113-1116.

469. Brown, S. A.; Parker, M.-C.; Turner, N. J. *Tetrahedron: Asymm.* **2000**, *11*, 1687-1690.

470. Seebach, D.; Jaeschke, G.; Gottwald, K.; Matsuda, K.; Formisano, R.; Chaplin, D. A.; Breuning, M.; Bringmann, G. *Tetrahedron* **1997**, *53*, 7539-7556.

471. Xie, L.; Hua, W.; Chan, A. S. C.; Leung, Y.-C. *Tetrahedron: Asymm.* **1999**, *10*, 4715-4728.

472. Liang, J.; Ruble, J. C.; Fu, G. C. *J. Org. Chem.* **1998**, *63*, 3154-3155.

473. Berkessel, A.; Cleemann, F.; Mukherjee, S.; Müller, T. N.; Lex. J. *Angew. Chem., Int. Ed.* **2005**, *44*, 807-811.

474. Berkessel, A.; Mukherjee, S.; Cleemann, F.; Müller, T. N.; Lex. J. *Chem. Commun.* **2005**, 1898-1900.

475. Tang, L.; Deng, L. *J. Am. Chem. Soc.* **2002**, *124*, 2870-2871.

476. Hang, J.; Li, H.; Deng, L. *Org. Lett.* **2002**, *4*, 3321-3324.

477. Chan, V.; Kim, J. G.; Jimeno, C.; Carroll, P. J.; Walsh, P. J. *Org. Lett.* **2004**, *6*, 2051-2053.

478. Tunge, J. A.; Gately, D. A.; Norton, J. R. *J. Am. Chem. Soc.* **1999**, *121*, 4520-4521.

479. Chen, J.-X.; Tunge, J. A.; Norton, J. R. *J. Org. Chem.* **2002**, *67*, 4366-4369.

480. Calmes, M.; Glot, C.; Michel, T.; Rolland, M.; Martinez, J. *Tetrahedron: Asymm.* **2000**, *11*, 737-741.

481. Yamada, S.; Noguchi, E. *Tetrahedron Lett.* **2001**, *42*, 3621-3624.

482. Mikami, K.; Yoshida, A. *Tetrahedron* **2001**, *57*, 889-898.

483. Rein, T.; Kreuder, R.; von Zezschwitz, P.; Wulff, C.; Reiser, O. *Angew. Chem., Int. Ed. Engl.* **1995**, *34*, 1023-1025.

484. Evans, J. W.; Fierman, M. B.; Miller, S. J.; Ellman, J. A. *J. Am. Chem. Soc.* **2004**, *126*, 8134-8135.

485. Shibata, N.; Matsunaga, M.; Nakagawa, M.; Fukuzumi, T.; Nakamura, S.; Toru, T. *J. Am. Chem. Soc.* **2005**, *127*, 1374-1375.

486. Khiar, N.; Alcudia, F.; Espartero, J.-L.; Rodriguez, L.; Fernandez, I. *J. Am. Chem. Soc.* **2000**, *122*, 7598-7599.

487. Khiar, N.; Araujo, C. S.; Alcudia, F.; Fernandez, I. *J. Org. Chem.* **2002**, *67*, 345-356.

488. Yamada, S.-I.; Mori, Y.; Morimatsu, K.; Ishizu, Y.; Ozaki, Y.; Yoshioka, R.; Nakatani, T.; Seko, H. *J. Org. Chem.* **1996**, *61*, 8586-8590.

489. Perlikowska, W.; Gouygou, M.; Mikolajczyk, M.; Daran, J.-C. *Tetrahedron: Asymm.* **2004**, *15*, 3519-3529.

490. Trost, B. M.; Krische, M. J.; Radinov, R.; Zanoni, G. *J. Am. Chem. Soc.* **1996**, *118*, 6297-6298.

491. Trost, B. M.; Fandrick, D. R.; Dinh, D. C. *J. Am. Chem. Soc.* **2005**, *127*, 14186-14187.

492. Trost, B. M.; Toste, F. D. *J. Am. Chem. Soc.* **1999**, *121*, 3543-3544.

493. Trost, B. M.; Toste, F. D. *J. Am. Chem. Soc.* **2003**, *125*, 3090-3100.

494. Trost, B. M.; Patterson, D. E.; Hembre, E. J. *J. Am. Chem. Soc.* **1999**, *121*, 10834-10835.

495. Trost, B. M.; Patterson, D. E.; Hembre, E. J. *Chem. Eur. J.* **2001**, *7*, 3768-3775.

496. Trost, B. M.; Crawley, M. L. *J. Am. Chem. Soc.* **2002**, *124*, 9328-9329.

497. Trost, B. M.; McEachern, E. J.; Toste, F. D. *J. Am. Chem. Soc.* **1998**, *120*, 12702-12703.

498. Trost, B. M.; Brown, B. S.; McEachern, E. J.; Kuhn, O. *Chem. Eur. J.* **2003**, *9*, 4442-4451.

499. Trost, B. M.; McEachern, E. J. *J. Am. Chem. Soc.* **1999**, *121*, 8649-8650.

500. Trost, B. M.; Tang, W.; Schulte, J. L. *Org. Lett.* **2000**, *2*, 4013-4015.

501. Trost, B. M.; Bunt, R. C.; Lemoine, R. C.; Calkins, T. L. *J. Am. Chem. Soc.* **2000**, *122*, 5968-5976.

502. Trost, B. M.; Tang, W. *Org. Lett.* **2001**, *3*, 3409-3411.

503. Trost, B. M.; Jiang, C. *Org. Lett.* **2003**, *5*, 1563-1565.

504. Cheeseman, N.; Fox, M.; Jackson, M.; Lennon, I. C.; Meek, G. *Proc. Natl. Acad. Sci.* **2004**, *101*, 5396-5399.

505. Trost, B. M.; Fandrick, D. R. *J. Am. Chem. Soc.* **2003**, *125*, 11836-11837.

506. Trost, B. M.; Fandrick, D. R. *Org. Lett.* **2005**, *7*, 823-826.

507. Trost, B. M.; Morgan, M. G. *J. Am. Chem. Soc.* **1994**, *116*, 10320-10321.

508. Hughes, D. L.; Palucki, M.; Yasuda, N.; Reamer, R. A.; Reider, P. J. *J. Org. Chem.* **2002**, *67*, 2762-2768.

509. Trost, B. M.; Andersen, N. G. *J. Am. Chem. Soc.* **2002**, *124*, 14320-14321.

510. Trost, B. M.; Sorum, M. T. *Org. Proc. Res. Dev.* **2003**, *7*, 432-435.

511. Braun, M.; Kotter, W. *Angew. Chem., Int. Ed.* **2004**, *43*, 514-517.

512. Trost, B. M.; Tsui, H.-C.; Toste, F. D. *J. Am. Chem. Soc.* **2000**, *122*, 3534-3535.

513. Trost, B. M.; Thiel, O. R.; Tsui, H.-C. *J. Am. Chem. Soc.* **2002**, *124*, 11616-11617.

514. Trost, B. M.; Thiel, O. R.; Tsui, H.-C. *J. Am. Chem. Soc.* **2003**, *125*, 13155-13164.

515. Trost, B. M.; Machacek, M. R.; Tsui, H.-C. *J. Am. Chem. Soc.* **2005**, *127*, 7014-7024.

516. Persson, B. A.; Huerta, F. F.; Bäckvall, J. E. *J. Org. Chem.* **1999**, *64*, 5237-5240.

517. Edin, M.; Steinreiber, J.; Bäckvall, J.-E. *Proc. Natl. Acad. Sci.* **2004**, *101*, 5761-5766.

518. Martin-Matute, B. J.; Bäckvall, J.-E. *J. Org. Chem.* **2004**, *69*, 9191-9195.

519. Amat, M.; Canto, M.; Llor, N.; Escolano, C.; Molins, E.; Espinosa, E.; Bosch, J. *J. Org. Chem.* **2002**, *67*, 5343-5351.

520. Amat, M.; Canto, M.; Llor, N.; Ponzo, V.; Perez, M.; Bosch, J. *Angew. Chem., Int. Ed.* **2002**, *41*, 335-338.

521. Amat, M.; Bassas, O.; Pericas, M. A.; Pasto, M.; Bosch, J. *Chem. Commun.* **2005**, 1327-1329.

522. Kubo, A.; Kubota, H.; Takahashi, M.; Nunami, K.-I. *J. Org. Chem.* **1997**, *62*, 5830-5837.

523. Caddick, S.; Jenkins, K.; Treweeke, N.; Candeias, S. X.; Afonso, C. A. M. *Tetrahedron Lett.* **1998**, *39*, 2203-2206.

524. Caddick, S.; Afonso, C. A. M.; Candeias, S. X.; Hitchcock, P. B.; Jenkins, K.; Murtagh, L.; Pardoe, D.; Santos, A. G.; Treweeke, N.; Weaving, R. *Tetrahedron* **2001**, *57*, 6589-6605.

525. Santos, A. G.; Pereira, J.; Afonso, C. A. M.; Frenking, G. *Chem. Eur. J.* **2005**, *11*, 330-343.

526. O'Meara, J. A.; Gardee, N.; Jung, M.; Ben, R. N.; Durst, T. *J. Org. Chem.* **1998**, *63*, 3117-3119.

527. Camps, P.; Perez, F.; Soldevilla, N. *Tetrahedron: Asymm.* **1997**, *8*, 1877-1894.

528. Camps, P.; Perez, F.; Soldevilla, N. *Tetrahedron: Asymm.* **1998**, *9*, 2065-2079.

529. Camps, P.; Perez, F.; Soldevilla, N.; Borrego, M. A. *Tetrahedron: Asymm.* **1999**, *10*, 493-509.

530. Amoroso, R.; Bettoni, G.; de Filippis, B.; Tricca, M. L. *Chirality* **1999**, *11*, 483-486.

531. Cook, G. R.; Shanker, P. S.; Pararajasingham, K. *Angew. Chem., Int. Ed.* **1999**, *38*, 110-113.

532. Enright, A.; Alexandre, F.-R.; Roff, G.; Fotheringham, I. G.; Dawson, M. J.; Turner, N. J. *Chem. Commun.* **2003**, 2636-2637.

533. Enders, D.; Maaßen, R.; Runsink, J. *Tetrahedron: Asymm.* **1998**, *9*, 2155-2180.

534. Lavergne, D.; Mordant, C.; Ratovelomanana-Vidal, V.; Genet, J.-P. *Org. Lett.* **2001**, *3*, 1909-1912.

535. Taniguchi, T.; Ogasawara, K. *Chem. Commun.* **1997**, 1399-1400.

536. Alayrac, C.; Metzner, P. *Tetrahedron. Lett.* **2000**, *41*, 2537-2539.

537. Xu, K.; Lalic, G.; Sheehan, S. M.; Shair, M. D. *Angew. Chem., Int. Ed.* **2005**, *44*, 2259-2261.

538. Beak, P.; Anderson, D. R.; Curtis, M. D.; Laumer, J. M.; Pippel, D. J.; Weisenburger, G. A. *Acc. Chem. Res.* **2000**, *33*, 715-727.

539. Cohen, T.; Lin, M.-T. *J. Am. Chem. Soc.* **1984**, *106*, 1130-1131.

540. McDougal, P. G.; Condon, B. D.; Laffosse Jr., M. D.; Lauro, A. M.; VanDerveer, D. *Tetrahedron Lett.* **1988**, *29*, 2547-2550.

541. Reich, H. J.; Bowe, M. D. *J. Am. Chem. Soc.* **1990**, *112*, 8994-8995.

542. Rychnovsky, S. D.; Buckmelter, A. J.; Dahanakur, V. H.; Skalitzky, D. J. *J. Org. Chem.* **1999**, *64*, 6849-6860.

543. Hoppe, D.; Hense, T. *Angew. Chem., Int. Ed. Engl.* **1997**, *36*, 2282-2316.

544. Basu, A.; Beak, P. *J. Am. Chem. Soc.* **1996**, *118*, 1575-1576.

545. Thayumananvan, S.; Basu, A.; Beak, P. *J. Am. Chem. Soc.* **1997**, *119*, 8209-8216.

546. Basu, A.; Gallagher, D. J.; Beak, P. *J. Org. Chem.* **1996**, *61*, 5718-5719.

547. Gallagher, D. J.; Du, H.; Long, S. A. Beak, P. *J. Am. Chem. Soc.* **1996**, *118*, 11391-11398.

548. Weisenburger, G. A.; Faibish, N. C.; Pippel, D. J.; Beak, P. *J. Am. Chem. Soc.* **1999**, *121*, 9522-9530.

549. Laumer, J. M.; Kim, D. D.; Beak, P. *J. Org. Chem.* **2002**, *67*, 6797-6804.

550. Hoppe, D.; Zschage, O. *Angew. Chem., Int. Ed. Engl.* **1989**, *28*, 69-70.

551. Hoppe, D.; Zschage, O. *Tetrahedron* **1992**, *48*, 8389-8392.

552. Kaiser, B.; Hoppe, D. *Angew. Chem., Int. Ed. Engl.* **1995**, *34*, 323-325.

553. Behrens, K.; Fröhlich, R.; Meyer, O.; Hoppe, D. *Eur. J. Org. Chem.* **1998**, 2397-2403.

554. Hoppe, I.; Marsch, M.; Harms, K.; Boche, G.; Hoppe, D. *Angew. Chem., Int. Ed. Engl.* **1995**, *34*, 2158-2160.

555. Wang, L.; Nakamura, S.; Toru, T. *Org. Biomol. Chem.* **2004**, *2*, 2168-2169.

556. Wang, L.; Nakamura, S.; Ito, Y.; Toru, T. *Tetrahedron: Asymm.* **2004**, *15*, 3059-3072.

557. Nakamura, S.; Ito, Y.; Wang, L.; Toru, T. *J. Org. Chem.* **2004**, *69*, 1581-1589.

558. Nakamura, S.; Furutani, A.; Toru, T. *Eur. J. Org. Chem.* **2002**, 1690-1695.

559. Nakamura, S.; Aoki, T.; Ogura, T.; Wang, L.; Toru, T. *J. Org. Chem.* **2004**, *69*, 8916-8923.

560. Nakamura, S.; Nakagawa, R.; Watanabe, Y.; Toru, T. *J. Am. Chem. Soc.* **2000**, *122*, 11340-11347.

561. Wolfe, B.; Livinghouse, T. *J. Am. Chem. Soc.* **1998**, *120*, 5116-5117.

562. Heath, H.; Wolfe, B.; Livinghouse, T.; Bae, S. K. *Synthesis* **2001**, 2341-2347.

563. Schlosser, M.; Limat, D. *J. Am. Chem. Soc.* **1995**, *117*, 12342-12343.

564. Coldham, I.; Dufour, S.; Haxell, T. F. N.; Howard, S.; Vennal, G. P. *Angew. Chem., Int. Ed.* **2002**, *41*, 3887-3889.

565. Coldham, I.; Dufour, S.; Haxell, T. F. N.; Vennal, G. P. *Tetrahedron* **2005**, *61*, 3205-3220.

566. Carlier, P. P.; Weldon, W.-F. L.; Wan, N. C.; Willimas, I. D. *Angew. Chem., Int. Ed.* **1998**, *37*, 2262-2254.

567. Gately, D. A.; Norton, J. R. *J. Am. Chem. Soc.* **1996**, *118*, 3479-3489.

568. Nam, J.; Lee, S.-K.; Park, Y. S. *Tetrahedron* **2003**, *59*, 2397-2401.

569. Komine, N.; Wang, L.-F.; Tommoka, K.; Nakai, T. *Tetrahedron Lett.* **1999**, *40*, 6809-6812.

570. Tommoka, K.; Wang, L.-F.; Komine, N.; Nakai, T. *Tetrahedron Lett.* **1999**, *40*, 6813-6816.

571. Clayden, J.; Lai, L. W. *Angew. Chem., Int. Ed.* **1999**, *38*, 2556-2558.

572. Clayden, J.; Johnson, P.; Pink, J. H.; Helliwell, M. *J. Org. Chem.* **2000**, *65*, 7033-7040.

573. Clayden, J.; Lai, L. W. *Tetrahedron Lett.* **2001**, *42*, 3163-3166.

574. Clayden, J.; Mitjans, D.; Youssef, L. H. *J. Am. Chem. Soc.* **2002**, *124*, 5266-5267.

575. Clayden, J.; Kubinski, P. M.; Sammicelli, F.; Helliwell, M.; Diorazio, L. *Tetrahedron* **2004**, *60*, 4387-4397.

576. Clayden, J.; Lai, L. W.; Helliwell, M. *Tetrahedron* **2004**, *60*, 4399-4412.

CHAPTER 8

From Chiral Propellers to Unidirectional Motors

The discovery of coordinated locomotion in muscle fibers, flagella, cilia, and other biological systems has inspired the development of molecular motors and other biomimetic devices.[1–3] The intriguing design of compounds resembling gears, switches, scissors, brakes, shuttles, turnstiles, and propellers illustrates the conceptual linkage between macroscopic mechanical tools and molecular analogs, Figure 8.1. In particular, the unique stereodynamic properties of chiral compounds provide ample opportunities to control molecular motion and have paved the way for artificial machines that play a crucial role at the interface between chemistry, engineering, physics, and molecular biology.[4]

Figure 8.1   Illustration of a four-bladed propeller, brake, bevel gear, and ratchet.

8.1 STABILITY AND REACTIVITY OF STEREODYNAMIC GEARS

Organic compounds typically fluctuate between rapidly equilibrating stereoisomeric structures at room temperature. The relative stability and energy barrier to interconversion of configurational and conformational isomers depend on intramolecular and intermolecular interactions. The rotational and vibrational freedom experienced by an individual substituent is not solely controlled by its connectivity (bond order and strength), shape and size, but also by neighbor effects. The sum of these interactions determines the structure and relative population of coexisting configurations and conformations. One can distinguish between three major classes of substituents: halides, nitriles, and other groups with a local C_∞-axis (1); planar structures including aryls (2); and polyhedral moieties such as alkyl groups (3). Both static and dynamic stereochemical properties of sterically crowded molecules result from bonding and nonbonding interactions between these substituents. Highly congested molecules possess rigid structures with limited conformational

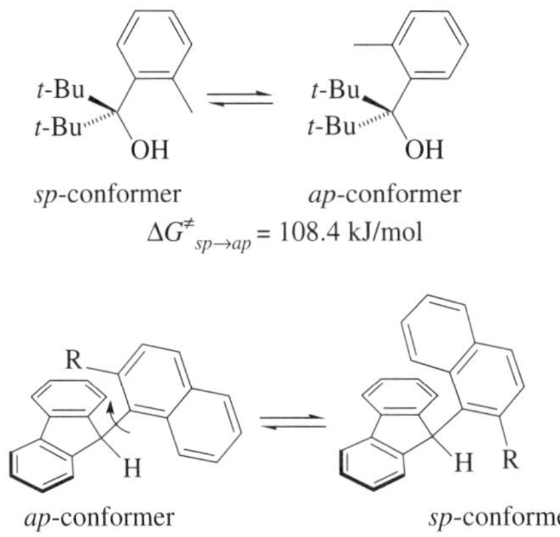

$\Delta G^{\neq}_{sp\rightarrow ap}$ = 108.4 kJ/mol

R=Me: $\Delta G^{\neq}_{ap\rightarrow sp}$ = 68.2 kJ/mol, $\Delta G^{\neq}_{sp\rightarrow ap}$ = 69.5 kJ/mol (0.0 °C)
i-Pr: $\Delta G^{\neq}_{ap\rightarrow sp}$ = 75.3 kJ/mol, $\Delta G^{\neq}_{sp\rightarrow ap}$ = 76.6 kJ/mol (0.0 °C)
OH: $\Delta G^{\neq}_{ap\rightarrow sp}$ = 110.1 kJ/mol, $\Delta G^{\neq}_{sp\rightarrow ap}$ = 108.0 kJ/mol (56.3 °C)
CO_2H: $\Delta G^{\neq}_{ap\rightarrow sp}$ = 99.6 kJ/mol, $\Delta G^{\neq}_{sp\rightarrow ap}$ = 107.6 kJ/mol (55.0 °C)

Scheme 8.1 Structures and conformational barriers of di-*tert*-butyl-*o*-tolylcarbinol and 9-(1-naphthyl)fluorenes.

flexibility and sometimes adopt unexpected ground state structures, for example nonalternating Newman projections, to balance attractive and repulsive forces.[5] Many propellers, gears, motors and other technomimetic compounds possess a rigid chiral structure which often facilitates kinetic and thermodynamic control of molecular stereodynamics.

The advance of molecular gears and propellers has been fueled by the pioneering studies of atropisomerism about sp^2–sp^3 and sp^3–sp^3 bonds in 9-arylfluorenes and triptycenes reported by Kessler, Oki, Siddall, and others.[6–12] Substituted 9-arylfluorenes and their analogs exist as a mixture of diastereoisomeric *ap*- and *sp*-conformers that can possess a high energy barrier to interconversion and remarkably different chemical properties, Scheme 8.1.[13–15] The conformers of some 2-substituted 9-(1-naphthyl)fluorenes are stable to isomerization at room temperature and show distinct conformational effects on reactivity. For example, *sp*-9-(2-formyl-1-naphthyl)fluorene can easily be oxidized to the corresponding carboxylic acid while the *ap*-conformer resists oxidation with chromium(VI) oxide.[16,17] Similarly, reaction between *sp*-9-(2-methoxy-1-naphthyl)fluorene and butyllithium is 1000 times faster than lithiation of the corresponding *ap*-isomer, and dehydration of di-*tert*-butyl-*o*-tolylcarbinol proceeds more rapidly when it occupies an *ap*-conformation, Scheme 8.2.[18,19]

Yamamoto and Oki prepared a series of separable *ap*- and *sc*-diastereoisomers of dibenzobicyclo[2.2.2]octadienes and 9-substituted triptycenes by Diels–Alder reaction of various anthracenes with dimethyl acetylenedicarboxylate and benzyne derivatives, respectively.[20,21] Because of the steric hindrance to rotation about the pivotal sp^3–sp^3 bond, which has been determined as 139.0 kJ/mol, they were able to separate the diastereomeric conformers of the octadiene atropisomers shown in Scheme 8.3. Selective hydrolysis of one ester group of the racemic mixture containing both synclinal conformers and conversion to diastereomeric menthyl derivatives allowed resolution and determination of the absolute configuration of individual stereoisomers.[22–24] Comparison of crystallographic data and conformational stability within the 9-tertiary alkyltriptycene series shown in Scheme 8.4 indicates that introduction of *peri*-substituents *X* into one arene ring results in considerable strain in the ground state and favors population of the *ap*-isomer.[25] Interestingly, the isomerization barrier increases with decreasing size of *X*. It is assumed that intramolecular repulsion involving a *peri*-substituent has a stronger destabilizing

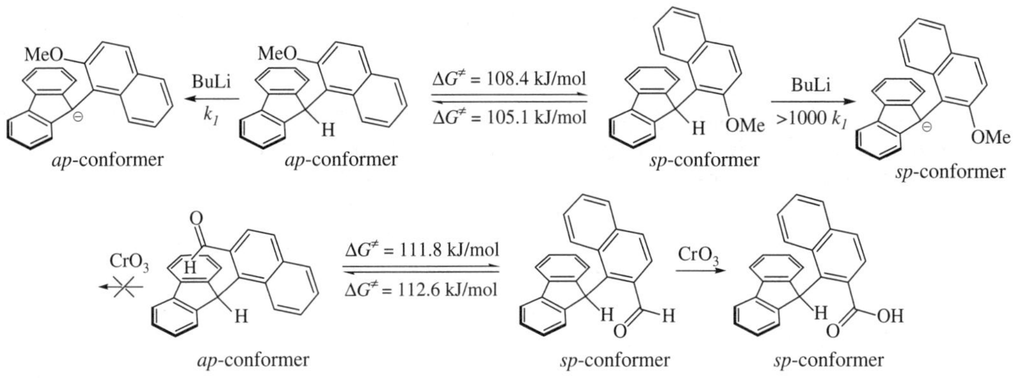

Scheme 8.2 Structure, rotational energy barrier and atropisomerselective reactivity of 9-(2-methoxy-1-naphthyl)fluorene and 9-(2-formyl-1-naphthyl)fluorene.

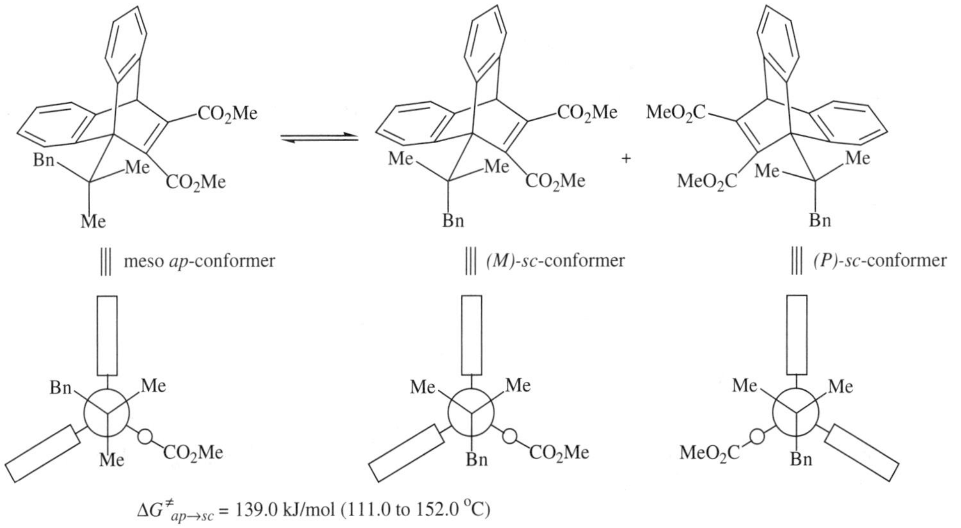

$\Delta G^{\neq}_{ap \to sc} = 139.0$ kJ/mol (111.0 to 152.0 °C)

Scheme 8.3 Atropisomerization of *ap*- and *sc*-diastereoisomers of a dibenzobicyclo[2.2.2]octadiene.

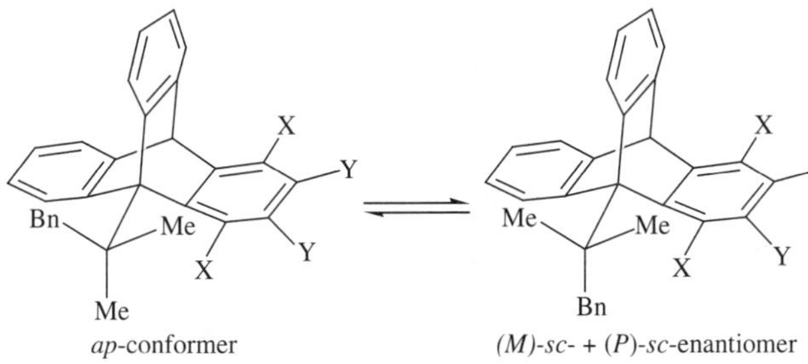

X=H, Y=Cl: $\Delta G^{\neq}_{ap \to sc} = 169.1$ kJ/mol (227.0 °C)

X=CH$_3$, Y=H: $\Delta G^{\neq}_{ap \to sc} = 161.6$ kJ/mol (227.0 °C)

X=OMe, Y=H: $\Delta G^{\neq}_{ap \to sc} = 177.5$ kJ/mol (227.0 °C)

X=Y=F: $\Delta G^{\neq}_{ap \to sc} = 185.4$ kJ/mol (227.0 °C)

Scheme 8.4 Atropisomerization of *ap*- and *sc*-diastereoisomers of some 9-substituted triptycenes.

effect on the congested ground state of these triptycenes than on the transition state and thus facilitates atropisomerization.

Employing variable-temperature NMR studies to elucidate the isomerization kinetics of atropisomeric aromatic amides, Clayden and coworkers found that the conformers of tertiary 2-methyl-1-naphthamides interconvert via correlated rotation about the aryl–carbonyl and the nitrogen–carbonyl bond. A closer look revealed that the geared motion competes with a two-step isomerization sequence which can be considered a gear-slippage process. However, the interdependent rotations about two molecular axes constitute a simple, albeit not perfect, microscopic gear, Scheme 8.5.[26–28] Anisotropic substituent interactions and geared bond rotation as observed with atropisomeric naphthalene amides can have pronounced stereochemical consequences and may be exploited for long-range conformational control and stereoselective synthesis.[29] Based on the observation that the amide groups in arenedicarboxamides preferably point in opposite

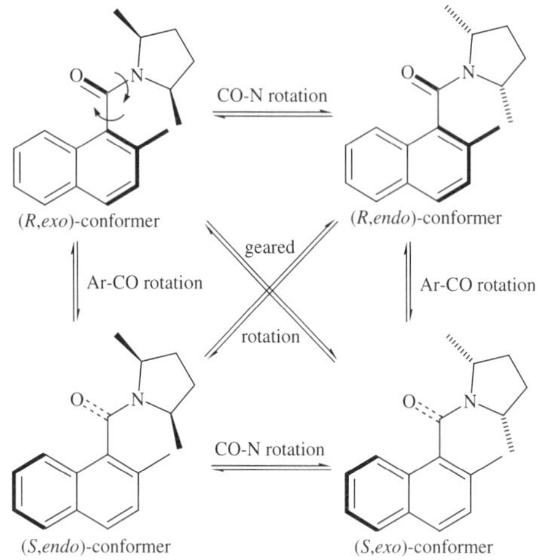

Scheme 8.5 Concerted rotation and sequential isomerization of a tertiary 2-methyl-1-naphthamide gear.

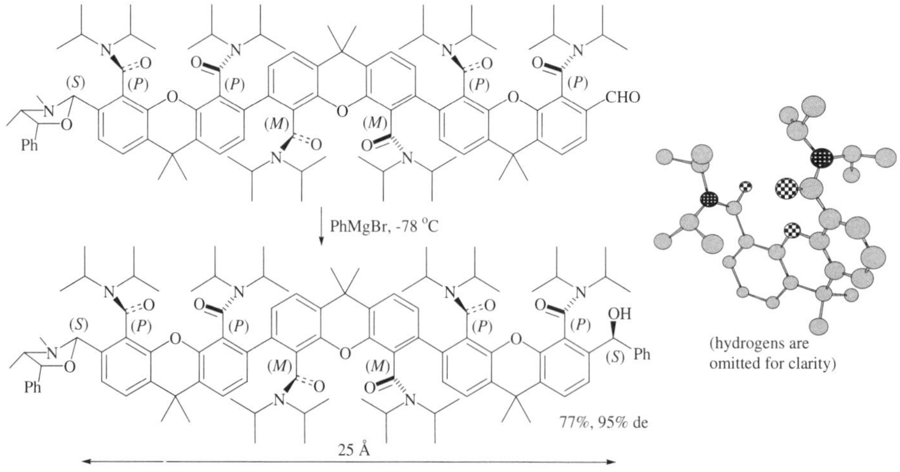

Scheme 8.6 Remote diastereoselectivity obtained by long-range conformational control along a tris(xanthenedicarboxamide) relay (left) and preferred *anti*-conformation of one arenedicarboxamide unit (right).

directions to minimize steric repulsion and overall dipole moment, Clayden designed a sterically crowded tris(xanthenedicarboxamide) relay exhibiting remote stereochemical communication, Scheme 8.6. Introduction of an ephedrine-derived (*S*)-oxazolidine residue at one terminus induces a right-handed twist (*P*-helicity) in the two proximate amide groups attached to the first arenedicarboxamide unit. This local conformational induction then propagates with a domino effect through the other xanthenedicarboxamide units populating *M*- and *P*-helical arrangements, respectively. Because each of the six amide groups adopts an *anti*-orientation relative to the previous one, the (*S*)-oxazolidine unit controls the stereochemistry of the whole atropisomeric array. As a result of [1,23]-asymmetric induction, Grignard addition of phenylmagnesium bromide to the prochiral aldehyde group located at the other terminus of the relay affords the secondary (*S*)-alcohol in 95% de. In other words, stereoselectivity is conformationally transmitted through 22 bonds spanning a distance of 2.5 nm.

8.2   STRUCTURE AND RING FLIPPING OF MOLECULAR PROPELLERS

In the early 1980s, Mislow and Iwamura independently introduced the concept of so-called static and dynamic gearing of proximate substituents in crowded molecules to design molecular analogs of propellers and bevel gears. Their findings represent the first step towards molecular switches, motors and nanomachines. By analogy with boat and airplane propellers, molecular propellers usually consist of two, three or four rigid aryl rings (blades) that are either connected to a central atom (hub) residing in the propeller axis or to a group of atoms residing in a plane perpendicular to the central rotation axis. The aryl rings are twisted, and adopt clockwise and counterclockwise arrangements that give rise to enantiomeric and diastereomeric helical conformations. This class of compounds must not be confused with propellanes which have a propeller-like structure consisting of three fused rings and, unlike molecular propellers, can exhibit a plane of symmetry, Figure 8.2.[i]

The aryl blades in molecular propellers may have the same or different substituents in *ortho*-, *meta*- or *para*-positions and the central hub usually consists of a boron, carbon, nitrogen, silicon or sulfur atom. Incorporation of aryl groups into ethylene and its derivatives yields vinyl propellers having the double bond perpendicular to the propeller axis. Molecular propellers bearing identical aryl blades with a local C_2-axis exist in two enantiomeric forms. Introduction of different groups (*X*, *Y*, *Z*), or incorporation of identical substituents into the *ortho*- or *meta*-position, but not along the local C_2-axes, generates diastereomeric conformations that display the aryl substituents directed to the same or opposite side of the propeller plane.[ii] Each diastereomeric conformation affords two enantiomeric forms, and the total amount of coexisting stereoisomers may be doubled if the central hub is an asymmetric atom. Accordingly, a triarylboron-derived propeller with three different *ortho*-substituents can access 2^3 stereoisomeric forms, *i.e.*, four racemic diastereoisomers,

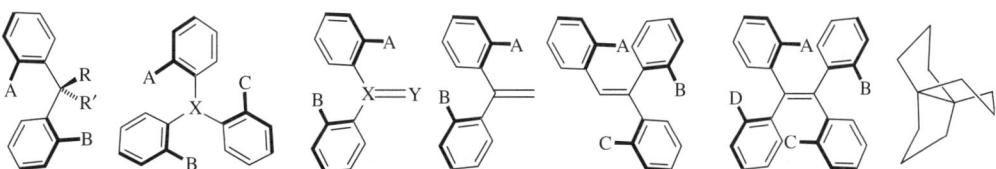

Figure 8.2   Structures of two-, three- and four-bladed molecular propellers, and a propellane (right).

---

[i] In contrast to propellanes, propellers are inherently chiral.
[ii] The propeller plane is sometimes referred to as the reference plane and is the origin of planar chirality.

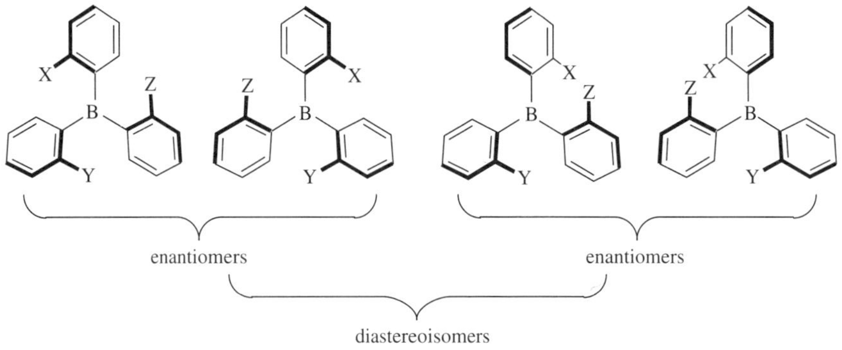

Figure 8.3   Stereoisomers of *ortho*-substituted arylboron propellers.

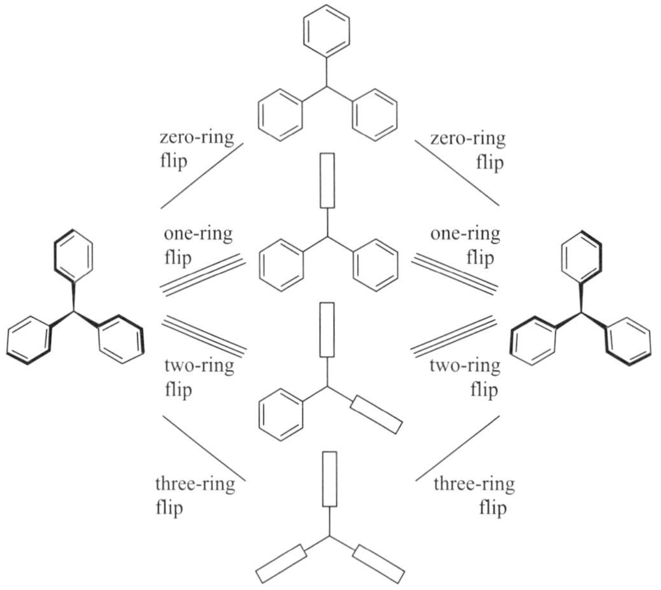

Scheme 8.7   Possible isomerization pathways of triaryl propellers. Flipping rings are shown from the side.

Figure 8.3. In addition, each propeller can possess right- or left-handed helicity with clockwise or counterclockwise blade orientation, resulting in a total amount of $2^4 = 16$ stereoisomers. The replacement of boron with a *CH* group generates a chiral center and consequently increases the number of possible stereoisomers to $2^5 = 32$. Propeller molecules may thus display three different elements of chirality, namely, a chiral center, a chiral plane and helicity.[iii] The intriguing stereochemistry of molecular propellers and the diversity of disrotatory or conrotatory correlated ring rotations have been studied extensively by Iwamura, Mislow, Rappoport, Biali, Lunazzi, Mazzanti, Casarini and others.[30–34]

As a consequence of significant steric interference between aryl rings, molecular propellers show gearing or cog-wheeling, *i.e.*, the blades do not rotate independently about the stereogenic axes but undergo energetically favored correlated movements. A discussion of the course of stereoisomer interconversion of molecular propellers requires analysis of competing ring flipping mechanisms. Except for so-called zero-ring flips, at least one ring passes through the plane containing the

---

[iii] The helical blade arrangement is often referred to as propeller chirality.

propeller axis and the pivotal bond about which the ring in question rotates.[iv] At the same time, all nonflipping rings, if any, pass through the propeller plane, Scheme 8.7. One can distinguish between zero-, one-, two-, three-, and four-ring flip mechanisms that reverse helicity and interconvert enantiomers or diastereoisomers. A zero-ring flip has a planar transition state in which all rings rotate through the reference plane. In an *N*-ring flip, *N* ring(s) undertake a conrotatory rotation while the nonflipping ring(s) rotate(s) in the opposite direction through the reference plane. In the case of the triarylboron propeller bearing three different *ortho*-substituents, discussed above, each of the 16 stereoisomers can in principle experience one zero-ring flip, three one-ring flips, three two-ring flips, and one three-ring flip, which corresponds to 64 individual interconversion pathways.[35] The rate of isomerization is determined by the threshold mechanism exhibiting the lowest activation energy. Deconvolution of competing isomerization reactions and determination of the corresponding energy barriers to rotation often entail variable-temperature NMR measurements in conjunction with molecular mechanical computations.

8.3 DYNAMIC GEARING IN BIARYL-, TRIARYL- AND TETRAARYL PROPELLERS

Introduction of two naphthyl or mesityl groups to a tetrahedral carbon atom yields stereolabile diaryl methane propellers, for example 2,2-dimesitylethanal,[36] dinaphthylmethane-derived crown ethers[37,38] and 1,1-dimesitylethanol.[39] Grilli and coworkers employed X-ray diffraction, variable-temperature NMR spectroscopy and molecular mechanics to confirm the two-blade propeller structure and disrotatory one-ring flipping of 1,1-dimesitylethyl methyl ether and its structural analogs, Scheme 8.8. Analysis of ^{1}H and ^{13}C NMR spectra showed that the mesityl methyl groups are diastereotopic, indicating that this propeller is devoid of any symmetry and belongs to the C_1-point group. Complete line shape analysis revealed that interconversion of the two helical

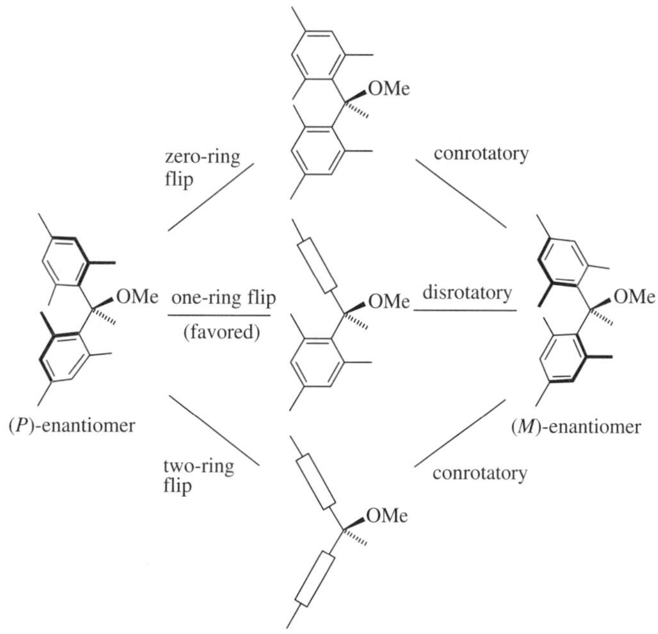

Scheme 8.8 Enantiomerization pathways of 1,1-dimesitylethyl methyl ether.

[iv] This plane is always perpendicular to the propeller plane.

enantiomers is fast and has an activation energy of 31.2 kJ/mol. This finding is in excellent agreement with theoretical calculations for a one-ring flip mechanism involving a correlated disrotatory arene motion. Accordingly, conrotatory zero-ring and two-ring flipping pathways have been ruled out for these biaryl propellers.

Other two-bladed propellers include diaryl ethers,[40] sulfides,[41,42] sulfoxides and sulfones,[43] sulfines,[44] boranes,[45] ketones,[46] ketenes,[47] and diarylcarbonium ions.[48] Many of these propellers prefer a disrotatory one-ring flip but this is not always the case. Solid-state and solution studies performed by Grilli and coworkers proved that dimesityl sulfine exhibits C_1-symmetric propeller-like conformations with distinguishable (*E*)- and (*Z*)-aryl rings.[44] In principle, the propeller conformations of dimesityl sulfine can undergo enantiomerization through four energetically different flipping circuits: a zero-ring flip, a (*Z*)-one-ring flip, an (*E*)-one-ring flip, and a two-ring flip. In the (*Z*)-one-ring flip, the mesityl ring that is closer to the sulfine moiety flips while the (*E*)-mesityl rotates through the reference plane. In contrast, the (*E*)-one-ring flip pathway has the (*Z*)-arene rotating through the reference plane. This is expected to result in increased steric repulsion with the planar sulfine group and can therefore be ruled out as the threshold mechanism. The zero-ring flip has two coplanar mesityl rings and thus affords an overcrowded transition state with a computed activation energy of at least 170 kJ/mol. In fact, NMR studies and molecular mechanics suggest that two competing interconversion mechanism are operative, and it is generally assumed that the helical enantiomers of dimesityl sulfine interconvert via (*Z*)-one-ring flip ($\Delta G^{\neq} = 24.7$ kJ/mol) and two-ring flip routes ($\Delta G^{\neq} = 57.8$ kJ/mol), Scheme 8.9. Computations and dynamic NMR measurements suggest that enantiomerization of dimesitylketone proceeds via

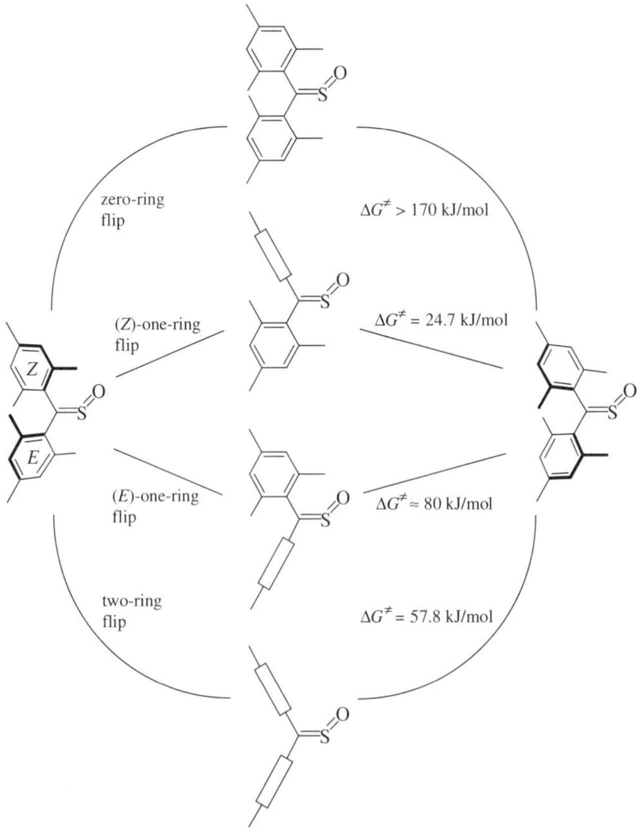

zero-ring flip $\Delta G^{\neq} > 170$ kJ/mol

(*Z*)-one-ring flip $\Delta G^{\neq} = 24.7$ kJ/mol

(*E*)-one-ring flip $\Delta G^{\neq} \approx 80$ kJ/mol

two-ring flip $\Delta G^{\neq} = 57.8$ kJ/mol

Scheme 8.9 Possible ring flip circuits of dimesityl sulfine.

disrotatory one-ring flipping with an activation energy of 19.3 kJ/mol, whereas the helical conformers of dimesitylethylene interconvert by a conrotatory two-ring flip mechanism with an interconversion barrier of 38.5 kJ/mol.[49,v] The conformers of propeller-like molecules are significantly more stable in the solid state. Force field calculations suggest an energy barrier to rotation about the phenyl–sulfoxide and naphthyl–sulfoxide axis in naphthylphenyl sulfoxide of 7.5 and 8.0 kJ/mol, respectively, but the barrier to phenyl rotation increases to 61.5 kJ/mol in the solid state, according to ^{13}C NMR CP-MAS (cross polarization magic angle spinning) analysis, Figure 8.4.[50]

Introduction of three aryl rings to one central hub or two adjacent atoms generates three-bladed propellers. Mislow *et al.* investigated the stereoisomerism of 1-(2-methoxynaphthyl)-1-(2-methylnaphthyl)-1-(3-methyl-2,4,6-trimethoxyphenyl)methane and other triarylmethanes, Figure 8.5. This compound displays a chiral carbon center and is devoid of a local C_2-axis due to the presence of different *ortho*- or *meta*-substituents within each aryl ring.[51–54] As discussed earlier, this class of propeller molecules exists in the form of 32 stereoisomers. Interestingly, two-ring flipping proceeds rapidly at room temperature and gives rise to residual diastereoisomerism, *i.e.*, separable diastereoisomeric propeller conformations that can only interconvert via one-ring flipping with an activation barrier of approximately 128 kJ/mol. Replacement of the 3-methyl-2,4,6-trimethoxyphenyl group

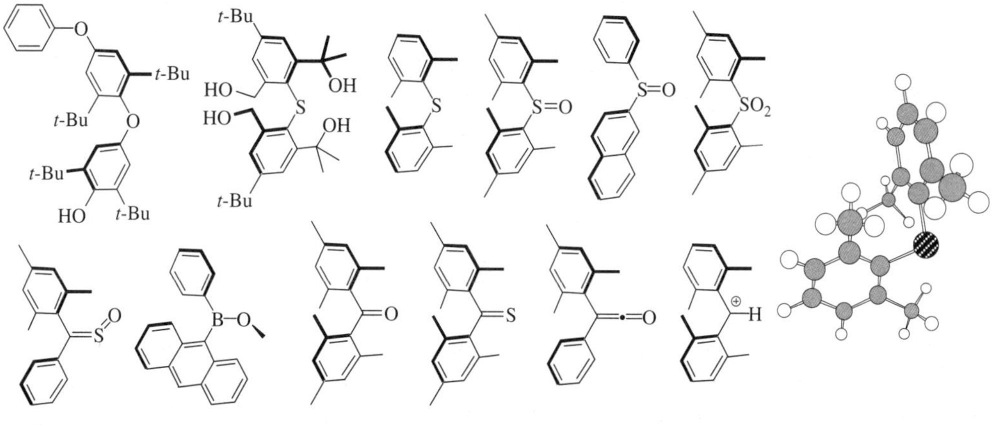

Figure 8.4 Two-bladed propellers and PM3-calculated structure of bis(2,6-dimethylphenyl)sulfide.

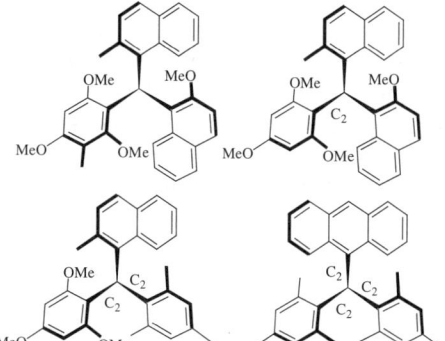

Figure 8.5 Structures of triarylmethane propellers with 0, 1, 2, and 3 local C_2-axes, reducing the amount of possible stereoisomers to 32, 16, 8, and 4.

---

[v] Stereomutations of diaryl-and triarylvinyl propellers are separately discussed in Chapter 8.5.

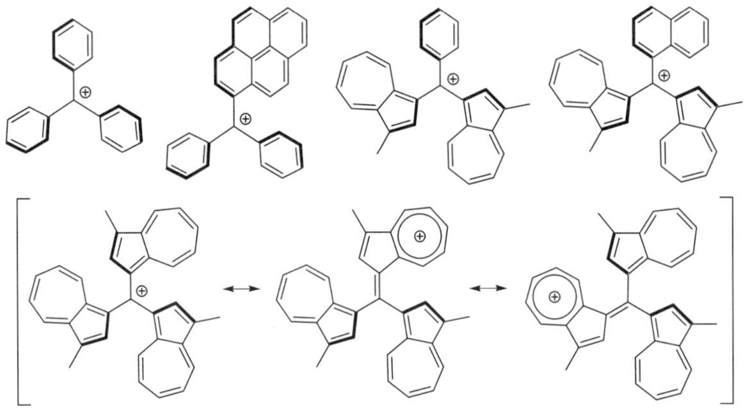

Figure 8.6   Three-bladed triarylcarbonium ions and resonance stabilization facilitating rotation of aryl rings through the reference plane.

with a 2,4,6-trimethoxyphenyl ring introduces a local C_2-axis and thus results in degeneracy, reducing the number of stereoisomers to 16. Since the remaining stereoisomers rapidly interconvert via two-ring flips, residual diastereoisomerism is not observed with 1-(2-methoxynaphthyl)-1-(2-methylnaphthyl)-1-(2,4,6-trimethoxyphenyl)methane.[55] For the same reason, mesityl-1-(2-methyl-naphthyl)-1-(2,4,6-trimethoxyphenyl)methane has two blades with a local C_2-axis and exists as a mixture of only eight propeller conformers that rapidly isomerize via two-ring flipping.

The fascinating stereodynamics of molecular propellers can be exploited in several ways. For example, triarylmethane moieties have been incorporated into crown ethers to control the three-dimensional structure and metal complexation properties of the macrocycle.[56–58] The remarkable stability and propeller-shape geometry of the triphenylmethyl cation has led to the development of other sterically crowded triarylcarbonium ions, Figure 8.6.[59–61] Asao *et al.* investigated the ring flipping mechanism of triazulenylmethyl hexafluorophosphate salts.[62–65] They rationalized that the transition state of the one-ring flip of tri(3-methyl-1-azulenyl)methyl hexafluorophosphate is energetically favored due to resonance stabilization. By contrast, the tri(2-methyl-1-azulenyl)methyl cation experiences increased steric hindrance impeding both resonance and the one-ring flip pathway. This triarylcarbonium ion therefore shows enhanced stability to isomerization and exhibits a two-ring flip threshold mechanism.

A variety of heteroatomic molecular propellers including triarylboranes,[66] triarylamine[67–69] and triarylsilanes[70–73] have been prepared, and even triaryl derivatives of germanium,[74] phosphorus and arsenic[75] are known. Conformational analysis of stereolabile triarylboranes such as tris(2-methyl-1-naphthyl)borane by Mislow *et al.* demonstrated that enantiomeric and diastereomeric conformers of these compounds undergo rapid isomerization with activation energies between 50 and 70 kJ/mol.[76] Variable-temperature NMR measurements with tris(2-methyl-1-naphthyl)borane uncovered that consecutive two-ring flips serve as the lowest energy pathways, whereas zero- and one-ring flips have been ruled out on steric grounds, Figure 8.7. Accordingly, enantiomers *B* and *B'* are slightly more stable than enantiomers *A* and *A'*. The diastereoisomers interconvert via two-ring flipping with an activation energy of 66.6 and 67.8 kJ/mol, respectively. The energy barrier to enantiomerization between *B* and *B'* is 61.1 kJ/mol. Similar stereomutations have been observed with tris[2-(trifluoromethyl)phenyl]borane.[77] Due to fewer steric interactions in the transition states, the stereoisomers of tris[2-(trifluoromethyl)phenyl]borane undergo two-ring flip enantiomerization and diastereomerization with activation energies of only 25.0 and 34.3 kJ/mol, respectively. Stereochemical analysis revealed that only the enantiomeric C_3-conformers crystallize in the form of CD-active conglomerates, while a mixture of rapidly interconverting diastereomeric pairs of

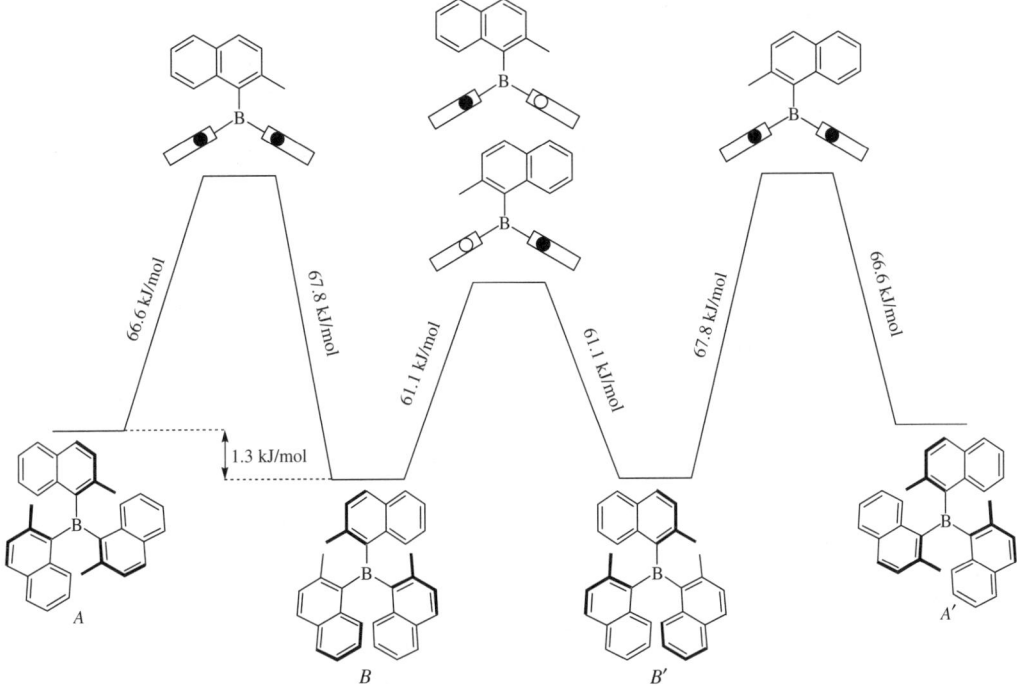

Figure 8.7 Stereoisomerization and free activation energies of two-ring flips of tris(2-methyl-1-naph-thyl)borane. Solid black (white) circles represent a methyl group pointing towards (away from) the viewer.

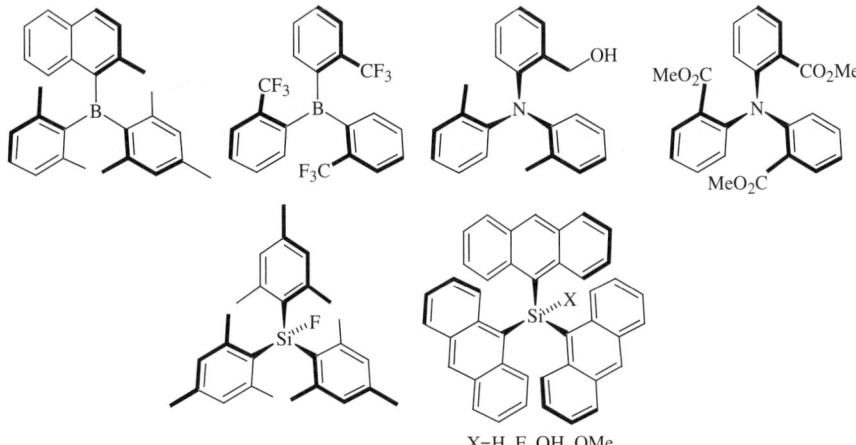

X=H, F, OH, OMe

Figure 8.8 Structures of triarylbladed boron, nitrogen and silane propellers.

enantiomers with C_3- and C_1-symmetry exists in solution. Crystallographic analysis confirmed that propeller-like conformations are stabilized by intramolecular $CF\cdots B$ interactions in addition to arene gearing, Figure 8.8.

Mislow's group introduced 1,1,2,2-tetraarylethanes that adopt C_2-symmetric ground state conformations having the vicinal methine protons in antiperiplanar orientation.[78,vi] Since each of the

---

vi The C_2 symmetry and chirality is conserved even in the presence of four identical aryl rings.

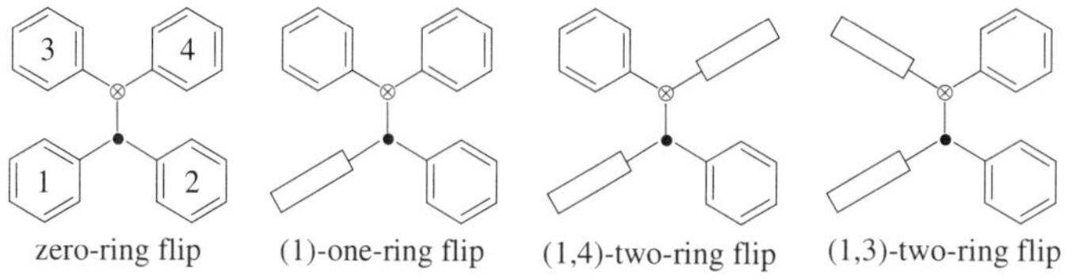

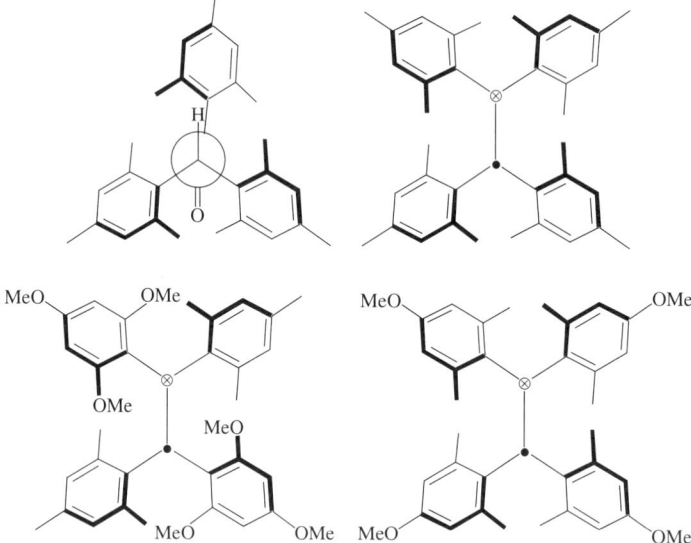

Figure 8.9   Transition states of the seven classes of flip mechanisms of *anti*-tetraarylethanes.

zero-ring flip   (1)-one-ring flip   (1,4)-two-ring flip   (1,3)-two-ring flip

(1,2)-two-ring flip   (1,2,3)-three-ring flip   (1,2,3,4)-four-ring flip

Figure 8.10   Structures of 1,2,2-trimesitylethanone and tetraarylethanes having antiperiplanar methine protons.

adjacent tetrahedral carbon atoms bears two aryl groups, the molecule displays the structure of a stretched four-bladed propeller. Helical inversion may, in principle, occur via seven classes of correlated blade rotation: one zero-ring-, one one-ring-, three two-ring-, one three-ring-, and one four-ring flip. Flipping rings rotate about the 0° dihedral angle ($H_{methine}$–C_{ethane}–C_{aryl}–$C_{ortho-aryl}$) while nonflipping rings rotate about a 90° dihedral angle in the opposite direction, Figure 8.9. Due to the presence of four aryl rings, the seven classes of flip mechanisms give rise to a total of 16 possible isomerization pathways: one four-ring, four one-ring, six two-ring, four three-ring, and one four-ring flip. Variable-temperature NMR spectroscopy of 1,1,2,2-tetramesitylethane and 1,2-dimesityl-1,2-bis(2,4,6-trimethoxyphenyl)ethane indicate that the threshold mechanism (the isomerization pathway with the lowest activation energy) is the four-ring flip.[79] The threshold barrier of tetramesitylethane was determined as 95.4 kJ/mol which is close to the activation energy of the two-ring flip of triarylmethanes carrying aryl rings of similar size. It has therefore been postulated that the four-ring flip may in fact be a composite of two two-ring flips. Because of the significant hindrance to isomerization in four-bladed propellers, Schlögl *et al.* were able to separate the enantiomers of 1,1,2,2-tetrakis(2,6-dimethyl-4-methoxyphenyl)ethane by chiral HPLC on triacetyl cellulose at 10 °C. They determined the barrier to racemization as 93.0 kJ/mol using CD spectroscopy. This value is in good agreement with DNMR results obtained with the structurally similar tetramesitylethane.[80] Biali and Rappoport prepared 1,2,2-trimesitylethanone which also has aryl blades attached to different atoms. According to X-ray crystallography and UV, IR and NMR spectroscopy, the blades form a propeller bisected by the carbonyl group, Figure 8.10.[81] The two β-mesityl blades participate in a two-ring flip mechanism with a free activation energy of 39.8 kJ/mol, and the energy barrier to rotation of the α-mesityl ring about the aryl–carbonyl bond is 57.8 kJ/mol.

8.4 MOLECULAR BEVEL GEARS

Introduction of two bridgehead-substituted triptycenes to a methylene group or an oxygen atom provides a molecular device bearing two tightly intermeshed rotors that undergo correlated rotation reminiscent of a pair of three-toothed bevel gears. Bis(9-triptycyl)methane experiences fast topomerization due to simultaneous disrotatory motion (viewed from the methylene group) of the intertwined triptycyl moieties, Figure 8.11. The phase relationships between the arene rings does not change during rotation, for example, arene *A* always remains between arenes *D* and *F* of the opposite rotor.

Diels–Alder reaction of substituted benzynes and bis(9-anthryl)methane or thermolysis of 9-triptycyl 9′-triptyceneperoxycarboxylates provides access to pairs of sterically crowded bevel gears. The stereomutations of selectively substituted bis(9-triptycyl)methanes and corresponding ether derivatives have been examined by Mislow and Iwamura using NMR spectroscopy, crystallography and empirical force-field calculations.[82–84] Bis(1,4-dimethyl-9-triptycyl)methane comprises three phase isomers: a meso isomer with a symmetry plane, and two enantiomeric forms that interconvert via correlated disrotatory cogwheeling motions and gear slippage. The 1-methyl group of one triptycyl rotor is located in the notch between two aryl rings of the other triptycyl moiety and *vice versa*, Figure 8.12. The tightly meshed triptycyl blades behave like interlocked wheel cogs which gives rise to dynamic gearing. As a consequence of the high barrier to gear

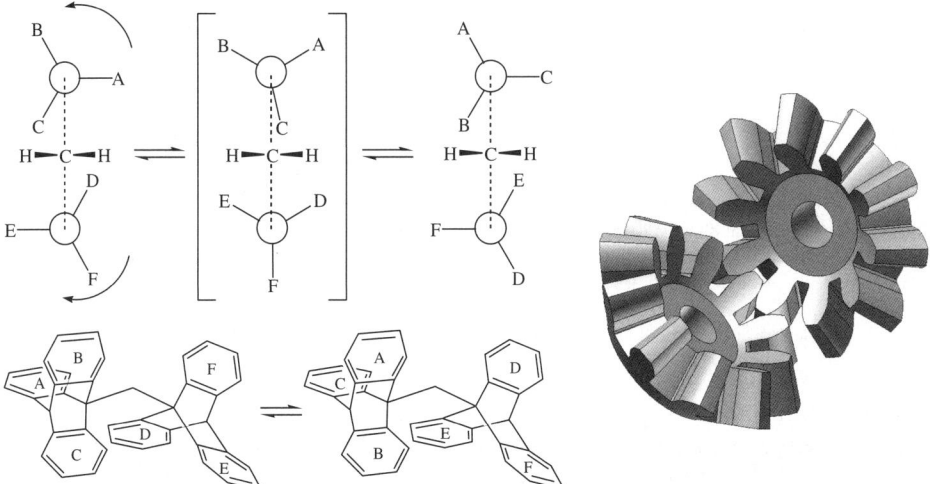

Figure 8.11 Schematic representation of cogwheel gearing of intermeshed triptycyl units in bis(triptycyl)methane (the gearing is viewed from the methylene hub). The large circles refer to the triptycyl bridgehead carbons connected to the methylene group and the propeller blades are represented by radiating lines. The arrows indicate the disrotatory motion.

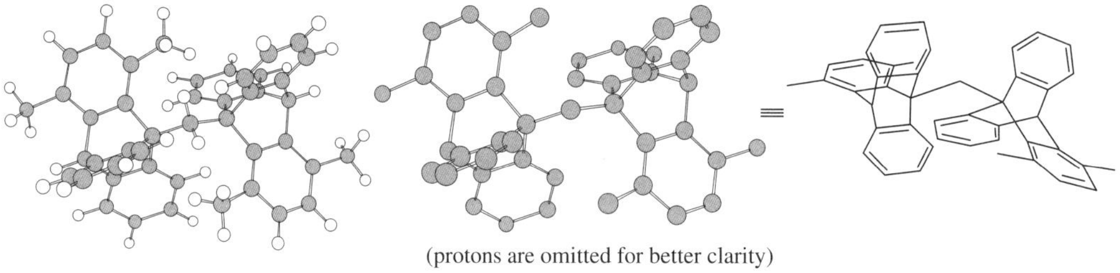

(protons are omitted for better clarity)

Figure 8.12 Computed structure of meso bis(1,4-dimethyl-9-triptycyl)methane bearing intermeshed trip-tycyl groups.

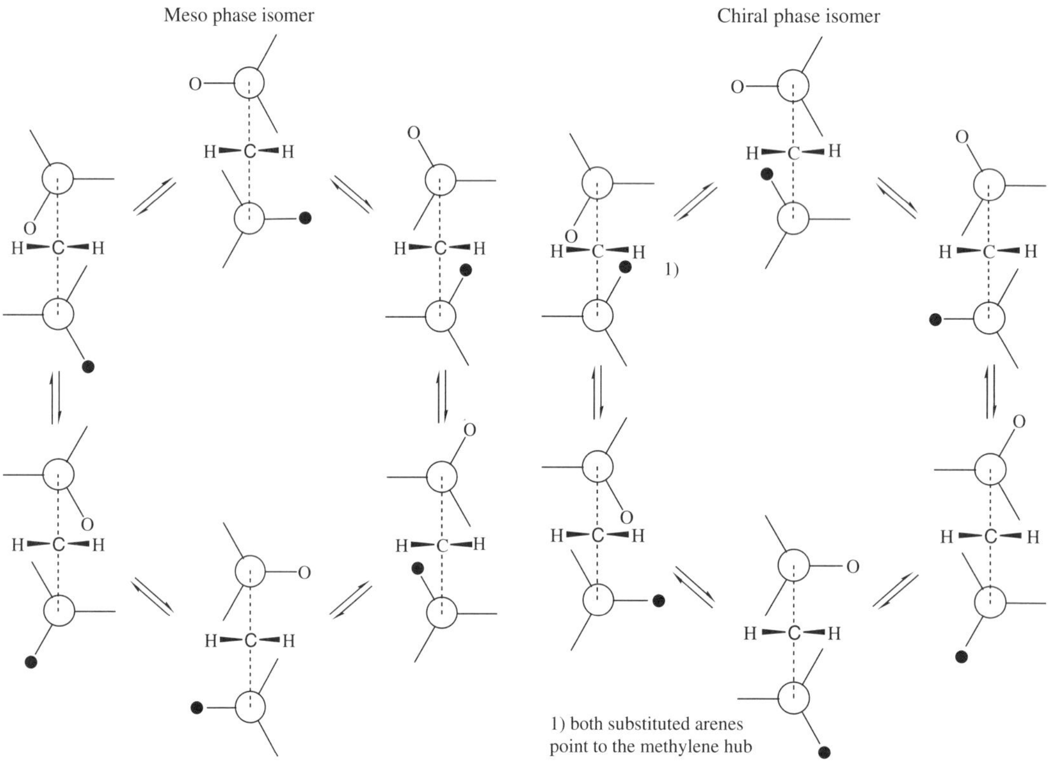

Meso phase isomer   Chiral phase isomer

1) both substituted arenes
point to the methylene hub

Scheme 8.10 Schematic representation of the geared conformers of the meso form (left) and one enantiomer (right) of selectively substituted bis(triptycyl)methane phase isomers, viewed from the methylene hub. The large circles are the triptycyl bridgehead carbons attached to the methylene hub. The propeller blades are represented by radiating lines; ● and ○ denote substituted benzene rings in each triptycyl moiety. The gearing course of the other enantiomeric phase isomer is the mirror image of the scheme shown at the right.

slippage in bis(1,4-dimethyl-9-triptycyl)methane, the meso form and the two enantiomers can be isolated at room temperature. Both the meso form and the chiral isomers show conformational gearing, *i.e.*, correlated disrotatory rotation of the two triptycyl units. Interestingly, the gearing motion of the meso isomer of substituted bis(9-triptycyl)methanes and similar bevel gear-type molecules does not bring the substituted aryl groups into close proximity. On the other hand, this is observed during gearing of the chiral phase isomers since one of the six interconverting conformers has both substituted arenes pointing to the methylene plane, Scheme 8.10.[85]

Many achiral molecules such as meso 2,3-diaminobutane exist as a complex mixture of rapidly interconverting achiral and chiral conformations, see Chapter 2.2. Similarly, the meso isomer of bis(1,4-dimethyl-9-triptycyl)methane comprises one achiral conformer and a pair of enantiomers that interconvert with an activation energy of 59.4 kJ/mol via disrotatory motion of the triptycyl moieties.[86] Each of the two chiral phase isomers consists of a mixture of three dynamically geared diastereomeric conformations. Interconversion of the three phase isomers is only observed at very high temperatures because it requires gear slippage which has an activation energy of 164.9 kJ/mol at 215 °C. Each phase isomer of bis(1,4-dimethyl-9-triptycyl)methane thus resembles a pair of three-toothed bevel gears, Scheme 8.11.

Cozzi *et al.* prepared bis(2,3-dimethyl-9-triptycyl)methane and isolated the diastereoisomeric phase isomers by chromatography, Figure 8.13. By analogy with bis(1,4-dimethyl-9-triptycyl)methane, one would expect nine conformers that undergo rapid geared interconversion at room temperature.[87] Indeed, three residual isomers were identified: a meso isomer exhibiting one achiral and two chiral rotamers, and a racemic mixture of two enantiomers, each comprising three rapidly

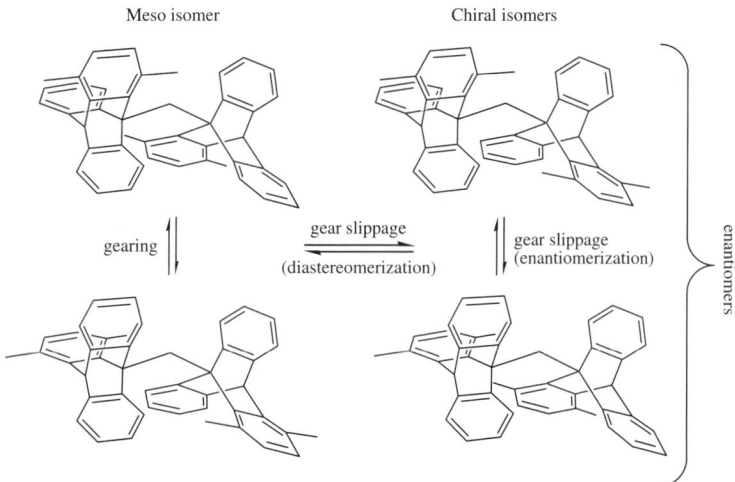

Scheme 8.11 Stereomutation via disrotatory gearing in the meso isomer and gear slippage resulting in interconversion of the phase isomers of bis(1,4-dimethyl-9-triptycyl)methane. Gearing of the chiral phase isomers is not shown.

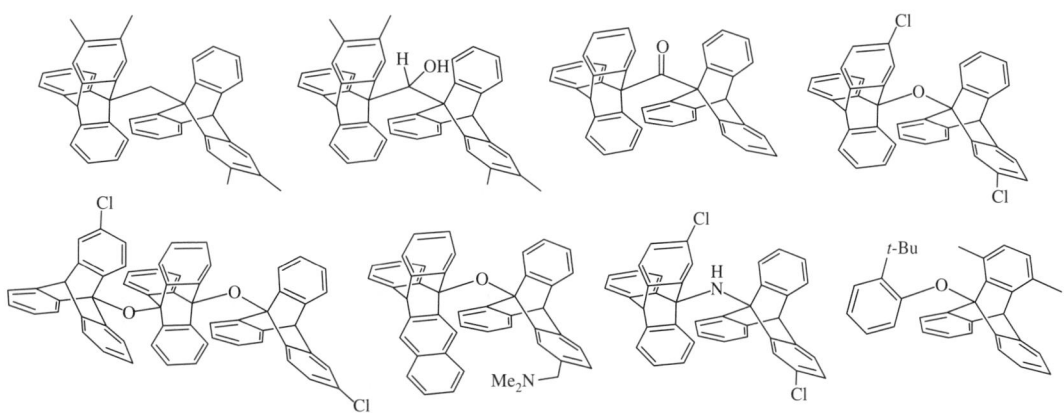

Figure 8.13 Structures of triptycyl-derived molecular bevel gears.

equilibrating diastereomeric conformations. Because the methyl groups do not penetrate as deeply into the adjacent triptycyl moiety as in the 1,4-dimethyl derivative discussed above, the energy barrier to gear slippage, which leads to interconversion of meso and chiral phase isomers, is reduced to approximately 142 kJ/mol. Covalently connected molecular pairs of toothed bevel gears showing phase isomerism and disrotatory gearing of intermeshed rotors have also been realized with bis(9-triptycyl)ethers,[88–93] bis(9-triptycyl)amines[94] and two-toothed analogs,[95] Figure 8.13. A closer look at phase isomerization rates and activation energies reveals that replacement of the methylene group in bis(9-triptycyl)methane by nitrogen or oxygen increases the barrier to gear slippage. This trend correlates well with the decreasing bond length and increasing stretching and bending force constants found in propane, dimethyl amine and dimethyl ether.[vii]

8.5   VINYL PROPELLERS

The three-dimensional structure and the stereodynamics of crowded styrenes carrying two or more aryl groups arise from a compromise between steric repulsion, which is minimized when the aryl rings are perpendicular to the vinyl plane, and π-conjugation between the aryl groups and the double bond, which is at a maximum in a coplanar orientation. Diarylethylenes can have either a geminal or an (E)- and (Z)-vicinal arene arrangement. Most two-bladed diarylvinyl propellers are enols with geminal aryl groups. The fast racemization of 1,1-diphenylethylene and the absence of an appropriate NMR probe such as an isopropyl group preclude experimental analysis of the flipping mechanism by dynamic NMR spectroscopy. However, molecular mechanics are in agreement with a threshold one-ring flip mechanism having an activation energy of approximately 5.0 kJ/mol, and the energy barriers to a zero- and two-ring flip have been calculated as 54.0 and 12.6 kJ/mol, respectively, Scheme 8.12.[96] Mazzanti *et al.* chose 1,1'-(2-isopropylphenyl)ethylene as a representative model to investigate stereomutations of diaryl ethylene propellers by NMR spectroscopy.[97] Stereochemical analysis revealed six coexisting conformations, *i.e.*, three racemic diastereomeric conformers belonging to C_1- and C_2-point groups.[viii] The C_2-symmetric *syn*-isomer

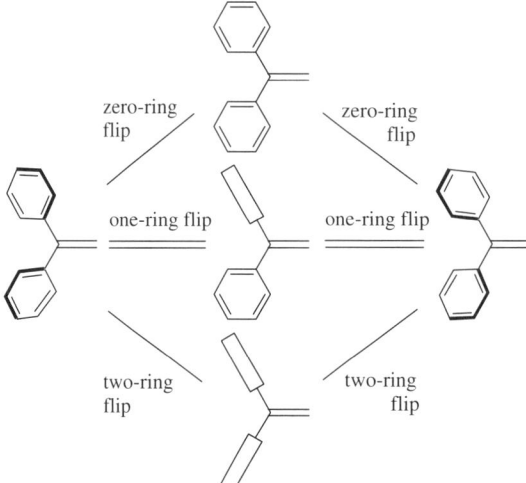

Scheme 8.12   Possible transition states of 1,1-diarylvinyl propellers. Flipping rings are viewed from the side.

[vii] It is generally assumed that nitrogen inversion does not play a role in gear slippage of bis(9-triptycyl)amines.

[viii] Objects belonging to the C_1-point group do not have any symmetry elements apart from the identity operation E. Objects that belong to the C_2-point group possess a C_2-symmetric rotation axis in addition to E. Compounds belonging to these point groups are chiral.

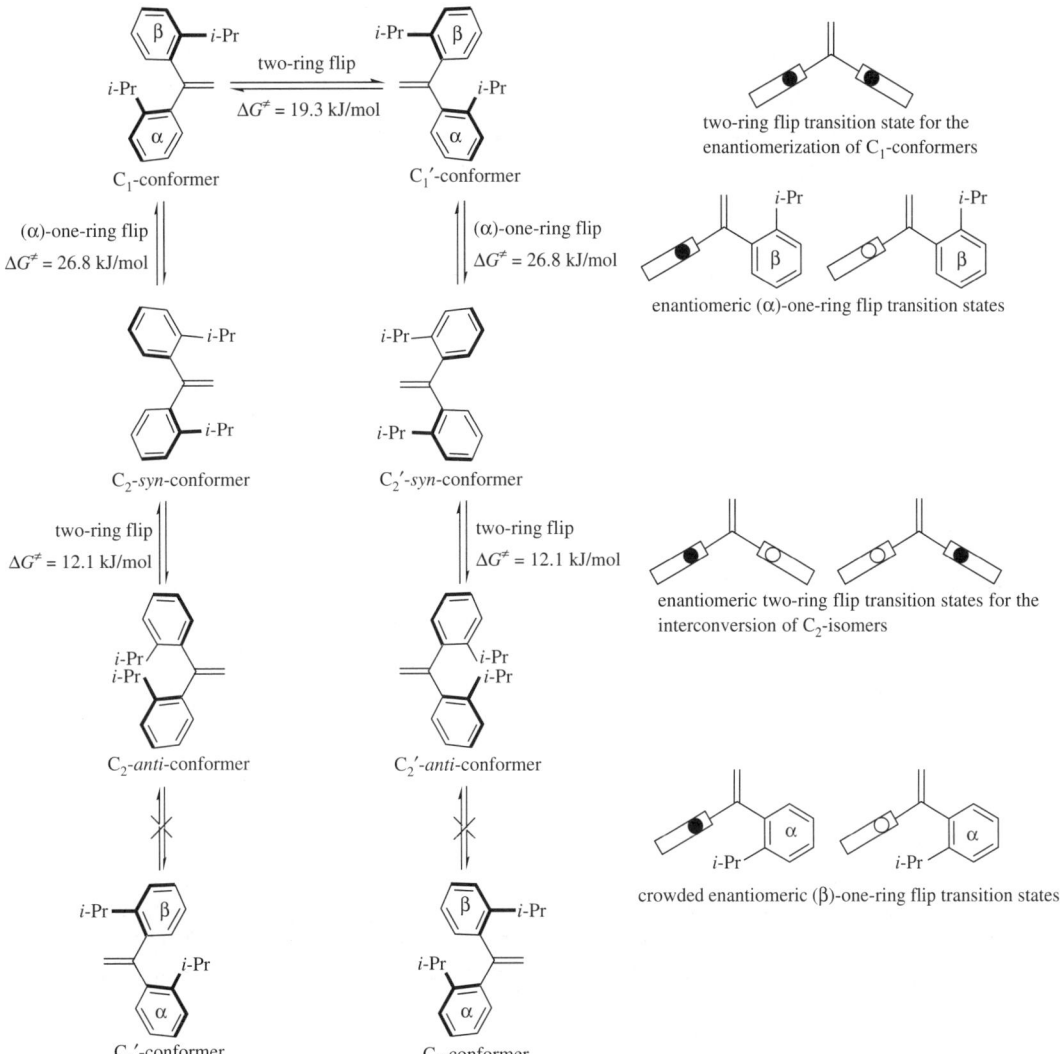

Scheme 8.13 Stereomutations of 1,1′-(2-isopropylphenyl)ethylene.

has both isopropyl groups pointing towards the methylene moiety, and the slightly more stable C_2-*anti*-isomer has them directed away from this group, Scheme 8.13. *Ab initio* calculations suggest that the C_1-conformer is the most stable isomer. According to cryogenic NMR measurements, interconversion of the C_1-conformer to the C_2-symmetric *syn*-diastereoisomers via (α)-one-ring flipping has a barrier of 26.8 kJ/mol, and enantiomerization of the C_1-conformer occurs via two-ring flipping with an activation energy of 19.3 kJ/mol. A two-ring-flip interconversion between the two diastereomeric C_2-conformers with an energy barrier of 12.1 kJ/mol has been predicted based on computational studies. Direct conversion of the enantiomeric C_1-conformers to the C_2-symmetric *anti*-diastereoisomers via (β)-one-ring flip has a relatively high energy barrier of 43.1 kJ/mol due to enhanced steric repulsion in the transition state.

Rappoport *et al.* investigated the structure and stereodynamics of 2,2-dimesitylethenol propellers, Figure 8.14. They observed that 2,2-dimesitylethenol prefers (β')-one-ring flipping which has an energy barrier of 43.5 kJ/mol. A less favorable process with an energy barrier of 59.4 kJ/mol was attributed to a (β,β')-two ring circuit.[98] Incorporation of a neighboring alkyl group results in a

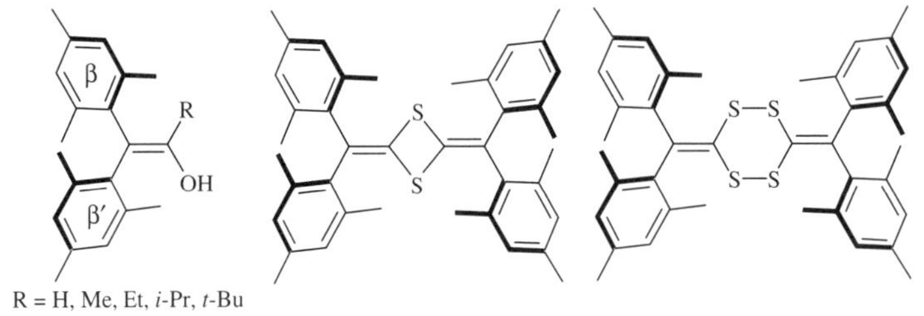

R = H, Me, Et, *i*-Pr, *t*-Bu

Figure 8.14 Structures of 2,2-dimesitylethenols, and dithiethane- and tetrathiane-derived multipropellers.

two-ring flip threshold mechanism with an activation energy of 52.7, 50.2, 49.0, and 43.5 kJ/mol in 1-methyl-, 1-ethyl-, 1-isopropyl-, and 1-*tert*-butyl-2,2-dimesitylethenol, respectively.[99] The change from a one- to a two-ring flip is probably a consequence of increased steric interactions during arene rotation. Accordingly, 2,2-dimesitylethenol can be expected to favor a (β′)-one-ring flip mechanism involving a transition state in which the β-group experiences little steric repulsion by the adjacent hydrogen, and therefore undergoes facile rotation through the reference plane when the β′-ring flips. Introduction of adjacent alkyl groups impedes rotation of the β-ring through the double bond plane and thus disfavors one-ring flipping, while the energy barrier to a two-ring flip is reduced due to enhanced torsion angles and increased ground state energies of crowded 1-alkyl-2,2-dimesitylethenols. Destabilization of the ground state conformations of these geminal two-bladed propellers by sterically demanding alkyl groups therefore facilitates isomerization via two-ring flipping, which explains why 1-*tert*-butyl-2,2-dimesitylethenol has the lowest conformational stability in this series. For the same reason, one would expect that incorporation of a bulky triptycyl unit would further destabilize the dimesitylethenol propeller. In contrast, 1-(9-triptycyl)-2,2-dimesitylethenol acetate has a helical inversion barrier of 71.2 kJ/mol, which has been attributed to correlated rotation of the three-toothed triptycyl rotor and both two-toothed mesityl rotors.[100] Replacement of the geminal mesityl rings in diarylethenol by bulky 2,4,6-triisopropylphenyl (tipyl) groups increases the energy barrier to ring flipping but does not change the threshold mechanism.[101] Both conformation and energy barrier to isomerization of 2,2-ditipylethenols are controlled by solvent-dependent intramolecular hydrogen bonding between the hydroxyl group and the *cis*-arene ring.

Introduction of two remote geminal propellers into one molecule has been accomplished through preparation of dithiethanes and tetrathianes from dimesityl thioketene. Crystallographic analysis and NMR studies proved that these multipropellers have two independent subunits that interconvert via the two-ring flip mechanism. This stereomutation is accompanied by simultaneous ring inversion in tetrathiane derivatives, Figure 8.14.[102] Vicinal diarylethylene propellers are generally less stable to isomerization than the geminal analogs discussed above. The structure and threshold mechanism of cyclic (Z)-1,2-diarylvinyl propellers depend on the ring size. 1,2-Diarylcyclopropenes are almost planar but larger diarylcycloalkenes adopt propeller conformations.[103] It is assumed that 1,2-diarylcyclobutenes behave like (Z)-1,2-diphenylethylene and enantiomerize via one-ring flipping. 1,2-Diarylcyclopentenes and cyclohexene derivatives seem to favor a two-ring flip mechanism.

Triarylvinyl propellers are more crowded than diarylalkenes and experience significantly higher helical inversion barriers. For example, interconversion of the four stereoisomeric forms of (E)- and (Z)-2-*meta*-methoxymesityl-1,2-dimesitylvinyl acetate is based on (α,β′)-two-ring flip enantiomerization in conjunction with a threshold (α,β,β′)-three-ring flip diastereomerization process, Scheme 8.14.[104,105] Comparison with other trimesitylethylenes confirms that the non-aromatic vinyl substituent has a strong influence on the relative stability of the transition states of the competing α,β-, α,β′-, β,β′- and α,β,β′-flip circuits. In general, trimesitylvinyl propellers with a

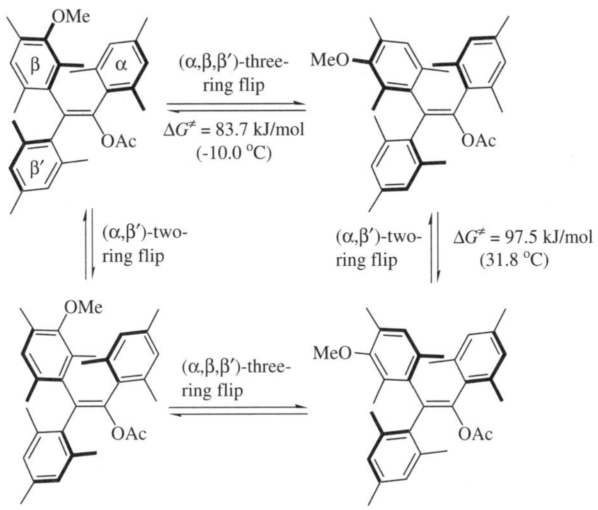

Scheme 8.14 Rotational energy barriers to α,β,β'-ring flip diastereomerization and α,β'-ring flip enantiomerization of (*E*)-2-*meta*-methoxymesityl-1,2-dimesitylvinyl acetate.

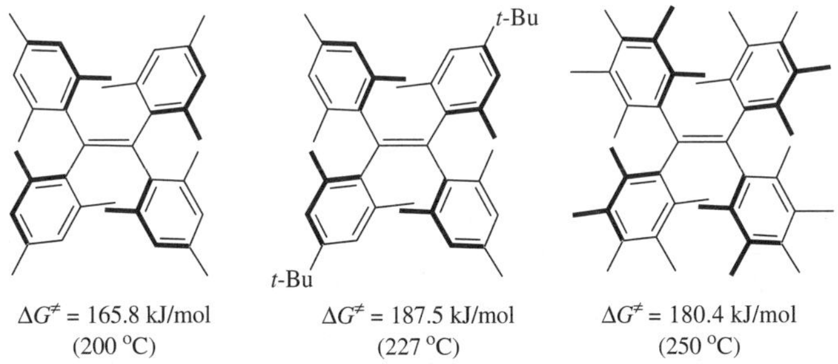

$\Delta G^{\neq} = 165.8$ kJ/mol (200 °C) $\Delta G^{\neq} = 187.5$ kJ/mol (227 °C) $\Delta G^{\neq} = 180.4$ kJ/mol (250 °C)

Figure 8.15 Structures of four-bladed vinyl propellers.

fourth vinylic substituent other than hydrogen have a three-ring flip threshold mechanism.[106–108] Trimesitylethylene prefers the (α,β)-two-ring circuit with an activation barrier of 70.3 kJ/mol over a higher energy process with an energy barrier of 85.8 kJ/mol, which is probably the three-ring flip.[109] Of course, the substitution pattern of the individual aryl blades also affects the stereodynamics of triarylvinyl propellers. This has been verified with several 2,2-dimesityl-1-arylethenol derivatives.[110,111] For example, 1-phenyl-2,2-dimesityl ethenol does not undergo isomerization via the three-ring flip route, but prefers a combination of independent low-energy rotation of the phenyl ring and correlated β,β'-two-ring flipping of the geminal mesityl blades.

Tetraarylethylenes such as tetra-*ortho*-tolylethylene exhibit the highest conformational stability among all vinyl propellers and isomerize via four-ring flipping, Figure 8.15.[112,113] Because of severe steric hindrance to blade rotation, the enantiomers of tetramesitylethylene can be isolated at room temperature. Helical inversion of this crowded tetraarylvinyl propeller occurs at very high temperatures and has a free activation energy of 165.8 kJ/mol.[114] Tetrakis(pentamethylphenyl)ethylene represents one of the most crowded four-bladed vinyl propellers reported to date and possesses an inversion barrier of 180.4 kJ/mol.[115]

8.6 PROPELLER-LIKE COORDINATION COMPLEXES WITH HELICITY CONTROL

Metal coordination of tripodal tris(2-pyridylmethyl)amines generates propeller-like coordination complexes that display a fluxional helical arrangement of three pyridyl rings. The helicity can be controlled by incorporation of a chiral center into one of the ligand tethers, Scheme 8.15.[116,117] Optical rotation, CD measurements and crystallographic analysis of trigonal bipyramidal Zn(II) and Cu(II) complexes of enantiopure *N,N*-bis[(2-pyridyl)methyl]-1-(2-pyridyl)ethylamine and its derivatives revealed pronounced molecular helicity. This has been attributed to intramolecular chiral induction. The chiral center controls the spatial arrangement of the chelating ligand and consequently the whole propeller-like coordination sphere of the complex. The methyl substituent attached to the chiral center determines the magnitude of propeller twisting and the equilibrium between two major diastereomeric conformers that display opposite helicity. The *anti*-conformer having the methyl group pointing away from the pyridylmethyl tethers is at least 4 kJ/mol more stable than the *syn*-isomer which experiences steric repulsion between the methyl group and a proximate pyridylmethyl moiety.[118,119] Some of these propeller-shaped transition metal complexes have been successfully used as chiral solvating agents to differentiate between enantiomers of sulfoxides by ^{1}H NMR spectroscopy. More importantly, trigonal bipyramidal copper complexes behave like molecular one-electron redox switches.[120,121] Canary's group observed that the propeller-like copper(II) complex of (*S*)-*N,N*-bis[(2-quinolyl)methyl]-1-(2-quinolyl)ethylamine has a strong exciton-coupled circular dichroism (ECCD) signal that almost disappears upon reduction ($\Delta\Delta\varepsilon_{240} = 430$). Subsequent oxidation of the ECCD-silent Cu(I) complex regenerates the helical Cu(II) complex, and is accompanied by the appearance of a strong ECCD signal. Voltammetric studies revealed that this reversible switch has a short response time and shows no sign of decreasing ECCD signal intensity after five redox cycles. The pronounced difference in the chiroptical properties of the two redox states has been explained by a structural change to a tetrahedral coordination sphere upon formation of the Cu(I) complex. The Cu(II) complex has all three quinolyl rings in close proximity and thus affords three couplets that contribute to a large CD amplitude. In contrast, the reduced Cu(I) center is coordinated by only two quinolyl rings, generating only one couplet, Scheme 8.16. The stereodynamics and redox properties of this class of copper complexes provide an on-off switch that controls molecular helicity and exciton-coupled circular dichroism. Other propeller-like copper complexes bearing enantiopure methylcysteine, methionine or methioninol in addition to two quinolylmethyl ligands undergo similar redox-induced inversion of helicity.[122–124] For example, the (*S*)-methionine-derived Cu(II) complex depicted in Scheme 8.17

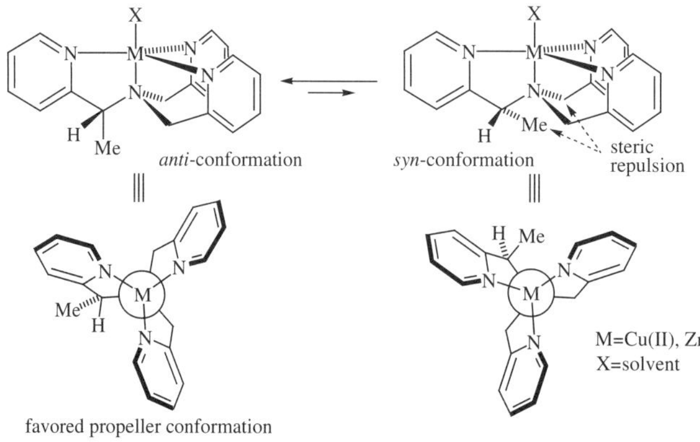

Scheme 8.15 Propeller conformations of trigonal bipyramidal metal complexes of (*R*)-*N,N*-bis[(2-pyridyl)methyl]-1-(2-pyridyl)ethylamine.

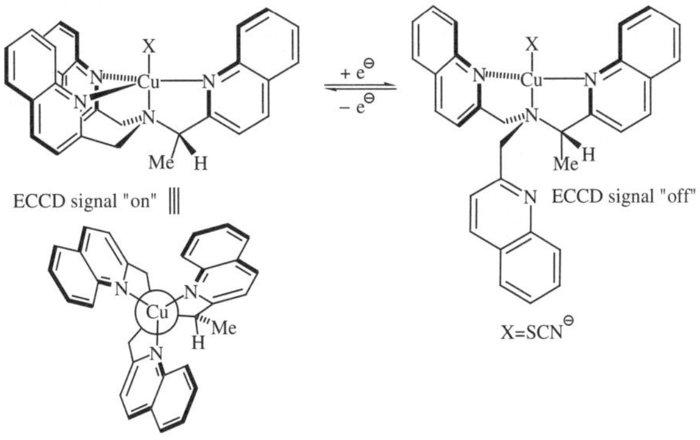

Scheme 8.16 Redox switching between a pentacoordinate propeller-like Cu(II) complex having a large ECCD signal and a tetracoordinate Cu(I) complex with low chiroptical activity.

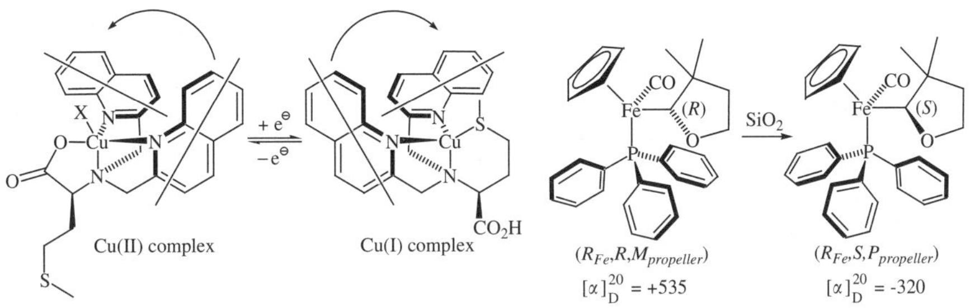

Scheme 8.17 Structural basis for exciton-coupled circular dichroism of an (*S*)-methionine-derived chiroptical molecular switch with redox-controlled reversible propeller helicity developed by Canary's group (left) and diastereomerization of Davies' [Fe(η^5-C$_5$H$_5$)(PPh$_3$)(CO)L] complex (right).

favors a trigonal bipyramidal pentacoordinate structure that is controlled by the absolute configuration of the chelating amino acid. Reduction to the tetracoordinate Cu(I) complex causes spontaneous reorganization of the metal coordination sphere in which a helical complex structure is maintained but the carboxylate group is replaced by the thioether function. The corresponding Cu(I) and Cu(II) complexes display reversed helicity and give ECCD couplets of opposite sign.

Davies *et al.* observed epimerization at the α-center in the chiral ligand *L* and concomitant helical inversion of the adjacent propeller-like triphenylphosphine group in [Fe(η^5-C$_5$H$_5$)(PPh$_3$)(CO)L], Scheme 8.17.[125] The SiO$_2$-mediated diastereomerization is accompanied by a large change in the specific rotation due to flipping of the triphenylphosphine propeller. Again, molecular helicity and chiroptical properties of a metal complex are controlled by a single chiral element.

8.7 STATIC GEARING AND CYCLOSTEREOISOMERISM

Adjacent isopropyl groups can adopt a bisected orientation in which the methine hydrogen of one isopropyl unit points between the two methyl groups of another. This can give rise to static gearing if the neighboring isopropyl groups are attached to a rigid planar framework, for example a benzene ring, Figure 8.16. The isopropyl and dichloromethyl groups in hexaisopropylbenzene and decakis(dichloromethyl)biphenyl are tightly intermeshed and can not rotate freely about the sp^2–sp^3 bonds. As a result of static gearing, the isopropyl groups in hexaisopropylbenzene favor a conformation that has all methine protons pointing in the same direction. Based on NMR analysis of

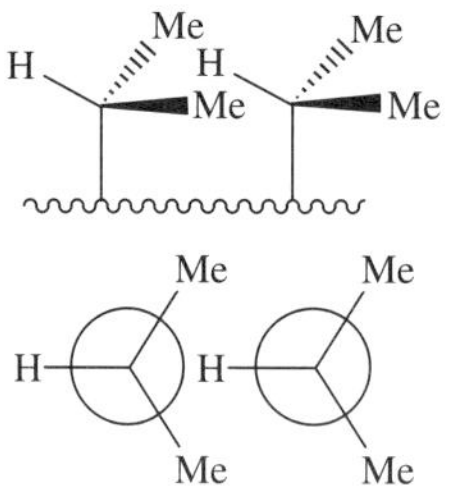

Figure 8.16   Static gearing of isopropyl groups.

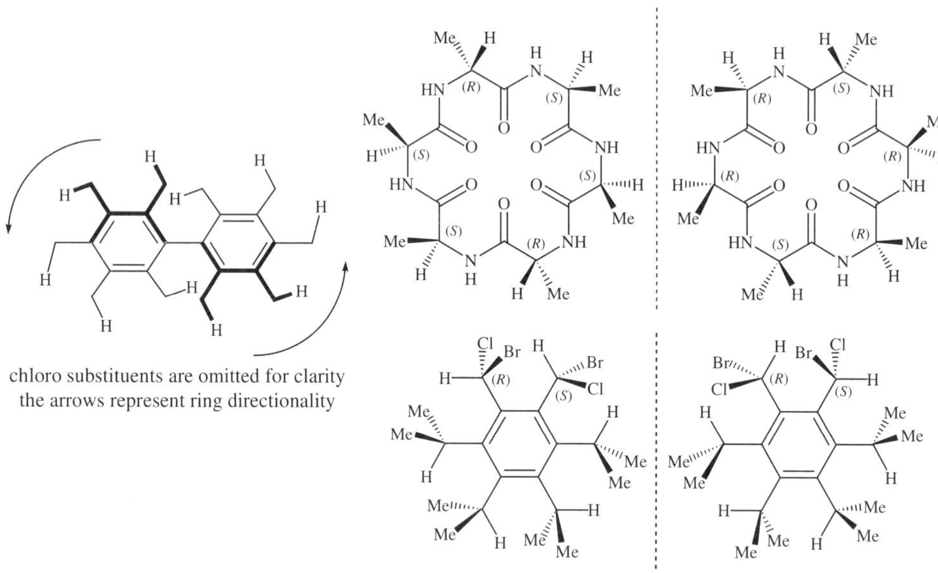

chloro substituents are omitted for clarity
the arrows represent ring directionality

Figure 8.17   Structure of decakis(dichloromethyl)biphenyl (left), and cycloenantiomeric cyclohexaalanine and 1,2-bis(bromochloromethyl)-3,4,5,6-tetraisopropylbenzene (right). Note that the two molecules of cyclohexaalanine differ in the sense of the peptide bonds: $(R \rightarrow S \rightarrow S \rightarrow R \rightarrow S \rightarrow S)$ versus $(S \leftarrow R \leftarrow R \leftarrow S \leftarrow R \leftarrow R)$.

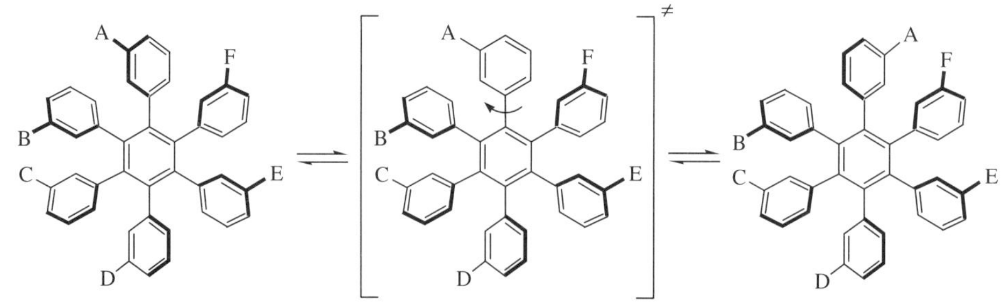

Scheme 8.18   Diastereomerization pathway of a hexaarylbenzene based on uncorrelated rotation of one aryl blade.

selectively deuterated hexaisopropylbenzene and force field calculations, Siegel and Mislow proved that the conformational orientation of interlocked isopropyl groups affords cyclic directionality and cyclostereoisomerism.[126,127] They postulated that substituent directionality can produce conformational cycloenantiomers in the case of appropriately hexasubstituted benzenes, and this has been confirmed in the case of 1,2-bis(1-bromoethyl)-3,4,5,6-tetraisopropylbenzene and 1,2-bis(bromochloromethyl)-3,4,5,6-tetraisopropylbenzene.[128,129] In general, cyclostereoisomerism is observed when constitutionally identical building blocks that possess a stereogenic element attain a directional arrangement around a ring system. These building blocks may be part of the ring system, for instance in a cyclic polypeptide such as cyclohexaalanine, or they may lie outside the ring. The latter is observed with 1,2-bis(bromochloromethyl)-3,4,5,6-tetraisopropylbenzene exhibiting static gearing of isopropyl and bromochloromethyl groups that are directionally oriented around the benzene core, Figure 8.17. Decakis(dichloromethyl)biphenyl provides another intriguing example of cycloenantiomerism, due to clockwise or counterclockwise directionality of geared dichloromethyl substituents. The racemization mechanism of this chiral biaryl is discussed in Chapter 3.3.1.[130]

Cogwheeling is by no means the only interconversion mechanism available to overcrowded molecules. Enantiomerization of tetraisopropylethylene, a highly strained alkene that adopts a twisted chiral conformation, was originally attributed to cogwheeling of intermeshed isopropyl groups.[131–133] It is now believed that the enantiomers of tetraisopropylethylene interconvert via a stepwise inversion process in which the substituents rotate consecutively but not through correlated motion. Similarly, sterically crowded polyarylbenzene and polyarylcyclopentadiene propellers undergo isomerization by entropically favored uncorrelated ring rotation with a threshold mechanism that has one ring rotating at a time, Scheme 8.18.[134,135]

8.8 MOLECULAR BRAKES, TURNSTILES AND SCISSORS

The intriguing stereodynamics of molecular propellers and the feasibility of redox-induced helicity switching in propeller-like trigonal bipyramidal transition metal complexes have drawn increasing attention to other chiral molecular devices that exploit molecular motion about a single bond. Triptycene-derived helicenes have been prepared in search of a molecular ratchet consisting of a toothed wheel, a pawl that allows rotary motion of the wheel in a single direction, and a spring that connects the two components. Kelly and coworkers envisioned that the triptycene moiety in 4-methyl-1-triptycylbenzo[c]phenanthrene represents the wheel of a ratchet, while the proximate helicene unit introduces rotational bias and thus operates like a simple pawl, Figure 8.18.[136,137] The covalent bond between wheel and pawl was expected to function like a spring that holds the pawl against the wheel. Molecular modeling indeed indicated that the helical shape of the pawl could favor unidirectional rotation of the triptycene unit, but spin polarization transfer NMR spectroscopy uncovered that the wheel rotates in both directions with an energy barrier of 105 kJ/mol.[ix]

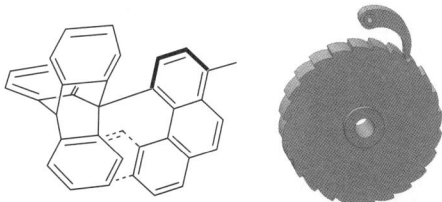

Figure 8.18   Structure of 4-methyl-1-triptycylbenzo[c]phenanthrene and a mechanical ratchet.

[ix] This molecule lacks the directional control of a ratchet, but the design of this triptycyl[4]helicene led to the development of a chemically-driven motor, see Chapter 8.11.

Molecular ratchets are still elusive but similar triptycyls have been incorporated into reversible brakes.[138,139] For example, attachment of a bipyridyl group to triptycene affords a stereodynamic device that can be controlled by an external stimulus. Variable-temperature NMR studies proved that the three-toothed gear shown in Scheme 8.19 rapidly rotates about the triptycyl-pyridyl bond. However, addition of Hg(II) or other metal ions locks the bipyridyl atropisomer into a rigid coplanar conformation that sterically impedes gear rotation. Metal coordination thus serves as a switch that halts triptycyl rotation. Bates designed a simple redox-mediated molecular brake that is derived from atropisomeric 2-[(2-methylthio)phenyl]isoindolin-1-one, Scheme 8.20.[140] The operation of this brake is based on an externally controlled, reversible sulfide–sulfoxide redox cycle. Transformation of the sulfide to the sulfoxide with *m*-chloroperbenzoic acid increases the steric hindrance to rotation about the chiral axis from 35.1 to 56.9 kJ/mol. Oxidation thus dramatically reduces the rate of atropisomerization by a factor of approximately 10^4 but the sulfoxide still undergoes rapid rotation about the biaryl axis at room temperature.[x] Despite the inherently low conformational stability, manipulation of the motion of this atropisomer is possible under cryogenic conditions.

A molecular mimic of a macroscopic pair of scissors has been realized with a light-driven chiral azobenzene-derived ferrocene complex, Scheme 8.21. The metallocene unit resembles the hinge,

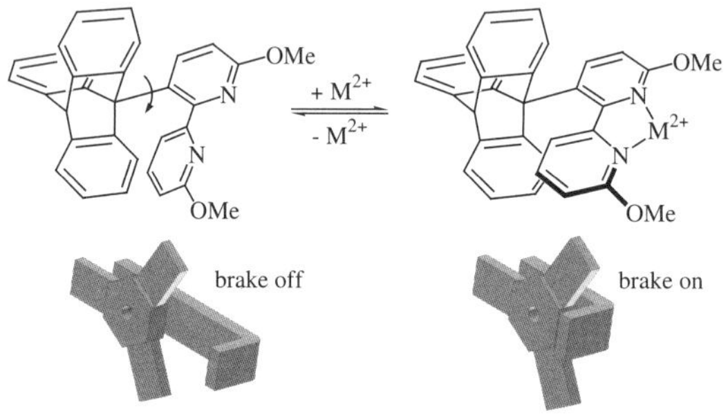

Scheme 8.19   Structure of a molecular brake consisting of a triptycyl wheel and a bipyridyl brake that impedes rotation of the three-toothed gear upon metal coordination.

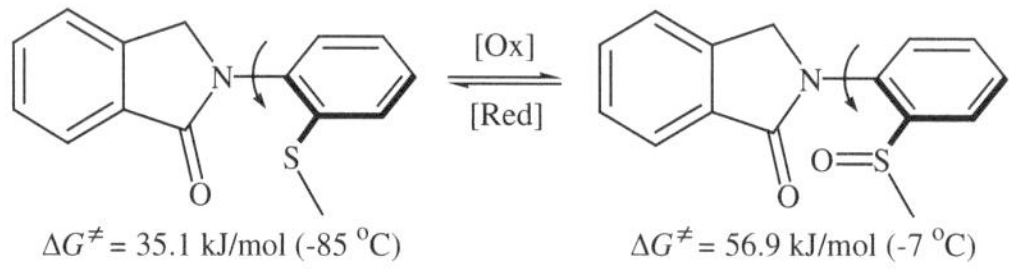

$\Delta G^{\neq} = 35.1$ kJ/mol (-85 °C) $\Delta G^{\neq} = 56.9$ kJ/mol (-7 °C)

Scheme 8.20   Structure of a redox-controlled atropisomeric brake.

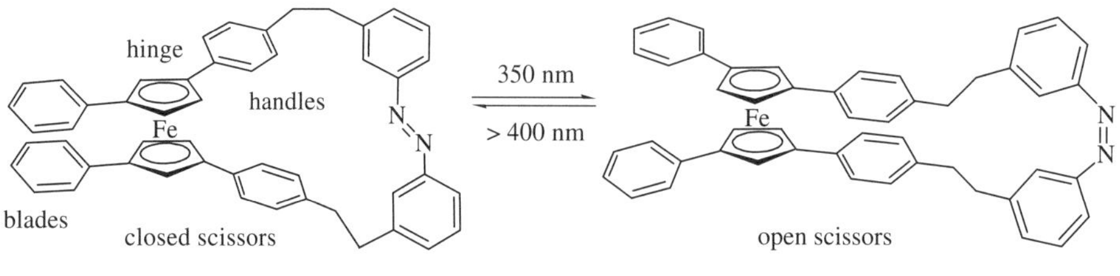

Scheme 8.21   Light-driven molecular scissors.

---

[x] The low conformational stability of this redox brake is a consequence of simultaneous *N*-pyramidalization which decreases steric repulsion in the transition state and thus facilitates rotation.

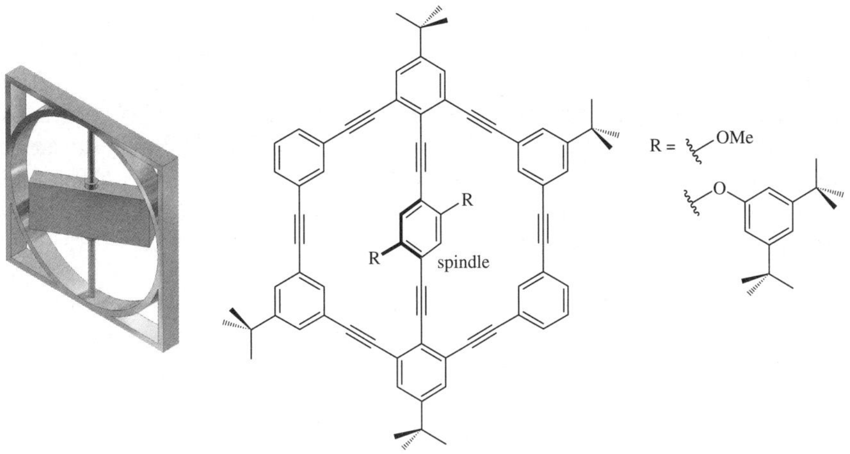

Figure 8.19 Illustration and structure of a molecular turnstile.

consisting of two easily rotating cofacial cyclopentadienyl planes, while the aryl substituents represent the blades and handles.[141] The azobenzene moiety provides the means for reversible opening and closing of the scissors through light-controlled *cis/trans*-photoisomerization; the blades are open when azobenzene adopts a *cis*-configuration and formation of the *trans*-isomer closes the blades. Irradiation of the *trans*-isomer with UV light at 350 nm for 3 minutes gives 89% of the *cis*-isomer and results in angular motion of the cyclopentadienyl rings. The molecular movement is reversed upon irradiation of visible light, establishing a *trans/cis* ratio of 46:54 within 15 seconds.

Moore *et al.* fabricated a molecular turnstile by assembling a rigid macrocyclic frame that holds a diethynylarene bridge.[142,xi] Incorporation of symmetry-breaking *tert*-butyl groups into the frame and two substituents into the spindle arene enabled them to monitor rotation about the bridging axis by variable-temperature NMR spectroscopy. The stereodynamics of the turnstile depend on the size of the substituents attached to the central aryl ring, but the direction of the rotation can not be controlled, Figure 8.19.

8.9 CHIRAL MOLECULAR SWITCHES

The desire to control and manipulate molecular structures by an external stimulus such as light has been inspired by the photomovements and photomorphogenesis frequently encountered in nature. A prominent example is the visual excitation in retina based on light-induced *cis/trans*-isomerization of 11-*cis*-retinal to all-*trans*-retinal. This process essentially converts a single photon into atomic motion through conformational changes and conversion of rhodopsin to metarhodopsin II, which triggers an enzymatic cascade and subsequently a nerve pulse. Retinal isomerase then regenerates 11-*cis*-retinal which basically serves as a photoresponsive biomolecular switch. Important properties of a molecular switch are bistability[xii] and fast, effective and reproducible responsiveness to a photochemical, thermal, electrochemical or chemical stimulus.[143] Other criteria for the usefulness of a switch are detectability and nondestructive read-out, stability to photochemical and thermal degradation, and stability to interconversion and concomitant loss of information over a wide temperature range. Many molecular switches are based on reversible redox

---

[xi] The development of molecular turnstiles exhibiting molecular bistability may lead to interesting nanomaterials, for example ferroelectric materials with short switching times.

[xii] A bistable system consists of two stable states that undergo reversible interconversion controlled by an external signal.

properties, analogous to Canary's propeller-like complexes discussed in Chapter 8.6. The photochromism of fulgides, azobenzenes, sterically overcrowded stilbenes, spiropyrans, and diarylethenes that undergo photochemically controlled *cis/trans*-isomerization, cyclization, electron transfer, and tautomerization has been exploited for the same purpose.[xiii] Photoreversible switching relies on selective interconversion of distinct isomers that absorb at different wavelengths. Chiral photochromic compounds are susceptible to selective diastereomerization and enantiomerization induced by irradiation at different wavelengths and circularly polarized light, respectively. Attractive features of chiroptical switches commonly include facile nondestructive read-out and detectability by ORD or CD spectroscopy at nonabsorbed wavelengths.

The investigation of photochromic and thermochromic properties of overcrowded alkenes by Feringa and coworkers has greatly contributed to the development of chiroptical molecular switches.[144–148] Benzoannulated bithioxanthylidenes exhibiting two stereogenic elements (helicity due to out-of-plane distortion of the molecular framework and a double bond giving rise to *cis/trans*-isomers) undergo stereospecific thermal and photochemical isomerization reactions, Scheme 8.22. Racemic synthesis of 12*H*-benzo[*a*]thioxanthenyl-12-(2′-methyl-9′*H*-thioxanthen-9′-ylidene) and HPLC isolation of all four stereoisomers using (+)-poly(triphenylmethylmethacrylate) as chiral stationary phase allowed systematic analysis of the stereodynamics of this crowded alkene. Thermal and photochemical isomerization of the (*P*)-*trans*-isomer generates the (*M*)-*cis*-diastereoisomer with an energy barrier of 119.7 kJ/mol but formation of the corresponding (*M*)-*trans*- and (*P*)-*cis*-isomers is not observed. Introduction of the (*P*)-*cis*-alkene to the same isomerization experiments results in exclusive formation of the (*M*)-*trans*-form. The stereospecific reaction outcome proves that *cis/trans*-isomerization and reversal of helicity are coupled processes in benzoannulated bithioxanthylidenes. Octahydrobiphenanthrylidenes represent another class of highly congested alkenes with a helically shaped π-electron system that gives rise to intense Cotton effects. Harada and Feringa applied enantioselective HPLC in conjunction with CD and NMR spectroscopy and crystallographic analysis to characterize the unique structure of these

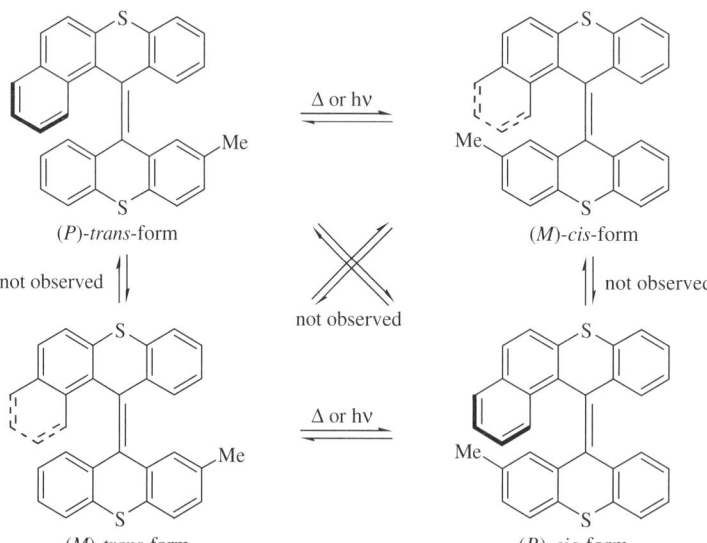

Scheme 8.22   Structure and stereospecific isomerization of 12*H*-benzo[*a*]thioxanthenyl-12-(2′-methyl-9′*H*-thioxanthen-9′-ylidene).

[xiii] Photochromism refers to photochemically induced interconversion of two or more isomers with different absorption spectra.

inherently chiral compounds.[149,150] Thermal racemization of (Z)-1,1',2,2',3,3',4,4'-octahydro-4,4'-biphenanthrylidene occurs at room temperature but the (E)-olefin does not racemize below 55 °C.[151] Interestingly, enantioconversion of the (E)- and (Z)-olefins does not concur with *cis/trans*-isomerization. The Gibbs activation energy for racemization of the (Z)- and the (E)-alkene is 97.7 and 114.3 kJ/mol, respectively, Scheme 8.23.

The unique thermochromic and photochromic properties of overcrowded alkenes originate from significant ground state distortion, molecular twisting and bistability of stereoisomeric states. Having developed the necessary tools to control the stereodynamics of helical alkenes exhibiting remarkable quantum yields of up to 0.72, Feringa's group introduced a thioxanthene-derived chiroptical switch.[152] The enantiomers of *cis*- and *trans*-4-[9'-(2'-methoxythioxanthylidene)]-7-methyl-1,2,3,4-tetrahydrophenanthrene are relatively stable to both racemization and *cis/trans*-isomerization at room temperature. However, stereospecific *cis/trans*-isomerization is observed upon irradiation of the (M)-*cis*-isomer with UV light, yielding a photostationary state consisting of 64% of the (M)-*cis*- and 36% of the (P)-*trans*-form. The equilibrium of *cis*- and *trans*-isomers at the photostationary state is determined by the ratio of the molar absorption coefficients of the two diastereoisomers at the wavelength used and the ratio of the quantum yields of each diastereomerization reaction. Since the energy barrier to thermal racemization of the (M)-*cis*- to the (P)-*cis*-isomer is only 110.5 kJ/mol, photochemical switching is accompanied by 10% racemization after 20 cycles, Scheme 8.24.

Further optimization of the thermal stability to racemization and fine-tuning of UV absorptivity of the photochromic states has led to a second generation of chiroptical switches with superior stereocontrol.[153–156] Incorporation of donor/acceptor-substituted thioxanthene units into the inherently chiral distorted alkene framework combines high racemization barriers with a pronounced difference in the absorption spectra of the *cis/trans*-isomers, which permits effective

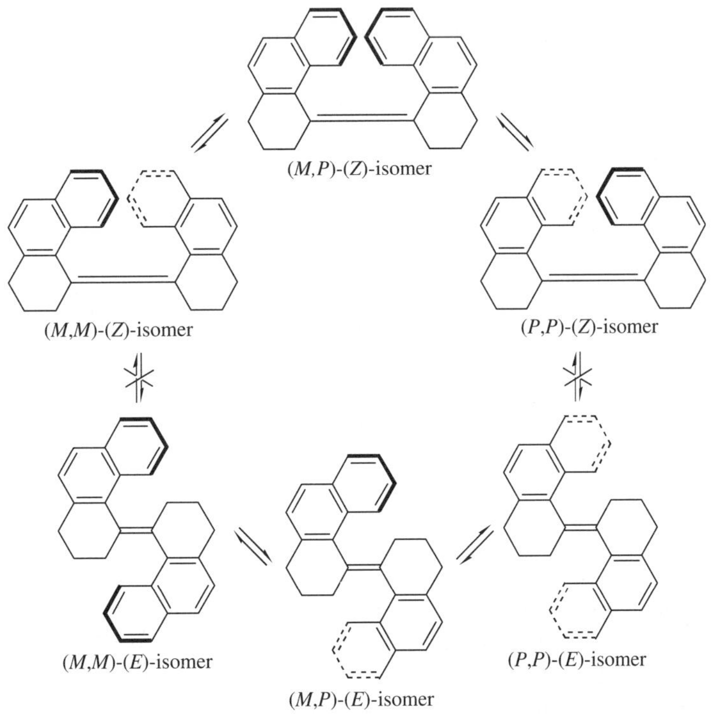

(M,P)-(Z)-isomer

(M,M)-(Z)-isomer

(P,P)-(Z)-isomer

(M,M)-(E)-isomer

(P,P)-(E)-isomer

(M,P)-(E)-isomer

Scheme 8.23 Enantiomerization pathways of (E)- and (Z)-isomers of 1,1',2,2',3,3',4,4'-octahydro-4,4'-biphenanthrylidene.

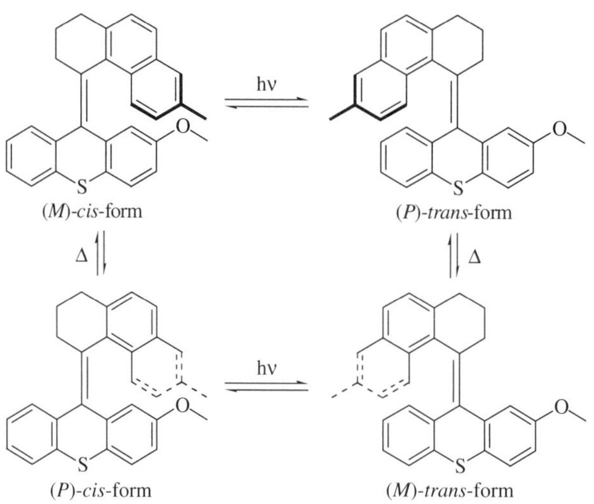

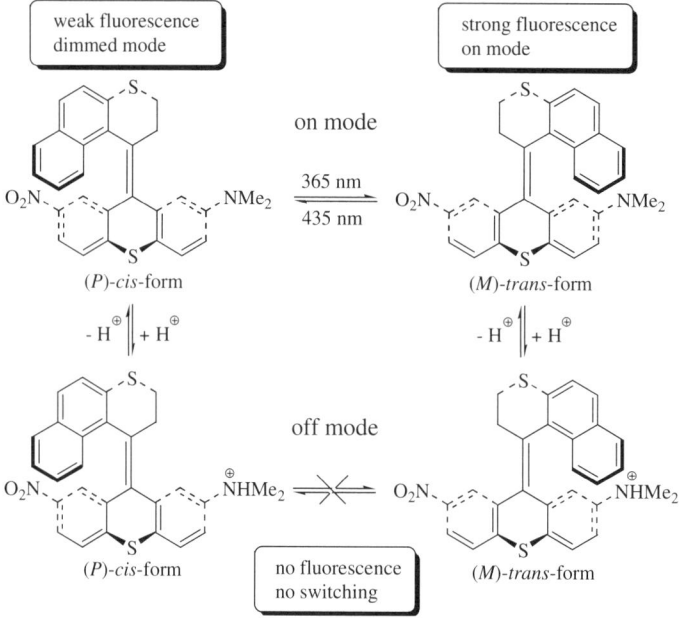

Scheme 8.24   A photoswitchable molecular system based on the isomers of [9′-(2′-methoxythioxanthyli-
dene)]-7-methyl-1,2,3,4-tetrahydrophenanthrene.

Scheme 8.25   Dual-mode photoswitching using a second generation chiroptical switch with a donor/
acceptor-substituted thioxanthene unit.

photoswitching based on diastereoselective excitation, Scheme 8.25. Polarimetric studies of isolated
enantiomers of a second generation switch derived from a donor/acceptor-substituted thioxanthene
unit revealed a racemization barrier of 122.2 kJ/mol, and no sign of racemization was observed
after 60 cycles. Alternating irradiation of 365 and 435 nm light showed excellent reversibility
between two photostationary states having (*P*)-*cis*/(*M*)-*trans* ratios of 30:70 and 90:10, respectively. It is noteworthy that reversible protonation with trifluoroacetic acid yields nonfluores-
cent ammonium salts that do not undergo photo-induced isomerization. This proton-regulated

photomodulation establishes a dual-mode switching system with three distinctive fluorescence states: dimmed, on and off. A change in the pH can therefore be used as an additional control element to effectively lock information stored in this chiroptical switch.[157,158] Incorporation of a configurationally stable chiral auxiliary into benzoannulated bithioxanthylidenes generates an additional means for fine-tuning of switching selectivity, and allows directional inversion of helicity.[159] In contrast to the chiroptical switches consisting of a diastereomeric pair of enantiomers discussed above, the presence of a chiral pyrrolidine unit changes the scenario to four inter-converting diastereomeric stations. The auxiliary renders photoequilibria distinguishable, and thermal helical inversion no longer affords a racemate but diastereoisomers with inherently different stability. One can therefore anticipate unidirectional interconversion of the four diastereo-isomeric states through a sequence of photochemically and thermally controlled isomerization steps, Scheme 8.26.

A remaining shortcoming of the switches shown above is the unsatisfactory diastereomeric ratio at the photostationary states. This limitation has been resolved through careful optimization of the stereodynamics and photochromic properties of sterically overcrowded alkenes bearing two chiral centers in addition to a helical structure.[160] Chiroptical switching between (3S,3'S)-(M,M)-*cis*- and (3S,3'S)-(P,P)-*trans*-1,1',2,2',3,3',4,4'-octahydro-3,3'-dimethyl-7,7'-dimethyl-4,4'-biphenanthrylidene proceeds with more than 99% stereoselectivity in both directions. The remarkable switching selectivity at 303 and 376 nm has been attributed to a dramatic red shift in the absorption spectrum of the *trans*-isomer, and to a significant difference in the thermodynamic stability of the diastereomers carrying the two methyl groups at carbons 3 and 3' either in pseudoequatorial or pseudoaxial position, Scheme 8.27.

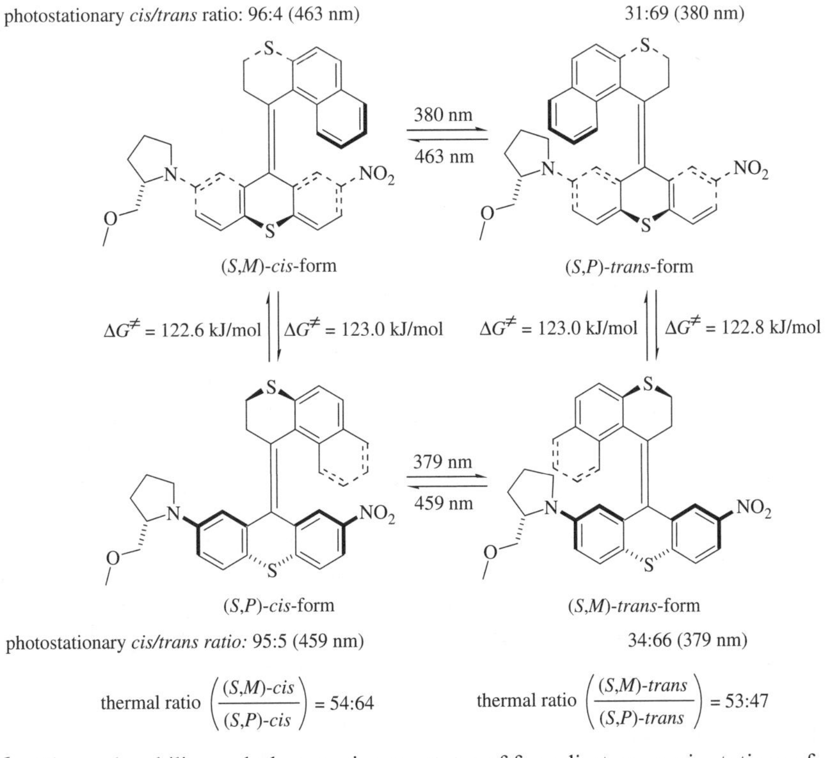

Scheme 8.26 Thermal stability and photostationary states of four diastereomeric stations of a benzoannulated bithioxanthylidene switch.

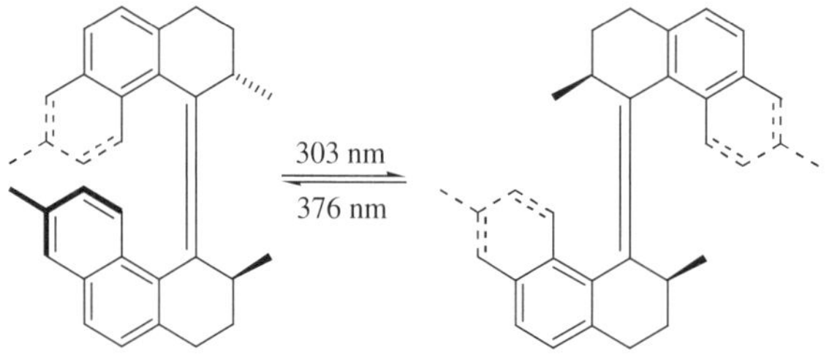

(3*S*,3'*S*)-(*M*,*M*)-*cis*-form   (3*S*,3'*S*)-(*P*,*P*)-*trans*-form
          > 99%   > 99%

Scheme 8.27   A chiroptical switch providing more than 99% stereoselectivity in both directions.

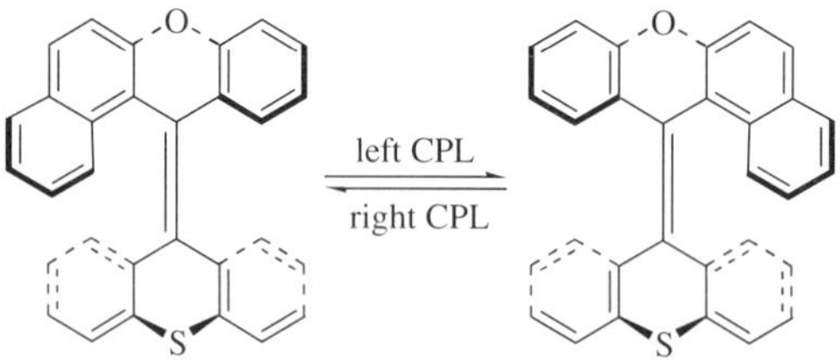

(*P*)-enantiomer   (*M*)-enantiomer

Scheme 8.28   Photochemical interconversion between the enantiomers of 12-(9'*H*-thioxanthen-9'-ylidene)-12*H*-benzo[*a*]xanthene using circularly polarized light.

Chiroptical switching with alkenes is also feasible by irradiation of circularly polarized light (CPL). The efficacy of dynamic switching between enantiomers and the enantiomeric excess that can be expected at a photostationary state, ee_{pss}, is determined by the quantum efficiency for photoracemization, Φ_{rac}, because the rate of photoresolution increases exponentially with Φ_{rac} and the anisotropy factor, g, which is defined as the ratio of the CD signal, $\Delta\varepsilon$, and the absorption, ε, at a fixed wavelength:

$$ee_{pss} = \frac{g_\lambda}{2} = \frac{\Delta\varepsilon}{2\varepsilon} \tag{8.1}$$

The inherently low g-factor of helical alkenes limits the enantiomeric excess that can be obtained by irradiation of CPL to less than 1.0%. Nevertheless, the inherently large CD absorptions and optical rotations of these compounds allow very sensitive detection of enantiospecific photoresolution. For instance, alternating irradiation of left- and right-circularly polarized light at 313 nm and 400 nm results in deracemization of thermally stable 12-(9'*H*-thioxanthen-9'-ylidene)-12*H*-benzo[*a*]xanthene and gives rise to an ee of 0.07 and –0.07%, respectively, Scheme 8.28.[161] Bridged diarylethenes that undergo photochochemically induced conrotatory ring closure and opening upon irradiation of monochromatic light afford another class of molecular switches with excellent thermal stability and fatigue resistance.[162–169] Irie *et al.* reported diastereoselective photocyclization of a chiral dibenzo[*b*]thienyl maleimide switch exhibiting a (1*R*,2*S*,5*R*)-menthyl auxiliary, Scheme 8.29.[170] Irradiation of 450 nm light induces ring closure, yielding a photostationary state with 43% of the menthyl-derived (*S*,*S*)- and (*R*,*R*)-diastereoisomers and 87% diastereomeric excess at –40 °C. A change in the wavelength of the irradiated light to 570 nm favors ring opening and regenerates the original state. The diastereoselectivity of the cyclization is highly sensitive to solvent and temperature, which somewhat limits the scope of this switch.[xiv]

[xiv] Bridging of the ethylene double bond is necessary to avoid concomitant *cis/trans*-isomerization.

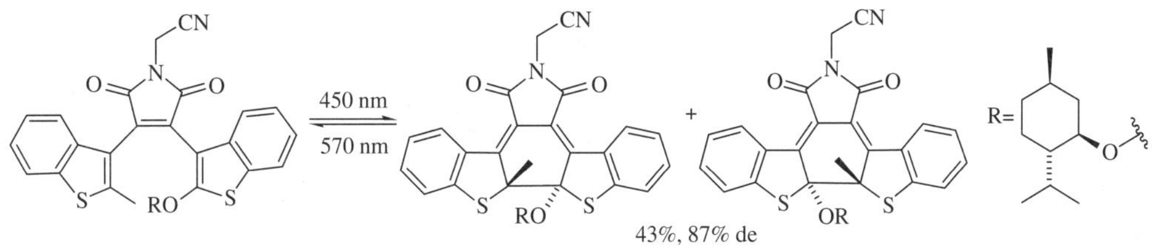

Scheme 8.29 Diastereoselective ring closure and opening of a chiral diarylethene switch.

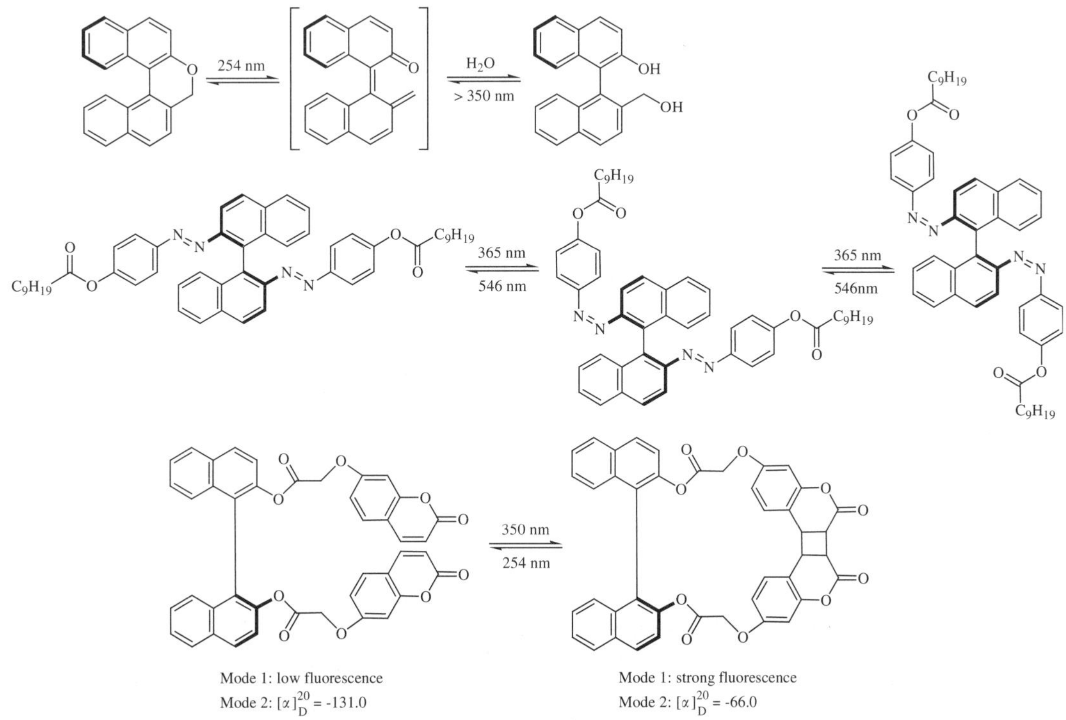

Scheme 8.30 Photochromic binaphthyl-derived chiral switches.

Several chiroptical binaphthyl switches are known. Schuster investigated photoracemization and photochromism of thermally stable 1,1′-binaphthylpyran and 2-hydroxy-2′-hydroxymethyl-1,1′-binaphthylene, Scheme 8.30.[171] Irradiation of enantiomerically enriched pyran above 390 nm causes formation of the diol accompanied by undesirable racemization but photochemical interconversion of the diol to the pyran proceeds without measurable loss of enantiopurity. Time-resolved absorption spectroscopy (nanosecond laser flash photolysis) revealed a 1,1′-binaphthyl quinine methide intermediate in both the photochromic and the photoracemization reaction. Similar to Feringa's overcrowded alkenes, the low quantum yield for photoracemization ($\Phi_{rac} = 0.0025$) and the low g-factor ($g_{max} \approx 0.002$) render this system unsuitable for chiroptical switching. A bistable dual mode coumarin-derived binaphthol system[172] generating a macrocycle upon irradiation and a diazobinaphthyl switch[173] that shows reversible *E/Z*-isomerization altering its skewed chiral conformation have been reported, Scheme 8.30.[xv]

[xv] The diazobinaphthyl switch has also been incorporated into nematic phases and used as a photoresponsive chiral dopant for induction of cholesteric liquid crystals with strong helical twisting power.

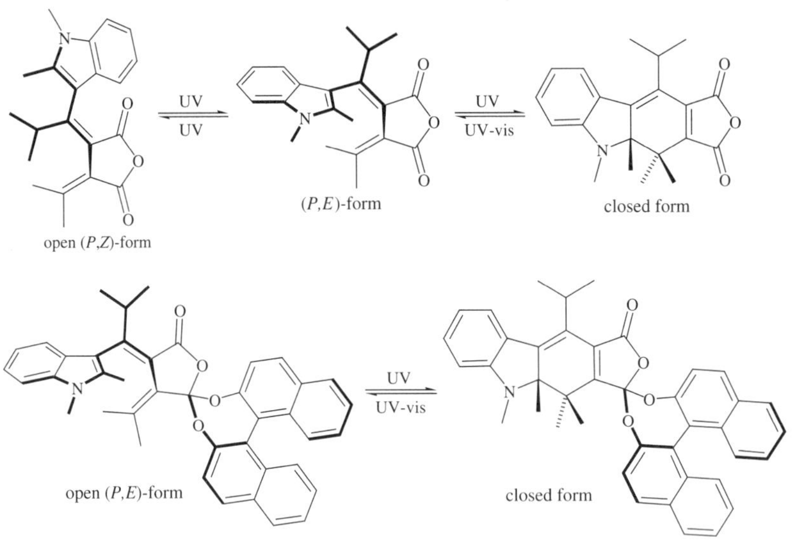

Scheme 8.31   Interconversion of photochromic indolyl fulgides.

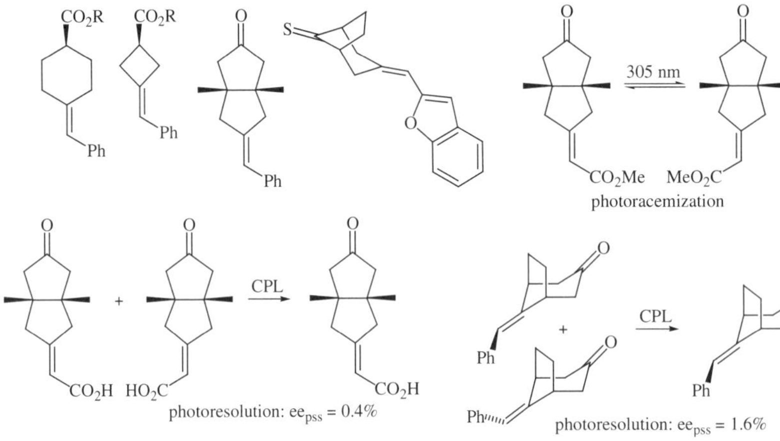

Figure 8.20   Axially chiral alkylidenecycloalkane switches.

Yokoyama's group developed photochromic helical fulgide switches that undergo interconversion of colorless (*E*,*Z*)-isomers and subsequent conrotatory electrocyclization of the (*E*)-form to a colorful rigid derivative, Scheme 8.31.[174–178] Repetitive photoswitching between the open (*E*)-isomer and the closed form is feasible but of limited use due to gradual photoracemization of small amounts of the (*Z*)-form produced upon prolonged irradiation at 405 nm. Combination of the indolyl fulgide structure with an (*R*)-binaphthyl moiety significantly increases thermal stability and fatigue resistance during iterative photochromic interconversions which are accompanied by a dramatic change of specific rotation values.[xvi]

Photoracemization of axially chiral alkylidenecycloalkane ketone and ester switches including bicyclo[3.3.0]- and [3.2.1]octan-3-one derivatives involves rotation about the exocyclic double bond, Figure 8.20.[179–181] As a consequence of the high *g*-factor ($g_{313} = 0.0502$) of

---

[xvi] Incorporation of substituents into indolyl fulgides is necessary to avoid sigmatropic hydrogen migration in the (*E*)-isomer.

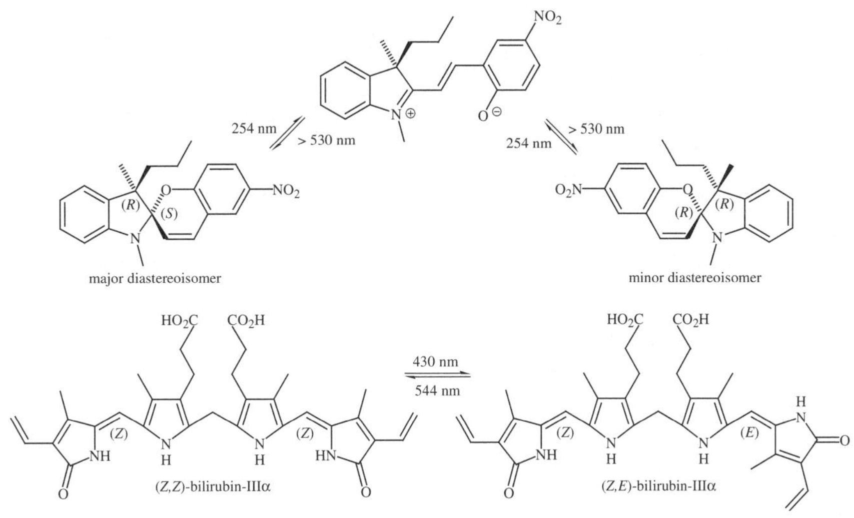

Scheme 8.32 Diastereoselective photoswitching with a spiroindolinopyran (top) and bilirubin-IIIα (bottom).

8-(phenylmethylene)bicyclo[3.2.1]octan-3-one, photoresolution resulting in 1.6% ee at the photo-stationary state after irradiation of CPL at 313 nm for 47 hours is possible and has been confirmed by CD spectroscopy.[182] The photochemically induced rotation proceeds through excitation of the ketone group, subsequent intersystem crossing to a ketone triplet state, and energy transfer forming a styrene triplet that readily rotates about the axially chiral double bond. The remarkable properties of this class of switches include high stability to photochemical decomposition and selective photoracemization with high quantum yields. For example, 7-[(methoxycarbonyl)methylene]-*cis*-bicyclo[3.3.0]octan-3-one has a quantum efficiency for photoracemization of 0.45 at wavelengths above 305 nm.[xvii] Unfortunately, effective photoresolution of this alkylidenecycloalkane ester is not feasible due to a very low *g*-factor, but partial deracemization (ee$_{pss}$ = 0.4%) has been observed with the corresponding carboxylic acid after irradiation of CPL at 305 nm for 400 minutes.

Spiropyrans are prone to fast thermal and photochemical interconversion of colorless enantiomers via a colorful achiral merocyanine intermediate.[183] Incorporation of a chiral center into close proximity of the stereogenic spiro carbon atom is necessary to prevent racemization.[184,185] A spiroindolinopyran-derived chiroptical switch that operates through formation of a chiral merocyanine is shown in Scheme 8.32.[186] Irradiation of UV light (254 nm) transforms the thermodynamically more stable spiro structures to the colored merocyanine form, whereas visible light (>530 nm) favors formation of the former. Spiropyran-derived switches have also been immobilized on cysteine residues of proteins including bovine serum albumin and used as biomolecular photochromic probes.[187] Site-selective incorporation of chiral photochromic switches into proteins provides intriguing opportunities for chiroptical control of biomolecular interactions in cells. Optical switching of a complex consisting of human serum albumin (HSA) and achiral bichromophoric bilirubin-IIIα, an isomer of the naturally occurring yellow pigment bilirubin-IXα, has been reported.[188] Photochemically controlled interconversion of two diastereoisomers of a water-soluble stoichiometric 1:1 bilirubin-IIIα/HSA adduct yields significant exciton coupling due to strong interactions between the chiral protein structure and closely embedded bilirubin chromophores. Photoexcitation of the (Z,Z)- and the (Z,E)-form, respectively, results in selective E/Z-isomerization accompanied by substantial changes of chiroptical properties. Irradiation of blue light (430 nm)

[xvii] The maximum theoretical value for Φ_{rac} is 0.5 because the excited state can form either enantiomer.

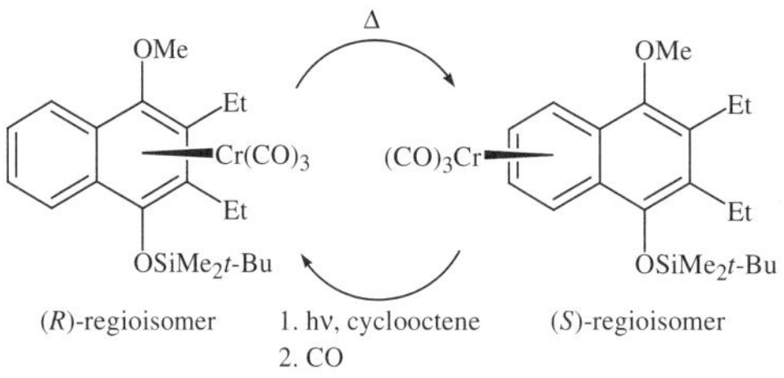

Scheme 8.33 Switching between chiral chromium complexes via haptotropic migration.

produces a photostationary state containing 32% of (Z,E)-bilirubin-IIIα/HSA, and the use of green light (544 nm) reduces the amount of the (Z,E)-isomer to 20%. Photoresponsive cycling by alternating one-minute irradiation of blue and green light coincides with less than 1.5% of photooxidative pigment loss after 10 repetitions.

A chiral metallorganic switch based on controlled migration of a tricarbonyl chromium moiety along a naphthohydroquinone skeleton due to haptotropic rearrangement has been described by Dötz and coworkers.[189] They discovered a cyclooctene-assisted photo-induced intramolecular metal shift along the fused aromatic system of a tricarbonyl naphthalene chromium complex, generating the thermodynamically disfavored (R)-isomer. This reaction reverses the thermally controlled haptotropic metal migration to the more stable (S)-regioisomeric complex. Haptotropic switching between the two chromium complexes exhibiting a chiral plane proceeds without any sign of racemization, Scheme 8.33.

8.10 STEREODYNAMIC SENSORS

As is described in the preceding chapter, irradiation of monochromatic and circularly polarized light allows selective interconversion of distinctive states of chiroptical switches. In some cases, incorporation of additional chiral elements into the switching device proved to enhance stereocontrol. For example, chiral centers complement the helical structure in some of Feringa's sterically overcrowded alkenes, and the spiroindolinopyran-derived chiroptical switch shown in Scheme 8.32 has a chiral center in close proximity to the stereogenic spiro carbon to exclude undesirable racemization. Similarly to irradiation of CPL, the presence of molecular chirality can induce chiroptical switching which can be exploited for stereoselective sensing. Inoue and coworkers observed enantioselective Z/E-photoisomerization of cyclooctene during singlet photosensitization with chiral menthyl benzenepolycarboxylates, β-cyclodextrin-derived monobenzoate and (R)- or (S)-1-methylheptylbenzoates immobilized on the surface of zeolites.[190–193] Formation of a singlet exciplex from the chiral sensitizer and (Z)-cyclooctene gives rise to enantiodifferentiating rotational relaxation towards chiral (E)-cyclooctene in up to 53% ee. Replacement of all (R)-1-methylheptyl units attached to the benzenepolycarboxylate unit with the corresponding (S)-alkyl groups results in reversed enantiomeric composition of the (E)-cyclooctene. Covalent linkage of the cyclooctene to a methyl arylcarboxylate sensitizer with a $(2R,4R)$-2,4-pentanediol tether provides intramolecular control of the photoisomerization and allows diastereoselective conversion of the achiral (E)-cyclooctene moiety to the chiral (Z)-isomer with 44% de, Scheme 8.34.[194]

Since the structure of the chiral benzenepolycarboxylate determines the stereochemical outcome of the Z/E-photoisomerization of cyclooctene, the enantiomeric distribution of the olefin product can be used to gain information about the chirality of the sensitizer. A molecular switch applicable to sensitive and selective recognition of stereoisomeric analytes must interconvert between photochromic states with distinct measurable output, while the ratio of these states is controlled by diastereomeric interactions with the chiral substrate. Tumambac and Wolf used axially chiral

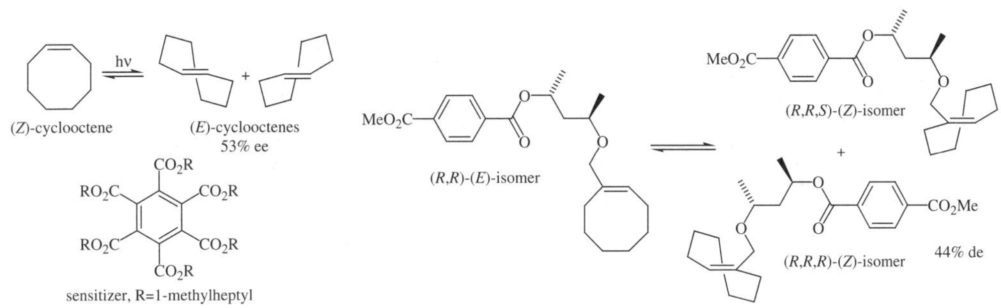

Scheme 8.34 Inter- and intramolecular photosensitization of cyclooctene.

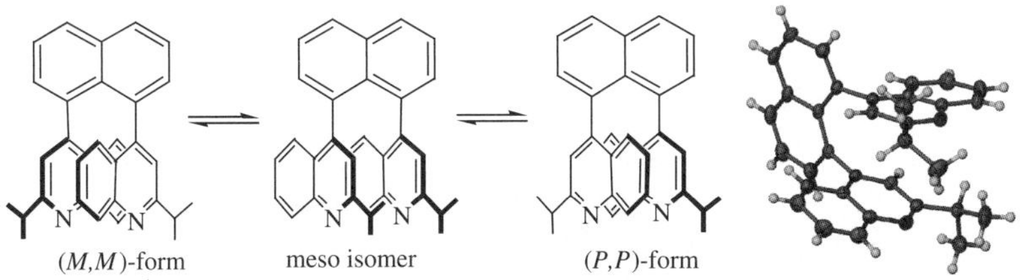

Scheme 8.35 Interconversion of *anti*- and *syn*-isomers and single crystal structure of 1,8-bis(2′-isopropyl-4′-quinolyl)naphthalene. [Reproduced with permission from *Eur. J. Org. Chem.* **2004**, 3850-3856.]

diquinolylnaphthalenes for stereodynamic fluorescence sensing of the *syn*- and *anti*-isomers of diaminocyclohexane, Scheme 8.35.[195] The C_2-symmetric *anti*-isomers of 1,8-bis(2′-isopropyl-4′-quinolyl)naphthalene are thermodynamically more stable than the meso *syn*-isomer. The atropisomers are conformationally stable at 25 °C but interconvert at higher temperatures. Importantly, the *syn*- and *anti*-isomers possess strikingly different fluorescence intensities at 380 nm and afford a well-defined binding environment for stereoselective recognition of hydrogen bond donating substrates.[xviii] Titration experiments revealed little fluorescence quenching upon addition of the *cis*- or *trans*-isomers of 1,2-diaminocyclohexane to a solution of the sensor. However, significant quenching is observed when the diquinolylnaphthalene is heated to 80 °C in the presence of *cis*-1,2-diaminocyclohexane, which has been attributed to substrate-controlled *anti*-to-*syn*-interconversion of the stereodynamic fluorosensor. The complementary geometry of the meso *cis*-diamine and the meso *syn*-isomer of the sensor thus favors formation of the latter through stabilizing hydrogen bonding. In contrast, the *trans*-1,2-diamine stabilizes the more fluorescent *anti*-isomer of the diquinolylnaphthalene-derived switch. This is a consequence of effective hydrogen bonding between the C_2-symmetric substrate and sensor structures, Scheme 8.36. The ability of *cis*- and *trans*-1,2-diaminocyclohexane to switch between the diastereomeric conformers of 1,8-bis(2′-isopropyl-4′-quinolyl)naphthalene has been verified by NMR spectroscopy. The use of this photochromic sensor for analysis of *cis*- and *trans*-isomers of diaminocyclohexane indicates the potential of stereodynamic recognition which complements applications of static binaphthol-, tartrate- and diacridylnaphthalene-derived stereoselective fluorosensors.[196–200]

Dynamic molecular recognition involving substrate-induced switching between atropisomeric conformations of a catecholamine receptor has been reported by Kawai *et al.*[201] Incorporation

---

[xviii] Kinetic studies revealed a Gibbs activation energy, $\Delta G^{\neq}$, of 115.2 and 111.1 kJ/mol for *anti*-to-*syn*- and *syn*-to-*anti*-isomerization of 1,8-bis(2′-isopropyl-4′-quinolyl)naphthalene at 66.2 °C. The fluorescence quantum yield of the *syn*- and the *anti*-isomers is 2.0 and 11.6%, respectively.

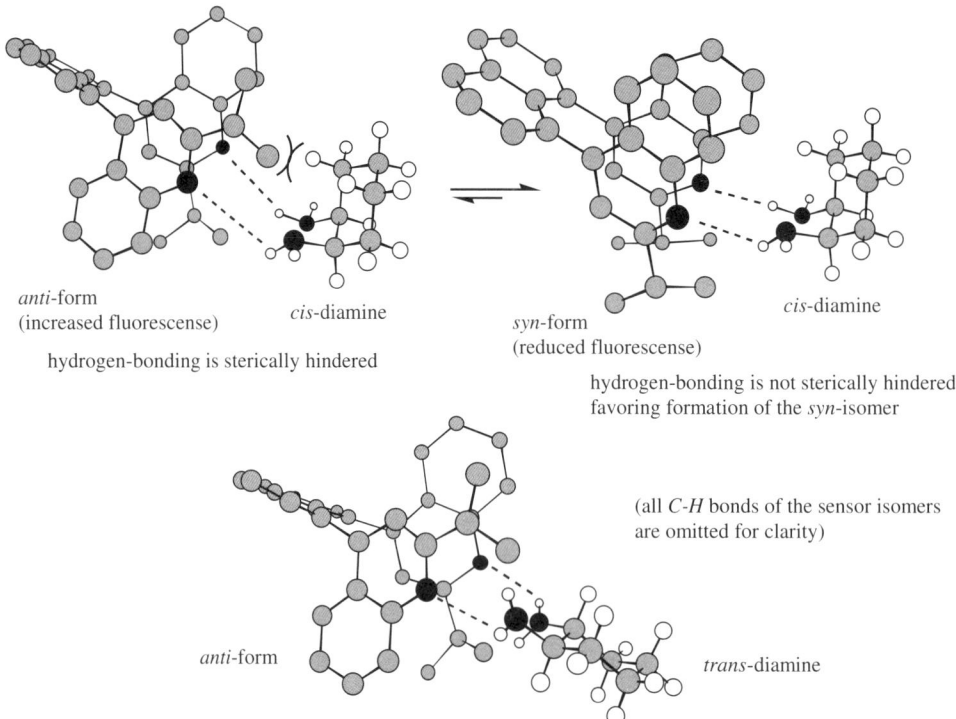

Scheme 8.36   *Anti*-to-*syn*-switching of the sensor in the presence of *cis*-1,2-diaminocyclohexane (top) and stabilization of the *anti*-isomer through hydrogen bonding with *trans*-1,2-diaminocyclohexane (bottom). [Reproduced with permission from *Eur. J. Org. Chem.* **2004**, 3850-3856.]

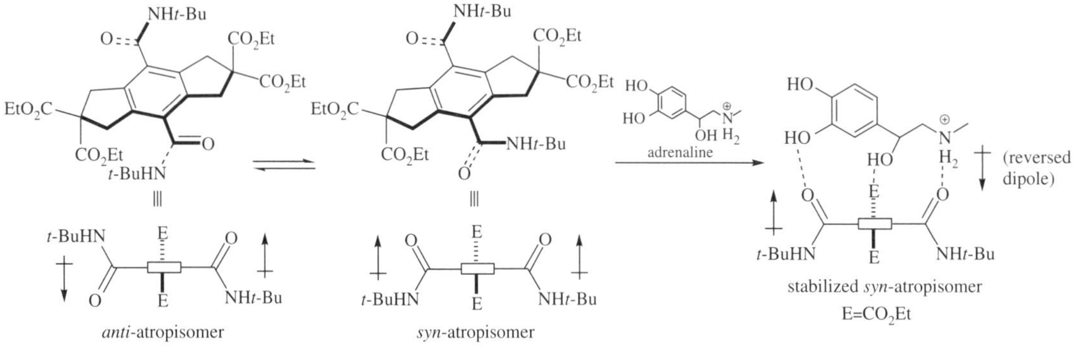

Scheme 8.37   *Anti*-to-*syn*-switching of an atropisomeric hydrindacene-derived receptor.

of two amide groups into a hydrindacene platform produces a mixture of rapidly interconverting *syn*- and *anti*-atropisomers, Scheme 8.37. The *anti*-isomer is more stable than the *syn*-form which has a larger net dipole moment. Binding studies with various catecholamines showed that the atropisomeric ratio of the receptor changes significantly upon addition of adrenaline and the *syn*-conformer becomes the dominant species due to dipole reversal and multiple hydrogen bonding. The formation of a stoichiometric adrenaline complex thus stabilizes the *syn*-conformation of the receptor. Artificial sensors and receptors that mimic natural switching processes are important for several reasons. They provide an entry to biological sensing and afford a mechanistic probe for analysis of allosteric binding and regulation of enzymes.[202,203]

8.11   CHIRAL MOLECULAR MOTORS

The discovery of biological locomotion has triggered increasing interest in the design of artificial motors.[204] Fueled by hydrolysis of adenosine triphosphate (ATP), myosins, kinesins, dyneins, and bacterial flagella serve as biomolecular machines and motors. These proteins operate in response to a biological stimulus and play a key role in biomolecular transport and biomechanical motion. A molecular motor generally consists of a stationary component (stator) that does not move, and a smaller rotating unit (rotor).[205] The development of bevel gears, ratchets, brakes, scissors, turnstiles, and switches has been possible through careful design and combination of molecular functionality and dynamic stereochemistry of chiral compounds. The ability to manipulate mechanical motion at the molecular level has culminated in a search for molecular motors that are able to convert energy into coordinated movement with directional control.[206–209]

Kelly's group designed a chemically powered 9-triptycyl[4]helicene-derived molecular motor using phosgene to fuel a unidirectional movement.[210,211] The triptycyl unit resembles a three-bladed rotor which is connected to the helicene stator by a single bond representing the axle of the motor, Scheme 8.38. Introduction of an amino group to the triptycyl rotor and a tethered alcohol to the helicene unit allows preparation of a carbamoyl derivative that undergoes chemically fueled unidirectional motion. In the first step, phosgene is used to convert the amine to the corresponding isocyanate which then rotates clockwise to a conformation having the isocyanate and the alcohol group in close proximity. At this point, the motor can also rotate in a counterclockwise direction, but the initial clockwise movement is rendered irreversible by formation of an intermediate carbamate. This reaction traps the rotor in a relatively high energy conformation. Ambient thermal energy then drives this intermediate towards a more stable conformation by continuation of the initial clockwise movement. Finally, the carbamate is hydrolyzed to complete the unidirectional motion. Shortening of the tether length has been found to accelerate rotation and provides a tool to vary the frequency of the motor.

The preceding discussion of chiroptical switches underscores the feasibility of light-driven unidirectional motion. In order to complete a full 360° rotary motion, Feringa and coworkers developed light-driven octahydro-4,4′-biphenanthrylidene-derived motors exhibiting selective

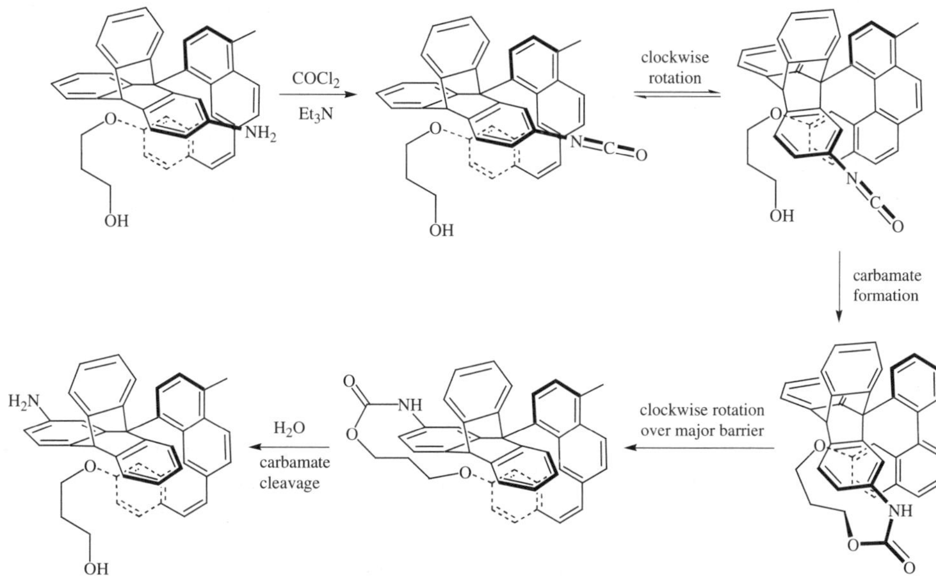

Scheme 8.38   Unidirectional rotation of a phosgene-driven motor.

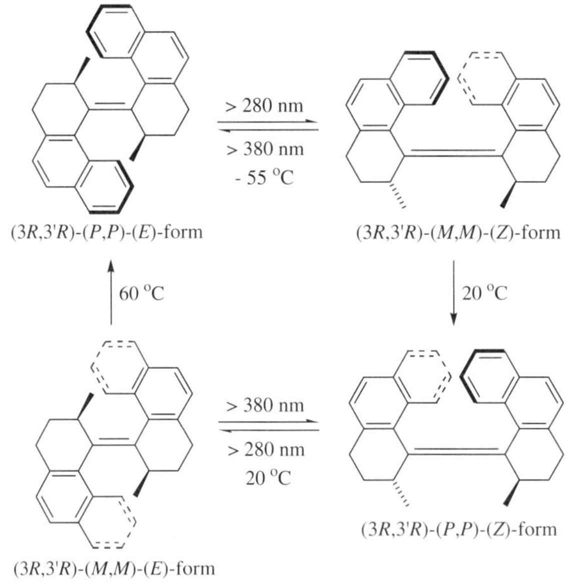

> 280 nm
> 380 nm
- 55 °C

(3R,3'R)-(P,P)-(E)-form

(3R,3'R)-(M,M)-(Z)-form

60 °C

20 °C

> 380 nm
> 280 nm
20 °C

(3R,3'R)-(M,M)-(E)-form

(3R,3'R)-(P,P)-(Z)-form

Scheme 8.39 Four-step directional rotation of an overcrowded alkene.

stereomutations, thermal and photochemical bistability, and nondestructive read-out.[212] The methyl substituents at *C*-3 in each phenanthrylidene unit introduce two chiral centers to the intrinsically helical geometry of these overcrowded alkenes, Scheme 8.39.[213,214] The presence of central chirality allows control of rotation about the motor axle (the carbon–carbon double bond) by thermally and photochemically initiated isomerization steps. Enantiomerically pure (3R,3'R)-(P,P)-(E)-1,1',2,2',3,3',4,4'-octahydro-3,3'-dimethyl-4,4'-biphenanthrylidene isomerizes to the (Z)-isomer upon irradiation of light at 280 nm. This reaction includes intermediate formation of the (3R,3'R)-(M,M)-(Z)-alkene. One can therefore photochemically convert the (P,P)-(E)-form to the (M,M)-(Z)-diastereosiomer under cryogenic conditions, which is followed by irreversible isomerization of the latter to the (P,P)-(Z)-form at 20 °C. The directional movement continues upon irradiation of light at 280 nm. This produces the (M,M)-(E)-isomer, and irreversible thermal isomerization at 60 °C finally regenerates the enantiomerically pure (3R,3'R)-(P,P)-(E)-alkene. The four isomerization steps result in a controlled 360° rotation and can be conveniently monitored by CD spectroscopy. It is noteworthy that the unidirectionality of this system is governed by the methyl groups which adopt pseudoaxial positions in the more stable (P,P)-isomers and pseudoequatorial orientations in the less stable (M,M)-forms.[xix] Accordingly, the first step resembles a photochemical E/Z-isomerization to a less stable form having the methyl groups in sterically congested pseudoequatorial positions. The strain is then released by thermal interconversion to a more stable (P,P)-isomer bearing pseudoaxial methyl groups. Another energetically uphill photoisomerization step followed by a downhill thermal diastereomerization completes the rotation. The communication between different chiral elements, *i.e.*, central and helical chirality, and the unique stereochemical dynamics of these alkenes are crucial for unidirectional motion of the rotor around the stator.[215]

The combination of a 2-methyl-2,3-dihydronaphthopyran rotor and a thioxanthene stator constitutes a second generation molecular motor that uses a single chiral center to govern unidirectional motion, Scheme 8.40.[216,217] The first isomerization step of this motor requires irradiation of UV light to a solution of the stable (3'S)-(P)-*trans*-form. This establishes a

---

xix Molecular modeling confirmed that pseudoequatorial methyl groups increase the strain in these crowded alkenes and consequently destabilize the (M,M)-isomers.

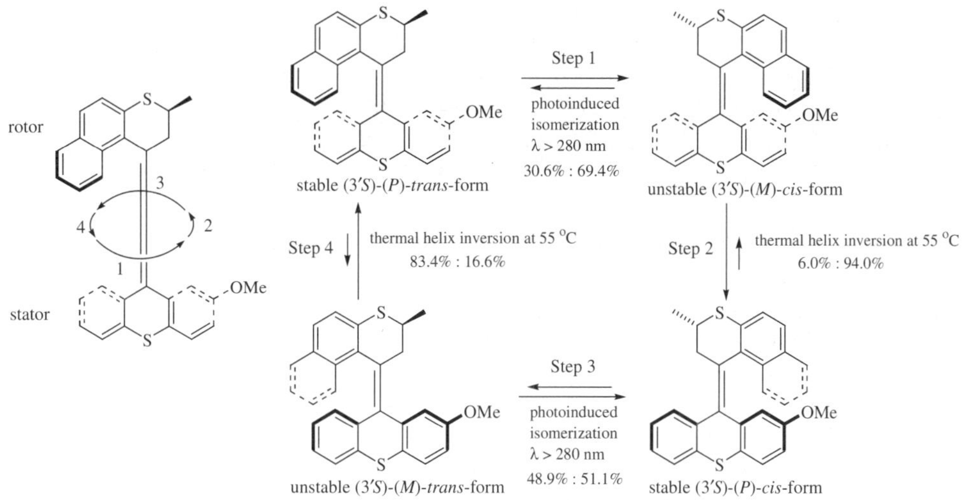

Scheme 8.40 Four isomerization steps of a second generation alkene motor.

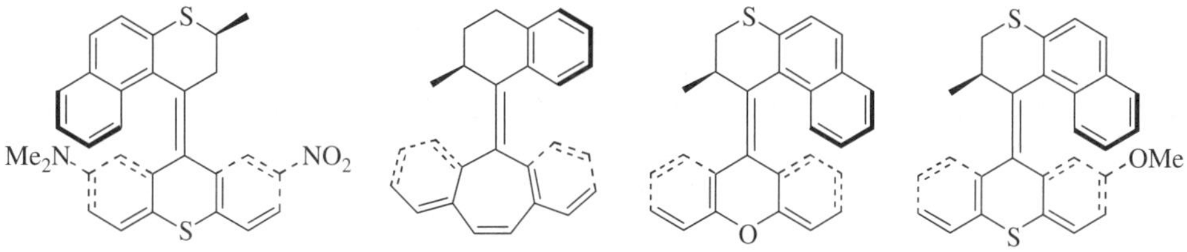

Figure 8.21 Structures of unidirectional motors derived from overcrowded alkenes.

photostationary state favoring 69.4% of the unstable $(3'S)$-(M)-*cis*-form which thermally relaxes to 94% of the stable $(3'S)$-(P)-*cis*-isomer in the second step. The same principle is exploited for photoisomerization of the $(3'S)$-(P)-*cis*-alkene to 48.9% of the $(3'S)$-(M)-*trans*-isomer in the third step, and successive thermal helix inversion regenerates 83.4% of the $(3'S)$-(P)-*trans*-isomer in the final step. Since the interconverting isomers exist as mixtures obeying Boltzmann distributions and reversible reaction kinetics, the molecular motor is not rotating in an exclusively monodirectional sense. However, the discrete and synergetic photochemical and thermal isomerization reactions accomplish an overall clockwise motion observed from the stator.

A variety of other second generation alkene motors that require only one chiral center to control unidirectional rotation has been developed. The structures of these motors have been modified to increase the conversion at the photostationary state of each photoisomerization step and to adjust the speed of rotation, which is ultimately determined by the half-lives of the two thermal isomerization steps, Figure 8.21.[218–223] Motors and shuttles based on translational isomerism are discussed in Chapter 9.

REFERENCES

1. Kelly, T. R.; de Silva, H.; Silva, R. A. *Nature* **1999**, *401*, 150-152.
2. Koumura, N.; Zijlstra, R. W. J.; van Delden, R. A.; Harada, N.; Feringa, B. L. *Nature* **1999**, *401*, 152-155.
3. Leigh, D. A.; Wong, J. K. Y.; Dehez, F.; Zerbetto, F. *Nature* **2003**, *424*, 174-179.

4. Schliwa, M. *Molecular Motors*, Wiley-VCH, Weinheim, 2003.
5. Hounshell, W. D.; Dougherty, D. A.; Mislow, K. *J. Am. Chem. Soc.* **1978**, *100*, 3149-3156.
6. Rieker, A.; Kessler, H. *Tetrahedron Lett.* **1969**, *10*, 1227-1230.
7. Siddall, T. H.; Stewart, W. E. *J. Org. Chem.* **1969**, *34*, 233-237.
8. Nakamura, M.; Oki, M. *Tetrahedron Lett.* **1974**, *13*, 505-508.
9. Ford, W. T.; Thompson, T. B.; Snoble, K. A. J.; Timko, J. M. *J. Am. Chem. Soc.* **1975**, *97*, 95-101.
10. Nakamura, M.; Oki, M. *Bull. Chem. Soc. Jpn.* **1980**, *53*, 2977-2980.
11. Mori, T.; Nakamura, N.; Oki, M. *Bull. Chem. Soc. Jpn.* **1981**, *54*, 1199-1202.
12. Oki, M. *Top. Stereochem.* **1983**, *14*, 1-81.
13. Lomas, J. S.; Dubois, J.-E. *J. Org. Chem.* **1976**, *41*, 3033-3034.
14. Nakamura, M.; Nakamura, N.; Oki, M. *Bull. Chem. Soc. Jpn.* **1977**, *50*, 2986-2990.
15. Nakamura, M.; Oki, M. *Bull. Chem. Soc. Jpn.* **1980**, *53*, 3248-3251.
16. Saito, R.; Oki, M. *Chem. Lett.* **1981**, 649-652.
17. Saito, R.; Oki, M. *Bull. Chem. Soc. Jpn.* **1982**, *55*, 3273-3276.
18. Nakamura, M.; Oki, M. *Chem. Lett.* **1975**, 671-674.
19. Lomas, J. S.; Dubois, J.-E. *Tetrahedron* **1978**, *34*, 1597-1604.
20. Oki, M.; Suda, M. *Bull. Chem. Soc. Jpn.* **1971**, *44*, 1876-1880.
21. Yamamoto, G.; Suzuki, M.; Oki, M. *Angew. Chem., Int. Ed. Engl.* **1981**, *20*, 607-608.
22. Yamamoto, G.; Oki, M. *Chem. Lett.* **1972**, 45-48.
23. Yamamoto, G.; Oki, M. *Bull. Chem. Soc. Jpn.* **1975**, *48*, 2592-2596.
24. Yamamoto, G.; Oki, M. *Bull. Chem. Soc. Jpn.* **1975**, *48*, 3686-3690.
25. Mikami, M.; Toriumi, K.; Kondo, M.; Saito, Y. *Acta Cryst.* **1975**, *B31*, 2474-2478.
26. Clayden, J.; Pink, J. H. *Angew. Chem., Int. Ed.* **1998**, *37*, 1937-1939.
27. Bragg, R. A.; Clayden, J. *Org. Lett.* **2000**, *2*, 3351-3354.
28. Bragg, R. A.; Clayden, J.; Morris, G. A.; Pink, J. L. *Chem. Eur. J.* **2002**, *8*, 1279-1289.
29. Clayden, J.; Lund, A.; Vallverdu, L.; Helliwell, M. *Nature* **2004**, *431*, 966-971.
30. Mislow, K. *Acc. Chem. Res.* **1976**, *9*, 26-33.
31. Iwamura, H.; Mislow, K. *Acc. Chem. Res.* **1988**, *21*, 175-182.
32. Rappoport, Z.; Biali, S. E. *Acc. Chem. Res.* **1997**, *30*, 307-314.
33. Berg, U.; Liljefors, T.; Roussel, C.; Sandström, J. *Acc. Chem. Res.* **1985**, *18*, 80-86.
34. Sedó, J.; Ventosa, N.; Molins, M. A.; Pons, M.; Rovira, C.; Veciana, J. *J. Org. Chem.* **2001**, *66*, 1579-1589.
35. Gust, D.; Mislow, K. *J. Am. Chem. Soc.* **1973**, *95*, 1535-1547.
36. Biali, S. E.; Rappoport, Z. *J. Am. Chem. Soc.* **1985**, *107*, 1007-1015.
37. Clegg, W.; Lockhart, J. C.; McDonnell, M. B. *J. Chem. Soc., Perkin Trans. 1* **1985**, 1019-1023.
38. Lockhart, J. C.; McDonnell, M. B.; Clegg, W. *J. Chem. Soc., Perkin Trans. 1* **1983**, 2153-2159.
39. Grilli, S.; Lunazzi, L.; Mazzanti, A. *J. Org. Chem.* **2001**, *66*, 5853-5858.
40. Bolon, D. A. *J. Am. Chem. Soc.* **1966**, *88*, 3148-3149.
41. Lam, W. Y.; Martin, J. C. *J. Org. Chem.* **1981**, *46*, 4458-4462.
42. Grilli, S.; Lunazzi, L.; Mazzanti, A. *J. Org. Chem.* **2001**, *66*, 4444-4446.
43. Casarini, D.; Grilli, S.; Lunazzi, L.; Mazzanti, A. *J. Org. Chem.* **2001**, *66*, 2757-2763.
44. Grilli, S.; Lunazzi, L.; Mazzanti, A.; Mazzanti, G. *J. Org. Chem.* **2001**, *66*, 748-754.
45. Finocchiaro, P.; Gust, D.; Mislow, K. *J. Am. Chem. Soc.* **1973**, *95*, 7029-7036.
46. Rappoport, Z.; Biali, S. E.; Kaftory, M. *J. Am. Chem. Soc.* **1990**, *112*, 7742-7748.
47. Yamataka, H.; Aleksiuk, O.; Biali, S. E.; Rappoport, Z. *J. Am. Chem. Soc.* **1996**, *118*, 12580-12587.
48. Kwart, H.; Alekman, A. *J. Am. Chem. Soc.* **1968**, *90*, 4482-4483.
49. Grilli, S.; Lunazzi, L.; Mazzanti, A.; Casarini, D.; Femoni, C. *J. Org. Chem.* **2001**, *66*, 488-495.

50. Casarini, D.; Lunazzi, L.; Mazzanti, A.; Mercandelli, P.; Sironi, A. *J. Org. Chem.* **2004**, *69*, 3574-3577.

51. Finocchiaro, P.; Gust, D.; Mislow, K. *J. Am. Chem. Soc.* **1974**, *96*, 2165-2167.

52. Andose, J. D.; Mislow, K. *J. Am. Chem. Soc.* **1974**, *96*, 2168-2176.

53. Finocchiaro, P.; Gust, D.; Mislow, K. *J. Am. Chem. Soc.* **1974**, *96*, 2176-2182.

54. Finocchiaro, P.; Gust, D.; Mislow, K. *J. Am. Chem. Soc.* **1974**, *96*, 3198-3205.

55. Finocchiaro, P.; Gust, D.; Mislow, K. *J. Am. Chem. Soc.* **1974**, *96*, 3205-3213.

56. Lockhart, J. C.; McDonnell, M. B.; Clegg, W.; Hill, M. N. S. *J. Chem. Soc., Perkin Trans. 2* **1987**, 639-649.

57. Clegg, W.; Lockhart, J. C. *J. Chem. Soc., Perkin Trans. 2* **1987**, 1621-1627.

58. Cooper, P. J.; Hill, M. N. S.; Lockhart, J. C. *J. Chem. Soc., Perkin Trans. 2* **1994**, 2145-2148.

59. Rakshys Jr., J. W.; McKindley, S. V.; Freedman, H. H. *J. Am. Chem. Soc.* **1970**, *92*, 3518-3520.

60. Rakshys Jr., J. W.; McKindley, S. V.; Freedman, H. H. *J. Am. Chem. Soc.* **1970**, *92*, 6522-6529.

61. Hansen, P. E.; Spanget-Lansen, J.; Laali, K. K. *J. Org. Chem.* **1998**, *63*, 1827-1835.

62. Ito, S.; Morita, N.; Asao, T. *Tetrahedron Lett.* **1992**, *33*, 6669-6672.

63. Ito, S.; Morita, N.; Asao, T. *Tetrahedron Lett.* **1994**, *35*, 3723-3726.

64. Ito, S.; Morita, N.; Asao, T. *Bull. Chem. Soc. Jpn.* **1995**, *68*, 1409-1436.

65. Ito, S.; Fujita, M.; Morita, N.; Asao, T. *Bull. Chem. Soc. Jpn.* **1995**, *68*, 3611-3620.

66. Hummel, J. P.; Gust, D.; Mislow, K. *J. Am. Chem. Soc.* **1974**, *96*, 3679-3681.

67. Hellwinkel, D.; Melan, M.; Degel, C. R. *Tetrahedron* **1973**, *29*, 1895-1907.

68. Hellwinkel, D.; Melan, M.; Egan, W.; Degel, C. R. *Chem. Ber.* **1975**, *108*, 2219-2231.

69. Glaser, R.; Blount, J. F.; Mislow, K. *J. Am. Chem. Soc.* **1980**, *102*, 2777-2786.

70. Boettcher, R. J.; Gust, D.; Mislow, K. *J. Am. Chem. Soc.* **1973**, *95*, 7157-7158.

71. Turnblom, E. W.; Boettcher, R. J.; Mislow, K. *J. Am. Chem. Soc.* **1975**, *97*, 1766-1772.

72. Kates, M. R.; Andose, J. D.; Finocchiaro, P.; Gust, D.; Mislow, K. *J. Am. Chem. Soc.* **1975**, *97*, 1772-1778.

73. Yamaguchi, S.; Akiyama, S.; Tamao, K. *Organometallics* **1998**, *17*, 4347-4352.

74. Chance, J. M.; Geiger, J. H.; Okamoto, Y.; Aburatani, R.; Mislow, K. *J. Am. Chem. Soc.* **1990**, *112*, 3540-3547.

75. Wille, E. E.; Stephenson, D. S.; Capriel, P.; Binsch, G. *J. Am. Chem. Soc.* **1982**, *104*, 405-415.

76. Blount, J. F.; Finocchiaro, P.; Gust, D.; Mislow, K. *J. Am. Chem. Soc.* **1973**, *95*, 7019-7029.

77. Toyota, S.; Asakura, M.; Oki, M.; Toda, F. *Bull. Chem. Soc. Jpn.* **2000**, *73*, 2357-2362.

78. Finocchiaro, P.; Gust, D.; Hounshell, W. D.; Hummel, J. P.; Maravigna, P.; Mislow, K. *J. Am. Chem. Soc.* **1976**, *98*, 4945-4952.

79. Finocchiaro, P.; Hounshell, W. D.; Mislow, K. *J. Am. Chem. Soc.* **1976**, *98*, 4952-4963.

80. Schlögl, K.; Weissensteiner, W.; Widhalm, M. *J. Org. Chem.* **1982**, *47*, 5025-5027.

81. Biali, S. E.; Rappoport, Z. *J. Am. Chem. Soc.* **1985**, *107*, 1007-1015.

82. Johnson, C. A.; Guenzi, A.; Nachbar, R. B.; Blount, J. F.; Wennerström, O.; Mislow, K. *J. Am. Chem. Soc.* **1982**, *104*, 5163-5168.

83. Bürgi, H.-B.; Hounshell, W. D.; Nachbar Jr., R. B.; Mislow, K. *J. Am. Chem. Soc.* **1983**, *105*, 1427-1438.

84. Kawada, Y.; Iwamura, H. *J. Am. Chem. Soc.* **1983**, *105*, 1449-1459.

85. Guenzi, A.; Johnson, C. A.; Cozzi, F.; Mislow, K. *J. Am. Chem. Soc.* **1983**, *105*, 1438-1448.

86. Johnson, C. A.; Guenzi, A.; Mislow, K. *J. Am. Chem. Soc.* **1981**, *103*, 6240-6242.

87. Cozzi, F.; Guenzi, A.; Johnson, C. A.; Mislow, K.; Hounshell, W. D.; Blount, J. F. *J. Am. Chem. Soc.* **1981**, *103*, 957-958.

88. Kawada, Y.; Iwamura, H. *J. Org. Chem.* **1980**, *45*, 2547-2548.

89. Kawada, Y.; Iwamura, H. *J. Am. Chem. Soc.* **1981**, *103*, 958-960.

90. Koga, N.; Kawada, Y.; Iwamura, H. *J. Am. Chem. Soc.* **1983**, *105*, 5498-5499.

91. Iwamura, H.; Ito, T.; Ito, H.; Toriumi, K.; Kawada, Y.; Osawa, E.; Fujiyoshi, T.; Jaime, C. *J. Am. Chem. Soc.* **1984**, *106*, 4712-4717.

92. Koga, N.; Iwamura, H. *J. Am. Chem. Soc.* **1985**, *107*, 1426-1427.

93. Yamamoto, G.; Oki, M. *Bull. Chem. Soc. Jpn.* **1985**, *58*, 1953-1961.

94. Kawada, Y.; Yamazaki, H.; Koga, N.; Murata, S.; Iwamura, H. *J. Org. Chem.* **1986**, *51*, 1472-1477.

95. Fuji, K.; Oka, T.; Kawabata, T.; Kinosahita, T. *Tetrahedron Lett.* **1998**, *39*, 1373-1376.

96. Kaftory, M.; Nugiel, D. A.; Biali, S.; Rappoport, Z. *J. Am. Chem. Soc.* **1989**, *111*, 8181-1891.

97. Lunazzi, L.; Mazzanti, A.; Minzoni, M. *J. Org. Chem.* **2005**, *70*, 456-462.

98. Nugiel, D. A.; Biali, S.; Rappoport, Z. *J. Am. Chem. Soc.* **1984**, *106*, 3357-3359.

99. Nugiel, D. A.; Biali, S.; Nugiel, D. A.; Rappoport, Z. *J. Am. Chem. Soc.* **1989**, *111*, 846-852.

100. Schottland, E.; Frey, J.; Rappoport, Z. *J. Org. Chem.* **1994**, *59*, 1663-1671.

101. Frey, J.; Rappoport, Z. *J. Org. Chem.* **1997**, *62*, 8372-8386.

102. Selzer, T.; Rappoport, Z. *J. Org. Chem.* **1996**, *61*, 7326-7334.

103. Gur, E.; Kaftory, M.; Bialy, S. E.; Rappoport, Z. *J. Org. Chem.* **1999**, *64*, 8144-8148.

104. Biali, S. E.; Rappoport, Z.; Mannschreck, A.; Pustet, N. *Angew. Chem., Int. Ed. Engl.* **1989**, *28*, 199-201.

105. Rochlin, E.; Rappoport, Z.; Kastner, F.; Pustet, N.; Mannschreck, A. *J. Org. Chem.* **1999**, *64*, 8840-8845.

106. Biali, S. E.; Rappoport, Z. *J. Org. Chem.* **1984**, *49*, 477-496.

107. Rochlin, E.; Rappoport, Z. *J. Org. Chem.* **1994**, *59*, 3857-3870.

108. Rappoport, Z.; Frey, J.; Sigalov, M.; Rochlin, E. *Pure Appl. Chem.* **1997**, *69*, 1933-1940.

109. Biali, S. E.; Rappoport, Z. *J. Org. Chem.* **1986**, *51*, 2245-2250.

110. Schmittel, M.; Keller, M.; Burghart, A.; Rappaport, Z.; Langels, A. *J. Chem. Soc., Perkin Trans. 2* **1998**, 869-875.

111. Rochlin, E.; Rappoport, Z. *J. Org. Chem.* **2003**, *68*, 216-226.

112. Willem, R.; Pepermans, H.; Hallenga, K.; Gielen, M.; Dams, R.; Geise, H. J. *J. Org. Chem.* **1983**, *48*, 1890-1898.

113. Maeda, K.; Okamoto, Y.; Morlender, N.; Haddad, N.; Eventova, I.; Biali, S.; Rappoport, Z. *J. Am. Chem. Soc.* **1995**, *117*, 9686-9689.

114. Gur, E.; Kaida, Y.; Okamoto, Y.; Biali, S.; Rappoport, Z. *J. Org. Chem.* **1992**, *57*, 3689-3693.

115. Maeda, K.; Okamoto, Y.; Toledano, O.; Becker, D.; Biali. S. E.; Rappoport, Z. *J. Org. Chem.* **1994**, *59*, 5473-5475.

116. Canary, J. W.; Allen, C. S.; Castagnetto, J. M.; Chiu, Y.-H.; Toscano, P. J.; Wang, Y. *J. Inorg. Chem.* **1998**, *37*, 6255-6262.

117. Canary, J. W.; Allen, C. S.; Castagnetto, J. M.; Wang, Y. *J. Am. Chem. Soc.* **1995**, *117*, 8484-8485.

118. Chiu, Y.-H.; dos Santos, O.; Canary, J. W. *Tetrahedron* **1999**, *55*, 12069-12078.

119. Xu, X.; Maresca, K. J.; Das, D.; Zahn, S.; Zubieta, J.; Canary, J. W. *Chem. Eur. J.* **2002**, *8*, 5679-5683.

120. Zahn, S.; Canary, J. W. *Angew. Chem., Int. Ed.* **1998**, *37*, 305-307.

121. Zahn, S.; Canary, J. W. *J. Am. Chem. Soc.* **2002**, *124*, 9204-9211.

122. Zahn, S.; Canary, J. W. *Science* **2000**, *288*, 1404-1407.

123. Zahn, S.; Canary, J. W. *Org. Lett.* **1999**, *1*, 861-864.

124. Barcena, H. S.; Holmes, A. E.; Zahn, S.; Canary, J. W. *Org. Lett.* **2003**, *5*, 709-711.

125. Ayscough, A. P.; Costello, J. F.; Davies, S. G. *Tetrahedron: Asymm.* **2001**, *12*, 1621-1624.

126. Siegel, J.; Mislow, K. *J. Am. Chem. Soc.* **1983**, *105*, 7763-7764.

127. Siegel, J.; Gutierrez, A.; Schweizer, W. B.; Ermer, O.; Mislow, K. *J. Am. Chem. Soc.* **1986**, *108*, 1569-1575.
128. Singh, M. D.; Siegel, J.; Biali, S. E.; Mislow, K. *J. Am. Chem. Soc.* **1987**, *109*, 3397-3402.
129. Biali, S. E.; Mislow, K. *J. Org. Chem.* **1988**, *53*, 1318-1320.
130. Biali, S. E.; Kahr, B.; Okamoto, Y.; Aburatani, R.; Mislow, K. *J. Am. Chem. Soc.* **1988**, *110*, 1917-1922.
131. Langler, R. F.; Tidwell, T. T. *Tetrahedron Lett.* **1975**, *16*, 777-780.
132. Bomse, D. S.; Morton, T. H. *Tetrahedron Lett.* **1975**, *16*, 781-784.
133. Ermer, O. *Angew. Chem., Int. Ed. Engl.* **1983**, *22*, 998-1000.
134. Gust, D.; Patton, A. *J. Am. Chem. Soc.* **1978**, *100*, 8175-8181.
135. Brydges, S.; McGlinchey, M. J. *J. Org. Chem.* **2002**, *67*, 7688-7698.
136. Kelly, T. R.; Tellitu, I.; Sestelo, J. P. *Angew. Chem., Int. Ed. Engl.* **1997**, *36*, 1866-1868.
137. Kelly, T. R.; Sestelo, J. P.; Tellitu, I. *J. Org. Chem.* **1998**, *63*, 3655-3665.
138. Kelly, T. R.; Bowyer, M. C.; Bhaskar, K. V.; Bebbington, D.; Garcia, A.; Lang, F.; Kim, M. H.; Jette, M. P. *J. Am. Chem. Soc.* **1994**, *116*, 3657-3658.
139. Harrington, L. E.; Cahill, L. S.; McGlinchey, M. J. *Organometallics* **2004**, *23*, 2884-2891.
140. Jog, P. V.; Brown, R. E.; Bates, D. K. *J. Org. Chem.* **2003**, *68*, 8240-8243.
141. Muraoka, T.; Kinbara, K.; Kobayashi, Y.; Aida, T. *J. Am. Chem. Soc.* **2003**, *125*, 5612-5613.
142. Bedard, T. C.; Moore, J. S. *J. Am. Chem. Soc.* **1995**, *117*, 10662-10671.
143. Feringa, B. L.; van Delden, R. A.; Koumura, N.; Geertsma, E. M. *Chem. Rev.* **2000**, *100*, 1789-1816.
144. Feringa, B. L.; Wynberg, H. *J. Am. Chem. Soc.* **1977**, *99*, 602-603.
145. Feringa, B. L.; Jager, W. F.; de Lange, B. *Tetrahedron Lett.* **1992**, *33*, 2887-2890.
146. Jager, W. F.; de Lange, B.; Schoevaars, A. M.; van Bolhuis, F.; Feringa, B. L. *Tetrahedron: Asymm.* **1993**, *4*, 1481-1497.
147. Feringa, B. L.; Jager, W. F.; de Lange, B. *J. Chem. Soc., Chem. Commun.* **1993**, 288-290.
148. Smid, W. I.; Schoevaars, A. M.; Kruizinga, W.; Veldman, N.; Smeets, W. J. J.; Spek, A. L.; Feringa, B. L. *Chem. Commun.* **1996**, 2265-2266.
149. Harada, N.; Saito, A.; Koumura, N.; Uda, H.; de Lange, B.; Jager, W. F.; Wynberg, H.; Feringa, B. L. *J. Am. Chem. Soc.* **1997**, *119*, 7241-7248.
150. Harada, N.; Koumura, N.; Feringa, B. L. *J. Am. Chem. Soc.* **1997**, *119*, 7256-7264.
151. Harada, N.; Saito, A.; Koumura, N.; Roe, D. C.; Jager, W. F.; Zijlstra, R. W. J.; de Lange, B.; Feringa, B. L. *J. Am. Chem. Soc.* **1997**, *119*, 7249-7255.
152. Feringa, B. L.; Jager, W. F.; de Lange, B. *J. Am. Chem. Soc.* **1991**, *113*, 5468-5470.
153. Jager, W. F.; de Jong, J. C.; de Lange, B.; Huck, N. P. M.; Meetsma, A.; Feringa, B. L. *Angew. Chem., Int. Ed. Engl.* **1995**, *34*, 348-350.
154. Feringa, B. L.; Huck, N. P. M.; Schoevaars, A. M. *Adv. Mater.* **1996**, *8*, 681-684.
155. Schoevaars, A. M.; Kruizinga, W.; Zijlstra, R. W. J.; Veldman, N.; Spek, A. L.; Feringa, B. L. *J. Org. Chem.* **1997**, *62*, 4943-4948.
156. Van Delden, R. A.; Huck, N. P. M.; Piet, J. J.; Warman, J. M.; Meskers, S. C. J.; Dekkers, H. P. J. M.; Feringa, B. L. *J. Am. Chem. Soc.* **2003**, *125*, 15659-15665.
157. Feringa, B. L.; Jager, W. F.; de Lange, B. *Tetrahedron* **1993**, *49*, 8267-8310.
158. Huck, N. P. M.; Feringa, B. L. *J. Chem. Soc., Chem. Commun.* **1995**, 1095-1096.
159. Van Delden, R. A.; Hurenkamp, J. H.; Feringa, B. L. *Chem. Eur. J.* **2003**, *9*, 2845-2853.
160. Van Delden, R. A.; ter Wiel, M. K. J.; Feringa, B. L. *Chem. Commun.* **2004**, 200-201.
161. Huck, N. P. M.; Jager, W. F.; de Lange, B.; Feringa, B. L. *Science* **1996**, *273*, 1686-1688.
162. Gilat, S. L.; Kawai, S. H.; Lehn, J.-M. *J. Chem. Soc., Chem. Commun.* **1993**, 1439-1442.
163. Gilat, S. L.; Kawai, S. H.; Lehn, J.-M. *Chem. Eur. J.* **1995**, *1*, 275-284.
164. Tsivgoulis, G. M.; Lehn, J.-M. *Angew. Chem., Int. Ed. Engl.* **1995**, *34*, 1119-1122.
165. Irie, M.; Sakemura, K.; Okinaka, M.; Uchida, K. *J. Org. Chem.* **1995**, *60*, 8305-8309.

166. Irie, M. *Chem. Rev.* **2000**, *100*, 1685-1716.

167. Lucas, L. N.; van Esch, J.; Kellog, R. M.; Feringa, B. L. *Chem. Commun.* **2001**, 759-760.

168. Lucas, L. N.; de Jong, J. J. D.; van Esch, J.; Kellog, R. M.; Feringa, B. L. *Eur. J. Org. Chem.* **2003**, 155-166.

169. De Jong, J. J. D.; Lucas, L. N.; Hania, R.; Pugzlys, A.; Kellog, R. M.; Feringa, B. L.; Duppen, K.; van Esch, J. H. *Eur. J. Org. Chem.* **2003**, 1887-1893.

170. Yamaguchi, T.; Uchida, K.; Irie, M. *J. Am. Chem. Soc.* **1997**, *119*, 6066-6071.

171. Burnham, K. S.; Schuster, G. B. *J. Am. Chem. Soc.* **1998**, *120*, 12619-12625.

172. Birau, M. M.; Wang, Z. Y. *Tetrahedron Lett.* **2000**, *41*, 4025-4028.

173. Pieraccini, S.; Masiero, S.; Spada, G. P.; Gottarelli, G. *Chem. Commun.* **2003**, 598-599.

174. Matsui, F.; Taniguchi, H.; Yokoyama, Y.; Sugiyama, K.; Kurita, Y. *Chem. Lett.* **1994**, 1869-1872.

175. Yokoyama, Y.; Iwai, T.; Yokoyama, Y.; Kurita, Y. *Chem. Lett.* **1994**, 225-226.

176. Yokoyama, Y.; Shimizu, Y.; Uchida, S.; Yokoyama, Y. *J. Chem. Soc., Chem. Commun.* **1995**, 785-786.

177. Yokoyama, Y.; Uchida, S.; Yokoyama, Y.; Sugawara, Y.; Kurita, Y. *J. Am. Chem. Soc.* **1996**, *118*, 3100-3107.

178. Yokoyama, Y. *Chem. Rev.* **2000**, *100*, 1717-1740.

179. Lemieux, R. P.; Schuster, G. B. *J. Org. Chem.* **1993**, *58*, 100-110.

180. Suarez, M.; Schuster, G. B. *J. Am. Chem. Soc.* **1995**, *117*, 6732-6738.

181. Fisher Bradford, R.; Schuster, G. B. *J. Org. Chem.* **2003**, *68*, 1075-1080.

182. Zhang, Y.; Schuster, G. B. *J. Org. Chem.* **1995**, *60*, 7192-7197.

183. Berkovic, G.; Krongauz, V.; Weiss, V. *Chem. Rev.* **2000**, *100*, 1741-1754.

184. Gruda, I.; Leblanc, R. M. *Can. J. Chem.* **1976**, *54*, 576-580.

185. Gruda, I.; Leblanc, R. M.; Sochanski, J. *Can. J. Chem.* **1978**, *56*, 1296-1301.

186. Eggers, L.; Buss, V. *Angew. Chem., Int. Ed. Engl.* **1997**, *36*, 881-883.

187. Sakata, T.; Yan, Y.; Marriott, G. *J. Org. Chem.* **2005**, *70*, 2009-2013.

188. Agati, G.; McDonagh, A. F. *J. Am. Chem. Soc.* **1995**, *117*, 4425-4426.

189. Jahr, H. C.; Nieger, M.; Dötz, K. H. *Chem. Commun.* **2003**, 2866-2867.

190. Inoue, Y.; Yamasaki, N.; Yokoyama, T.; Tai, A. *J. Org. Chem.* **1992**, *57*, 1332-1345.

191. Inoue, Y.; Yamasaki, N.; Yokoyama, T.; Tai, A. *J. Org. Chem.* **1993**, *58*, 1011-1018.

192. Inoue, Y.; Dong, F.; Yamamoto, K.; Tong, L.-H.; Tsuneishi, H.; Hakushi, T.; Tai, A. *J. Am. Chem. Soc.* **1995**, *117*, 11033-11034.

193. Wada, T.; Shikimi, M.; Inoue, Y.; Lem, G.; Turro, N. J. *Chem. Commun.* **2001**, 1864-1865.

194. Sugimura, T.; Shimizu, H.; Umemoto, S.; Tsuneishi, H.; Hakushi, T.; Inoue, Y.; Tai, A. *Chem. Lett.* **1998**, 323-324.

195. Tumambac, G. E.; Mei, X.; Wolf, C. *Eur. J. Org. Chem.* **2004**, 3850-3856.

196. Zhu, L.; Anslyn, E. V. *J. Am. Chem. Soc.* **2004**, *126*, 3676-3677.

197. Pu, L. *Chem. Rev.* **2004**, *104*, 1687-1716.

198. Zhao, J.; Fyles, T. M.; James, T. D. *Angew. Chem., Int. Ed.* **2004**, *43*, 3461-3464.

199. Mei, X.; Wolf, C. *Chem. Commun.* **2004**, 2078-2079.

200. Mei, X.; Wolf, C. *J. Am. Chem. Soc.* **2004**, *126*, 14736-14737.

201. Kawai, H.; Katoono, R.; Fujiwara, K.; Tsuji, T.; Suzuki, T. *Chem. Eur. J.* **2005**, *11*, 815-824.

202. Rebek Jr., J. *Acc. Chem. Res.* **1984**, *17*, 258-264.

203. Takeuchi, M.; Ikeda, M.; Sugasaki, A.; Shinkai, S. *Acc. Chem. Res.* **2001**, *34*, 865-873.

204. Kinbara, K.; Aida, T. *Chem. Rev.* **2005**, *105*, 1377-1400.

205. Kottas, G. S.; Clarke, L. I.; Horinek, D.; Michl, J. *Chem. Rev.* **2005**, *105*, 1281-1376.

206. Davis, A. P. *Nature* **1999**, *401*, 120-121.

207. Feringa, B. L. *Nature* **2000**, *408*, 151-154.

208. Dahl, B. J.; Branchaud, B. P. *Tetrahedron Lett.* **2004**, *45*, 9599-9602.

209. Lin, Y.; Dahl, B. J.; Branchaud, B. P. *Tetrahedron Lett.* **2005**, *46*, 8359-8362.

210. Kelly, T. R.; Silva, R. A.; de Silva, H.; Jasmin, S.; Zhao, Y. *J. Am. Chem. Soc.* **2000**, *122*, 6935-6949.

211. Kelly, T. R. *Acc. Chem. Res.* **2001**, *34*, 514-522.

212. Feringa, B. L. *Acc. Chem. Res.* **2001**, *34*, 504-513.

213. Koumura, N.; Zijlstra, R. W. J.; van Delden, R. A.; Harada, N.; Feringa, B. L. *Nature* **1999**, *401*, 152-155.

214. Van Delden, R. A.; Koumura, N.; Harada, N.; Feringa, B. L. *Proc. Natl. Acad. Sci.* **2002**, *99*, 4945-4949.

215. Ter Wiel, M. K. J.; van Delden, R. A.; Meetsma, A.; Feringa, B. L. *J. Am. Chem. Soc.* **2005**, *127*, 14208-14222.

216. Koumura, N.; Geertsema, E. M.; Meetsma, A.; Feringa, B. L. *J. Am. Chem. Soc.* **2000**, *122*, 12005-12006.

217. Van Delden, R. A.; ter Wiel, M. K. J.; de Jong, H.; Meetsma, A.; Feringa, B. L. *Org. Biomol. Chem.* **2004**, *2*, 1531-1541.

218. Koumura, N.; Geertsema, E. M.; van Gelder, M. B.; Meetsma, A.; Feringa, B. L. *J. Am. Chem. Soc.* **2002**, *124*, 5037-5051.

219. Geertsema, E. M.; Koumura, N.; ter Wiel, M. K. J.; Meetsma, A.; Feringa, B. L. *Chem. Commun.* **2002**, 2962-2963.

220. Van Delden, R. A.; Koumura, N.; Schoevaars, A.; Meetsma, A.; Feringa, B. L. *Org. Biomol. Chem.* **2003**, *1*, 33-35.

221. Ter Wiel, M. K. J.; van Delden, R. A.; Meetsma, A.; Feringa, B. L. *J. Am. Chem. Soc.* **2003**, *125*, 15076-15086.

222. Van Delden, R. A.; ter Wiel, M. K. J.; Pollard, M. M.; Vicario, J.; Koumura, N.; Feringa, B. L. *Nature* **2005**, *437*, 1337-1340.

223. Ter Wiel, M. K. J.; van Delden, R. A.; Meetsma, A.; Feringa, B. L. *Org. Biomol. Chem.* **2005**, *3*, 4071-4076.

Topological Isomerism and Chirality

The concept of topological isomerism and chirality of interlocked molecules was first mentioned by Frisch and Wasserman in 1961.[1] Catenanes and rotaxanes are prominent examples of mechanically interlocked molecular assemblies consisting of two or more individual components that can not dissociate without cleavage of a covalent bond, Figure 9.1. A catenane is derived from two or more interlocked rings. Homocircuit catenanes contain identical rings and heterocircuit catenanes display different ring components. A rotaxane consists of a linear component that is threaded through a macrocycle which is trapped on the rod by two bulky terminal groups, generally referred to as stoppers. In contrast to catenanes and rotaxanes, the so-called Möbius strip and molecular knots consist of a single mechanically intertwined entity. A Möbius strip is obtained through a half-twist of a rectangular strip and connection of the two ends. The structure of molecular knots resembles the topology of macroscopic analogs such as ties, shoelaces and sailors' knots. Catenanes, rotaxanes, Möbius strips, and knotanes give rise to topological isomerism and chirality. For example, a trefoil knot and a macrocycle constitute a pair of topological isomers. Similarly, a [2]catenane is topologically isomeric to two separate macrocycles, Figure 9.2. Topological isomers have identical constitution and atom connectivity but different topology. In most cases, inter-conversion of topological isomers requires a sequence of bond cleavage, reorientation and bond formation.[i]

The unique stereochemistry and properties of intertwined molecular assemblies is reminiscent of nontrivial tangles with distinctive biochemical functions in DNA, RNA and proteins.[2–7] The

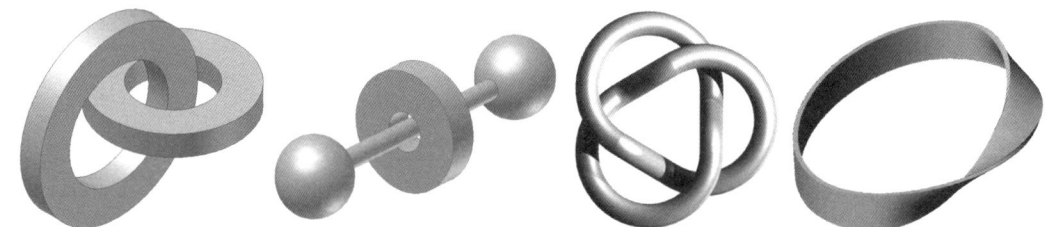

Figure 9.1 Illustration of the architectures of a catenane, rotaxane, trefoil knot, and Möbius strip (from left to right).

---

[i]Isomerization of rotaxanes due to slippage of the threaded ring over one of the stoppers is also possible, see Chapter 9.1.3.

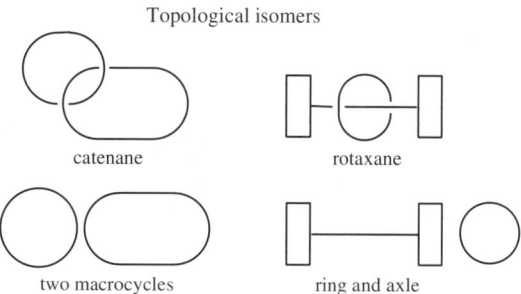

Figure 9.2 Topological isomers.

development of synthetic strategies towards interlocked assemblies exhibiting controlled movements of noncovalently interacting individual units has paved the way to a variety of important applications. Topological isomerism of mechanically interlocked systems has become a common motif in the design of molecular amplifiers, sensors, shuttles, switches, and machineries.[8–16] The intriguing structures and properties of topological assemblies have been utilized in material sciences, and numerous supramolecular networks with entangled architectures of polythreaded, polycatenated and polyknotted species are known, including infinite Borromean rings.[17–19]

9.1 SYNTHESIS OF CATENANES AND ROTAXANES

The biosynthesis of naturally occurring knotted and catenated structures of DNA and RNA is governed by topoisomerases and gyrases. These enzymes catalyze a multi-step process involving transient cleavage of a nucleic acid strand, ATP-driven interconversion of a relaxed strand into a new topological structure by passing a segment through the break, and finally resealing of the two ends of the previously cut strand. The pioneering work of Wasserman and subsequent discoveries of natural topological isomers have inspired the synthesis of mechanically interlocked assemblies. This has been accomplished by three different strategies: statistical synthesis, template-directed assembly and topological isomerization based on slippage of preformed units.

9.1.1 Statistical Methods

The synthesis of a catenane can be accomplished by formation of a large ring from a sufficiently long acyclic molecule with appropriate terminal functionalities in the presence of another large ring, Scheme 9.1. Statistically, some acyclic molecules will be threaded through the other macrocycle upon cyclization and thus form a catenane. Statistical threading of a linear molecule through another cyclic compound can also be used to prepare a rotaxane by trapping the rod with bulky groups at each end. A major drawback of this approach is that it relies on an unfavorable equilibrium between individual components and the desired intertwined assembly during ring formation or incorporation of terminal stoppers. Yields obtained by statistical synthesis are often lower than 1%.[20] To overcome low yields of statistical methods, one can utilize Merrifield's solid phase synthesis approach. This has been demonstrated with a resin-bound macrocyclic α-alkoxy ketone. As expected, treatment of the immobilized macrocycle with a solution of 1,10-decanediol in the presence of triphenylmethyl chloride gave the ditrityl ether of 1,10-decanediol as the major product and only traces of the threaded α-alkoxy ketone endcapped by bulky trityl groups were obtained. However, by-products and residual reagents were conveniently washed away at the end of the reaction, providing a mixture of immobilized unreacted macrocycle and small amounts of the

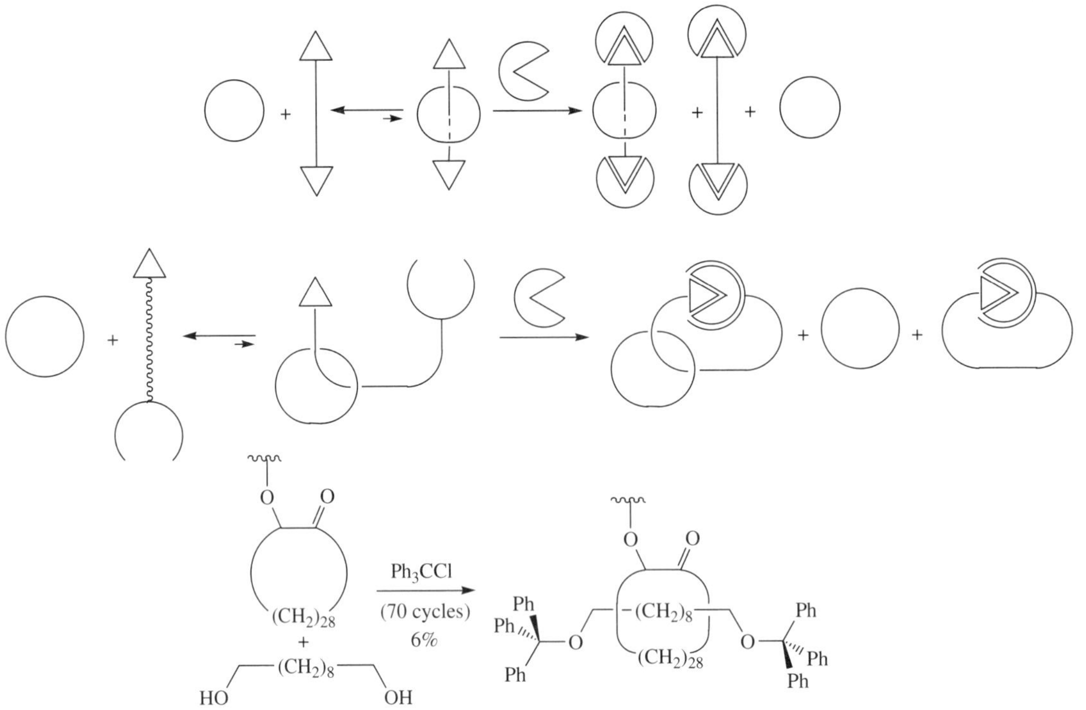

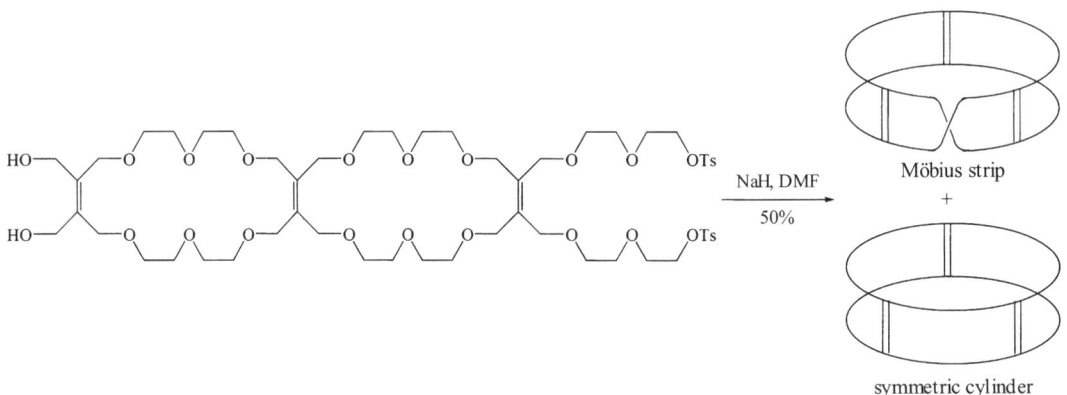

Scheme 9.1   Schematic illustration of statistical syntheses of rotaxanes and catenanes (top). Threading of an immobilized α-alkoxy ketone by 1,10-decanediol and trapping of the rod with trityl groups (bottom).

Scheme 9.2   Synthesis of a Möbius strip and its cylindrical topological isomer by statistical ring closure.

rotaxane. This procedure was repeated 70 times to afford 6% of the desired rotaxane which was finally detached from the resin, Scheme 9.1.[21]

In 1982, Walba *et al.* reported the first molecular Möbius strip. Cyclization of an array of crown ethers at high dilution provided a twisted ring structure in remarkable 24% yield, Scheme 9.2.[22,23,ii] The chemical and physical properties of molecular Möbius strips are closely related to their unique

---

[ii] In addition to the inherently chiral Möbius strip, the cylindrical topological isomer was formed in similar amounts.

topology. This is probably most obvious in the case of annulenes. According to the Hückel rule, planar annulenes (cyclic molecules with alternating single and double bonds) are aromatic if they contain $(4N + 2)$ π-electrons but are antiaromatic and unstable if they have $4N$ π-electrons. In 1964, Heilbronner predicted that annulenes with $4N$ π-electrons would be aromatic and stable if they possess Möbius topology.[24] The synthesis of an aromatic Möbius compound proved to be quite difficult since the twist results in considerable ring strain and reduced *p*-orbital overlap, which ultimately limits π-electron delocalization. It took almost 40 years before a stable aromatic Möbius compound containing $4N$ π-electrons was prepared by Herges to confirm Heilbronner's prediction.[25]

9.1.2 Template-assisted Assembly

With the advance of supramolecular chemistry, template-assisted strategies that significantly enhance the chance of threading prior to ring closure or endcapping have become available.[26] Ingenious methods developed by the groups of Stoddart, Vögtle, Sauvage, Leigh, and others have produced mechanically interlocked architectures in high yields. The template approach relies on preorganization of individual components prior to formation of covalent bonds and trapping of a threading rod or ring.[27-44] Predictable preorganization of appropriately functionalized precursors has been accomplished through hydrogen bonding, metal coordination, π-π-interactions, and hydrophobic forces. For example, Sauvage *et al.* demonstrated the use of copper(I) salts in metal ion-assisted synthesis of 1,10-phenanthroline-derived catenanes, Scheme 9.3.[45] Apparently, two phenanthroline ligands coordinate to the tetrahedral copper center to form an entwined perpendicular arrangement.

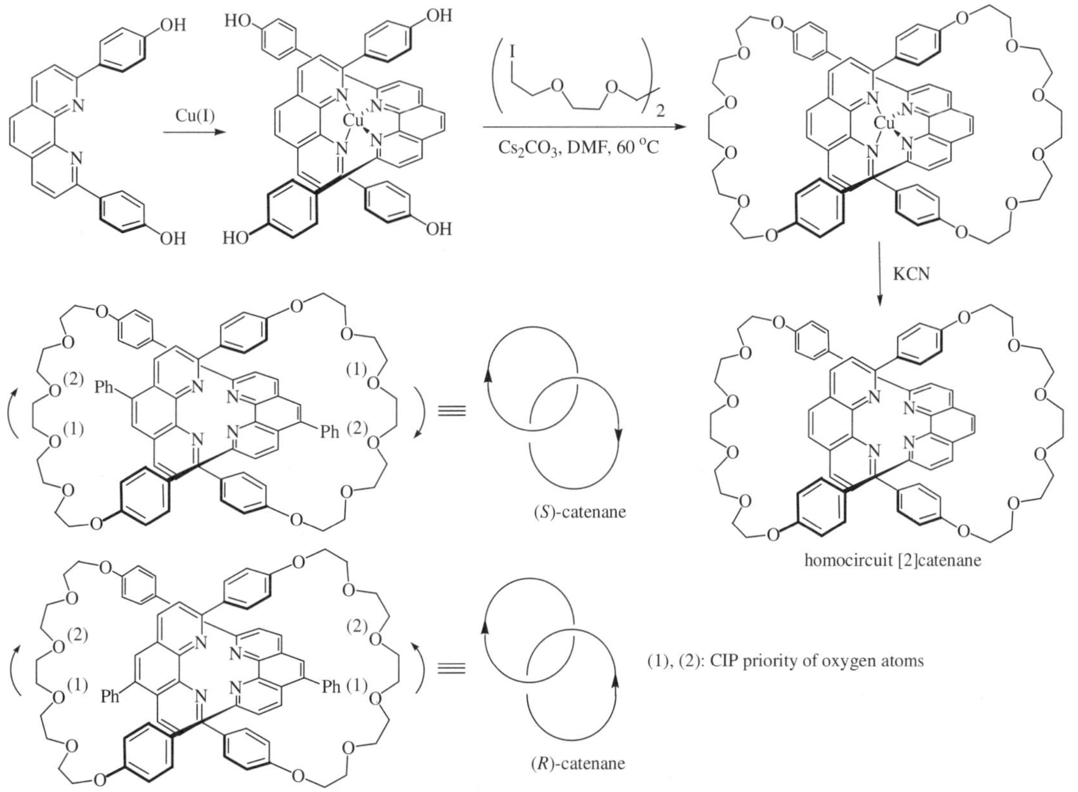

Scheme 9.3 Template-directed synthesis of a [2]catenane using metal coordination to preorganize intertwining components prior to ring closure, and structures of topologically chiral catenanes.

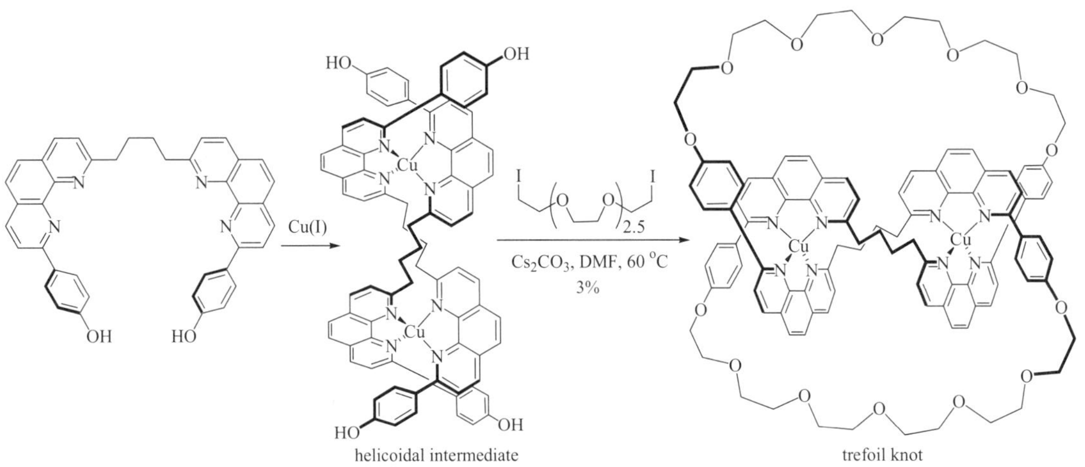

Scheme 9.4 Synthesis of a trefoil knot using the metal ion template method.

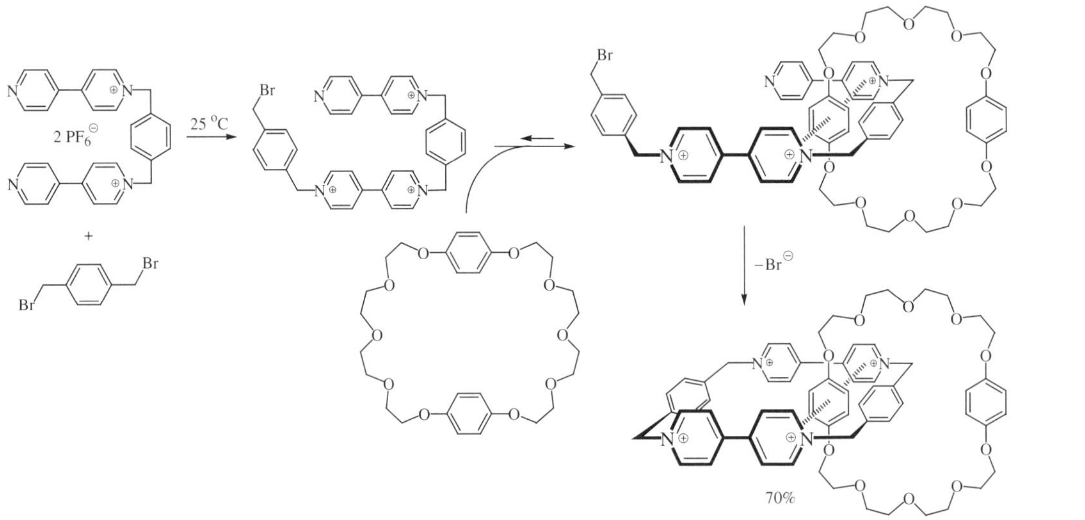

Scheme 9.5 Synthesis of a heterocircuit [2]catenane using π-π-interactions and Coulomb attraction to preorganize intertwining components prior to ring closure.

Reaction of the terminal phenol groups with α,ω-diiodopentaethylenegylcol at high dilution and removal of Cu(I) with potassium cyanide produced a homocircuit [2]catenane in 20% yield. Importantly, incorporation of substituted phenanthrolines gives rise to topologically chiral catenanes exhibiting oriented rings, see Chapter 9.2.[46,47] Preorganization of reactive precursors based on metal-directed self-assembly has successfully been applied to the synthesis of a trefoil knot.[48] In this case, two metal ions are used to entangle molecular threads consisting of two 1,10-phenanthroline units and peripheral phenol groups. Cyclization of the helicoidal intermediate with α,ω-di-iodohexaethylenegylcol gave the tetraphenanthroline-derived trefoil knot in 3% yield, Scheme 9.4. NMR spectroscopy revealed that the helicoidal intermediate is formed in approximately 15% yield, which explains the low yield of the trefoil knot and substantial formation of topological isomers.

Stoddart's group introduced predictable self-assembly of stereoelectronically matched components as a highly efficient route to catenanes and rotaxanes. This approach has furnished heterocircuit [2]catenanes from crown ethers containing electron-rich hydroquinone or

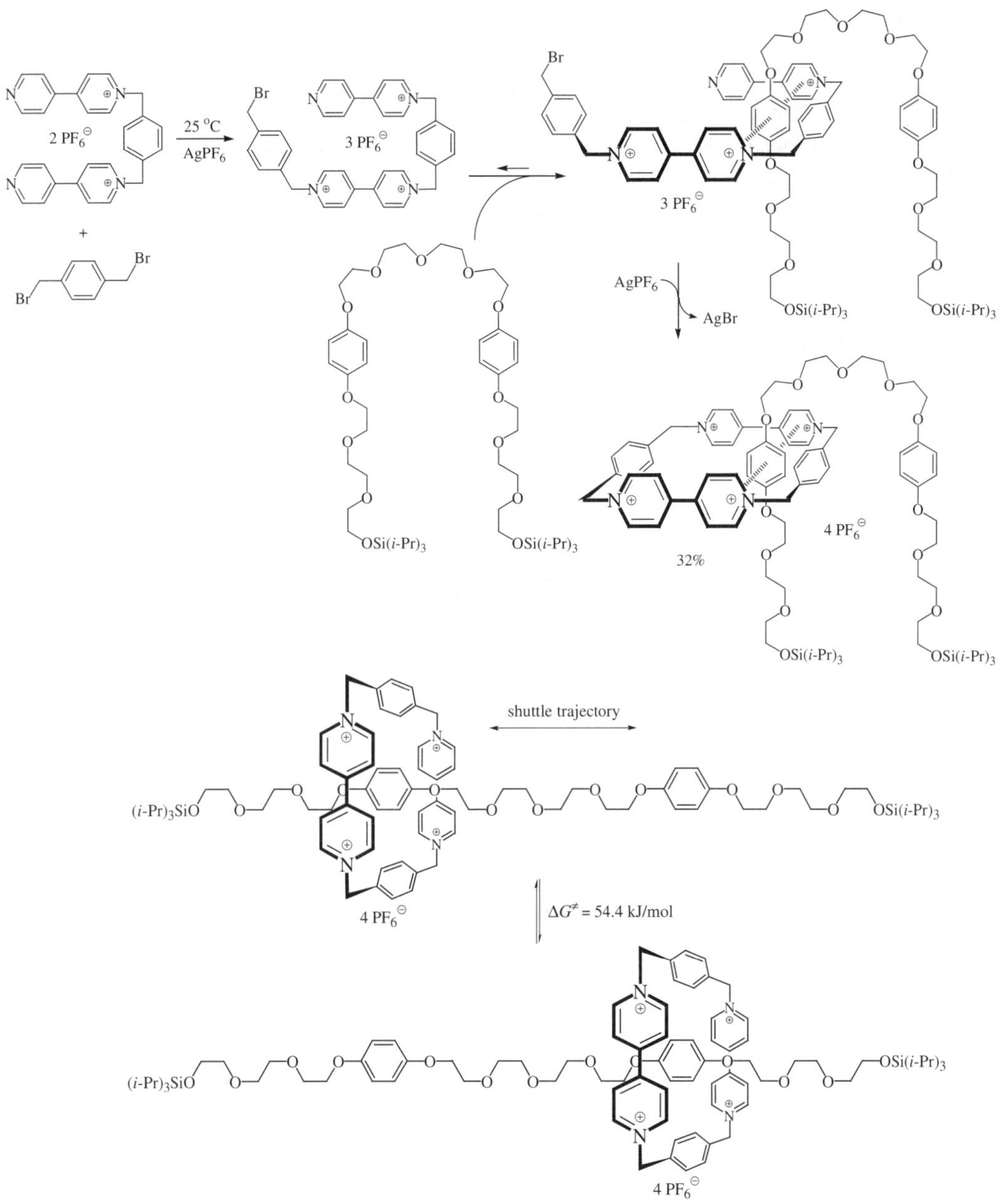

Scheme 9.6 Template-directed synthesis of a [2]rotaxane (top) and topomerization (bottom).

dioxynaphthalene moieties and 4,4'-bipyridinium-derived rods, Scheme 9.5.[49,50] Threading of bis-*p*-biphenylene-34-crown-10, exhibiting two electron-rich hydroquinone portions, occurs upon formation of a tricationic electron-deficient bis-4,4'-bipyridinium rod. Thus, π-π-interactions and Coulomb attraction are utilized as directing forces to preorganize an entangled structure which finally undergoes ring closure to the tetracationic [2]catenane in remarkable 70% yield. The use of a *linear* hydroquinone-derived polyether carrying bulky stoppers in combination with the positively charged bipyridinium-dixylylene motif allows one to apply the same concept in the synthesis of [2]rotaxanes. Cyclization of the intermediate tricationic ring precursor in the presence of an electron-rich aromatic polyether rod bearing two terminal triisopropylsilyl groups gives the corresponding rotaxane in 32% yield, Scheme 9.6.[51] This rotaxane shows degenerate translational movement of the trapped ring along the rod and affords two topomers. The barrier to topomerization has been determined as 54.4 kJ/mol by NMR spectroscopy. Translational isomerism due to relative movements of intertwined components in topological assemblies is a common feature of catenane- and rotaxane-derived molecular shuttles and switches, see Chapter 9.5.

9.1.3   Topological Isomerization

A [2]rotaxane and a mixture of the corresponding unthreaded ring and rod components establish a pair of topological isomers. Topological isomerization based on the so-called slippage approach affords another synthetic entry to rotaxanes.[52–57] For example, heating of a solution containing a macrocyclic electron-rich bis-*p*-phenylene-34-crown-10 and a dumbbell-shaped electron-deficient bipyridinium unit produces the topologically isomeric [2]rotaxane in 52% yield, Scheme 9.7.[58] Absorption UV–Vis and NMR spectroscopy, as well as computational studies, revealed that the energy barrier to slippage of the ring onto the rod is mainly controlled by the size of the macrocycle and the bulkiness of the stoppers. Stabilizing π-π-interactions between the two rotaxane components decrease the activation barrier for the slipping-on process relative to that of the slipping-off process. Since the passage requires elevated temperatures, the rotaxane is kinetically stable at room temperature.

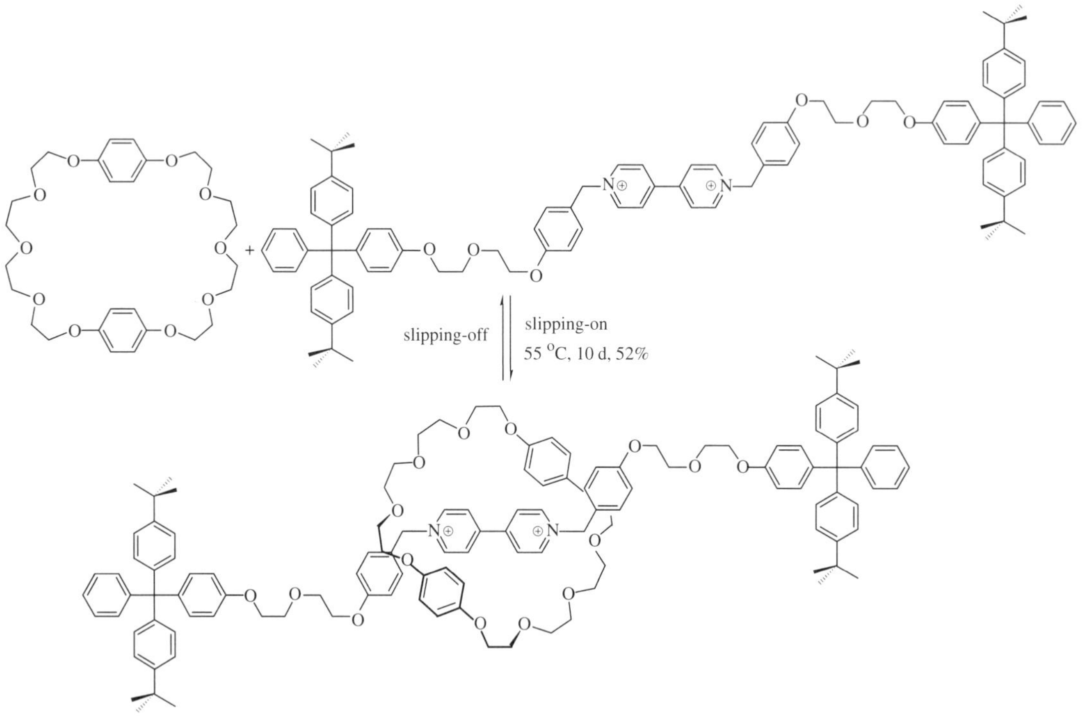

Scheme 9.7   Topological isomerization of a rotaxane.

9.2 CHIRAL CATENANES

The chirality of catenanes can originate either from the presence of classical chiral elements (chiral center, axis, plane or helix) or from directionality in both rings.[59–61] The former scenario affords so-called Euclidian chirality and the latter establishes topological chirality. In contrast to stereoisomers possessing classical elements of chirality, interconversion of enantiomers and diastereoisomers of topologically chiral molecules generally requires bond cleavage.[iii] The combination of two rings devoid of chiral elements and directionality generates an achiral catenane. Chiral assemblies can be prepared either from chiral precursors (macrocycles with at least one element of Euclidian chirality) or from achiral rings that possess cyclic directionality, compare (a) to (d) in Figure 9.3. Interlocking of homochiral rings or combination of one chiral macrocycle and one achiral ring always produces a chiral assembly, see (c) and (d). However, catenanes generated from a racemic mixture of chiral rings or ring precursors may either be chiral or achiral because meso isomers can also be formed, (e) in Figure 9.3. When enantiomeric macrocycles are threaded by an achiral ring only chiral catenanes are obtained (f). The stereochemistry is further complicated when (R,R)-, (S,S)- and meso (R,S)-macrocycles are used as starting materials (g). In this case, six topological isomers of homocircuit [2]catenanes can be expected.

Interestingly, interlocking of achiral macrocycles can generate topological chirality. Puddephatt and coworkers demonstrated that the combination of two meso rings can give a chiral [2]catenane devoid of the symmetry elements of the individual components.[62] They found that *trans*-[PdCl$_2$(SMe)$_2$] and racemic bis(amidopyridyl)binaphthyl ligands selectively form tetrameric meso (M,P)-macrocycles that assemble into a C$_2$-symmetric catenane, Scheme 9.8.[iv] The formation of the intermediate macrocycle could in principle generate a mixture of three stereoisomers, consisting of (M,M)-, (M,P)- and (P,P)-bis(binaphthyl) units, but highly diastereoselective [2 + 2]-assembly favors the meso form. Catenation of two meso macrocycles to an intertwined [4 + 4]-assembly involves temporary ring opening of the kinetically labile pyridylpalladium(II) complex and yields an equimolar mixture of topological enantiomers.

As shown in Scheme 9.3, formation of copper-coordinating [2]catenanes from substituted phenanthrolines gives rise to topologically chiral assemblies due to cyclic directionality in both interlocked rings.[v] Since the individual ring components are achiral the observed chirality is solely of topological origin and the enantiomers are produced in equal amounts. Topologically chiral catenates are generally stable to racemization and have been separated into enantiomers by HPLC, Figure 9.4.[63] Sauvage's group synthesized a chiral (P)-1,1'-binaphthyl-derived [2]catenate in the presence of a templating metal ion, Scheme 9.9.[64] To improve overall yields, a preformed ring coordinating to copper(I) was employed in the catenation process and a homocircuit assembly was prepared through two nucleophilic substitutions. Threading of the four-coordinate metal complex by 2,9-bis(4-hydroxyphenyl)-1,10-phenanthroline and subsequent ring closure with a complementary α,ω-diiodo polyether gave the corresponding (P,P)-[2]catenate in 21% yield.

Puddephatt *et al.* prepared an organometallic catenane through self-assembly of prochiral digold(I) diacetylides and bis(diphenylphosphino)propane.[65,66] Variation of the length of the diphosphine spacer gave access to a series of topologically chiral [2]catenanes that are stabilized by aurophilic interactions, Scheme 9.10.[67] Vögtle's group utilized two consecutive cyclizations for one-pot syntheses of sulfonamide-derived catenanes that are chiral by virtue of directionality in both ring components. Regioselective threading of a macrocyclic lactam formed *in situ* from a

---

[iii] An exception to this rule is possible with topologically chiral rotaxanes that undergo racemization or diastereomerization through topological isomerization, *i.e.*, reversible on- and off-slipping and formation of intermediate unthreaded ring and rod components.

[iv] The individual meso macrocycles display an inversion center and are achiral, but the symmetry is lost in the corresponding [2]catenane.

[v] By analogy with organometallic ate complexes, metal-derived catenanes are sometimes called catenates.

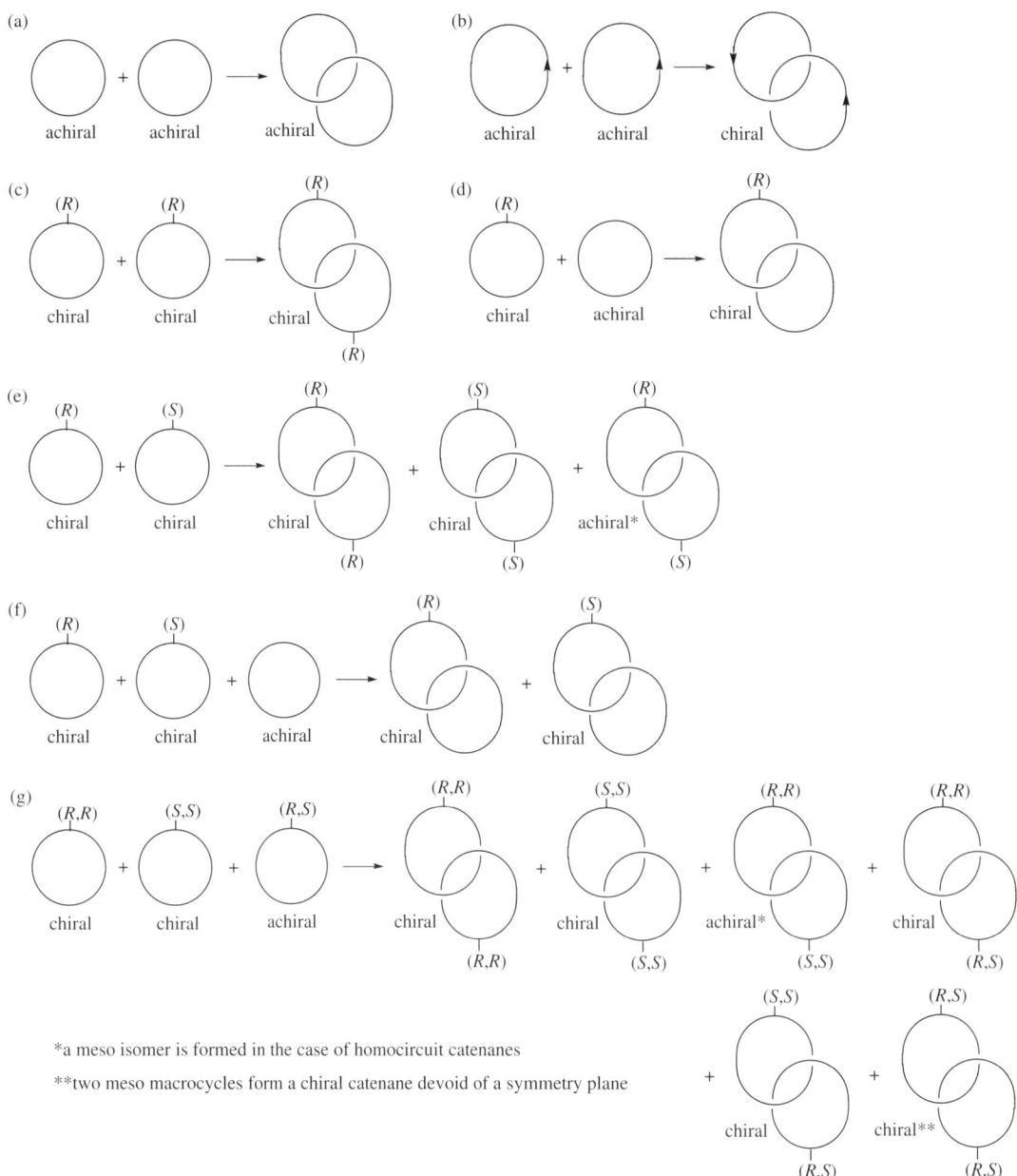

Figure 9.3 Catenanes derived from achiral and chiral macrocycles.

diamine and 5-methoxyisophthaloyl dichloride, and subsequent ring closure furnished a racemic homocircuit sulfonamide [2]catenane in 19% yield, Scheme 9.11.[68] According to NMR titration experiments, the catenation process is controlled by hydrogen bonding between the intertwining components.[69] The chiral sulfonamide catenanes are stable to racemization and can be separated into enantiomers by liquid chromatography on Chiralcel OD.[70] Introduction of dendritic branches to the sulfonamide moieties affords chiral dendro[2]catenanes that are also resolvable by HPLC.[71,vi]

[vi] Dendrocatenanes are dendrimers derived from a (topologically chiral) catenane core.

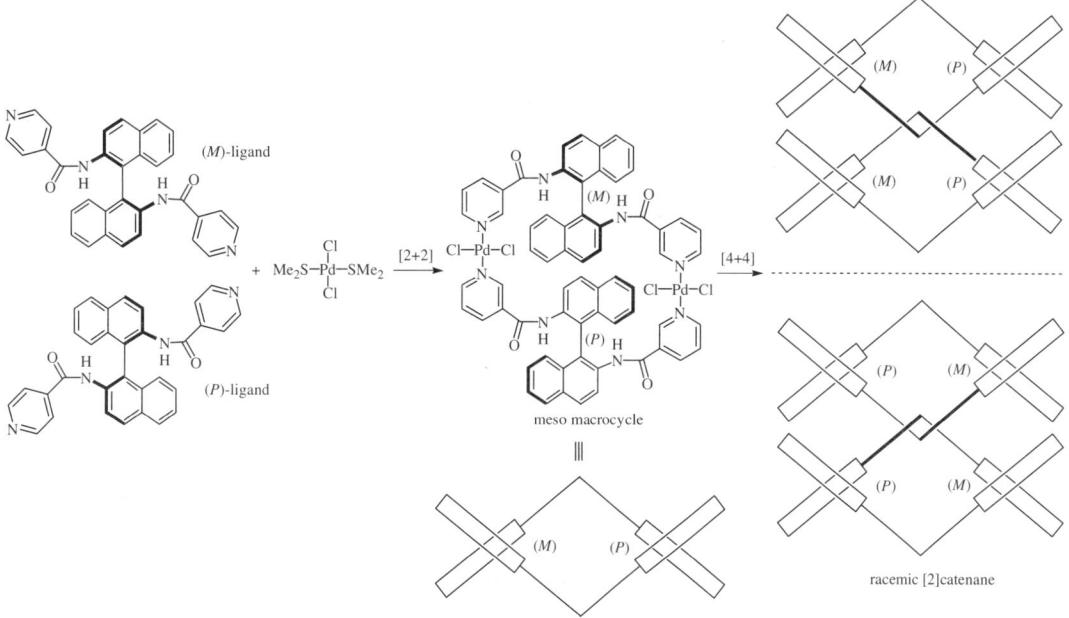

Scheme 9.8 Formation of C_2-symmetric [2]catenanes from a *trans*-palladium(II) complex and racemic binaphthyl ligands.

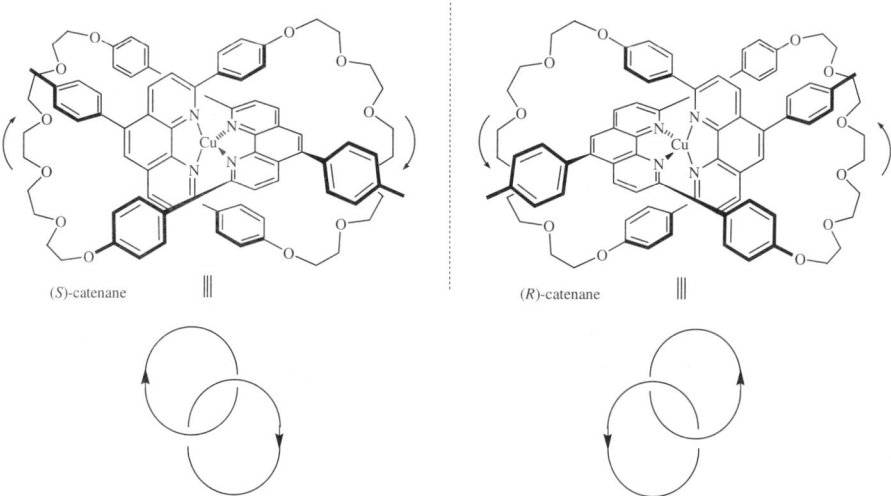

Figure 9.4 Enantiomers of topologically chiral [2]catenates. The arrows illustrate the cyclic directionality based on CIP rules.

Incorporation of covalent linkages of varying length between two interlocked rings reduces the translational mobility of the individual units.[72] Because of the pretzel-like structure, this class of compounds has been named pretzelanes. Stoddart prepared a pretzelane consisting of a tetra-cationic cyclophane intertwined with a crown ether that is covalently connected to the former by a phthalimido tether, Figure 9.5.[73,74] By analogy with ansa compounds and cyclophanes, bridging of

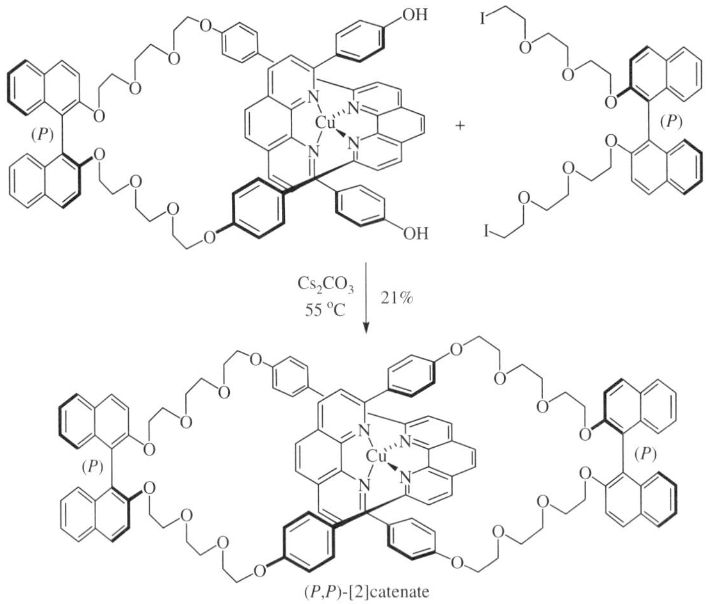

Scheme 9.9   Preparation of a binaphthyl-derived copper catenate.

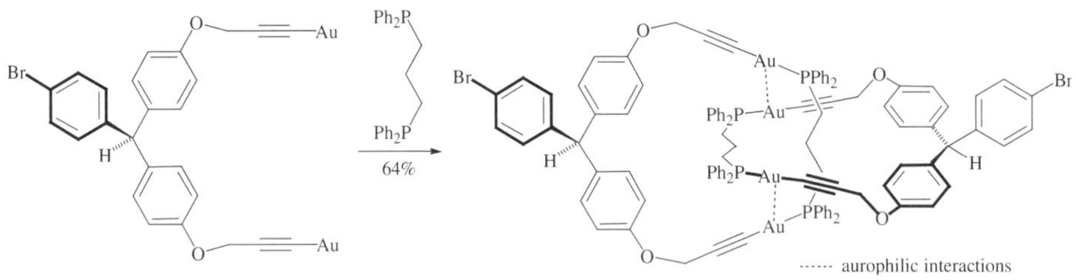

Scheme 9.10   Synthesis of a topologically chiral organometallic catenane.

the 1,5-dioxynaphthalene ring generates planar chirality, while the relative positioning of the interlocked rings in the pretzelane gives rise to helical chirality. As a result, the pretzelane exhibits two elements of chirality and exists as a mixture of two diastereomeric pairs of enantiomers. Variable-temperature NMR spectroscopy proved that the enantiomers can interconvert via a tandem process involving reorientation of the 1,5-dioxynaphthalene moiety outside the cyclophane cavity and subsequent flipping of the phthalimido group around its –CH$_2$ArCH$_2$– axis. The first process inverts the chiral plane and generates a diastereomeric intermediate that undergoes fast 180° rotation of the phthalimido linker, thus reversing helical chirality to afford the opposite enantiomer. Alternatively, a pirouetting process in which the macrocyclic polyether ring migrates around the paracyclophane from one bipyridinium portion to the other has been proposed. Spin-saturation transfer NMR experiments revealed activation barriers of 73.3 and 73.7 kJ/mol for these two possible enantiomerization pathways. Because of the striking similarity of the activation energies, it is assumed that reorientation of the dioxynaphthalene moiety is the rate determining step and is followed either by phthalimido linker flipping or pirouetting. The architecture of pretzelanes emphasizes the unique stereochemistry of mechanically interlocked molecules that may display central, axial, planar, and helical chirality. The stereochemical diversity and dynamic behavior of catenanes comprise translational isomerism as well as ring rocking and controllable

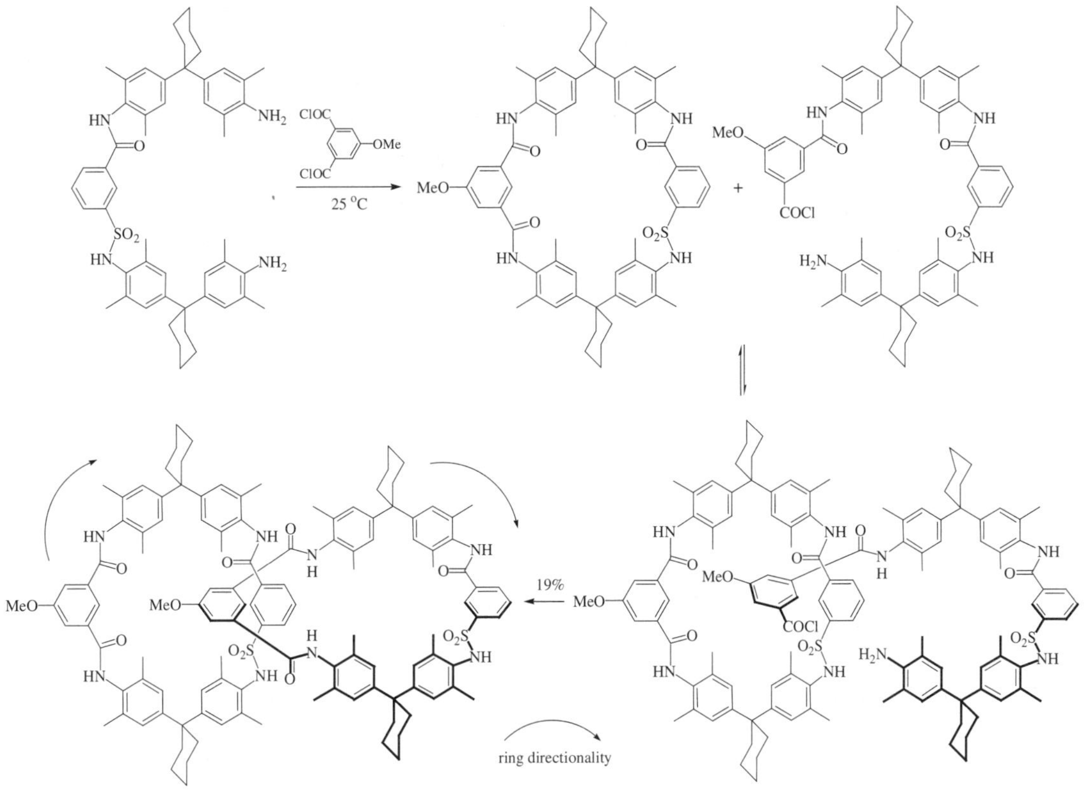

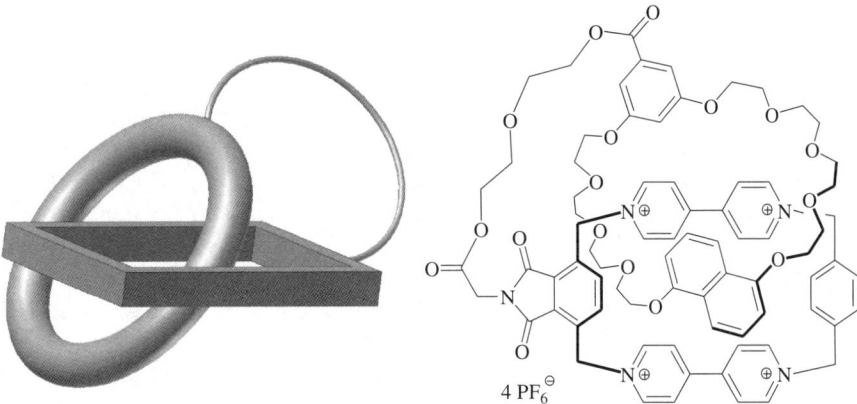

Scheme 9.11 Racemic one-pot synthesis of a topologically chiral sulfonamide catenane. Only the (*R*)-enantiomer is shown.

Figure 9.5 Illustration of the structure of a pretzelane (left) and Stoddart's pretzelane (right).

circumrotation processes, which have been exploited for the development of molecular shuttles and switches.

Extending the concept of molecular recognition between ring precursors bearing π-electron-rich and electron-deficient aromatic units, respectively, to the synthesis of chiral [2]catenanes,

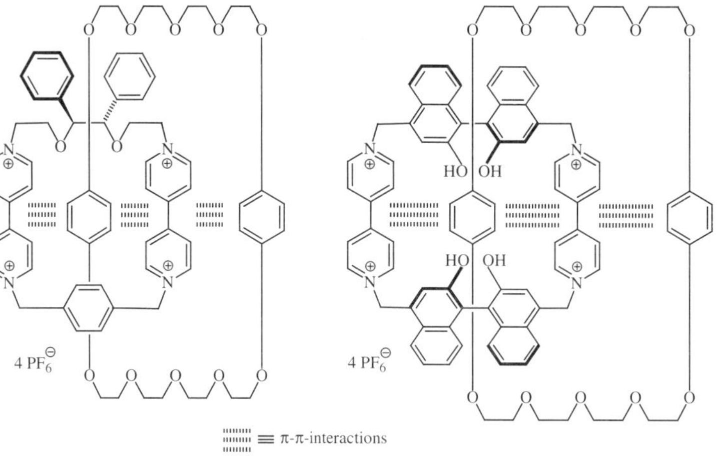

Figure 9.6   Structures of (*S,S*)-hydrobenzoin- and (*M,M*)-binaphthol-derived [2]catenanes.

Stoddart *et al.* observed that cyclodextrins, hydrobenzoin, mannitol, and binaphthols can be incorporated into at least one ring component without impeding stereoelectronic matching and self-assembly prior to catenation, Figure 9.6.[75–78,vii] Crystallographic analysis of a binaphthyl-derived catenane revealed that one electron-rich hydroquinone moiety is sandwiched by two electron-deficient bipyridinium units in the solid state. Additional variable-temperature NMR spectroscopic measurements uncovered two degenerate circumrotation processes, in which either the tetracationic cyclophane ring or the dihydroquinone ring migrates through the cavity of the other macrocycle. Accordingly, both hydroquinone moieties can reside inside the region of the cyclophane ring and both bipyridiniums can be sandwiched by π-donating hydroquinones in solution, Scheme 9.12.

   Template-directed self-assembly of an electron-deficient cyclobis(paraquat-1,5-naphthalene) component in the presence of electron-rich bis-1,5-dinaphtho[38]-crown-10 gives a mixture of stereoisomeric [2]catenanes in 66% yield, Scheme 9.13.[79] The high yield is probably a consequence of favorable preorganization due to charge transfer between the crown ether and the bipyridinium ring precursor. Each 1,5-disubstituted naphthalene moiety introduces a chiral plane. The presence of four planes of chirality, and the local reflection symmetry of tetracationic cyclophanes carrying heterochiral naphthalene planes, yields a total of twelve stereoisomers.[viii] Flipping of one naphthalene unit causes diastereomerization whereas enantioconversion requires reversal of all four elements of chirality.[ix] The latter is associated with a very high activation barrier but flipping of the alongside dioxynaphthalene occurs rapidly on the NMR time scale. In addition to isomerization of one naphthalene unit, circumrotation of the crown ether through the tetracationic cyclophane causes diastereomerization, if the exchanging 1,5-dioxynaphthalene residues possess an opposite sense of chirality (one has (*R*)- and the other one has (*S*)-configuration). In this case the tetracationic cyclophane ring remains unchanged, but the cavity hosts either an (*R*)- or (*S*)-1,5-dioxynaphthalene plane which gives rise to ($R_{inside}S_{alongside}$)- and ($S_{inside}R_{alongside}$)-translational isomers. In contrast, circumrotation of the tetracationic cyclophane ring exhibiting two 1,5-disubstituted naphthalene groups of opposite chirality is a degenerate process and yields the original isomer. The tetracationic cyclophane ring and the macrocyclic aromatic polyether undergo

---

[vii] Incorporation of a chiral group into one of the intertwined macrocycles provides a useful probe for NMR spectroscopic analysis of catenane stereomutations.

[viii] In principle, the combination of four chiral planes can afford a total of $2^4 = 16$ stereoisomers, but only six diastereoisomeric pairs of enantiomers are possible due to the local reflection symmetry present in catenanes with a heterochiral tetracationic ring.

[ix] See Chapter 3.3.3 for atropisomerization of cyclophanes.

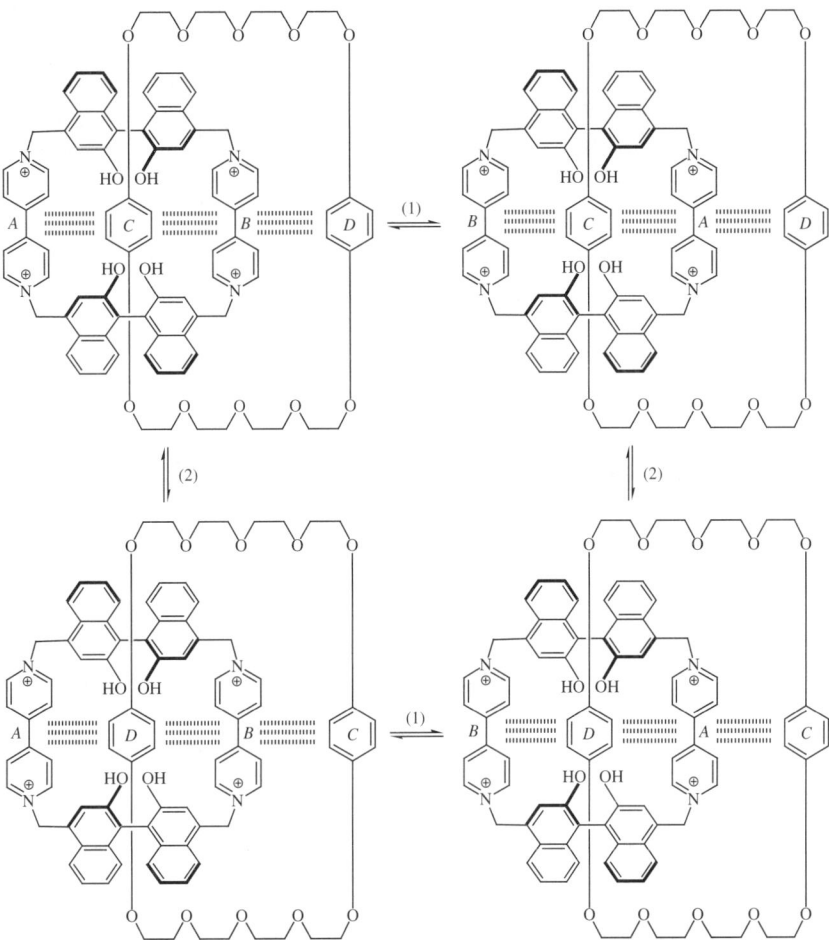

Scheme 9.12 Translational isomers of an (*M,M*)-binaphthol-derived [2]catenane. Circumrotation processes: translation of the tetracationic cyclophane ring through the crown ether exchanging bipyridinium units *A* and *B* (1); translation of the crown ether through the tetracationic cyclophane ring exchanging hydroquinones *C* and *D* (2).

CH···O- and *CH/π*-interactions in addition to π-stacking. The relative stability of translational isomers and the activation barrier to translational isomerization of Stoddart's catenanes is controlled by these forces. Circumrotation (interconversion of translational isomers) requires disruption of all three bonding interactions. In addition to circumrotation and reversal of local planar chirality due to ring flipping of a 1,5-disubstituted naphthalene, catenanes show so-called rocking processes. Rocking refers to interconversion of helical arrangements of the two ring components. As illustrated in Figure 9.7, the helicity originates from tilting of the crown ether with respect to the plane of the tetracationic ring.[80] The relative orientation of the interlocked rings is mostly determined by *CH···O-* and *CH/π*-interactions which are maximized when the rings are tilted by approximately 45°. Rocking requires breaking of the *CH···O-* and *CH/π*-bonds but does not affect π-π-interactions. Accordingly, ring rocking proceeds more easily than circumrotation. Since ring rocking results in interconversion of enantiomeric ring arrangements (if other chiral elements are not present), Stoddart's group resorted to chiral shift reagents and incorporation of chiral probes into one ring to investigate the dynamic chirality, relative movements and switching

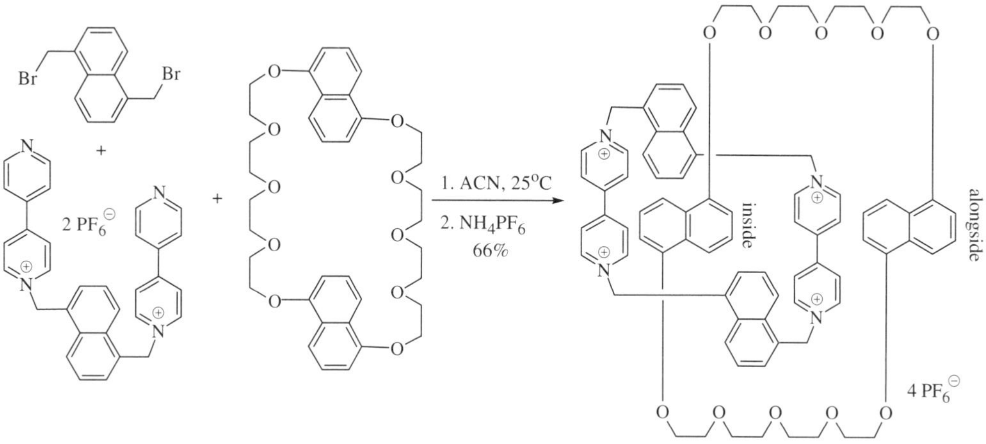

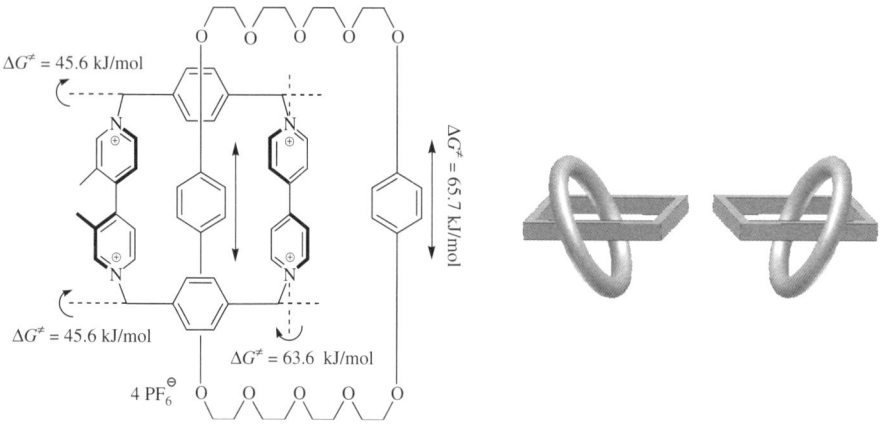

Scheme 9.13   Synthesis of a [2]catenane with four chiral 1,5-disubstituted naphthalene units.

Figure 9.7   Phenylene rotation, pyridinium rotation and crown ether circumrotation of a [2]catenane (left), and schematic illustration of the helical arrangement of the two tilted rings that interconvert via ring rocking (right).

properties of catenanes by NMR spectroscopy.[81,82] Stereodynamic analysis of a catenane derived from a bisparaphenylene-[34]crown-10 and a cyclobis(paraquat-*p*-phenylene) ring, in which one bipyridinium unit was replaced by an axially chiral bipicolinium motif, revealed an activation barrier to crown ether circumrotation of 65.7 kJ/mol (–17 °C) and the activation energy for the ring rocking process was determined as 42.7 kJ/mol (–91 °C). Phenylene rotation and pyridinium rotation were also monitored, and have activation barriers of 45.6 kJ/mol (–91 °C) and 63.6 kJ/mol (–26 °C), respectively.

9.3   CHIRAL ROTAXANES

The stereochemistry of rotaxanes is closely related to that of catenanes and other mechanically interlocked molecules.[83] Rotaxanes can be rendered chiral by incorporation of either a chiral thread or a chiral ring component. Alternatively, combination of achiral rings and rods having a sense of directionality establishes topological chirality. A variety of rotaxanes derived from binaphthol-derived crown ethers, amino acid-derived calix[4]arenes or other inherently chiral rings,[84–86] and

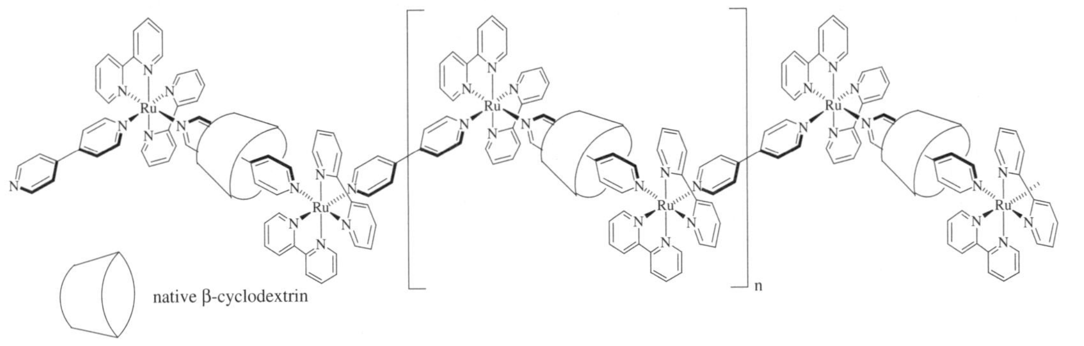

Figure 9.8 Structure of a polymeric chiral rotaxane.

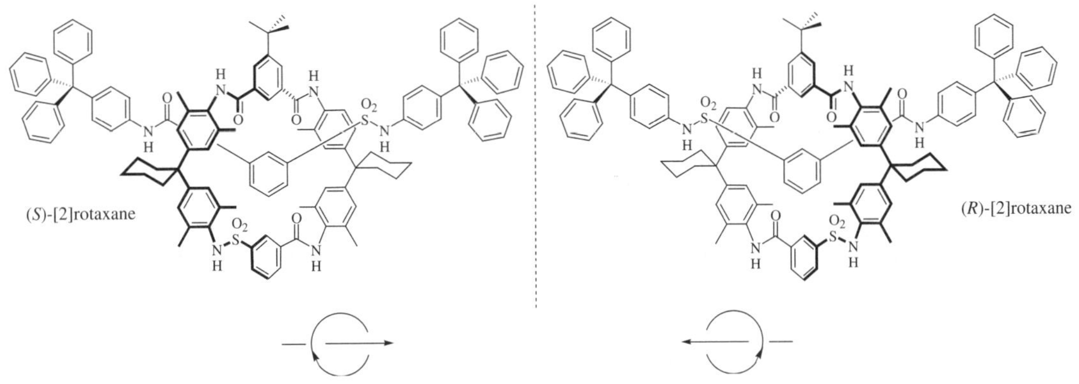

Figure 9.9 Cycloenantiomeric [2]rotaxanes.

chiral rods[87] has been reported. For example, Liu *et al.* constructed polymeric chiral rotaxanes via multi-component self-assembly of nickel(II)- and ruthenium(II) coordination complexes carrying bipyridine linkers threaded through β-cyclodextrin cavities.[88,89] Scanning tunneling microscopy and transmission electron microscopy proved that the polymers adopt a zigzag chain with an approximate length of 250 nm, Figure 9.8. By analogy with catenanes, introduction of directionality into both axle and wheel of a rotaxane generates topological chirality. The wheel affords either a clockwise or a counterclockwise orientation with respect to the axle, which results in so-called cycloenantiomerism.[90] For instance, incorporation of amide and sulfonamide groups into both interlocked components establishes directionality and yields chiral [2]rotaxanes that can often be separated into enantiomers by chiral HPLC, Figure 9.9.

The concept of cyclostereoisomerism originating from ring directionality and distinctive distribution of chiral but otherwise identical building blocks within a ring structure was first mentioned by Prelog and Gerlach. They demonstrated that the amino acid sequence and connectivity of stereogenic centers exhibiting both (*R*)- and (*S*)-configuration in cyclopeptides such as cyclohexaalanine gives rise to cycloenantiomers and cyclodiastereoisomers. It should be noted that the presence of ring directionality in cyclopeptides consisting of both enantiomers of one amino acid does not change the overall number of possible stereoisomers: a cyclic peptide exists in as many stereoisomeric forms as the open chain analog. Mislow therefore suggested that the term cyclostereoisomerism should be restricted to cases in which ring directionality increases the number of stereoisomers, and proved that this can be achieved with appropriately substituted hexaisopropylbenzene derivatives. In these cases, directionality due to static gearing increases the number

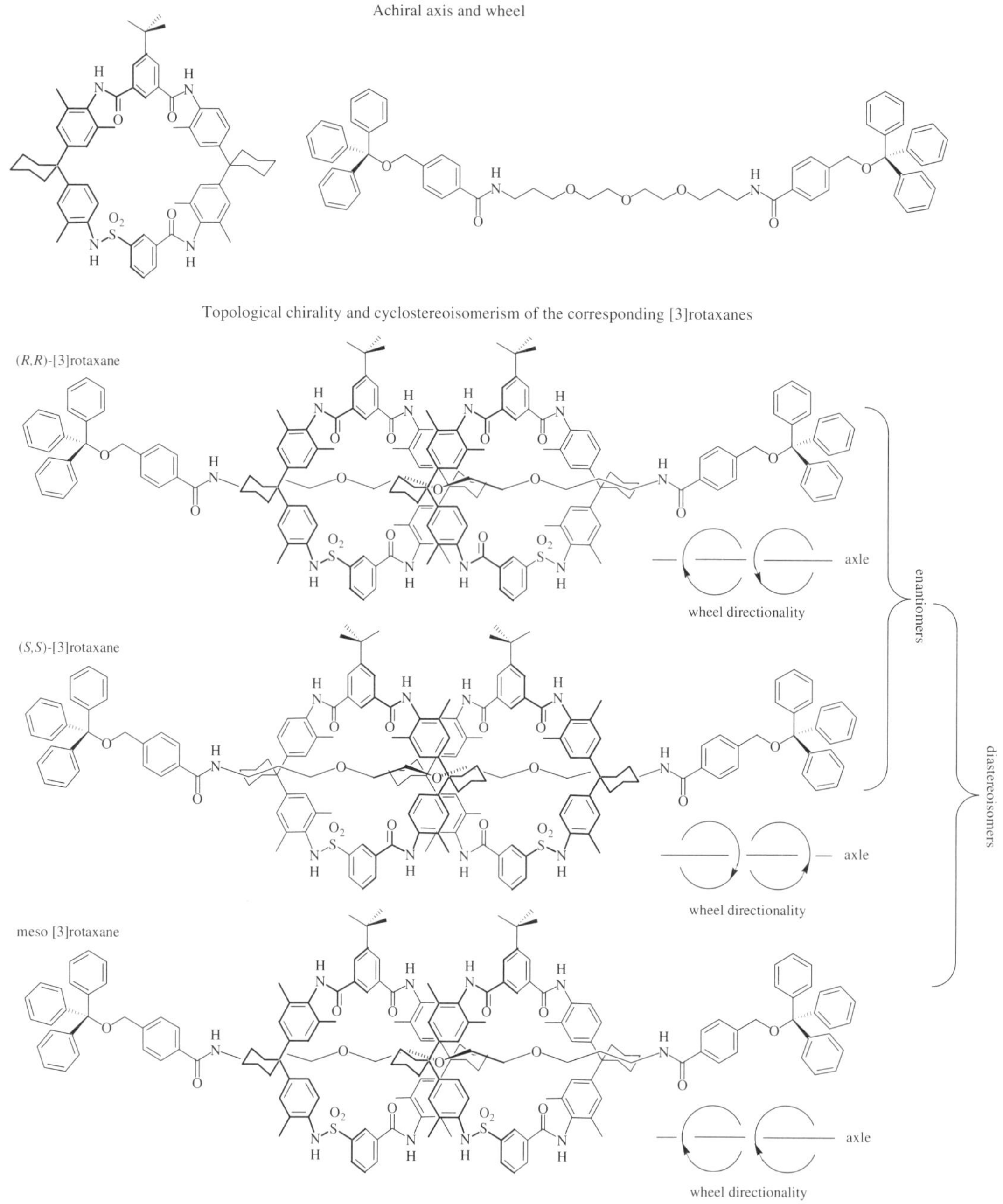

Figure 9.10   Structure of an achiral directional wheel and a symmetric axis (top). Topological chirality and cyclostereoisomerism of the corresponding [3]rotaxanes (bottom).

of possible stereoisomers, see Chapter 8.7.[91–95] The same situation arises with cyclostereoisomerism of rotaxanes. Removal of the directionality of either the axle or the wheel in the chiral rotaxane shown in Figure 9.9 would eliminate topological chirality, and thus reduce the number of stereoisomers: one would observe a single achiral [2]rotaxane instead of two enantiomeric forms.

Cyclodiastereoisomeric [3]rotaxanes carrying two wheels on one axle have been prepared by Vögtle's group, Figure 9.10.[96] Because of the directionality of the two threaded macrocycles, which

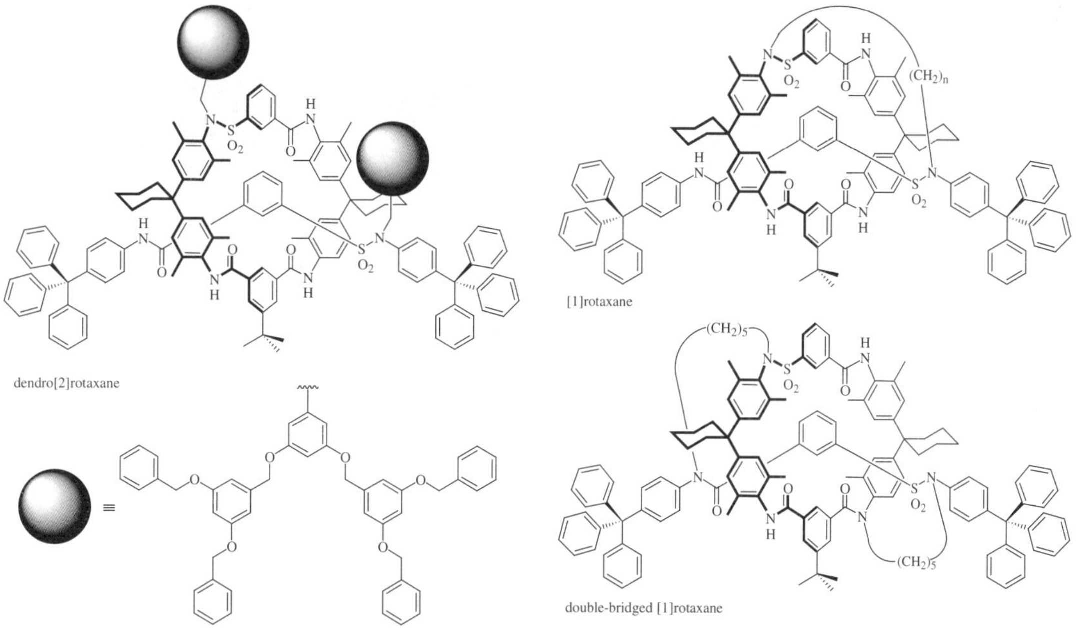

Figure 9.11 Structures of a chiral dendro[2]rotaxane, [1]⟨*n*⟩rotaxanes and a double-bridged [1]⟨*5,5*⟩rotaxane.

is a consequence of the presence of the sulfonamide and amide groups, the two wheels can possess the same or opposite orientation. Although the symmetric thread lacks directionality, this [3]rotaxane exists in the form of three stereoisomers: one meso and two enantiomeric assemblies. Evidence for the coexistence of these cyclodiastereoisomeric catenanes was obtained by HPLC separation of the three isomers on a Chiralpak AD column and chiroptical measurements. Cycloenantiomeric dendro[2]rotaxanes have been obtained by incorporation of dendritic branches into a topologically chiral [2]rotaxane core, Figure 9.11. Similar to dendro[2]catenanes, dendro[2]rotaxanes can be separated into enantiomers by chiral HPLC.[97] Introduction of a tether with *N* atoms between the wheel and the axis of a chiral rotaxane generates cycloenantiomeric [1]⟨*n*⟩rotaxanes that display cyclochirality and restricted mobility of the covalently linked wheel.[98–100] The presence of a second linkage establishes a chiral double-bridged [1]rotaxane reminiscent of a self-intertwining molecular figure of eight.[101,x] The chiral information of rotaxanes can also be contained in the dumbbell. Both Euclidian and topological chirality have been introduced to nondirectional axes in the form of terminal glucose units[102] and knot stoppers.[103,104] The combination of rotaxane topology with inherently chiral knots at both termini of the axle affords so-called knotaxanes.

Stoddart's group was among the first to recognize the potential of rotaxanes and pseudo-rotaxanes[105,106,xi] as molecular shuttles,[107] and systematically incorporated elements of chirality into thread and ring components to study the dynamic stereochemistry of these compounds by NMR spectroscopy and chiroptical methods.[108–113] The synthesis of [2]rotaxanes exhibiting an element of planar chirality enabled them to monitor enantioconversion and diastereomerization in addition to degenerate site exchange processes, Scheme 9.14.[114] They observed that three degenerate dynamic processes occur rapidly at room temperature with activation barriers between 59.0 an 63.2 kJ/mol. These stereomutations involve temporary dislodgement of the 1,5-dioxynaphthalene

---

[x] The molecular 8 displays cycloenantiomerism and chiral helicity.
[xi] Pseudorotaxanes are rotaxanes with stoppers that are small enough to allow de-threading.

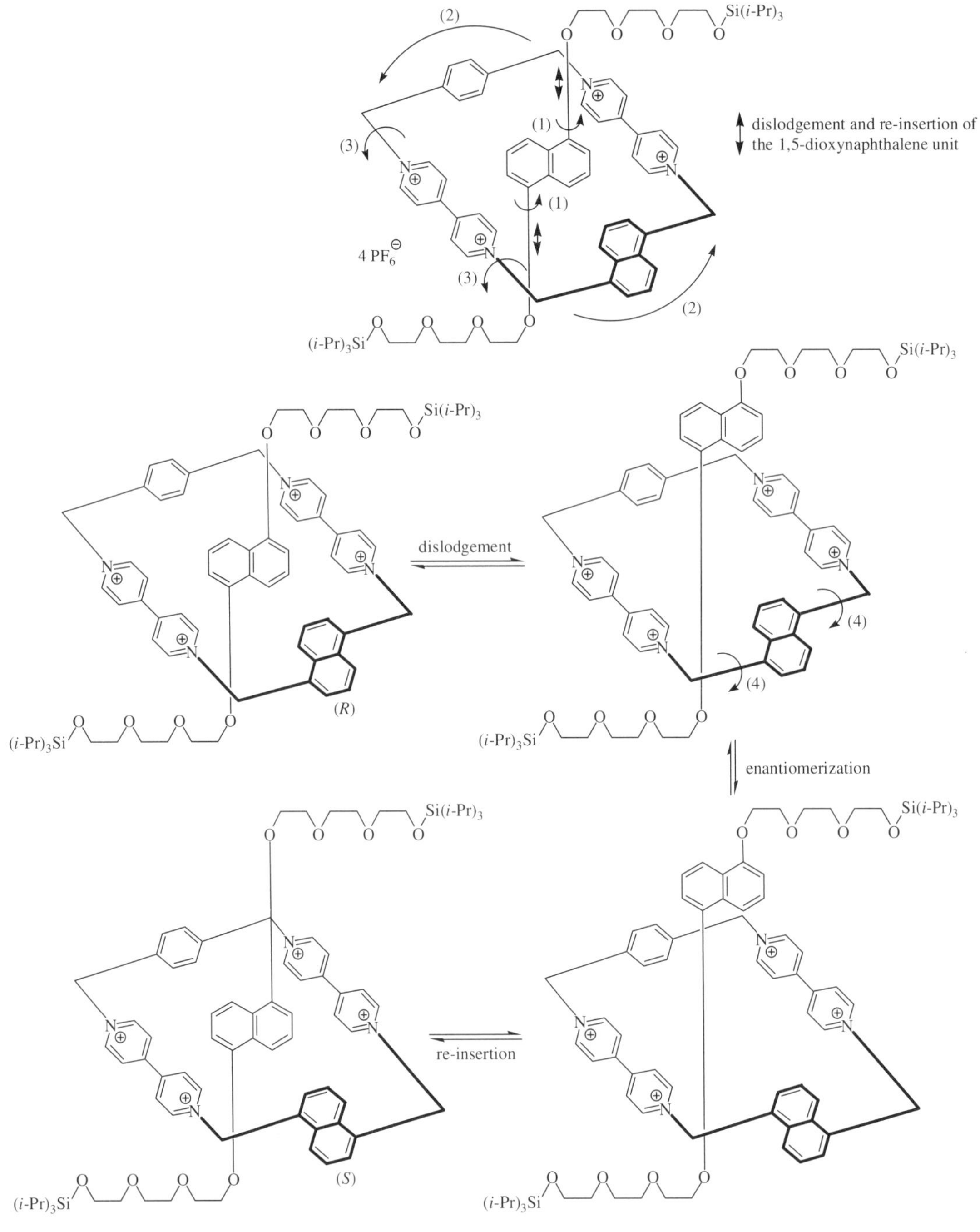

Scheme 9.14   Degenerate and nondegenerate stereomutations of a [2]catenane.

unit from the ring cavity in conjunction with either 180° rotation of the dioxynaphthalene portion about its central carbon–carbon bond axis, (1) in Scheme 9.14, translation through 180° rotation of the tetracationic cyclophane about an axis that is perpendicular to its plane and passing through its inversion center (2), or 180° rotation of one of the bipyridinium moieties about an axis passing

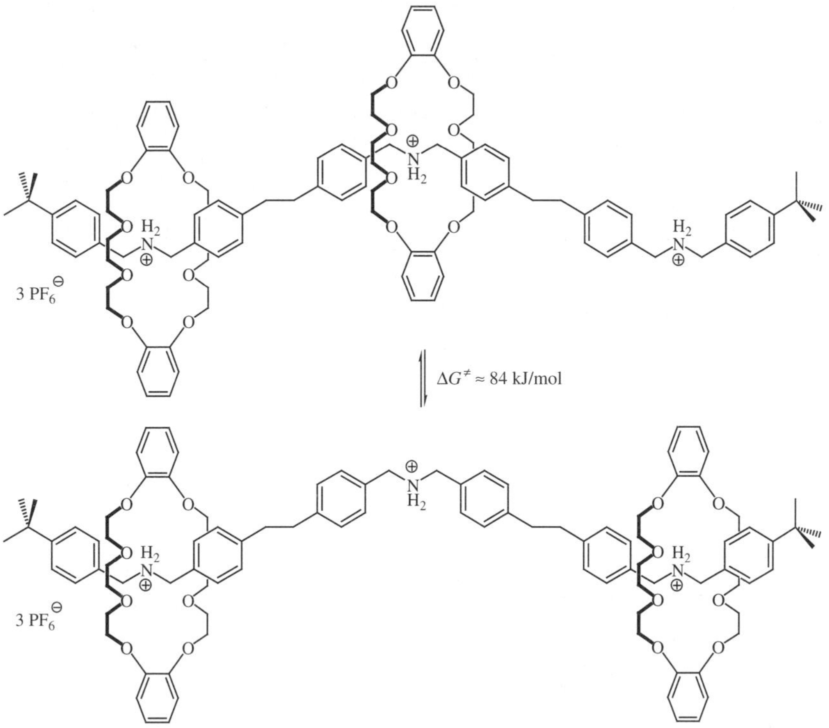

Scheme 9.15 Translational isomerization of a [3]rotaxane.

through both nitrogen atoms (3). In contrast, enantiomerization resulting from dislodgement of 1,5-dioxynaphthalene followed by nondegenerate 180° rotation of one of the 1,5-disubstituted naphthalene units located in the tetracationic paracyclophane macrocycle (4) and re-insertion of the 1,5-dioxynaphthalene into the ring cavity does not proceed within the ^{1}H NMR time scale, even at high temperatures.

The ring components of rotaxanes often undergo lateral movements that are easily observed or even externally controlled. For example, threading of two macrocyclic crown ethers by a tricationic rod provides a [3]rotaxane that exists as a mixture of translational isomers, Scheme 9.15.[115] The two rings shuttle along the thread between the stoppers and reside at two of the three available ammonium stations. The relative movement interconverts two distinguishable translational isomers and has an activation energy of approximately 84 kJ/mol according to variable-temperature NMR studies. Leigh *et al.* prepared chiral peptido[2]rotaxanes consisting of an achiral tetraamide ring and a chiral dipeptide rod derived from Gly-(S)-Ala, Gly-(S)-Leu, Gly-(S)-Met, or Gly-(S)-Phe in 32–45% yield via hydrogen bond-directed template synthesis, Scheme 9.16.[116] The circular dichroism signal of these rotaxanes can be selectively switched on and off through manipulation of intercomponent interactions that control translational isomerism. Chiroptical studies in conjunction with molecular modeling suggest that the ring glides almost freely along the thread when interactions between the two mechanically intertwined components are diminished by the presence of polar solvents or at increased temperatures. Under these conditions, CD measurements of the fluxional rotaxanes show little positive elliptical polarization. However, at lower temperatures and in the absence of polar solvents, hydrogen bonding between the thread and the macrocycle places the latter close to the asymmetric carbon center at the *C*-terminal amino acid. In this arrangement, the ring effectively relays chiral information from the rod to the diphenylmethine stopper at the *C*-terminus, which induces a strong negative CD response with molar ellipticities, θ,

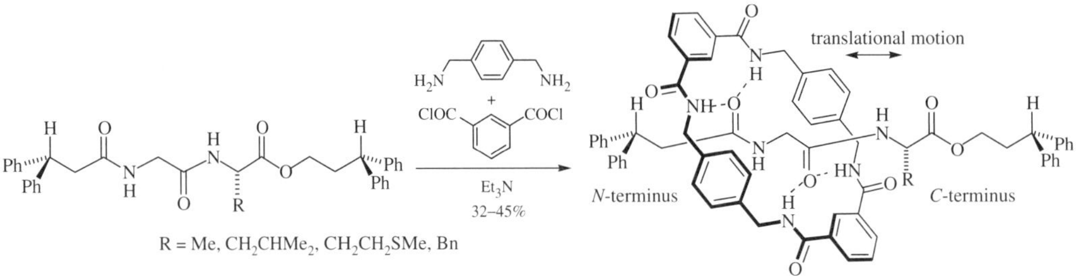

Scheme 9.16   Hydrogen bond-directed synthesis of dipeptide rotaxanes.

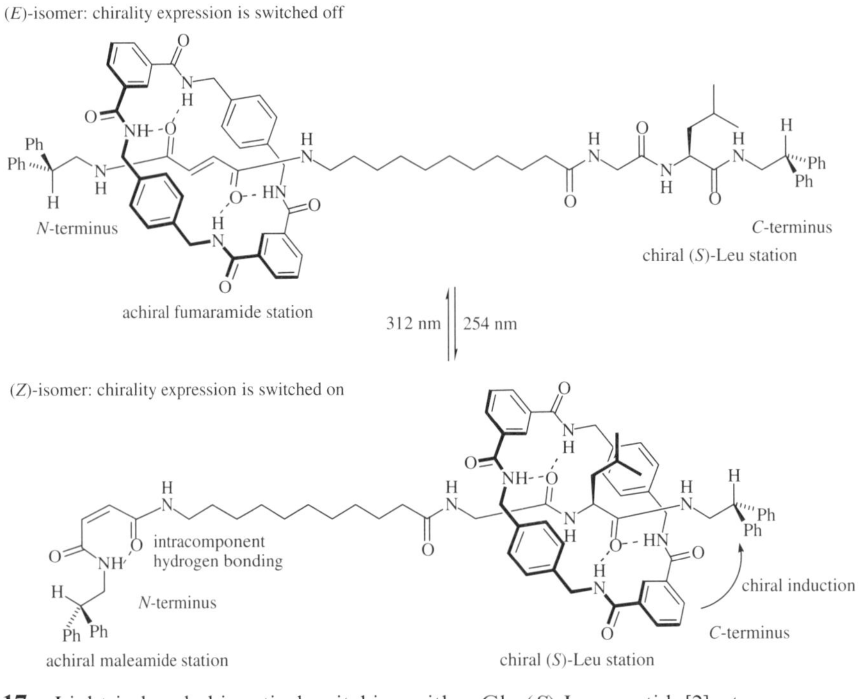

Scheme 9.17   Light-induced chiroptical switching with a Gly-(*S*)-Leu peptido[2]rotaxane.

up to –20,000 deg cm^2/dmol. Accordingly, these peptido[2]rotaxanes emit a chiral signal in response to externally controlled translational isomerization. A Gly-(*S*)-Leu-derived chiroptical switch exhibiting controlled expression of chirality based on light-induced *E*/*Z*-isomerization of the rod has also been developed, Scheme 9.17.[117] Reversible photoisomerization of this peptidorotaxane allows interconversion of diastereoisomeric states that have remarkably different binding affinities to the threaded macrocycle. The major translational isomer of the CD-inactive (*E*)-isomer holds the ring at the achiral *N*-terminus of the peptide chain due to strong hydrogen bonding between the interlocked components at the fumaramide station. Formation of the (*Z*)-isomer favors *intra*component hydrogen bonding within the isomeric maleamide unit and prevents strong *inter*component interactions at the olefin station. The macrocycle therefore shuttles to the remote station at the *C*-terminus where it adopts a rigid conformation in the chiral environment of (*S*)-leucine. In this state, the ring component can transmit chirality to the *C*-terminal stopper, resulting in strong negative CD induction. The photoisomerizations occur upon irradiation at 254 and 312 nm and allow control of the translational motion of the ring shuttle and selective on and off switching of the CD signal.

9.4 KNOTS AND BORROMEAN RINGS

Knotanes are topological isomers of macrocycles and resemble macroscopic knots, Figure 9.12. A molecular knot can be obtained by tying a knot in a flexible strand and subsequent formation of a covalent bond between the loose ends. In contrast to catenanes and rotaxanes, knotanes do not have two or more mechanically interlocked components and consist of only one molecular entity. Catenanes and rotaxanes can be rendered chiral through incorporation of classical elements of chirality or by introduction of directionality into the individual ring and rod components, but trefoil knots are inherently chiral.

The first molecular knot was prepared by metal template-directed synthesis in 1989.[118] Racemization of knots requires bond cleavage followed by reorientation of the loose ends and ring closure. Knotanes are therefore stable to racemization and can be separated into enantiomers at room temperature. For example, diastereoselective crystallization of the racemic trefoil knot shown in Figure 9.13 in the presence of (S)-(+)-1,1′-binaphthyl-2,2′-diyl phosphate furnishes a cocrystal containing the dextrorotatory knot and two equivalents of the chiral auxiliary, while the levo-rotatory enantiomer remains in the mother liquor.[119] Several molecular knots have been prepared

Figure 9.12 Illustration of the structures of enantiomeric trefoil knots and the topologically isomeric ring.

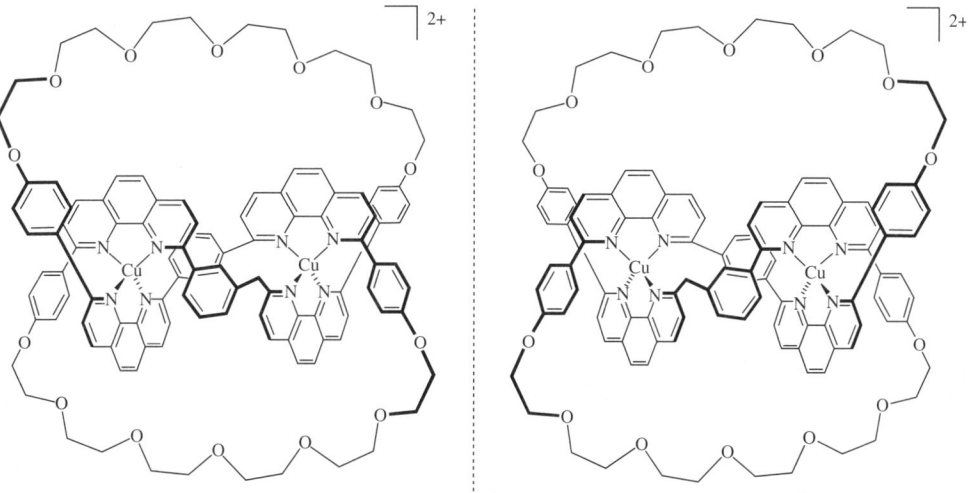

Figure 9.13 Structures of the enantiomers of a dicationic trefoil knot.

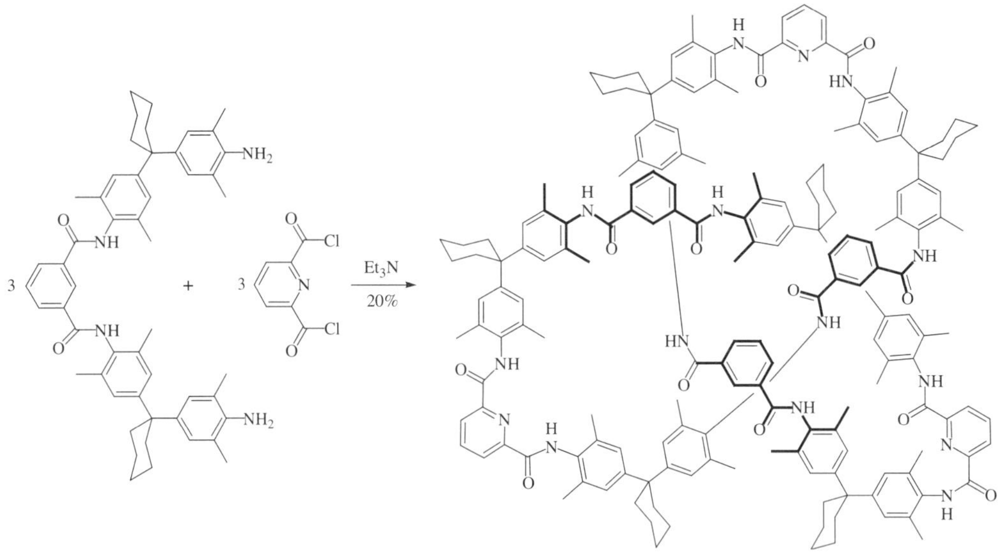

Scheme 9.18 Self-assembly of an amidoknotane.

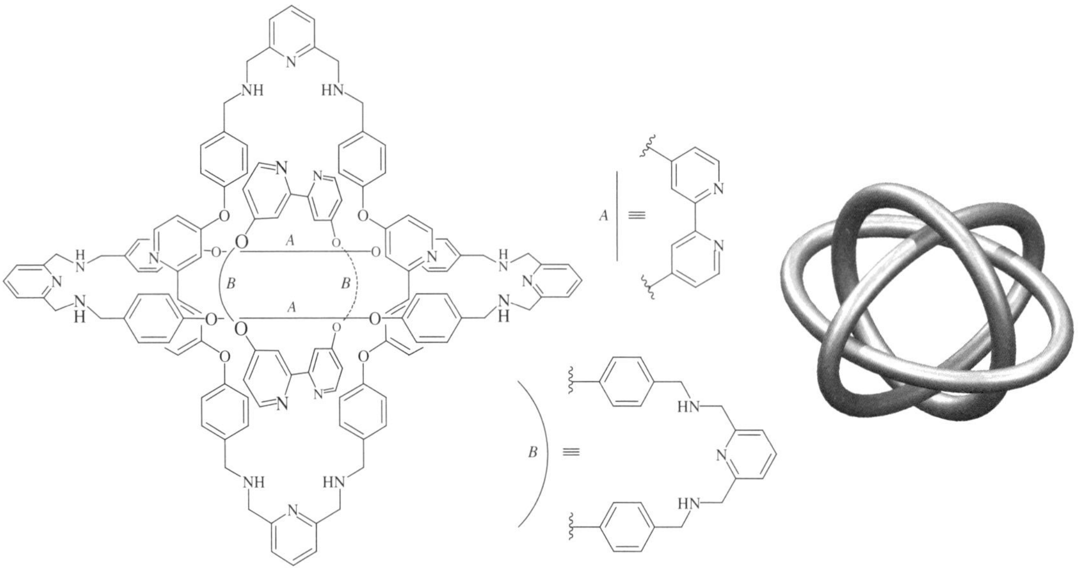

Figure 9.14 Borromean rings.

by Sauvage,[120–123] Stoddart,[124] Hunter,[125] and Vögtle, Scheme 9.18.[126–131] Large-amplitude movements are typical in many catenanes and rotaxanes but knotanes possess a more rigid structure, and this lack of translational mobility limits the potential of these molecules as shuttles, switches, sensors, and nanomachines. Metal coordination of the phenanthroline rings in the knotane depicted in Figure 9.13 further reduces the conformational flexibility of the intertwined strand.[132,133] Despite their congested structure, knotanes are by no means static molecules. Variable-temperature NMR studies of amidoknotanes uncovered coexisting conformations that interconvert with an activation energy of approximately 67 kJ/mol.[134]

Borromean rings consist of three inseparable macrocycles that are intertwined in such a way that no two of the three ring components are concatenated.[135] As has been realized with a variety of catenanes and rotaxanes, incorporation of ring directionality can render Borromean rings topologically chiral. An important difference between [3]catenanes and Borromean rings is that in the latter removal of one of the three intertwined rings unlocks the whole assembly and releases each ring. The advance of supramolecular chemistry, dynamic combinatorial methods and template-driven assembly has generated the necessary tools for the synthesis of these challenging targets, and the first molecular Borromean rings have been reported, Figure 9.14.[136–140] Even more entangled topologies have been constructed from single-stranded DNA.[141]

9.5 TOPOLOGICAL ISOMERISM OF SHUTTLES, SWITCHES, SENSORS, AND ROTORS

The chiral dipeptidorotaxanes described in Chapter 9.3 prove that the making and breaking of noncovalent interactions between mechanically interlocked components of rotaxanes and catenanes can be controlled by external stimuli. Topological isomerism provides an entry to the development of positionally bistable molecular switches, shuttles and elevators.[142] Interlocked structures can therefore be associated with other technomimetic compounds such as molecular brakes, bevel gears, propellers, and rotors that simulate mechanical-like movements of macroscopic devices and machines, see Chapter 8.[143] Translational isomerization of numerous mechanically intertwined assemblies induced by chemical,[144–147] thermal,[148] photochemical,[149–151] and electrochemical[152,153] changes have been reported to date.[154] Various stimuli, including changes in pH,[155,156] redox potential,[157–159] addition of ions or other additives,[160–162] irradiation of photons,[163,164] and application of alternating electric fields,[165] can be used to trigger intercomponent movements that alter structure and molecular properties of interlocked assemblies. In some cases, large-amplitude translational motions result in remarkable changes in molecular size, shape, color, and conductivity.[166,167] Alternatively, switchable rotaxanes and catenanes can be utilized for sensing and detection of pH and other parameters.

Sauvage *et al.* developed an electrochemically-driven pirouetting device that exploits distinct stereoelectronic requirements of copper(I) and copper(II) complexes.[168] They prepared a bistable copper rotaxane from a 2,2′-bipyridine thread and a bifunctional macrocycle that undergoes metal coordination using either its bidentate 1,10-phenanthroline unit or its tridentate tripyridyl moiety. The Cu(I)-derived rotaxane prefers coordination to the 1,10-phenanthroline moiety of the ring in a tetrahedral complex. However, oxidation to the corresponding Cu(II) complex is followed by facile ligand exchange to a five-coordinate species. Cyclic voltammetry measurements proved that the electrochemically-driven circumrotations are reversible and proceed with rate constants of 5 s^{-1} and above 500 s^{-1}, respectively, Scheme 9.19.

Numerous examples of switchable rotaxanes and catenanes have been reported by Stoddart's group. In particular, [2]catenanes exhibiting a cyclophane with two electron-deficient paraquat units and a ring bearing a tetrathiafulvalene and a 1,5-dioxynaphthalene group show reproducible circumrotation triggered by chemical and electrochemical stimuli, Scheme 9.20.[169] Both UV–Vis and NMR spectroscopy in conjunction with crystallographic analysis confirmed that the tetrathiafulvalene (TTF) core is located in close proximity to the dicationic bipyridinium portions due to strong π-π-interactions and *C-H···O* hydrogen bonding between α-bipyridinium hydrogen atoms and nearby oxygen atoms in the polyether chain. One- or two-electron oxidation of TTF to the corresponding cationic radical or dication results in spontaneous circumrotation to the translational isomer having the dioxynaphthalene portion in the cyclobis(paraquat-*p*-phenylene) cavity. Reduction of TTF to its neutral form regenerates the resting state. Reversible redox switching between the two isomers can be accomplished by electrochemical or chemical means, *e.g.*, through

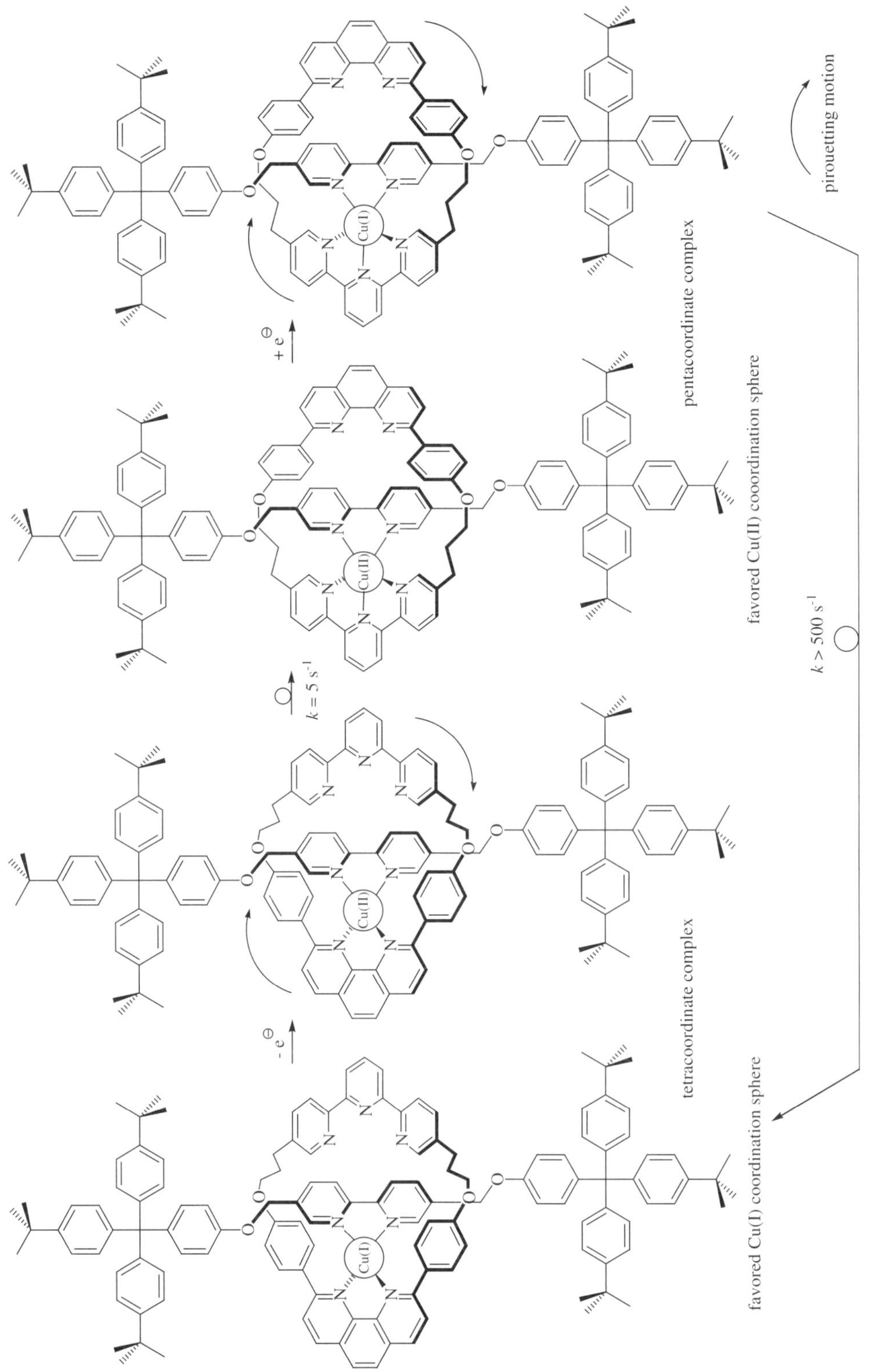

Scheme 9.19 Electrochemically-induced circumrotation of a copper rotaxane complex.

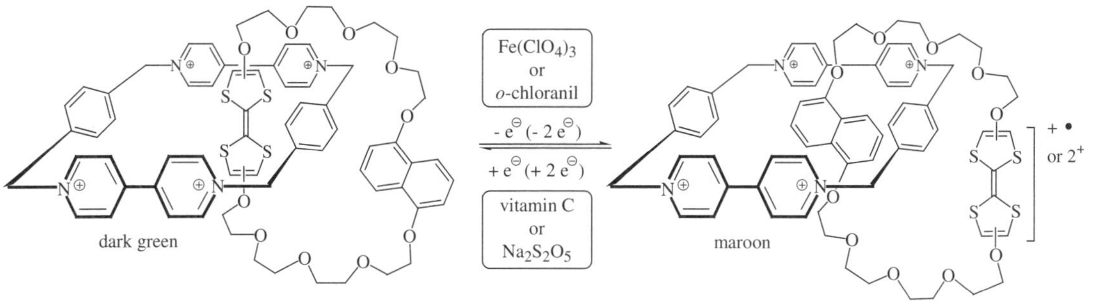

Scheme 9.20 Redox switching of a catenane.

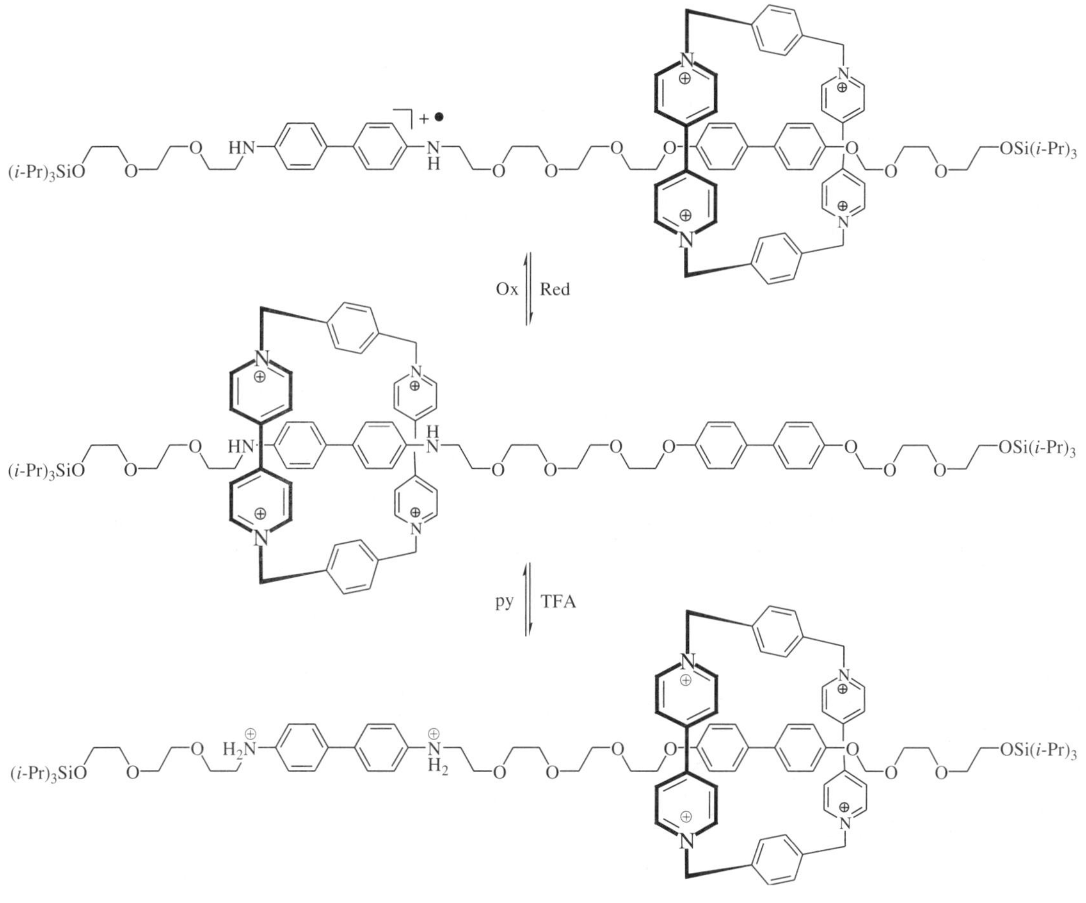

Scheme 9.21 Electrochemically and chemically induced translational isomerization of a rotaxane.

addition of *o*-chloranil and vitamin C to a solution of the [2]catenane. The oxidation and subsequent circumrotation is accompanied by a characteristic color change from dark green to maroon. A similar design is realized with rotaxanes carrying an axle with two complementary stations and a tetracationic cyclobis(paraquat-*p*-phenylene) wheel, Scheme 9.21.[170] In the reduced state of this rotaxane switch, the electron-deficient wheel experiences strong π-π-interactions with the electron-rich benzidine unit. Disruption of these intercomponent interactions by oxidation or

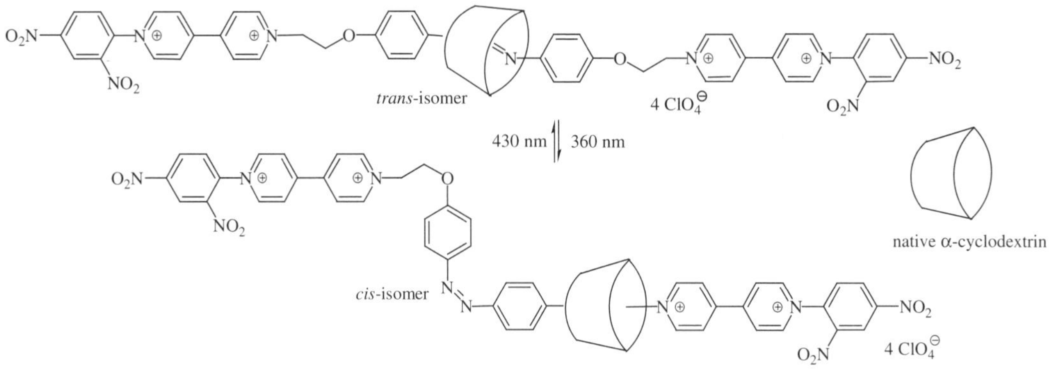

Scheme 9.22   A light-driven cyclodextrin shuttle.

protonation of the benzidine system with trifluoroacetic acid triggers a translational movement. Because of Coulomb repulsion with the benzidinium portion, the cyclophane ring glides to the biphenol moiety. Both processes are reversible and reduction or addition of pyridine or another base regenerates the original isomer.

Murakami *et al.* prepared a light-driven chiral molecular shuttle from native α-cyclodextrin and an azobenzene-derived rod, Scheme 9.22.[171] The azobenzene moiety undergoes *trans*-to-*cis*-photoisomerization upon UV light irradiation (360 nm) and the *trans*-isomer is regenerated by irradiation of visible light (430 nm). The photostationary states of the two rotaxane diastereo-isomers are obtained within 15 minutes.[xii] Nuclear Overhauser NMR spectroscopy proved that the cyclodextrin resides at the *trans*-azobenzene station but it moves into close proximity of the bipyridinium group upon formation of the *cis*-isomer. The *trans*-isomer gives a pronounced induced circular dichroism signal and CD spectroscopy has been used to demonstrate that alternating photoirradiation of UV and visible light allows reproducible switching between the two states.

The feasibility of large-amplitude translational motion in both achiral and chiral interlocked molecules introduces an opportunity for the development of rotors and motors. Leigh's group introduced a [3]catenane that shows repetitive unidirectional rotation of two small rings threaded by a large macrocycle, providing cyclic directionality and three distinct stations.[172] They first studied sequential translational movements in a related [2]catenane consisting of a small tetraamide ring and a macrocycle with three binding sites (stations) of which two can be externally manipulated to change the binding affinity to the small ring, Figure 9.15. The three stations of this technomimetic device are a secondary fumaramide *A*, a tertiary fumaramide *B* and a succinic amide ester group *C*. Each station participates in hydrogen bonding with the tetraamide ring, albeit with different affinity. The secondary (*E*)-fumaramide moiety can form the strongest hydrogen bonds with the tetraamide shuttle. Conversion of this station to the *(Z)*-isomeric state greatly diminishes intercomponent interactions and thus switches off the high affinity to the small ring. Fine-tuning of a second station with slightly less affinity was accomplished by introduction of an additional methyl group to the fumaramide group. Because steric repulsion impedes hydrogen bonding, the tertiary fumaramide in station *B* binds less strongly to the tetraamide shuttle than does station A. Finally, replacement of one amide group by an ester function and incorporation of the more flexible succinic acid spacer establishes only weak hydrogen bonding between station *C* and the amide groups of the shuttle. The ingenious design of this [2]catenane provides control of the affinity of fumaramide stations *A* and *B* to the tetraamide cycle by *cis/trans*-isomerization. In order to enable

---

[xii] Thermal isomerization has a half-life time of 13 hours at 5 °C and does not interfere with the photoswitching process.

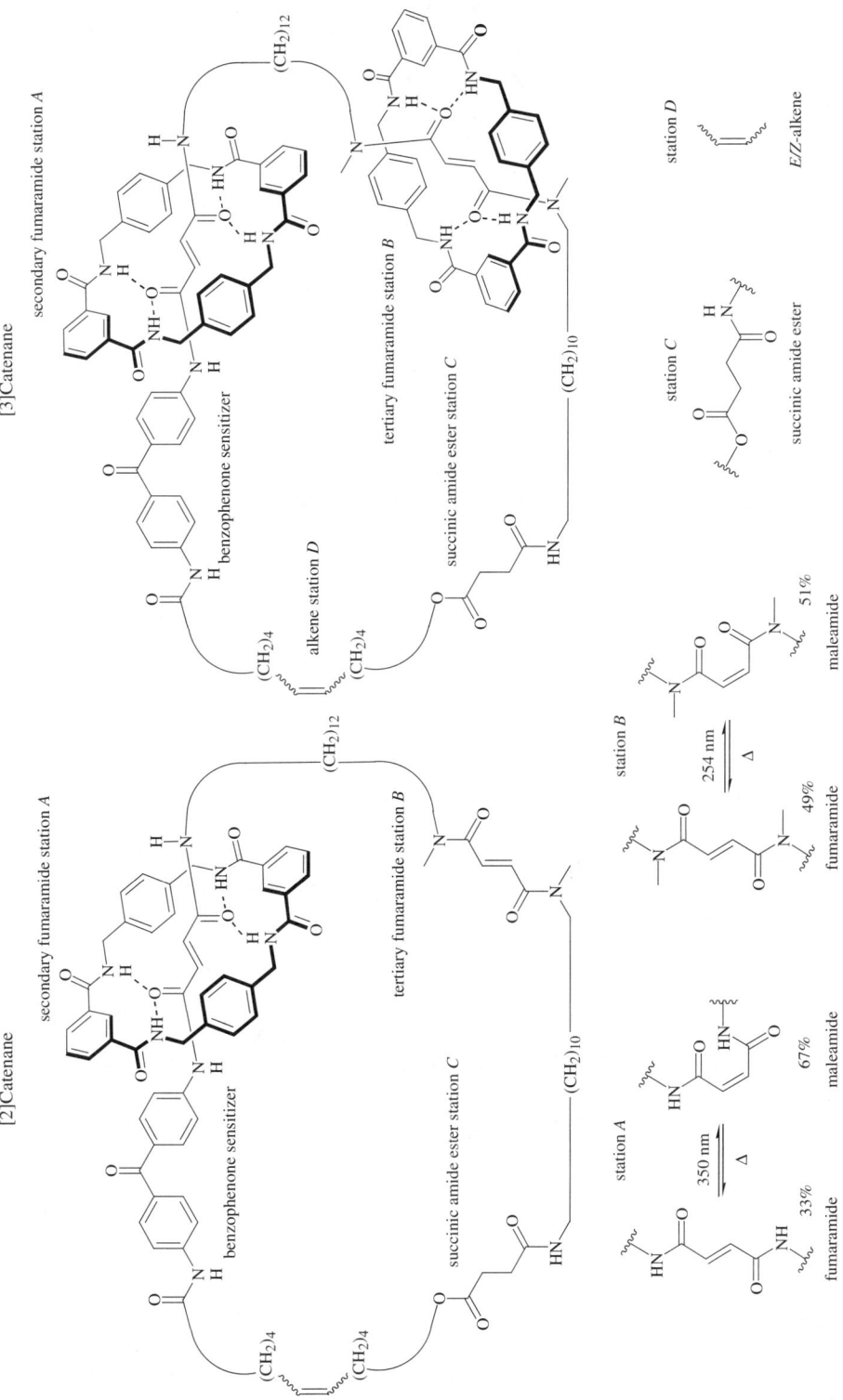

Figure 9.15 Structures of Leigh's molecular catenane rotors exhibiting stations *A*, *B*, *C*, and *D*. The yields of the (*E*)- and (*Z*)-isomers of stations *A* and *B* refer to photostationary states.

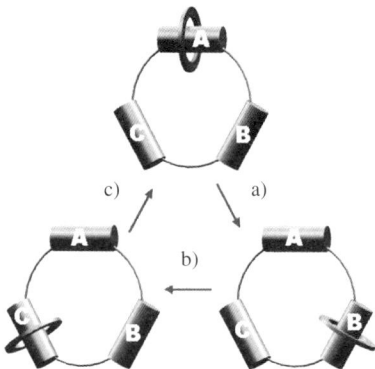

Scheme 9.23   Illustration of sequential translational movement of the tetraamide ring around the oriented
macrocycle of Leigh's [2]catenane. [Modified with permission from *Nature* **2003**, *424*,
174–179.]

selective switching of these two stations, a benzophenone unit is located in close proximity to
station A. The benzophenone serves as a photosensitizer, and allows selective isomerization of the
fumaramide group having the highest affinity to the tetraamide shuttle. According to NMR
spectroscopic analysis, the small ring resides preferentially at station A when both fumaramides
have (E)-configuration, Scheme 9.23. Irradiation at 350 nm for 5 minutes causes diastereomerization of station A with 67% yield, thus interconverting the bulk of the (E_A,E_B)-catenane to
the (Z_A,E_B)-isomer. As expected, isomerization of station A significantly reduces intercomponent
hydrogen bonding with the small ring and initiates translational movement of the latter to station
B. In the next switching step, the (Z,Z)-catenane is produced in 51% yield by irradiation at 254 nm
for 20 minutes, which is followed by gliding of the small ring to station C. Finally, thermal
isomerization at 100 °C for 24 hours regenerates the (E,E)-macrocycle in quantitative amounts and
the small ring returns to station A. The mechanically interlocked rotor is thus fueled by thermal and
photochemical energy.[xiii]

Since the translational movement in the [2]catenane is not directionally biased,[xiv] Leigh's group
developed a [3]catenane by incorporating a second tetraamide ring into the rotor to achieve
unidirectional movement around the macrocycle, Figure 9.15. The second tetraamide ring blocks
Brownian motion and allows a net shuttle movement in one direction. Completion of two
sequences of the three isomerization steps outlined for the [2]catenane results in unidirectional
360° rotation, Scheme 9.24. Selective photoisomerization of station A of the (E_A,E_B)-[3]catenane
carrying both small rings at stations A and B gives the corresponding (Z_A,E_B)-isomer. The
cis/trans-isomerization induces translational movement of one ring from station A to C. Subsequent formation of the (Z_A,Z_B)-catenane then renders hydrogen bonding with both fumaramides
unfavorable and the other ring moves from station B to station D where it experiences the least
steric repulsion. Importantly, this is only possible through a counterclockwise movement because
the ring on station C blocks the passage to station D from the other direction. In the third step,
thermal isomerization regenerates the initial (E_A,E_B)-[3]catenane, which favors movement of both
tetraamide rings to stations A and B. Overall, both shuttles have experienced a net counterclockwise motion but they accomplished only a 90° and a 270° turn, respectively. A full rotation of both

---

[xiii] Thermal isomerization at 100 °C generates fumaramide structures at both stations A and B in quantitative yields.
Alternatively, maleamide-to-fumaramide-isomerization can be accomplished at –78 °C by irradiation at 400–670 nm in the
presence of bromine.

[xiv] For example, the tetraamide ring can reach station A during the final thermal isomerization step directly from station C
(clockwise motion), or by gliding through station B (counterclockwise motion).

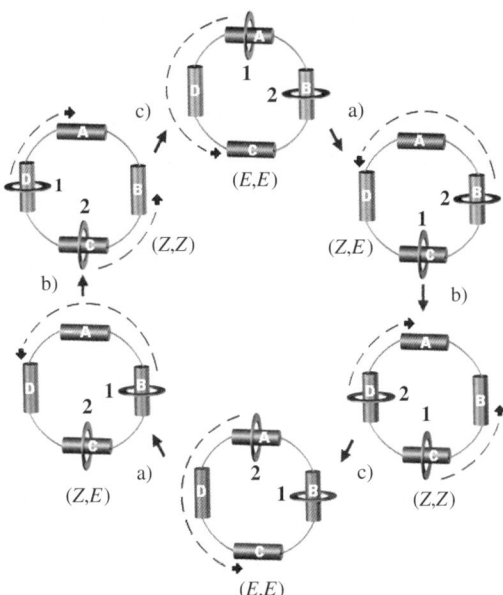

Scheme 9.24 Illustration of the unidirectional translational movement of the tetraamide shuttles around the [3]catenane. The two small rings are labeled for differentiation of individual translocations which are indicated with dashed arrows. [Modified with permission from *Nature* **2003**, *424*, 174–179.]

rings can be realized with another sequence of photochemically and thermally induced isomerization steps. Spectroscopic studies revealed that the ring translocations proceed rapidly even under cryogenic conditions. The activation energy for the translational movement of the small rings from station *C* to *B* was determined as 47.3 kJ/mol, and all other translational barriers are less than 34 kJ/mol. In contrast, circumrotation of the macrocycle is relatively slow and has an activation barrier above 96 kJ/mol. To minimize random background rotation and macrocycle circumrotation, conditions for the thermal maleamide-to-fumaramide isomerization were optimized. Addition of one equivalent of bromine and irradiation at 400–670 nm allows quantitative interconversion of the (Z_A,Z_B)-catenane to the (E_A,E_B)-isomer at −78 °C. Since the photoisomerization steps proceed within a few minutes under these conditions, undesirable slippage of the unidirectional rotor due to thermal background ring translation is negligible during operation at low temperatures.[xv] It is noteworthy that switchable catenanes and rotaxanes have been incorporated into Langmuir–Blodgett films,[173–175] self-assembled monolayers[176] and solid-state molecular switch tunnel junctions,[177,178] which underscores the potential of mechanically interlocked molecules for the development of nanoscale electronic devices and machines.[179,180]

REFERENCES

1. Frisch, H. L.; Wassermann, E. *J. Am. Chem. Soc.* **1961**, *83*, 3789-3795.
2. Wassermann, S. A.; Cozzarelli, N. R. *Proc. Natl. Acad. Sci.* **1985**, *82*, 1079-1083.
3. Griffith, J. D. *Proc. Natl. Acad. Sci.* **1985**, *82*, 3124-3128.

[xv] In contrast to the unidirectional rotors and motors discussed in Chapter 8.11, Leigh's rotor is achiral but still effectively favors translational movement into one direction. This proves that chirality is not a strict prerequisite for unidirectionality in molecular motors and machines.

4. Liang, C.; Mislow, K. *J. Am. Chem. Soc.* **1994**, *116*, 11189-11190.

5. Taylor, W. R. *Nature* **2000**, *406*, 916-919.

6. Taylor, W. R.; Lin, K. *Nature* **2003**, *421*, 25.

7. Zhou, H.-X. *J. Am. Chem. Soc.* **2003**, *125*, 9280-9281.

8. Balzani, V.; Gomez-Lopez, M.; Stoddart, J. F. *Acc. Chem. Res.* **1998**, *31*, 405-414.

9. Sauvage, J.-P. *Acc. Chem. Res.* **1998**, *31*, 611-619.

10. Balzani, V.; Credi, A.; Raymo, F. M.; Stoddart, J. F. *Angew. Chem., Int. Ed.* **2000**, *39*, 3348-3391.

11. Pease, A. R.; Jeppesen, J. O.; Stoddart, J. F.; Luo, Y.; Collier, C. P.; Heath, J. R. *Acc. Chem. Res.* **2001**, *34*, 433-444.

12. Ballardini, R.; Balzani, V.; Credi, A.; Gandolfi, M. T.; Venturi, M. *Acc. Chem. Res.* **2001**, *34*, 445-455.

13. Schalley, C. A.; Beizai, K.; Vögtle, F. *Acc. Chem. Res.* **2001**, *34*, 465-476.

14. Collin, J.-P.; Dietrich-Buchecker, C.; Gavina, P.; Jimenez-Molero, M. C.; Sauvage, J.-P. *Acc. Chem. Res.* **2001**, *34*, 477-487.

15. Luo, Y.; Collier, C. P.; Jeppesen, J. O.; Nielsen, K. A.; DeIonno, E.; Ho, G.; Perkins, J.; Tseng, H.-R.; Yamamoto, T.; Stoddart, J. F.; Heath, J. R. *ChemPhysChem* **2002**, *3*, 519-525.

16. Mandl, C. P.; König, B. *Angew. Chem., Int. Ed.* **2004**, *43*, 1622-1624.

17. Carlucci, L.; Ciani, G.; Proserpio, D. M. *Coord. Chem. Rev.* **2003**, *246*, 247-289.

18. Carlucci, L.; Ciani, G.; Proserpio, D. M. *CrystEngComm* **2003**, *5*, 269-279.

19. Loeb, S. J. *Chem. Commun.* **2005**, 1511-1518.

20. Wassermann, S. A. *J. Am. Chem. Soc.* **1960**, *82*, 4433-4434.

21. Harrison, I. T.; Harrison, S. *J. Am. Chem. Soc.* **1967**, *89*, 5723-5734.

22. Walba, D. M.; Richards, R. M.; Haltiwanger, R. C. *J. Am. Chem. Soc.* **1982**, *104*, 3219-3221.

23. Walba, D. M.; Homan, T. C.; Richards, R. M.; Haltiwanger, R. C. *New J. Chem.* **1993**, *17*, 661-681.

24. Heilbronner, E. *Tetrahedron Lett.* **1964**, *29*, 1923-1928.

25. Ajami, D.; Oeckler, O.; Simon, A.; Herges, R. *Nature* **2003**, *426*, 819-821.

26. Lehn, J.-M. *Angew. Chem., Int. Ed. Engl.* **1990**, *29*, 1304-1319.

27. Dietrich-Buchecker, C.; Sauvage, J.-P. *Chem. Rev.* **1987**, *87*, 795-810.

28. Anderson, S.; Anderson, H. L.; Sanders, J. K. M. *Acc. Chem. Res.* **1993**, *26*, 469-475.

29. Johnston, A. G.; Leigh, D. A.; Pritchard, R. J.; Deegan, M. D. *Angew. Chem., Int. Ed. Engl.* **1995**, *34*, 1212-1216.

30. Fyfe, M. C. T.; Stoddart, J. F. *Acc. Chem. Res.* **1997**, *30*, 393-401.

31. Breault, G. A.; Hunter, C. A.; Mayers, P. C. *Tetrahedron* **1999**, *55*, 5265-5293.

32. Amabilino, D. B.; Ashton, P. R.; Bravo, J. A.; Raymo, F. M.; Stoddart, J. F.; White, A. J. P.; Williams, D. J. *Eur. J. Org. Chem.* **1999**, 1295-1302.

33. Seel, C.; Vögtle, F. *Chem. Eur. J.* **2000**, *6*, 21-24.

34. Hansen, J. G.; Feeder, N.; Hamilton, D. G.; Gunter, M. J.; Becher, J.; Sanders, J. K. M. *Org. Lett.* **2000**, *2*, 449-452.

35. Ashton, P. R.; Baldoni, V.; Balzani, V.; Claessens, C. G.; Credi, A.; Hoffmann, H. D. A.; Raymo, F. M.; Stoddart, J. F.; Venturi, M.; White, A. J. P.; Williams, D. J. *Eur. J. Org. Chem.* **2000**, 1121-1130.

36. Perez-Alvarez, M.; Raymo, F. M.; Rowan, S. J.; Schiraldi, D.; Stoddart, J. F.; Wang, Z.-H.; White, A. J. P.; Williams, D. J. *Tetrahedron* **2001**, *57*, 3799-3808.

37. Gatti, F. G.; Leigh, D. A.; Nepogodiev, S. A.; Slawin, A. M. Z.; Teat, S. J.; Wong, J. K. Y. *J. Am. Chem. Soc.* **2001**, *123*, 5983-5989.

38. Hubin, T. J.; Busch, D. H. *Coord. Chem. Rev.* **2000**, *200-202*, 5-52.

39. Stoddart, J. F.; Tseng, H. R. *Proc. Natl. Acad. Sci.* **2002**, *99*, 4797-4800.

40. Dietrich-Buchecker, C.; Colasson, B.; Fujita, M.; Hori, A.; Geum, N.; Sakamoto, S.; Yamaguchi, K.; Sauvage, J.-P. *J. Am. Chem. Soc.* **2003**, *125*, 5717-5725.

41. Kaiser, G.; Jarrosson, T.; Otto, S.; Ng, Y.-F.; Bond, A. D.; Sanders, J. K. M. *Angew. Chem., Int. Ed.* **2004**, *43*, 1959-1962.

42. Schalley, C. A.; Reckien, W.; Peyerimhoff, S.; Baytekin, B.; Vögtle, F. *Chem. Eur. J.* **2004**, *10*, 4777-4789.

43. Sambrook, M. R.; Beer, P. D.; Wisner, J. A.; Paul, R. L.; Cowley, A. R. *J. Am. Chem. Soc.* **2004**, *126*, 15364-15365.

44. Loren, J. C.; Gantzel, P.; Linden, A.; Siegel, J. S. *Org. Biomol. Chem.* **2005**, *3*, 3105-3116.

45. Sauvage, J.-P. *Acc. Chem. Res.* **1990**, *23*, 319-327.

46. Mitchell, D. K.; Sauvage, J.-P. *Angew. Chem., Int. Ed. Engl.* **1988**, *27*, 930-931.

47. Chambron, J.-C.; Mitchell, D. K.; Sauvage, J.-P. *J. Am. Chem. Soc.* **1992**, *114*, 4625-4631.

48. Dietrich-Buchecker, C.; Sauvage, J.-P. *Angew. Chem., Int. Ed. Engl.* **1989**, *28*, 189-192.

49. Ashton, P. R.; Goodnow, T. T.; Kaifer, A.; Reddington, M. V.; Slawin, A. M. Z.; Spencer, M.; Stoddart, J. F.; Vicent, C.; Willimas, D. J. *Angew. Chem., Int. Ed. Engl.* **1989**, *28*, 1396-1399.

50. Anelli, P. L.; Ashton, P. R.; Ballardini, R.; Balzani, V.; Delgado, M.; Gandolfi, M. T.; Goodnow, T. T.; Kaifer, A. E.; Philp, D.; Pietraszkiewicz, M.; Prodi, L.; Reddington, M. V.; Slawin, A. M. Z.; Spencer, N.; Stoddart, J. F.; Vicent, C.; Williams, D. J. *J. Am. Chem. Soc.* **1992**, *114*, 193-218.

51. Anelli, P. L.; Spencer, N.; Stoddart, J. F. *J. Am. Chem. Soc.* **1991**, *113*, 5131-5133.

52. Ashton, P. R.; Belohradsky, M.; Philp, M.; Stoddart, J. F. *J. Chem. Soc., Chem. Commun.* **1993**, 1269-1274.

53. Ashton, P. R.; Belohradsky, M.; Philp, D.; Spencer, N.; Stoddart, J. F. *J. Chem. Soc., Chem. Commun.* **1993**, 1274-1277.

54. Amabilino, D. B.; Ashton, P. R.; Belohradsky, M.; Raymo, F. M.; Stoddart, J. F. *J. Chem. Soc., Chem. Commun.* **1995**, 747-750.

55. Amabilino, D. B.; Ashton, P. R.; Belohradsky, M.; Raymo, F. M.; Stoddart, J. F. *J. Chem. Soc., Chem. Commun.* **1995**, 751-753.

56. Ashton, P. R.; Ballardini, R.; Balzani, V.; Belohradsky, M.; Gandolfi, M. T.; Philp, D.; Prodi, L.; Raymo, F. M.; Reddington, M. V.; Spencer, N.; Stoddart, J. F.; Venturi, M.; Williams, D. J. *J. Am. Chem. Soc.* **1996**, *118*, 4931-4951.

57. Asakawa, M.; Ashton, P. R.; Ballardini, R.; Balzani, V.; Belohradsky, M.; Gandolfi, M. T.; Kocian, O.; Prodi, L.; Raymo, F. M.; Stoddart, J. F.; Venturi, M. *J. Am. Chem. Soc.* **1997**, *119*, 302-310.

58. Raymo, F. M.; Houk, K. N.; Stoddart, J. F. *J. Am. Chem. Soc.* **1998**, *120*, 9318-9322.

59. Nierengarten, J.-F.; Dietrich-Buchecker, C. O.; Sauvage, J.-P. *J. Am. Chem. Soc.* **1994**, *116*, 375-376.

60. Chambron, J.-C.; Sauvage, J.-P. *J. Am. Chem. Soc.* **1997**, *119*, 9558-9559.

61. Hori, A.; Akasaka, A.; Biradha, K.; Sakamoto, S.; Yamaguchi, K.; Fujita, M. *Angew. Chem., Int. Ed.* **2002**, *41*, 3269-3271.

62. Burchell, T. J.; Eisler, D. J.; Puddephatt, R. J. *J. Chem. Soc., Dalton Trans.* **2005**, 268-272.

63. Kaida, Y.; Okamoto, Y.; Chambron, J. C.; Mitchel, D. K.; Sauvage, J.-P. *Tetrahedron Lett.* **1993**, *28*, 1019-1022.

64. Koizumi, M.; Dietrich-Buchecker, C.; Sauvage, J.-P. *Eur. J. Org. Chem.* **2004**, 770-775.

65. McArdle, C. P.; Irwin, M. J.; Jennings, M. C.; Puddephatt, R. J. *Angew. Chem., Int. Ed.* **1999**, *38*, 3376-3378.

66. McArdle, C. P.; Vittal, J. J.; Puddephatt, R. J. *Angew. Chem., Int. Ed.* **2000**, *39*, 3819-3822.

67. McArdle, C. P.; Van, S.; Jennings, M. C.; Puddephatt, R. J. *J. Am. Chem. Soc.* **2002**, *124*, 3959-3965.

68. Ottens-Hildebrandt, S.; Schmidt, T.; Harren, J.; Vögtle, F. *Liebigs. Ann.* **1995**, 1855-1860.

69. Mohry, A.; Vögtle, F.; Nieger, M.; Hupfer, H. *Chirality* **2000**, *12*, 76-83.

70. Yamamoto, C.; Okamoto, Y.; Schmidt, T.; Jäger, R.; Vögtle, F. *J. Am. Chem. Soc.* **1997**, *119*, 10547-10548.

71. Reuter, C.; Pawlitzki, G.; Wörsdörfer, U.; Plevots, M.; Mohry, A.; Kubota, T.; Okamoto, Y.; Vögtle, F. *Eur. J. Org. Chem.* **2000**, 3059-3067.

72. Reuter, C.; Mohry, A.; Sobanski, A.; Vögtle, F. *Chem. Eur. J.* **2000**, *6*, 1674-1682.

73. Liu, Y.; Bonvallet, P. A.; Vignon, S. A.; Khan, S. I.; Stoddart, J. F. *Angew. Chem., Int. Ed.* **2005**, *44*, 3050-3055.

74. Liu, Y.; Vignon, S. A.; Zhang, X.; Bonvallet, P. A.; Khan, S. I.; Houk, K. N.; Stoddart, J. F. *J. Org. Chem.* **2005**, *70*, 9334-9344.

75. Armspach, D.; Ashton, P. R.; Moore, C. P.; Spencer, N.; Stoddart, J. F.; Wear, T. J.; Williams, D. J. *Angew. Chem., Int. Ed. Engl.* **1993**, *32*, 854-858.

76. Ashton, P. R.; Iriepa, I.; Reddington, M. V.; Spencer, N.; Slawin, A. M. Z.; Stoddart, J. F.; Williams, D. J. *Tetrahedron Lett.* **1994**, *35*, 4835-4838.

77. Asakawa, M.; Ashton, P. R.; Boyd, S. E.; Brown, C. L.; Menzer, S.; Pasini, D.; Stoddart, J. F.; Tolley, M. S.; White, A. J. P.; Williams, D. J.; Wyatt, P. G. *Chem. Eur. J.* **1997**, *3*, 463-481.

78. Ashton, P. R.; Heiss, A. M.; Pasini, D.; Raymo, F. M.; Shipway, A. N.; Stoddart, J. F.; Spencer, N. *Eur. J. Org. Chem.* **1999**, 995-1004.

79. Ashton, P. R.; Boyd, S. E.; Menzer, S.; Pasini, D.; Raymo, F. M.; Spencer, N.; Stoddart, J. F.; White, A. J. P.; Williams, D. J.; Wyatt, P. G. *Chem. Eur. J.* **1998**, *4*, 299-310.

80. Ashton, P. R.; Preece, J. A.; Stoddart, J. F.; Tolley, M. S.; White, A. J. P.; Williams, D. J. *Synthesis* **1994**, 1344-1352.

81. Tseng, H.-R.; Vignon, S. A.; Celestre, P. C.; Stoddart, J. F.; White, A. J. P.; Williams, D. J. *Chem. Eur. J.* **2003**, *9*, 543-556.

82. Vignon, S. A.; Wong, J.; Tseng, H.-R.; Stoddart, J. F. *Org. Lett.* **2004**, *6*, 1095-1098.

83. Chambron, J.-C.; Sauvage, J.-P.; Mislow, K.; De Cian, A.; Fischer, J. *Chem. Eur. J.* **2001**, *7*, 4086-4096.

84. Tachibana, Y.; Kihara, N.; Ohga, Y.; Takata, T. *Chem. Lett.* **2000**, 806-807.

85. Kameta, N.; Hiratani, K.; Nagawa, Y. *Chem. Commun.* **2004**, 466-467.

86. Smutske, I.; House, B. E.; Smithrud, D. B. *J. Org. Chem.* **2003**, *68*, 2559-2571.

87. Smutske, I.; Smithrud, D. B. *J. Org. Chem.* **2003**, *68*, 2547-2558.

88. Liu, Y.-L.; Zhao, H.-Y.; Song, H.-B. *Angew. Chem., Int. Ed.* **2003**, *42*, 3260-3263.

89. Liu, Y.-L.; Song, H.-B.; Chen, Y.; Zhao, Y.-L.; Yang, Y.-W. *Chem. Commun.* **2005**, 1702-1704.

90. Jäger, R.; Händel, M.; Harren, J.; Rissanen, K.; Vögtle, F. *Liebigs Ann.* **1996**, 1201-1207.

91. Prelog, V.; Gerlach, H. *Helv. Chim. Acta* **1964**, *47*, 2288-2294.

92. Gerlach, H.; Owtschinnikow, J. A.; Prelog, V. *Helv. Chim. Acta* **1964**, *47*, 2294-2302.

93. Mislow, K. *Chimia* **1986**, *40*, 395-402.

94. Chorev, M.; Goodman, M. *Acc. Chem. Res.* **1992**, *25*, 266-272.

95. Eliel, E. L.; Wilen, S. H. *Stereochemistry of Organic Compounds*, Wiley, New York, 1994, pp. 1176-1181.

96. Schmieder, R.; Hübner, G.; Seel, C.; Vögtle, F. *Angew. Chem., Int. Ed.* **1999**, *38*, 3528-3530.

97. Reuter, C.; Pawlitzki, G.; Wörsdörfer, U.; Plevots, M.; Mohry, A.; Kubota, T.; Okamoto, Y.; Vögtle, F. *Eur. J. Org. Chem.* **2000**, 3059-3067.

98. Reuter, C.; Mohry, A.; Sobanski, A.; Vögtle, F. *Chem. Eur. J.* **2000**, *6*, 1674-1682.

99. Vögtle, F.; Safarowsky, O.; Heim, C.; Affeld, A.; Braun, O.; Mohry, A. *Pure Appl. Chem.* **1999**, *71*, 247-251.

100. Reuter, C.; Schmieder, R.; Vögtle, F. *Pure Appl. Chem.* **2000**, *72*, 2233-2241.

101. Reuter, C.; Wienand, W.; Schmuck, C.; Vögtle, F. *Chem. Eur. J.* **2001**, *7*, 1728-1733.

102. Schmidt, T.; Schmieder, R.; Müller, W. M.; Kiupel, B.; Vögtle, F. *Eur. J. Org. Chem.* **1998**, 2003-2007.

103. Lukin, O.; Recker, J. Böhmer, A.; Müller, W. M.; Kubota, T.; Okamoto, Y.; Nieger, M.; Fröhlich, R.; Vögtle, F. *Angew. Chem., Int. Ed.* **2003**, *42*, 442-445.

104. Lukin, O.; Kubota, T.; Okamoto, Y.; Schelhase, F.; Yoneva, A.; Müller, W. M.; Müller, U.; Vögtle, F. *Angew. Chem., Int. Ed.* **2003**, *42*, 4542-4545.

105. Ashton, P. R.; Philp, D.; Reddington, M. V.; Slawin, A. M. Z.; Stoddart, J. F.; Williams, D. *J. Chem. Soc., Chem. Commun.* **1991**, 1680-1683.

106. Ashton, P. R.; Philp, D.; Spencer, N.; Stoddart, J. F.; Williams, D. *J. Chem. Soc., Chem. Commun.* **1994**, 181-184.

107. Anelli, P. L.; Asakawa, M.; Ashton, P. R.; Bissell, R. A.; Clavier, G.; Gorski, G.; Kaifer, A. E.; Langford, S. J.; Mattersteig, G.; Menzer, S.; Philp, S.; Slawin, A. M. Z.; Spencer, N.; Stoddart, J. F.; Tolley, M. S.; Williams, D. J. *J. Chem. Eur. J.* **1997**, *3*, 1113-1135.

108. Ashton, P. R.; Everitt, S. R. L.; Gómez-López, M.; Jayaraman, N.; Stoddart, J. F. *Tetrahedron Lett.* **1997**, *38*, 5691-5694.

109. Asakawa, M.; Ashton, P. R.; Hayes, W.; Janssen, H. M.; Meijer, E. W.; Menzer, S.; Pasini, D.; Stoddart, J. F.; White, A. J. P.; Williams, D. J. *J. Am. Chem. Soc.* **1998**, *120*, 920-931.

110. Asakawa, M.; Janssen, H. M.; Meijer, E. W.; Pasini, D.; Stoddart, J. F.; White, A. J. P.; Williams, D. J. *Eur. J. Org. Chem.* **1998**, 983-986.

111. Fuchs, B.; Nelson, A.; Star, A.; Stoddart, J. F.; Vidal, S. *Angew. Chem., Int. Ed.* **2003**, *42*, 4220-4224.

112. Jeppesen, J. O.; Vignon, S. A.; Stoddart, J. F. *Chem. Eur. J.* **2003**, *9*, 4611-4625.

113. Laursen, B. W.; Nygaard, S.; Jeppesen, J. O.; Stoddart, J. F. *Org. Lett.* **2004**, *6*, 4167-4170.

114. Ashton, P. R.; Bravo, J. A.; Raymo, F. M.; Stoddart, J. F.; White, A. J. P.; Williams, D. *Eur. J. Org. Chem.* **1999**, 899-908.

115. Chiu, S. H.; Elizarov, A. M.; Glink, P. T.; Stoddart, J. F. *Org. Lett.* **2002**, *4*, 3561-3564.

116. Asakawa, M.; Brancato, G.; Fanti, M.; Leigh, D. A.; Shimizu, T.; Slawin, A. M. Z.; Wong, J. K. Y.; Zerbetto, F.; Zhang, S. *J. Am. Chem. Soc.* **2002**, *124*, 2939-2950.

117. Bottari, G.; Leigh, D. A.; Pérez, E. M. *J. Am. Chem. Soc.* **2003**, *125*, 13360-13361.

118. Dietrich-Buchecker, C. O.; Sauvage, J.-P. *Angew. Chem., Int. Ed. Engl.* **1989**, *28*, 189-192.

119. Chambron, J. C.; Dietrich-Buchecker, C.; Rapenne, G.; Sauvage, J.-P. *Chirality* **1998**, *19*, 125-133.

120. Dietrich-Buchecker, C. O.; Nierengarten, J. F.; Sauvage, J.-P.; Armaroli, N.; Balzani, V.; de Cola, L. *J. Am. Chem. Soc.* **1993**, *115*, 11234-11237.

121. Carina, R. F.; Dietrich-Buchecker, C. O.; Sauvage, J. P. *J. Am. Chem. Soc.* **1996**, *118*, 9110-9116.

122. Dietrich-Buchecker, C. O.; Rapenne, G.; Sauvage, J.-P. *Chem. Commun.* **1997**, 2053-2054.

123. Rapenne, G.; Dietrich-Buchecker, C.; Sauvage, J.-P. *J. Am. Chem. Soc.* **1999**, *121*, 994-1001.

124. Ashton, P. R.; Matthews, O. A.; Menzer, S.; Raymo, F. M.; Spencer, N.; Stoddart, J. F.; Williams, D. J. *Liebigs Ann.* **1997**, 2485-2494.

125. Adams, H.; Ashworth, E.; Breault, G. A.; Guo, J.; Hunter, C. A. Mayers, P. C. *Nature* **2001**, *411*, 763.

126. Safarowsky, O.; Nieger, M.; Fröhlich, R.; Vögtle, F. *Angew. Chem., Int. Ed.* **2000**, *39*, 1616-1618.

127. Vögtle, F.; Hünten, A.; Vogel, E.; Buschbeck, S.; Safarowsky, O.; Recker, J.; Parham, A.-H.; Knott, M.; Müller, W. M.; Müller, U.; Okamoto, Y.; Kubota, T.; Lindner, W.; Francotte, E.; Grimme, S. *Angew. Chem., Int. Ed.* **2001**, *40*, 2468-2471.

128. Recker, J.; Müller, W. M.; Müller, U.; Kubota, T.; Okamoto, Y.; Nieger, M.; Vögtle, F. *Chem. Eur. J.* **2002**, *8*, 4434-4442.

129. Lukin, O.; Recker, J.; Böhmer, A.; Müller, W. M.; Okamoto, Y.; Nieger, M.; Vögtle, F. *Angew. Chem., Int. Ed.* **2003**, *42*, 442-445.

130. Lukin, O.; Müller, W. M.; Müller, U.; Kaufmann, A.; Schmidt, C.; Leszcynski, J.; Vögtle, F. *Chem. Eur. J.* **2003**, *9*, 3507-3517.

131. Lukin, O.; Kubota, T.; Okamoto, Y.; Kaufmann, A.; Vögtle, F. *Chem. Eur. J.* **2004**, *10*, 2804-2810.

132. Meyer, M.; Albrecht-Gary, A.-M.; Dietrich-Buchecker, C. O.; Sauvage, J. P. *J. Am. Chem. Soc.* **1997**, *119*, 4599-4607.

133. Dietrich-Buchecker, C. O.; Sauvage, J. P.; Armaroli, N.; Ceroni, P.; Balzani, V. *Angew. Chem., Int. Ed. Engl.* **1996**, *35*, 1119-1121.

134. Lukin, O.; Vögtle, F. *Angew. Chem., Int. Ed.* **2005**, *44*, 1456-1477.

135. Siegel, J. S. *Science* **2004**, *304*, 1256-1258.

136. Loren, J. C.; Yoshizawa, M.; Haldimann, R. F.; Linden, A.; Siegel, J. S. *Angew. Chem., Int. Ed.* **2003**, *42*, 5702-5705.

137. Chichak, K. S.; Cantrill, S. J.; Pease, A. R.; Chiu, S.-H.; Cave, G. W. V.; Atwood, J. L.; Stoddart, J. F. *Science* **2004**, *304*, 1308-1312.

138. Cantrill, S. J.; Chichak, K. S.; Peters, A. J.; Stoddart, J. F. *Acc. Chem. Res.* **2005**, *38*, 1-9.

139. Chichak, K. S.; Cantrill, S. J.; Stoddart, J. F. *Chem. Commun.* **2005**, 3391-3393.

140. Peters, A. J.; Chichak, K. S.; Cantrill, S. J.; Stoddart, J. F. *Chem. Commun.* **2005**, 3394-3396.

141. Mao, C.; Sun, W.; Seeman, N. C. *Science* **1997**, *386*, 137-138.

142. Badjic, J. D.; Balzani, V.; Credi, A.; Silvi, S.; Stoddart, J. F. *Science* **2004**, *303*, 1845-1849.

143. Rapenne, G. *Org. Biomol. Chem.* **2005**, *3*, 1165-1169.

144. Martinez-Diaz, M.-V.; Spencer, N.; Stoddart, J. F. *Angew. Chem., Int. Ed. Engl.* **1997**, *36*, 1904-1907.

145. Ashton, P. R.; Ballardini, R.; Balzani, V.; Baxter, V.; Credi, A.; Fyfe, M. C. T.; Gandolfi, M. T.; Gomez-Lopez, M.; Martinez-Diaz, M.-V.; Piersanti, A.; Spencer, N.; Stoddart, J. F.; Venturi, M.; White, A. J. P.; Williams, D. J. *J. Am. Chem. Soc.* **1998**, *120*, 11932-11942.

146. Badjic, J. D.; Balzani, V.; Credi, A.; Silvi, S.; Stoddart, J. F. *Science* **2004**, *303*, 1845-1849.

147. Tseng, H.-R.; Vignon, S. A.; Stoddart, J. F. *Angew. Chem., Int. Ed.* **2003**, *42*, 1491-1495.

148. Anelli, P. L.; Spencer, N.; Stoddart, J. F. *J. Am. Chem. Soc.* **1991**, *113*, 5131-5133.

149. Gatti, F. G.; Leon, S.; Wong, J. K. Y.; Bottari, G.; Allieri, A.; Morales, M. A. F.; Teat, S. J.; Frochet, C.; Leigh, D. A.; Brouwer, A. M.; Zerbetto, F. *Proc. Natl. Acad. Sci.* **2003**, *100*, 10-14.

150. Ashton, P. R.; Ballardini, R.; Balzani, V.; Credi, A.; Dress, R.; Ishow, E.; Kocian, O.; Preece, J. A.; Spencer, N.; Stoddart, J. F.; Venturi, M.; Wenger, S. *Chem. Eur. J.* **2000**, *6*, 3558-3574.

151. Poleschak, I.; Kern, J.-M.; Sauvage, J.-P. *Chem. Commun.* **2004**, 474-476.

152. Cardenas, D. J.; Livoreil, A.; Sauvage, J.-P. *J. Am. Chem. Soc.* **1996**, *118*, 11980-11981.

153. Balzani, V.; Credi, A.; Mattersteig, G.; Matthews, O. A.; Raymo, F. M.; Stoddart, J. F.; Venturi, M.; White, A. J. P.; Williams, D. J. *J. Org. Chem.* **2000**, *65*, 1924-1936.

154. Balzani, V.; Credi, A.; Raymo, F. M.; Stoddart, J. F. *Angew. Chem., Int. Ed.* **2000**, *39*, 3348-3391.

155. Lee, J. W.; Kim, K.; Kim, K. *Chem. Commun.* **2001**, 1042-1043.

156. Elizarov, A. M.; Chiu, S.-H.; Stoddart, J. F. *J. Org. Chem.* **2002**, *67*, 9175-9181.

157. Armaroli, N.; Balzani, V.; Collin, J.-P.; Gavina, P.; Sauvage, J.-P.; Ventura, B. *J. Am. Chem. Soc.* **1999**, *121*, 4397-4408.

158. Korybut-Daszkiewicz, B.; Wieckowska, A.; Bilewicz, R.; Domagala, S.; Wozniak, K. *Angew. Chem., Int. Ed.* **2004**, *43*, 1668-1672.

159. Tseng, H.-R.; Vignon, S. A.; Celestre, P. C.; Perkins, J.; Jeppesen, J. O.; Di Fabio, A.; Ballardini, R.; Gandolfi, M. T.; Venturi, M.; Balzani, V.; Stoddart, J. F. *Chem. Eur. J.* **2004**, *10*, 155-172.

160. Lane, A. S.; Leigh, D. A.; Murphy, A. *J. Am. Chem. Soc.* **1997**, *119*, 11092-11093.
161. Balzani, V.; Credi, A.; Langford, S. J.; Raymo, F. M.; Stoddart, J. F.; Venturi, M. *J. Am. Chem. Soc.* **2000**, *122*, 3542-3543.
162. Vignon, S. A.; Jarrosson, T.; Iijima, T.; Tseng, H.-R.; Sanders, J. K. M.; Stoddart, J. F. *J. Am. Chem. Soc.* **2004**, *126*, 9884-9885.
163. Brouwer, A. M.; Frochot, C.; Gatti, F. G.; Leigh, D. A.; Mottier, L.; Paolucci, F.; Roffia, S.; Wurpel, G. W. H. *Science* **2001**, *291*, 2124-2128.
164. Perez, E. M.; Dryden, D. T. F.; Leigh, D. A.; Teobaldi, G.; Zerbetto, F. *J. Am. Chem. Soc.* **2004**, *126*, 12210-12211.
165. Bermudez, V.; Capron, N.; Gase, T.; Gatti, F. G.; Kajzar, F.; Leigh, D. A.; Zerbetto, F.; Zhang, S. *Nature* **2000**, *406*, 608-611.
166. Jiminez-Molero, M. C.; Dietrich-Buchecker, C.; Sauvage, J. P. *Chem. Eur. J.* **2002**, *8*, 1456-1466.
167. Collier, C. P.; Wong, E. W.; Belohradsky, M.; Raymo, F.; Stoddart, J. F.; Kuekes, P. J.; Williams, R. S.; Heath, J. R. *Science* **1999**, *285*, 391-394.
168. Sauvage, J.-P. *Chem. Commun.* **2005**, 1507-1510.
169. Asakawa, M.; Ashton, P. R.; Balzani, V.; Credi, A.; Hamers, C.; Mattersteig, G.; Montalti, M.; Shipway, A. N.; Spencer, N.; Stoddart, J. F.; Tolley, M. S.; Venturi, M.; White, A. J. P.; Williams, D. J. *Angew. Chem., Int. Ed.* **1998**, *37*, 333-337.
170. Bissell, R. A.; Córdova, E.; Kaifer, A. E.; Stoddart, J. F. *Nature* **1994**, *369*, 133-137.
171. Murakami, H.; Kawabuchi, A.; Kotoo, K.; Kunitake, M.; Nakashima, N. *J. Am. Chem. Soc.* **1997**, *119*, 7605-7606.
172. Leigh, D. A.; Wong, J. K. Y.; Dehez, F.; Zerbetto, F. *Nature* **2003**, *424*, 174-179.
173. Asakawa, M.; Higuchi, M.; Mattersteig, G.; Nakamura, T.; Pease, A. R.; Raymo, F. M.; Shimizu, T.; Stoddart, J. F. *Adv. Mater.* **2000**, *12*, 1099-1102.
174. Brown, C. L.; Jonas, U.; Preece, J. A.; Ringsdorf, H.; Seitz, M.; Stoddart, J. F. *Langmuir* **2000**, *16*, 1924-1930.
175. Lee, I. C.; Frank, C. W.; Yamamoto, T.; Tseng, H.-R.; Flood, A. H.; Stoddart, J. F. *Langmuir* **2004**, *20*, 5809-5828.
176. Tseng, H.-R.; Wu, D.; Fang, N. X.; Zhang, X.; Stoddart, J. F. *ChemPhysChem* **2004**, *5*, 111-116.
177. Collier, C. P.; Mattersteig, G.; Wong, E. W.; Luo, Y.; Beverly, K.; Sampaio, J.; Raymo, F. M.; Stoddart, J. F.; Heath, J. R. *Science* **2000**, *289*, 1172-1175.
178. Collier, C. P.; Jeppesen, J. O.; Luo, Y.; Perkins, J.; Wong, E. W.; Heath, J. R.; Stoddart, J. F. *J. Am. Chem. Soc.* **2001**, *123*, 12632-12641.
179. Tseng, H.-S.; Vignon, S. A.; Stoddart, J. F. *Angew. Chem., Int. Ed.* **2003**, *42*, 1491-1495.
180. Steuermann, D. W.; Tseng, H.-R.; Peters, A. J.; Flood, A. H.; Jeppesen, J. O.; Nielsen, K. A.; Stoddart, J. F.; Heath, J. R. *Angew. Chem., Int. Ed.* **2004**, *43*, 6486-6491.

Stereochemical Definitions and Terms

Absolute configuration

The spatial arrangement of substituents around an element of chirality and the corresponding stereochemical description, for example R, S, P, M. Absolute configuration is reflection variant.

Achiral

A molecule is achiral if it is superimposable on its mirror image.

Allylic strain

Allylic [1,2]-strain ($A^{1,2}$) and allylic [1,3]-strain ($A^{1,3}$) originate from van der Waals repulsion between vinylic and allylic substituents.

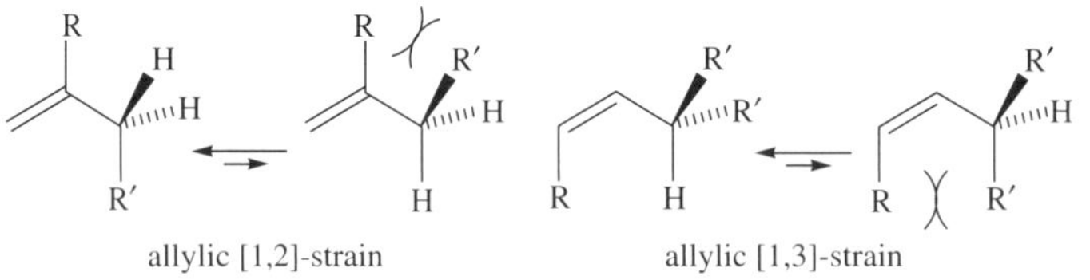

allylic [1,2]-strain   allylic [1,3]-strain

α (alpha), β (beta)

Stereodescriptors used in carbohydrate and steroid nomenclature to describe relative configuration. In carbohydrates, the configuration at the anomeric carbon is often related to the reference atom which also defines D or L configuration. In the Fischer projection, an α-anomer has the hydroxyl group attached to C-1 on the same side as the hydroxyl group that is involved in acetal or ketal formation. For example, in α-D-glucopyranose the reference atom is C-5 and the hydroxyl group at C-1 is on the same side as the hydroxyl group at C-5. The two groups are on opposite sides in β-D-glucopyranose.

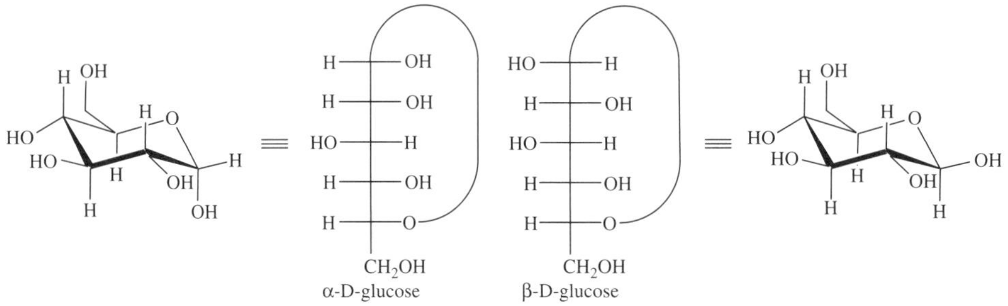

α-D-glucose   β-D-glucose

In steroids, β-substituents are on the same side of the fused ring plane as the 3-hydroxyl group in cholesterol, whereas α-substituents are on the opposite side.

Alternating symmetry axis (S_n)
See rotation-reflection axis.

Angle strain
Strain due to deviation in bond angles from "normal" values such as the tetrahedral angle of 109° 28'. For example, epoxides can not accommodate a perfect tetrahedral bond angle and possess a ring strain of approximately 105 kJ/mol. Angle strain is synonymous with Baeyer strain.

Anomeric effect
Describes the thermodynamic preference of a polar group attached to the anomeric carbon of a pyranosyl derivative for the axial position.

α-anomer (more stable isomer)   β-anomer

The anomeric effect observed with carbohydrates is a special case of a general preference of heteroatoms for synclinal (gauche) conformations in a system $C–X–C–Y$, where X and Y are heteroatoms with nonbonding electron pairs such as oxygen, nitrogen or a halide. The anomeric effect has been attributed to resonance involving a nonbonding orbital of atom X and an antibonding σ*-orbital of the adjacent $C–Y$ bond.

synclinal (gauche)   antiperiplanar
(more stable)   (less stable)

Similarly, electronegative substituents (oxygen, nitrogen, halides) occupying an allylic position in cyclohexenes prefer a pseudoaxial orientation. In accordance to the anomeric effect, the preference for the pseudoaxial position has been explained by stabilizing resonance between the π-system of the double bond and an antibonding orbital of the allylic σ-bond.

Y=Cl, Br, NHR, OH, OR

Anomers
Diastereomeric glycosides or related compounds with different configuration only at the anomeric carbon, *e.g.*, at *C*-1 in an aldose and *C*-2 in a 2-ketose.

Antarafacial
Describes the stereochemical course of pericyclic reactions (cycloadditions, electrocyclic reactions, sigmatropic rearrangements, and cheletropic reactions). In an antarafacial process, bonds are formed and cleaved on opposite sides of a π-system. See also suprafacial.

Anti
Describes the relative configuration in a molecule with two or more chiral centers. If the molecule is shown in a planar zigzag projection and the substituents are on opposite sides, the orientation is called *anti*. If they are on the same side, the relative configuration is *syn*.

Anticlinal (ac)
See torsion angle.

Antiperiplanar (ap)
See torsion angle.

Apical
See Berry pseudorotation.

Asymmetric
Describes the absence of symmetry elements other than identity E and C_1. See dissymmetric.

Asymmetric carbon atom
Van't Hoff's traditional definition of a carbon atom that bears four different substituents or groups. See chiral center.

Asymmetric induction
Describes the preferential formation of one enantiomer or diastereoisomer over another as a result of the influence of a chiral element present in the substrate, reagent, catalyst or solvent.

Asymmetric synthesis
An asymmetric reaction selectively introduces at least one chiral element to a substrate in such a way that one chiral compound is formed in excess of other stereoisomeric products.

Asymmetric transformation of the first and second kind
The thermodynamically controlled conversion of a 1:1 mixture of enantiomers or diastereo-isomers into a pure enantiomer or diastereoisomer or into a mixture in which one stereoisomer is enriched. A conversion that proceeds without separation of stereoisomers is called an asymmetric transformation of the first kind. If it coincides with separation of stereoisomers, for example crystallization, it is considered an asymmetric transformation of the second kind. See also stereoconvergent.

Atropisomer
Kuhn introduced the term atropisomer for stereosiomers that result from restricted rotation about a single bond and that can be separated at room temperature. See Chapter 3.3.

Axial, equatorial
Relative positions of substituents in a six-membered ring are described as axial (a) or equatorial (e) according to whether the bonds are approximately parallel to the C_3-axis or parallel to two of the six ring bonds. The corresponding orientations at the allylic carbons in unsaturated six-membered rings are pseudoaxial (a') and pseudoequatorial (e').

Axial chirality
Chirality resulting from a nonplanar arrangement of four groups about an axis. The absolute configuration is specified by stereodescriptors R_a and S_a or by P and M. Common examples are allenes, biaryls and cyclohexylidenes. See Chapter 2.1.

Baeyer strain
See angle strain.

Berry pseudorotation
Mechanism for the interconversion of bipyramidal structures via a tetragonal pyramidal intermediate or transition state. While one equatorial ligand is not affected (ligand 3) the positions of the other two equatorial ligands 4 and 5 are exchanged with apical ligands 1 and 2.

Boat
The C_{2v}-conformation of cyclohexane and similar six-membered rings. See chair and half-chair.

Boltzmann equation
Describes the ratio of the population of two states A and B with different Gibbs energy, G_A and G_B, at equilibrium.

$$\frac{n_A}{n_B} = e^{-\Delta G/RT}$$

where R = universal gas constant ($8.3144 \, \mathrm{J\,K^{-1}\,mol^{-1}}$); T = temperature [K]; n_A and n_B describe the population of states A and B; and $\Delta G = G_A - G_B$.

Bowsprit, flagpole

Positions on the two carbon atoms that are not in the same plane as the other four in the boat form of cyclohexane and similar six-membered rings. Exocyclic bonds to these two atoms are called bowsprit, the other two are called flagpole.

b = bowsprit
f = flagpole

Bürgi–Dunitz trajectory

Preferred angle of approach (109°) of a nucleophile attacking a carbonyl group.

CDA

Chiral derivatizing agent. An enantiopure reagent that is used to convert enantiomers to diastereoisomers for chromatographic or spectroscopic resolution.

Center of symmetry (i)

A molecule or object has a center of symmetry or inversion center if point reflection in the origin affords a superimposable image.

Chair, boat, twist-boat

The chair conformation of cyclohexanes and other six-membered rings exhibits carbon atoms 1, 2, 4, and 5 in one plane while carbon atoms 3 and 6 are on opposite sides of the plane (symmetry group D_{3d}). In the boat conformation, carbons 3 and 6 are on the same side of the plane (symmetry group C_{2v}). For cyclohexane and most analogs, the chair form is the most stable conformation. Interconversion of boat conformers proceeds via a twist-boat conformation also known as skew form (symmetry group D_2). See also axial and half-chair.

chair   boat   twist-boat

Chiral

A compound is chiral if it is not superimposable on its mirror image. This constitutes a necessary and sufficient condition for chirality that may be fulfilled by the presence of an element of chirality or directionality in interlocked components of topological isomers.

Chiral axis

See axial chirality.

Chiral center

A tetrahedral (*Cabcd*) or trigonal pyramidal (*Xabc*, $X = N$, P, S, etc.) arrangement which is reflection variant. The term chirality center is an extension of the concept of van't Hoff's asymmetric carbon atom. See Chapter 2.1.

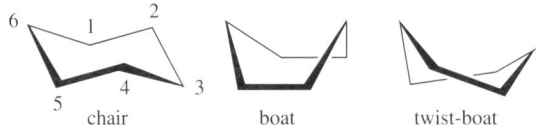

Chiral element
Refers to a chiral center, axis, plane or helix. Synonymous with element of chirality.

Chiral plane
Originates from a planar moiety connected to a group that can not lie in the same plane and thus destroys molecular symmetry. Examples are paracyclophanes and metallocenes. See Chapter 2.1.

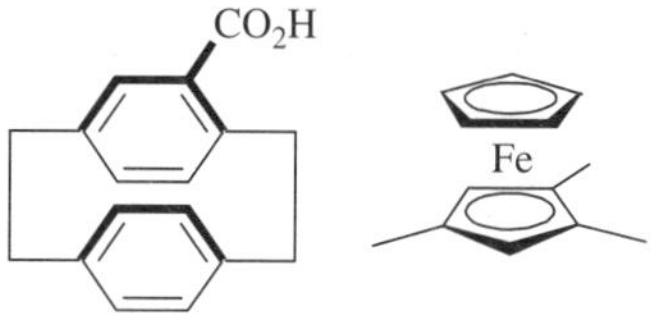

Chirality
The geometric property of an object or molecule that is not superimposable on its mirror image. A chiral object is devoid of symmetry elements of the second kind (mirror plane, $\sigma = S_1$, center of inversion, $i = S_2$, and rotation–reflection axis, S_n). Chirality is a molecular property but it is common practice to refer to a chiral center, axis, etc.

Chiroptical
Refers to optical properties of chiral compounds including optical rotation, optical rotatory dispersion (ORD), circular dichroism (CD), vibrational CD, and circular polarization of luminescence (CPL).

Chirotopic
Describes a molecular atom or group that is located in a chiral environment. The whole molecule must not be chiral. Chirotopic atoms or groups may be enantiotopic or diastereotopic. If the atom or group lies in a symmetry plane, in an inversion center or at the core of a rotation–reflection axis, it is achirotopic. See stereogenic.

CIP rules
Short for Cahn–Ingold–Prelog system. See Chapter 2.1.

Cis, trans
Descriptors for the relative configuration of two substituents attached to a double bond or ring structure. Two groups afford *cis*-configuration if they are located on the same side of the plane. The corresponding *trans*-isomer has the two substituents on opposite sides. For alkenes, the *cis/trans*-nomenclature has largely been replaced by the E/Z-convention which is based on CIP rules.

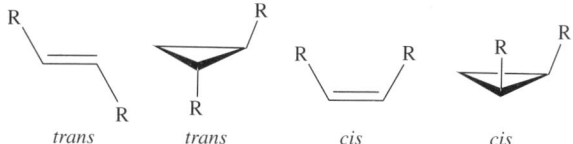

trans   *trans*   *cis*   *cis*

Cisoid, transoid
Obsolescent nomenclature. See *s-cis, s-trans*.

Configuration
Describes the spatial arrangement of atoms in a molecule and the corresponding stereoisomerism. Configurational isomers have bond angles of different amplitude or sign. Stereosiomers that are not

configurational isomers are called conformational isomers. See also absolute configuration, relative configuration, and Chapter 2.

Conformation
Describes the spatial arrangement of atoms in a molecule that can afford different stereoisomers varying by torsion angles (amplitude or sign). See configuration and Chapter 2.

Conformers, conformational isomers
Stereoisomers with identical configuration but different torsion angles and distinct potential energy minima.

Conglomerate
An equimolar mechanical mixture of enantiopure crystals. The process of conglomerate formation by crystallization of a racemate is called spontaneous resolution. Sorting of a conglomerate yields pure enantiomers.

Conrotatory
Descriptor for the relative movement of bonds or groups within one molecule. Used to describe the stereochemistry of pericyclic reactions, such as cycloadditions and electrocyclic reactions, and to illustrate the mechanism of ring flipping processes. Interconversion of molecular propellers may proceed via conrotatory (in the same direction, *i.e.*, all clockwise or all counterclockwise) or disrotatory (in opposite directions, *e.g.*, one clockwise and one counterclockwise) correlated ring rotations.

Constitution
The description of the connectivity of atoms in a molecule, disregarding spatial orientation.

Constitutional isomers
Isomers of different constitution, *e.g.*, CH_3OCH_3 and CH_3CH_2OH.

Cram's rule
A model for the prediction of the stereochemical outcome of diastereofacial nucleophilic additions to aldehydes, ketones and derivatives thereof bearing a chiral center in α-position. See Chapter 2.2.

CSA
Chiral solvating agent. A nonracemic chiral additive employed in NMR spectroscopy to induce diastereotopic and hence anisochronous signals.

CSP
Chiral stationary phase. Used for chromatographic separation of enantiomers.

CSR
Chiral shift reagent. A paramagnetic chiral lanthanide complex that renders enantiotopic signals diastereotopic for NMR analysis.

Cyclostereoisomerism
Cyclostereoisomerism is observed when constitutionally identical building blocks that possess a stereogenic element afford a directional arrangement within a ring system. The stereogenic elements may be part of the ring system which can be a cyclic polypeptide such as cycloenantiomeric

cyclohexaalanines. Note that the two molecules of cyclohexaalanine differ in the sense of the peptide bonds, *i.e.*, $(R \rightarrow S \rightarrow S \rightarrow R \rightarrow S \rightarrow S)$ versus $(S \leftarrow R \leftarrow R \leftarrow S \leftarrow R \leftarrow R)$. Alternatively, the stereogenic elements may reside outside the ring as in 1,2-bis(bromochloromethyl)-3,4,5,6-tetra-isopropylbenzene exhibiting static gearing and directional orientation of isopropyl and bromo-chloromethyl groups.

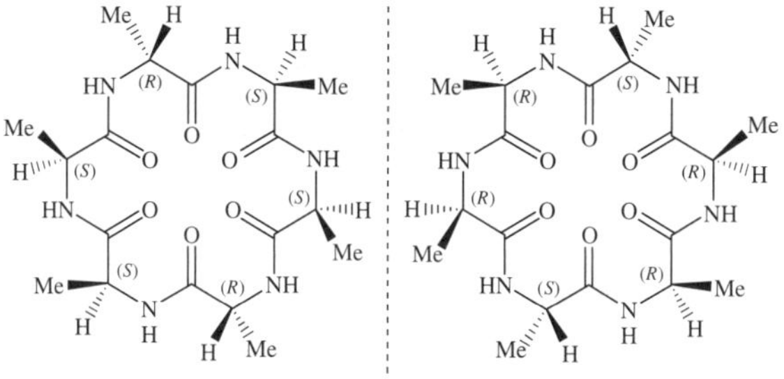

cycloenantiomeric cyclohexaalanines

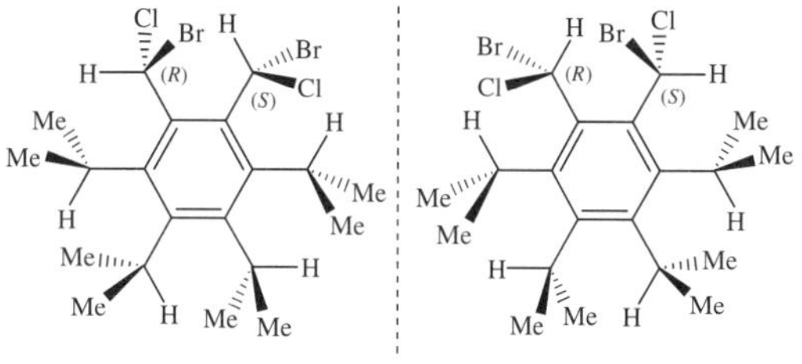

cycloenantiomeric
1,2-bis(bromochloromethyl)-3,4,5,6-tetraisopropylbenzene

Introduction of directionality into both axle and wheel of a rotaxane generates cycloenantiomerism since the wheel can either afford a clockwise or a counterclockwise orientation with respect to the axle. See topological chirality.

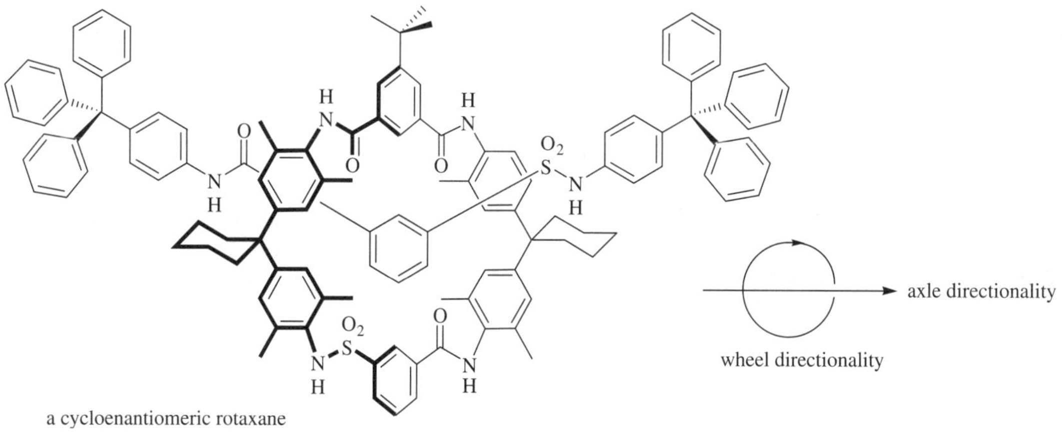

a cycloenantiomeric rotaxane

axle directionality

wheel directionality

D, L
See Fischer–Rosanoff convention.

Δ (delta), Λ (lambda)
Descriptors that denote the chirality of tris(bidentate) metal complexes and other octahedral complexes having left-handed (Λ) or right-handed (Δ) structural helicity. See helicity.

Diastereofacial
The presence of a chiral element in a molecule exhibiting a prochiral trigonal group renders the two faces of this trigonal system diastereotopic. Addition of a new atom or group to either heterotopic face can afford unequal amounts of diastereoisomers. A reaction proceeds with diastereofacial selectivity if attack on one heterotopic face is favored. For example, diastereofacial addition of ethylmagnesium bromide to the *Re*- or *Si*-face of (*R*)-3-phenyl-2-butanone produces the

corresponding (2*R*,3*R*)- and (2*S*,3*R*)-alcohol, respectively. The two faces of the trigonal system are designated by *Re* and *Si* stereodescriptors. See also diastereotopic.

Diastereomeric excess (% de)

By analogy with enantiomeric excess, % ee, diastereoisomeric excess is defined as:

$$\% \ de = \left(\frac{A - B}{A + B}\right) 100 = \% \ A - \% \ B, \ \text{with} \ A > B$$

where *A* and *B* are the fractions of the two diastereoisomers ($A + B = 1$). The term is not applicable to mixtures of more than two diastereoisomers.

Diastereomeric ratio (dr)

By analogy with enantiomeric ratio, er, the diastereomeric ratio is defined as the ratio of the amount of one diastereoisomer to that of another.

Diastereoselectivity (% ds)

By analogy with enantioselectivity, % es, diastereoselectivity is defined as the mol fraction *A* of one diastereomer in percent:

$$\% \ ds = \left(\frac{A}{A + B + C + \ldots}\right) 100$$

where *A*, *B*, *C*, etc. describe the amount of each diastereomer.

Diastereoisomerization (diastereomerization)

The interconversion of diastereoisomers. See Chapter 3.1.

Diastereoisomers (diastereomers)

Stereoisomers are either enantiomers or diastereoisomers. Chiral molecules that have the relationship of mirror images are called enantiomers. All other stereoisomers are diastereoisomers including *E*/*Z*-isomers.

Diastereotopic

Descriptor for constitutionally equivalent atoms or groups of a molecule that are not symmetry related. Replacement of one or the other of two diastereotopic substituents by a new atom or group gives rise to diastereoisomers. For example, replacement of the β-hydrogens in 2-hydroxybutanoic acid by a bromo substituent produces two diastereoisomers. The two hydrogen atoms of the methylene group at *C*-3 are diastereotopic. See also prochirality, enantiotopic, heterotopic, diastereofacial.

diastereotopic hydrogens   diastereoisomers

Dihedral angle
The angle between two intersecting planes. Commonly used to describe the angle between two vicinal substituents.

Disrotatory
Descriptor for the relative movement of bonds or groups within one molecule. Used to describe the stereochemistry of pericyclic reactions such as cycloadditions and electrocyclic reactions, and to illustrate the mechanism of ring flipping processes. Interconversion of molecular propellers may proceed via conrotatory (in the same direction) or disrotatory (in opposite directions) correlated ring rotations.

Dissymmetric
Describes an object that has no inversion center, symmetry plane and rotation–reflection axis, while symmetry axes may be present.

E, Z
Stereodescriptors for isomeric alkenes. The groups of highest priority on either end of a double bond are identified based on CIP rules. The diastereoisomer is designated as *Z* (German *zusammen*, together) if the groups are on the same side and as *E* (German *entgegen*, opposite) if they reside on opposite sides.

Eclipsed conformation
Two atoms or groups attached to adjacent atoms are eclipsed if the torsion angle is zero or almost zero. See synperiplanar, torsion angle.

Enantiofacial
A prochiral trigonal moiety of an achiral molecule has two enantiotopic faces. Asymmetric addition of a new atom or group to either enantiotopic face, for example in the presence of a chiral catalyst, produces a nonracemic mixture of enantiomeric products. A reaction proceeds with enantiofacial selectivity if attack on one enantiotopic face is favored. For example, enantiofacial addition of diethylzinc to the *Re-* or *Si*-face of benzaldehyde catalyzed by a nonracemic chiral amino alcohol such as *N,N*-dimethyl aminoisoborneol generates the corresponding (*R*)- and (*S*)-alcohol in unequal amounts. The two faces of the trigonal system are designated by *Re* and *Si* stereodescriptors. See also enantiotopic.

Enantiomer
Chiral molecules that have the relationship of mirror images are called enantiomers. All other stereoisomers are diastereoisomers.

Enantiomeric excess (% ee)
The enantiomeric excess, % ee, is defined as:

$$\% \ ee = \left(\frac{A - B}{A + B}\right) 100 = \% \ A - \% \ B, \text{ with } A > B$$

where A and B are the fractions of the enantiomers ($A + B = 1$).

Enantiomeric ratio (er)
The enantiomeric ratio is defined as the ratio of the amount of one enantiomer to that of the other.

Enantiomerically enriched (enantioenriched)
A chiral compound with an enantiomeric excess between 0 and 100%.

Enantiomerically pure (enantiopure)
A chiral compound with 100% ee (within limits of detection of the minor enantiomer).

Enantiomerization
The interconversion of enantiomers. See racemization and Chapter 3.1.

Enantiomorph
One of a pair of chiral objects. For example, crystals that are nonsuperimposable mirror images.

Enantioselectivity (% es)
Enantioselectivity, % es, is defined as the mol fraction A of one enantiomer in percent:

$$\% \ es = \left(\frac{A}{A + B}\right) 100$$

where A and B describe the amount of each enantiomer.

Enantiotopic
Descriptor for constitutionally equivalent atoms or groups of a molecule that are related by symmetry elements of the second kind only (symmetry plane, inversion center, rotation-reflection axis). Replacement of one or the other of two enantiotopic substituents by a new atom or group gives enantiomers. For example, replacement of the α-hydrogens in propanoic acid by a bromo substituent produces enantiomers. The two hydrogen atoms of the methylene group at *C*-2 are enantiotopic. See also prochirality, diastereotopic, heterotopic, enantiofacial.

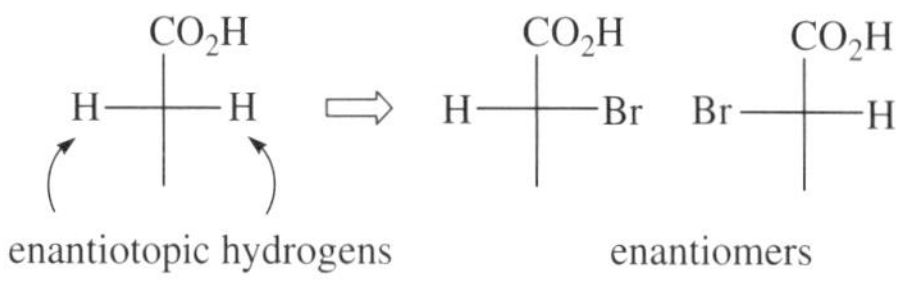

enantiotopic hydrogens   enantiomers

Endo, exo
Descriptors of the relative configuration of groups attached to a nonbridgehead atom in a bicyclic compound. If the group points towards the larger bridge it is *endo*. If it is directed towards the smaller bridge it is *exo*.

smaller bridge

larger bridge

(*R*)-2-*exo*-bicyclo[2.2.1]heptane   (*S*)-2-*endo*-bicyclo[2.2.1]heptane

Envelope conformation
The conformation of a five-membered ring (symmetry group C_s), in which four atoms are coplanar while the fifth atom (the flap) resides outside the plane.

Epimerization
Epimerization is a special case of diastereomerization and refers to the interconversion of epimers. The macroscopic analog to epimerization is mutarotation. See Chapter 3.1.4.

Epimers
Diastereoisomers that differ in only one configuration of two or more chiral elements.

Equatorial
See axial.

Erythro, threo
Obsolescent descriptors for the relative configuration of diastereoisomers with two adjacent chiral centers. This notation is derived from the two carbohydrates erythrose and threose. The threo diastereoisomer has similar or identical substituents on opposite sides of the vertical chain when it is drawn in the Fischer projection while the erythro isomer has the corresponding substituents on the same side of the chain.

Exo
See *endo*.

Eyring equation
Correlation of the rate constant, k, and the Gibbs activation energy, $\Delta G^{\neq}$, of a reaction based on the activated complex theory.

$$k = \kappa \, \frac{k_B \, T}{h} \, e^{-\Delta G^{\neq}/RT}$$

where k = rate constant at temperature T measured in Kelvin; h = Planck's constant $(6.6261 \cdot 10^{-34} \, \text{J s})$; k_B = Boltzmann's constant $(1.3807 \cdot 10^{-23} \, \text{J K}^{-1})$; $\Delta G^{\neq}$ = Gibbs activation energy [J mol^{-1}]; R = gas constant $(8.3144 \, \text{J K}^{-1} \text{mol}^{-1})$; and κ is the transmission coefficient (usually assumed to be 1).

Felkin–Anh model
Refinement of Cram's rule. See Chapter 2.2.

Fischer projection (or Fischer–Tollens projection)
A two-dimensional projection in which vertically drawn bonds are considered to be behind the plane while horizontal bonds lie above that plane. Commonly used for carbohydrate structures.

Fischer–Rosanoff convention
A convention often used to describe the absolute configuration at one stereocenter in carbohydrates and amino acids. The molecule is drawn in the Fischer projection with the carbon having the highest oxidation state at the top and the carbon skeleton in a vertical line. In this projection D-sugars have the hydroxyl group at the highest numbered chiral center on the right side, whereas L-sugars have it on the left side. The same principle is applied to the chiral center of α-amino acids. Naturally occurring amino acids have (S)-configuration which is identical with L-configuration. See CIP rules.

D-glucose   L-glucose

D-alanine   L-alanine

Flagpole
See bowsprit, flagpole

Gauche
Vicinal substituents occupy a gauche conformation if the torsion angle is $+60$ or $-60°$. See also torsion angle.

Gauche effect
Often observed preference of a molecule with vicinal electronegative substituents (N, O, F) for a gauche conformation, *e.g.*, 1,2-difluoroethane. A gauche conformation can also be stabilized through intramolecular hydrogen bonding.

Gibbs–Helmholtz equation
Fundamental thermodynamic equation describing the relationship between the change in the standard Gibbs free energy, ΔG^o, that accompanies a chemical reaction under standard conditions (1 atm, $T = 298.15\,\text{K}$) and the corresponding change in standard enthalpy, ΔH^o, and standard entropy, ΔS^o.

$$\Delta G^o = \Delta H^o - T\,\Delta S^o$$

Half-chair
The conformation of cyclohexenes or cyclopentanes exhibiting four and three atoms, respectively, in one plane while the other two atoms lie on opposite sides of this plane. See also chair.

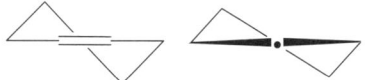

Helicity
Chirality of a helical or propeller-shaped molecule. A right-handed helix is described as P and a left-handed one as M. Similarly, helical chirality in tris(bidentate) metal complexes is described as Δ (delta) and Λ (lambda). See Chapter 2.1.

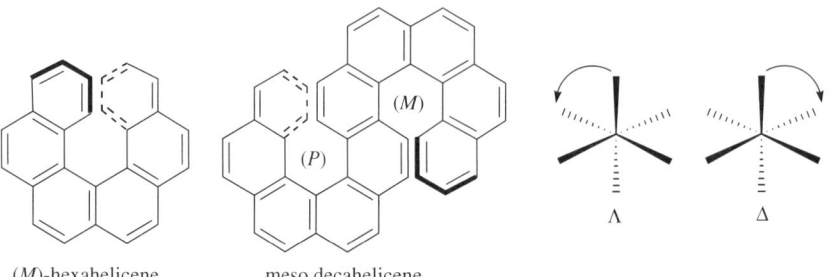

(*M*)-hexahelicene   meso decahelicene

Heterochiral
Objects are heterochiral if they have the opposite sense of chirality.

Heterofacial
Either enantiofacial or diastereofacial. See *Re*, *Si*.

Heterotopic
Either enantiotopic or diastereotopic. See *pro-R*, *pro-S*, *Re*, *Si*.

Homochiral
Objects are homochiral if they have the same sense of chirality.

Homotopic
Atoms or groups of a molecule that are related by an *n*-fold rotation axis and occupy configurationally equivalent positions upon rotation by $(360/n)°$. Replacement of one or the other of two homotopic substituents by a new atom or group gives identical products.

Inversion
See Walden inversion, pyramidal inversion and ring inversion.

Inversion center
See center of symmetry.

Isomers
Compounds with the same molecular formula but different structure due to variation in constitution (constitutional isomers), configuration (configurational isomers) or conformation (conformational isomers).

Kinetic resolution
Kinetically controlled process that leads to partial or complete resolution of enantiomers exhibiting unequal reaction rates in the presence of a nonracemic chiral reagent or catalyst. See Chapter 7.4.

L, D
See Fischer–Rosanoff convention.

l, u
Stereodescriptors for the relative configuration of diastereoisomers with two chirality elements. When the two chirality elements are similar, for example *RR*, *SS*, *MM*, or *PP*, they have an *l*-relationship (*l* = like). Combinations such as *RS*, *MP*, etc. are considered *u* (*u* = unlike).

lk, ul
Description of topicity in stereoselective reactions based on the *l/u*-nomenclature. A preferential approach of an achiral reagent to the *Re*-face of a heterofacial group of an (*R*)-enantiomer has *lk*-topicity (*lk* = like). The combination *Si/S* has also *lk*-topicity. A preferential approach of an achiral reagent to the *Re*-face of a heterofacial group of an (*S*)-enantiomer is considered *ul*-topicity (*ul* = unlike). Similarly, the combination *Si/R* has *ul*-topicity. Also used to describe the preferential approach of a chiral reagent having (*R*)- or (*S*)-configuration to a *Re*-face and *Si*-face, respectively, of a substrate (*lk*-topicity). Again, the combinations *Si/R* and *Re/S* have *ul*-topicity.

Λ (lambda)
See Δ (delta).

M, P
Stereodescriptors for axially chiral and helical molecules. See Chapter 2.1.

Meso
A meso compound contains more than one chiral element but is achiral because of the existence of a symmetry plane, rotation–reflection axis or inversion center, for example (R,S)-tartaric acid.

Mutarotation
The macroscopic analog to epimerization is mutarotation which refers to the irreversible change of the optical rotation of an epimeric mixture until equilibrium is reached. See Chapter 3.1.4.

Newman projection
Projection illustrating the torsion of vicinal groups attached to two adjacent atoms. The viewer looks along the pivotal bond and the bonds pointing from the front atom are drawn to the center of a circle. The bonds to the rear atom are drawn to the periphery of the circle.

Newman projection   Sawhorse projection

Optical activity
The property of a chiral molecule to rotate the plane of linearly polarized light.

Optical purity (% op)
The ratio of the optical rotation of a sample, $[\alpha]$, to the optical rotation of the pure enantiomer, $[\alpha_{max}]$.

$$\% \text{ op} = \frac{[\alpha]}{[\alpha_{max}]} \ 100$$

Optical purity is not identical to enantiomeric excess. The use of optical purity to describe the enantiopurity of a sample based on polarimetric measurements is often inaccurate and therefore discouraged, see Chapter 4.1.

Optical rotation

Rotation of the plane of linearly polarized light caused by a chiral sample. The specific rotation $[\alpha]_\lambda^T$ is defined as:

$$[\alpha]_\lambda^T = \frac{[\alpha]}{c\ l}$$

where $[\alpha]$ is the measured optical rotation in degrees; l is the length of the cuvette in dm; and c is the sample concentration in g/mL.

P, M

See *M, P*.

Periplanar

See torsion angle.

Phase isomers

Stereoisomers of molecular propellers exhibiting a pair of bevel gears such as 1,4-substituted bis(9-triptycyl)methanes. Phase isomers differ in the phase relationship of their mechanically interlocked wheel cogs and interconvert via sterically hindered gear slippage. See Chapter 8.4.

Pitzer strain

See torsion strain.

Planar chirality

See chiral plane.

Point group

Classification of the symmetry elements of an object based on the Schoenflies notation (C_3, D_2, T_d, etc.).

Prochirality

Prochirality originates from the existence of heterotopic substituents or faces in a molecule. The replacement of a heterotopic substituent by a new group or the selective addition to one side of a heterofacial plane in a prochiral molecule generates a chiral product. In principle, prochirality (like chirality) is a molecular property but it is commonly used to refer to an atom or group within a molecule. A more general term is prostereoisomerism which also comprises formation of achiral diastereoisomers containing stereogenic but not necessarily chiral elements as a result of replacement of a heterotopic ligand or selective heterofacial addition. See enantiotopic, diastereotopic, enantiofacial, diastereofacial.

Pro-R, pro-S

Descriptors for differentiation between heterotopic groups such as C in $XABC_2$. An arbitrarily selected heterotopic group C is called *pro-R* if its replacement by a ligand with higher CIP priority than the other heterotopic group C results in the formation of a chiral center with (R)-configuration. The other group C is then described as *pro-S*.

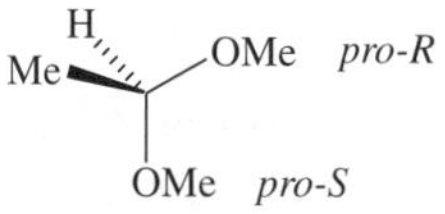

Prostereoisomerism
See prochirality.

Pseudoasymmetric
Discouraged term for an achirotopic but stereogenic center that bears two enantiomorphic ligands (same constitution but opposite sense of chirality) and two other groups. As a result, molecules with a pseudoasymmetric atom (better described as a stereogenic center) are achiral but exchange of any set of two ligands produces a diastereoisomer. The corresponding CIP descriptors r and s are reflection invariant because the mirror images are identical.

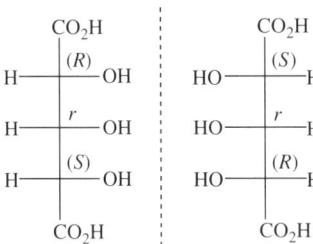

Pseudoaxial, pseudoequatorial
See axial, equatorial.

Pyramidal inversion
A change in the spatial orientation of substituents in a pyramidal or tripodal arrangement due to flipping of the central atom to the other side of the pyramidal base. If the ligands attached to the central atom are different, pyramidal inversion results in enantioconversion.

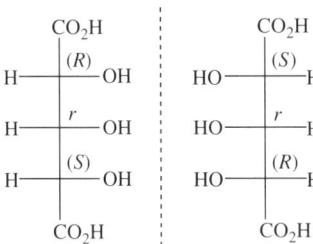

R, S
Stereodescriptors for chiral molecules. See Chapter 2.1.

r, s
See pseudoasymmetric.

R_a, S_a, R_p, S_p
CIP stereodescriptors R and S for a chiral axis and plane, respectively.

Racemate
An equimolar mixture of enantiomers, usually described by the prefix ($\pm$)- or *rac*-.

Racemization
Irreversible conversion of a pure or enriched enantiomer to the corresponding racemate. See Chapter 3.1.

Re, Si
Stereodescriptors for heterotopic faces. A heterotopic face of a trigonal unit is described as *Re* if the ligands of the trigonal atom afford a clockwise orientation based on CIP rules. The opposite side is

designated *Si* and shows a counterclockwise priority sense of the ligands. See also enantiofacial, diastereofacial.

Reflection invariant, reflection variant
An object that is identical with its mirror image is reflection invariant. If the object and its mirror image are enantiomorphous, they are reflection variant.

Relative configuration
The configuration of a stereogenic element with respect to another in the same molecule. In contrast to absolute configuration, relative configuration (*syn*, *anti*, *cis*, *trans*, *E*, *Z*, etc.) is reflection invariant.

Ring inversion
Interconversion of the chair conformation of cyclohexane and similar ring structures (ring flipping). Axial substituents occupy equatorial positions after ring inversion and vice versa.

Rotamer
Conformational isomers due to hindered rotation about a single bond.

Rotational energy barrier
Free activation energy for the interconversion of rotamers.

Rotation-reflection axis (S$_n$)
An axis that goes through a molecule or object such that rotation by an angle of $360°/n$ about the axis followed by reflection across a perpendicular plane affords a superimposable image.

S, R
See *R*.

Sawhorse projection
A perspective drawing showing the spatial orientation of all bonds of two adjacent tetrahedral atoms. See Newman projection.

s-Cis, s-trans
Stereodescriptors for conformational isomers with synperiplanar (*s-cis*) or antiperiplanar (*s-trans*) conjugated double bonds.

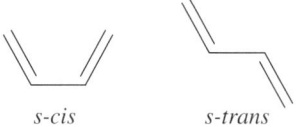

s-cis   s-trans

Si
See *Re*.

Specific rotation
See optical rotation.

Spontaneous resolution
See conglomerate.

Staggered conformation
Vicinal substituents attached to adjacent atoms are staggered if the torsion angle is 60° or close to this. See torsion angle.

Stereoconvergent
A reaction or reaction sequence that results in the formation of one stereoisomer from stereochemically different starting materials. If the process affords one enantiomer (diastereoisomer) it is described as enantioconvergent (diastereoconvergent).

Stereodescriptor
A prefix to specify the configuration or conformation of a molecule. For example, *R, S, r, s, P, M, E, Z, s-cis*.

Stereogenic
The exchange of two atoms or groups attached to a stereogenic element results in the formation of a new stereoisomer. The presence of a stereogenic element does not necessarily render a molecule chiral, *i.e.*, a stereogenic atom must not be asymmetric. Exchange of the geminal vinyl hydrogen and chloride substituents at either carbon atom in (*Z*)-1,2-dichloroethene generates the corresponding (*E*)-isomer. The two carbon atoms in achiral 1,2-dichloroethene are therefore stereogenic. If a stereogenic element is reflection variant, it constitutes an element of chirality. Chirotopicity and stereogenicity are different concepts. For example, the halogens in CHFBrCl are chirotopic but not stereogenic, whereas the carbon atoms in (*E*)- and (*Z*)-1,2-dichloroethene are stereogenic but not chirotopic.

Stereoisomers
Stereoisomers are compounds that exhibit the same constitution (the same molecular formula and connectivity of atoms) but different spatial arrangements. Stereoisomers can be further classified as enantiomers and diastereoisomers. Chiral molecules that have the relationship of mirror images are called enantiomers. All other stereoisomers are diastereoisomers. See configuration, conformation and Chapter 2.

Stereomutation
General term for isomerization reactions such as racemization, epimerization, etc.

Stereoselectivity
Preferential formation of one stereoisomer over another during a reaction. When the stereoisomeric products are enantiomers (diastereoisomers) the reaction is called enantioselective (diastereoselective). See Chapter 5.

Stereospecificity
A stereospecific reaction converts stereoisomeric starting materials to stereoisomeric products. In other words, starting materials with different configuration are converted to different stereoisomers. This applies to stereochemically constrained reactions such as S_N2, epoxidation and *trans*-additions. For example, *trans*-addition of bromine to (*E*)-2-butene gives meso 2,3-dibromobutane, whereas (*Z*)-2-butene affords racemic 2,3-dibromobutane. A stereospecific process is intrinsically stereoselective, but stereoselective reactions are not necessarily stereospecific. Note that even a perfectly stereoselective reaction (100% ee or de) is not necessarily stereospecific. See Chapter 5.1.

Superimposable
Superimposable objects can be brought into coincidence through translation and rotation.

Suprafacial
Describes the stereochemical course of pericyclic reactions (cycloadditions, electrocyclic reactions, sigmatropic rearrangements, and cheletropic reactions). In a suprafacial process, bonds are formed and cleaved at the same side of a π-system. See antarafacial.

Symmetry axis (C_n)
An axis that goes through a molecule or object in such a way that rotation by an angle of $360°/n$ about this axis generates a superimposable image.

Symmetry plane (σ)
A plane that goes through a molecule or object in such a way that reflection across this plane affords a superimposable image.

Syn
Describes the relative configuration in a molecule with two or more chiral centers. If the molecule is shown in a planar zigzag projection and the substituents are on opposite sides, the orientation is called *anti*. If they are on the same side the relative configuration is called *syn*.

Synclinal (sc)
See torsion angle.

Synperiplanar (sp)
See torsion angle.

Threo
See erythro.

Topological chirality
Nonclassical chirality of intertwined and interlocked molecules. Introduction of directionality to mechanically interlocked components of catenanes and rotaxanes can result in topological chirality and thus formation of enantiomers and diastereoisomers including meso compounds. Molecular knots and Möbius strips are inherently chiral. See cyclostereoisomerism and Chapters 2 and 9.

the arrows describe the
directionality of the rings

enantiomers of a topologically chiral [2]catenane

Topological isomers
Intertwined and interlocked molecules with identical atom connectivity but different topology. A cyclic molecule and a trefoil knot are an example of a pair of topological isomers. A [2]catenane consists of two interlocked rings and is topologically isomeric to the two separate macrocycles. Examples of three-component topological isomers are three individual rings, a [3]catenane and the corresponding Borromean ring. Usually, topological isomerization requires breaking of a covalent bond but slippage of a threaded wheel over one of the stoppers has been observed with some rotaxanes. See Chapter 9.

Topomers
Degenerate isomers. See topomerization.

Topomerization
A degenerate stereomutation caused by exchange of the positions of identical ligands of a molecule. The result is the formation of a superimposable product (topomer).

Torsion angle
The dihedral angle between two planes, *A–B–C* and *B–C–D*, in a system *A–B–C–D*. Conformations exhibiting torsion angles between $0°$ and $\pm30°$ are synperiplanar (sp), whereas synclinal (sc) conformations have angles between $\pm30°$ and $\pm90°$, anticlinal (ac) conformations have angles between $\pm90°$ and $\pm150°$, and antiperiplanar (ap) conformations have angles between $\pm150°$ and $180°$. The synclinal arrangement is often called gauche conformation. See Newman projection.

Torsion strain

Destabilizing interactions between vicinal substituents as a result of deviation from the optimal torsion angle of 60°. Synonymous with Pitzer strain.

Trans

See *cis*.

Transannular strain

Repulsive interactions between substituents attached to nonadjacent atoms in a ring structure.

Translational isomerism

Relative movement of interlocked components, for example gliding of a threaded ring in a [2]rotaxane from one station to another.

Translational isomers

Isomers of interlocked molecules that interconvert through translational movement of one component and without breaking of a covalent bond. See translational isomerism.

Transoid

See cisoid.

Twist form

See chair.

u

See *l*.

ul

See *lk*.

Walden inversion

Inversion of configuration, for example during S_N2 attack at an asymmetric carbon atom.

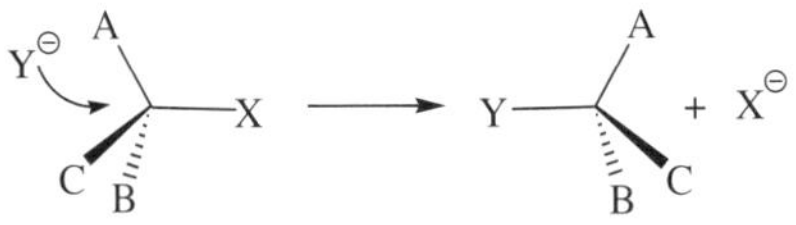

Woodward–Hoffmann Rule

A thermally initiated pericyclic reaction is symmetry-allowed if the total number of $(4q+2)$ suprafacial and $4r$ antarafacial components is uneven ($q, r = 0, 1, 2, 3$, etc.).

Z

See *E*.

Zigzag projection

Projection showing the main chain of an acyclic molecule in a planar zigzag line and the substituents above or below the plane.

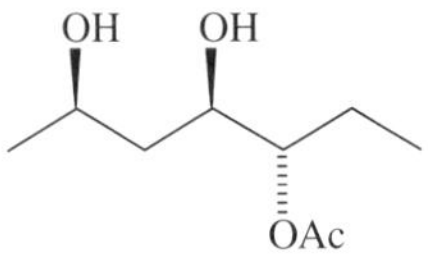

Subject Index